Fourth Edition
INTERMEDIATE ALGEBRA
A TEXT/WORKBOOK

Charles D. Miller

Margaret L. Lial
American River College

E. John Hornsby, Jr.
University of New Orleans

HarperCollins*Publishers*

TO THE STUDENT

If you need further help with algebra, you may want to obtain a copy of the *Student's Solutions Manual* that goes with this book. It contains solutions to the odd-numbered exercises that are not already solved at the back of the textbook, plus solutions to all chapter review exercises and chapter tests. Your college bookstore either has this book or can order it for you.

Sponsoring Editor:	Bill Poole
Developmental Editor:	Linda Youngman
Project Editor:	Janet Tilden/Kristin Syverson
Art Direction:	Julie Anderson
Text and Cover Design:	Lucy Lesiak Design/Cynthia Crampton
Cover Photo:	Front cover left and right, M. Angelo/West Light; Front illustration reprinted with permission from: Peitgen, H.-O./Richter, P.H.: *The Beauty of Fractals*. 1986 © 1986 Springer-Verlag Berlin-Heidelberg.
Photo Research:	Carol Parden/Judy Ladendorf
Director of Production:	Jeanie A. Berke
Production Assistant:	Linda Murray
Compositor:	The Clarinda Company
Printer and Binder:	Courier Corporation
Cover Printer:	Lehigh Press Lithographers

Intermediate Algebra: A Text/Workbook, Fourth Edition

Copyright © 1991 by HarperCollins Publishers Inc.

All rights reserved. Printed in the United States of America. No part of this book may be used or reproduced in any manner whatsoever without written permission, except in the case of brief quotations embodied in critical articles and reviews. For information address HarperCollins Publishers Inc., 10 East 53rd Street, New York, NY 10022.

ISBN 0-673-46272-2

The patterns in the small inserts on the cover of this text are examples of the types of structures that can be computer generated in accordance with the principles of fractal geometry. Fractal patterns also exist in nature; examples include the shapes of mountain ranges and the winding of coastlines.

PREFACE

Intermediate Algebra: A Text/Workbook, Fourth Edition, has been written and designed to give a thorough and complete treatment of those topics in algebra necessary for success in later courses. In particular, students will be prepared for such courses as college algebra, trigonometry, statistics, finite mathematics, analytic geometry, precalculus, and applied calculus. While we assume that most students using this book will have had a previous course in algebra, all necessary ideas are introduced or reviewed as needed.

The book's clear explanations, precise learning objectives, more than 700 detailed examples, carefully graded exercise sets referenced to examples, and open, accessible design make this an ideal text for either a traditional lecture class or individualized instruction.

NEW CONTENT FEATURES

Changes in the fourth edition include the following:

- New examples and exercises have been added, and the applications have been extensively rewritten and updated. The integrity of the applications in previous editions has been retained, but we have provided a freshness for instructors who have taught from those editions.

- At appropriate points in the text we discuss estimation and ways of determining whether or not an answer is reasonable.

- New intuitive introductions to some topics have been included to help students make connections with the real world. For example, the operations of subtraction and division in Section 1.3 and the concept of scientific notation in Section 3.2 are presented intuitively.

- In many sections, summaries of procedures have been moved earlier in the section, so students can refer to the summary while working through the examples.

- Interval notation has been introduced in Chapter 1 and used consistently throughout in talking about solutions to linear inequalities in one variable.

- Chapter 1 has been condensed from nine sections to five sections, to allow instructors to move through the review material more quickly. The operations on real numbers have been combined into one section, and the basic definitions of exponents and roots have been combined with the section on order of operations.

- The material on combining terms has been moved from Section 2.1 to Section 1.5, where the properties of real numbers are used to justify the steps.

- Chapter 2 now has an additional section on word problems, to break up the material and make it easier for students to absorb. Absolute value equations and inequalities have been combined into one section with a parallel development.

- Since students typically have trouble with exponents, two sections on the rules for exponents now introduce Chapter 3 on polynomials.

PREFACE

- Rational exponents are now introduced in Section 5.2 after a review of radicals, thus going from the familiar to the new. Rational exponents are then referred to and used elsewhere in the chapter, giving students additional exposure to this new concept.
- In Chapter 6, the material on the discriminant has been combined with that on the quadratic formula. The proof of the quadratic formula is now given at the end of the section, allowing students to see how it is used before considering the proof.
- Functions are introduced in Chapter 7, after straight-line equations and inequalities have been discussed. This is followed by the section on variation, where the variation formulas provide further examples of functions.
- The two sections on applications of linear systems in Chapter 8 have been combined into one section to reduce the amount of material that must be covered.
- A new section on second-degree inequalities and systems of inequalities has been added to Chapter 9 for those who wish to cover these topics.
- Inverse functions now open Chapter 10 on exponential and logarithmic functions, to help students understand the relationship between those functions. The section on using logarithms for calculations has been replaced with a section on using the calculator to evaluate common logarithms, natural logarithms, and logarithms to other bases. The use of tables of logarithms is explained in an appendix.

NEW FEATURES

All the successful features of the workbook format in the previous edition are carried over in the new edition: Learning objectives for each section, careful exposition and fully developed examples, sample problems in the margins for immediate feedback, carefully graded exercises with work space, and a clear, functional design guide students through the text. Screened boxes that set off important definitions, formulas, rules, and procedures further aid in learning and reviewing the course material.

In addition, the following new features have been developed to enhance the pedagogy and usefulness of the text.

Example Titles
Each example now has a title to help students see the purpose of the example. The titles also facilitate working the exercises and studying for examinations.

Cautionary Remarks
Common student errors and misconceptions, or difficulties students typically encounter, are highlighted graphically and identified with the headings "Caution" or "Note."

Skill Sharpeners
These short sets of review exercises at the end of each exercise set beginning with Chapter 2 provide students with a quick review of skills that will be needed in the next section.

Chapter Summaries
This new study aid at the end of each chapter provides a glossary of key terms with section references, a list of new symbols introduced in the chapter, and a "Quick Review" featuring section-referenced capsule summaries of key ideas accompanied by worked-out examples.

Chapter Review Exercises
A thorough set of chapter review exercises follows each chapter summary. Most of these exercises are organized by section, but a concluding section, titled "Mixed Review Exercises," includes randomly mixed exercises from the entire chapter. This helps students practice identifying problems by type. As in the previous edition, a sample test concludes each chapter.

Historical Reflections
These brief items, designed to emphasize the human dimension of mathematics, introduce eminent mathematicians and the important ideas they advanced. They are included as enrichment, to interest and motivate students.

Success in Algebra
This foreword to the students provides additional support by offering suggestions for successful study of the course material.

Section References in Answer Section
As another study aid, section references are given in the answers to the chapter tests and final examination at the back of the textbook. Students then can readily go back to study any section that gave them difficulty when they took the sample chapter test or final examination.

Glossary
A comprehensive glossary is placed at the end of the book. Each term in the glossary is defined and then referenced to the appropriate section in the text, where students may find more detailed explanations of the term.

SUPPLEMENTS

Our extensive supplemental package includes an annotated instructor's edition, testing materials, solutions, software, videotapes, and audiotapes.

Annotated Instructor's Edition
This edition provides instructors with immediate access to the answers to every exercise in the text; each answer is printed in color next to the corresponding text exercise. In addition, exercises that will require most students to stretch beyond the concepts discussed in the text are marked with the symbol **C**, to indicate *challenging*. Instructors may wish to use discretion in assigning these problems.

Instructor's Testing Manual
The Instructor's Testing Manual includes suggestions for using the textbook in a mathematics laboratory; short-answer and multiple-choice versions of a placement test; six forms of chapter tests for each chapter, including four open-response and two multiple-choice forms; two forms of a final examination; and an extensive set of additional exercises, providing 10 to 20 exercises for each textbook objective, which can be used as an additional source of questions for tests, quizzes, or student review of difficult topics.

Student's Solutions Manual
This book contains solutions to half of the odd-numbered section exercises (those not included at the back of the textbook) as well as solutions to all chapter review exercises and chapter tests.

Instructor's Solutions Manual

Available at no charge to instructors, this book includes solutions to all the margin problems in the textbook and solutions to the even-numbered section exercises. The two solutions manuals plus the solutions given at the back of the textbook provide detailed, worked-out solutions to each exercise and margin problem in the book.

HarperCollins Test Generator for Mathematics

Available in Apple, IBM, and Macintosh versions, the test generator enables instructors to select questions by objective, section, or chapter, or to use a ready-made test for each chapter. Instructors may generate tests in multiple-choice or open-response formats, scramble the order of questions while printing, and produce multiple versions of each test (up to 9 with Apple, up to 25 with IBM and Macintosh). The system features printed graphics and accurate mathematics symbols. It also features a preview option that allows instructors to view questions before printing, to regenerate variables, and to replace or skip questions if desired. The IBM version includes an editor that allows instructors to add their own problems to existing data disks.

Interactive Tutorial Software

This innovative package is also available in Apple, IBM, and Macintosh versions. It offers interactive modular units, specifically linked to the text, for reinforcement of selected topics. The tutorial is self-paced and provides unlimited opportunities to review lessons and to practice problem solving. When students give a wrong answer, they can request to see the problem worked out. The program is menu-driven for ease of use, and on-screen help can be obtained at any time with a single keystroke. Students' scores are automatically recorded and can be printed for a permanent record.

Audiotapes

A set of audiotapes, one per chapter, is available for the text. The tapes guide students through each topic, allowing individualized study and additional practice for troublesome areas. They are especially helpful for visually impaired students.

Videotapes

A new videotape series, *Algebra Connection: The Intermediate Algebra Course*, has been developed to accompany *Intermediate Algebra*, Fourth Edition. Produced by an Emmy Award-winning team in consultation with a task force of academicians from both two-year and four-year colleges, the tapes cover all objectives, topics, and problem-solving techniques within the text. In addition, each lesson is preceded by motivational "launchers" that connect classroom activity to real-world applications.

ACKNOWLEDGMENTS

We wish to thank the many users of the third edition for their insightful suggestions on improvements for this book.

We also wish to thank our reviewers for their contributions: Robin Alexander, Ball State University; Donna Brann, Moraine Valley Community College; Roselynn Cacy, University of Alaska; Clifford B. Capp, Fort Lewis College; Robert Christie, Miami-Dade Community College–South Campus; Helen D. Darcey, Cleveland State Community College; Robert L. Davidson, East Tennessee State University; William Edgar, American River College; Yvonne E. Fockler, Tidewater Community College; Jill Hanson, University of North Carolina at Wilmington; Karen Ann Holmes, Eastern Michigan University; Katherine J. Huppler, Saint Cloud State University; Nancy A. Kitt, Ball State University; Joanne F. Korsmo, New Mexico State University; F. Arnold Lowry III, East Tennessee State University; Gael Mericle, Mankato State University; Kathy

Monaghan, American River College; Marcia B. Murray, Texas Tech University; Alan Natapoff, University of Massachusetts; Patricia J. Newell, Edison Community College; Laura R. Perez, Bowling Green State University; Catherine H. Pirri, Northern Essex Community College; Jack Preston, Canada College; Sharon Taylor Riley, Wayne County Community College; Anthony Soychak, University of Southern Maine; John S. Szwec, Jr., Wayne County Community College; Mary T. Teegarden, San Diego Mesa College; Lucy C. Thrower, Francis Marion College; Paul Van Erden, American River College; David Zerangue, Nicholls State University.

We wish to thank Janet Krantz of American River College, who did an excellent job of checking all the answers for us.

Paul J. Eldersveld, College of DuPage, deserves heartfelt thanks for undertaking the enormous job of coordinating all of the print ancillaries for us.

We also want to thank Tommy Thompson, Seminole Community College, for his suggestions about the essay to the students on studying algebra at the beginning of the book.

Our special thanks go to our editors, Bill Poole, Linda Youngman, and Janet Tilden. As always, they have done a wonderful job under difficult circumstances.

Margaret L. Lial
E. John Hornsby, Jr.

CONTENTS

To the Student: Success in Algebra xii

Diagnostic Pretest xiii

1 THE REAL NUMBERS 1

   1.1 Basic Terms 1

   1.2 Inequality 9

   1.3 Operations on Real Numbers 15

   1.4 Exponents and Roots; Order of Operations 25

   1.5 Properties of Real Numbers 35

   Chapter 1 Summary 43

   Chapter 1 Review Exercises 47

   Chapter 1 Test 51

2 LINEAR EQUATIONS AND INEQUALITIES 53

   2.1 Linear Equations in One Variable 53

   Historical Reflections 58

   2.2 Formulas 63

   2.3 Applications of Linear Equations 71

   2.4 More Applications of Linear Equations 83

   Historical Reflections 86

   2.5 Linear Inequalities in One Variable 91

   2.6 Set Operations and Compound Inequalities 101

   2.7 Absolute Value Equations and Inequalities 109

   Summary on Solving Linear Equations and Inequalities 119

   Chapter 2 Summary 121

   Chapter 2 Review Exercises 125

   Chapter 2 Test 131

3 POLYNOMIALS 133

   3.1 Integer Exponents 133

   3.2 Further Properties of Exponents 141

   Historical Reflections 146

   3.3 Addition and Subtraction of Polynomials 149

   3.4 Multiplication of Polynomials 159

   3.5 Division of Polynomials 169

3.6 Synthetic Division (Optional)   175

3.7 Greatest Common Factors; Factoring by Grouping   181

3.8 Factoring Trinomials   187

3.9 Special Factoring   195

Summary of Factoring Methods   201

3.10 Solving Equations by Factoring   205

Chapter 3 Summary   211

Chapter 3 Review Exercises   215

Chapter 3 Test   221

4 RATIONAL EXPRESSIONS   223

4.1 Basics of Rational Expressions   223

4.2 Multiplication and Division of Rational Expressions   229

4.3 Addition and Subtraction of Rational Expressions   233

4.4 Complex Fractions   243

4.5 Equations with Rational Expressions   247

Summary of Operations Involving Rational Expressions   253

4.6 Formulas with Rational Expressions   255

4.7 Applications   259

Historical Reflections   267

Chapter 4 Summary   268

Chapter 4 Review Exercises   271

Chapter 4 Test   275

5 ROOTS AND RADICALS   277

5.1 Radicals   277

5.2 Rational Exponents   283

5.3 Simplifying Radicals   289

5.4 Addition and Subtraction of Radical Expressions   297

5.5 Multiplication and Division of Radical Expressions   301

5.6 Equations with Radicals   311

Historical Reflections   316

5.7 Complex Numbers   321

Chapter 5 Summary   331

Chapter 5 Review Exercises   335

Chapter 5 Test   339

CONTENTS

6 QUADRATIC EQUATIONS AND INEQUALITIES   341

   6.1 Completing the Square   341

   6.2 The Quadratic Formula   349

   6.3 Equations Quadratic in Form   359

   6.4 Formulas and Applications   369

   6.5 Nonlinear Inequalities   377

   Historical Reflections   382

   Chapter 6 Summary   385

   Chapter 6 Review Exercises   387

   Chapter 6 Test   393

7 THE STRAIGHT LINE   395

   7.1 The Rectangular Coordinate System   395

   Historical Reflections   400

   7.2 The Slope of a Line   405

   7.3 Linear Equations in Two Variables   415

   Historical Reflections   420

   7.4 Linear Inequalities in Two Variables   425

   Historical Reflections   430

   7.5 Functions   435

   7.6 Variation (Optional)   445

   Historical Reflections   450

   Historical Reflections   453

   Chapter 7 Summary   454

   Chapter 7 Review Exercises   457

   Chapter 7 Test   463

8 SYSTEMS OF LINEAR EQUATIONS   465

   8.1 Linear Systems of Equations in Two Variables   465

   8.2 Linear Systems of Equations in Three Variables   477

   Historical Reflections   482

   8.3 Applications of Linear Systems   487

   8.4 Determinants   499

   8.5 Solution of Linear Systems of Equations by Determinants—Cramer's Rule (Optional)   505

   Historical Reflections   511

   Chapter 8 Summary   512

   Chapter 8 Review Exercises   517

   Chapter 8 Test   521

9 GRAPHS OF FUNCTIONS AND CONIC SECTIONS 523

9.1 Graphing Parabolas 523

9.2 More About Parabolas 533

9.3 The Distance Formula and the Circle 543

9.4 The Ellipse and the Hyperbola 551

9.5 Nonlinear Systems of Equations 561

9.6 Second-Degree Inequalities; Systems of Inequalities (Optional) 569

Historical Reflections 572

Chapter 9 Summary 577

Chapter 9 Review Exercises 581

Chapter 9 Test 587

10 EXPONENTIAL AND LOGARITHMIC FUNCTIONS 591

10.1 Inverse Functions 591

Historical Reflections 596

10.2 Exponential Functions 601

Historical Reflections 606

10.3 Logarithmic Functions 609

Historical Reflections 614

10.4 Properties of Logarithms 619

Historical Reflections 624

10.5 Evaluating Logarithms 627

10.6 Exponential and Logarithmic Equations 637

Historical Reflections 642

Historical Reflections 647

Chapter 10 Summary 648

Chapter 10 Review Exercises 651

Chapter 10 Test 655

Final Examination 659

Appendices 665

Appendix A: Using the Table of Common Logarithms 665

Table of Common Logarithms 670

Appendix B: Formulas 672

Answers to Selected Exercises A-1

Solutions to Selected Exercises A-29

Glossary A-84

Index A-89

TO THE STUDENT: SUCCESS IN ALGEBRA

The main reason students have difficulty with mathematics is that they don't know how to study it. Studying mathematics *is* different from studying subjects like English or history. The key to success is regular practice.

This should not be surprising. After all, can you learn to play the piano or to ski well without a lot of regular practice? The same thing is true for learning mathematics. Working problems nearly every day is the key to becoming successful. Here is a list of things you can do to help you succeed in studying algebra.

1. Pay attention in class to what your instructor says and does, and make careful notes. In particular, note the problems the instructor works on the board and copy the complete solutions. Keep these notes separate from your homework to avoid confusion when you read them over later.

2. Don't hesitate to ask questions in class. It is not a sign of weakness, but of strength. There are always other students with the same question who are too shy to ask.

3. Before you start on your homework assignment, rework the problems the instructor worked in class. This will reinforce what you have learned. Many students say, "I understand it perfectly when you do it, but I get stuck when I try to work the problem myself."

4. *Read your text carefully.* Many students read only enough to get by, usually only the examples. Reading the complete section will help you to be successful with the homework problems. As a bonus you will be able to do the problems more quickly if you have read the text first. As you read the text, work the sample problems given in the margin and check the answers to see if you have worked them correctly. This will test your understanding of what you have read.

5. Do your homework assignment only *after* reading the text and reviewing your notes from class. Check your work with the answers in the back of the book. If you get a problem wrong and are unable to see why, mark that problem and ask your instructor about it.

6. Work as neatly as you can. Write your symbols clearly, and make sure the problems are clearly separated from each other. Working neatly will help you to think clearly and also make it easier to review the homework before a test.

7. After you have completed a homework assignment, look over the text again. Try to decide what the main ideas are in the lesson. Often they are clearly highlighted or boxed in the text.

8. Keep any quizzes and tests that are returned to you and use them when you study for future tests and the final exam. These quizzes and tests indicate what your instructor considers most important. Be sure to correct any problems on these tests that you missed, so you will have the corrected work to study.

9. Don't worry if you do not understand a new topic right away. As you read more about it and work through the problems, you will gain understanding. Each time you look back at a topic you will understand it a little better. No one understands each topic completely right from the start.

10. What a great feeling you will experience when, after a lot of time and hard work, you can say, "I really do understand this now."

DIAGNOSTIC PRETEST

Ask your instructor whether or not you should work this pretest. It is designed to tell what material in the course may already be familiar to you. The actual course begins on page 1.

1. Complete the following: $-15 - 8 =$

2. Evaluate $\dfrac{3ab + c}{b - a}$ if $a = 5$, $b = -6$, and $c = -1$.

Choose the smaller of each pair of numbers.

3. $-5, 2$

4. $-3, -8$

5. $|-6|, -|-2|$

Simplify the following.

6. $-5x + 3x - 8x$

7. $4(2x + 6) - 2(3x - 1)$

Solve the equations or inequalities in Exercises 8–10.

8. $3x - 4 = 12$

9. $2m - (5m + 3) = m$

DIAGNOSTIC PRETEST

10. _____

10. $3y - 2 < 4y$

11. _____

11. Multiply: $6xy(2x^4y^2 - 3x^3y^3)$.

12. _____

12. Add: $(k^2 - 3) + (k^3 + 2k^2 - k + 4)$.

13. _____

13. Multiply: $(3x - 2)(2x + 7)$.

14. _____

14. Multiply: $(5x - 2y)^2$.

15. _____

15. Divide $x^3 - 3x^2 - 3x - 35$ by $x - 5$.

Factor as completely as possible in Exercises 16–18.

16. _____

16. $p^2 - 4q^2$

17. _____

17. $x^2 + 2x - 15$

18. $6y^2 + 7y - 3$

18. _____

19. Multiply: $\dfrac{4m^2(m + 3)}{m - 2} \cdot \dfrac{3(m - 2)}{12m^3}$.

19. _____

20. Divide: $\dfrac{3a^3b^2}{a^2 - 2ab + b^2} \div \dfrac{9ab^3}{a^2 - b^2}$.

20. _____

21. Write as a single fraction: $\dfrac{1}{k} + \dfrac{2}{r} - \dfrac{3}{rk}$.

21. _____

Solve the equations in Exercises 22 and 23.

22. $z^2 + 14z + 48 = 0$

22. _____

23. $\dfrac{5}{b} + 2 = \dfrac{7}{b^2}$

23. _____

24. Solve the system: $x + 3y = 7$
$2x - y = 0$.

24. _____

DIAGNOSTIC PRETEST

25. _____

25. Solve $p = \dfrac{rt}{D}$ for r.

26. _____

26. What are the square roots of $36k^{10}$?

27. _____

27. Multiply: $\sqrt{7} \cdot 4\sqrt{2}$.

28. _____

28. Subtract: $3\sqrt{18} - 6\sqrt{8}$.

29.

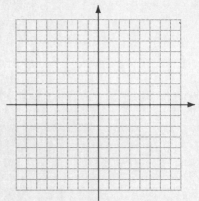

29. Graph $y = 3x + 1$.

30. _____

30. A lot is in the shape of a rectangle with a perimeter of 70 meters. The length is 5 meters more than the width. Find the length and width of the lot.

THE REAL NUMBERS

1.1 BASIC TERMS

The study of algebra extends and generalizes the rules of arithmetic. In algebra, symbols are used to convey ideas in an economical way. Many of the symbols used throughout the book are introduced in this chapter.

1 A **set** is a collection of objects. These objects are called the **elements** or **members** of the set. In algebra, the elements in a set are usually numbers. Sets are written with braces used to enclose the elements. For example, 2 is an element of the set {1, 2, 3}. In our study of algebra we will find it useful to refer to certain sets of numbers by name. The set

$$N = \{1, 2, 3, 4, 5, 6, \ldots\}$$

is called the **natural numbers** or the **counting numbers.** The three dots show that the list continues in the same pattern indefinitely. A set with no elements is called the **empty set** and is written as ∅.

WORK PROBLEM 1 AT THE SIDE.

Letters called **variables** are often used to represent numbers. Variables can also be used to define sets of numbers. For example,

$$\{x|x \text{ is a natural number between 3 and 15}\}$$

(read "the set of all elements x such that x is a natural number between 3 and 15") defines the set

$$\{4, 5, 6, 7, \ldots, 14\}.$$

The notation $\{x|x$ is a natural number between 3 and 15$\}$ is an example of **set-builder notation,** read as follows.

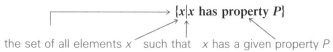

The property P describes the elements to be included in the set, as illustrated in the next example.

EXAMPLE 1 Listing the Elements in a Set
List the elements in each set.

(a) $\{x|x$ is a natural number less than 4$\}$

The natural numbers less than 4 are 1, 2, and 3. The given set is written as {1, 2, 3}.

OBJECTIVES

1 Write sets of numbers.
2 Use number lines.
3 Know the common sets of numbers.
4 Find additive inverses.
5 Use absolute value.

1. What are the elements of the set $S = \{\text{red, yellow, blue}\}$?

ANSWER
1. There are three elements in the set S: red, yellow, and blue.

2. List the elements in each set.

   (a) {x|x is a natural number less than 5}

   (b) {x|x is a natural number greater than 12}

3. Graph each set.

   (a) {−4, −2, 0, 2, 4, 6}

   (b) $\left\{-1, 0, \frac{2}{3}, \frac{5}{2}\right\}$

   (c) $\left\{5, \frac{16}{3}, 6, \frac{13}{2}, 7, \frac{29}{4}\right\}$

   (b) {y|y is one of the first five even natural numbers}

   This set is written as {2, 4, 6, 8, 10}. ∎

   WORK PROBLEM 2 AT THE SIDE.

   2 A good way to get a picture of sets of numbers is to use a diagram called a **number line.** To construct a number line, choose any point on a horizontal line and label it 0. Next, choose a point to the right of 0 and label it 1. The distance from 0 to 1 establishes a scale that can be used to locate more points, with positive numbers to the right of 0 and negative numbers to the left of 0. The number 0 is neither positive nor negative. A number line is shown in Figure 1.

FIGURE 1

Each number is called the **coordinate** of the point that it labels, while the point is the **graph** of the number. Figure 2 shows a number line with several selected points graphed on it.

FIGURE 2

Numbers written with a positive or negative sign, such as +4, +8, −9, −5, and so on, are called **signed numbers.** Positive numbers can be called signed numbers even if the plus sign is left off (as it usually is).

WORK PROBLEM 3 AT THE SIDE.

3 The sets of numbers listed below will be used throughout the rest of the book.

Real numbers	{x|x is a coordinate of a point on a number line}*
Natural numbers or counting numbers	{1, 2, 3, 4, 5, 6, 7, 8, . . .}
Whole numbers	{0, 1, 2, 3, 4, 5, 6, . . .}
Integers	{. . ., −3, −2, −1, 0, 1, 2, 3, . . .}
Rational numbers	$\left\{\frac{p}{q} \mid p \text{ and } q \text{ are integers}, q \neq 0\right\}$ Examples: $\frac{2}{3}, -\frac{9}{2}, \frac{16}{8}$
Irrational numbers	{x|x is a real number that is not rational} Examples: $\sqrt{3}, -\sqrt{2}, \pi$

*An example of a number that is not a coordinate of a point on a number line is $\sqrt{-1}$. This number, called an *imaginary* number, is discussed in Chapter 5.

ANSWERS
2. (a) {1, 2, 3, 4}
   (b) {13, 14, 15, . . .}

3. (a)

   (b)

   (c)

Every point on a number line corresponds to a real number, and every real number corresponds to a point on the line. Real numbers can be written in decimal form, either as a terminating decimal, such as .6 or .125; a repeating decimal, such as .33333... or .127127...; or as a decimal that neither repeats nor terminates, such as .12534875.... Repeating and terminating decimal numbers are rational numbers; they *can* be written as fractions in which the numerator and denominator are integers. Decimal numbers that do not repeat or terminate are *not* rational, and thus are called irrational numbers. Many square roots are irrational numbers; some examples are $\sqrt{7}$, $\sqrt{11}$, $\sqrt{2}$, and $-\sqrt{5}$. (Some square roots *are* rational: $\sqrt{16} = 4$, $\sqrt{100} = 10$, and so on.) Another irrational number is π, the ratio of the circumference of a circle to its diameter. All irrational numbers are real numbers.

The relationships among these various sets of numbers are shown in Figure 3; in particular, the figure shows that the set of real numbers includes both the rational and irrational numbers. Every real number is either rational or irrational. Also, notice that the integers are elements of the set of rational numbers, and the whole numbers and the natural numbers are elements of the set of integers.

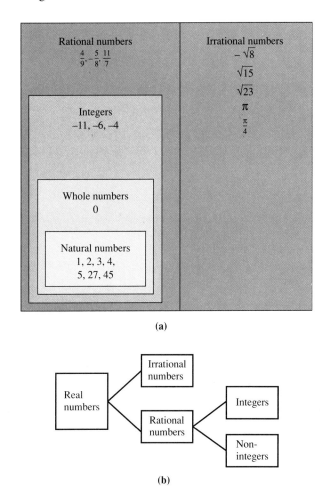

FIGURE 3 The Real Numbers

CHAPTER 1 THE REAL NUMBERS

4. Select all the words from the following list that apply to each number.

> Whole number
> Rational number
> Irrational number
> Real number
> Undefined

(a) -6

(b) 12

(c) $.333...$

(d) $\dfrac{8}{2}$ (*Hint:* Simplify.)

(e) $-\sqrt{15}$

(f) $\sqrt{4}$

(g) $-\dfrac{6}{0}$

ANSWERS
4. (a) rational, real
(b) whole, rational, real
(c) rational, real
(d) $\left(\dfrac{8}{2} = 4\right)$ whole, rational, real
(e) irrational, real
(f) ($\sqrt{4} = 2$) whole, rational, real
(g) undefined

EXAMPLE 2 Identifying Examples of Number Sets

(a) $0, \dfrac{2}{3}, -\dfrac{8}{1}$ (or -8), $-\dfrac{9}{64}, \dfrac{28}{7}$ (or 4), .5, and 1.1212... are rational numbers.

(b) $\sqrt{3}, \pi, -\sqrt{2}$, and $\sqrt{7} + \sqrt{3}$ are irrational numbers.

(c) $-8, \dfrac{4}{2}$ (or 2), $-\dfrac{3}{1}$ (or -3), and $\dfrac{25}{5}$ (or 5) are integers.

(d) All the numbers in parts (a), (b), and (c) above are real numbers.

(e) $\dfrac{4}{0}$ is undefined since the definition requires the denominator of a rational number to be nonzero. ∎

■ WORK PROBLEM 4 AT THE SIDE.

4 Now look again at the number line in Figure 1. For each positive number, there is a negative number that lies the same distance from zero. These pairs of numbers are called **additive inverses, negatives,** or **opposites** of each other. For example, 5 is the additive inverse of -5, and -5 is the additive inverse of 5.

For any real number a, the number $-a$ is the additive inverse of a.

The sum of a number and its additive inverse is always zero.

The symbol "$-$" can be used to indicate any of the following things:

(a) a negative number, such as -9 or -15;
(b) the additive inverse of a number, as in "-4 is the additive inverse of 4";
(c) subtraction, as in $12 - 3$.

In the expression $-(-5)$, the symbol "$-$" is being used in two ways: the first $-$ indicates the additive inverse of -5, and the second indicates a negative number, -5. The additive inverse of -5 is 5, so $-(-5) = 5$.

For any real number a, $\quad -(-a) = a.$

1.1 BASIC TERMS

EXAMPLE 3 Finding Additive Inverses
The following list shows several signed numbers and the additive inverse of each.

Number	Additive inverse
6	−6
−4	4
$\frac{2}{3}$	$-\frac{2}{3}$
0	0

WORK PROBLEM 5 AT THE SIDE.

5 Geometrically, the **absolute value** of a number a, written $|a|$, is the distance on the number line from 0 to a. For example, the absolute value of 5 is the same as the absolute value of −5, since each number lies five units from 0 (see Figure 4). That is, both

$$|5| = 5 \quad \text{and} \quad |-5| = 5.$$

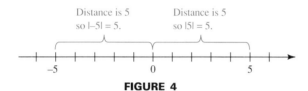

FIGURE 4

CAUTION Since absolute value is used to represent distance, and since distance is never negative, *the absolute value of a number is never negative.* However, the absolute value of 0 is 0.

The algebraic definition of absolute value is written as follows.

$$|a| = \begin{cases} a & \text{if } a \text{ is positive or zero} \\ -a & \text{if } a \text{ is negative} \end{cases}$$

The second part of this definition, $|a| = -a$ if a is negative, requires careful thought. If a is a *negative* number, then $-a$, the additive inverse or opposite of a, is a positive number, so that $|a| = -a$ is positive. For example, if $a = -3$,

$$|a| = |-3| = -(-3) = 3.$$

5. Give the additive inverse of each number.

(a) 9

(b) 17

(c) −12

(d) $-\frac{6}{5}$

(e) −0

ANSWERS
5. (a) −9 (b) −17 (c) 12 (d) $\frac{6}{5}$
   (e) 0

CHAPTER 1 · THE REAL NUMBERS

6. Find the value of each expression.

(a) $|6|$

(b) $|-3|$

(c) $-|5|$

(d) $-|-2|$

(e) $|-6| + |-3|$

(f) $|-9| - |-4|$

EXAMPLE 4 Evaluating Absolute Value Expressions

Find the value of each expression.

(a) $|13| = 13$

(b) $|-2| = -(-2) = 2$

(c) $|0| = 0$

(d) $-|8|$

Evaluate the absolute value first. Then find the additive inverse.
$$-|8| = -(8) = -8$$

(e) $-|-8|$

Work as in part (d): $|-8| = 8$, so
$$-|-8| = -(8) = -8.$$

(f) $|-2| + |5|$

Evaluate each absolute value first, then add.
$$|-2| + |5| = 2 + 5 = 7$$

▬ **WORK PROBLEM 6 AT THE SIDE.**

ANSWERS

6. (a) 6 (b) 3 (c) −5 (d) −2
   (e) 9 (f) 5

name   date   hour

1.1 EXERCISES

Graph the elements of each set on a number line.

1. $\{2, 3, 4, 5\}$

2. $\{0, 2, 4, 6, 8\}$

3. $\{-4, -3, -2, -1, 0, 1\}$

4. $\{-6, -5, -4, -3, -2\}$

5. $\left\{-\dfrac{1}{2}, \dfrac{3}{4}, \dfrac{5}{3}, \dfrac{7}{2}\right\}$

6. $\left\{-\dfrac{3}{5}, -\dfrac{1}{10}, \dfrac{9}{8}, \dfrac{12}{5}, \dfrac{13}{4}\right\}$

List the elements in each set. See Example 1.

7. $\{x \mid x \text{ is a natural number less than } 7\}$

8. $\{m \mid m \text{ is a whole number less than } 9\}$

9. $\{z \mid z \text{ is an integer greater than } 11\}$

10. $\{y \mid y \text{ is a natural number greater than } 8\}$

11. $\{a \mid a \text{ is an even integer greater than } 10\}$

12. $\{k \mid k \text{ is a counting number less than } 1\}$

13. $\{x \mid x \text{ is an irrational number that is also rational}\}$

14. $\{r \mid r \text{ is a negative integer greater than } -1\}$

15. $\{p \mid p \text{ is a number whose absolute value is } 3\}$

16. $\{w \mid w \text{ is a number whose absolute value is } 12\}$

17. $\{z \mid z \text{ is a whole number and a multiple of } 5\}$

18. $\{n \mid n \text{ is a counting number and a multiple of } 3\}$

Find the value of each expression. See Example 4.

19. $|-9|$ 20. $|12|$ 21. $-|6|$ 22. $-|15|$ 23. $-|-3|$

24. $-|-7|$ 25. $-|16|$ 26. $-|1|$ 27. $|-2| + |3|$ 28. $|-12| + |2|$

29. $|-8| - |3|$
30. $|-10| - |5|$
31. $|-6| - |-2|$
32. $|-3| - |-2|$
33. $|-12| + |-3|$
34. $|-15| + |-10|$
35. $|-9| - |-1|$
36. $|8| + |-11|$

Give the additive inverse for each number. See Examples 3 and 4.

37. 8
38. 15
39. -9
40. -12
41. 0
42. -0
43. $-(-3)$
44. $-(-5)$
45. $|10|$
46. $|17|$
47. $-|8|$
48. $-|21|$
49. $-|-1|$
50. $-|-9|$
51. $-|-18|$
52. $-|-4|$
53. $|-2| + |-3|$
54. $|-9| - |-4|$

Which elements of the set $S = \left\{-6, -\sqrt{3}, -.5, 0, \dfrac{2}{3}, \sqrt{2}, 2, 3, \dfrac{7}{2}, 4.5, \dfrac{10}{2}\right\}$
are elements of the following sets? See Example 2.

55. Natural numbers
56. Whole numbers
57. Integers
58. Rational numbers
59. Irrational numbers
60. Real numbers

Mark each statement as true *or* false.

61. Every rational number is an integer.
62. Every natural number is an integer.
63. Every irrational number is an integer.
64. Every integer is a rational number.
65. Every whole number is a real number.
66. Every natural number is a whole number.
67. Some rational numbers are irrational.
68. Some natural numbers are whole numbers.
69. Some rational numbers are integers.
70. Some real numbers are integers.
71. The absolute value of any number is positive.
72. The absolute value of any nonzero number is positive.

For what values of x are the following statements true?

73. $|x| = -|x|$
74. $|-x| = |x|$
75. $|x| = |x + 2|$
76. $|x - 1| = |x + 1|$

1.2 INEQUALITY

The statement $4 + 2 = 6$ is an **equation**; it states that two quantities are equal. The statement $4 \neq 6$ (read "4 is not equal to 6") is an **inequality**, a statement that two expressions are *not* equal.

OBJECTIVES
1. Use the symbols $<$ and $>$.
2. Use other inequality symbols.
3. Graph sets of real numbers.
4. Graph sets with "between."

1 When two numbers are not equal, one must be less than the other. The symbol $<$ means "is less than." For example,

$$8 < 9, \quad -6 < 15, \quad \text{and} \quad 0 < \frac{4}{3}.$$

"Is greater than" is written with the symbol $>$. For example,

$$12 > 5, \quad 9 > -2, \quad \text{and} \quad \frac{6}{5} > 0.$$

CAUTION Notice that in either case the symbol "points" toward the smaller number.

The number line in Figure 5 shows the numbers 4 and 9, and we know that $4 < 9$. On the graph, 4 is to the left of 9. The smaller of two numbers is always to the left of the other on a number line.

FIGURE 5

The geometric definitions of $<$ and $>$ are as follows.

$a < b$ if a is to the left of b on the number line.
$b > a$ if b is to the right of a on the number line.

EXAMPLE 1 Using a Number Line to Determine Order

(a) As the number line in Figure 6 shows,

$$-6 < 1.$$

Also, $1 > -6$.

(b) From the same number line, $-5 < -2$, or $-2 > -5$. ∎

FIGURE 6

CHAPTER 1 THE REAL NUMBERS

1. Insert $<$ or $>$ in each blank.

(a) $3 \_\_\_ 7$

(b) $9 \_\_\_ 2$

(c) $-4 \_\_\_ -8$

(d) $-2 \_\_\_ -1$

(e) $0 \_\_\_ -3$

The following table summarizes results about positive and negative numbers. The same statement is given both in words and in symbols.

Words	Symbols
Every negative number is less than 0.	If a is negative, then $a < 0$.
Every positive number is greater than 0.	If a is positive, then $a > 0$.
0 is neither positive nor negative.	

■ **WORK PROBLEM 1 AT THE SIDE.**

2 In addition to the symbols $<$ and $>$, several other inequality symbols are often used. Here is a list of all these symbols.

INEQUALITY SYMBOLS

Symbol	Meaning	Example
$\neq$	is not equal to	$3 \neq 7$
$<$	is less than	$-4 < -1$
$>$	is greater than	$3 > -2$
$\leq$	is less than or equal to	$6 \leq 6$
$\geq$	is greater than or equal to	$-8 \geq -10$
$\not<$	is not less than	$5 \not< 2$
$\not>$	is not greater than	$-3 \not> -1$
$\not\leq$	is not less than or equal to	$5 \not\leq 2$
$\not\geq$	is not greater than or equal to	$3 \not\geq 4$

2. Answer *true* or *false*.

(a) $-2 \leq -3$

(b) $8 \leq 8$

(c) $-12 \not< -3$

(d) $-9 \geq -1$

(e) $5 \cdot 8 \leq 7 \cdot 7$

(f) $3(4) > 2(6)$

EXAMPLE 2 Interpreting Inequality Symbols

The following table shows several statements and the reason that each is true.

Statement	Reason
$6 \leq 8$	$6 < 8$
$-2 \leq -2$	$-2 = -2$
$-9 \geq -12$	$-9 > -12$
$-3 \geq -3$	$-3 = -3$
$8 \not> 10$	$8 < 10$
$6 \cdot 4 \leq 5(5)$	$24 < 25$

In the last line, the dot in $6 \cdot 4$ indicates the product 6×4, or 24. Also, $5(5)$ means 5×5, or 25. The statement $6 \cdot 4 \leq 5(5)$ becomes $24 \leq 25$, which is true. ■

■ **WORK PROBLEM 2 AT THE SIDE.**

ANSWERS
1. (a) $<$ (b) $>$ (c) $>$ (d) $<$ (e) $>$
2. (a) false (b) true (c) false
   (d) false (e) true (f) false

3 Inequality symbols can be used to write sets of real numbers. For example, the set of all real numbers greater than -2 can be written as $\{x \mid x > -2\}$. To show the elements of this set on a number line, draw a line from -2 to the right. Place a parenthesis at -2 to show that -2 is not an element of the set. The result, shown in Figure 7, is called the **graph** of the set $\{x \mid x > -2\}$.

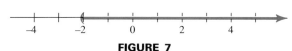

FIGURE 7

The set of numbers greater than -2 is an example of an **interval** on the number line. A simplified notation, called **interval notation,** is used for writing intervals. For example, using this notation, the interval of numbers greater than -2 is written as $(-2, \infty)$. The infinity symbol ∞ does not indicate a number; it is used to show that the interval includes all real numbers greater than -2. The left parenthesis indicates that -2 is not included. A parenthesis is always used next to the infinity symbol in interval notation.

EXAMPLE 3 Graphing an Inequality Written in Interval Notation
Write $\{x \mid x < 4\}$ in interval notation and graph the interval.

In interval notation, $\{x \mid x < 4\}$ is written as $(-\infty, 4)$. The graph is shown in Figure 8. Since the elements of the set are all the numbers *less* than 4, the graph extends to the left. ■

FIGURE 8

WORK PROBLEM 3 AT THE SIDE.

The elements of the set $\{x \mid x \leq -6\}$ are all the real numbers less than or equal to -6. To show that -6 is part of the set, we use a square bracket on the graph at -6, as shown in Figure 9. The square bracket is also used in the interval notation $(-\infty, -6]$ to show that -6 is included in the set.

FIGURE 9

EXAMPLE 4 Graphing an Inequality Written in Interval Notation
Write $\{x \mid x \geq -4\}$ in interval notation and graph the interval.

This interval is written as $[-4, \infty)$. The square bracket indicates that -4 is included in the set. The graph is shown in Figure 10. A square bracket is also used on the graph at -4 to show that -4 is part of the set. ■

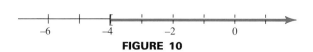

FIGURE 10

WORK PROBLEM 4 AT THE SIDE.

3. Write in interval notation and graph.

   (a) $\{x \mid x < -1\}$

   (b) $\{x \mid x > 0\}$

4. Write in interval notation and graph.

   (a) $\{x \mid x \geq -3\}$

   (b) $\{x \mid x \leq 5\}$

ANSWERS
3. (a) $(-\infty, -1)$
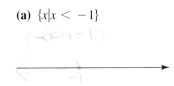
   (b) $(0, \infty)$
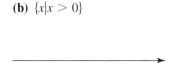
4. (a) $[-3, \infty)$

   (b) $(-\infty, 5]$
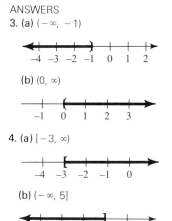

CHAPTER 1 THE REAL NUMBERS

5. Write in interval notation and graph.

(a) $\{x | -1 \leq x \leq 2\}$

(b) $\{x | 8 < x < 12\}$

(c) $\{x | -4 \leq x < 2\}$

4 The next example shows a set whose elements are *between* two numbers.

EXAMPLE 5 Graphing a Three-Part Inequality
Write in interval notation and graph $\{x | -2 < x < 4\}$.

The inequality $-2 < x < 4$ is read "x is greater than -2 and less than 4," or "x is between -2 and 4." The set $\{x | -2 < x < 4\}$ includes all real numbers between but not including -2 and 4. In interval notation, the set is written as $(-2, 4)$. The graph of this set goes from -2 to 4, with parentheses at -2 and 4. See Figure 11. ∎

FIGURE 11

EXAMPLE 6 Graphing a Three-Part Inequality
Write in interval notation and graph $\{x | 3 < x \leq 10\}$.

In interval notation the set is written as $(3, 10]$. As shown in Figure 12, the graph has a parenthesis at 3 and a square bracket at 10. ∎

FIGURE 12

■ WORK PROBLEM 5 AT THE SIDE.

ANSWERS
5. (a) $[-1, 2]$

(b) $(8, 12)$
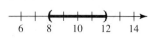

(c) $[-4, 2)$
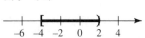

1.2 EXERCISES

Use inequality symbols to rewrite each of the following statements.

1. 6 is less than 10.

2. -4 is less than 2.

3. 2 is greater than x.

4. 5 is greater than $2p$.

5. r is not equal to 4.

6. -6 is not equal to a.

7. $2p + 1$ is less than or equal to 9.

8. -5 is less than or equal to $a - 6$.

9. 3 is greater than or equal to $7y$.

10. 12 is greater than or equal to $8z + 3$.

11. -6 is less than or equal to -6.

12. -8 is greater than or equal to -8.

13. x is between 5 and 9.

14. x is between -4 and 3.

15. $3k$ is between -8 and 4, including -8 and 4.

16. $7y$ is between 9 and 12, including 9 and 12.

17. a is between 1 and 5, including 1 and excluding 5.

18. $4m$ is between -4 and 0, excluding -4 and including 0.

Identify each inequality as true or false. See Example 2.

19. $6 < -12$

20. $2 \geq -5$

21. $4 \leq 4$

22. $-8 \geq -8$

23. $4 \not< 12$

24. $8 \not> 3$

Using your knowledge of arithmetic, first simplify on each side of the following inequalities. Then identify each inequality as true or false.

25. $6 \neq 5 + 2$

26. $4 \neq 8 + 1$

27. $9 \neq 5 + 4$

28. $11 \neq 15 - 4$

29. $-6 \leq 5 + 2$

30. $-12 > 4 + 10$

31. $2 \cdot 4 > 1 + 5$

32. $3 \cdot 8 \leq 10 + 12$

33. $3 \cdot 10 > 6 - 5$

34. $2 \cdot 5 \geq 8 + 2$

35. $8 + 7 \leq 4 \cdot 5$

36. $3 + 2 < 8 - 2$

37. $|-5| \leq |5|$

38. $-|2| \leq |-2|$

39. $4 + |-4| \not> |4|$

40. $-|-6| < 0$

41. $9 \cdot |-6| < 18 + |36|$

42. $|-3| \cdot 9 \geq |-24| + 3$

CHAPTER 1 THE REAL NUMBERS

Write in interval notation and graph each of the following sets of real numbers. See Examples 3–6.

43. $\{x \mid x > -1\}$

44. $\{x \mid x < 3\}$

45. $\{x \mid x \leq 4\}$

46. $\{x \mid x \geq -2\}$

47. $\{x \mid 0 < x < 3\}$

48. $\{x \mid -4 < x < 4\}$

49. $\{x \mid 2 \leq x \leq 5\}$

50. $\{x \mid -3 \leq x \leq -1\}$

51. $\{x \mid -4 < x \leq 2\}$

52. $\{x \mid 3 \leq x < 5\}$

53. $\{x \mid 0 < x \leq 2\}$

54. $\{x \mid -1 \leq x < 3\}$

55. Suppose $x^2 \leq 16$. Is it always true, then, that $x \leq 4$?

56. If $x^2 \geq 16$, must it be true that $x \geq 4$?

57. For what values of x is $\dfrac{1}{x} < x$? Assume $x \neq 0$.

58. For what values of x is $x < x^2$?

1.3 OPERATIONS ON REAL NUMBERS

In this section we will review the rules for the four operations with signed numbers: addition, subtraction, multiplication, and division.

1 To find the sum of two real numbers, we add them using the following rules.

To add two numbers with the *same* sign, first add their absolute values. The sign of the answer (either + or −) is the same as the sign of the two numbers.

To add two numbers with *different* signs, first subtract their absolute values. The answer is positive if the positive number has the larger absolute value. The answer is negative if the negative number has the larger absolute value.

OBJECTIVES

1. Add signed numbers.
2. Subtract signed numbers.
3. Multiply signed numbers.
4. Find the reciprocal of a number.
5. Divide signed numbers.

EXAMPLE 1 Adding Two Negative Numbers
Find the following sums.

(a) $-12 + (-8)$

First find the absolute values.

$$|-12| = 12 \quad \text{and} \quad |-8| = 8$$

Since these numbers have the same sign, add their absolute values. Both numbers are negative, so the answer is negative.

$$-12 + (-8) = -(12 + 8)$$
$$= -(20) = -20$$

(b) $-6 + (-3) = -(6 + 3) = -9$

(c) $-12 + (-4) = -(12 + 4) = -16$ ∎

WORK PROBLEM 1 AT THE SIDE.

EXAMPLE 2 Adding Numbers with Different Signs
Add -17 and 11.

To find $-17 + 11$, first find the absolute value of each number:

$$|-17| = 17 \quad \text{and} \quad |11| = 11.$$

Since -17 and 11 have *different* signs, subtract these absolute values:

$$17 - 11 = 6.$$

The number -17 has a larger absolute value than 11 has, so the answer should be negative.

$$-17 + 11 = -6$$
↑
Negative, since $|-17| > |11|$ ∎

1. Add.

(a) $-4 + (-2)$

(b) $-2 + (-7)$

(c) $-15 + (-6)$

(d) $-8 + (-9)$

(e) $-11 + (-12)$

ANSWERS
1. (a) -6 (b) -9 (c) -21
   (d) -17 (e) -23

CHAPTER 1 THE REAL NUMBERS

2. Add.

(a) $12 + (-1)$

(b) $3 + (-7)$

(c) $-6 + 9$

(d) $-8 + 2$

(e) $-17 + 5$

ANSWERS
2. (a) 11 (b) -4 (c) 3 (d) -6
   (e) -12

EXAMPLE 3 Adding Numbers with Different Signs
Find the following sums.

(a) $4 + (-1)$

The absolute values are 4 and 1. Since 4 is positive, the sum must be positive.
$$4 + (-1) = 4 - 1 = 3$$

(b) $-9 + 17 = 17 - 9 = 8$

(c) $-16 + 12 = -(16 - 12) = -4$ ∎

WORK PROBLEM 2 AT THE SIDE.

It is important to be able to add numbers mentally and quickly.

2 The result of subtraction is called the **difference.** Thus, the difference between 7 and 5 is 2. To see how subtraction should be defined, compare the two statements shown below.

$$7 - 5 = 2$$
$$7 + (-5) = 2$$

Also notice that

$$9 - 3 = 6 \quad \text{and} \quad 9 + (-3) = 6$$

so that

$$9 - 3 = 9 + (-3).$$

These examples suggest the following rule for subtraction of signed numbers.

For all real numbers a and b,
$$a - b = a + (-b).$$
(Change the sign of the second number and add.)

This method of observing patterns and similarities and generalizing from them is used often in mathematics. Looking at many examples strengthens our confidence in such generalizations; if possible, though, mathematicians prefer to prove the results using previously established facts.

EXAMPLE 4 Subtracting Signed Numbers
Find each difference.

(a) $6 - 8 = 6 + (-8) = -2$

(Change to addition. Change sign of second number.)

1.3 OPERATIONS ON REAL NUMBERS

(b) $-12 - 4 = -12 + (-4) = -16$

(Changed / Sign changed)

(c) $-10 - (-7) = -10 + [-(-7)]$ This step is often omitted.
$$= -10 + 7$$
$$= -3$$

(d) $-2 - (-8) = -2 + [-(-8)] = -2 + 8 = 6$

(e) $8 - (-5) = 8 + [-(-5)] = 8 + 5 = 13$ ■

WORK PROBLEM 3 AT THE SIDE.

When working a problem that involves both addition and subtraction, perform the additions and subtractions in order from left to right, as in the following example. Do not forget to work inside the brackets or parentheses first.

EXAMPLE 5 Adding and Subtracting Signed Numbers
Perform the indicated operations.

(a) $-8 + 5 - 6 = (-8 + 5) - 6$
$$= -3 - 6$$
$$= -3 + (-6) = -9$$

(b) $15 - (-3) - 5 - 12 = (15 + 3) - 5 - 12$
$$= 18 - 5 - 12$$
$$= 13 - 12 = 1$$

(c) $-4 - (-6) + 7 - 1 = (-4 + 6) + 7 - 1$
$$= 2 + 7 - 1$$
$$= 9 - 1 = 8$$

(d) $-9 - [-8 - (-4)] + 6 = -9 - [-8 + 4] + 6$
$$= -9 - [-4] + 6$$
$$= -9 + 4 + 6$$
$$= -5 + 6 = 1$$ ■

WORK PROBLEM 4 AT THE SIDE.

3 A **product** is the result of multiplying two or more numbers. For example, 24 is the product of 8 and 3. The rules for finding products of signed numbers are given below.

The product of two numbers with the *same* sign is positive.
The product of two numbers with *different* signs is negative.

3. Subtract.

(a) $9 - 12$

(b) $-7 - 2$

(c) $-8 - (-2)$

(d) $-6 - (-11)$

(e) $12 - (-5)$

4. Simplify.

(a) $-6 + 9 - 2$

(b) $12 - (-4) + 8$

(c) $-6 - (-2) - 8 - 1$

(d) $-3 - (-7) + 15 + 6$

ANSWERS
3. (a) -3 (b) -9 (c) -6 (d) 5
   (e) 17
4. (a) 1 (b) 24 (c) -13 (d) 25

CHAPTER 1 THE REAL NUMBERS

5. Multiply.

(a) $(-7)(-5)$

(b) $(-9)(-15)$

(c) $\left(-\dfrac{4}{7}\right)\left(-\dfrac{14}{3}\right)$

(d) $7(-2)$

(e) $(-8)6$

(f) $\dfrac{5}{8}(-16)$

(g) $\left(-\dfrac{2}{3}\right)(12)$

ANSWERS
5. (a) 35 (b) 135 (c) $\dfrac{8}{3}$
   (d) -14 (e) -48
   (f) -10 (g) -8

EXAMPLE 6 Finding Products of Signed Numbers
Find each product.

(a) $(-3)(-9) = 27$

(b) $(-5)(-4) = 20$

(c) $\left(-\dfrac{3}{4}\right)\left(-\dfrac{5}{3}\right) = \dfrac{5}{4}$

(d) $6(-9) = -54$

(e) $(-5)3 = -15$

(f) $\dfrac{2}{3}(-3) = -2$

(g) $-\dfrac{5}{8}\left(\dfrac{12}{13}\right) = -\dfrac{15}{26}$ ■

■ WORK PROBLEM 5 AT THE SIDE.

4 Earlier, subtraction was defined in terms of addition. Now division is defined in terms of multiplication. This definition depends on the idea of reciprocals; two numbers are **reciprocals** if they have a product of 1.

The **reciprocal** of a nonzero number a is $\dfrac{1}{a}$.

EXAMPLE 7 Finding the Reciprocal of a Number
The following chart gives several numbers and their reciprocals.

Number	Reciprocal
3	$\dfrac{1}{3}$
$-\dfrac{2}{5}$	$-\dfrac{5}{2}$
-6	$-\dfrac{1}{6}$
$\dfrac{7}{11}$	$\dfrac{11}{7}$
0	None

There is no reciprocal for 0, since there is no number that can be multiplied by 0 to give a product of 1. ■

1.3 OPERATIONS ON REAL NUMBERS

WORK PROBLEM 6 AT THE SIDE.

6. Give the reciprocal of each number.

(a) 15

(b) -7

(c) $\dfrac{8}{9}$

CAUTION A number and its additive inverse have *opposite* signs; however, a number and its reciprocal always have the same sign.

5 The result of dividing one number by another is called the **quotient.** For example, when 45 is divided by 3, the quotient is 15. Now division of signed numbers can be defined. To see how, first write 15 as $\dfrac{45}{3}$, the quotient of 45 and 3. The same answer will be obtained if 45 and $\dfrac{1}{3}$ are multiplied, as follows.

$$\dfrac{45}{3} = 45 \cdot \dfrac{1}{3}$$
$$= 15$$

Looking at many other examples of this type suggests the definition of division of signed numbers.

For all real numbers a and b (where $b \neq 0$),

$$a \div b = \dfrac{a}{b} = a \cdot \dfrac{1}{b}.$$

(Multiply the first number by the reciprocal of the second number.)

NOTE There is no reciprocal for the number 0, so division by 0 is not defined. (If a problem in this book has a zero denominator, the answer will say "undefined.")

Since division is defined as multiplication by the reciprocal, the rules for quotients resemble those for products.

The quotient of two nonzero real numbers with the same sign is positive.

The quotient of two nonzero real numbers with different signs is negative.

ANSWERS

6. (a) $\dfrac{1}{15}$ (b) $-\dfrac{1}{7}$ (c) $\dfrac{9}{8}$

CHAPTER 1 THE REAL NUMBERS

7. Divide.

(a) $\dfrac{-16}{4}$

(b) $\dfrac{-120}{10}$

(c) $\dfrac{8}{-2}$

(d) $\dfrac{-12}{-6}$

(e) $\dfrac{-15}{-3}$

(f) $\dfrac{\frac{3}{8}}{\frac{11}{16}}$

ANSWERS
7. (a) -4 (b) -12 (c) -4 (d) 2
 (e) 5 (f) $\dfrac{6}{11}$

EXAMPLE 8 Finding the Quotient of Two Signed Numbers
Find each quotient.

(a) $\dfrac{-12}{4} = -12 \cdot \dfrac{1}{4} = -3$

(b) $\dfrac{6}{-3} = 6\left(-\dfrac{1}{3}\right) = -2$

(c) $\dfrac{-30}{-2} = -30\left(-\dfrac{1}{2}\right) = 15$

(d) $\dfrac{\frac{2}{3}}{\frac{5}{9}} = \dfrac{2}{3} \cdot \dfrac{1}{\frac{5}{9}} = \dfrac{2}{3} \cdot \dfrac{9}{5} = \dfrac{6}{5}$ ∎

Recall that the product of a number and its reciprocal is 1. The reciprocal of $\dfrac{5}{9}$, written

$$\dfrac{1}{\frac{5}{9}} \quad \text{or} \quad 1 \div \dfrac{5}{9}$$

is equal to $\dfrac{9}{5}$ since

$$\dfrac{5}{9} \cdot \dfrac{9}{5} = 1.$$

This example shows the origin of the usual rule for dividing by a fraction: "invert the divisor fraction and multiply."

■ **WORK PROBLEM 7 AT THE SIDE.**

The rules for multiplication and division suggest the results given below.

The fractions $\dfrac{-x}{y}$, $-\dfrac{x}{y}$, and $\dfrac{x}{-y}$ are equal.

Also, the fractions $\dfrac{x}{y}$ and $\dfrac{-x}{-y}$ are equal. (Assume $y \neq 0$.)

The forms $\dfrac{x}{-y}$ and $\dfrac{-x}{-y}$ are not used very often.

Every fraction has three signs: the sign of the numerator, the sign of the denominator, and the sign of the fraction itself. As shown above, changing any two of these three signs does not change the value of the fraction. (Changing only one sign, or changing all three, does change the value.)

name   date   hour

1.3 EXERCISES

Add or subtract as indicated. See Examples 1–5.

1. $12 + (-3)$
2. $6 + (-2)$
3. $-8 + 3$
4. $-9 + 7$

5. $-12 + (-8)$
6. $-5 + (-2)$
7. $9 + (-15)$
8. $\dfrac{5}{8} + \left(-\dfrac{2}{3}\right)$

9. $-\dfrac{7}{3} + \dfrac{3}{4}$
10. $-\dfrac{5}{6} + \dfrac{3}{8}$
11. $13 - 15$
12. $-6 - 17$

13. $-3 - 9$
14. $-8 - (-3)$
15. $-12 - (-1)$
16. $-3 - (-14)$

17. $-\dfrac{3}{8} - \left(-\dfrac{1}{6}\right)$
18. $\dfrac{9}{10} - \left(-\dfrac{4}{3}\right)$
19. $-\dfrac{3}{4} - \dfrac{1}{3}$
20. $\dfrac{7}{5} - \dfrac{9}{2}$

21. $-\dfrac{9}{7} + \dfrac{3}{4}$
22. $\dfrac{3}{14} - \left(-\dfrac{1}{4}\right)$
23. $8.9 - 3.3$
24. $-7.6 + 9.8$

25. $-18.31 - 7.51$*
26. $14.72 - 8.25$
27. $-12.3 + (-45.3)$
28. $-2 + |-3|$

29. $|-5| + 8$
30. $12 + |-5|$
31. $-4 - |-8|$
32. $|-19| - |-12|$

33. $|-20| + |-9|$
34. $-|-4| + |-3|$
35. $-6 + |-13|$
36. $-11 + |-40|$

---

*Exercises that can be solved more easily using a calculator are identified with colored exercise numbers (as well as a calculator symbol next to the first exercise in each group) throughout the book.

CHAPTER 1 THE REAL NUMBERS

Multiply. See Example 6.

37. $6(-7)$ **38.** $5(-9)$ **39.** $3(-11)$ **40.** $-8(3)$

41. $-9(7)$ **42.** $(-12)(-2)$ **43.** $(-2)(-4)$ **44.** $(-7)(-8)$

45. $(-3)(-12)(6)$ **46.** $5(-4)(-2)$ **47.** $7(-4)(6)$ **48.** $8(-5)(3)$

49. $-8\left(\dfrac{1}{4}\right)$ **50.** $-\dfrac{1}{2}(12)$ **51.** $\dfrac{2}{3}(-9)$ **52.** $\dfrac{3}{4}(-16)$

53. $\dfrac{5}{8}(-40)$ **54.** $\dfrac{9}{16}(-80)$ **55.** $\dfrac{3}{5}(-20)$ **56.** $-\dfrac{9}{8}(-24)$

57. $-\dfrac{5}{2}\left(-\dfrac{12}{25}\right)$ **58.** $-\dfrac{9}{7}\left(-\dfrac{35}{36}\right)$ **59.** $-\dfrac{3}{8}\left(-\dfrac{24}{9}\right)$ **60.** $-\dfrac{2}{11}\left(-\dfrac{99}{4}\right)$

61. $-4.7(3.98)$ **62.** $(-11.8)(-2.1)$ **63.** $(7.69)(-.14)$

1.3 EXERCISES

Give the reciprocal of each number. See Example 7.

64. 2

65. -8

66. -7

67. $\dfrac{5}{8}$

68. $-\dfrac{9}{19}$

69. $-\dfrac{4}{13}$

Divide where possible. See Example 8.

70. $\dfrac{-12}{2}$

71. $\dfrac{-8}{4}$

72. $\dfrac{-18}{-3}$

73. $\dfrac{18}{-2}$

74. $\dfrac{36}{-4}$

75. $\dfrac{-120}{-10}$

76. $\dfrac{-480}{-24}$

77. $\dfrac{-360}{-18}$

78. $\dfrac{100}{-4}$

79. $\dfrac{9}{0}$

80. $\dfrac{-15}{0}$

81. $-\dfrac{3}{4} \div \dfrac{6}{7}$

82. $-\dfrac{10}{17} \div \left(-\dfrac{12}{5}\right)$

83. $-\dfrac{22}{23} \div \left(-\dfrac{33}{4}\right)$

84. $\dfrac{\dfrac{7}{10}}{-\dfrac{8}{15}}$

85. $\dfrac{-1.71}{.9}$

86. $\dfrac{-.96}{.8}$

87. $\dfrac{.171}{-1.9}$

By replacing a, b, and c with various integers, decide whether the statements below are always true. *If they are not, mark them* false. *For each false statement, give an example showing that it is false.*

88. $a - b = b - a$

89. $a - (b - c) = (a - b) - c$

90. $|a + b| = |a| + |b|$

91. $|a - b| = |a| - |b|$

1.4 EXPONENTS AND ROOTS; ORDER OF OPERATIONS

Two or more integers whose product is a third number are **factors** of that third number. For example, 2 and 6 are factors of 12 since $2 \cdot 6 = 12$. Other factors of 12 are 1, 3, 4, 12, -1, -2, -3, -4, -6, and -12.

OBJECTIVES

1. Use exponents.
2. Identify exponents and bases.
3. Find square roots and higher roots.
4. Use the order of operations.
5. Substitute numbers for variables.

1 In algebra, *exponents* are used as a way of writing products of repeated factors. For example, the product $2 \cdot 2 \cdot 2 \cdot 2 \cdot 2$ is written

$$2 \cdot 2 \cdot 2 \cdot 2 \cdot 2 = \mathbf{2^5}.$$

The number 5 shows that 2 appears as a factor 5 times. The number 5 is the **exponent,** and 2 is the **base.**

2^5 ← Exponent
↑— Base

Multiplying out the five 2s gives

$$2^5 = 2 \cdot 2 \cdot 2 \cdot 2 \cdot 2 = 32.$$

If a is a real number and n is a natural number,

$$a^n = \underbrace{a \cdot a \cdot a \cdots a}_{n \text{ factors of } a}$$

where n is the **exponent**, a is the **base**, and a^n is an **exponential** or a **power**.

EXAMPLE 1 Evaluating an Exponential
Write each number without exponents.

(a) $5^2 = 5 \cdot 5 = 25$

Read 5^2 as "5 squared."

(b) $9^3 = 9 \cdot 9 \cdot 9 = 729$

Read 9^3 as "9 cubed."

(c) $2^6 = 2 \cdot 2 \cdot 2 \cdot 2 \cdot 2 \cdot 2 = 64$

Read 2^6 as "2 to the sixth."

(d) $(-2)^4 = (-2)(-2)(-2)(-2) = 16$

(e) $(-3)^5 = (-3)(-3)(-3)(-3)(-3) = -243$ ∎

Parts (d) and (e) of Example 1 suggest the following generalization.

The product of an *even* number of negative factors is positive.

The product of an *odd* number of negative factors is negative.

CHAPTER 1 THE REAL NUMBERS

1. Write without exponents.

   (a) 2^3

   (b) 5^3

   (c) 3^4

   (d) $(-4)^5$

   (e) $(-3)^3$

   (f) $\left(\dfrac{3}{4}\right)^3$

   (g) $\left(\dfrac{2}{5}\right)^4$

2. Identify the exponent and the base in each exponential.

   (a) 7^5

   (b) m^3

   (c) $(-5)^7$

   (d) -12^4

   (e) $(-3)^2$

   (f) -3^2

ANSWERS
1. (a) 8 (b) 125 (c) 81 (d) -1024
   (e) -27 (f) $\dfrac{27}{64}$ (g) $\dfrac{16}{625}$
2. (a) 5; 7 (b) 3; m (c) 7; -5
   (d) 4; 12 (e) 2; -3 (f) 2; 3

In parts (a) and (b) of Example 1, we used the terms "squared" and "cubed" to refer to powers of 2 and 3, respectively. The term "squared" comes from the figure of a square, which has the same measure for both length and width, as shown in Figure 13(a). Similarly, the term "cubed" comes from the figure of a cube. As shown in Figure 13(b), the length, width, and height of a cube have the same measure.

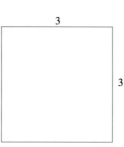
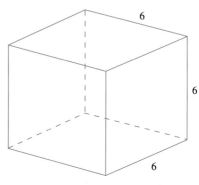

(a) $3 \cdot 3 = 3$ squared, or 3^2 (b) $6 \cdot 6 \cdot 6 = 6$ cubed, or 6^3

FIGURE 13

■ WORK PROBLEM 1 AT THE SIDE.

The following example shows how to identify exponents and bases.

EXAMPLE 2 Identifying Exponents and Bases
Identify the base and the exponent.

(a) 3^6

The base is 3 and the exponent is 6.

(b) 5^4

The base is 5 and the exponent is 4.

(c) $(-2)^6$

The exponent of 6 refers to the number -2, so the base is -2. Evaluating $(-2)^6$ gives

$$(-2)^6 = (-2)(-2)(-2)(-2)(-2)(-2) = 64.$$

(d) -2^6

Here the lack of parentheses shows that the exponent 6 refers *only* to the number 2, and not to -2, so the base is 2. Evaluate -2^6 as follows:

$$-2^6 = -(2 \cdot 2 \cdot 2 \cdot 2 \cdot 2 \cdot 2) = -64. \blacksquare$$

CAUTION As shown by parts (c) and (d) of Example 2, it is important to be careful to distinguish between $-a^n$ and $(-a)^n$.

■ WORK PROBLEM 2 AT THE SIDE.

1.4 EXPONENTS AND ROOTS; ORDER OF OPERATIONS

3 We know that $5^2 = 5 \cdot 5 = 25$, so that 5 squared is 25. The opposite of squaring a number is called taking its **square root**. For example, a square root of 25 is 5. Another square root of 25 is -5, since $(-5)^2 = 25$; thus, 25 has two square roots, 5 and -5.

Write the positive square root of a number with the symbol $\sqrt{}$. For example, the positive square root of 25 is written $\sqrt{25} = 5$. The negative square root of 25 is written $-\sqrt{25} = -5$. Since the square of any real number is positive, the square root of a negative number, such as $\sqrt{-4}$, is not a real number.

EXAMPLE 3 Finding Square Roots
Find each square root that is a real number.

(a) $\sqrt{36} = 6$ since 6 is positive and $6^2 = 36$.

(b) $\sqrt{144} = 12$ since $12^2 = 144$.

(c) $\sqrt{4} = 2$

(d) $-\sqrt{100} = -10$

(e) $\sqrt{0} = 0$ since $0^2 = 0$.

(f) $\sqrt{\dfrac{9}{16}} = \dfrac{3}{4}$

(g) $\sqrt{-16}$ is not a real number. ■

CAUTION The symbol $\sqrt{}$ is used only for the *positive* square root, except that $\sqrt{0} = 0$.

WORK PROBLEM 3 AT THE SIDE.

Since 6 cubed is $6^3 = 6 \cdot 6 \cdot 6 = 216$, the **cube root** of 216 is 6. Write this as
$$\sqrt[3]{216} = 6.$$
In the same way, the **fourth root** of 81 is 3, written
$$\sqrt[4]{81} = 3.$$
The number -3 is also a fourth root of 81, but the symbol $\sqrt[4]{}$ is reserved for roots that are not negative. We discuss negative roots in Chapter 5.

EXAMPLE 4 Finding Higher Roots
Find each root that is a real number.

(a) $\sqrt[3]{27} = 3$ since $3^3 = 27$.

(b) $\sqrt[3]{125} = 5$ since $5^3 = 125$.

(c) $\sqrt[3]{8} = 2$ since $2^3 = 8$.

(d) $\sqrt[4]{16} = 2$ since $2^4 = 16$.

(e) $\sqrt[5]{32} = 2$ since $2^5 = 32$.

(f) $\sqrt[7]{128} = 2$ since $2^7 = 128$.

---

3. Find the following.

(a) $\sqrt{9}$

(b) $\sqrt{49}$

(c) $-\sqrt{81}$

(d) $-\sqrt{169}$

(e) $\sqrt{\dfrac{121}{81}}$

(f) $\sqrt{-9}$

ANSWERS
3. (a) 3 (b) 7 (c) -9 (d) -13
   (e) $\dfrac{11}{9}$ (f) not a real number

CHAPTER 1 THE REAL NUMBERS

4. Find each root.

(a) $\sqrt[3]{64}$

(b) $\sqrt[3]{\dfrac{27}{8}}$

(c) $\sqrt[3]{0}$

(d) $\sqrt[4]{625}$

(e) $\sqrt[4]{\dfrac{16}{625}}$

(f) $\sqrt[4]{-16}$

ANSWERS

4. (a) 4 (b) $\dfrac{3}{2}$ (c) 0 (d) 5 (e) $\dfrac{2}{5}$
   (f) not a real number

(g) $\sqrt[4]{-81}$ is not a real number, since there is no real number whose fourth power is negative. ■

WORK PROBLEM 4 AT THE SIDE.

For reference, we now give a portion of a table of squares and a table of selected powers of numbers, both of which will be helpful in the exercises for this section.

TABLE OF SQUARES (PORTION)

$11^2 = 121$	$16^2 = 256$	$21^2 = 441$	$26^2 = 676$
$12^2 = 144$	$17^2 = 289$	$22^2 = 484$	$27^2 = 729$
$13^2 = 169$	$18^2 = 324$	$23^2 = 529$	$28^2 = 784$
$14^2 = 196$	$19^2 = 361$	$24^2 = 576$	$29^2 = 841$
$15^2 = 225$	$20^2 = 400$	$25^2 = 625$	$30^2 = 900$

SELECTED POWERS OF NUMBERS

Number n	Cube n^3	Fourth power n^4	Fifth power n^5	Sixth power n^6
2	8	16	32	64
3	27	81	243	729
4	64	256	1024	4096
5	125	625	3125	15,625
6	216	1296	7776	46,656
7	343	2401	16,807	117,649
8	512	4096	32,768	262,144
9	729	6561	59,049	531,441
10	1000	10,000	100,000	1,000,000
11	1331	14,641	161,051	1,771,561
12	1728	20,736	248,832	2,985,984

4 Given a problem such as $5 + 2 \cdot 3$, should 5 and 2 be added first, or should 2 and 3 be multiplied first? When a problem involves more than one operation symbol, use the following **order of operations.**

ORDER OF OPERATIONS

1. Work separately above and below any fraction bar.
2. Use the rules below within each set of parentheses or square brackets. Start with the innermost set and work outward.
3. Evaluate all powers and roots.
4. Do any multiplications or divisions in the order in which they occur, working from left to right.
5. Do any additions or subtractions in the order in which they occur, working from left to right.

1.4 EXPONENTS AND ROOTS; ORDER OF OPERATIONS

EXAMPLE 5 Using Order of Operations
To simplify $5 + 2 \cdot 3$, first multiply and then add.

$$5 + 2 \cdot 3 = 5 + 6 \quad \text{Multiply.}$$
$$= 11 \quad \text{Add.} \blacksquare$$

WORK PROBLEM 5 AT THE SIDE.

EXAMPLE 6 Using Order of Operations
Simplify $4 \cdot 3^2 + 7 - (2 + 8)$.
   Work inside the parentheses.

$$4 \cdot 3^2 + 7 - (2 + 8) = 4 \cdot 3^2 + 7 - 10$$

Simplify powers and roots. Since $3^2 = 3 \cdot 3 = 9$,

$$4 \cdot 3^2 + 7 - 10 = 4 \cdot 9 + 7 - 10.$$

Do all multiplications or divisions, working from left to right.

$$4 \cdot 9 + 7 - 10 = 36 + 7 - 10 \quad \text{Multiply.}$$

Finally, do all additions or subtractions, working from left to right.

$$36 + 7 - 10 = 43 - 10 \quad \text{Add.}$$
$$= 33 \quad \text{Subtract.} \blacksquare$$

WORK PROBLEM 6 AT THE SIDE.

EXAMPLE 7 Using Order of Operations
Simplify $\frac{1}{2} \cdot 4 + (6 \div 3 - 7)$.
   Work inside the parentheses, doing the division before the subtraction.

$$\frac{1}{2} \cdot 4 + (6 \div 3 - 7) = \frac{1}{2} \cdot 4 + (2 - 7) \quad \text{Divide.}$$
$$= \frac{1}{2} \cdot 4 + (-5) \quad \text{Subtract.}$$
$$= 2 + (-5) \quad \text{Multiply.}$$
$$= -3 \quad \text{Add.} \blacksquare$$

WORK PROBLEM 7 AT THE SIDE.

EXAMPLE 8 Using Order of Operations

Simplify $\dfrac{\frac{2}{3} \cdot 3 + 1}{\frac{1}{2} \cdot 8 - 6}$.

   Go through the steps given above. Work above and below the fraction bar separately.

$$\frac{\frac{2}{3} \cdot 3 + 1}{\frac{1}{2} \cdot 8 - 6} = \frac{2 + 1}{4 - 6} \quad \text{Multiply.}$$
$$= \frac{3}{-2} \quad \text{Add and subtract.}$$
$$= -\frac{3}{2} \quad \blacksquare$$

5. Simplify.

(a) $4 + 6 \cdot 9$

(b) $5 \cdot 9 + 2 \cdot 4$

6. Simplify.

(a) $9 \cdot 6 - 4 + (2 + 3)$

(b) $(4 + 2) - 3^2 - (8 - 3)$

7. Simplify.

(a) $6 + \dfrac{2}{3}(-9) - \dfrac{5}{8} \cdot 16$

(b) $-\dfrac{4}{7}(-14) - 6\left(-\dfrac{2}{3}\right) + \dfrac{1}{2}(-12)$

ANSWERS
5. (a) 58 (b) 53
6. (a) 55 (b) −8
7. (a) −10 (b) 6

CHAPTER 1 THE REAL NUMBERS

8. Simplify.

(a) $\dfrac{\frac{1}{2} \cdot 10 - 6 + \sqrt{9}}{\frac{5}{6} \cdot 12 - 3(2)^2}$

(b) $\dfrac{4 - 3^2 \cdot 2 + 7 \cdot 6}{(-6)(-2) - 3 \cdot \sqrt{16}}$

9. Evaluate if $x = -12$, $y = 64$, and $z = -3$.

(a) $5x - 2 \cdot \sqrt{y}$

(b) $-6(x - \sqrt[3]{y})$

(c) $\dfrac{5x - 3 \cdot \sqrt{y}}{x - 1}$

(d) $x^2 + 2z^3$

EXAMPLE 9 Using Order of Operations

Simplify $\dfrac{5 + 2^4}{6\sqrt{9} - 9 \cdot 2}$.

$$\dfrac{5 + 2^4}{6\sqrt{9} - 9 \cdot 2} = \dfrac{5 + 16}{6 \cdot 3 - 9 \cdot 2} \quad \text{Evaluate powers and roots.}$$

$$= \dfrac{5 + 16}{18 - 18} \quad \text{Multiply.}$$

$$= \dfrac{21}{0} \quad \text{Add and subtract.}$$

Since division by zero is undefined, the given expression is undefined. ■

WORK PROBLEM 8 AT THE SIDE.

5 The final example in this section shows how to evaluate expressions for given values of the variables.

EXAMPLE 10 Evaluating Expressions for Given Values of the Variables

Evaluate the following expressions if $m = -4$, $n = 5$, $p = -6$, and $q = \sqrt{25}$.

(a) $5m - 9n$

Replace m with -4 and n with 5.

$$5m - 9n = 5(-4) - 9(5) = -20 - 45 = -65$$

(b) $-2p + 7n = -2(-6) + 7(5) = 47$

(c) $\dfrac{m + 2n}{4p} = \dfrac{-4 + 2(5)}{4(-6)} = \dfrac{-4 + 10}{-24} = \dfrac{6}{-24} = -\dfrac{1}{4}$

(d) $-3m^3 - n^2q$

$$-3m^3 - n^2q = -3(-4)^3 - (5)^2(\sqrt{25}) \quad \text{Substitute.}$$

$$= -3(-64) - 25(5) \quad \text{Evaluate powers and roots.}$$

$$= 192 - 125 \quad \text{Multiply.}$$

$$= 67 \quad \text{■ Subtract.}$$

WORK PROBLEM 9 AT THE SIDE.

ANSWERS
8. (a) -1 (b) undefined
9. (a) -76 (b) 96 (c) $\dfrac{84}{13}$ (d) 90

30

1.4 EXERCISES

Evaluate. See Example 1.

1. 2^3
2. 4^2
3. 10^3
4. 3^5
5. $\left(\dfrac{1}{5}\right)^2$

6. $\left(\dfrac{2}{3}\right)^2$
7. $\left(\dfrac{9}{10}\right)^3$
8. $\left(\dfrac{5}{8}\right)^4$
9. $(-3)^3$
10. $(-5)^5$

11. $(-2)^8$
12. $(-4)^2$
13. -2^3
14. -3^2
15. -5^3

16. -4^4
17. $(.4)^4$
18. $(1.2)^5$
19. $(-3.04)^2$
20. $(-.72)^3$

Identify the exponent and the base. Do not evaluate. See Example 2.

	Exponent	Base		Exponent	Base		Exponent	Base
21. 5^7			22. 9^4			23. -12^5		
24. -7^8			25. $(-9)^4$			26. $(-2)^{11}$		

Find the given square roots. See Example 3.

27. $\sqrt{256}$
28. $\sqrt{529}$
29. $\sqrt{784}$
30. $\sqrt{121}$

31. $\sqrt{324}$
32. $\sqrt{900}$
33. $\sqrt{361}$
34. $\sqrt{841}$

CHAPTER 1 THE REAL NUMBERS

Use the table of powers in the text to find the given roots. See Example 4.

35. $\sqrt[3]{125}$

36. $\sqrt[3]{216}$

37. $\sqrt[3]{1000}$

38. $\sqrt[3]{1728}$

39. $\sqrt[3]{27}$

40. $\sqrt[3]{343}$

41. $\sqrt[4]{625}$

42. $\sqrt[4]{10,000}$

43. $\sqrt[5]{243}$

44. $\sqrt[5]{1024}$

45. $\sqrt[6]{729}$

46. $\sqrt[6]{4096}$

Simplify, using the order of operations given in the text. See Examples 5 and 6.

47. $-6[2 + (-5)]$

48. $-8[4 + (-12)]$

49. $2[-5 - (-7)]$

50. $5(-6) + 7(-3)$

51. $-12\left(-\dfrac{3}{4}\right) - (-5)$

52. $-7\left(\dfrac{6}{14}\right) - (-8)$

53. $-6 - 5(-8) + 3^2$

54. $(-9)(2)^2 - (-3)(-2)$

55. $(-7)(\sqrt{36}) - (-2)(-3)$

56. $(-6 - 3)(-2 - \sqrt{9})$

57. $(-8 - 5)(-\sqrt[3]{8} - 1)$

58. $(-3 - 1)(-2 + 5) - 3^2$

1.4 EXERCISES

Perform the given operations. See Examples 5–9.

59. $12 \div 3 \cdot 4 \div 8$

60. $25 \div 5 \cdot 2 \cdot 3 \div 5$

61. $\dfrac{-8 + (-7)}{-3}$

62. $\dfrac{-9 + (-11)}{-4^2}$

63. $\dfrac{(-6 + 3)(-2^2)}{-5 - 1}$

64. $\dfrac{-4[\sqrt{25} + (-6)]}{2^3 - 6}$

65. $\dfrac{2(-5) + (-3)(-4)}{-6 + 5 + 1}$

66. $\dfrac{3(-4) + (-5)(-2)}{8 - 2 - 6}$

67. $\dfrac{4(-3) - (2^3)(-5)}{5 - \sqrt{16} + 3}$

68. $\dfrac{-3(\sqrt{4}) + (-3^2)5}{9 - 2 - 4}$

69. $\dfrac{-3(-4) - 5(2^2)}{2 - 4 + 10}$

70. $\dfrac{5(-1) + 3(\sqrt{4})}{\sqrt{9} - 1 - 3}$

71. $-\dfrac{4}{5}[6(-4) + (-5)(-5)]$

72. $-\dfrac{1}{4}[3(-5) + 7(-5) + 1(-2)]$

73. $5\left[1 + \dfrac{3}{4}(-12) - 8 \cdot \dfrac{3}{2}\right]$

74. $-7\left[6 - \dfrac{5}{8}(24) + 3\left(\dfrac{8}{3}\right)\right]$

Evaluate if a = −4, b = 8, and c = −7. See Example 10.

75. $8a + 9b$

76. $3c - 2a$

77. $-2(a^2 + 4c)$

78. $\dfrac{3b - c^2}{a}$

79. $\dfrac{2c + a^3}{4b + 6a}$

80. $\dfrac{6(\sqrt[3]{b}) - a}{7a - 4c}$

81. $-6a^2 + b$

82. $2c^3 - a^2$

83. $-a^2 + b^3$

84. Replace x with 5 to show that $2 + 6x \neq 8x$. What order of operations rule does this illustrate?

85. Replace x with 5 to show that $4x - x \neq 4$. What order of operations rule does this illustrate?

1.5 PROPERTIES OF REAL NUMBERS

The study of any object is simplified when we know the properties of the object. The property that water boils when heated to 100°C helps us to predict the behavior of water. The study of numbers is no different. The basic properties of addition and multiplication of real numbers studied in this section will be used in later work in algebra. The properties are results that have been observed to occur consistently in work with numbers, so they have been generalized to apply to expressions with variables as well.

OBJECTIVES

1. Use the distributive property.
2. Use the inverse properties.
3. Use the identity properties.
4. Use the commutative and associative properties.
5. Use the multiplication property of zero.

1 The properties we discuss in this section are used in simplifying algebraic expressions. For example, notice that

$$2(3 + 5) = 2 \cdot 8 = \boxed{16}$$

and

$$2 \cdot 3 + 2 \cdot 5 = 6 + 10 = \boxed{16}$$

so that

$$2(3 + 5) = 2 \cdot 3 + 2 \cdot 5.$$

This idea is illustrated by the divided rectangle in Figure 14. Similarly,

$$-4[5 + (-3)] = -4(2) = \boxed{-8}$$

and

$$-4(5) + (-4)(-3) = -20 + 12 = \boxed{-8}$$

so

$$-4[5 + (-3)] = -4(5) + (-4)(-3).$$

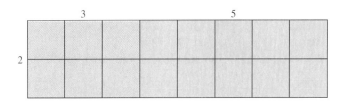

Area of left part is $2 \cdot 3 = 6$.
Area of right part is $2 \cdot 5 = 10$.
Area of total rectangle is $2(3 + 5) = 16$.

FIGURE 14

These examples suggest the **distributive property.**

For any real numbers a, b, and c,

$$a(b + c) = ab + ac \quad \text{and} \quad (b + c)a = ba + ca.$$

The distributive property can also be written

$$ab + ac = a(b + c).$$

CHAPTER 1 THE REAL NUMBERS

1. Use the distributive property to rewrite each expression.

   (a) $8(m + n)$

   (b) $-4(p - 5)$

   (c) $3k + 6k$

   (d) $-6m + 2m$

   (e) $2r + 3s$

2. Complete the following statements.

   (a) $4 + \underline{} = 0$

   (b) $-7 + \underline{} = 0$

   (c) $-9 + 9 = \underline{}$

   (d) $5 \cdot \underline{} = 1$

   (e) $-\dfrac{3}{4} \cdot \underline{} = 1$

   (f) $7 \cdot \dfrac{1}{7} = \underline{}$

EXAMPLE 1 Using the Distributive Property
Use the distributive property to rewrite each expression.

(a) $3(x + y)$

In the statement of the property, let $a = 3$, $b = x$ and $c = y$. Then
$$3(x + y) = 3x + 3y.$$

(b) $-2(5 + k) = -2(5) + (-2)(k)$
$$= -10 - 2k$$

(c) $4x + 8x$

Use the second form of the property.
$$4x + 8x = (4 + 8)x = 12x$$

(d) $3r - 7r = 3r + (-7)r$ Definition of subtraction
$$= [3 + (-7)]r$$
$$= -4r$$

(e) $5p + 7q$

Since there is no common factor here, we cannot use the distributive property to simplify the expression. ■

As illustrated in Example 1(d), the distributive property can also be used for subtraction, so that
$$a(b - c) = ab - ac.$$

■ WORK PROBLEM 1 AT THE SIDE.

2 In Section 1.1 we saw that the additive inverse of a number a is $-a$ and that the sum of a number and its additive inverse is 0. For example, 3 and -3 are additive inverses, as are -8 and 8. The number 0 is its own additive inverse. In Section 1.3, we saw that two numbers with a product of 1 are reciprocals. Another name for a reciprocal is **multiplicative inverse**. This is similar to the idea of an additive inverse. Thus, 4 and $\dfrac{1}{4}$ are multiplicative inverses, and so are $-\dfrac{2}{3}$ and $-\dfrac{3}{2}$. (Note again that a pair of reciprocals has the same sign.) These properties are called the **inverse properties** of addition and multiplication.

For any real number a, there is a single real number $-a$, such that
$$a + (-a) = 0 \quad \text{and} \quad -a + a = 0.$$

For any nonzero real number a, there is a single real number $\dfrac{1}{a}$ such that
$$a \cdot \dfrac{1}{a} = 1 \quad \text{and} \quad \dfrac{1}{a} \cdot a = 1.$$

■ WORK PROBLEM 2 AT THE SIDE.

ANSWERS
1. (a) $8m + 8n$
   (b) $-4p + 20$
   (c) $9k$ (d) $-4m$
   (e) cannot be simplified
2. (a) -4 (b) 7 (c) 0
   (d) $\dfrac{1}{5}$ (e) $-\dfrac{4}{3}$ (f) 1

1.5 PROPERTIES OF REAL NUMBERS

3 The numbers 0 and 1 each have a special property. Zero is the only number that can be added to any number to get that number. That is, adding 0 leaves the identity of a number unchanged. For this reason, 0 is called the **identity element for addition.** In a similar way, multiplying by 1 leaves the identity of any number unchanged, so 1 is the **identity element for multiplication.** The following **identity properties** summarize this discussion.

For any real number a,

$$a + 0 = 0 + a = a$$

and

$$a \cdot 1 = 1 \cdot a = a.$$

The identity property for 1 is especially useful in simplifying algebraic expressions.

EXAMPLE 2 Using the Identity Property $1 \cdot a = a$
Simplify each expression.

(a) $12m + m$

$$\begin{aligned} 12m + m &= 12m + 1m & \text{Identity property} \\ &= (12 + 1)m & \text{Distributive property} \\ &= 13m \end{aligned}$$

(b) $y + y = 1y + 1y$ Identity property
$ = (1 + 1)y$ Distributive property
$ = 2y$

(c) $-(m - 5n) = -1(m - 5n)$ Identity property
$ = -1 \cdot m + (-1)(-5n)$ Distributive property
$ = -m + 5n$ ■

WORK PROBLEM 3 AT THE SIDE.

Expressions such as $12m$ and $5n$ from Example 2 are examples of *terms*. A **term** is a number or the product of a number and one or more variables. Terms with exactly the same variables raised to exactly the same powers are called **like terms.** The number in the product is called the **numerical coefficient** or just the **coefficient.** For example, in the term $5p$, the coefficient is 5.

4 Simplifying expressions as in parts (a) and (b) of Example 2 is called **combining like terms.** Only like terms may be combined.

3. Use the properties to rewrite each expression.

(a) $p - 3p$

(b) $r + r + r$

(c) $-(3 + 4p)$

(d) $-(k - 2)$

ANSWERS
3. (a) $-2p$ (b) $3r$
   (c) $-3 - 4p$ (d) $-k + 2$

To combine like terms in an expression such as

$$-2m + 5m + 3 - 6m + 8$$

we need two more properties. Note that

$$3 + 9 = 12 \quad \text{and} \quad 9 + 3 = 12.$$

Also,

$$3 \cdot 9 = 27 \quad \text{and} \quad 9 \cdot 3 = 27.$$

Furthermore,

$$(5 + 7) + (-2) = 12 + (-2) = \boxed{10}$$

and

$$5 + (7 + (-2)) = 5 + 5 = \boxed{10}.$$

Also,

$$(5 \cdot 7)(-2) = 35(-2) = \boxed{-70}$$

and

$$(5)(7 \cdot -2) = 5(-14) = \boxed{-70}.$$

These observations suggest the following properties.

For any real numbers a, b, and c,

$$a + b = b + a$$
$$ab = ba$$

Commutative properties

and

$$a + (b + c) = (a + b) + c$$
$$a(bc) = (ab)c.$$

Associative properties

The associative properties are used to *regroup* the terms of an expression. The commutative properties are used to change the *order* of the terms in an expression.

EXAMPLE 3 Using the Commutative and Associative Properties
Simplify $-2m + 5m + 3 - 6m + 8$.

$$-2m + 5m + 3 - 6m + 8$$
$$= (-2m + 5m) + 3 - 6m + 8 \qquad \text{Order of operations}$$
$$= 3m + 3 - 6m + 8 \qquad \text{Distributive property}$$

1.5 PROPERTIES OF REAL NUMBERS

By the order of operations, the next step would be to add $3m$ and 3, but they are unlike terms. To get $3m$ and $-6m$ together, use the associative and commutative properties. Begin by putting in parentheses and brackets, as shown.

$[(3m + 3) - 6m] + 8$
$= [3m + (3 - 6m)] + 8$ Associative property
$= [3m + (-6m + 3)] + 8$ Commutative property
$= [(3m + [-6m]) + 3] + 8$ Associative property
$= (-3m + 3) + 8$ Combine like terms.
$= -3m + (3 + 8)$ Associative property
$= -3m + 11$ Add. ∎

In practice, many of the steps are not written down, but you should realize that the commutative and associative properties are used whenever the terms in an expression are rearranged in order to combine like terms.

EXAMPLE 4 Using the Properties of Real Numbers
Simplify each expression.

(a) $5y - 8y - 6y + 11y$

$5y - 8y - 6y + 11y = (5 - 8 - 6 + 11)y = 2y$

(b) $-2(m - 3)$

$-2(m - 3) = -2(m) - (-2)(3) = -2m + 6$

(c) $3x + 4 - 5(x + 1) - 8$

First use the distributive property to eliminate the parentheses.

$3x + 4 - 5(x + 1) - 8 = 3x + 4 - 5x - 5 - 8$

Next use the commutative and associative properties to rearrange the terms; then combine like terms.

$= 3x - 5x + 4 - 5 - 8$
$= -2x - 9$

(d) $8 - (3m + 2)$

Think of $8 - (3m + 2)$ as $8 - 1 \cdot (3m + 2)$.

$8 - 1 \cdot (3m + 2) = 8 - 3m - 2 = 6 - 3m$

(e) $(3x)(5)(y) = [(3x)(5)]y$ Order of operations
$= [3(x \cdot 5)]y$ Associative property
$= [3(5x)]y$ Commutative property
$= [(3 \cdot 5)x]y$ Associative property
$= (15x)y$
$= 15(xy)$ Associative property
$= 15xy$

As mentioned above, many of these steps usually are not written out. ∎

WORK PROBLEM 4 AT THE SIDE.

4. Simplify each expression.

(a) $12b - 9b + 4b - 7b + b$

(b) $-3w + 7 - 8w - 2$

(c) $-3(6 + 2t)$

(d) $9 - 2(a - 3) + 4 - a$

(e) $(4m)(2n)$

ANSWERS
4. (a) b (b) $-11w + 5$
(c) $-18 - 6t$ (d) $19 - 3a$
(e) $8mn$

CHAPTER 1 THE REAL NUMBERS

5. Complete the following.

(a) $197 \cdot 0 =$ _____

(b) $(0)(-8) =$ _____

(c) $\left(-\dfrac{8}{15}\right)(0) =$ _____

(d) $0 \cdot$ _____ $= 0$

---

5 The additive identity property gives a special property of zero, namely that $a + 0 = a$ for any real number a. The **multiplication property of zero** gives a special property of zero that involves multiplication: The product of any real number and zero is zero.

For all real numbers a,
$$a \cdot 0 = 0$$
and
$$0 \cdot a = 0.$$

This property just extends to all real numbers what is true for positive numbers multiplied by 0.

WORK PROBLEM 5 AT THE SIDE.

ANSWERS
5. (a) 0 (b) 0 (c) 0
   (d) any real number

1.5 EXERCISES

Use the properties of real numbers to simplify each expression. See Examples 1 and 2.

1. $5k + 3k$
2. $6a + 5a$
3. $9r + 7r$
4. $-4n + 6n$

5. $-8z + 3w$
6. $-12k + 2h$
7. $-a + 7a$
8. $7t + t$

9. $14c + c$
10. $2(m + p)$
11. $3(a + b)$
12. $12(x - y)$

13. $-5(2d + f)$
14. $-2(3m + n)$
15. $-4(5k - 3)$

16. $-(8p - 1)$
17. $-(3k + 7)$
18. $-(p - 3q)$

Simplify each of the following expressions by removing parentheses and combining terms. See Examples 1–4.

19. $4x + 3x + 7 + 9$
20. $5m + 9m + 8 + 4$
21. $-12y + 4y - 3 + 2y$

22. $5r - 9r + 8r - 5$
23. $-6p + 11p - 4p + 6 - 5$
24. $8x - 5x + 3x - 12 + 9$

25. $3(k + 2) - 5k + 6 - 3$
26. $5(r - 3) + 6r - 2r + 1$
27. $-2(m + 1) + 3(m - 4)$

28. $6(a - 5) - 4(a + 6)$
29. $3(2y + 3) - 4(3y - 5)$
30. $-2(5r - 2) + 4(5r - 1)$

31. $-(2p + 5) + 3(2p + 4) + p$
32. $-(7m - 12) - 2(4m + 7) - 3m$

33. $2 + 3(2z - 5) - 3(4z + 6) - 8$
34. $-4 + 4(4k - 3) - 6(2k + 8) + 7$

35. $m + 2(-m + 4) - 3(m + 6)$
36. $-k + 2(k - 5) + 5(2k - 1)$

37. $4(-2x - 5) + 6(2 - 3x) - (4 - 2x)$
38. $3(-4z + 5) - 2(4 - 5z) - (2 - 9z)$

CHAPTER 1 THE REAL NUMBERS

Each of the following exercises shows half of a statement illustrating the indicated property. Complete each statement. Simplify all answers.

39. $2x + 3x = $ _____
(distributive property)

40. $-4 \cdot 1 = $ _____
(identity property)

41. $2(4x) = $ _____
(associative property)

42. $-3 + 3 = $ _____
(commutative property)

43. $-3 + 3 = $ _____
(inverse property)

44. $4y + 4z = $ _____
(distributive property)

45. $0 + 7 = $ _____
(identity property)

46. $8 \cdot \dfrac{1}{8} = $ _____
(inverse property)

47. $3a + 5a + 6a = $ _____
(distributive property)

48. $\dfrac{9}{28} \cdot 0 = $ _____
(multiplication property of 0)

49. $0 = 2 + ($ _____ $)$
(inverse property)

50. $-15 = -15 ($ _____ $)$
(identity property)

51. $8(2 + 3) = ($ _____ $)2 + ($ _____ $)3$
(distributive property)

52. $6 + (2 + x) = $ _____
(associative property)

53. $0(3 - 8) = $ _____
(multiplication property of 0)

54. $3(2 \cdot 5) = 3($ _____ $)$
(commutative property)

55. By the distributive property, $a(b + c) = ab + ac$. If we exchange the operations of multiplication and addition in this statement, we get $a + (b \cdot c) = (a + b)(a + c)$. Is this new statement true for all values of a, b, and c? To find out, try various sample values for a, b, and c.

56. Do *any* different numbers satisfy the statement $a - b = b - a$? Give an example if your answer is yes.

57. Are there *any* two different numbers a and b for which $a/b = b/a$? Give an example if your answer is yes.

CHAPTER 1 SUMMARY

KEY TERMS

1.1	**set**	A set is a collection of objects.
	empty set	The set with no elements is called the empty set.
	variable	A variable is a letter used to represent a number or a set of numbers.
	set-builder notation	Set-builder notation is used to describe a set of numbers without listing them.
	number line	A number line is a line with a scale to indicate the set of real numbers.
	coordinate	The number that corresponds to a point on the number line is its coordinate.
	graph	The point on the number line that corresponds to a number is its graph.
	signed numbers	Positive and negative numbers are signed numbers.
	additive inverse	The additive inverse (**negative, opposite**) of a number a is $-a$.
	absolute value	The absolute value of a number is its distance from 0 or its magnitude without regard to sign.
1.2	**inequality**	An inequality is a mathematical statement that two quantities are not equal.
	interval	An interval is a portion of a number line.
	interval notation	Interval notation uses symbols to describe an interval on the number line.
1.3	**sum**	The result of addition is called the sum.
	difference	The result of subtraction is called the difference.
	product	The result of multiplication is called the product.
	reciprocals	Two numbers whose product is 1 are reciprocals.
	quotient	The result of division is called the quotient.
1.4	**factors**	Two numbers whose product is a third number are factors of that third number.
	exponent	An exponent is a number that shows how many times a factor is repeated in a product.
	base	The base is a number that is a repeated factor in a product.
	exponential	A base with an exponent is called an exponential or a **power**.
	square root	A square root of a number r is a number that can be squared to get r.
	cube root	The cube root of a number r is the number that can be cubed to get r.
1.5	**term**	A term is a number or the product of a number and one or more variables.
	like terms	Like terms are terms with the same variables raised to the same powers.
	coefficient	A coefficient (**numerical coefficient**) is the numerical factor of a term.
	combining like terms	Combining like terms is a method of adding or subtracting like terms by using the properties of real numbers.

CHAPTER 1 THE REAL NUMBERS

NEW SYMBOLS

{*a*, *b*}	set containing the elements *a* and *b*
∅	the empty set
{*x*|*x* has property *P*}	set-builder notation
|*x*|	the absolute value of *x*
≠	is not equal to
<	is less than
>	is greater than
(*a*, ∞)	the interval $\{x \mid x > a\}$
(−∞, *a*)	the interval $\{x \mid x < a\}$
(*a*, *b*]	the interval $\{x \mid a < x \leq b\}$
a^m	*m* factors of *a*
$\sqrt{a}$	the square root of *a*
$\sqrt[n]{a}$	the *n*th root of *a*

QUICK REVIEW

Section	Concepts	Examples
1.1 Basic Terms	**Sets of Numbers**	
	Real Numbers {*x*|*x* is a coordinate of a point on a number line}	$-3, .7, \pi, -\dfrac{2}{3}$
	Natural Numbers {1, 2, 3, 4, . . .}	10, 25, 143
	Whole Numbers {0, 1, 2, 3, 4, . . .}	0, 8, 47
	Integers {..., −2, −1, 0, 1, 2, ...}	−22, −7, 0, 4, 9
	Rational Numbers $\left\{\dfrac{p}{q} \mid p \text{ and } q \text{ are integers}, q \neq 0\right\}$, or all terminating or repeating decimals	$-\dfrac{2}{3}, -.14, 0, 6, \dfrac{5}{8}, .33333...$
	Irrational Numbers {*x*|*x* is a real number that is not rational} or all non-terminating, non-repeating decimals	$\pi, .125469..., \sqrt{3}, -\sqrt{22}$

CHAPTER 1 SUMMARY

Section	Concepts	Examples
1.2 Inequality	**Inequality Symbols** 	

Symbol	Meaning	
$\neq$	is not equal to	$-3 \neq 3$
$<$	is less than	$-4 < -1$
$>$	is greater than	$3 > -2$
$\leq$	is less than or equal to	$6 \leq 6$
$\geq$	is greater than or equal to	$-8 \geq -10$
$\not<$	is not less than	$5 \not< 2$
$\not>$	is not greater than	$-3 \not> -1$
$\not\leq$	is not less than or equal to	$5 \not\leq 2$
$\not\geq$	is not greater than or equal to	$3 \not\geq 4$

Section	Concepts	Examples
1.3 Operations on Real Numbers	**Addition** *Same sign:* Add the absolute values. The answer has the same sign as the numbers.	$-2 + (-7) = -(2 + 7)$ $= -9$
	Different signs: Subtract the absolute values. The answer has the sign of the number with the larger absolute value.	$-5 + 8 = 8 - 5 = 3$ $-12 + 4 = -(12 - 4) = -8$
	Subtraction Change the sign of the second number and add.	$-5 - (-3) = -5 + 3 = -2$
	Multiplication *Same sign:* The answer is positive. *Different signs:* The answer is negative.	$(-3)(-8) = 24$ $(-7)(5) = -35$
	Division *Same sign:* The answer is positive. *Different signs:* The answer is negative.	$\dfrac{-15}{-5} = 3$ $\dfrac{-24}{12} = -2$

CHAPTER 1 THE REAL NUMBERS

Section	Concepts	Examples
1.4 Exponents and Roots; Order of Operations	The product of an even number of negative factors is positive. The product of an odd number of negative factors is negative.	$(-5)^6$ is positive. $(-5)^7$ is negative.
	Order of Operations 1. Work separately above and below any fraction bar. 2. Use the rules below within each set of parentheses or square brackets. Start with the innermost set and work outward. 3. Evaluate all powers and roots. 4. Do any multiplications or divisions in the order in which they occur, working from left to right. 5. Do any additions or subtractions in the order in which they occur, working from left to right.	$\dfrac{12 + 3}{5 \cdot 2} = \dfrac{15}{10} = \dfrac{3}{2}$ $(-6)[2^2 - (3 + 4)] + 3$ $= (-6)[2^2 - 7] + 3$ $= (-6)[4 - 7] + 3$ $= (-6)[-3] + 3$ $= 18 + 3$ $= 21$
1.5 Properties of Real Numbers	For any real numbers a, b, and c: **Distributive Property** $a(b + c) = ab + ac$	$12(4 + 2) = 12 \cdot 4 + 12 \cdot 2$
	Inverse Properties $a + (-a) = 0$ $-a + a = 0$	$5 + (-5) = 0$
	$a \cdot \dfrac{1}{a} = 1$ and $\dfrac{1}{a} \cdot a = 1$ $(a \neq 0)$	$-\dfrac{1}{3} \cdot -3 = 1$
	Identity Properties $a + 0 = a$ and $0 + a = a$ $a \cdot 1 = a$ and $1 \cdot a = a$	$-32 + 0 = -32$ $17.5 \cdot 1 = 17.5$
	Associative Properties $a + (b + c) = (a + b) + c$ $a(bc) = (ab)c$	$7 + (5 + 3) = (7 + 5) + 3$ $-4(6 \cdot 3) = (-4 \cdot 6)3$
	Commutative Properties $a + b = b + a$ $ab = ba$	$9 + (-3) = -3 + 9$ $6(-4) = (-4)6$

CHAPTER 1 REVIEW EXERCISES

[1.1] *Graph each set on a number line.**

1. $\left\{-3, -5, 1, \dfrac{1}{4}, \dfrac{9}{4}\right\}$

2. $\left\{-6, -\dfrac{11}{4}, -1, 0, 3, \dfrac{12}{5}\right\}$

Find the value of each expression.

3. $|-12|$

4. $|31|$

5. $-|4|$

6. $-|-7| + |-2|$

Let set $S = \left\{-9, -\dfrac{4}{3}, -\sqrt{10}, 0, \dfrac{5}{3}, \sqrt{7}, \dfrac{12}{3}\right\}$. *Simplify the elements of S as necessary and then list the elements that belong to each set listed below.*

7. Natural numbers

8. Integers

9. Rational numbers

10. Irrational numbers

[1.2] *Write each statement using inequality symbols.*

11. x is between -2 and 1.

12. x is between 0 and 3, excluding 0 and including 3.

_____

*For help with any of these exercises, look in the section given in brackets.

CHAPTER 1 THE REAL NUMBERS

Write **true** *or* **false** *for each inequality.*

13. $2 \geq -5$

14. $-3 < -3$

15. $6 \not> |-2|$

16. $3 \cdot 2 \leq |12 - 4|$

17. $2 + |-2| \geq 3$

18. $4(3 + 7) > -|-41|$

Write in interval notation and graph.

19. $\{x|x < -4\}$

20. $\{x|-2 < x \leq 5\}$

⟶   ⟶

[1.3] *Add or subtract, as indicated.*

21. $-\dfrac{5}{8} - \left(-\dfrac{7}{3}\right)$

22. $-\dfrac{4}{5} - \left(-\dfrac{3}{10}\right)$

23. $-5 + (-11) + 20$

24. $-9.42 + 1.83 - 7.6 - 1.9$

25. $-15 + (-13) + (-11)$

26. $-1 - 3 - (-10)$

27. $\dfrac{3}{4} - \left(\dfrac{1}{2} - \dfrac{9}{10}\right)$

28. $-\dfrac{2}{3} - \left(\dfrac{1}{6} - \dfrac{5}{9}\right)$

29. $-|-12| - |-9| + (-4) - 10$

Find each product.

30. $2(-5)(-3)$

31. $-\dfrac{3}{7}\left(-\dfrac{14}{9}\right)$

32. $-4.6(-3.9)$

CHAPTER 1 REVIEW EXERCISES

Divide.

33. $\dfrac{-15}{-3}$

34. $\dfrac{-7}{0}$

35. $\dfrac{9.28}{-3.2}$

[1.4] *Evaluate.*

36. 3^2

37. 10^4

38. $\left(\dfrac{3}{7}\right)^3$

39. $(-5)^3$

40. -5^3

41. $(1.7)^2$

Find each root.

42. $\sqrt{400}$

43. $\sqrt[3]{27}$

44. $\sqrt[3]{343}$

45. $\sqrt[4]{81}$

46. $\sqrt[6]{64}$

Use the order of operations to simplify.

47. $-14\left(\dfrac{3}{7}\right) + 6 \div 2$

48. $-\dfrac{2}{3}[5(-2) + 8 - 4^2]$

49. $\dfrac{-4(\sqrt{25}) - (-3)(-5)}{3 + (-6)(\sqrt{9})}$

50. $\dfrac{-5(3^2) + 9(\sqrt{4}) - 5}{6 - 5(\sqrt[3]{-8})}$

Evaluate. Assume that $k = -2$, $m = 3$, and $n = 16$.

51. $4k - 7m$

52. $-3(\sqrt{16}) + m + 5k$

53. $-2(3k^2 + 5m)$

54. $\dfrac{4m^3 - 3n}{7k^2 - 10}$

55. $\dfrac{4k - 3k(\sqrt{n} + 3)}{k - m}$

56. $\dfrac{k + \sqrt{n} - m(2 + k)}{m - 5 - 2k}$

CHAPTER 1 THE REAL NUMBERS

[1.5] *Use the properties of real numbers to simplify each expression.*

57. $2q + 9q$

58. $3z - 17z$

59. $-m + 6m$

60. $5p - p$

61. $-2(k - 3)$

62. $6(r + 3)$

63. $9(2m + 3n)$

64. $-(3k - 4h)$

65. $-(-p + 6q)$

66. $2x + 5 - 4x + 1$

67. $-3y + 6 - 5 - 4y$

68. $2a + 3 - a - 1$

69. $2(k - 1) + 3k - k$

70. $-3(4m - 2) + 2(3m - 1)$

71. $-(5p + 2) - 3p - 8 + p$

72. $-k + 3(k - 1) - 2(4k + 3)$

Complete each statement to illustrate the indicated property.

73. $-2(5y) = $ _____
(associative property)

74. $3 \cdot 9 = $ _____
(commutative property)

75. _____ $+ z = z$
(identity property)

76. $0 \cdot z = $ _____
(multiplication property of 0)

MIXED REVIEW EXERCISES*
Perform the indicated operations.

77. $\left(-\dfrac{4}{5}\right)^4$

78. $-\dfrac{5}{8}(-32)$

79. $-25\left(-\dfrac{4}{5}\right) + 3^3 - 32 \div \sqrt{64}$

80. $-8 + |-14| + |-3|$

81. $\dfrac{6 \cdot \sqrt{4} - 3 \cdot \sqrt{16}}{-2 \cdot 5 + 7(-3) - 1}$

82. $\sqrt[5]{32}$

83. $-\dfrac{10}{21} \div \dfrac{5}{14}$

84. $.8 - 4.9 - 3.2 + 1.14$

85. -3^2

---

*The order of exercises in this final group does not correspond to the order in which topics occur in the chapter. This random ordering should help you prepare for the chapter test in yet another way.

CHAPTER 1 TEST

Graph each set on a number line.

1. $\left\{-3, \dfrac{3}{4}, -2, |-1|, 3\right\}$

2. $\left\{2, -\dfrac{5}{8}, -1, \dfrac{5}{2}, 0\right\}$

Let $A = \left\{-\sqrt{7}, -2, -\dfrac{1}{2}, \dfrac{0}{2}, 7.5, 3, \dfrac{24}{2}, \sqrt{11}\right\}$. *First simplify each element in set A as needed, then list the elements from A that belong to each set below.*

3. Irrational numbers

4. Rational numbers

5. Integers

Write in interval notation and graph.

6. $\{x \mid x > 2\}$

7. $\{x \mid -2 \leq x < 5\}$

Evaluate.

8. $-5 + 12 + (-11)$

9. $-2 + (-3) - |-5| + |-12 + 2|$

10. $6 - 3 \cdot 3 + 5(-4) - \dfrac{8}{-2}$

11. $6 - 3^3 + 2(5) + (-4)^2$

12. $\dfrac{10 - 24 + (-6)}{\sqrt{16}(-5)}$

13. $\dfrac{-2[3 - (-1 - 2) + 2]}{\sqrt{9}(-3) - (-2)}$

14. $\dfrac{8 \cdot 4 - 3^2 \cdot 5 - 2(-1)}{-3 \cdot 2^3 + 1}$

15. Name the exponent and the base in $-8y^4$.

1. _____
2. _____
3. _____
4. _____
5. _____
6. _____
7. _____
8. _____
9. _____
10. _____
11. _____
12. _____
13. _____
14. _____
15. _____

CHAPTER 1 THE REAL NUMBERS

Find each of the following.

16. $(-2)^3$

17. $\left(\dfrac{3}{2}\right)^4$

18. $(-12)^2$

19. $-4^2 + (-4)^2$

20. $\sqrt[3]{64}$

21. $\sqrt{64}$

Evaluate each expression if $k = -4$, $m = -3$, and $r = 5$.

22. $8r^2 + m^3$

23. $\dfrac{5k + 6r}{k + r - 1}$

24. $-6(2m - 3k)$

25. $\dfrac{8k + 2m^2}{r - 2}$

Use the properties of real numbers to simplify each of the following.

26. $3(2x - y)$

27. $4k - 2 + k - 8 - 3k$

28. $5p - 3(p + 2) + 7$

Complete each statement using the indicated property.

29. $\dfrac{2}{3} \cdot \underline{\quad} = 1$ (multiplicative inverse)

30. $32 + \underline{\quad} = 32$ (additive identity)

2 LINEAR EQUATIONS AND INEQUALITIES

2.1 LINEAR EQUATIONS IN ONE VARIABLE

An **algebraic expression** is an expression indicating any combination of the following operations: addition, subtraction, multiplication, division (except by 0), and taking roots on any collection of variables and numbers. Some examples of algebraic expressions are shown below.

$$8x + 9 \qquad y + 4 \qquad \frac{x^3 y^8}{z}$$

Applications of mathematics often lead to **equations,** statements that two algebraic expressions are equal. A *linear equation in one variable* involves only real numbers and one variable. Examples include the following.

$$x + 1 = -2 \qquad y - 3 = 5 \qquad 2k + 5 = 10$$

It is very important to be able to distinguish between algebraic expressions and equations. An equation always includes an equals sign, while an expression does not.

An equation in the variable x is **linear** if it can be written in the form

$$ax + b = c$$

where a, b, and c are real numbers, with $a \neq 0$.

A linear equation is also called a **first-degree** equation, since the highest power on the variable is one.

1 If the variable in an equation can be replaced by a real number that makes the statement true, then that number is a **solution** of the equation. For example, 8 is a solution of the equation $y - 3 = 5$, since replacing y with 8 gives a true statement. An equation is **solved** by finding its **solution set,** the set of all solutions. The solution set of the equation $y - 3 = 5$ is $\{8\}$.

WORK PROBLEM 1 AT THE SIDE.

Equivalent equations are equations that have the same solution set. To solve an equation we usually start with the given equation and replace it with a series of simpler equivalent equations. For example, $5x + 2 = 17$, $5x = 15$, and $x = 3$ are all equivalent, since each has the solution set $\{3\}$.

OBJECTIVES

1. Determine whether a number is a solution of a linear equation.
2. Solve linear equations using the addition and multiplication properties of equality.
3. Solve linear equations using the distributive property.
4. Solve linear equations with fractions.
5. Identify conditional equations, contradictions, and identities.

---

1. Are the given numbers solutions for the given equations?

(a) $3k = 15$; 5

(b) $r + 5 = 4$; 1

(c) $-8m = 12$; $\dfrac{3}{2}$

ANSWERS
1. (a) yes (b) no (c) no

2.1 LINEAR EQUATIONS IN ONE VARIABLE   53

CHAPTER 2 LINEAR EQUATIONS AND INEQUALITIES

2. Solve.

(a) $6x - 4 = 14$

(b) $3p + 2p + 1 = -24$

(c) $3p = 2p + 4p + 5$

(d) $3m - 4 + 2m = 10 - 6m + 4m$

(e) $4a + 8a = 17a - 9 - 1$

(f) $-7 + 3y - 9y = 12y - 5$

ANSWERS

2. (a) $\{3\}$ (b) $\{-5\}$ (c) $\left\{-\dfrac{5}{3}\right\}$

   (d) $\{2\}$ (e) $\{2\}$ (f) $\left\{-\dfrac{1}{9}\right\}$

Two important properties that are used in producing equivalent equations are the **addition and multiplication properties of equality.**

ADDITION PROPERTY OF EQUALITY

For all real numbers a, b, and c, the equations

$$a = b \quad \text{and} \quad a + c = b + c$$

are equivalent equations.

MULTIPLICATION PROPERTY OF EQUALITY

For all real numbers a, b, and c, where $c \neq 0$, the equations

$$a = b \quad \text{and} \quad ac = bc$$

are equivalent equations.

By the addition property, the same number may be added to both sides of an equation without affecting the solution set. By the multiplication property, both sides of an equation may be multiplied by the same nonzero number to produce an equivalent equation. Because subtraction and division are defined in terms of addition and multiplication, respectively, these properties can be extended: The same number may be subtracted from both sides of an equation, and both sides may be divided by the same nonzero number.

EXAMPLE 1 Solving a Linear Equation

Solve $4y - 2y - 5 = 4 + 6y + 3$.

First, combine terms separately on both sides of the equation to get

$$2y - 5 = 7 + 6y.$$

Next, use the addition property to get the terms with y on the same side of the equation and the remaining terms (the numbers) on the other side. One way to do this is first to add 5 to both sides.

$$2y - 5 \boxed{+ 5} = 7 + 6y \boxed{+ 5} \qquad \text{Add 5.}$$
$$2y = 12 + 6y$$

Now subtract $6y$ from both sides.

$$2y \boxed{- 6y} = 12 + 6y \boxed{- 6y} \qquad \text{Subtract } 6y.$$
$$-4y = 12$$

Finally, divide both sides by -4 to get just y on the left.

$$\frac{-4y}{\boxed{-4}} = \frac{12}{\boxed{-4}} \qquad \text{Divide by } -4.$$
$$y = -3$$

The solution set is $\{-3\}$. Check by substituting -3 for y in the original equation. ∎

■ **WORK PROBLEM 2 AT THE SIDE.** ■

2.1 LINEAR EQUATIONS IN ONE VARIABLE

The steps needed to solve a linear equation in one variable are as follows. (Some equations may not require all of these steps.)

SOLVING A LINEAR EQUATION IN ONE VARIABLE

Step 1 Eliminate any fractions by multiplying both sides of the equation by a common denominator.

Step 2 Simplify each side of the equation as much as possible by using the distributive property as needed to eliminate any parentheses and to combine like terms.

Step 3 Use the addition property of equality to simplify further so that the variable term is on one side of the equation and a number is on the other side.

Step 4 Use the multiplication property of equality to make the coefficient of the variable equal to 1.

Step 5 Check the solution by substituting back in the *original* equation.

3 In Example 1 we did not use Steps 1 and 2. Solving many other equations will require one or both of these steps, however, as shown in the next examples.

EXAMPLE 2 Using the Distributive Property to Solve a Linear Equation

Solve $2(k - 5) + 3k = k + 6$.

Step 1 does not apply here, since there are no fractions. Begin with Step 2, using the distributive property to simplify the left side, and then combine like terms.

$$2(k - 5) + 3k = k + 6$$
$$\boxed{2k - 10} + 3k = k + 6 \qquad \text{Distributive property}$$
$$5k - 10 = k + 6 \qquad \text{Combine terms.}$$

Now apply Step 3 to get the variable term on the left side and a number on the right side.

$$5k \boxed{-k} = k + 16 \boxed{-k} \qquad \text{Subtract } k; \text{ add 10.}$$
$$4k = 16$$

Use Step 4 to make the coefficient of the variable equal to 1.

$$\frac{4k}{4} = \frac{16}{4} \qquad \text{Divide by 4.}$$

Finally, apply Step 5 to check that the solution set is {4} by substituting 4 for k in the original equation. ■

WORK PROBLEM 3 AT THE SIDE.

3. Solve.

(a) $5p + 4(3 - 2p) = 2 + p - 10$

(b) $3(z - 2) + 5z = 2$

(c) $-2 + 3(y + 4) = 8y$

(d) $6 - (4 + m) = 8m - 2(3m + 5)$

ANSWERS
3. (a) {5} (b) {1} (c) {2} (d) {4}

CHAPTER 2 LINEAR EQUATIONS AND INEQUALITIES

4. Solve.

(a) $\dfrac{2p}{7} - \dfrac{p}{2} = -3$

(b) $\dfrac{3r}{5} - \dfrac{2r}{7} = -\dfrac{11}{35}$

5. Solve.

$\dfrac{k+1}{2} + \dfrac{k+3}{4} = \dfrac{1}{2}$

When fractions appear in an equation, Step 1 requires that we eliminate them by multiplying both sides of the equation by a common denominator. We ususally choose the least common denominator. This is shown in the next example.

EXAMPLE 3 Solving a Linear Equation with Fractions

Solve $\dfrac{y}{2} - \dfrac{3y}{5} = -1$.

Apply Step 1 by multiplying both sides of the equation by 10, the least common denominator of 2 and 5.

$$10\left(\dfrac{y}{2} - \dfrac{3y}{5}\right) = 10(-1) \qquad \text{Multiply by 10.}$$

Use the distributive property and combine like terms (Step 2):

$$10\left(\dfrac{y}{2}\right) - 10\left(\dfrac{3y}{5}\right) = -10 \qquad \text{Distributive property}$$
$$5y - 6y = -10$$
$$-y = -10$$

Step 3 does not apply, so go on to Step 4, multiplying both sides by -1 to get

$$-1(-y) = -1(-10)$$
$$y = 10.$$

Now apply Step 5 to check that the solution set is $\{10\}$. ■

WORK PROBLEM 4 AT THE SIDE.

(In the remaining examples, the various steps are not identified by number.)

EXAMPLE 4 Solving a Linear Equation with Fractions

Solve $\dfrac{x+7}{6} + \dfrac{2x-8}{2} = -4$.

Start by eliminating the fractions. Multiply both sides by 6.

$$6\left[\dfrac{x+7}{6} + \dfrac{2x-8}{2}\right] = 6 \cdot (-4)$$
$$6\left(\dfrac{x+7}{6}\right) + 6\left(\dfrac{2x-8}{2}\right) = 6(-4)$$
$$x + 7 + 3(2x - 8) = -24$$
$$x + 7 + 6x - 24 = -24 \qquad \text{Distributive property}$$
$$7x - 17 = -24 \qquad \text{Combine terms.}$$
$$7x - 17 + 17 = -24 + 17 \qquad \text{Add 17.}$$
$$7x = -7$$
$$\dfrac{7x}{7} = \dfrac{-7}{7} \qquad \text{Divide by 7.}$$
$$x = -1$$

Check to see that $\{-1\}$ is the solution set. ■

WORK PROBLEM 5 AT THE SIDE.

ANSWERS
4. (a) $\{14\}$ (b) $\{-1\}$
5. $\{-1\}$

5 Each of the equations above had a solution set containing one element; for example, $\frac{y}{2} - \frac{3y}{5} = -1$ has solution set {10}, containing only the single number 10. An equation that has a finite number of elements in its solution set is a **conditional equation.** Sometimes an equation has no solutions. Such an equation is a **contradiction** and its solution set is ∅. It is also possible for an equation to have an infinite number of solutions. An equation that is satisfied by every number for which both sides are defined is called an **identity.** The next example shows how to recognize these types of equations.

EXAMPLE 5 Recognizing Conditional Equations, Identities, and Contradictions

Solve each equation. Decide whether it is a conditional equation, an identity, or a contradiction.

(a) $5x - 9 = 4(x - 3)$

Work as in the previous examples.

$5x - 9 = 4(x - 3)$	
$5x - 9 = 4x - 12$	Distributive property
$5x - 9 - 4x = 4x - 12 - 4x$	Subtract $4x$.
$x - 9 = -12$	Combine terms.
$x - 9 + 9 = -12 + 9$	Add 9.
$x = -3$	

The solution set, $\{-3\}$, has only one element, so $5x - 9 = 4(x - 3)$ is a *conditional equation.*

(b) $5x - 15 = 5(x - 3)$

Use the distributive property on the right side.

$$5x - 15 = 5x - 15$$

Both sides of the equation are *exactly the same,* so any real number would make the equation true. For this reason, the solution set is {all real numbers}, and the equation $5x - 15 = 5(x - 3)$ is an *identity.*

(c) $5x - 15 = 5(x - 4)$

Use the distributive property.

$5x - 15 = 5x - 20$	Distributive property
$5x - 15 - 5x = 5x - 20 - 5x$	Subtract $5x$ from each side.
$-15 = -20$	False

Since the result, $-15 = -20$, is *false,* the equation has no solution. The solution set is ∅, so the equation $5x - 15 = 5(x - 4)$ is a *contradiction.* ∎

WORK PROBLEM 6 AT THE SIDE.

6. Solve each equation. Decide whether it is a conditional equation, an identity, or a contradiction. Give the solution set.

(a) $5(x + 2) - 2(x + 1) = 3x + 1$

(b) $\dfrac{x+1}{3} + \dfrac{2x}{3} = x + \dfrac{1}{3}$

(c) $5(3x + 1) = x + 5$

ANSWERS
6. (a) contradiction; ∅
 (b) identity; {all real numbers}
 (c) conditional; {0}

Historical Reflections

Archimedes (ca. 287 – ca. 212 B.C.)

Archimedes, who was one of the greatest mathematicians of all time, was surely the greatest of the early Greek period. He wrote many treatises, and in his *Method* he used a theory that anticipated the ideas of integral calculus—ideas that were not to surface for nearly 2000 years. He was also an inventor, and his water screw is still used in Egypt today.

There are numerous colorful stories about the life of Archimedes. The best known of these illustrates his reaction to one of his discoveries. While taking a bath, he noticed that the water level rose as he lay in the water. Realizing that from this observation he could state a property of buoyancy, he became so excited that he ran through the streets shouting "Eureka" ("I have found it"), without bothering to clothe himself.

Archimedes met his death during the pillage of Syracuse at the hand of a Roman soldier. As was his custom, he was using a sand tray to draw his geometric figures when the soldier came upon him. He ordered the soldier to move clear of his "circles," but the soldier refused and killed him.

Art: Culver Pictures

name　　　　　　　　　　　　　　　　date　　　　　　　　hour

2.1 EXERCISES

Solve each of the following equations. See Examples 1–3.

1. $2k + 6 = 12$
2. $5m - 4 = 16$
3. $5 - 8k = -19$

4. $4 - 2m = 10$
5. $3 - 2r = 9$
6. $-5z + 2 = 7$

7. $7z - 5z + 3 = z - 4$
8. $2x + 7 - x = 4x - 2$

9. $7m - 2m + 4 - 5 = 3m - 5 + 6$
10. $12a + 7 = 7a - 2$

11. $11p - 9 + 8p - 7 + 14p = 12p + 9p + 11 - 7$
12. $12z - 15z - 8 + 6 = 4z + 6 - 1$

13. $2(k - 4) = 5k + 2$
14. $3(x + 5) = 2x - 1$

15. $3(2t + 1) - 2(t - 2) = 5$
16. $4(r - 1) + 2(r + 3) = 6$

17. $2y + 3(y - 4) = 2(y - 3)$
18. $6s - 3(5s + 2) = 4 - 5s$

19. $4k - 3(4 - 2k) = 2(k - 3) + 6k + 2$
20. $6x - 4(3 - 2x) = 5(x - 4) - 10$

21. $-[z - (4z + 2)] = 2 + (2z + 7)$

22. $5y - (8 - y) = 2[-4 - (3 + 5y) - 13]$

23. $-9m - (4 + 3m) = -(2m - 1) - 5$

24. $-11r - (5 - 6r) = -(6 - 3r) + 1$

25. $-(9 - 3a) - (4 + 2a) - 3 = -(2 - 5a) + (-a) + 1$

26. $-(-2 + 4x) - (3 - 4x) + 5 = -(-3 + 6x) + x + 1$

Solve each of the following equations. See Example 4. (In Exercises 38 and 39, be especially careful with signs.)

27. $-\dfrac{5}{9}k = 2$

28. $\dfrac{3}{11}z = -5$

29. $\dfrac{6x}{5} = -1$

30. $\dfrac{m}{2} + \dfrac{m}{3} = 5$

31. $\dfrac{y}{5} - \dfrac{y}{4} = 1$

32. $\dfrac{3k}{5} - \dfrac{2k}{3} = 1$

33. $\dfrac{8r}{3} - \dfrac{6r}{5} = 22$

34. $\dfrac{m-2}{3} + \dfrac{m}{4} = \dfrac{1}{2}$

35. $\dfrac{y-8}{5} + \dfrac{y}{3} = -\dfrac{8}{5}$

36. $\dfrac{a+8}{6} + \dfrac{4a}{9} = \dfrac{2a+12}{9}$

37. $\dfrac{2p-3}{4} + \dfrac{p-1}{4} = \dfrac{p-4}{2}$

38. $\dfrac{4x+1}{3} - \dfrac{x-3}{6} = \dfrac{x+5}{6}$

39. $\dfrac{2x+5}{5} - \dfrac{3x+1}{2} = \dfrac{7-x}{2}$

2.1 EXERCISES

Decide whether the following equations are conditional, identities, or contradictions. Give the solution set of each. See Example 5.

40. $9k + 4 - 3k = 2(3k + 4) - 4$

41. $-7m + 8 + 4m = -3(m - 3) - 1$

42. $-2p + 5p - 9 = 3(p - 4) - 5$

43. $-6k + 2k - 11 = -2(2k - 3) + 4$

44. $-11m + 4(m - 3) + 6m = 4m - 12$

45. $3p - 5(p + 4) + 9 = -11 + 15p$

46. $7[2 - (3 + 4r)] - 2r = -9 + 2(1 - 15r)$

47. $4[6 - (1 + 2m)] + 10m = 2(10 - 3m) + 8m$

48. $-4[6 - (-2 + 3q)] = 3(7 + 4q)$

49. $-3[-5 - (-9 + 2r)] = 2(3r - 1)$

Find a value of k so that the following pairs of equations are equivalent equations.

50. $6x - 5 = k$ and $2x + 1 = 15$

51. $3y + k = 11$ and $5y - 8 = 22$

Find a value of k so that each of the following equations has solution set $\{-4\}$.

52. $3x + 6 = k + 8$

53. $8x - 9 = k - 7$

54. $\dfrac{2x + 1}{3x - 4} = k$

55. $\dfrac{9 - 7x}{3 + 2x} = k$

61

CHAPTER 2 LINEAR EQUATIONS AND INEQUALITIES

Decide which of the following pairs of equations are equivalent.

56. $5x = 10$
$\dfrac{5x}{x+2} = \dfrac{10}{x+2}$

57. $x + 1 = 9$
$\dfrac{x+1}{8} = \dfrac{9}{8}$

58. $y = -3$
$\dfrac{y}{y+3} = \dfrac{-3}{y+3}$

59. $m = 1$
$\dfrac{m+1}{m-1} = \dfrac{2}{m-1}$

60. $k = 3$
$k^2 = 9$

61. $p^2 = 25$
$p = 5$

SKILL SHARPENERS
Most of the exercise sets in the rest of this book end with a few "skill sharpeners." These problems are designed to help you review ideas needed for the next section. If you need help with these, look in the section indicated each time.

Evaluate each expression, using the given values. See Section 1.4.

62. $2L + 2W$; $L = 10, W = 6$

63. rt; $r = 15, t = 9$

64. $\dfrac{1}{3}Bh$; $B = 27, h = 10$

65. prt; $p = 8000, r = .12, t = 2$

66. $\dfrac{5}{9}(F - 32)$; $F = 212$

67. $\dfrac{9}{5}C + 32$; $C = 100$

68. $\dfrac{1}{2}(B + b)h$; $B = 9, b = 4, h = 2$

69. $\dfrac{1}{2}bh$; $b = 21, h = 5$

2.2 FORMULAS

Solving a problem in algebra often involves working with a mathematical statement or **formula** in which more than one letter is used to express a relationship. Examples of formulas are

$$d = rt, \quad I = prt, \quad \text{and} \quad P = 2L + 2W.$$

A list of the formulas used in this book is given in an appendix. Metric units are used in some of the exercises for this section, but no knowledge of the metric system is needed to work the problems.

1 We say a formula is *solved for a specified variable* when that variable is isolated on one side of the equation. In some applications, a particular formula may be useful, but the formula, as given, does not isolate the "right" variable. The following examples show how to solve a formula for any one of its variables. Notice that the steps used in these examples are very similar to those used in solving a linear equation.

EXAMPLE 1 Solving for a Specified Variable
Solve the formula $I = prt$ for t.

This formula gives the amount of simple interest, I, in terms of the principal (the amount of money deposited), p, the yearly rate of interest, r, and time in years, t. To solve this formula for t, assume that I, p, and r are constants (having a fixed value) and that t is the variable. Then use properties from the previous section as follows.

$$I = prt$$
$$I = (pr)t \quad \text{Associative property}$$
$$\frac{I}{pr} = \frac{(pr)t}{pr} \quad \text{Divide by } pr.$$
$$\frac{I}{pr} = t$$

This result is a formula for t, time in years. ∎

WORK PROBLEM 1 AT THE SIDE.

The process of solving for a specified variable uses the same steps as the ones in Section 2.1 for solving a linear equation, but the following additional suggestions may be helpful.

SOLVING FOR A SPECIFIED VARIABLE

Step 1 Use the addition or multiplication properties as necessary to get all terms containing the specified variable on one side of the equals sign.

Step 2 All terms not containing the specified variable should be on the other side of the equals sign.

Step 3 If necessary, use the distributive property to write the side with the specified variable as the product of that variable and a sum of terms.

Step 4 Complete the solution using the steps listed in Section 2.1.

OBJECTIVES

1 Solve a formula for a specified variable.

2 Solve word problems using formulas.

3 Work with percent and percentages.

1. Solve $I = prt$ for each of the following.

   (a) p

   (b) r

ANSWERS
1. (a) $p = \dfrac{I}{rt}$ (b) $r = \dfrac{I}{pt}$

2. Solve the formula

$$m = 2k + 3b$$

for each of the following.

(a) k

(b) b

3. Solve the formula

$$P = a + b + c$$

for each of the following.

(a) b

(b) c

ANSWERS
2. (a) $k = \dfrac{m - 3b}{2}$ (b) $b = \dfrac{m - 2k}{3}$
3. (a) $b = P - a - c$
    (b) $c = P - a - b$

EXAMPLE 2 Solving for a Specified Variable
Solve the formula $P = 2L + 2W$ for W.

This formula gives the relationship between the perimeter (P) of a rectangle, and the length (L) and width (W) of the rectangle. See Figure 1.

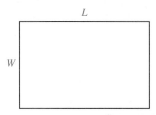

Perimeter, P, distance around the rectangle, is given by
$P = 2L + 2W$.

FIGURE 1

Solve the formula for W by getting W alone on one side of the equals sign. Start by subtracting $2L$ from both sides.

$$P = 2L + 2W$$
$$P - 2L = 2L + 2W - 2L \quad \text{Subtract } 2L.$$
$$P - 2L = 2W$$
$$\frac{P - 2L}{2} = \frac{2W}{2} \quad \text{Divide by 2.}$$
$$\frac{P - 2L}{2} = W \quad \blacksquare$$

■ **WORK PROBLEMS 2 AND 3 AT THE SIDE.**

EXAMPLE 3 Solving for a Specified Variable
Given the volume and height of a right pyramid, write a formula for the area of the base.

Look in the appendix to find the formula

$$V = \frac{1}{3}Bh.$$

A sketch is shown in Figure 2.

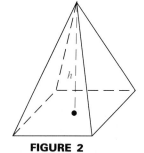

FIGURE 2

In the formula, V is the volume, h is the height, and B is the area of the base. To solve for the area of the base, B, first multiply both sides by 3.

$$3V = 3\left(\frac{1}{3}Bh\right)$$
$$= \left(3 \cdot \frac{1}{3}\right) Bh$$
$$= Bh$$

Now divide both sides of $3V = Bh$ by h. This will leave just B on the right, as required.

$$\frac{3V}{h} = \frac{Bh}{h} \quad \text{Divide by } h.$$

$$\frac{3V}{h} = B \quad \text{or} \quad B = \frac{3V}{h} \quad \blacksquare$$

WORK PROBLEM 4 AT THE SIDE.

2 The next examples show how to solve word problems using formulas.

EXAMPLE 4 Finding Average Speed

Ben Whitney found that on the average it took $\frac{3}{4}$ hour each day to drive a distance of 15 miles to work. What was his average speed?

This problem requires the formula for distance, $d = rt$, where d represents distance traveled; r, rate of speed; and t, time elapsed. Find the rate of speed, r, by solving the formula for r.

$$d = rt$$

$$\frac{d}{t} = \frac{rt}{t} \quad \text{Divide by } t.$$

$$\frac{d}{t} = r$$

Now find r by substituting the given values of d and t into this formula.

$$r = \frac{15}{\frac{3}{4}} = 20$$

Whitney's average speed was 20 miles per hour. $\blacksquare$

WORK PROBLEM 5 AT THE SIDE.

3 A important everyday use of mathematics involves the concept of percent. The word **percent** means "per one hundred." Percent is written with the symbol %. One percent means "one per one hundred" or "one one-hundredth."

$$1\% = .01 \quad \text{or} \quad 1\% = \frac{1}{100}$$

A **percentage** is part of a whole. For example, since

$$25\% = \frac{25}{100} = \frac{1}{4}$$

of a whole, 25% of 400 is $\frac{1}{4}$ of 400, or 100.

4. (a) Given the perimeter and sides a and c of a triangle, solve the appropriate formula for side b.

(b) Given the area and base of a triangle, solve the appropriate formula for the height.

5. (a) A triangle has an area of 36 and a base of 12. Find its height.

(b) The distance is 500 miles and the rate is 25 miles per hour. Find the time.

ANSWERS

4. (a) $b = P - a - c$ (b) $h = \frac{2A}{b}$

5. (a) 6 (b) 20 hours

CHAPTER 2 LINEAR EQUATIONS AND INEQUALITIES

6. (a) A mixture of gasoline and oil contains 20 ounces, 1 of which is oil. What percent of the mixture is oil?

(b) An automobile salesman earns an 8% commission on every car he sells. How much does he earn on a car that sells for $12,000?

ANSWERS
6. (a) 5%   (b) $960

EXAMPLE 5 Finding a Percent and a Percentage

(a) Fifty liters of a mixture of acid and water contains 10 liters of acid. What is the percent of acid in the mixture?

The given amount of the mixture is 50 liters and the part that is acid (percentage) is 10 liters. Thus, the percent of acid in the mixture is

$$\frac{10}{50} = \frac{20}{100} = 20\%.$$

(b) If a savings account balance of $3550 earns 8% interest in one year, how much interest is earned?

Here we must find percentage; that is, what part of the whole is earned as interest? To do this, multiply the percent interest by the account balance.

$$8\% \text{ of } 3550 = (.08)(3550) = 284.00$$

The amount of interest is $284.00. ■

The two types of problems in Example 5 are summarized as follows.

Divide to find percent:

$$\text{Percent} = \frac{\text{Percentage}}{\text{Amount}}.$$

Multiply to find percentage:

$$\text{Percentage} = \text{Percent} \times \text{Amount}.$$

■ **WORK PROBLEM 6 AT THE SIDE.**

2.2 EXERCISES

Solve each formula for the specified variable. See Examples 1 and 2.

1. $d = rt$ for r (distance)

2. $I = prt$ for r (simple interest)

3. $A = bh$ for b
(area of a parallelogram)

4. $P = 2L + 2W$ for L
(perimeter of a rectangle)

5. $P = a + b + c$ for a
(perimeter of a triangle)

6. $V = LWH$ for W
(volume of a rectangular solid)

7. $A = \frac{1}{2}bh$ for h (area of a triangle)

8. $C = 2\pi r$ for r
(circumference of a circle)

9. $S = 2\pi rh + 2\pi r^2$ for h
(surface area of a right circular cylinder)

10. $A = \frac{1}{2}(B + b)h$ for B
(area of a trapezoid)

11. $C = \frac{5}{9}(F - 32)$ for F
(Fahrenheit to Celsius)

12. $F = \frac{9}{5}C + 32$ for C
(Celsius to Fahrenheit)

For each of the following, solve the appropriate formula for the required variable. See Example 3.

13. If the measures of the area and height of a triangle are known, write a formula for the base.

14. A rectangle has a known length and a known area. Write a formula for the width.

15. Given the volume, length, and height of a box, write a formula for the width.

16. If the rate, time, and simple interest for an investment are known, write a formula for the principal.

17. If the height, area, and short base of a trapezoid are known, write a formula for the long base.

18. If the surface area and radius of a right circular cylinder are known, write a formula for the height.

For each of the following, select the appropriate formula, solve the formula for the required variable, and then find the value of the required variable. See Example 4.

19. Faye travels 80 kilometers at a rate of 20 kilometers per hour. What is her travel time?

20. Ed travels 120 miles in 5 hours. What is his rate?

21. The perimeter of a rectangle is 36 meters and the width is 3.5 meters. Find the length.

22. Two sides of a triangle measure 4.6 and 12.2 centimeters. The perimeter is 28.4 centimeters. Find the measure of the third side.

23. An incinerator in the shape of a box is 1.5 meters wide and 5 meters long. If the volume of the incinerator is 75 cubic meters, find the height.

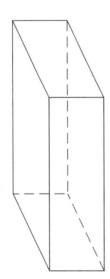

24. The area of a trapezoid is 26 square feet. The short base measures 5 feet and the height 4 feet. What is the length of the long base?

2.2 EXERCISES

25. The Fahrenheit temperature is 41°. Find the Celsius temperature.

26. The Fahrenheit temperature is 212°. Find the Celsius temperature.

27. The Celsius temperature is 40°. Find the Fahrenheit temperature.

28. The Celsius temperature is $-40°$. Find the Fahrenheit temperature.

29. The circumference of a circle is 240π inches. Find the radius of the circle.

30. The circumference of a circle is 60π meters. Find the diameter of the circle.

31. A can has surface area of 12π square inches. The radius of the can is 1 inch. Find the height of the can.

32. The surface area of a can is 32π square inches. The radius of the can is 2 inches. Find the height of the can.

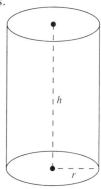

Find the required percent or percentage. See Example 5.

33. Alcohol and water are mixed to give 20 liters of liquid. The mixture contains 2 liters of alcohol. What is the percent of alcohol in the mixture?

34. In 15 liters of a mixture of acid and water, there are 3 liters of pure acid. What percent of the mixture is acid?

35. In a class of 32 students, 24 are boys. What percent of the class is made up of girls?

36. In a class of freshmen and sophomores only, there are 85 students. If 34 are freshmen, what percent of the class are sophomores?

37. A real estate agent earns a 10% commission of an owner's monthly rent while managing the owner's rental house. If the monthly rent is $450, how much does the agent earn each month?

38. Twelve percent of a college student body has a grade point average of 3.0 or better. If there are 1250 students enrolled in the college, how many have a grade point average of 3.0 or better?

39. A savings account balance of $1650 earns 5% annual interest. How much interest is earned in one year?

40. The purchase of a home requires a 20% down payment. If the purchase price is $140,000, how much is the down payment?

Solve each equation for the specified variable.

41. $4x = 7y - bx + 3$ for x

42. $-kx + 9 = 2m - 4x$ for x

43. $\dfrac{x + 5}{4} = 8y - 5$ for x

44. $\dfrac{9 + y}{x - 2} = 7y$ for x

45. $r = \dfrac{k - y}{k}$ for k

46. $y = \dfrac{ay - r}{r}$ for y

SKILL SHARPENERS
Solve each of the following equations. See Section 2.1.

47. $2x + 2(x + 7) = 34$

48. $2y + 2(3y - 4) = 32$

49. $x + .10x = 220$

50. $k - .15k = 680$

51. $.20x + .10(8000 - x) = 1000$

52. $.07x + .05(9000 - x) = 480$

2.3 APPLICATIONS OF LINEAR EQUATIONS

When algebra is used to solve practical problems, it is first necessary to translate the verbal statements of problems into mathematical sentences. Specific methods for making this translation are shown in this section.

1 Word problems tend to have key phrases that translate into mathematical expressions involving the operations of addition, subtraction, multiplication, and division. Translations of some commonly used expressions are listed in the following table.

OBJECTIVES

1. Translate from words to mathematical expressions.
2. Write equations from given information.
3. Distinguish between expressions and equations.
4. Solve geometry problems.
5. Solve percent problems.
6. Solve investment problems.
7. Solve mixture problems.

Verbal expression	Mathematical expression
Addition	
The sum of a number and 7	$x + 7$
6 more than a number	$x + 6$
3 plus 8	$3 + 8$
24 added to a number	$x + 24$
A number increased by 5	$x + 5$
The sum of two numbers	$x + y$
Subtraction	
2 less than a number	$x - 2$
12 minus a number	$12 - x$
A number decreased by 12	$x - 12$
The difference between two numbers	$x - y$
A number subtracted from 10	$10 - x$
Multiplication	
16 times a number	$16x$
Some number multiplied by 6	$6x$
$\frac{2}{3}$ of some number (used only with fractions and percent)	$\left(\frac{2}{3}\right)x$
13% of a number	$.13x$
Twice (2 times) a number	$2x$
The product of two numbers	xy
Division	
The quotient of 8 and some nonzero number	$\frac{8}{x}$
A number divided by 13	$\frac{x}{13}$
The ratio of two nonzero numbers or The quotient of two nonzero numbers	$\frac{x}{y}$

CHAPTER 2 LINEAR EQUATIONS AND INEQUALITIES

1. Write these verbal expressions as mathematical expressions. Use x as the variable.

    (a) 9 added to a number

    (b) The difference between 7 and a number

    (c) Four times a number

    (d) The product of -6 and a number

    (e) The quotient of 7 and a nonzero number

2. Translate each verbal sentence into an equation.

    (a) The sum of a number and 6 is 28.

    (b) If twice a number is decreased by 3, the result is 17.

    (c) The product of a number and 7 is twice the number plus 12.

    (d) The quotient of a number and 6, added to twice the number, is 7.

    (e) The quotient of a number and 3 is 8.

    (f) The sum of a number and twice the number is 3.

ANSWERS
1. (a) $9 + x$ or $x + 9$ (b) $7 - x$
   (c) $4x$ (d) $-6x$ (e) $\dfrac{7}{x}$
2. (a) $x + 6 = 28$ (b) $2x - 3 = 17$
   (c) $7x = 2x + 12$ (d) $\dfrac{x}{6} + 2x = 7$
   (e) $\dfrac{x}{3} = 8$ (f) $x + 2x = 3$

CAUTION Because subtraction and division are not commutative operations, it is important to correctly translate expressions involving them. For example, "2 less than a number" is translated as $x - 2$, *not* $2 - x$. "A number subtracted from 10" is expressed as $10 - x$, not $x - 10$. For division, it is understood that the number doing the dividing is the denominator, and the number that is divided is the numerator. For example, "a number divided by 13" and "13 dividing x" both translate as $\dfrac{x}{13}$.

■ WORK PROBLEM 1 AT THE SIDE.

2 The symbol for equality, =, is often indicated by the word "is." In fact, since equal mathematical expressions represent names for the same number, any words that indicate the idea of "sameness" indicate translation to =.

EXAMPLE 1 Translating Words into Equations
Translate the following verbal sentences into equations.

Verbal sentence	Equation
Twice a number, **decreased by** 3, **is** 42.	$2x - 3 = 42$
If the **product of a number and 12** is decreased by 7, the result is 105.	$12x - 7 = 105$
The **quotient of a number and the number plus 4** is 28.	$\dfrac{x}{x + 4} = 28$
The **quotient of a number and 4,** plus the number, is 10.	$\dfrac{x}{4} + x = 10$ ■

■ WORK PROBLEM 2 AT THE SIDE.

3 It is important to be able to distinguish between algebraic expressions and equations. An expression translates as a phrase; an equation includes the = symbol and translates as a sentence.

EXAMPLE 2 Distinguishing Between Equations and Expressions
For each of the following, decide whether it is an equation or an expression.

(a) $2(3 + x) - 4x + 7$

There is no equals sign, so this is an expression.

2.3 APPLICATIONS OF LINEAR EQUATIONS

(b) $2(3 + x) - 4x + 7 = -1$

Because of the equals sign, this is an equation. Verify that the expression in part (a) simplifies to $-2x + 13$, and the equation in part (b) has solution set $\{7\}$. ■

WORK PROBLEM 3 AT THE SIDE.

The rest of this section consists of examples of some common types of word problems. The following steps are helpful in solving word problems, and are used in Examples 3–6.

3. For each of the following, decide whether it is an equation or an expression.

(a) $5x - 3(x + 2) = 7$

(b) $5x - 3(x + 2)$

SOLVING A WORD PROBLEM

Step 1 Read the problem carefully. Decide what is given and what must be found. Use sketches as necessary.

Step 2 Choose a variable and write down exactly what it represents. (Do not make the mistake of omitting this step.)

Step 3 Write an equation using the information given in the problem.

Step 4 Solve the equation.

Step 5 Make sure you have answered the question of the problem.

Step 6 Check your answer *back in the words of the original problem,* not in the equation you obtained from the words.

4 The first example comes from geometry.

EXAMPLE 3 Solving a Geometry Problem
The length of a rectangle is 1 centimeter more than twice the width. The perimeter of the rectangle is 110 centimeters. Find the length and the width of the rectangle.

Step 1 What must be found?   The length and width of the rectangle

What is given?   The length is 1 more than twice the width; the perimeter is 110.

Make a sketch, as in Figure 3.

FIGURE 3

Step 2 Let W = the width; then $1 + 2W$ = the length.

Step 3 The perimeter of a rectangle is given by the formula $P = 2L + 2W$.

$P = 2L + 2W$

$110 = 2(1 + 2W) + 2W.$ Let $L = 1 + 2W$ and $P = 110$.

ANSWERS
3. (a) equation
   (b) expression

CHAPTER 2 LINEAR EQUATIONS AND INEQUALITIES

4. The length of a rectangle is 5 centimeters more than its width. The perimeter is five times the width. What are the dimensions of the rectangle?

Step 4 Solve the equation obtained in Step 3.

$$110 = 2(1 + 2W) + 2W$$
$$110 = 2 + 4W + 2W \qquad \text{Distributive property}$$
$$110 = 2 + 6W \qquad \text{Combine terms.}$$
$$110 - 2 = 2 + 6W - 2 \qquad \text{Subtract 2.}$$
$$108 = 6W$$
$$18 = W \qquad \text{Divide by 6.}$$

Step 5 The width of the rectangle is 18 centimeters and the length is $1 + 2(18) = 37$ centimeters.

Step 6 Check the answer by substituting these dimensions back into the words of the original problem. ∎

■ WORK PROBLEM 4 AT THE SIDE.

5 The next example involves percent. Recall from the previous section that percent means "per one hundred," so 5% means .05, 14% means .14, and so on.

EXAMPLE 4 Solving a Percent Problem

Karen Wilson paid $46.20 for a cassette tape player, including 5% sales tax. What was the price of the tape player?

Find: the price of the tape player.
Given: the price, increased by 5% tax, was $46.20.

Let x = the price of the tape player;
$.05x$ = the sales tax.

Since the price plus the tax equals $46.20, the equation is

$$x + .05x = 46.20.$$

Solve the equation.

$$\mathbf{1}x + \mathbf{.05}x = 46.20 \qquad x = 1x$$
$$\mathbf{1.05}x = 46.20 \qquad \text{Combine terms.}$$
$$x = \frac{46.20}{1.05} = 44 \qquad \text{Divide by 1.05.}$$

The price of the tape player was $44. ∎

CAUTION Watch out for two common errors that occur in solving problems like the one in Example 4. First, do not try to find 5% of the total price ($46.20) and then subtract from the total. The 5% tax is applied to the price of the *tape player,* and *not to the amount paid.* Second, avoid writing the equation incorrectly. It would be *wrong* to write the equation as

$$x + .05 = 46.20.$$

.05 *must* be multiplied by *x*, as shown in the example.

ANSWERS
4. 10 centimeters by 15 centimeters

2.3 APPLICATIONS OF LINEAR EQUATIONS

WORK PROBLEM 5 AT THE SIDE.

6 The next example shows how we can use linear equations to solve certain types of investment problems.

EXAMPLE 5 Solving an Investment Problem

After winning the state lottery, Grey Thornton had $40,000 to invest. He put part of the money in an account paying 14% interest, with the remainder going into stocks paying 12%. The total annual income from the investments is $5160. Find the amount invested at each rate.

Find: the amount invested at each rate.
Given: $40,000 invested, part at 14% and part at 12%; total annual income of $5160.

Let x = the amount invested at 14%;
$40,000 - x$ = the amount invested at 12%.

The formula for interest is $I = prt$. The interest earned in one year at 14% is

interest at 14% $= x(.14)(1) = $ **.14x.**

The interest at 12% is

interest at 12% $= (40,000 - x)(.12)(1) = $ **.12(40,000 − x).**

Since the total interest is $5160,

interest at 14% + interest at 12% = total interest
.14x + .12(40,000 − x) = 5160.

Solve the equation.

$.14x + 4800 - .12x = 5160$
$.02x + 4800 = 5160$ Combine terms.
$.02x = 360$ Subtract 4800.
$x = 18,000$ Divide by .02.

Thornton has $18,000 invested at 14% and $40,000 − $18,000 = $22,000 at 12%. Check by finding the annual interest at each rate; they should total $5160. ■

WORK PROBLEM 6 AT THE SIDE.

---

5. (a) A number increased by 15% is 287.5. Find the number.

(b) Michael Raymond was paid $162 for a week's work at his part-time job after 10% deductions for taxes. How much did he earn before the deductions were made?

6. (a) A woman invests $72,000 in two ways—some at 15% and some at 18%. Her total annual interest income is $11,460. Find the amount invested at each rate.

(b) A man has $34,000 to invest. He invests some at 17% and the balance at 20%. His total annual interest income is $6245. Find the amount invested at each rate.

ANSWERS
5. (a) 250   (b) $180
6. (a) $50,000 at 15%; $22,000 at 18%
   (b) $18,500 at 17%; $15,500 at 20%

7. (a) How many liters of a 10% solution should be mixed with 60 liters of a 25% solution to get a 15% solution?

(b) How many pounds of candy worth $8 per pound should be mixed with 100 pounds of candy worth $4 per pound to get a mixture that can be sold for $7 per pound?

7 Mixture problems involving rates of concentration may be solved with linear equations.

EXAMPLE 6 Solving a Mixture Problem

A chemist must mix 8 liters of a 40% solution of acid with some 70% solution to get a mixture that is a 50% solution. How much of the 70% solution should be used?

The information in the problem is illustrated in Figure 4.

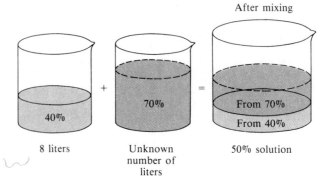

FIGURE 4

Find: the amount of 70% solution that should be used.
Given: 8 liters of 40% solution to mix with 70% solution to get a 50% solution.

Let x = the number of liters of 70% solution to be used.

Use the given information to complete the following table.

Strength	Liters of solution	Liters of pure acid
40%	8	.40(8) = 3.2
70%	x	.70x
50%	8 + x	.50(8 + x)

Sum must equal

The numbers in the right-hand column were found by multiplying the strengths and the numbers of liters. The number of liters of pure acid in the 40% solution plus the number of liters in the 70% solution must equal the number of liters in the 50% solution.

$$3.2 + .70x = .50(8 + x)$$
$$3.2 + .70x = 4 + .50x \quad \text{Distributive property}$$
$$.20x = .8 \quad \text{Subtract 3.2 and .50x.}$$
$$x = 4 \quad \text{Divide by .20.}$$

The chemist must use 4 liters of the 70% solution. ■

■■■ **WORK PROBLEM 7 AT THE SIDE.**

ANSWERS
7. (a) 120 liters **(b)** 300 pounds

2.3 EXERCISES

Translate the following verbal phrases into mathematical expressions. Use x to represent the unknown number.

1. A number decreased by 9

2. A number increased by 15

3. 13 increased by a number

4. 7 decreased by a number

5. The product of -2 and a number

6. The product of a number and 10

7. 7 less than a number

8. 4 more than a number

9. -1 increased by 6 times a number

10. 13 decreased by 8 times a number

11. The ratio of a number and -9

12. The ratio of a number and 5

13. The quotient of a number and 12

14. The quotient of 12 and a nonzero number

15. $\frac{2}{3}$ of a number

16. $\frac{8}{9}$ of a number

17. The quotient of 3 more than a number, and 6

18. The quotient of 12 less than a number, and 15

For each of the following, select a variable for the unknown. Next, write an equation and use it to solve the problem. See Example 1.

19. The sum of a number and -4 is -23. Find the number.

20. The sum of a number and 25 is -18. Find the number.

21. When $\frac{5}{12}$ of a number is subtracted from 7, the result is 17. Find the number.

22. If $\frac{2}{3}$ of a number is added to $\frac{5}{3}$, the result is 2. Find the number.

For each of the following, decide whether it is an equation or an expression. See Example 2.

23. $6(2x + 7) = 5(3x - 9)$

24. $-5(7y + 2) = y + 4$

25. $6(2x + 7) + 5(3x - 9)$

26. $-5(7y + 2) - y + 4$

27. $\dfrac{k}{2} - \dfrac{3k + 7}{3} = 6$

28. $\dfrac{k}{2} - \dfrac{3k + 7}{3} + 6$

Solve each of the following problems from geometry. See Example 3.

29. The length of a rectangle is 2 meters less than twice the width. The perimeter is 38 meters. Find the length and the width.

30. One side of a triangle is 7 meters shorter than the longest side. The third side is 3 meters shorter than the longest side. The perimeter is 26 meters. Find the lengths of the three sides of the triangle.

31. In a triangle with two sides of equal length, the third side is 4 feet less than two times the length of the equal sides. If the perimeter of the triangle is 36 feet, find the lengths of the equal sides.

32. The perimeter of a rectangle is 56 inches more than the length of the rectangle. The width is 10 inches. Find the length of the rectangle.

2.3 EXERCISES

Solve each of the following problems. See Examples 4, 5, and 6.

33. A number is decreased by 12% of the number, giving 44. Find the number.

34. After a number is increased by 15% of the number, the result is 69. Find the number.

35. An electronics store offered a VCR at 40% off. The sale price was $336. Find the regular price.

36. Kala Wanersdorfer bought a book that had been reduced by 30%. She paid $17.50. What was the original price?

37. At the end of a day, Jeff Hornsby found that the total cash register receipts at the motel where he works were $286.20. This included sales tax of 8%. Find the amount of the tax.
(*Hint:* Let x = total due the motel
$.08x$ = amount of the tax.)

38. Dongming Wei wants to sell his house. He would like to receive $136,000 after deducting the 8% commission due the real estate agent. How much should he ask for the house? (Round the answer to the nearest dollar.)

39. Mark Fong earned $12,000 last year by giving golf lessons. He invested part at 8% simple interest and the rest at 9%. He made a total of $1050 in interest. How much did he invest at each rate?

40. Kim Falgout won $60,000 on a slot machine in Las Vegas. She invested part at 8% simple interest and the rest at 12%. She earned a total of $6200 in interest. How much was invested at each rate?

41. Evelyn Pourciau invested some money at 8% simple interest and $1000 less than twice this amount at 14%. Her total annual income from the interest was $580. How much was invested at each rate?

42. Mabel Johnston invested some money at 9% simple interest, and $5000 more than 3 times this amount at 10%. She earned $2840 in interest. How much did she invest at each rate?

43. Bill Poole has $29,000 invested in bonds paying 11%. How much additional money should he invest in mortgages paying 15% so that the average return on the two investments is 14%?

44. Dina Abo-zahrah placed $14,000 in an account paying 12%. How much additional money should she deposit at 16% so that the average return on the two investments is 15%?

45. Five liters of a 4% acid solution must be mixed with a 10% solution to get a 6% solution. How many liters of the 10% solution are needed?

46. How many liters of a 14% alcohol solution must be mixed with 20 liters of a 50% solution to get a 20% solution?

47. How much pure dye must be added to 2 liters of a 25% dye solution to increase the solution to 40%? (*Hint:* Pure dye is 100% dye.)

48. How much water must be added to 3 gallons of a 4% insecticide solution to reduce the concentration to 3%? (*Hint:* Water is 0% insecticide.)

49. In a chemistry class, 6 liters of a 12% alcohol solution must be mixed with a 20% solution to get a 14% solution. How many liters of the 20% solution are needed?

50. How many liters of a 10% alcohol solution must be mixed with 40 liters of a 50% solution to get a 20% solution?

Solve each of the following problems.

51. A farmer wishes to enclose a rectangular area with 105 meters of fencing in such a way that the length is twice the width and the area is cut into two equal parts. (See the figure below.) What length and width should be used?

52. Pat Kelley has a sheet of tin 12 centimeters by 16 centimeters. He plans to make a box by cutting squares out of each of the four corners and folding up the remaining edges. How large a square should he cut so that the finished box will have a length that is three times its width?

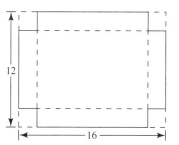

53. It is necessary to have a 40% antifreeze solution in the radiator of a certain car. The radiator now holds 20 liters of 20% solution. How many liters of this should be drained and replaced with 100% antifreeze to get the desired strength?
(*Hint:* The number of liters drained is equal to the number of liters replaced.)

54. A tank holds 80 liters of a chemical solution. Currently, the solution has a strength of 30%. How much of this should be drained and replaced with a 70% solution to get a final strength of 40%?

CHAPTER 2 LINEAR EQUATIONS AND INEQUALITIES

55. Prem Kythe invests $3000 at 8% annual interest. He would like to earn $740 per year in interest. How much should he invest in a second account paying 10% interest in order to accomplish this?

56. Rita Lieux invested some money at 7% interest and the same amount at 10%. Her total interest for the year was $150 less than one-tenth of the total amount she invested. How much did she invest at each rate?

57. The monthly phone bill includes a monthly charge of $10 for local calls plus an additional charge of 50¢ for each toll call in a certain area. A federal tax of 5% is added to the total bill. If all calls were local or within the 50¢ area and the total bill was $17.85, find the number of toll calls made.

58. A grocer buys lettuce for $5.20 a crate. Of the lettuce he buys, 10% cannot be sold. If he charges 40¢ for each head he sells and makes a profit of 10¢ on each head he buys, how many heads of lettuce are in a crate?

SKILL SHARPENERS
Solve each problem. See Section 1.4.

59. Find d if $r = 50$ and $t = 3$. Use $d = rt$.

60. Find r if $d = 75$ and $t = 5$. Use $d = rt$.

61. Find t if $d = 1000$ and $r = 20$. Use $d = rt$.

62. Find p if $b = 12$ and $r = .25$. Use $p = br$.

63. Find b if $p = 4.00$ and $r = .10$. Use $p = br$.

64. Find r if $b = 12$ and $p = .60$. Use $p = br$.

2.4 MORE APPLICATIONS OF LINEAR EQUATIONS

There are two common applications of linear equations that we did not discuss in Section 2.3: money problems and uniform motion problems. Money problems are very similar to the investment problems seen in Section 2.3; uniform motion problems use the formula

$$\text{Distance} = \text{Rate} \times \text{Time}.$$

OBJECTIVES

1. Solve problems involving money.
2. Solve problems involving uniform motion.

1 In problems involving money, use the basic fact that

$$\begin{bmatrix} \text{Number of} \\ \text{monetary units} \\ \text{of the same kind} \end{bmatrix} \times [\text{Denomination}] = \begin{bmatrix} \text{Total monetary} \\ \text{value.} \end{bmatrix}$$

For example, 30 dimes have a monetary value of $30(.10) = 3.00$ dollars. Fifteen five-dollar bills have a value of $15(5) = 75$ dollars.

EXAMPLE 1 Solving a Money Problem

For a bill totaling $5.65, a cashier received 25 coins consisting of nickels and quarters. How many of each type of coin did the cashier receive?

Let x represent the number of nickels;
then $25 - x$ represents the number of quarters.

The information for this problem may be arranged as shown in Figure 5.

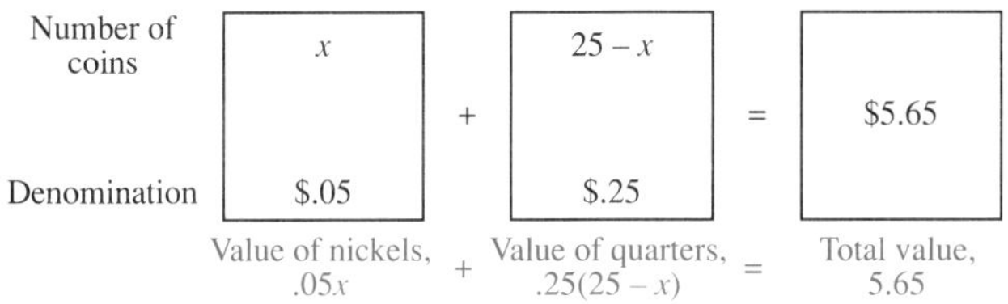

Number of coins	x		$25 - x$		
		+		=	$5.65
Denomination	$.05		$.25		

Value of nickels, $.05x$ + Value of quarters, $.25(25 - x)$ = Total value, $5.65

FIGURE 5

Multiply the numbers of coins by the denominations, and add the results to get 5.65.

$$.05x + .25(25 - x) = 5.65$$
$$5x + 25(25 - x) = 565 \quad \text{Multiply by 100.}$$
$$5x + 625 - 25x = 565 \quad \text{Distributive property}$$
$$-20x = -60 \quad \text{Subtract 625.}$$
$$x = 3 \quad \text{Divide by } -20.$$

The cashier has 3 nickels and $25 - 3 = 22$ quarters. Check to see that the total value of these coins is $5.65. ■

CAUTION Be sure that your answer is reasonable when working problems like Example 1. Since you are dealing with a number of coins, an answer can neither be negative nor a fraction.

1. Mohammed has a box of coins containing only dimes and half-dollars. There are 26 coins, and the total value is $8.60. How many of each type does he have?

WORK PROBLEM 1 AT THE SIDE.

ANSWERS
1. 11 dimes, 15 half-dollars

2. Two cars leave the same town at the same time. One travels north at 60 miles per hour and the other south at 45 miles per hour. In how many hours will they be 420 miles apart?

---

2 Uniform motion problems use the formula Distance = Rate × Time. In this formula, when rate (or speed) is given in miles per hour, time must be given in hours. When solving such problems, draw a sketch to illustrate what is happening in the problem, and make a chart to summarize the information of the problem.

EXAMPLE 2 Solving a Motion Problem

Two cars leave the same place at the same time, one going east and the other west. The eastbound car averages 40 miles per hour, while the westbound car averages 50 miles per hour. In how many hours will they be 300 miles apart?

A sketch shows what is happening in the problem: The cars are going in *opposite* directions. See Figure 6.

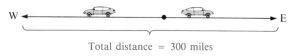

Total distance = 300 miles

FIGURE 6

Let x represent the time traveled by each car. Summarize the information of the problem in a chart.

	Rate	Time	Distance
Eastbound car	40	x	**$40x$**
Westbound car	50	x	**$50x$**

The distances traveled by the cars, $40x$ and $50x$, are obtained from the formula $d = rt$. When the expressions for rate and time are entered in the chart, *fill in the distance expression by multiplying rate by time*.

From the sketch in Figure 6, the sum of the two distances is 300.

$$40x + 50x = 300$$
$$90x = 300 \quad \text{Combine terms.}$$
$$x = \frac{300}{90} \quad \text{Divide by 90.}$$
$$x = \frac{10}{3}$$

The cars travel $\frac{10}{3} = 3\frac{1}{3}$ hours, or 3 hours and 20 minutes. ∎

> **CAUTION** It is a common error to write 300 as the distance for each car in Example 2. Three hundred miles is the *total* distance traveled.

■ **WORK PROBLEM 2 AT THE SIDE.**

ANSWERS
2. 4 hours

Example 2 involved motion in opposite directions. The next example deals with motion in the same direction.

EXAMPLE 3 Solving a Motion Problem

Train A leaves a town at 8:00 A.M., heading east. At 9:00 A.M., Train B leaves on a parallel track, also heading east. The rate of Train A is 80 kilometers per hour, and the rate of Train B is 100 kilometers per hour. How long will it take for Train B to catch up with Train A?

Begin by drawing a sketch that illustrates what is happening in the problem. See Figure 7.

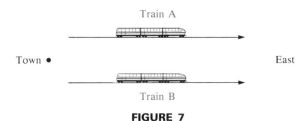

FIGURE 7

Let x = the time Train B will travel before it catches up with Train A (in hours);
then $x + 1$ = time that train A will travel (in hours),

since Train A left one hour earlier. The information is summarized in the following chart. Use $d = rt$ to get the expressions in the last column.

	Rate	Time	Distance
Train A	80	$x+1$	**$80(x+1)$**
Train B	100	x	**$100x$**

As seen in Figure 7, the distances traveled will be the same. Therefore, to get an equation to solve the problem, set the distance expressions equal to each other.

$$80(x + 1) = 100x$$

Now solve the equation.

$80x + 80 = 100x$ Distributive property
$80 = 20x$ Subtract $80x$.
$x = 4$ Divide by 20.

Train B will travel 4 hours before it catches up with Train A. (This will happen at 1:00 P.M. Do you see why?) ∎

WORK PROBLEM 3 AT THE SIDE.

CAUTION Difficulties often arise in uniform motion problems because the step of drawing a sketch is neglected. *The sketch will tell you how to set up the equation.*

---

3. Elayn begins jogging at 5:00 A.M., averaging 3 miles per hour. Clay leaves at 5:30 A.M., following her, averaging 5 miles per hour. How long will it take him to catch up to her? (*Hint:* 30 minutes = $\frac{1}{2}$ hour.)

ANSWERS

3. $\frac{3}{4}$ hour or 45 minutes

Historical Reflections

George Polya (1887–1985)

While there is no shortcut to proficiency at solving word problems, there is a method that provides an excellent general outline for problem solving. It was proposed by George Polya, in his classic book *How to Solve It*.

1. Understand the problem.
2. Devise a plan.
3. Carry out the plan.
4. Look back.

Polya, a native of Budapest, Hungary, wrote more than 250 papers in many languages, as well as a number of books. He was a brilliant lecturer and teacher, and numerous mathematical properties and theorems bear his name. He once was asked why so many good mathematicians came out of Hungary at the turn of the century. He theorized that it was because mathematics was the cheapest science: It did not require any expensive equipment, only pencil and paper. He also was quoted as saying, "I was not smart enough to become a physicist and too smart to be a philosopher, so I chose mathematics."

Art: Courtesy Stanford University

2.4 EXERCISES

Solve each of the following problems. See Example 1.

1. Mary Ann Parke has 30 coins in her change purse, consisting of pennies and nickels. The total value of the money is $.94. How many of each type of coin does she have?

2. Meredith Many has 28 coins in her pocket, consisting of nickels and dimes. The total value of the money is $2.70. How many of each type of coin does she have?

3. Leslie Cobar's piggy bank has 45 coins. Some are quarters and the rest are half-dollars. If the total value of the coins is $17.50, how many of each type does she have?

4. Charles Rees has a jar in his office that contains 39 coins. Some are pennies and the rest are dimes. If the total value of the coins is $2.55, how many of each type does he have?

5. Gary Gundersen has a box of coins that he uses when playing poker with his friends. The box currently contains 40 coins, consisting of pennies, dimes, and quarters. The number of pennies is equal to the number of dimes, and the total value is $8.05. How many of each type of coin does he have in the box?

6. Shannon Mulkey found some coins while looking under her sofa pillows. There were equal numbers of nickels and quarters, and twice as many half-dollars as quarters. If she found $19.50 in all, how many of each type of coin did she find?

7. In the nineteenth century, the United States minted two-cent and three-cent pieces. Charles Steib has three times as many three-cent pieces as two-cent pieces, and the face value of these coins is $1.10. How many of each type does he have?

8. Gerry Moore collects U.S. gold coins. He has a collection of 30 coins. Some are $10 coins and the rest are $20 coins. If the face value of the coins is $540, how many of each type does he have?

9. The school production of *The Music Man* was a big success. For opening night, 300 tickets were sold. Students paid $1.50 each, while non-students paid $3.50 each. If a total of $810 was collected, how many students and how many non-students attended?

10. A total of 550 people attended a Frankie Valli and the Four Seasons concert last night. Floor tickets cost $10 each, while balcony tickets cost $7 each. If a total of $5200 was collected, how many of each type of ticket were sold?

11. The matinee showing of *The Wizard of Oz* was attended by 54 people. Children three years old or younger were admitted free, and children between four and twelve paid $2. Anyone else paid $3. If the same number of children three or younger and children between four and twelve attended, and $82 was collected, how many of each type of ticket were sold?

12. On one day, 660 people visited the local zoo. Senior citizens attended free, children paid $3, and adults paid $5. If twice as many children visited as senior citizens, and $2292 was collected, how many of each age group visited the zoo that day?

Solve each of the following problems. See Examples 2 and 3.

13. Two cars leave a town at the same time. One travels east at 50 miles per hour, and the other travels west at 55 miles per hour. In how many hours will they be 315 miles apart?

r	t	d

14. Two planes leave O'Hare Airport in Chicago at the same time. One travels east at 550 miles per hour, and the other travels west at 500 miles per hour. How long will it take for the planes to be 2100 miles apart?

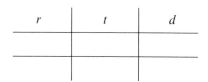

15. A train leaves Coon Rapids, Minnesota, and travels north at 85 kilometers per hour. Another train leaves at the same time and travels south at 95 kilometers per hour. How long will it take before they are 990 kilometers apart?

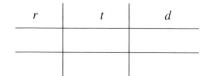

16. Two steamers leave a port on a river at the same time, traveling in opposite directions. Each is traveling 12 miles per hour. How long will it take for them to be 66 miles apart?

r	t	d

2.4 EXERCISES

17. A pleasure boat on the Mississippi River traveled from Baton Rouge to New Orleans with a stop at White Castle. On the first part of the trip, the boat traveled at an average speed of 10 miles per hour. From White Castle to New Orleans the average speed was 15 miles per hour. The entire trip covered 100 miles. How long did the entire trip take if the two parts each took the same number of hours?

r	t	d

18. A jet airliner flew across the United States at an average speed of 500 miles per hour. It then continued on across the Atlantic at an average speed of 530 miles per hour. If the entire trip covered 14,420 miles and both parts of the trip took the same number of hours, how many hours did the whole trip take?

r	t	d

19. Carl leaves his house on his bicycle at 9:30 A.M. and averages 5 miles per hour. His wife, Karen, leaves at 10:00 A.M., following the same path and averaging 8 miles per hour. How long will it take for Karen to catch up with Carl?

r	t	d

20. Joey and Liz commute to work, traveling in opposite directions. Joey leaves the house at 7:00 A.M. and averages 35 miles per hour. Liz leaves at 7:15 A.M. and averages 40 miles per hour. At what time will they be 65 miles apart?

r	t	d

21. Maria Gutierrez can get to school in $\frac{1}{4}$ hour if she rides her bike. It takes her $\frac{3}{4}$ hour if she walks. Her speed when walking is 10 miles per hour slower than her speed when riding. What is her speed when she rides?

r	t	d

22. On an automobile trip, Susan Blohm maintained a steady speed for the first two hours. Rush hour traffic slowed her speed by 25 miles per hour for the last part of the trip. The entire trip, a distance of 150 miles, took $2\frac{1}{2}$ hours. What was her speed during the first part of the trip?

r	t	d

89

23. When Jessie drives her car to work, the trip takes $\frac{1}{2}$ hour. When she rides the bus, it takes $\frac{3}{4}$ hour. The average speed of the bus is 12 miles per hour less than her speed when driving. Find the distance she travels to work.

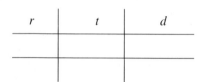

24. On a 100-mile trip to the mountains, the Nguyen family traveled at a steady speed for the first hour. Their speed was 16 miles per hour slower during the second hour of the trip. Find their speed during the first hour.

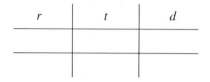

25. Steve leaves Nashville to visit his cousin David in Napa, 80 miles away. He travels at an average speed of 50 miles per hour. One-half hour later David leaves to visit Steve, traveling at an average speed of 60 miles per hour. How long after David leaves will it be before they meet?

26. In a run for charity Janet runs at a speed of 5 miles per hour. Paula leaves 10 minutes after Janet and runs at 6 miles per hour. How long will it take for Paula to catch up with Janet? (*Hint:* Change minutes to hours.)

SKILL SHARPENERS
Graph each of the following intervals. See Section 1.2.

27. $(-\infty, -2]$

28. $(4, \infty)$

29. $(3, \infty)$

30. $(-\infty, -1]$

31. $(-2, 5)$

32. $[0, 3]$

33. $[3, 9)$

34. $(-5, 0]$

2.5 LINEAR INEQUALITIES IN ONE VARIABLE

A *linear inequality in one variable* is an inequality such as

$$x + 5 < 2, \quad y - 3 \geq 5, \quad \text{or} \quad 2k + 5 \leq 10.$$

A linear inequality in one variable can be written in the form

$$ax + b < c$$

where a, b, and c are real numbers, with $a \neq 0$.

(Throughout this section the definitions and rules are given only for $<$, but they are also valid for $>$, $\leq$, and $\geq$.)

OBJECTIVES

1. Solve linear inequalities using the addition property.
2. Solve linear inequalities using the multiplication property.
3. Solve linear inequalities with three parts.
4. Solve word problems with inequalities.

1 We solve an inequality by finding all the numbers that make the inequality true. **Equivalent inequalities** are inequalities with the same solution set. Usually, an inequality has an infinite number of solutions. These solutions are found, as are solutions of equations, by producing a series of simpler equivalent inequalities. The inequalities in this chain of equivalent inequalities can be found with the **addition and multiplication properties of inequality.** We discuss the addition property first.

ADDITION PROPERTY OF INEQUALITY

For all real numbers a, b, and c, the inequalities

$$a < b \quad \text{and} \quad a + c < b + c$$

are equivalent. (The same number may be added to both sides of an inequality.)

As with equations, the addition property can be used to *subtract* the same number from both sides of an inequality.

EXAMPLE 1 Solving a Linear Inequality
Solve each inequality and graph the solution set.

(a) $x - 7 \leq -12$ Add 7 to both sides. $x - 7 + 7 \leq -12 + 7$
$$x \leq -5$$

The solution set, $(-\infty, -5]$, is graphed in Figure 8.

FIGURE 8

(b) $3m > 2m + 14$

First subtract $2m$ from each side.

$$3m - 2m > 2m + 14 - 2m \quad \text{Subtract } 2m.$$
$$m > 14$$

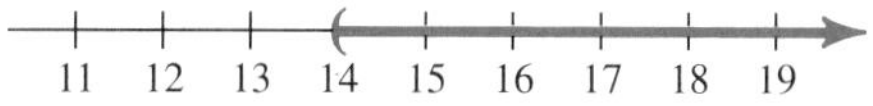

The solution set, $(14, \infty)$, is shown in Figure 9. ■

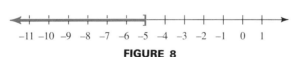

FIGURE 9

WORK PROBLEM 1 AT THE SIDE.

1. Find the solution set of each inequality and graph the solution set.

   (a) $p + 6 \leq 8$

   (b) $k - 5 \geq 1$

   (c) $8y < 7y - 6$

ANSWERS
1. (a) $(-\infty, 2]$

   (b) $[6, \infty)$

   (c) $(-\infty, -6)$

CHAPTER 2 LINEAR EQUATIONS AND INEQUALITIES

2. Multiply both sides of each of the following by -5. Then insert the correct symbol, either $<$ or $>$, in the middle blank.

(a) $7 < 8$

-35 _____ -40

(b) $-1 > -4$

5 _____ _____

2

Solving an inequality such as $3x \leq 15$ requires dividing both sides by 3. This is done with the multiplication property of inequality, which is a little more involved than the corresponding property for equality. To see how this property works, start with the true statement

$$-2 < 5.$$

Multiply both sides by, say, 8.

$-2(\mathbf{8}) < 5(\mathbf{8})$ Multiply by 8.
$-16 < 40$ True

This gives a true statement. Start again with $-2 < 5$, and this time multiply both sides by -8.

$-2(\mathbf{-8}) < 5(\mathbf{-8})$ Multiply by -8.
$16 < -40$ False

The result, $16 < -40$, is false. To make it true, change the direction of the inequality symbol to get

$$16 > -40.$$

■ **WORK PROBLEM 2 AT THE SIDE.**

As these examples suggest, multiplying both sides of an inequality by a negative number requires that the direction of the inequality symbol be reversed. Because division is defined in terms of multiplication, the direction of the inequality symbol must also be reversed when dividing by a negative number.

CAUTION   It is a common error to forget to reverse the direction of the inequality sign when multiplying or dividing by a negative number!

The results of multiplying inequalities by positive and negative numbers are summarized here.

MULTIPLICATION PROPERTY OF INEQUALITY

For all real numbers a, b, and c, with $c \neq 0$,

the inequalities

$$a < b \quad \text{and} \quad ac < bc$$

are equivalent if $c > 0$;

and the inequalities

$$a < b \quad \text{and} \quad ac > bc$$

are equivalent if $c < 0$.

(Both sides of an inequality may be multiplied by a *positive* number without changing the direction of the inequality symbol. Multiplying [or dividing] by a *negative* number requires that the inequality symbol be reversed.)

ANSWERS
2. (a) $>$ (b) $<$ 20

2.5 LINEAR INEQUALITIES IN ONE VARIABLE

EXAMPLE 2 Solving a Linear Inequality

Solve each inequality and graph the solution set.

(a) $5m \leq -30$

Use the multiplication property to divide both sides by 5. Since $5 > 0$, do *not* reverse the inequality symbol.

$$5m \leq -30$$
$$\frac{5m}{5} \leq \frac{-30}{5} \quad \text{Divide by 5.}$$
$$m \leq -6$$

The solution set, $(-\infty, -6]$ is graphed in Figure 10.

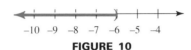

FIGURE 10

(b) $-4k \geq 32$

Divide both sides by -4. Since $-4 < 0$, the inequality symbol must be reversed.

$$-4k \geq 32$$
$$\frac{-4k}{-4} \leq \frac{32}{-4} \quad \text{Change} \geq \text{to} \leq.$$
$$k \leq -8$$

Figure 11 shows the graph of the solution set, $(-\infty, -8]$. ∎

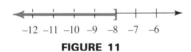

FIGURE 11

WORK PROBLEM 3 AT THE SIDE.

The steps required for solving a linear inequality are given below.

SOLVING A LINEAR INEQUALITY

Step 1 Simplify each side of the inequality separately by combining like terms and clearing parentheses as needed.

Step 2 Use the addition property of inequality to change the inequality so that the variable is on one side.

Step 3 Use the multiplication property to change the inequality to the form $x < k$ or $x > k$.

Remember to change the direction of the inequality symbol **only** when multiplying or dividing both sides of an inequality by a **negative** number, and never otherwise.

---

3. Graph the solution set of each inequality.

(a) $2y < -10$

(b) $7k \geq 8$

(c) $-9m < -81$

ANSWERS

3. (a)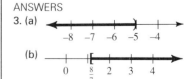

(b)

(c)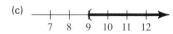

CHAPTER 2 LINEAR EQUATIONS AND INEQUALITIES

4. Graph the solution set of each inequality.

   (a) $-y \leq 2$

   (b) $-z > -11$

   (c) $4(x + 2) \geq 6x - 8$

   (d) $5 - 3(m - 1) \leq 2(m + 3) + 1$

EXAMPLE 3 Solving a Linear Inequality

Solve $-3(x + 4) + 2 \geq 8 - x$. Graph the solution set.

To begin, use the distributive property on the left side of the inequality.

$-3x - 12 + 2 \geq 8 - x$ Distributive property

$-3x - 10 \geq 8 - x$ Combine terms.

Next, use the addition property. First add x to both sides, then add 10 to both sides.

$-3x - 10 + x \geq 8 - x + x$ Add x.

$-2x - 10 \geq 8$

$-2x - 10 + 10 \geq 8 + 10$ Add 10.

$-2x \geq 18$

Finally, use the multiplication property and divide both sides of the inequality by the negative number -2. Dividing by a negative number means that $\geq$ will change to $\leq$.

$$\frac{-2x}{-2} \leq \frac{18}{-2}$$ Divide by -2.

$x \leq -9$

Figure 12 shows the graph of the solution set, $(-\infty, -9]$. ∎

FIGURE 12

EXAMPLE 4 Solving a Linear Inequality

Solve $2 - 4(r - 3) < 3(5 - r) + 5$. Graph the solution set.

First, simplify both sides of the inequality as much as possible.

$2 - 4(r - 3) < 3(5 - r) + 5$

$2 - 4r + 12 < 15 - 3r + 5$ Distributive property

$14 - 4r < 20 - 3r$ Combine terms.

$14 - 4r + 3r < 20 - 3r + 3r$ Add $3r$.

$14 - r < 20$

$14 - r - 14 < 20 - 14$ Subtract 14.

$-r < 6$

We want r, not $-r$. To get r, multiply both sides of the inequality by -1. Since -1 is negative, change the direction of the inequality symbol.

$-r < 6$

$(-1)(-r) > (-1)(6)$ Multiply by -1 and change $<$ to $>$.

$r > -6$

The solution set, $(-6, \infty)$, is graphed in Figure 13. ∎

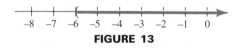

FIGURE 13

▪ **WORK PROBLEM 4 AT THE SIDE.**

ANSWERS

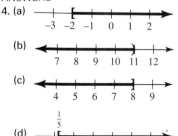

94

2.5 LINEAR INEQUALITIES IN ONE VARIABLE

3 In further work in mathematics, it is sometimes necessary to work with an inequality such as $3 < x + 2 < 8$ where $x + 2$ is *between* 3 and 8. To solve this inequality, subtract 2 from each of the three parts of the inequality, giving

$$3 - 2 < x + 2 - 2 < 8 - 2 \quad \text{Subtract 2.}$$
$$1 < x < 6.$$

The solution set, $(1, 6)$, is graphed in Figure 14.

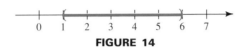

FIGURE 14

EXAMPLE 5 Solving a Three-Part Inequality

Solve the inequality $-2 \leq 3k - 1 \leq 5$ and graph the solution set.

Begin by adding 1 to each of the three parts.

$$-2 + 1 \leq 3k - 1 + 1 \leq 5 + 1 \quad \text{Add 1.}$$
$$-1 \leq 3k \leq 6$$

Now divide each of the three parts by the positive number 3.

$$\frac{-1}{3} \leq \frac{3k}{3} \leq \frac{6}{3} \quad \text{Divide by 3.}$$
$$-\frac{1}{3} \leq k \leq 2$$

A graph of the solution, $\left[-\frac{1}{3}, 2\right]$, is shown in Figure 15. ■

FIGURE 15

WORK PROBLEM 5 AT THE SIDE.

The types of solutions to be expected from solving linear equations or linear inequalities are shown below.

SOLUTIONS OF LINEAR EQUATIONS AND INEQUALITIES

Equation or inequality	Typical solution set	Graph of solution set
Linear equation $ax + b = c$	$\{p\}$	——•—— p
Linear inequality $ax + b < c$	$(-\infty, p)$ or (p, ∞)	←——— p
Three-part inequality $c < ax + b < d$	(p, q)	( ) $p \quad q$

5. Graph the solution set of each inequality.

(a) $-3 \leq x - 1 \leq 7$

(b) $5 < 3x - 4 < 9$

(c) $-6 \leq 5m + 8 \leq 0$

ANSWERS

5. (a) [number line with bracket from -2 to 8, marks at -4, -2, 0, 2, 4, 6, 8, 10]

(b)

(c)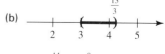

CHAPTER 2 LINEAR EQUATIONS AND INEQUALITIES

6. Teresa has been saving dimes and nickels. She has three times as many nickels as dimes and has at least 48 coins. What is the smallest number of nickels she might have?

4 There are several phrases that denote inequality. Some of them were discussed in Chapter 1. In addition to the familiar "is less than" and "is greater than" (which are examples of **strict** inequalities), the expressions "is no more than," "is at least," and others also denote inequalities. (These are called **non-strict**). Expressions like these sometimes appear in word problems that are solved using inequalities. The chart below shows how these expressions are interpreted.

Word expression	Interpretation	Word expression	Interpretation
a is at least b	$a \geq b$	a is at most b	$a \leq b$
a is no less than b	$a \geq b$	a is no more than b	$a \leq b$

In Examples 6 and 7 word problems are solved with inequalities.

EXAMPLE 6 Solving a Nutrition Problem
A dietician must include in a diet three foods, A, B, and C. He must include twice as many grams of food A as food C, and 5 grams of food B. The three foods must total at most 24 grams. What is the largest amount of food C that he can include?

Let x represent the amount of food C. Then the number of grams of food A will be $2x$. Since the total number of grams of food must not exceed 24,

$$\underset{\downarrow}{\text{Food A}} + \underset{\downarrow}{\text{Food B}} + \underset{\downarrow}{\text{Food C}} \;\; \underset{\downarrow}{\text{At most}} \;\; \underset{\downarrow}{24}$$
$$2x \;\; + \;\; 5 \;\; + \;\; x \;\; \leq \;\; 24.$$

$$3x + 5 \leq 24 \qquad \text{Combine terms.}$$
$$3x \leq 19 \qquad \text{Subtract 5.}$$
$$x \leq \frac{19}{3} \text{ or } 6\frac{1}{3} \qquad \text{Divide by 3.}$$

The diet can include no more than $6\frac{1}{3}$ grams of food C. ∎

■ WORK PROBLEM 6 AT THE SIDE.

7. A student has grades of 92, 90, and 84 on his first three tests. What grade must the student make on his fourth test in order to keep an average of 90 or greater?

EXAMPLE 7 Solving a Grade-Averaging Problem
A student has grades of 88, 86, and 90 on her first three algebra tests. An average of 90 or above will earn an A in the class. What grade must the student make on her fourth test in order to have an A average?

Let x represent the score on the fourth test. Her average must be at least 90. To find the average of four numbers, add them and divide by 4.

$$\frac{88 + 86 + 90 + x}{4} \geq 90$$
$$\frac{264 + x}{4} \geq 90 \qquad \text{Add the scores.}$$
$$264 + x \geq 360 \qquad \text{Multiply by 4.}$$
$$x \geq 96 \qquad \text{Subtract 264.}$$

She must score at least 96 on her fourth test to keep an A average. ∎

ANSWERS
6. 36
7. at least 94

■ WORK PROBLEM 7 AT THE SIDE.

2.5 EXERCISES

Solve each inequality. Graph each solution set. See Examples 1–4.

1. $4x > 8$

2. $6y > 18$

3. $2m \leq -6$

4. $5k \leq -15$

5. $3r + 1 \geq 16$

6. $2m - 5 \geq 15$

7. $\dfrac{3k - 1}{4} > 2$

8. $\dfrac{5z - 6}{8} < 3$

9. $-4x < 12$

10. $-2m \geq 6$

11. $-\dfrac{3}{4}r \geq 21$

12. $-\dfrac{2}{3}y < -10$

13. $-\dfrac{3}{2}y \leq -\dfrac{9}{2}$

14. $-\dfrac{2}{5}x \geq -4$

15. $-1.3m > 3.9$

16. $-2.5y \leq -7.5$

17. $\dfrac{2k - 5}{-4} > 1$

18. $\dfrac{3z - 2}{-5} \leq 4$

19. $3(x + 2) - 5x < x$

20. $2(3k - 5) + 7 > k + 12$

21. $y + 4(2y - 1) \geq 5y$

22. $-2(m - 4) \leq -3(m + 1)$

23. $-(4 + r) + 2 - 3r < -10$

24. $-(9 + k) - 5 + 4k \geq 1$

25. $-3(z - 6) > 2z - 5$

26. $-2(y + 4) \leq 6y + 8$

27. $\dfrac{2}{3}(3k - 1) \geq \dfrac{1}{2}(2k - 3)$

28. $\dfrac{7}{5}(10m - 1) < \dfrac{2}{3}(6m - 5)$

29. $-\dfrac{1}{4}(p + 6) + \dfrac{3}{2}(2p - 5) < 0$

30. $\dfrac{3}{5}(k - 2) - \dfrac{1}{4}(2k - 7) < 3$

Solve each inequality. Graph each solution set. See Example 5.

31. $-3 < x - 5 < 6$

32. $-6 < x + 1 < 8$

33. $9 < k + 5 < 13$

34. $-4 \leq m + 3 \leq 6$

35. $-6 \leq 2z + 4 \leq 12$

36. $-15 < 3p + 6 < -9$

2.5 EXERCISES

37. $-19 \leq 3x - 5 \leq -9$

38. $-16 < 3t + 2 < -11$

39. $-4 \leq \dfrac{2x - 5}{6} \leq 5$

40. $-8 \leq \dfrac{3m + 1}{4} \leq 3$

Solve each inequality. Write its solution set using interval notation. See Examples 1–5.

41. $3x - 8 > -5$

42. $4x + 12 > 0$

43. $5(2x + 3) - 1 < 12$

44. $8(6 + 4x) + 9 < -4$

45. $3 - 2x \geq -7$

46. $5 - 4x \geq 13$

47. $-4 < 6x + 3 < 8$

48. $-7 < 4x + 2 < -3$

49. $-1 \leq 5 - 9x \leq 0$

50. $4 \leq 3 - 2x \leq 10$

Solve each of the following word problems. See Examples 6 and 7.

51. Louise Siedelmann gets scores of 88 and 78 on her first two tests. What score must she make on her third test to keep an average of 80 or greater?

52. Simon Cohen scored 94 and 82 on his first two tests. What score must he make on his third test to keep an average of 90 or greater?

53. A Christmas tree has twice as many green lights as red lights, and it has at least 15 lights. At least how many red lights does it have?

54. Samantha has a total of 815 points so far in her algebra class. At the end of the course she must have 82% of the 1100 points possible in order to get a B. What is the lowest score she can earn on the 100-point final to get a B in the class?

55. A nurse must make sure that Ms. Carlson receives at least 30 units of a certain drug each day. This drug comes from red pills or green pills, each of which provides three units of the drug. The patient must have twice as many red pills as green pills. Find the smallest number of green pills that will satisfy the requirement.

56. A nearby pharmacy school charges a tuition of $6440 annually. Tom makes no more than $1610 per year in his summer job. What is the least number of summers that he must work in order to make enough for one year's tuition?

Find the solution set of each of the following inequalities.

57. $3(2x - 4) - 4x < 2x + 3$

58. $7(4 - x) + 5x < 2(16 - x)$

59. $8(\frac{1}{2}x + 3) < 8(\frac{1}{2}x - 1)$

60. $10x + 2(x - 4) < 12x - 10$

SKILL SHARPENERS

Graph each of the following sets. See Section 1.2.

61. $[-2, \infty)$

62. $(-\infty, 4)$

63. $(-\infty, -4)$

64. $[3, \infty)$

2.6 SET OPERATIONS AND COMPOUND INEQUALITIES

The words *and* and *or* are very important in interpreting certain kinds of equations and inequalities in algebra. They are encountered when studying sets as well. In this section we study the use of these two words as they relate to sets and inequalities.

1 We start by looking at the use of the word "and" with sets. The intersection of sets is defined below.

OBJECTIVES

1. Find the intersection of two sets.
2. Solve compound inequalities with *and*.
3. Find the union of two sets.
4. Solve compound inequalities with *or*.

For any two sets A and B, the **intersection** of A and B, symbolized $A \cap B$, is defined as follows:

$A \cap B = \{x | x$ is an element of A **and** x is an element of $B\}$.

1. Let $A = \{3, 4, 5, 6\}$ and $B = \{5, 6, 7\}$. Find $A \cap B$.

EXAMPLE 1 Finding the Intersection of Two Sets
Let $A = \{1, 2, 3, 4\}$ and $B = \{2, 4, 6\}$. Find $A \cap B$.

The set $A \cap B$ is made up of those elements that belong to both A and B at the same time: the numbers 2 and 4. Therefore,

$$A \cap B = \{1, 2, 3, 4\} \cap \{2, 4, 6\}$$
$$= \{2, 4\}. \blacksquare$$

WORK PROBLEM 1 AT THE SIDE.

2 A **compound inequality** consists of two inequalities linked by a connective word such as *and* or *or*. Examples of compound inequalities are

$$x + 1 \leq 9 \quad \textbf{and} \quad x - 2 \geq 3$$

and

$$2x > 4 \quad \textbf{or} \quad 3x - 6 < 5.$$

To solve a compound inequality with the word *and*, we use the following steps.

SOLVING A COMPOUND INEQUALITY WITH *AND*

Step 1   Solve each inequality in the compound inequality individually.

Step 2   Since the inequalities are joined with *and*, the solution will include all numbers that satisfy both solutions in Step 1 at the same time (the intersection of the solutions).

The next example shows how a compound inequality with *and* is solved.

ANSWER
1. $\{5, 6\}$

2. Graph the solution set of each compound inequality.

   (a) $x < 10$ and $x > 2$

   (b) $x \geq 6$ and $x \leq 12$

   (c) $x + 3 < 1$ and $x - 4 > -12$

3. Solve and graph.

   (a) $2x \geq x - 1$ and $3x \geq 3 + 2x$

   (b) $2y \leq 4y + 8$ and $3y \geq y - 6$

EXAMPLE 2 Solving a Compound Inequality with *and*
Solve the compound inequality

$$x + 1 \leq 9 \quad \text{and} \quad x - 2 \geq 3.$$

Step 1 directs that we solve each inequality in the compound inequality individually.

$$\begin{array}{ccc} x + 1 \leq 9 & \text{and} & x - 2 \geq 3 \\ x + 1 - 1 \leq 9 - 1 & \text{and} & x - 2 + 2 \geq 3 + 2 \\ x \leq 8 & \text{and} & x \geq 5 \end{array}$$

Now we apply Step 2. Since the inequalities are joined with the word *and*, the solution will include all numbers that satisfy both solutions in Step 1 at the same time. Thus, the compound inequality is true whenever $x \leq 8$ and $x \geq 5$ are both true. The top graph in Figure 16 shows $x \leq 8$, and the middle graph shows $x \geq 5$. The bottom graph shows the numbers common to the first two graphs. As shown by this third graph, the solution consists of all numbers between 5 and 8, including both 5 and 8. This is the intersection of the two graphs, and we write it as [5, 8]. ∎

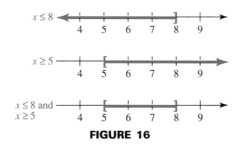

FIGURE 16

▬ WORK PROBLEM 2 AT THE SIDE.

EXAMPLE 3 Solving a Compound Inequality with *and*
Solve the compound inequality

$$-3x - 2 > 4 \quad \text{and} \quad 5x - 1 \leq -21.$$

Begin by solving $-3x - 2 > 4$ and $5x - 1 \leq -21$ separately.

$$\begin{array}{ccc} -3x - 2 > 4 & \text{and} & 5x - 1 \leq -21 \\ -3x > 6 & \text{and} & 5x \leq -20 \\ x < -2 & \text{and} & x \leq -4 \end{array}$$

Now find all values of x that satisfy both conditions; that is, the real numbers that are less than -2 and also less than or equal to -4. As shown by the graphs in Figure 17, the solution set is $(-\infty, -4]$. ∎

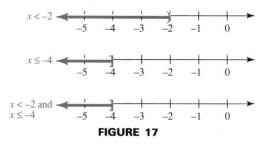

FIGURE 17

▬ WORK PROBLEM 3 AT THE SIDE.

ANSWERS

2. (a)

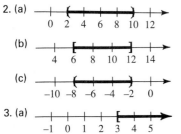

3. (a)

   (b)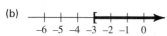

2.6 SET OPERATIONS AND COMPOUND INEQUALITIES

EXAMPLE 4 Solving a Compound Inequality with *and*
Solve $x + 2 < 5$ and $x - 10 > 2$.

First solve each inequality separately.

$$x + 2 < 5 \quad \text{and} \quad x - 10 > 2$$
$$x < 3 \quad \text{and} \quad x > 12$$

There is no number that is both less than 3 and greater than 12, so the given compound sentence has no solution (solution set is ∅). See Figure 18. ∎

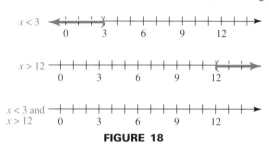

FIGURE 18

WORK PROBLEM 4 AT THE SIDE.

3 We now discuss the union of two sets, which involves the use of the word "or."

For any two sets A and B, the **union** of A and B, symbolized A ∪ B, is defined as follows:

$A \cup B = \{x | x$ is an element of A **or** x is an element of B$\}$.

EXAMPLE 5 Finding the Union of Two Sets
Find the union of the sets $A = \{1, 2, 3, 4\}$ and $B = \{2, 4, 6\}$.

Begin by listing all the elements of set A: 1, 2, 3, 4. Then list any additional elements from set B. In this case the elements 2 and 4 are already listed, so the only additional element is 6. Therefore,

$$A \cup B = \{1, 2, 3, 4\} \cup \{2, 4, 6\} = \{1, 2, 3, 4, 6\}.$$

The union consists of all elements in either *A or B* (or both). ∎

Notice in Example 5, that even though the elements 2 and 4 appeared in both sets A and B, they are only written once in A ∪ B. It is not necessary to write them more than once in the union.

WORK PROBLEM 5 AT THE SIDE.

4 To solve compound inequalities with the word *or*, we use the following steps.

SOLVING A COMPOUND INEQUALITY WITH OR

Step 1 Solve each inequality in the compound inequality individually.

Step 2 Since the inequalities are joined with *or*, the solution will include all numbers that satisfy either one of the solutions in Step 1 (the union of the solutions).

4. Solve $x + 2 > 3$ and $2x + 1 < -3$.

5. Let $A = \{3, 4, 5, 6\}$ and $B = \{5, 6, 7\}$. Find $A \cup B$.

ANSWERS
4. ∅
5. {3, 4, 5, 6, 7}

6. Graph each solution set.

   (a) $x + 2 > 3$ or
   $2x + 1 < -3$

   (b) $y - 1 > 2$ or
   $3y + 5 < 2y + 6$

   (c) $5a + 6 \leq 3a$ or
   $2a - 1 \geq 9$

7. Solve
   $3x - 2 \leq 13$ or
   $x + 5 \geq 7$.

ANSWERS
6. (a), (b), (c) — graphs shown
7. {all real numbers}

The next examples show how to solve a compound inequality with *or*.

EXAMPLE 6 Solving a Compound Inequality with *or*

Solve $6x - 4 < 2x$ or $-3x \leq -9$.

Solve each inequality separately:

$$6x - 4 < 2x \quad \text{or} \quad -3x \leq -9$$
$$4x < 4$$
$$x < 1 \quad \text{or} \quad x \geq 3.$$

The graphs of these results are shown in Figure 19. The third graph gives the combination of these two solutions. This final solution set is written

$$(-\infty, 1) \cup [3, \infty).$$

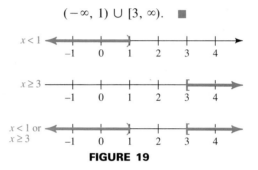

FIGURE 19

CAUTION When interval notation is used to write the solution of Example 6, it *must* be written as

$$(-\infty, 1) \cup [3, \infty).$$

There is no short-cut way to write this solution.

■ **WORK PROBLEM 6 AT THE SIDE.**

EXAMPLE 7 Solving a Compound Inequality with *or*

Solve $-4x + 1 \geq 9$ or $5x + 3 \geq -12$.

Solve each inequality separately.

$$-4x + 1 \geq 9 \quad \text{or} \quad 5x + 3 \geq -12$$
$$-4x \geq 8 \quad \text{or} \quad 5x \geq -15$$
$$x \leq -2 \quad \text{or} \quad x \geq -3$$

The graphs of these solutions are shown in Figure 20. As shown in the figure, every real number is a solution of the compound sentence, since every real number satisfies at least one of the two inequalities. ■

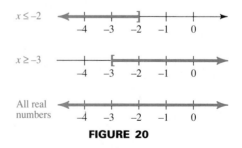

FIGURE 20

■ **WORK PROBLEM 7 AT THE SIDE.**

2.6 EXERCISES

Let $A = \{a, b, c, d, e, f\}$, $B = \{a, c, e\}$, $C = \{a, f\}$, and $D = \{d\}$. List the elements in each of the following sets. See Examples 1 and 5.

1. $A \cap B$
2. $B \cap A$
3. $A \cap D$
4. $B \cap D$

5. $B \cap C$
6. $A \cap \emptyset$
7. $A \cup B$
8. $B \cup D$

9. $B \cup C$
10. $C \cup B$
11. $C \cup D$
12. $D \cup C$

Solve each compound inequality. Graph each solution set. See Examples 2, 3, and 4.

13. $x < 2$ and $x > -1$
14. $x < 4$ and $x > 0$

15. $x \leq 3$ and $x \leq 5$
16. $x \geq 3$ and $x \geq 5$

17. $x \leq 3$ and $x \geq 5$
18. $x \geq 3$ and $x \leq 5$

19. $x \geq -1$ and $x \geq 2$
20. $x \leq -1$ and $x \leq 2$

21. $x \geq -1$ and $x \leq 2$

22. $x \leq -1$ and $x \geq 2$

23. $x - 3 \leq 5$ and $x + 2 \geq 6$

24. $x + 5 \leq 9$ and $x - 3 \geq -2$

25. $3x < -3$ and $x + 2 > 0$

26. $3x > -3$ and $x + 2 < 2$

27. $2x - 1 < 3$ and $2x - 8 > -4$

28. $5x + 2 > 2$ and $5x + 2 \leq 7$

29. $6x - 8 \leq 16$ and $4x - 1 \leq 15$

30. $7x + 6 \leq 48$ and $-2x \geq -6$

Solve each compound inequality. Graph each solution set. See Examples 6 and 7.

31. $x > 3$ or $x < -2$

32. $x < 5$ or $x > 0$

2.6 EXERCISES

33. $x \leq 1$ or $x \geq 4$

34. $x \leq -5$ or $x \geq 4$

35. $x \leq 1$ or $x \leq 7$

36. $x \geq 1$ or $x \geq 7$

37. $x \leq 1$ or $x \geq 7$

38. $x \geq 1$ or $x \leq 7$

39. $x \geq -2$ or $x \geq 3$

40. $x \leq -2$ or $x \leq 3$

41. $x \geq -2$ or $x \leq 3$

42. $x \leq -2$ or $x \geq 3$

43. $x + 2 > 6$ or $x - 1 < -5$

44. $x - 3 > 2$ or $x + 4 < 3$

45. $2x + 2 > 6$ or $4x - 1 < 3$

46. $3x < x + 12$ or $x - 1 > 5$

CHAPTER 2 LINEAR EQUATIONS AND INEQUALITIES

Solve each compound inequality. Graph each solution set. See Examples 2–4, 6, and 7.

47. $x < -1$ and $x > -4$

48. $x > -1$ and $x < 9$

49. $x < 3$ or $x < -2$

50. $x < 5$ or $x < -1$

51. $x + 1 \geq 5$ and $x - 1 \leq 11$

52. $2x - 6 \leq -18$ and $3x \geq -27$

53. $-3x \leq -6$ or $-5x \geq 0$

54. $-8x \leq -24$ or $-2x \geq 6$

SKILL SHARPENERS
Evaluate each of the following. See Section 1.1.

55. $|-3| + |-2|$

56. $|-5| - |-9|$

57. $|-8| - |4| - 3$

58. $-|6| - |-11| + (-4)$

59. $(-5) - (-9) + |-7|$

60. $|-4| - (-6) - (-3)$

2.7 ABSOLUTE VALUE EQUATIONS AND INEQUALITIES

OBJECTIVES

1. Use the distance definition of absolute value.

For $k > 0$,

2. Solve $|ax + b| = k$.

3. Solve $|ax + b| < k$ and solve $|ax + b| > k$.

4. Solve absolute value equations that involve rewriting.

5. Solve absolute value equations of the form
$$|ax + b| = |cx + d|.$$

6. Solve absolute value equations and inequalities that have only nonpositive constants on one side.

1 In Chapter 1 it was shown that the absolute value of a number x, written $|x|$, represents the distance from x to 0 on the number line. For example, the solutions of $|x| = 4$ are 4 and -4, as shown in Figure 21.

$x = -4$ or $x = 4$

FIGURE 21

Since absolute value represents distance from 0, it is reasonable to interpret the solutions of $|x| > 4$ to be all numbers that are *more* than 4 units from 0. The set $(-\infty, -4) \cup (4, \infty)$ fits this description. Figure 22 shows the graph of the solution set of $|x| > 4$.

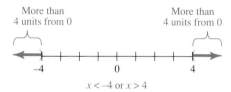

$x < -4$ or $x > 4$

FIGURE 22

The solution set of $|x| < 4$ consists of all numbers that are *less* than 4 units from 0 on the number line. Another way of thinking of this is to think of all numbers between -4 and 4. This set of numbers is given by $(-4, 4)$, as shown in Figure 23.

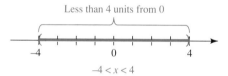

$-4 < x < 4$

FIGURE 23

WORK PROBLEM 1 AT THE SIDE.

The equation and inequalities just described are examples of **absolute value equations and inequalities.** These are equations and inequalities that involve the absolute value of a variable expression. These generally take the form

$$|ax + b| = k, \quad |ax + b| > k, \quad \text{or} \quad |ax + b| < k$$

where k is a positive number. We may solve them by rewriting them as compound equations or inequalities. The various methods of solving them are explained in the following examples.

1. Graph the solution set of each equation or inequality.

(a) $|x| = 3$

(b) $|x| > 3$

(c) $|x| < 3$

ANSWERS

1.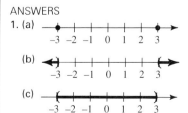

CHAPTER 2 LINEAR EQUATIONS AND INEQUALITIES

2. Solve each equation and graph the solution set.

   (a) $|x + 2| = 3$

   (b) $|3x - 4| = 11$

3. Solve each inequality and graph the solution set.

   (a) $|x + 2| > 3$

   (b) $|3x - 4| \geq 11$

2 The first example shows how to solve a typical absolute value equation. Remember that since absolute value refers to distance from the origin, each absolute value equation will have two cases.

EXAMPLE 1 Solving an Absolute Value Equation
Solve $|2x + 1| = 7$.

For $2x + 1$ to equal 7, $2x + 1$ must be 7 units from 0 on the number line. This can happen only when $2x + 1 = 7$ or $2x + 1 = -7$. Solve this compound equation as follows.

$$2x + 1 = 7 \quad \text{or} \quad 2x + 1 = -7$$
$$2x = 6 \quad \text{or} \quad 2x = -8$$
$$x = 3 \quad \text{or} \quad x = -4$$

The solution set is $\{-4, 3\}$. Its graph is shown in Figure 24. ■

FIGURE 24

■ WORK PROBLEM 2 AT THE SIDE.

3 We now discuss how to solve absolute value inequalities.

EXAMPLE 2 Solving an Absolute Value Inequality with >
Solve $|2x + 1| > 7$.

This absolute value inequality must be rewritten as

$$2x + 1 > 7 \quad \text{or} \quad 2x + 1 < -7,$$

because $2x + 1$ must represent a number that is *more* than 7 units from 0 on the number line. Now solve the compound inequality.

$$2x + 1 > 7 \quad \text{or} \quad 2x + 1 < -7$$
$$2x > 6 \quad \text{or} \quad 2x < -8$$
$$x > 3 \quad \text{or} \quad x < -4$$

The solution set, $(-\infty, -4) \cup (3, \infty)$, is graphed in Figure 25. Notice that the graph consists of two intervals. ■

FIGURE 25

■ WORK PROBLEM 3 AT THE SIDE.

ANSWERS
2. (a) $\{-5, 1\}$

(b) $\left\{-\dfrac{7}{3}, 5\right\}$

3. (a) $(-\infty, -5) \cup (1, \infty)$

(b) $\left(-\infty, -\dfrac{7}{3}\right] \cup \left[5, \infty\right)$

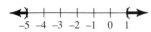

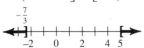

2.7 ABSOLUTE VALUE EQUATIONS AND INEQUALITIES

EXAMPLE 3 Solving an Absolute Value Inequality with $<$
Solve $|2x + 1| < 7$.

The expression $2x + 1$ must represent a number that is less than 7 units from 0 on the number line. Another way of thinking of this is to realize that $2x + 1$ must be between -7 and 7. This is written as the three-part inequality

$$-7 < 2x + 1 < 7.$$

We solved such inequalities in Section 2.5 by working with all three parts at the same time.

$-7 < 2x + 1 < 7$
$-8 < 2x < 6$ Subtract 1 from each part.
$-4 < x < 3$ Divide each part by 2.

The solution set is $(-4, 3)$, and the graph consists of a single interval, as shown in Figure 26. ∎

FIGURE 26

WORK PROBLEM 4 AT THE SIDE.

Look back at Figures 24, 25, and 26. These are the graphs of $|2x + 1| = 7$, $|2x + 1| > 7$, and $|2x + 1| < 7$. If we find the union of the three sets, we get the set of all real numbers. This is because for any value of x, $|2x + 1|$ will satisfy one and only one of the following: it is equal to 7, greater than 7, or less than 7.

When solving absolute value equations and inequalities of the types in Examples 1, 2, and 3, be sure to remember the following.

1. The methods described apply when the constant is alone on one side of the equation or inequality and is *positive*.

2. Absolute value equations and absolute value inequalities in the form $|ax + b| > k$ translate into "or" compound statements.

3. Absolute value inequalities in the form $|ax + b| < k$ translate into "and" compound statements, which may be written as three-part inequalities.

4. An "or" statement *cannot* be written in three parts. It would be *incorrect* to use

$$-7 > 2x + 1 > 7$$

in Example 2, since this would imply that $-7 > 7$, which is false.

---

4. Solve each inequality and graph the solution set.

   (a) $|x + 2| < 3$

   (b) $|3x - 4| \le 11$

ANSWERS
4. (a) $(-5, 1)$

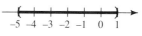

(b) $\left[-\dfrac{7}{3}, 5\right]$

CHAPTER 2 LINEAR EQUATIONS AND INEQUALITIES

5. (a) Solve $|5a + 2| - 9 = -7$.

(b) Solve and graph the solution set.
$|m + 2| - 3 > 2$

(c) Solve and graph the solution set.
$|3a + 2| + 4 \leq 15$

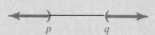

ANSWERS

5. (a) $\left\{-\dfrac{4}{5}, 0\right\}$

(b) $(-\infty, -7) \cup (3, \infty)$

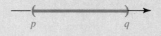

(c) $\left[-\dfrac{13}{3}, 3\right]$

---

4 Sometimes an absolute value equation or inequality is given in a form that requires some rewriting before it can be set up as a compound statement. The next example illustrates this process for an absolute value equation.

EXAMPLE 4 Solving an Absolute Value Equation Requiring Rewriting

Solve the equation $|x + 3| + 5 = 12$.

First get the absolute value alone on one side of the equals sign. Do this by subtracting 5 on each side.

$$|x + 3| + 5 - 5 = 12 - 5 \qquad \text{Subtract 5.}$$
$$|x + 3| = 7$$

Then use the method shown in Example 1.

$$x + 3 = 7 \quad \text{or} \quad x + 3 = -7$$
$$x = 4 \quad \text{or} \quad x = -10$$

Check that the solution set is $\{4, -10\}$ by substituting in the original equation. ■

Solving an absolute value *inequality* requiring rewriting is done in a similar manner.

▬ **WORK PROBLEM 5 AT THE SIDE.**

The methods used in Examples 1, 2, and 3 are summarized below.

SOLVING ABSOLUTE VALUE EQUATIONS AND INEQUALITIES

Let k be a positive number, and p and q be two numbers.

1. To solve $|ax + b| = k$, solve the compound equation

$$ax + b = k \quad \text{or} \quad ax + b = -k.$$

The solution set is usually of the form $\{p, q\}$, with two numbers.

2. To solve $|ax + b| > k$, solve the compound inequality

$$ax + b > k \quad \text{or} \quad ax + b < -k.$$

The solution set is of the form $(-\infty, p) \cup (q, \infty)$, which consists of two separate intervals.

3. To solve $|ax + b| < k$, solve the compound inequality

$$-k < ax + b < k.$$

The solution set is of the form (p, q), a single interval.

2.7 ABSOLUTE VALUE EQUATIONS AND INEQUALITIES

5 The next example shows how to solve an absolute value equation with two absolute value expressions. For two expressions to have the same absolute value, they must either be equal or be negatives of each other.

To solve an absolute value equation of the form

$$|ax + b| = |cx + d|$$

solve the compound equation

$$ax + b = cx + d \quad \textbf{or} \quad ax + b = -(cx + d).$$

EXAMPLE 5 Solving an Equation with Two Absolute Values
Solve the equation $|z + 6| = |2z - 3|$.

This equation is satisfied either if $z + 6$ and $2z - 3$ are equal to each other, or if $z + 6$ and $2z - 3$ are negatives of each other. Thus, we have

$$z + 6 = \boxed{2z - 3} \quad \text{or} \quad z + 6 = \boxed{-(2z - 3)}.$$

Solve each equation.

$$9 = z \quad \text{or} \quad z + 6 = -2z + 3$$
$$3z = -3$$
$$z = -1$$

The solution set is $\{9, -1\}$. ∎

6. Solve each equation.

(a) $|k - 1| = |5k + 7|$

(b) $|4r - 1| = |3r + 5|$

WORK PROBLEM 6 AT THE SIDE.

6 When a typical absolute value equation or inequality involves a *negative* constant or *zero* alone on one side, simply use the properties of absolute value to solve. Keep in mind the following.

1. The absolute value of an expression can never be negative: $|a| \geq 0$ for all real numbers a.

2. The absolute value of an expression equals 0 only when the expression is equal to 0.

The next two examples illustrate these special cases.

EXAMPLE 6 Solving Special Cases of Absolute Value Equations
Solve each equation.

(a) $|5r - 3| = -4$

Since the absolute value of an expression can never be negative, there are no solutions for this equation. The solution set is ∅.

(b) $|7x - 3| = 0$

The expression $7x - 3$ will equal 0 *only* for the solution of the equation

$$7x - 3 = 0.$$

The solution of this equation is $\frac{3}{7}$. The solution set is $\left\{\frac{3}{7}\right\}$. It consists of only one element. ∎

WORK PROBLEM 7 AT THE SIDE.

7. Solve.

(a) $|6x + 7| = -5$

(b) $\left|\frac{1}{4}x - 3\right| = 0$

ANSWERS
6. (a) $\{-1, -2\}$
   (b) $\left\{-\frac{4}{7}, 6\right\}$
7. (a) ∅   (b) $\{12\}$

CHAPTER 2 LINEAR EQUATIONS AND INEQUALITIES

8. Solve.

(a) $|x| > -1$

(b) $|y| < -5$

(c) $|k + 2| \leq 0$

EXAMPLE 7 Solving Special Cases of Absolute Value Inequalities
Solve each of the following inequalities.

(a) $|x| \geq -4$

The absolute value of a number is never negative. For this reason, $|x| \geq -4$ is true for *all* real numbers. The solution set is {all real numbers}.

(b) $|k + 6| < -2$

There is no number whose absolute value is less than -2, so this inequality has no solution. The solution set is $\emptyset$.

(c) $|m - 7| \leq 0$

The value of $|m - 7|$ will never be less than 0. However, $|m - 7|$ will equal 0 when $m = 7$. Therefore, the solution set is {7}.

■ **WORK PROBLEM 8 AT THE SIDE.**

ANSWERS
8. (a) {all real numbers}
   (b) $\emptyset$ (c) $\{-2\}$

2.7 EXERCISES

Solve each equation. See Example 1.

1. $|x| = 8$
2. $|k| = 11$
3. $|4x| = 12$

4. $|5x| = 10$
5. $|y - 3| = 8$
6. $|p - 5| = 11$

7. $|2x + 1| = 9$
8. $|2y + 3| = 17$
9. $|4r - 5| = 13$

10. $|5t - 1| = 26$
11. $|2y + 5| = 12$
12. $|2x - 9| = 16$

13. $\left|\dfrac{1}{2}x + 3\right| = 4$
14. $\left|\dfrac{2}{3}q - 1\right| = 2$
15. $\left|1 - \dfrac{3}{4}k\right| = 3$

16. $\left|-4 + \dfrac{5}{2}m\right| = 14$
17. $|2(3x + 1)| = 4$
18. $|-3(x - 1)| = 7$

Solve each inequality and graph the solution set. See Example 2.

19. $|x| > 2$

20. $|y| > 4$

21. $|k| \geq 3$

22. $|r| \geq 5$

23. $|t + 2| > 8$

24. $|r + 5| > 15$

25. $|3x - 1| \geq 5$

26. $|4x + 1| \geq 17$

27. $|3 - x| > 4$

28. $|5 - x| > 9$

CHAPTER 2 LINEAR EQUATIONS AND INEQUALITIES

Solve each inequality and graph the solution set. See Example 3. (Hint: Compare your answers to those in Exercises 19–28.)

29. $|x| \leq 2$ **30.** $|y| \leq 4$ **31.** $|k| < 3$

32. $|r| < 5$ **33.** $|t + 2| \leq 8$ **34.** $|r + 5| \leq 15$

35. $|3x - 1| < 5$ **36.** $|4x + 1| < 17$

37. $|3 - x| \leq 4$ **38.** $|5 - x| \leq 9$

Solve each equation or inequality. Graph the solution set of each inequality. See Examples 1, 2, and 3.

39. $|4 + k| > 9$ **40.** $|-3 + t| > 5$ **41.** $|7 + 2z| = 3$

42. $|9 - 3p| = 6$ **43.** $|3r - 1| \leq 8$ **44.** $|2s - 6| \leq 4$

45. $|6(x + 1)| < 1$ **46.** $|-2(x + 3)| < 5$

2.7 EXERCISES

Solve each equation or inequality. Graph the solution set of each inequality. See Example 4.

47. $|x| - 1 = 7$ **48.** $|y| + 3 = 9$

49. $|x + 4| + 1 = 3$ **50.** $|y + 5| - 2 = 8$

51. $|2x + 1| + 3 > 5$ **52.** $|6x - 1| - 2 > 1$

53. $|x + 5| - 6 \leq -2$ **54.** $|r - 2| - 3 \leq 6$

Solve each equation. See Example 5.

55. $|3x + 1| = |2x - 7|$ **56.** $|7x + 12| = |x - 4|$

57. $\left|m - \dfrac{1}{2}\right| = \left|\dfrac{1}{2}m - 2\right|$ **58.** $\left|\dfrac{2}{3}r - 2\right| = \left|\dfrac{1}{3}r + 3\right|$

59. $|6x| = |9x + 5|$ **60.** $|13y| = |2y - 1|$

61. $|2p - 6| = |2p + 5|$ **62.** $|3x - 1| = |3x + 4|$

Solve each equation or inequality. See Examples 6 and 7.

63. $|2x - 7| = -4$ **64.** $|8x + 1| = -9$

65. $|12t - 3| = -4$ **66.** $|13w + 1| = -\dfrac{1}{4}$

CHAPTER 2 LINEAR EQUATIONS AND INEQUALITIES

67. $|4x - 1| = 0$

68. $|6r + 2| = 0$

69. $|2q - 1| < -5$

70. $|8n + 4| < -3$

71. $|x + 5| > -8$

72. $|x + 9| > -4$

73. $|7x - 3| \leq 0$

74. $|4x + 1| \leq 0$

75. $|5x + 2| \geq 0$

76. $|4 - 7x| \geq 0$

77. $|10z + 9| > 0$

78. $|-8w + 1| > 0$

Find the unknown numbers in Exercises 79–84.

79. If 5 is added to a number, the absolute value of the result is 8.

80. Half a number is added to 1. The absolute value of the result is 9.

81. If the absolute value of a number is found, the result is at least 4.

82. If 5 is added to twice a number, the absolute value of the result is at most 2.

83. The absolute value of the sum of a number and 4 is found. Six is subtracted from the absolute value. The final result is at most 2.

84. Twice a number is added to 12. The absolute value of this sum is found. Eight is added to the absolute value. The result is at least 11.

SKILL SHARPENERS
Evaluate each of the following. See Section 1.4.

85. 2^3

86. 5^2

87. $(-4)^3$

88. $(-5)^4$

89. -5^2

90. -3^4

91. $\left(\dfrac{2}{3}\right)^3$

92. $\left(\dfrac{3}{2}\right)^4$

SUMMARY ON SOLVING LINEAR EQUATIONS AND INEQUALITIES

Students often have difficulty distinguishing between the various equations and inequalities studied in this chapter. This section of miscellaneous equations and inequalities should help with this difficulty. Solve each of these. (See the boxes in this chapter that summarize these equations and inequalities.)

1. $4z + 1 = 53$

2. $|m - 1| = 4$

3. $6q - 9 = 12 - q$

4. $3p + 7 = 9 - 8p$

5. $|a + 3| = 7$

6. $2m - 1 \leq m$

7. $8r + 2 \geq 7r$

8. $4(a - 11) + 3a = 2a + 11$

9. $2q - 1 = -5$

10. $|3q - 7| = 4$

11. $6z - 5 \leq 3z + 7$

12. $|5z - 8| = 9$

13. $9y - 3(y + 1) = 8y - 3$

14. $|y| \geq 4$

15. $9y - 5 \geq 4y + 3$

16. $13p - 5 > 8p + 1$

17. $|q| > 1$

18. $4z - 1 = 12 + 3z$

19. $\frac{2}{3}y - 8 = \frac{1}{4}y$

20. $-\frac{5}{8}y \geq 10$

21. $\frac{1}{4}p < -3$

22. $7z - 3 + 2z = 9z + 5$

23. $\frac{3}{5}q - \frac{1}{10} = 1$

24. $|r - 1| < 5$

25. $3r + 9 + 5r = 4(3 + 2r) - 3$

26. $6 - 3(2 - p) < 2(1 + p) + 1$

27. $|2p - 3| > 1$

28. $\dfrac{x}{4} - \dfrac{2x}{3} = -5$

29. $|5a + 1| < 6$

30. $5z - (3 + z) \geq 6z + 2$

31. $-2 \leq 3x - 1 \leq 5$

32. $-1 \leq 4a + 3 \leq 5$

33. $|7z - 1| = |5z + 1|$

34. $|p + 2| = |p + 8|$

35. $|3r - 1| \geq 4$

36. $\dfrac{1}{2} \leq \dfrac{2}{3}r \leq \dfrac{3}{4}$

37. $-(m - 4) + 2 = 3m - 8$

38. $\dfrac{p}{6} - \dfrac{3p}{5} = -13$

39. $-6 \leq \dfrac{3}{2} - x \leq 2$

40. $|5 - y| < 3$

41. $|y - 1| \geq -2$

42. $|2r - 5| = |r + 3|$

43. $8q - (2 - q) = 3(1 + 3q) - 5$

44. $8y - (y + 3) = -(2y + 1) - 3$

45. $|r - 5| = |r + 7|$

46. $|r + 2| < -3$

47. $|2 + 5q| < 3$

48. $-1 \leq \dfrac{x}{4} + 2 \leq 1$

CHAPTER 2 SUMMARY

KEY TERMS

2.1

algebraic expression
An algebraic expression is an expression indicating any combination of the following operations: addition, subtraction, multiplication, division (except by 0) and taking roots on any collection of variables and numbers.

equation
An equation is a statement that two algebraic expressions are equal.

linear equation or first-degree equation in one variable
An equation is linear or first-degree in the variable x if it can be written in the form $ax + b = c$, where a, b, and c are real numbers, with $a \neq 0$.

solution
A solution of an equation is a number that makes the equation true when substituted for the variable.

solution set
The solution set of an equation is the set of all its solutions.

equivalent equations
Equivalent equations are equations that have the same solution set.

addition and multiplication properties of equality
These properties state that the same number may be added to (or subtracted from) both sides of an equation to obtain an equivalent equation; and the same nonzero number may be multiplied by or divided into both sides of an equation to obtain an equivalent equation.

conditional equation
An equation that has a finite number of elements in its solution set is called a conditional equation.

contradiction
An equation that has no solutions (that is, its solution set is $\emptyset$) is called a contradiction.

identity
An equation that is satisfied by every number for which both sides are defined is called an identity.

2.2

formula
A formula is a mathematical statement in which more than one letter is used to express a relationship.

2.5

linear inequality in one variable
An inequality is linear in the variable x if it can be written in the form $ax + b < c$, $ax + b \leq c$, $ax + b > c$, or $ax + b \geq c$, where a, b, and c are real numbers, with $a \neq 0$.

equivalent inequalities
Equivalent inequalities are inequalities with the same solution set.

addition and multiplication properties of inequality
The same number may be added to (or subtracted from) both sides of an inequality to obtain an equivalent inequality. Both sides of an inequality may be multiplied or divided by the same positive number. If both sides are multiplied by or divided by a negative number, the inequality symbol must be reversed.

strict inequality
An inequality that involves $>$ or $<$ is called a strict inequality.

non-strict inequality
An inequality that involves $\geq$ or $\leq$ is called a non-strict inequality.

2.6

intersection
The intersection of two sets A and B is the set of elements that belong to both A and B.

union
The union of two sets A and B is the set of elements that belong to either A or B (or both).

compound inequality
A compound inequality is formed by joining two inequalities with a connective word, such as *and* or *or*.

CHAPTER 2 LINEAR EQUATIONS AND INEQUALITIES

2.7 **absolute value equation; absolute value inequality** — Absolute value equations and inequalities are equations and inequalities that involve the absolute value of a variable expression.

NEW SYMBOLS

∩   set intersection

∪   set union

QUICK REVIEW

Section	Concepts	Examples
2.1 Linear Equations in One Variable	**Solving A Linear Equation** If necessary, eliminate fractions by multiplying both sides by the LCD. Simplify each side, and then use the addition property of equality to get the variables on one side and the numbers on the other. Combine terms if possible, and then use the multiplication property of equality to make the coefficient of the variable equal to 1. Check by substituting into the original equation.	Solve the equation $4(8 - 3t) = 32 - 8(t + 2)$. $$32 - 12t = 32 - 8t - 16$$ $$32 - 12t = 16 - 8t$$ $$32 - 12t + 12t = 16 - 8t + 12t$$ $$32 = 16 + 4t$$ $$32 - 16 = 16 + 4t - 16$$ $$16 = 4t$$ $$\frac{16}{4} = \frac{4t}{4}$$ $$4 = t \quad \text{or} \quad t = 4$$ Check: $$4(8 - 3t) = 32 - 8(t + 2)$$ $$4(8 - 3 \cdot 4) = 32 - 8(4 + 2)$$ $$4(8 - 12) = 32 - 8(6)$$ $$4(-4) = 32 - 48$$ $$-16 = -16$$ The solution set is $\{4\}$.
2.2 Formulas	**Solving for a Specified Variable** Use the addition or multiplication properties as necessary to get all terms with the specified variable on one side of the equals sign, and all other terms on the other side. If necessary, use the distributive property to write the terms with the specified variable as the product of that variable and a sum of terms. Complete the solution.	Solve for h: $A = \frac{1}{2}bh$. $$A = \frac{1}{2}bh$$ $$2A = 2(\frac{1}{2}bh)$$ $$2A = bh$$ $$\frac{2A}{b} = h$$

CHAPTER 2 SUMMARY

Section	Concepts	Examples
2.3 Applications of Linear Equations **2.4** More Applications of Linear Equations	**Solving a Word Problem** *Step 1* Read the problem carefully. Decide what is given and what must be found. Use sketches as necessary. *Step 2* Choose a variable and write down exactly what it represents. (Do not make the mistake of omitting this step.) *Step 3* Write an equation using the information given in the problem. *Step 4* Solve the equation. *Step 5* Make sure you have answered the question of the problem. *Step 6* Check your answer back in the words of the original problem, not in the equation you obtained from the words.	The perimeter of a triangle is 34 inches. The middle side is twice as long as the shortest side. The longest side is 2 inches less than three times the shortest side. Find the length of the shortest side. Let x represent the length of the shortest side; then $2x$ represents the length of the middle side; and $3x - 2$ represents the length of the longest side. Formula is $P = a + b + c$. $$34 = x + 2x + (3x - 2)$$ Solve this equation to get $x = 6$. Since $6 + 12 + 16 = 34$, it checks. The shortest side is 6 inches.
2.5 Linear Inequalities in One Variable	**Solving a Linear Inequality** Simplify each side separately, combining like terms and removing parentheses. Use the addition property of inequality to get the variables on one side of the inequality sign and the numbers on the other. Combine like terms, and then use the multiplication property to change the inequality to the form $x < k$ or $x > k$. If an inequality is multiplied or divided by a *negative* number, the inequality symbol *must be reversed*.	Solve $3(x + 2) - 5x \leq 12$. $$3x + 6 - 5x \leq 12$$ $$-2x + 6 \leq 12$$ $$-2x \leq 6$$ $$\frac{-2x}{-2} \geq \frac{6}{-2}$$ $$x \geq -3$$ The solution set is $[-3, \infty)$ and is graphed below.

Section	Concepts	Examples
2.6 Set Operations and Compound Inequalities	**Solving a Compound Inequality** Solve each inequality in the compound inequality individually. If the inequalities are joined with *and*, the solution is the intersection of the two individual solutions. If the inequalities are joined with *or*, the solution is the union of the two individual solutions.	Solve $x + 1 > 2$ and $2x < 6$. $$x + 1 > 2 \quad \text{and} \quad 2x < 6$$ $$x > 1 \quad \text{and} \quad x < 3$$ The solution set is $(1, 3)$. Solve $x \geq 4$ or $x \leq 0$. The solution set is $(-\infty, 0] \cup [4, \infty)$.
2.7 Absolute Value Equations and Inequalities	Let k be a positive number. To solve $\|ax + b\| = k$, solve the compound equation $$ax + b = k \quad \text{or} \quad ax + b = -k.$$	Solve $\|x - 7\| = 3$. $$x - 7 = 3 \quad \text{or} \quad x - 7 = -3$$ $$x = 10 \qquad \qquad x = 4$$ The solution set is $\{4, 10\}$.
	To solve $\|ax + b\| > k$, solve the compound inequality $$ax + b > k \quad \text{or} \quad ax + b < -k.$$	Solve $\|x - 7\| > 3$. $$x - 7 > 3 \quad \text{or} \quad x - 7 < -3$$ $$x > 10 \quad \text{or} \quad x < 4$$ The solution set is $(-\infty, 4) \cup (10, \infty)$.
	To solve $\|ax + b\| < k$, solve the compound inequality $$-k < ax + b < k.$$	Solve $\|x - 7\| < 3$. $$-3 < x - 7 < 3$$ $$4 < x < 10$$ The solution set is $(4, 10)$.
	To solve an absolute value equation of the form $$\|ax + b\| = \|cx + d\|$$ solve the compound equation $$ax + b = cx + d \quad \text{or}$$ $$ax + b = -(cx + d).$$	Solve $\|x + 2\| = \|2x - 6\|$. $$x + 2 = 2x - 6$$ $$x = 8$$ or $$x + 2 = -(2x - 6)$$ $$x = \frac{4}{3}$$ The solution set is $\left\{\frac{4}{3}, 8\right\}$.

CHAPTER 2 REVIEW EXERCISES

[2.1] *Solve each equation.*

1. $5y - 9 + 4y = 4y + 1$

2. $-6r + 2r - 5 = r$

3. $-(8 + 3y) + 5 = 2y + 3$

4. $-(r + 5) - (2 + 7r) + 8r = 3r - 5$

5. $\dfrac{m - 2}{4} + \dfrac{m + 1}{2} = 1$

6. $\dfrac{2q + 1}{3} - \dfrac{q - 1}{4} = -2$

7. $5(2x - 3) = 8(x - 1) + 2x$

8. $-3x + 2(4x + 5) = 6x - 3$

9. $\dfrac{1}{2}(x - 4) = 2x - 2$

10. $7x - 2x + 10 = 5(x + 2)$

[2.2] *Solve each equation for x.*

11. $4xyz = 9$

12. $Q = 3x + 2y + 4z$

13. $M = \dfrac{1}{4}(x + 2y)$

14. $P = \dfrac{2}{3}x - 6$

Find the unknown value.

15. Distance is 150 kilometers and time is 5 hours. Find the rate.

16. A triangle has a perimeter of 120 meters. Two equal sides of the triangle each have a length of 50 meters. Find the length of the third side.

17. A rectangle has a perimeter of 46 centimeters. One side is 8 centimeters long. Find the length of the other side.

18. How much principal must be deposited for 6 years at 8% per year simple interest to earn $720 in interest?

19. A circle has a circumference of 36π millimeters. Find its radius.

20. The Celsius temperature is 90°. Find the Fahrenheit temperature.

[2.3] *Write the following as mathematical expressions. Use x as the variable.*

21. Half a number, added to 5

22. The product of 4 and a number, subtracted from 8

Solve each of the following.

23. The length of a rectangle is 3 meters less than twice the width. The perimeter of the rectangle is 42 meters. Find the length and width of the rectangle.

24. In a triangle with two sides of equal length, the third side measures 15 inches less than the sum of the two equal sides. The perimeter of the triangle is 53 inches. Find the lengths of the three sides.

25. A candy clerk has three times as many kilograms of chocolate creams as peanut clusters. The clerk has 48 kilograms of the two candies altogether. How many kilograms of peanut clusters does the clerk have?

26. How many liters of a 20% solution of a chemical should be mixed with 15 liters of a 50% solution to get a 30% mixture?

CHAPTER 2 REVIEW EXERCISES

27. Chico Ruiz invested some money at 12% and $4000 less than this amount at 14%. Find the amount invested at each rate if his total annual interest income is $4120.

28. Lori Johns earned $42,000 in royalties on her book. She invested part at 15% and part at 12%, earning $5790 per year interest. How much did she invest at each rate?

[2.4] *Solve each of the following.*

29. A total of 1096 people attended the Beach Boys concert yesterday. Reserved seats cost $15 each, and general admission seats cost $12 each. If $15,702 was collected, how many of each type of seat were sold?

30. There were 311 tickets sold for a soccer game, some for students and some for non-students. Student tickets cost 25¢ each and non-student tickets cost 75¢ each. The total receipts were $108.75. How many of each type of ticket were sold?

31. A passenger train and a freight train leave a town at the same time and go in opposite directions. They travel at 60 miles per hour and 75 miles per hour, respectively. How long will it take for them to be 297 miles apart?

32. Two cars leave towns 230 kilometers apart at the same time, traveling directly toward one another. One car travels 15 kilometers per hour slower than the other. They pass one another 2 hours later. What are their speeds?

33. An automobile averaged 45 miles per hour for the first part of a trip and 50 miles per hour for the second part. If the entire trip took 4 hours and covered 195 miles, for how long was the rate of speed 45 miles per hour?

34. An 85-mile trip to the beach took the Rodriguez family 2 hours. During the second hour, a rainstorm caused them to average 7 miles per hour less than they traveled during the first hour. Find their average speed for the first hour.

[2.5] *Solve each inequality. Graph each solution set.*

35. $2y + 7 > -21$

36. $-6z \leq 72$

37. $-\dfrac{2}{3}k < 4$

38. $-5x - 4 \geq 10$

39. $\dfrac{6a + 3}{-4} < -2$

40. $\dfrac{9y + 5}{-3} > 2$

41. $5 - (6 - 4k) > 2k - 5$

42. $-6 \leq 2k \leq 12$

43. $8 \leq 3y - 1 < 14$

44. $-4 < 2k - 3 < 9$

45. $\dfrac{5}{3}(m - 2) + \dfrac{2}{5}(m + 1) > 1$

46. $\dfrac{3}{4}(a - 2) - \dfrac{1}{3}(5 - 2a) < -2$

[2.6] *In Exercises 47 and 48, let* $A = \{2, 4, 6, 8, 10, 12\}$ *and let* $B = \{4, 8, 12, 16\}$.

47. Find $A \cap B$.

48. Find $A \cup B$.

Solve each compound inequality. Graph each solution set.

49. $x > 6$ and $x < 8$

50. $x + 4 > 12$ and $x - 2 < 1$

CHAPTER 2 REVIEW EXERCISES

51. $x > 5$ or $x \leq -1$

52. $x \geq -2$ or $x < 4$

53. $x - 4 > 6$ or $x + 3 \leq 18$

54. $-5x + 1 \geq 11$ or $3x + 5 \geq 26$

[2.7] *Solve each absolute value equation.*

55. $|x| = 4$

56. $|y + 2| = 6$

57. $|3k - 7| = 4$

58. $|z - 4| = -4$

59. $|2k - 7| + 4 = 9$

60. $|4a + 2| + 7 = 3$

61. $|3p + 1| = |p + 7|$

62. $|m - 1| = |2m + 3|$

Solve each absolute value inequality. Graph each solution set.

63. $|p| < 4$

64. $|-y + 6| \leq 5$

65. $|2p + 5| \leq 3$

66. $|x + 1| \geq 9$

67. $|5r - 1| > 14$

68. $|3k + 9| \geq 0$

MIXED REVIEW EXERCISES*
Solve.

69. $(7 - 2k) + 3(5 - 3k) > k + 6$

70. $R = \dfrac{3}{4}x - 12$ for x

71. $x < 5$ and $x \geq -1$

72. $-5(6p + 4) - 2p = -6p$

73. A triangle has perimeter 49 inches. Two sides are of equal length, and the third side is 19 inches. Find the lengths of the equal sides.

74. $-5r \geq -30$

75. $|7x - 2| > 5$

76. $x \geq -3$ or $x < 6$

77. $|2x - 10| = 12$

78. $|m + 3| \leq 1$

79. In an election, one candidate received 135 more votes than the other. The total number of votes cast in the election was 1215. Find the number of votes received by each candidate.

80. A cashier has 34 bills worth $455. Some are five-dollar bills and the rest are twenty-dollar bills. How many of each type are there?

---

*The order of exercises in this final group does not correspond to the order in which topics occur in the chapter. This random ordering should help you prepare for the chapter test in yet another way.

CHAPTER 2 TEST

Solve each equation.

1. $6x - 4 + 2x = 3(x + 2) - 5$

2. $\dfrac{2x + 3}{2} - \dfrac{x + 12}{8} = -7$

3. $3x + 2(x - 4) = 5x - 8$

4. Solve for v: $S = vt - 16t^2$.

Solve each word problem.

5. Find the period of time for which $2000 must be invested at 7% to produce $560 simple interest.

6. The sale price of a compact disc player is $280. This represents 30% off the regular price. What is the regular price?

7. The perimeter of a rectangle is 28 meters. The length is 2 meters less than 3 times the width. Find the width of the rectangle.

8. Earl Karn invested some money at 7% and three times as much at 6%. His total annual interest was $375. How much did he invest at each rate?

Solve each inequality and graph the solution set.

9. $-\dfrac{5}{6}x > -25$

10. $4 - 7(p + 4) < -3p$

1. _____
2. _____
3. _____
4. _____
5. _____
6. _____
7. _____
8. _____
9. _____
10. _____

11. _____

11. $-3 \leq \dfrac{2}{3}x - 1 \leq 1$

12. _____

12. A student must have an average grade of at least 80 on the four tests in a course to get a grade of B. A student had grades of 79, 76, and 87 on the first three tests. What minimum score on the fourth test would earn the student a B?

Solve each compound inequality and graph the solution set.

13. _____

13. $3x - 2 < 10$ and $-2x < 10$

14. _____

14. $2k \geq 6$ and $k - 3 \geq 5$

15. _____

15. $-4x \leq -20$ or $4x - 2 < 10$

Solve each absolute value equation.

16. _____

16. $|3k - 2| = 9$

17. _____

17. $|5x - 3| = |2x + 6|$

18. _____

18. $|6t - 4| + 7 = 2$

Solve and graph each absolute value inequality.

19. _____

19. $|4r + 3| \leq 11$

20. _____

20. $|2r + 5| > 7$

3 POLYNOMIALS

3.1 INTEGER EXPONENTS

In Chapter 1 we showed that two integers whose product is a third number are factors of that number. For example, 2 and 5 are factors of 10, since $2 \cdot 5 = 10$.

1 Recall that exponents are used to write products of repeated factors. For example, 2^5 is defined as $2 \cdot 2 \cdot 2 \cdot 2 \cdot 2 = 32$. The number 5, the *exponent*, shows that the *base* 2 appears as a factor 5 times. The quantity 2^5 is called an *exponential* or a *power*. Read 2^5 as "2 to the fifth power" or "2 to the fifth."

EXAMPLE 1 Identifying Exponents and Bases and Evaluating Exponentials

Identify the exponent and the base, and evaluate each exponential.

(a) -2^4

The exponent is 4 and the base is 2 (not -2).
$$-2^4 = -(2^4) = -16$$

(b) $(-2)^4$

The exponent is 4, and the base is -2.
$$(-2)^4 = (-2)(-2)(-2)(-2) = 16$$

(c) $3z^7$

The exponent is 7. The base is z (not $3z$).
$$3z^7 = 3z \cdot z \cdot z \cdot z \cdot z \cdot z \cdot z$$

(d) $(-9y)^3$

Because of the parentheses, the exponent 3 refers to the base $-9y$, so
$$(-9y)^3 = (-9y)(-9y)(-9y) = -729y^3. \quad \blacksquare$$

CAUTION As shown in Example 1 parts (a) and (b), when no parentheses are used, the exponent refers only to the factor closest to it.

WORK PROBLEM 1 AT THE SIDE.

OBJECTIVES

1 Review identifying exponents and bases.

2 Use the product rule.

3 Define a negative exponent.

4 Use the quotient rule.

5 Define a zero exponent.

---

1. Identify the base of each exponential.

(a) $-2p^8$

(b) $(-2p)^8$

(c) $3y^7$

(d) $-m^{10}$

ANSWERS
1. (a) p (b) $-2p$ (c) y (d) m

3.1 INTEGER EXPONENTS   133

2. Find the product.

(a) $m^8 \cdot m^6$

(b) $r^7 \cdot r$

(c) $k^4 k^3 k^6$

(d) $(-4a^3)(6a^2)$

(e) $(-5p^4)(-9p^5)$

ANSWERS
2. (a) m^{14} (b) r^8 (c) k^{13}
(d) $-24a^5$ (e) $45p^9$

---

2 Several useful rules can be developed that simplify the use of exponents. For example, the product $2^5 \cdot 2^3$ can be simplified as follows.

$$2^5 \cdot 2^3 = (2 \cdot 2 \cdot 2 \cdot 2 \cdot 2)(2 \cdot 2 \cdot 2) = 2^8$$

$$5 + 3 = 8$$

This result, that products of exponentials with the same base are found by adding exponents, can be generalized by the product rule for exponents.

PRODUCT RULE FOR EXPONENTS

If m and n are integers and a is any nonzero real number, then
$$a^m \cdot a^n = a^{m+n}.$$

EXAMPLE 2 Using the Product Rule

(a) $3^4 \cdot 3^7 = 3^{4+7} = 3^{11}$

(b) $5^3 \cdot 5 = 5^3 \cdot 5^1 = 5^{3+1} = 5^4$

(c) $y^3 \cdot y^8 \cdot y^2 = y^{3+8+2} = y^{13}$ ∎

EXAMPLE 3 Using the Product Rule
Use the product rule for exponents to find each product.

(a) $(5y^2)(-3y^4)$

Use the associative and commutative properties as necessary to multiply the numbers and multiply the variables.

$$(5y^2)(-3y^4) = 5(-3)y^2 y^4$$
$$= -15y^{2+4}$$
$$= -15y^6$$

(b) $(7p^3q)(2p^5q^2) = 7(2)p^3 p^5 q q^2$
$$= 14p^8 q^3$$ ∎

■ WORK PROBLEM 2 AT THE SIDE.

3 A quotient such as $\dfrac{a^8}{a^3}$ can be simplified in much the same way as a product. (In all quotients of this type, we assume the denominator is not zero.) Using the definition of an exponent,

$$\frac{a^8}{a^3} = \frac{a \cdot a \cdot a \cdot a \cdot a \cdot a \cdot a \cdot a}{a \cdot a \cdot a}$$
$$= a \cdot a \cdot a \cdot a \cdot a$$
$$= a^5.$$

Notice that $8 - 3 = 5$. In the same way,

$$\frac{a^3}{a^8} = \frac{a \cdot a \cdot a}{a \cdot a \cdot a \cdot a \cdot a \cdot a \cdot a \cdot a}$$
$$\frac{a^3}{a^8} = \frac{1}{a^5}.$$

3.1 INTEGER EXPONENTS

Here again, $8 - 3 = 5$. In the first example, the exponent in the denominator was subtracted from the one in the numerator; the reverse was true in the second example, $\dfrac{a^3}{a^8}$. The order in which we subtracted the exponents depended on the location of the larger exponent. To simplify the rule for quotients, we first define a **negative exponent.**

DEFINITION OF NEGATIVE EXPONENT

If a is any nonzero number, and n is any whole number,
$$a^{-n} = \frac{1}{a^n}.$$

With this definition, the expression a^n is meaningful for any nonzero integer exponent n and any nonzero real number a.

EXAMPLE 4 Using Negative Exponents
Simplify the following expressions and evaluate, if possible.

(a) $2^{-3} = \dfrac{1}{2^3} = \dfrac{1}{8}$

(b) $3^{-2} = \dfrac{1}{3^2} = \dfrac{1}{9}$

(c) $6^{-1} = \dfrac{1}{6^1} = \dfrac{1}{6}$

(d) $5^{-3} = \dfrac{1}{5^3} = \dfrac{1}{125}$

(e) $k^{-7} = \dfrac{1}{k^7} \quad (k \neq 0)$

(f) $(5z^2)^{-3} = \dfrac{1}{(5z^2)^3} \quad (z \neq 0)$ ∎

CAUTION A negative exponent *does not indicate* a negative number; negative exponents lead to reciprocals, as shown below.

Expression	Example	
a^{-m}	$3^{-2} = \dfrac{1}{3^2} = \dfrac{1}{9}$	Not negative
$-a^{-m}$	$-3^{-2} = -\dfrac{1}{3^2} = -\dfrac{1}{9}$	Negative

WORK PROBLEM 3 AT THE SIDE.

3. Simplify, using the definition of negative exponents.

(a) 5^{-2}

(b) 8^{-1}

(c) 6^{-3}

(d) $y^{-3}, \quad y \neq 0$

(e) $(2m^4)^{-3}, \quad m \neq 0$

ANSWERS
3. (a) $\dfrac{1}{25}$ (b) $\dfrac{1}{8}$ (c) $\dfrac{1}{216}$
   (d) $\dfrac{1}{y^3}$ (e) $\dfrac{1}{(2m^4)^3}$

4. Evaluate each expression.

(a) $3^{-1} + 5^{-1}$

(b) $4^{-1} - 2^{-1}$

(c) $\dfrac{4^{-1}}{2^{-2}}$

(d) $\dfrac{3^{-3}}{9^{-1}}$

EXAMPLE 5 Using Negative Exponents

Evaluate the following expressions.

(a) $3^{-1} + 4^{-1}$

Since $3^{-1} = \dfrac{1}{3}$ and $4^{-1} = \dfrac{1}{4}$,

$$3^{-1} + 4^{-1} = \dfrac{1}{3} + \dfrac{1}{4} = \dfrac{4}{12} + \dfrac{3}{12} = \dfrac{7}{12}.$$

(b) $5^{-1} - 2^{-1} = \dfrac{1}{5} - \dfrac{1}{2} = \dfrac{2}{10} - \dfrac{5}{10} = -\dfrac{3}{10}$ ∎

CAUTION In Example 5, note that

$$3^{-1} + 4^{-1} \neq (3+4)^{-1}$$

because $3^{-1} + 4^{-1} = \dfrac{7}{12}$ but $(3+4)^{-1} = 7^{-1} = \dfrac{1}{7}$.

Also,

$$5^{-1} - 2^{-1} \neq (5-2)^{-1}$$

because $5^{-1} - 2^{-1} = -\dfrac{3}{10}$ but $(5-2)^{-1} = 3^{-1} = \dfrac{1}{3}$.

EXAMPLE 6 Using Negative Exponents

Evaluate each expression.

(a) $\dfrac{1}{2^{-3}} = \dfrac{1}{\frac{1}{2^3}} = 2^3 = 8$

(b) $\dfrac{1}{3^{-2}} = \dfrac{1}{\frac{1}{3^2}} = 3^2 = 9$

(c) $\dfrac{2^{-3}}{3^{-2}} = \dfrac{\frac{1}{2^3}}{\frac{1}{3^2}} = \dfrac{1}{2^3} \cdot \dfrac{3^2}{1} = \dfrac{3^2}{2^3} = \dfrac{9}{8}$ ∎

Example 6 suggests the following generalizations. If $a \neq 0$, $b \neq 0$,

$$\dfrac{1}{a^{-n}} = a^n \quad \text{and} \quad \dfrac{a^{-n}}{b^{-m}} = \dfrac{b^m}{a^n}.$$

■ **WORK PROBLEM 4 AT THE SIDE.**

4 As we saw at the beginning of this section, to divide exponentials with the same base, we should subtract the exponents.

ANSWERS

4. (a) $\dfrac{8}{15}$ (b) $-\dfrac{1}{4}$ (c) 1 (d) $\dfrac{1}{3}$

3.1 INTEGER EXPONENTS

QUOTIENT RULE FOR EXPONENTS

If a is any nonzero real number, and m and n are integers, then

$$\frac{a^m}{a^n} = a^{m-n}.$$

EXAMPLE 7 Using the Quotient Rule

(a) $\dfrac{3^7}{3^2} = 3^{7-2} = 3^5$

(b) $\dfrac{p^6}{p^2} = p^{6-2} = p^4 \quad (p \neq 0)$

(c) $\dfrac{12^{10}}{12^9} = 12^{10-9} = 12^1 = 12$

(d) $\dfrac{7^4}{7^6} = 7^{4-6} = 7^{-2} = \dfrac{1}{7^2}$

(e) $\dfrac{k^7}{k^{12}} = k^{7-12} = k^{-5} = \dfrac{1}{k^5} \quad (k \neq 0)$ ■

WORK PROBLEM 5 AT THE SIDE.

CAUTION Be very careful when working with quotients that involve negative exponents in the denominator. Always be sure to write the numerator exponent, then a minus sign, and then the denominator exponent.

EXAMPLE 8 Using the Quotient Rule

(a) $\dfrac{8^{-2}}{8^5} = 8^{-2-5} = 8^{-7} = \dfrac{1}{8^7}$
(Numerator exponent, Minus sign, Denominator exponent)

(b) $\dfrac{6^{-5}}{6^{-2}} = 6^{-5-(-2)} = 6^{-3} = \dfrac{1}{6^3}$
(Numerator exponent, Minus sign, Denominator exponent)

(c) $\dfrac{2^7}{2^{-3}} = 2^{7-(-3)} = 2^{10}$

(d) $\dfrac{6}{6^{-1}} = \dfrac{6^1}{6^{-1}} = 6^{1-(-1)} = 6^2$

(e) $\dfrac{z^{-5}}{z^{-8}} = z^{-5-(-8)} = z^3 \quad (z \neq 0)$ ■

WORK PROBLEM 6 AT THE SIDE.

---

5. Divide. Write all answers with only positive exponents.

(a) $\dfrac{6^9}{6^4}$

(b) $\dfrac{9^{12}}{9^7}$

(c) $\dfrac{15^7}{15^{10}}$

(d) $\dfrac{m^3}{m^8}, \quad m \neq 0$

6. Simplify. Write all answers with only positive exponents.

(a) $\dfrac{2^{-4}}{2^2}$

(b) $\dfrac{8^{-2}}{8^{-6}}$

(c) $\dfrac{9^{-5}}{9^{-2}}$

(d) $\dfrac{7^{-1}}{7}$

(e) $\dfrac{k^4}{k^{-5}}, \quad k \neq 0$

ANSWERS
5. (a) 6^5 (b) 9^5 (c) $\dfrac{1}{15^3}$ (d) $\dfrac{1}{m^5}$
6. (a) $\dfrac{1}{2^6}$ (b) 8^4 (c) $\dfrac{1}{9^3}$
(d) $\dfrac{1}{7^2}$ (e) k^9

POLYNOMIALS

7. Simplify.

(a) 46^0

(b) $(-29)^0$

(c) -29^0

(d) $8^0 + 15^0$

(e) $(-15p^5)^0$, $p \neq 0$

5 By the quotient rule, $\dfrac{a^m}{a^n} = a^{m-n}$. Suppose the exponents in the numerator and the denominator are the same. We would have, for example,

$$\frac{8^4}{8^4} = 8^{4-4} = 8^0.$$

The quotient of any nonzero number and itself is 1, so $\dfrac{8^4}{8^4} = 1$. Therefore,

$$1 = \frac{8^4}{8^4} = 8^0.$$

For this reason, a^0 is defined as follows for any nonzero real number a.

ZERO EXPONENT

For any nonzero real number a,

$$a^0 = 1.$$

0^0 is undefined.

EXAMPLE 9 Using Zero Exponents

(a) $12^0 = 1$

(b) $(-19)^0 = 1$

(c) $-19^0 = -(19^0) = -1$

(d) $5^0 + 12^0 = 1 + 1 = 2$

(e) $(8k)^0 = 1$, if $k \neq 0$ ∎

WORK PROBLEM 7 AT THE SIDE.

ANSWERS
7. (a) 1 (b) 1 (c) −1 (d) 2 (e) 1

3.1 EXERCISES

Identify the exponent and the base. Do not evaluate. See Example 1.

1. 5^7
2. 9^4
3. $(-9)^4$

4. $(-2)^{11}$
5. -9^4
6. -2^{11}

7. p^{-7}
8. m^{-9}
9. $-3q^{-4}$

10. $-9m^{-6}$
11. $(-m+z)^3$
12. $-(6a-5b)^5$

Evaluate. See Examples 1 and 4–6.

13. 5^4
14. 10^3
15. $\left(\dfrac{5}{3}\right)^2$
16. $\left(\dfrac{3}{5}\right)^3$

17. 7^{-2}
18. 4^{-1}
19. -2^{-3}
20. -3^{-2}

21. $-(-3)^{-4}$
22. $-(-5)^{-2}$
23. $(-2)^{-5}$
24. $(-5)^{-4}$

25. $\dfrac{1}{5^{-2}}$
26. $\dfrac{6}{3^{-3}}$
27. $\dfrac{2}{(-4)^{-3}}$
28. $\dfrac{-3}{(-5)^{-2}}$

29. $\dfrac{2^{-3}}{3^{-2}}$
30. $\dfrac{5^{-1}}{4^{-2}}$
31. $\left(\dfrac{1}{2}\right)^{-3}$
32. $\left(\dfrac{1}{5}\right)^{-3}$

33. $\left(\dfrac{2}{3}\right)^{-2}$
34. $\left(\dfrac{4}{5}\right)^{-2}$
35. $3^{-1}+2^{-1}$
36. $4^{-1}+5^{-1}$

Use the product rule or quotient rule to simplify. Write all answers with only positive exponents. Assume that all variables represent nonzero real numbers. See Examples 2, 3, 7, and 8.

37. $2^6 \cdot 2^{10}$ **38.** $3^5 \cdot 3^7$ **39.** $7^{12} \cdot 7^{-8}$ **40.** $10^{-4} \cdot 10^5$

41. $\dfrac{3^5}{3^2}$ **42.** $\dfrac{5^7}{5^2}$ **43.** $\dfrac{6^7}{6^9}$ **44.** $\dfrac{7^5}{7^9}$

45. $\dfrac{3^{-5}}{3^{-2}}$ **46.** $\dfrac{2^{-4}}{2^{-3}}$ **47.** $\dfrac{9^{-1}}{9}$ **48.** $\dfrac{12}{12^{-1}}$

49. $t^5 t^{-12}$ **50.** $p^5 p^{-6}$ **51.** $r^4 r$ **52.** $k \cdot k^6$

53. $a^{-3} a^2 a^{-4}$ **54.** $k^{-5} k^{-3} k^4$ **55.** $\dfrac{x^7}{x^{-4}}$ **56.** $\dfrac{p^{-3}}{p^5}$

57. $\dfrac{r^3 r^{-4}}{r^{-2} r^{-5}}$ **58.** $\dfrac{z^{-4} z^{-2}}{z^3 z^{-1}}$ **59.** $7k^2(-2k)(4k^{-5})$ **60.** $3a^2(-5a^{-6})(-2a)$

Evaluate each expression. Assume all variables are nonzero. See Example 9.

61. 8^0 **62.** 12^0 **63.** $(-23)^0$ **64.** $(-4)^0$

65. -2^0 **66.** -7^0 **67.** $3^0 - y^0$ **68.** $-8^0 - k^0$

Many students believe that the following statements are true; however, they are false. Show that each is false by replacing x with 2 and y with 3.

69. $(x+y)^{-1} = x^{-1} + y^{-1}$ **70.** $(x^{-1} + y^{-1})^{-1} = x + y$ **71.** $(x+y)^2 = x^2 + y^2$

SKILL SHARPENERS

Simplify each expression, writing answers in exponential form. See Section 1.4.

72. $5^2 \cdot 5^2 \cdot 5^2$ **73.** $(-2)^3 \cdot (-2)^3$ **74.** $-(2^3) \cdot (2^3)$ **75.** $\left(\dfrac{3}{4}\right)^2 \cdot \left(\dfrac{3}{4}\right)^2$

3.2 FURTHER PROPERTIES OF EXPONENTS

1 By the product rule for exponents, the expression $(3^4)^2$ can be simplified as

$$(3^4)^2 = 3^4 \cdot 3^4 = 3^{4+4} = 3^8$$

where $4 \cdot 2 = 8$. This example suggests the first of the **power rules for exponents**; the other two parts can be demonstrated with similar examples.

POWER RULES FOR EXPONENTS

If a and b are real numbers and m and n are integers, then

$$(a^m)^n = a^{mn}$$
$$(ab)^m = a^m b^m$$
$$\left(\frac{a}{b}\right)^m = \frac{a^m}{b^m} \quad (b \neq 0).$$

OBJECTIVES

1 Use the power rules for exponents.

2 Simplify exponential expressions.

3 Use the rules for exponents with scientific notation.

EXAMPLE 1 Using the Power Rules

(a) $(p^8)^3 = p^{8 \cdot 3} = p^{24}$

(b) $\left(\frac{2}{3}\right)^4 = \frac{2^4}{3^4} = \frac{16}{81}$

(c) $(3y)^4 = 3^4 y^4 = 81 y^4$

(d) $(6p^7)^2 = 6^2 p^{7 \cdot 2} = 6^2 p^{14} = 36 p^{14}$

(e) $\left(\frac{2m^5}{z}\right)^3 = \frac{2^3 m^{5 \cdot 3}}{z^3} = \frac{2^3 m^{15}}{z^3} = \frac{8 m^{15}}{z^3}$

WORK PROBLEM 1 AT THE SIDE.

In Section 3.1 we saw that

$$\frac{1}{a^{-n}} = a^n \quad \text{and} \quad \frac{a^{-n}}{b^{-m}} = \frac{b^m}{a^n}.$$

We can extend these ideas as follows.

The reciprocal of a^n is $\frac{1}{a^n} = \left(\frac{1}{a}\right)^n$. Also, by definition, a^n and a^{-n} are reciprocals, since

$$a^n \cdot a^{-n} = a^n \cdot \frac{1}{a^n} = 1.$$

Thus, since both are reciprocals of a^n, $a^{-n} = \left(\frac{1}{a}\right)^n$.

That is, any nonzero number raised to the negative nth power equals the *reciprocal* of that number raised to the nth power. For example, using this result,

$$6^{-3} = \left(\frac{1}{6}\right)^3 \quad \text{and} \quad \left(\frac{1}{3}\right)^{-2} = 3^2$$

since $\frac{1}{6}$ is the reciprocal of 6 and 3 is the reciprocal of $\frac{1}{3}$.

1. Simplify.

(a) $(m^5)^4$

(b) $(x^3)^9$

(c) $\left(\frac{3}{8}\right)^7$

(d) $(2r)^{10}$

(e) $(-3y^5)^2$

(f) $\left(\frac{5p^2}{r^3}\right)^4$, $r \neq 0$

ANSWERS

1. (a) m^{20} (b) x^{27} (c) $\frac{3^7}{8^7}$ (d) $2^{10} r^{10}$
(e) $(-3)^2 y^{10}$ or $9y^{10}$ (f) $\frac{5^4 p^8}{r^{12}}$

2. Write each expression with a positive exponent, and evaluate it.

   (a) 4^{-3}

   (b) $\left(\dfrac{2}{3}\right)^{-4}$

   (c) $\left(\dfrac{1}{6}\right)^{-3}$

EXAMPLE 2 Using Negative Exponents with Fractions

Write the following expressions with only positive exponents and then evaluate.

(a) $5^{-4} = \left(\dfrac{1}{5}\right)^4 = \dfrac{1}{625}$

(b) $\left(\dfrac{3}{7}\right)^{-2} = \left(\dfrac{7}{3}\right)^2 = \dfrac{49}{9}$

(c) $\left(\dfrac{4}{5}\right)^{-3} = \left(\dfrac{5}{4}\right)^3 = \dfrac{125}{64}$ ■

WORK PROBLEM 2 AT THE SIDE.

Example 2 suggests the following generalizations. If $a \neq 0$ and $b \neq 0$,

$$a^{-n} = \left(\dfrac{1}{a}\right)^n \quad \text{and} \quad \left(\dfrac{a}{b}\right)^{-n} = \left(\dfrac{b}{a}\right)^n.$$

In the previous section, we expanded the definition of an exponent to include *all* integers—positive, negative, or zero. This was done in such a way that all past rules and definitions for exponents are still valid. These rules and definitions are summarized below.

DEFINITIONS AND RULES FOR EXPONENTS

If a and b are real numbers, m and n are integers, and no denominators are zero,

Product rule $\quad a^m \cdot a^n = a^{m+n}$

Quotient rule $\quad \dfrac{a^m}{a^n} = a^{m-n}$

Power rules $\quad (a^m)^n = a^{mn}$

$\qquad\qquad\qquad (ab)^m = a^m b^m$

$\qquad\qquad\qquad \left(\dfrac{a}{b}\right)^m = \dfrac{a^m}{b^m}$

Zero exponent $\quad a^0 = 1$

Negative exponent $\quad a^{-n} = \dfrac{1}{a^n} = \left(\dfrac{1}{a}\right)^n$

$\qquad\qquad\qquad \dfrac{a^{-n}}{b^{-m}} = \dfrac{b^m}{a^n}$

$\qquad\qquad\qquad \left(\dfrac{a}{b}\right)^{-n} = \left(\dfrac{b}{a}\right)^n.$

ANSWERS

2. (a) $\dfrac{1}{64}$ (b) $\dfrac{81}{16}$ (c) 216

3.2 FURTHER PROPERTIES OF EXPONENTS

2 The next two examples show how these definitions and rules for exponents are used to simplify exponential expressions.

EXAMPLE 3 Using the Definitions and Rules for Exponents
Simplify.

(a) $3^2 \cdot 3^{-5} = 3^{2+(-5)} = 3^{-3} = \dfrac{1}{3^3}$

(b) $x^{-3} \cdot x^{-4} \cdot x^2 = x^{-3+(-4)+2} = x^{-5} = \dfrac{1}{x^5}$ $(x \neq 0)$

(c) $(2^5)^{-3} = 2^{5(-3)} = 2^{-15} = \dfrac{1}{2^{15}}$

(d) $(4^{-2})^{-5} = 4^{(-2)(-5)} = 4^{10}$

(e) $(x^{-4})^6 = x^{(-4)6} = x^{-24} = \dfrac{1}{x^{24}}$ $(x \neq 0)$

(f) $(xy)^3 = x^3 y^3$ ■

CAUTION As shown in part (f) of Example 3, $(xy)^3 = x^3 y^3$, so that $(xy)^3 \neq xy^3$. Remember that ab^m is *not* the same as $(ab)^m$.

3. Simplify each expression and write the answer with a positive exponent.

(a) $5^2 \cdot 5^{-4}$

(b) $p^{-5} \cdot p^{-3} \cdot p^4$ $(p \neq 0)$

(c) $(4^2)^{-5}$

(d) $(m^{-3})^{-4}$ $(m \neq 0)$

(e) $(4a)^5$

WORK PROBLEM 3 AT THE SIDE.

EXAMPLE 4 Using the Definitions and Rules for Exponents
Simplify. Assume that all variables represent nonzero real numbers.

(a) $\dfrac{x^{-4}y^2}{x^2 y^{-5}} = \dfrac{x^{-4}}{x^2} \cdot \dfrac{y^2}{y^{-5}}$

$= x^{-4-2} \cdot y^{2-(-5)}$

$= x^{-6} y^7$

$= \dfrac{y^7}{x^6}$

(b) $(2^3 x^{-2})^{-2} = (2^3)^{-2} \cdot (x^{-2})^{-2}$

$= 2^{-6} x^4$

$= \dfrac{x^4}{2^6}$ or $\dfrac{x^4}{64}$ ■

4. Simplify, writing answers with positive exponents. Assume all variables are nonzero.

(a) $\dfrac{a^{-3} b^5}{a^4 b^{-2}}$

(b) $(3^2 k^{-4})^{-1}$

(c) $\dfrac{(m^2 n)^{-2}}{m^{-3} n}$

WORK PROBLEM 4 AT THE SIDE.

3 Many numbers that occur in science are very large. For example, the number of one-celled organisms that will sustain a whale for a few hours is 400,000,000,000,000. Other numbers are very small, such as the wavelength of visible light, which is .0000004 meters. Writing these numbers is simpler when we use *scientific notation*.

ANSWERS

3. (a) $\dfrac{1}{5^2}$ or $\dfrac{1}{25}$ (b) $\dfrac{1}{p^4}$ (c) $\dfrac{1}{4^{10}}$
   (d) m^{12} (e) $4^5 a^5$

4. (a) $\dfrac{b^7}{a^7}$ (b) $\dfrac{k^4}{3^2}$ (c) $\dfrac{1}{mn^3}$

POLYNOMIALS

5. Write in scientific notation.

(a) 400,000

(b) 29,800,000

(c) −6083

(d) .00172

(e) .0000000503

In **scientific notation,** a number is written as the product of a number between 1 and 10 (or −1 and −10) and some integer power of 10.

For example, since $8000 = 8 \cdot 1000 = 8 \cdot 10^3$, the number 8000 is written in scientific notation as

$$8000 = \boxed{8 \times 10^3}.$$

(It is customary to use × instead of a dot to show multiplication.)

A number can be converted to scientific notation with the following steps. (If the number is negative, ignore the minus sign, go through these steps, and then attach a minus sign to the result.)

CONVERTING TO SCIENTIFIC NOTATION

Step 1 Place a caret, ∧, to the right of the first nonzero digit.

Step 2 Count the number of digits from the caret to the decimal point. This number gives the absolute value of the exponent on the ten.

Step 3 Decide whether multiplying by 10^n should make the number larger or smaller. The exponent should be positive to make the product larger; it should be negative to make the product smaller.

EXAMPLE 5 Writing a Number in Scientific Notation
Write in scientific notation.

(a) 820,000

Place a caret after the 8 (the first nonzero digit).

$$8_\wedge 20,000$$

Count from the caret to the decimal point, which is understood to be after the last 0.

$$8_\wedge 20,000. \leftarrow \text{Decimal point}$$
$$\underbrace{}_{\text{Count 5 places}}$$

Now decide whether the exponent on 10 should be 5 or −5. The number will be written in scientific notation as 8.2×10^n. Comparing 8.2×10^n with 820,000 shows that n must be 5 so the product 8.2×10^5 will equal 820,000.

(b) −.000072

$$-.00007_\wedge 2$$
$$\underbrace{}_{\text{5 places}}$$

Since .000072 is smaller than 7.2, use a *negative* exponent on 10.

$$-.000072 = -7.2 \times 10^{-5} \blacksquare$$

WORK PROBLEM 5 AT THE SIDE.

ANSWERS
5. (a) 4×10^5 (b) 2.98×10^7
   (c) -6.083×10^3 (d) 1.72×10^{-3}
   (e) 5.03×10^{-8}

3.2 FURTHER PROPERTIES OF EXPONENTS

To convert a number written in scientific notation to standard notation, work in reverse. Multiplying a number by a positive power of 10 makes the number larger; multiplying by a negative power of 10 makes the number smaller.

EXAMPLE 6 Converting from Scientific Notation to Standard Notation

Write the following numbers in standard notation.

(a) 6.93×10^5

$$6.93000. \quad \text{5 places}$$

Since the exponent is positive, we moved the decimal point 5 places to the right to get a larger number. (We had to attach 3 zeros.)

$$6.93 \times 10^5 = 693,000$$

(b) $3.52 \times 10^7 = 35,200,000$

(c) 4.7×10^{-6}

$$\text{6 places} \quad .000004.7$$

Because of the negative exponent, we moved the decimal point 6 places to the left to get a smaller number. ■

WORK PROBLEM 6 AT THE SIDE.

With scientific notation, rules for exponents can be used to simplify lengthy computation.

EXAMPLE 7 Using Scientific Notation in Computation

Find $\dfrac{1,920,000 \times .0015}{.000032 \times 45,000}$.

First, express all numbers in scientific notation.

$$\frac{1,920,000 \times .0015}{.000032 \times 45,000} = \frac{1.92 \times 10^6 \times 1.5 \times 10^{-3}}{3.2 \times 10^{-5} \times 4.5 \times 10^4}$$

Next, use the rules for exponents to simplify the expression.

$$\frac{1.92 \times 10^6 \times 1.5 \times 10^{-3}}{3.2 \times 10^{-5} \times 4.5 \times 10^4} = \frac{1.92 \times 1.5}{3.2 \times 4.5} \times \frac{10^6 \times 10^{-3}}{10^{-5} \times 10^4}$$

$$= \frac{1.92 \times 1.5}{3.2 \times 4.5} \times 10^4$$

Now perform the calculations and write the answer in standard notation.

$$= .2 \times 10^4$$
$$= 2000 \quad ■$$

WORK PROBLEM 7 AT THE SIDE.

6. Write in standard notation.

(a) 3.7×10^8

(b) 2.51×10^3

(c) 4.6×10^{-5}

(d) 9.372×10^{-6}

(e) 8.5×10^{-1}

7. Use the rules for exponents to find each value.

(a) $\dfrac{8 \times 10^3}{2 \times 10^2}$

(b) $\dfrac{9 \times 10^3}{3 \times 10^{-2}}$

(c) $\dfrac{.06}{.003}$

(d) $\dfrac{200,000 \times .0003}{.06}$

ANSWERS
6. (a) 370,000,000 (b) 2510
   (c) .000046 (d) .000009372
   (e) .85
7. (a) 40 (b) 300,000 (c) 20
   (d) 1000

Historical Reflections

WOMEN IN MATHEMATICS:
Emilie, Marquise du Châtelet (1706–1749)

This French mathematician participated in the scientific activity of the generation after Isaac Newton and G. W. Leibniz (who independently developed the ideas of calculus). She was educated in science, music, and literature, and she was studying mathematics in 1733 when she began a long relationship with the philosopher Voltaire (1694–1778). She spent the last four years of her life translating Newton's *Principia Mathematica*, and today it remains the only French translation of this work.

Art: Library of Congress

3.2 EXERCISES

Simplify each expression. Write with only positive exponents. Assume that variables represent nonzero real numbers. See Examples 1–4.

1. $\left(\dfrac{3}{4}\right)^{-2}$

2. $\left(\dfrac{2}{5}\right)^{-3}$

3. $\left(\dfrac{6}{5}\right)^{-1}$

4. $\left(\dfrac{8}{3}\right)^{-2}$

5. $(3^{-2} \cdot 4^{-1})^3$

6. $(5^{-3} \cdot 6^{-2})^2$

7. $(4^{-3} \cdot 7^{-5})^{-2}$

8. $(9^{-3} \cdot 8^{-2})^{-2}$

9. $(z^3)^{-2} z^2$

10. $(p^{-1})^3 p^{-4}$

11. $-3r^{-1}(r^{-3})^2$

12. $2(y^{-3})^4 (y^6)$

13. $(3a^{-2})^3 (a^3)^{-4}$

14. $(m^5)^{-2}(3m^{-2})^3$

15. $(x^{-5}y^2)^{-1}$

16. $(a^{-3}b^{-5})^2$

17. $(2p^2q^{-3})^2(4p^{-3}q)^2$

18. $(-5y^2z^{-4})^2(2yz^5)^{-3}$

19. $\dfrac{(p^{-2})^3}{5p^4}$

20. $\dfrac{(m^4)^{-1}}{9m^3}$

21. $\dfrac{4a^5(a^{-1})^3}{(a^{-2})^{-2}}$

22. $\dfrac{12k^{-2}(k^{-3})^{-4}}{6k^5}$

23. $\dfrac{(-y^{-4})^2}{6(y^{-5})^{-1}}$

24. $\dfrac{2(-m^{-1})^{-4}}{9(m^{-3})^2}$

25. $\dfrac{(2k)^2 m^{-5}}{(km)^{-3}}$

26. $\dfrac{(3rs)^{-2}}{3^2 r^2 s^{-4}}$

27. $\dfrac{2^{-1}y^{-3}z^2}{2y^2 z^{-3}}$

28. $\dfrac{2^{-2}m^{-5}p^{-3}}{2^{-4}m^3 p^2}$

29. $\dfrac{(2k)^2 k^3}{k^{-1} k^{-5}}(5k^{-2})^{-3}$

30. $\dfrac{(3r^2)^2 r^{-5}}{r^{-2} r^3}(2r^{-6})^2$

31. $\left(\dfrac{3k^{-2}}{k^4}\right)^{-1} \cdot \dfrac{2}{k}$

32. $\left(\dfrac{7m^{-2}}{m^{-3}}\right)^{-2} \cdot \dfrac{m^3}{4}$

33. $\left(\dfrac{2p}{q^2}\right)^3 \left(\dfrac{3p^4}{q^{-4}}\right)^{-1}$

34. $\left(\dfrac{5z^3}{2a^2}\right)^{-3} \left(\dfrac{8a^{-1}}{15z^{-2}}\right)^{-3}$

35. $\dfrac{2^2 y^4 (y^{-3})^{-1}}{2^5 y^{-2}}$

36. $\dfrac{3^{-1}m^4(m^2)^{-1}}{3^2 m^{-2}}$

Write each number in scientific notation. See Example 5.

37. 230

38. 46,500

39. .02

40. .0051

41. .00000327

42. 93,000,000

43. −47,200

44. −.93506

POLYNOMIALS

Write each of the following in standard notation. See Example 6.

45. 6.5×10^3
46. 2.317×10^5
47. 1.52×10^{-2}
48. 1.63×10^{-4}

49. 5×10^{-3}
50. 8×10^7
51. -5.68×10^{-4}
52. -9.87×10^{-3}

Use the rules for exponents to find each value. See Example 7.

53. $\dfrac{9 \times 10^2}{3 \times 10^6}$
54. $\dfrac{8 \times 10^6}{4 \times 10^5}$
55. $\dfrac{6 \times 10^{-3}}{2 \times 10^2}$

56. $\dfrac{5 \times 10^{-2}}{1 \times 10^{-4}}$
57. $\dfrac{.002 \times 3900}{.000013}$
58. $\dfrac{.009 \times 600}{.02}$

59. $\dfrac{210{,}000}{.07 \times 20{,}000}$
60. $\dfrac{.018 \times 20{,}000}{300 \times .0004}$
61. $\dfrac{840{,}000 \times .03}{.00021 \times 600}$

Use scientific notation to work the following problems.

62. The distance to the sun is 9.3×10^7 miles. How long would it take a space probe, traveling at 3.1×10^3 miles per hour, to reach the sun? (Recall the formula $d = rt$.)

63. A *light-year* is the distance that light travels in one year. Find the number of miles in a light-year if light travels 1.86×10^5 miles per second.

64. Use the information given in the previous two exercises to find the number of minutes necessary for light from the sun to reach the earth.

65. A computer can do one addition in 2×10^{-7} seconds. How long would it take the computer to do a trillion (10^{12}) calculations? Give the answer in seconds and then in hours.

SKILL SHARPENERS
Use the properties of real numbers to simplify. See Section 1.5.

66. $9x + 5x - x$

67. $3p + 2p - 8p$

68. $6 - 4(3 - y)$

69. $7x - (5 + 5x)$

3.3 ADDITION AND SUBTRACTION OF POLYNOMIALS

Just as whole numbers are the basis of arithmetic, *polynomials* (defined below) are the basis of algebra. Recall that a *term* is the product of a number and one or more variables raised to a power. Examples of terms include

$$4x, \quad \frac{1}{2}m^5, \quad -7z^9, \quad \text{and} \quad 5.$$

Any factor in a term is the **coefficient** of the product of the remaining factors. For example, $3x^2$ is the coefficient of y in the term $3x^2y$, and $3y$ is the coefficient of x^2 in $3x^2y$. However, "coefficient" is often used to mean "numerical coefficient," and from now on we will use it that way in this book. With this understanding, in the term $8x^3$, the coefficient is 8, and in the term $-4p^5$, it is -4. The coefficient of the term k is understood to be 1. The coefficient of $-r$ is -1.

WORK PROBLEM 1 AT THE SIDE.

1 Recall that any combination of variables or constants joined by the basic operations of addition, subtraction, multiplication, division (except by zero), or extraction of roots is called an *algebraic expression*. The simplest kind of algebraic expression is a **polynomial,** a finite sum of terms in which all variables have only whole number exponents and no variables appear in denominators. Examples of polynomials include

$$3x - 5, \quad 4m^3 - 5m^2p + 8, \quad \text{and} \quad -5t^2s^3.$$

On the other hand, the expressions

$$5x^2 + 2x + \frac{1}{x} \quad \text{and} \quad x^{-3} - 6x$$

are *not* polynomials since variables appear in a denominator in the first case and an exponent on a variable is negative in the second case. Even though the expression $3x - 5$ involves subtraction, it is still called a sum of terms, since it could be written as $3x + (-5)$. Also, $-5t^2s^3$ can be treated as a sum of terms by writing it as $0 + (-5t^2s^3)$.

Although polynomials may have more than one variable, most of the polynomials discussed in this chapter have only one variable. A polynomial containing only the variable x is called a **polynomial in x.** A polynomial in one variable is written in **descending powers** of the variable if the exponents on the terms of the polynomial decrease from left to right. For example,

$$x^5 - 6x^2 + 12x - 5$$

is a polynomial in descending powers of x (the term -5 can be thought of as $-5x^0$). This is usually the preferred way to write a polynomial.

OBJECTIVES

1. Know the basic definitions for polynomials.
2. Identify monomials, binomials, and trinomials.
3. Find the degree of a polynomial.
4. Add and subtract polynomials.
5. Use $P(x)$ notation.

1. Identify each coefficient.

(a) $-9m^5$

(b) $12y^2x$

(c) x

(d) $-y$

ANSWERS
1. (a) -9 (b) 12 (c) 1 (d) -1

POLYNOMIALS

2. Write each polynomial in descending powers.

(a) $-4 + 9y + y^3$

(b) $-3z^4 + 2z^3 + z^5 - 6z$

(c) $-12m^{10} + 8m^9 + 10m^{12}$

3. Identify each polynomial as a trinomial, binomial, monomial, or none of these.

(a) $12m^4 - 6m^2$

(b) $-6y^3 + 2y^2 - 8y$

(c) $3a^5$

(d) $-2k^{10} + 2k^9 - 8k^5 + 2k$

(e) -7

ANSWERS
2. (a) $y^3 + 9y - 4$
   (b) $z^5 - 3z^4 + 2z^3 - 6z$
   (c) $10m^{12} - 12m^{10} + 8m^9$
3. (a) binomial (b) trinomial
   (c) monomial (d) none
   (e) monomial

EXAMPLE 1 Writing a Polynomial in Descending Powers
Write each of the following in descending powers of the variable.

(a) $y - 6y^3 + 8y^5 - 9y^4 + 12$

Write the polynomial as
$$8y^5 - 9y^4 - 6y^3 + y + 12.$$

(b) $-2 + m + 6m^2 - 4m^3$ would be written as
$$-4m^3 + 6m^2 + m - 2. \blacksquare$$

WORK PROBLEM 2 AT THE SIDE.

2 Certain types of polynomials are so common that they are given special names based on the number of terms that they have. A polynomial with exactly three terms is a **trinomial**, while a polynomial with two terms is a **binomial**. A single-term polynomial is a **monomial**.

EXAMPLE 2 Identifying Types of Polynomials
The list below gives examples of monomials, binomials, and trinomials, as well as polynomials that are none of these.

Type of Polynomial	Examples
Monomial	$5x$, $7m^9$, -8
Binomial	$3x^2 - 6$, $11y + 8$, $5k + 15$
Trinomial	$y^2 + 11y + 6$, $8p^3 - 7p + 2m$, $-3 + 2k^5 + 9z^4$
None of the above	$p^3 - 5p^2 + 2p - 5$, $-9z^3 + 5c^3 + 2m^5 + 11r^2 - 7r$ $\blacksquare$

WORK PROBLEM 3 AT THE SIDE.

3 The **degree** of a term with one variable is the exponent on that variable. For example, the degree of $2x^3$ is 3, the degree of $-x^4$ is 4, and the degree of $17x$ is 1. The highest degree of any of the terms in a polynomial in one variable is called the **degree of the polynomial**.

EXAMPLE 3 Finding the Degree of a Polynomial
Find the degree of each polynomial.

(a) $9x^2 - 5x + 8$ — Largest exponent is 2.

The largest exponent is 2, so the polynomial is of degree 2.

(b) $17m^9 + 8m^{14} - 9m^3$

This polynomial is of degree 14.

3.3 ADDITION AND SUBTRACTION OF POLYNOMIALS

(c) $5x$

The degree is 1, since $5x = 5x^1$.

(d) -2

A constant term, other than 0, has degree 0. (This is because, for example, $-2 = -2 \cdot x^0$.) There is no degree for the number 0 itself since 0 times x to any power is 0. ∎

WORK PROBLEM 4 AT THE SIDE.

4 The distributive property can sometimes be used to simplify polynomials. For example, to simplify $4x^2 + 5x^2$ we use the distributive property to combine terms as follows.

$$4x^2 + 5x^2 = (4 + 5)x^2 = 9x^2$$

In the same way,

$$-6m + 4m = (-6 + 4)m = -2m.$$

It is not possible to combine the terms in the polynomial $4x + 5x^2$. As these examples suggest, only terms containing exactly the same variables to the same powers may be combined. As mentioned in Chapter 1, such terms are called like terms.

EXAMPLE 4 Combining Terms
Combine terms.

(a) $-5y^3 + 8y^3 - 6y^3$

These like terms may be combined by using the distributive property.

$$-5y^3 + 8y^3 - 6y^3 = (-5 + 8 - 6)y^3 = -3y^3$$

(b) $6x + 5y - 9x + 2y$

Use the associative and commutative properties to rewrite the expression with all the x's together and all the y's together.

$$6x + 5y - 9x + 2y = 6x - 9x + 5y + 2y$$

Now combine like terms: $\quad = -3x + 7y.$

It is not possible to go further, since $-3x$ and $7y$ are unlike terms.

(c) $5x^2y - 6xy^2 + 9x^2y + 13xy^2 = 5x^2y + 9x^2y - 6xy^2 + 13xy^2$
$$= 14x^2y + 7xy^2 \quad ∎$$

CAUTION Remember that only like terms can be combined.

WORK PROBLEM 5 AT THE SIDE.

---

4. Give the degree of each polynomial.

(a) $9y^4 + 8y^3 - 6$

(b) $-12m^7 + 11m^3 + m^9$

(c) $-2k$

(d) 10

(e) $3 + 2x$

5. Combine terms.

(a) $11x + 12x - 7x - 3x$

(b) $5m^3 + m^3 + 9m^3$

(c) $-9y^4 + 8y^4 + 2y^4$

(d) $11p^5 + 4p^5 - 6p^3 + 8p^3$

(e) $2z^4 + 3x^4 + 5z^4 - 9x^4$

ANSWERS
4. (a) 4 (b) 9 (c) 1 (d) 0 (e) 1
5. (a) $13x$ (b) $15m^3$ (c) y^4
   (d) $15p^5 + 2p^3$ (e) $7z^4 - 6x^4$

151

POLYNOMIALS

Add.

(a) $-4x^3 + 2x^2$
$8x^3 - 6x^2$

(b) $(-5p^3 + 6p^2) + (8p^3 - 12p^2)$

(c) $-6r^5 + 2r^3 - r^2$
$8r^5 - 2r^3 + 5r^2$

(d) $(12y^2 - 7y + 9) + (-4y^2 - 11y + 5)$

The following rule is used to add two polynomials.

ADDING POLYNOMIALS

To add two polynomials, combine like terms.

EXAMPLE 5 Adding Polynomials Horizontally
Add $4k^2 - 5k + 2$ and $-9k^2 + 3k - 7$.

Use the commutative and associative properties to rearrange the polynomials so that like terms are together.

$$(4k^2 - 5k + 2) + (-9k^2 + 3k - 7)$$
$$= 4k^2 - 9k^2 - 5k + 3k + 2 - 7$$
$$= -5k^2 - 2k - 5 \quad \blacksquare$$

EXAMPLE 6 Adding Polynomials Vertically
The two polynomials of Example 1 can also be added vertically by lining up like terms in columns, then adding by columns.

$$\begin{array}{r} 4k^2 - 5k + 2 \\ -9k^2 + 3k - 7 \\ \hline -5k^2 - 2k - 5 \end{array} \quad \text{Add columns.} \quad \blacksquare$$

EXAMPLE 7 Adding Polynomials
Add $3a^5 - 9a^3 + 4a^2$ and $-8a^5 + 8a^3 + 2$.

By the first method shown above,

$$(3a^5 - 9a^3 + 4a^2) + (-8a^5 + 8a^3 + 2)$$
$$= 3a^5 - 8a^5 - 9a^3 + 8a^3 + 4a^2 + 2$$
$$= -5a^5 - a^3 + 4a^2 + 2. \quad \text{Combine like terms.}$$

We can also add these two polynomials by placing them in columns, with like terms in the same columns.

$$\begin{array}{r} 3a^5 - 9a^3 + 4a^2 \\ -8a^5 + 8a^3 + 2 \\ \hline -5a^5 - a^3 + 4a^2 + 2 \end{array} \quad \text{Add like terms.}$$

For many people, there is less chance of error with vertical addition. $\blacksquare$

WORK PROBLEM 6 AT THE SIDE.

In Chapter 1, subtraction of real numbers was defined for real numbers a and b as

$$a - b = a + (-b).$$

That is, the first number and the negative of the second are added. The definition for subtraction of polynomials is similar. The negative of a polynomial is defined as that polynomial with every sign changed.

ANSWERS
6. (a) $4x^3 - 4x^2$ (b) $3p^3 - 6p^2$
 (c) $2r^5 + 4r^2$ (d) $8y^2 - 18y + 14$

3.3 ADDITION AND SUBTRACTION OF POLYNOMIALS

SUBTRACTING POLYNOMIALS

To subtract two polynomials, add the negative of the *second* polynomial to the first polynomial.

EXAMPLE 8 Subtracting Polynomials Horizontally
Subtract: $(-6m^2 - 8m + 5) - (-3m^2 + 7m - 8)$.

Change every sign in the second polynomial and add.

$$(-6m^2 - 8m + 5) - (-3m^2 + 7m - 8)$$
$$= -6m^2 - 8m + 5 + 3m^2 - 7m + 8$$

↑ ↑ ↑
Change every sign.

Now add by combining terms.

$$= -6m^2 + 3m^2 - 8m - 7m + 5 + 8$$
$$= -3m^2 - 15m + 13$$

Check by adding $-3m^2 - 15m + 13$ to the second polynomial. The result should be the first polynomial. ■

WORK PROBLEM 7 AT THE SIDE.

EXAMPLE 9 Subtracting Polynomials Vertically
Use the same polynomials as in Example 8, and subtract in columns.

Write the first polynomial over the second, lining up like terms in columns.

$$\begin{array}{r} -6m^2 - 8m + 5 \\ -3m^2 + 7m - 8 \end{array}$$

Change all the signs in the second polynomial, and add.

$$\begin{array}{r} -6m^2 - 8m + 5 \\ +3m^2 - 7m + 8 \\ \hline -3m^2 - 15m + 13 \end{array}$$ All signs changed
Add in columns. ■

WORK PROBLEM 8 AT THE SIDE.

5 Sometimes one problem will involve several polynomials. To keep track of these polynomials, we use capital letters to name the polynomials. For example, we might let $P(x)$ (read "P of x") represent the polynomial $3x^2 - 5x + 7$, with

$$P(x) = 3x^2 - 5x + 7.$$

The x in $P(x)$ is used to show that x is the variable in the polynomial. If $x = -2$, then $P(x) = 3x^2 - 5x + 7$ takes on the value

$$P(-2) = 3(-2)^2 - 5(-2) + 7 \quad \text{Let } x = -2.$$
$$= 3 \cdot 4 + 10 + 7$$
$$= 29.$$

7. Subtract.

(a) $(2a^2 - a) - (5a^2 + 8a)$

(b) $(6y^3 - 9y^2 + 8)$
    $- (2y^3 + y^2 + 5)$

(c) $(p^4 + p^3 + 5)$
    $- (3p^4 + 5p^3 + 2)$

8. Subtract.

(a) $m^4 - 2m^3$
    $5m^4 + 8m^3$

(b) $6y^3 - 2y^2 + 5y$
    $-2y^3 + 8y^2 - 11y$

(c) $2k^3 - 3k^2 - 2k + 5$
    $4k^3 + 6k^2 - 5k + 8$

ANSWERS
7. (a) $-3a^2 - 9a$
   (b) $4y^3 - 10y^2 + 3$
   (c) $-2p^4 - 4p^3 + 3$
8. (a) $-4m^4 - 10m^3$
   (b) $8y^3 - 10y^2 + 16y$
   (c) $-2k^3 - 9k^2 + 3k - 3$

CHAPTER 3 POLYNOMIALS

9. Let $P(x) = -x^2 + 5x - 11$. Find each of the following.

(a) $P(1)$

(b) $P(-4)$

(c) $P(5)$

(d) $P(0)$

CAUTION Note that $P(x)$ does *not* mean P times x. It is a special notation to name a polynomial in x.

EXAMPLE 10 Using $P(x)$ Notation

Let $P(x) = 4x^3 - x^2 + 5$. Find each of the following.

(a) $P(3)$

First, substitute 3 for x.

$$P(x) = 4x^3 - x^2 + 5$$
$$P(3) = 4 \cdot 3^3 - 3^2 + 5 \qquad \text{Let } x = 3.$$

Now use the order of operations from Chapter 1.

$$= 4 \cdot 27 - 9 + 5 \qquad \text{Evaluate exponentials.}$$
$$= 108 - 9 + 5 \qquad \text{Multiply.}$$
$$= 104 \qquad \text{Subtract and add.}$$

(b) $P(-4) = 4 \cdot (-4)^3 - (-4)^2 + 5 \qquad \text{Let } x = -4.$
$$= 4 \cdot (-64) - 16 + 5$$
$$= -267 \quad \blacksquare$$

WORK PROBLEM 9 AT THE SIDE.

ANSWERS

9. (a) -7 (b) -47 (c) -11 (d) -11

3.3 EXERCISES

Give the coefficient of each term.

1. $9k$
2. $-14z$
3. $-11p^7$
4. $9y^{12}$
5. $-y$
6. $-a$
7. z
8. x

Write each polynomial in descending powers of the variable. See Example 1.

9. $8x + 9x^2 + 3x^3$
10. $5m - 3m^4 + 8m^6$
11. $-6y^3 + 8y^2 + 4y^4 - 3$
12. $9 - 4z^3 + 8z - 12z^2$

Identify each polynomial as a monomial, binomial, trinomial, or none of these. Give the degree of each. See Examples 2 and 3.

13. $4k - 9$
14. $12z^5$
15. $-8a^2 - 9a + 2$
16. $7y^5 - 6y^{12}$
17. $a^3 - a^2 + a^4$
18. $-m^9 + m^7 - m^6 + m^2$
19. -2
20. $29k$
21. $-5x^3 + 8x^2 - 5x + 2$

Combine terms when possible. See Example 4.

22. $6y + 4y$
23. $9k - 5k$
24. $8m^3 - 5m^3$
25. $12z^5 + 8z^5$
26. $2y^3 + 4y^3 - y^3$
27. $-5p^5 + 2p^5 + 7p^5$
28. $k + k + k + k$
29. $z - z + z - z$
30. $6x - 5x + x^2$
31. $2a^2 + 3a - 5a + 6a^2$
32. $6p^2 - 11p + 8p^2 - 21p$
33. $8r + 3r^2 - 5r^2 - 2r$
34. $-5y + 2y - 3y^2 + 4y^2 - 6y$
35. $3x + 5x^2 - 3x^2 + 2x - 3x^2$
36. $4m + 2m^2 + 6m^2 - 3m - m$

For each polynomial, find **(a)** $P(-1)$ *and* **(b)** $P(2)$. *See Example 10.*

37. $P(x) = 3x + 3$
    (a)   **(b)**

38. $P(x) = x - 10$
    (a)   **(b)**

39. $P(x) = -2x^2 + 4x - 1$
    (a)   **(b)**

40. $P(x) = 2x^2 - 5x + 2$
    (a)   **(b)**

41. $P(x) = x^4 - 3x^2 + 2x$
    (a)   **(b)**

42. $P(x) = 2x^4 + 3x^2 - 5x$
    (a)   **(b)**

Add or subtract as indicated. See Examples 5–9.

43. Add.
$$-12p^2 + 4p - 1$$
$$3p^2 + 7p - 8$$

44. Add.
$$-6y^3 + 8y + 5$$
$$9y^3 + 4y - 6$$

45. Subtract.
$$12a + 15$$
$$7a - 3$$

46. Subtract.
$$-3b + 6$$
$$2b - 8$$

47. Subtract.
$$6m^2 - 11m + 5$$
$$-8m^2 + 2m - 1$$

48. Subtract.
$$-4z^2 + 2z - 1$$
$$3z^2 - 5z + 2$$

49. Add.
$$12z^2 - 11z + 8$$
$$5z^2 + 16z - 2$$
$$-4z^2 + 5z - 9$$

50. Add.
$$-6m^3 + 2m^2 + 5m$$
$$8m^3 + 4m^2 - 6m$$
$$-3m^3 + 2m^2 - 7m$$

51. Add.
$$14p^5 + 8p^3 - p$$
$$6p^5 - 4p^3 + 8p$$
$$7p^5 + 3p^3 - 5p$$

52. Add.
$$6y^3 - 9y^2 + 8$$
$$4y^3 + 2y^2 + 5y$$

53. Add.
$$12k^5 - 9k^4 $$
$$ 8k^4 + 5k - 6$$

54. Add.
$$-7r^8 + 2r^6 - r^5$$
$$ 3r^6 + 5$$

3.3 EXERCISES

55. Subtract.
$$-5a^4 \phantom{{}+{}} + 8a^2 - 9$$
$$\underline{\; 6a^3 - a^2 + 2}$$

56. Subtract.
$$- 2m^3 + 8m^2 \phantom{{}+ 2m}$$
$$\underline{m^4 - m^3 + 2m}$$

57. Subtract.
$$6x^5 - 9x^4 + x^3 \phantom{{}+ x}$$
$$\underline{ 2x^4 - 3x^3 + x}$$

58. $(6 + 3p) + (2p + 1)$

59. $(4x - 8) + (-1 + x)$

60. $(2k - 7) - (3k - 8)$

61. $(5z + 6) - (8z - 3)$

62. $(3x^2 + 2x - 5) + (7x^2 - 4x + 3)$

63. $(y^3 + 3y + 2) + (4y^3 - 9y + 5)$

64. $(3a^4 - 9a^2 + 8) - (4a^4 + a^2 - 9)$

65. $(-2m^3 + 5m^2 - 6) - (3m^3 - 2m^2 + 5)$

66. $(8k^5 - 2k^2 + k) - (4k^5 + 5k^2 - 3k)$

67. $(-4p^3 + 6p^2 - 9p) - (p^3 + 8p^2 + 7p)$

68. $(5x^3 + 2x^2 + 5) + (x^3 - 5x^2 + 7x)$

69. $(-2z^3 - 5z^2 - 8z) + (-4z^3 + 2z^2 - 4)$

70. $[-(4m^2 - 8m + 4m^3) - (3m^2 + 2m + 5m^3)] + m^2$

71. $[-(y^4 - y^2 + 1) - (y^4 + 2y^2 + 1)] + (3y^4 - 3y^2 - 2)$

CHAPTER 3 POLYNOMIALS

Perform each operation.

72. Add $3x^2 - 5x$ and $5x - 3x^2$.

73. Subtract $x^2 - 7x + 2$ from $-2x^2 + 5x - 4$.

74. Subtract $2k - 3k^2$ from the sum of $3k^2 + 2$ and $6k + 2k^2$.

75. Add $-5y + 3y^2$ to the sum of $2y - 3y^3$ and $4y + 6y^2$.

SKILL SHARPENERS
Find each product. See Section 3.1.

76. $9p^2(4p^3)$ **77.** $8r^2(6r^4)$ **78.** $10z(5z^4)$

79. $6xy^3(2xy^4)$ **80.** $-2a^4(3a^5b^2)$ **81.** $-9m(-4mn^4)$

3.4 MULTIPLICATION OF POLYNOMIALS

In the previous section we saw how polynomials are added and subtracted. Polynomial multiplication is discussed in this section.

1 A polynomial with just one term is a monomial. To see how to multiply two monomials, recall that the product of the two terms $3x^4$ and $5x^3$ is found by using the commutative and associative properties, along with the rules for exponents.

$$(3x^4)(5x^3) = 3 \cdot 5 \cdot x^4 \cdot x^3$$
$$= 15x^{4+3}$$
$$= 15x^7$$

EXAMPLE 1 Multiplying Monomials
(a) $(-4a^3)(3a^5) = (-4)(3)a^3 \cdot a^5 = -12a^8$
(b) $(2m^2z^4)(8m^3z^2) = (2)(8)m^2 \cdot m^3 \cdot z^4 \cdot z^2 = 16m^5z^6$ ∎

WORK PROBLEM 1 AT THE SIDE.

2 The distributive property can be used to extend this process to find the product of any two polynomials.

EXAMPLE 2 Multiplying a Monomial and a Polynomial
(a) Find the product of -2 and $8x^3 - 9x^2$.
Use the distributive property.

$$-2(8x^3 - 9x^2) = \mathbf{-2}(8x^3) \mathbf{-} \mathbf{2}(-9x^2)$$
$$= -16x^3 + 18x^2$$

(b) Find the product of $5x^2$ and $-4x^2 + 3x - 2$.

$$5x^2(-4x^2 + 3x - 2)$$
$$= 5x^2(-4x^2) + 5x^2(3x) + 5x^2(-2) \quad \text{Distributive property}$$
$$= -20x^4 + 15x^3 - 10x^2 \quad \blacksquare$$

WORK PROBLEM 2 AT THE SIDE.

EXAMPLE 3 Multiply Two Polynomials
Find the product of $3x - 4$ and $2x^2 + x$.
Use the distributive property to multiply each term of $2x^2 + x$ by $3x - 4$.

$$(3x - 4)(2x^2 + x) = \mathbf{(3x - 4)}(2x^2) + \mathbf{(3x - 4)}(x)$$

Here $3x - 4$ has been treated as a single expression so that the distributive property could be used. Now use the distributive property twice again.

$$(3x - 4)(2x^2 + x) = 3x(\mathbf{2x^2}) + (-4)(\mathbf{2x^2}) + (3x)(\mathbf{x}) + (-4)(\mathbf{x})$$
$$= 6x^3 - 8x^2 + 3x^2 - 4x$$
$$= 6x^3 - 5x^2 - 4x \quad \blacksquare$$

OBJECTIVES

1 Multiply monomials.

2 Multiply any two polynomials.

3 Multiply two binomials.

4 Find the product of the sum and difference of two terms.

5 Find the square of a binomial.

---

1. Find each product.

   (a) $8k(-4k)$

   (b) $-6m^5(2m^4)$

   (c) $3p^4(2p)$

   (d) $(8k^3y)(9ky^3)$

2. Find each product.

   (a) $8(6b + 5)$

   (b) $p(3p + 7)$

   (c) $-2r(9r - 5)$

   (d) $3p^2(5p^3 + 2p^2 - 7)$

ANSWERS
1. (a) $-32k^2$ (b) $-12m^9$ (c) $6p^5$
   (d) $72k^4y^4$
2. (a) $48b + 40$ (b) $3p^2 + 7p$
   (c) $-18r^2 + 10r$
   (d) $15p^5 + 6p^4 - 21p^2$

CHAPTER 3 POLYNOMIALS

3. Find each product.

(a) $2m - 5$
$\underline{3m + 4}$

(b) $(4a - 5)(3a + 6)$

(c) $(2k - 5m)(3k + 2m)$

4. Find each product.

(a) $p^2 - 4p + 6$
$\underline{3p - 2}$

(b) $(r^4 - 2r^3 + 6)(3r - 1)$

(c) $(5a^3 - 6a^2 + 2a - 3) \cdot (2a - 5)$

ANSWERS
3. (a) $6m^2 - 7m - 20$
(b) $12a^2 + 9a - 30$
(c) $6k^2 - 11km - 10m^2$
4. (a) $3p^3 - 14p^2 + 26p - 12$
(b) $3r^5 - 7r^4 + 2r^3 + 18r - 6$
(c) $10a^4 - 37a^3 + 34a^2 - 16a + 15$

It is often easier to multiply polynomials by writing them vertically, in much the same way that numbers are multiplied. Proceed as follows to find the result in Example 3 by this method. (Notice how similar this process is to finding the product of two numbers, such as 24×78.)

Step 1   Multiply x and $3x - 4$.

$$\begin{array}{r} 3x - 4 \\ 2x^2 + x \\ \hline x(3x-4) \rightarrow 3x^2 - 4x \end{array}$$

Step 2   Multiply $2x^2$ and $3x - 4$. Line up like terms of the products in columns.

Step 3   Combine like terms.

$$\begin{array}{r} 3x - 4 \\ 2x^2 + x \\ \hline 3x^2 - 4x \\ 2x^2(3x-4) \rightarrow 6x^3 - 8x^2 \\ \hline 6x^3 - 5x^2 - 4x \end{array}$$

EXAMPLE 4 Multiplying Polynomials Vertically

Find the product of $5a - 2b$ and $3a + b$.

$$\begin{array}{r} 5a - 2b \\ 3a + b \\ \hline 5ab - 2b^2 \\ 15a^2 - 6ab \\ \hline 15a^2 - ab - 2b^2 \end{array}$$ ∎

■ WORK PROBLEM 3 AT THE SIDE.

EXAMPLE 5 Multiplying Polynomials Vertically

Find the product of $3m - 5$ and $3m^3 - 2m^2 + 4$.

$$\begin{array}{r} 3m^3 - 2m^2 + 4 \\ 3m - 5 \\ \hline -15m^3 + 10m^2 - 20 \\ 9m^4 - 6m^3 + 12m \\ \hline 9m^4 - 21m^3 + 10m^2 + 12m - 20 \end{array}$$

-5 times $3m^3 - 2m^2 + 4$
$3m$ times $3m^3 - 2m^2 + 4$
Combine like terms. ∎

■ WORK PROBLEM 4 AT THE SIDE.

3 In work with polynomials a special kind of product comes up repeatedly—the product of two binomials. A method for finding these products is discussed in the rest of this section.

Recall that a binomial is a polynomial with just two terms, such as $3x - 4$ or $2x + 3$. The product of the binomials $3x - 4$ and $2x + 3$ can be found with the distributive property as follows.

160

3.4 MULTIPLICATION OF POLYNOMIALS

$$(3x - 4)(2x + 3) = (3x - 4)(2x) + (3x - 4)(3)$$
$$= (3x)(2x) + (-4)(2x) + (3x)(3) + (-4)(3)$$
$$= 6x^2 - 8x + 9x - 12$$

Before combining like terms to find the simplest form of the answer, let us check the origin of each of the four terms in the sum. First, $6x^2$ is the product of the two *first* terms.

$$(\mathbf{3x} - 4)(\mathbf{2x} + 3) \qquad (3x)(2x) = 6x^2 \qquad \text{First terms}$$

To get $9x$, multiply the *outside* terms.

$$(\mathbf{3x} - 4)(2x + \mathbf{3}) \qquad 3x(3) = 9x \qquad \text{Outside terms}$$

The term $-8x$ comes from the *inside terms*.

$$(3x - \mathbf{4})(\mathbf{2x} + 3) \qquad -4(2x) = -8x \qquad \text{Inside terms}$$

Finally, -12 comes from the *last terms*.

$$(3x - \mathbf{4})(2x + \mathbf{3}) \qquad -4(3) = -12 \qquad \text{Last terms}$$

The product is found by combining these four results.

$$(3x - 4)(2x + 3) = 6x^2 + 9x - 8x - 12$$
$$= 6x^2 + x - 12$$

To help keep track of the order of multiplying these terms, use the initials FOIL (First-Outside-Inside-Last). The steps of the FOIL method can be done quickly, as follows.

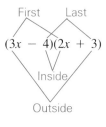

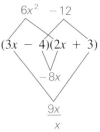

The FOIL method will be very helpful in factoring, which is discussed in the rest of this chapter.

CHAPTER 3 POLYNOMIALS

5. Find each product.

(a) $(3z - 2)(z + 1)$

(b) $(5r + 3)(2r - 5)$

(c) $(8r + 1)(8r - 1)$

(d) $(3p + 5)(3p + 5)$

6. Find each product.

(a) $(6k + y)(k - 3y)$

(b) $(4p + 5q)(3p - 2q)$

(c) $(4y - z)(2y + 3z)$

(d) $(2z + 3w)(7z - w)$

ANSWERS
5. (a) $3z^2 + z - 2$
   (b) $10r^2 - 19r - 15$
   (c) $64r^2 - 1$
   (d) $9p^2 + 30p + 25$
6. (a) $6k^2 - 17ky - 3y^2$
   (b) $12p^2 + 7pq - 10q^2$
   (c) $8y^2 + 10yz - 3z^2$
   (d) $14z^2 + 19zw - 3w^2$

EXAMPLE 6 Using the FOIL Method
Use the FOIL method to find $(4m - 5)(3m + 1)$.

Find the product of the first terms.

$(\mathbf{4m} - 5)(\mathbf{3m} + 1)$ $(4m)(3m) = \mathbf{12m^2}$

Multiply the outside terms.

$(\mathbf{4m} - 5)(3m + \mathbf{1})$ $(4m)(1) = \mathbf{4m}$

Find the product of the inside terms.

$(4m - \mathbf{5})(\mathbf{3m} + 1)$ $(-5)(3m) = \mathbf{-15m}$

Multiply the last terms.

$(4m - \mathbf{5})(3m + \mathbf{1})$ $(-5)(1) = \mathbf{-5}$

Combine the four terms obtained above, and simplify.

$$(4m - 5)(3m + 1) = 12m^2 + 4m - 15m - 5$$
$$= 12m^2 - 11m - 5$$

Practice adding the middle terms mentally so that you can write down the three terms of the answer directly, or use the shortcut notation shown below.

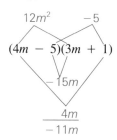

Combine these four results to get $12m^2 - 11m - 5$. ■

■ WORK PROBLEM 5 AT THE SIDE.

EXAMPLE 7 Using the FOIL Method
Find each product.

(a) $(6a - 5b)(3a + 4b) = 18a^2 + 24ab - 15ab - 20b^2$
  First   Outside   Inside   Last
  $= 18a^2 + 9ab - 20b^2$

(b) $(2k + 3z)(5k - 3z) = 10k^2 - 6kz + 15kz - 9z^2$
  $= 10k^2 + 9kz - 9z^2$ ■

■ WORK PROBLEM 6 AT THE SIDE.

3.4 MULTIPLICATION OF POLYNOMIALS

4 A special type of binomial product that occurs frequently is the sum and difference of the same two terms. By the FOIL method, the product of $(x + y)(x - y)$ is

$$(x + y)(x - y) = x^2 - xy + xy - y^2$$
$$= x^2 - y^2.$$

The **product of the sum and difference of two terms** is the difference of the squares of the terms, or

$$(x + y)(x - y) = x^2 - y^2.$$

EXAMPLE 8 Multiplying the Sum and Difference of Two Terms

(a) $(p + 7)(p - 7) = p^2 - 7^2 = p^2 - 49$

(b) $(2r + 5)(2r - 5) = (2r)^2 - 5^2$
$$= 2^2 r^2 - 25$$
$$= 4r^2 - 25$$

(c) $(6m + 5n)(6m - 5n) = (6m)^2 - (5n)^2$
$$= 36m^2 - 25n^2$$

(d) $(9y + 2z)(9y - 2z) = (9y)^2 - (2z)^2$
$$= 81y^2 - 4z^2$$

WORK PROBLEM 7 AT THE SIDE.

5 Another special binomial product is the *square of a binomial*.

To find the square of $x + y$ or $(x + y)^2$, multiply $x + y$ and $x + y$.

$$(x + y)^2 = (x + y)(x + y) = x^2 + xy + xy + y^2$$
$$= x^2 + 2xy + y^2$$

A similar result is true for the square of a difference, as shown below.

The **square of a binomial** is the sum of the square of the first term, twice the product of the two terms, and the square of the last term.

$$(x + y)^2 = x^2 + 2xy + y^2$$
$$(x - y)^2 = x^2 - 2xy + y^2$$

7. Find each product.

(a) $(m + 5)(m - 5)$

(b) $(x - 4y)(x + 4y)$

(c) $(7m - 2n)(7m + 2n)$

(d) $(6a + b)(6a - b)$

ANSWERS
7. (a) $m^2 - 25$ (b) $x^2 - 16y^2$
   (c) $49m^2 - 4n^2$
   (d) $36a^2 - b^2$

CHAPTER 3 POLYNOMIALS

8. Find each product.

(a) $(a + 2)^2$

(b) $(2m - 5)^2$

(c) $(y + 6z)^2$

(d) $(3k - 2n)^2$

9. Find each product.

(a) $[(m - 2n) - 3] \cdot [(m - 2n) + 3]$

(b) $(x - y + z)(x - y - z)$

(c) $[(k - 5h) + 2]^2$

EXAMPLE 9 Squaring a Binomial
(a) $(m + 7)^2 = \mathbf{m^2} + \mathbf{2 \cdot m \cdot 7} + \mathbf{7^2}$
$ = m^2 + 14m + 49$

(b) $(p - 5)^2 = p^2 - 2 \cdot p \cdot 5 + 5^2$
$ = p^2 - 10p + 25$

(c) $(2p + 3v)^2 = (2p)^2 + 2(2p)(3v) + (3v)^2$
$ = 4p^2 + 12pv + 9v^2$

(d) $(3r - 5s)^2 = (3r)^2 - 2(3r)(5s) + (5s)^2$
$ = 9r^2 - 30rs + 25s^2$ ■

■ **WORK PROBLEM 8 AT THE SIDE.**

CAUTION As the products in the definition of the square of a binomial show,
$$(x + y)^2 \neq x^2 + y^2.$$
Also, more generally,
$$(x + y)^n \neq x^n + y^n.$$

The patterns given above for the product of the sum and difference of two terms and for the square of a binomial can be used with more complicated binomial expressions.

EXAMPLE 10 Multiplying More Complicated Binomials
Find each product.

(a) $[(3p - 2) + 5q][(3p - 2) - 5q]$
$= (3p - 2)^2 - (5q)^2$ Product of sum and difference of terms
$= 9p^2 - 12p + 4 - 25q^2$ Square both quantities

(b) $[(2z + r) + 3]^2$
$= (2z + r)^2 + 2(2z + r)(3) + 3^2$ Square of a binomial
$= 4z^2 + 4zr + r^2 + 12z + 6r + 9$ Square $2z + r$. ■

■ **WORK PROBLEM 9 AT THE SIDE.**

ANSWERS
8. (a) $a^2 + 4a + 4$
 (b) $4m^2 - 20m + 25$
 (c) $y^2 + 12yz + 36z^2$
 (d) $9k^2 - 12kn + 4n^2$
9. (a) $m^2 - 4mn + 4n^2 - 9$
 (b) $x^2 - 2xy + y^2 - z^2$
 (c) $k^2 - 10kh + 25h^2 + 4k - 20h + 4$

3.4 EXERCISES

Find each product. See Examples 1–5.

1. $3p(5p)$

2. $4m(7m)$

3. $-9a^2(5a^3)$

4. $6y^3(-2y^3)$

5. $3(5a - 9)$

6. $-7(2k + 5)$

7. $2r(-5r + 2)$

8. $3y(2y - 9)$

9. $8a^2(a + 5)$

10. $-9p^3(2p - 7)$

11. $-6r^3(3r - 2)$

12. $2y^7(5y + 1)$

13. $-3z^2(2z^2 + 5z - 1)$

14. $2r^5(3r^2 - 5r + 2)$

15. $3b^3(2b^2 - b + 8)$

16. $-9k^4(-5k^3 + 7k^2 - 8)$

17. $2y^3(6y^2 - 4y^3 + 8y - 5)$

18. $3m^3(5m^3 - m^4 + 8m^2 - 2m)$

19. $3k - 2$
    $\underline{5k + 1}$

20. $2m - 5$
    $\underline{m + 7}$

21. $8a - 5$
    $\underline{2a + 1}$

22. $6m^2 + 2m - 1$
    $\underline{2m + 3}$

23. $-y^2 + 2y + 1$
    $\underline{3y - 5}$

24. $8r^2 - 2r + 5$
    $\underline{3r + 2}$

25. $2k - 5z$
    $3k + 2z$

26. $6m - 5n$
    $6m + 5n$

27. $7y + 11z$
    $7y - 11z$

28. $2z^3 - 5z^2 + 8z - 1$
    $ 4z - 3$

29. $5k^3 + 2k^2 - 3k - 4$
    $ 2k + 5$

30. $-4p^4 + 8p^3 - 9p^2 + 6p - 1$
    $ 3p - 2$

31. $8y^4 - 9y^3 + 3y^2 + 5y + 6$
    $ 2y + 5$

32. $-m^2 + 8m - 3$
    $ 2m^2 + 5m$

33. $2k^2 + 6k + 5$
    $ -k^2 + 2k$

Use the FOIL method to find each product. See Examples 6 and 7.

34. $(p + 4)(p - 6)$

35. $(m + 5)(m - 8)$

36. $(y + 7)(y + 5)$

37. $(3a - 2)(2a + 1)$

38. $(5m + 1)(m + 3)$

39. $(5y - 6)(2y + 7)$

40. $(2g - 3)(g - 5)$

41. $(6a - 5)(3a + 2)$

42. $(2p + 11)(2p + 1)$

43. $(7a + 5)(3a + 1)$

44. $(a + b)(3a - b)$

45. $(r - s)(2r + s)$

46. $(3p + q)(2p - q)$

47. $(8m - 3n)(3m - 5n)$

48. $(2p - 5q)(2p + 7q)$

3.4 EXERCISES

49. $\left(k - \dfrac{1}{2}\right)\left(k + \dfrac{2}{3}\right)$

50. $\left(r + \dfrac{5}{3}\right)\left(r - \dfrac{4}{5}\right)$

51. $\left(3w + \dfrac{1}{4}z\right)(w - 2z)$

52. $\left(5r - \dfrac{3}{5}y\right)(r + 5y)$

Find each product. See Example 8.

53. $(6a - 1)(6a + 1)$

54. $(3x + 2)(3x - 2)$

55. $(5x + 2y)(5x - 2y)$

56. $(2a - 5d)(2a + 5d)$

57. $(5r - 3w)(5r + 3w)$

58. $(10m - n)(10m + n)$

59. $(8x - y)(8x + y)$

60. $(7k - 6m)(7k + 6m)$

61. $\left(r - \dfrac{3}{4}\right)\left(r + \dfrac{3}{4}\right)$

62. $\left(p + \dfrac{1}{8}\right)\left(p - \dfrac{1}{8}\right)$

63. $(2y^3 + 1)(2y^3 - 1)$

64. $(3x^3 + 5)(3x^3 - 5)$

Find each of the following squares. See Example 9.

65. $(k - 8)^2$

66. $(z + 3)^2$

67. $(3r + t)^2$

68. $(10r - s)^2$

69. $(6p - 5q)^2$

70. $(2a - 7b)^2$

71. $\left(3k - \dfrac{1}{9}\right)^2$

72. $\left(z - \dfrac{3}{8}y\right)^2$

73. $\left(r + \dfrac{2}{3}s\right)^2$

CHAPTER 3 POLYNOMIALS

Find each of the following products. See Example 10.

74. $[(5x + 1) + 6y]^2$

75. $[(3m - 2) + p]^2$

76. $[(2a + b) - 3]^2$

77. $[(4k + h) - 4]^2$

78. $[(2a + b) - 3][(2a + b) + 3]$

79. $[(m + p) + 5][(m + p) - 5]$

80. $[(2h - k) + j][(2h - k) - j]$

81. $[(3m - y) + z][(3m - y) - z]$

Show that each of the following is false by replacing x with 2 and y with 3, and then rewrite each statement with the correct product.

82. $(x + y)^2 = x^2 + y^2$

83. $(x + y)^3 = x^3 + y^3$

84. $(x + y)^4 = x^4 + y^4$

SKILL SHARPENERS
Perform the indicated operations. See Sections 3.1 and 3.3.

85. $\dfrac{12p^7}{6p^3}$

86. $\dfrac{-8a^3b^7}{6a^5b}$

87. $\dfrac{20rs^5}{15rs^9}$

88. Subtract.

$\quad -3a^2 + 4a - 5$
$\quad \underline{5a^2 + 3a - 9}$

89. Subtract.

$\quad -4p^2 - 8p + 5$
$\quad \underline{3p^2 + 2p + 9}$

90. Subtract.

$\quad 10x^3 - 8x^2 + 4x$
$\quad \underline{10x^3 + 5x^2 - 7x}$

3.5 DIVISION OF POLYNOMIALS

1 Methods for adding, subtracting, and multiplying polynomials have now been shown. Polynomial division is discussed in this section, beginning with the division of a polynomial by a monomial. (Recall that a monomial is a single term, such as $8x$, $-9m^4$, or $11y^2$.)

DIVIDING A POLYNOMIAL BY A MONOMIAL

To divide a polynomial by a monomial, divide each term in the polynomial by the monomial and then write each quotient in lowest terms.

EXAMPLE 1 Dividing a Polynomial by a Monomial
(a) Divide $15x^2 - 12x + 6$ by 3.

Divide each term of the polynomial by 3; then write in lowest terms.

$$\frac{15x^2 - 12x + 6}{3} = \frac{15x^2}{3} - \frac{12x}{3} + \frac{6}{3}$$
$$= 5x^2 - 4x + 2$$

Check this answer by multiplying it by the divisor, 3. You should get $15x^2 - 12x + 6$ as the result.

$$\underset{\text{Divisor}}{3}(\underset{\text{Quotient}}{5x^2 - 4x + 2}) = \underset{\text{Original polynomial}}{15x^2 - 12x + 6}$$

(b) $\frac{5m^3 - 9m^2 + 10m}{5m^2} = \frac{5m^3}{5m^2} - \frac{9m^2}{5m^2} + \frac{10m}{5m^2}$ Divide each term by $5m^2$.

$= m - \frac{9}{5} + \frac{2}{m}$ Write in lowest terms.

This result is not a polynomial. The quotient of two polynomials need not be a polynomial.

(c) $\frac{8xy^2 - 9x^2y + 6x^2y^2}{x^2y^2} = \frac{8xy^2}{x^2y^2} - \frac{9x^2y}{x^2y^2} + \frac{6x^2y^2}{x^2y^2}$

$= \frac{8}{x} - \frac{9}{y} + 6$ ■

WORK PROBLEM 1 AT THE SIDE.

2 The process for dividing one polynomial by a polynomial that is not a monomial is very similar to that for dividing one whole number by another. The following examples show how this is done.

OBJECTIVES

1 Divide a polynomial by a monomial.

2 Divide a polynomial by a polynomial of two or more terms.

1. Divide.

(a) $\frac{12p + 30}{6}$

(b) $\frac{6z^2 + 12z + 18}{2z}$

(c) $\frac{9y^3 - 4y^2 + 8y}{2y^2}$

(d) $\frac{8a^2b^2 - 20ab^3}{4a^3b}$

ANSWERS
1. (a) $2p + 5$ (b) $3z + 6 + \frac{9}{z}$
   (c) $\frac{9y}{2} - 2 + \frac{4}{y}$ (d) $\frac{2b}{a} - \frac{5b^2}{a^2}$

EXAMPLE 2 Dividing One Polynomial by Another

Divide $2m^2 + m - 10$ by $m - 2$.

Write the problem, making sure that both polynomials are written in descending powers of the variables.

$$m - 2 \overline{\smash{)}2m^2 + m - 10}$$

Divide the first term of $2m^2 + m - 10$ by the first term of $m - 2$. Since $2m^2 \div m = 2m$, place this result above the division line.

$$\begin{array}{r} \mathbf{2m} \\ m - 2 \overline{\smash{)}2m^2 + m - 10} \end{array} \longleftarrow \text{Result of } \frac{2m^2}{m}$$

Multiply $m - 2$ and $2m$, and write the result below $2m^2 + m - 10$.

$$\begin{array}{r} 2m \\ m - 2 \overline{\smash{)}2m^2 + m - 10} \\ \mathbf{2m^2 - 4m} \end{array} \longleftarrow 2m(m - 2) = 2m^2 - 4m$$

Now subtract $2m^2 - 4m$ from $2m^2 + m$. Do this by changing the signs on $2m^2 - 4m$ and *adding*.

$$\begin{array}{r} 2m \\ m - 2 \overline{\smash{)}2m^2 + m - 10} \\ \underline{2m^2 - 4m} \\ \mathbf{5m} \end{array} \longleftarrow \text{Subtract.}$$

Bring down -10 and continue by dividing $5m$ by m.

$$\begin{array}{r} 2m \mathbf{+\ 5} \\ m - 2 \overline{\smash{)}2m^2 + m - 10} \\ \underline{2m^2 - 4m} \\ 5m - 10 \\ \underline{5m - 10} \\ 0 \end{array} \begin{array}{l} \longleftarrow \frac{5m}{m} = 5 \\ \\ \\ \longleftarrow 5(m - 2) = 5m - 10 \\ \longleftarrow \text{Subtract.} \end{array}$$

Finally, $(2m^2 + m - 10) \div (m - 2) = 2m + 5$. To check, multiply $m - 2$ and $2m + 5$. ∎

EXAMPLE 3 Dividing Polynomials

Divide $2p^3 + 7p^2 - 5p - 8$ by $2p + 1$.

Work as follows.

$$\begin{array}{r} p^2 + 3p - 4 \\ 2p + 1 \overline{\smash{)}2p^3 + 7p^2 - 5p - 8} \\ \underline{2p^3 + p^2} \\ 6p^2 - 5p \\ \underline{6p^2 + 3p} \\ -8p - 8 \\ \underline{-8p - 4} \\ -4 \end{array}$$

We write the remainder as a fraction: $\frac{-4}{2p + 1}$. The quotient is $p^2 + 3p - 4 + \frac{-4}{2p + 1}$. Check by multiplying $2p + 1$ and $p^2 + 3p - 4$ and adding -4 to the result. ∎

3.5 DIVISION OF POLYNOMIALS

WORK PROBLEM 2 AT THE SIDE.

EXAMPLE 4 Dividing a Polynomial with a Missing Term
Divide $3x^3 - 2x - 75$ by $x - 3$.

Make sure that $3x^3 - 2x - 75$ is in descending powers of the variable. Add a term with a 0 coefficient for the missing x^2 term.

$$x - 3 \overline{)3x^3 + 0x^2 - 2x - 75} \quad \leftarrow \text{Missing term}$$

Start with $3x^3 \div x = 3x^2$.

$$\begin{array}{r} 3x^2 \\ x - 3 \overline{)3x^3 + 0x^2 - 2x - 75} \\ 3x^3 - 9x^2 \end{array} \quad \begin{array}{l} \frac{3x^3}{x} = 3x^2 \\ 3x^2(x - 3) \end{array}$$

Subtract by changing the signs on $3x^3 - 9x^2$ and adding.

$$\begin{array}{r} 3x^2 \\ x - 3 \overline{)3x^3 + 0x^2 - 2x - 75} \\ \underline{3x^3 - 9x^2} \\ 9x^2 \quad \leftarrow \text{Subtract.} \end{array}$$

Bring down the next term.

$$\begin{array}{r} 3x^2 \\ x - 3 \overline{)3x^3 + 0x^2 - 2x - 75} \\ \underline{3x^3 - 9x^2} \\ 9x^2 - 2x \quad \leftarrow \text{Bring down } -2x. \end{array}$$

In the next step, $9x^2 \div x = 9x$.

$$\begin{array}{r} 3x^2 + 9x \\ x - 3 \overline{)3x^3 + 0x^2 - 2x - 75} \\ \underline{3x^3 - 9x^2} \\ 9x^2 - 2x \\ \underline{9x^2 - 27x} \quad \leftarrow 9x(x - 3) \\ 25x - 75 \leftarrow \text{Subtract and bring down } -75. \end{array}$$

Finally, $25x \div x = 25$.

$$\begin{array}{r} 3x^2 + 9x + 25 \\ x - 3 \overline{)3x^3 + 0x^2 - 2x - 75} \\ \underline{3x^3 - 9x^2} \\ 9x^2 - 2x \\ \underline{9x^2 - 27x} \\ 25x - 75 \\ \underline{25x - 75} \leftarrow 25(x - 3) \\ 0 \leftarrow \text{Subtract.} \end{array}$$

In summary,

$$\frac{3x^3 - 2x - 75}{x - 3} = 3x^2 + 9x + 25.$$

Check by multiplying $x - 3$ and $3x^2 + 9x + 25$. ■

WORK PROBLEM 3 AT THE SIDE.

---

2. Divide.

(a) $\dfrac{2r^2 + r - 21}{r - 3}$

(b) $\dfrac{2k^2 + 17k + 30}{2k + 5}$

(c) $\dfrac{10y^3 - 39y^2 + 41y - 10}{2y - 5}$

3. Divide.

(a) $\dfrac{2r^2 + 7r + 2}{2r - 1}$

(b) $\dfrac{3k^3 + 13k^2 + 8k - 12}{3k - 2}$

(c) $\dfrac{5x^4 - 13x^3 + 11x^2 - 33x + 10}{5x - 3}$

ANSWERS
2. (a) $2r + 7$ (b) $k + 6$
 (c) $5y^2 - 7y + 3 + \dfrac{5}{2y - 5}$
3. (a) $r + 4 + \dfrac{6}{2r - 1}$
 (b) $k^2 + 5k + 6$
 (c) $x^3 - 2x^2 + x - 6 + \dfrac{-8}{5x - 3}$

4. Divide.

(a) $\dfrac{4x^4 - 7x^2 + x + 5}{2x^2 - x}$

(b) $\dfrac{4m^4 - 23m^3 + 16m^2 - 4m - 1}{m^2 - 5m}$

(c) $\dfrac{3r^5 - 15r^4 - 2r^3 + 19r^2 - 7}{3r^2 - 2}$

EXAMPLE 5 Dividing Polynomials with Missing Terms
Divide $6r^4 + 5r^3 + r - 9$ by $3r^2 - 2r$.

The polynomial $6r^4 + 5r^3 + r - 9$ has a missing r^2 term. Insert the missing term with a 0 coefficient.

$$
\begin{array}{r}
2r^2 + 3r + 2 \\
3r^2 - 2r{\overline{\smash{\big)}\,6r^4 + 5r^3 + 0r^2 + r - 9}} \\
\underline{6r^4 - 4r^3} \\
9r^3 + 0r^2 \\
\underline{9r^3 - 6r^2} \\
6r^2 + r \\
\underline{6r^2 - 4r} \\
5r - 9
\end{array}
$$

Since the degree of the remainder, $5r - 9$, is less than the degree of the divisor, $3r^2 - 2r$, the division process is now finished. The result is written

$$2r^2 + 3r + 2 + \dfrac{5r - 9}{3r^2 - 2r}.\ \blacksquare$$

WORK PROBLEM 4 AT THE SIDE.

ANSWERS

4. (a) $2x^2 + x - 3 + \dfrac{-2x + 5}{2x^2 - x}$

    (b) $4m^2 - 3m + 1 + \dfrac{m - 1}{m^2 - 5m}$

    (c) $r^3 - 5r^2 + 3 + \dfrac{-1}{3r^2 - 2}$

3.5 EXERCISES

Divide. See Examples 1–5.

1. $\dfrac{5a^2 - 10a + 5}{5}$

2. $\dfrac{9y^2 - 12y + 27}{3}$

3. $\dfrac{8m^2 + 16m + 24}{8m}$

4. $\dfrac{22p^2 - 11p + 33}{11p}$

5. $\dfrac{12a^3 + 8a^2 - 10a}{4a^3}$

6. $\dfrac{8m^4 - 6m^3 + 4m^2}{8m^3}$

7. $\dfrac{9p^2q + 18pq^2 - 6pq}{12p^2q^2}$

8. $\dfrac{15m^2k^2 - 9mk^3 + 12m^2k}{6m^2k}$

9. $\dfrac{3abc^2 - 6ab^2c + 9a^2bc}{15abc}$

10. $\dfrac{27x^2y^2z - 18xy^2z^2 + 36xyz^2}{9x^2y^3z}$

11. $\dfrac{x^2 - 2x - 15}{x - 5}$

12. $\dfrac{p^2 - 3p - 18}{p - 6}$

13. $\dfrac{2r^2 + 13r + 21}{2r + 7}$

14. $\dfrac{10a^2 - 11a - 6}{5a + 2}$

15. $(6b^3 - 7b^2 - 3b + 1) \div (2b - 3)$

16. $(4a^3 + 5a^2 + 4a + 5) \div (4a + 5)$

17. $(a^3 - 8a^2 + 6a - 3) \div (a - 3)$

18. $(3r^3 - 22r^2 + 25r - 10) \div (3r - 1)$

19. $\dfrac{4x^3 - 18x^2 + 22x - 10}{2x^2 - 4x + 3}$

20. $\dfrac{12m^3 - 17m^2 + 30m - 10}{3m^2 - 2m + 5}$

21. $\dfrac{8p^3 + 2p^2 + p + 18}{2p^2 + 3}$

22. $\dfrac{12z^3 + 9z^2 - 10z + 21}{3z^2 - 2}$

23. $\dfrac{4y^4 + 6y^3 + 3y - 1}{2y^2 + 1}$

24. $\dfrac{15r^4 + 3r^3 + 4r^2 + 4}{3r^2 - 1}$

25. $(9t^4 - 13t^3 + 23t^2 - 9t + 2) \div (t^2 - t + 2)$

26. $(2m^4 + 5m^3 - 11m^2 + 11m - 20) \div (2m^2 - m + 2)$

27. $\left(2x^2 - \dfrac{7}{3}x - 1\right) \div (3x + 1)$

28. $\left(m^2 + \dfrac{7}{2}m + 3\right) \div (2m + 3)$

29. $\left(3a^2 - \dfrac{23}{4}a - 5\right) \div (4a + 3)$

30. $\left(3q^2 + \dfrac{19}{5}q - 3\right) \div (5q - 2)$

Solve each of the following word problems.

31. The area of a rectangle is $6k^2 - 5k - 6$. Find the length if the width is $2k - 3$.

32. The volume of a box is $2p^3 + 15p^2 + 28p$. The height is p and the length is $p + 4$; find the width.

33. Suppose a car goes $2m^3 + 15m^2 + 13m - 63$ kilometers in $2m + 9$ hours. Find its rate of speed.

34. A garden has $6z^3 - 17z^2 + 4z + 7$ flowers planted in $3z - 7$ rows, with an equal number of flowers in each row. Find the number of flowers in each row.

SKILL SHARPENERS

Evaluate each polynomial for the given value of x. See Section 1.4.

35. $x^2 + 2x - 9$; $x = 1$

36. $3x^2 - 5x + 7$; $x = 2$

37. $2x^3 + x^2 - 5x - 4$; $x = -2$

38. $3x^3 - x^2 + 4x - 8$; $x = -3$

3.6 SYNTHETIC DIVISION

OBJECTIVES

1. Use synthetic division to divide by a polynomial of the form $x - k$.
2. Use the remainder theorem to evaluate a polynomial.
3. Decide whether a given number is a solution of an equation.

1 Many times when one polynomial is divided by a second, the second polynomial is of the form $x - k$, where the coefficient of the x term is 1. There is a shortcut way for doing these divisions. To see how this shortcut works, look first at the left below, where the division of $3x^3 - 2x + 5$ by $x - 3$ is shown. (Notice that a 0 was inserted for the missing x^2 term.)

$$
\begin{array}{r}
3x^2 + 9x + 25 \\
x - 3{\overline{\smash{\big)}\,3x^3 + 0x^2 - 2x + 5}} \\
\underline{3x^3 - 9x^2} \\
9x^2 - 2x \\
\underline{9x^2 - 27x} \\
25x + 5 \\
\underline{25x - 75} \\
80
\end{array}
\qquad
\begin{array}{r}
3 \quad\; 9 \quad\; 25 \\
1 - 3{\overline{\smash{\big)}\,3 \quad\; 0 \;-2 \quad\;\; 5}} \\
(3) \;-9 \\
9 \;-2 \\
(9) \;-27 \\
25 \quad\; 5 \\
(25) \;-75 \\
80
\end{array}
$$

On the right, we show exactly the same division, written without the variables. All the numbers in parentheses on the right are repetitions of the numbers directly above them, so that they may be omitted, as shown below.

$$
\begin{array}{r}
3 \quad\; 9 \quad\; 25 \\
1 - 3{\overline{\smash{\big)}\,3 \quad\; 0 \;-2 \quad\;\; 5}} \\
-9 \\
9 \;\;(-2) \\
-27 \\
25 \quad (5) \\
-75 \\
80
\end{array}
$$

The numbers in parentheses are again repetitions of the numbers directly above them; they too can be omitted, as shown here.

$$
\begin{array}{r}
3 \quad\; 9 \quad\; 25 \\
1 - 3{\overline{\smash{\big)}\,3 \quad\; 0 \;-2 \quad\;\; 5}} \\
-9 \\
9 \\
-27 \\
25 \\
-75 \\
80
\end{array}
$$

Now the problem can be condensed. If the 3 in the top row is brought down to the beginning of the bottom row, the top row can be omitted, since it duplicates the bottom row.

$$
\begin{array}{r}
1 - 3{\overline{\smash{\big)}\,3 \quad\;\; 0 \;\;-2 \quad\;\; 5}} \\
-9 \;-27 \;-75 \\
\hline
3 \quad\; 9 \quad\;\; 25 \quad\; 80
\end{array}
$$

CHAPTER 3 POLYNOMIALS

1. Divide, using synthetic division.

(a) $\dfrac{3z^2 + 10z - 8}{z + 4}$

(b) $\dfrac{p^2 - 4p + 7}{p - 2}$

(c) $\dfrac{2x^3 + 13x^2 + 30x + 25}{x + 3}$

(d) $(2x^2 + 3x - 5) \div (x + 1)$

ANSWERS

1. (a) $3z - 2$ (b) $p - 2 + \dfrac{3}{p - 2}$
   (c) $2x^2 + 7x + 9 + \dfrac{-2}{x + 3}$
   (d) $2x + 1 + \dfrac{-6}{x + 1}$

Finally, the number at the upper left can be omitted since that number will always be a 1. Also, to simplify the arithmetic, subtraction in the second row is replaced by addition. Compensate for this by changing the -3 at upper left to its additive inverse, 3. The result of doing all this is shown below.

Additive inverse → $3 \overline{)\,3 \quad\; 0 \quad -2 \quad\;\; 5}$
$\qquad\qquad\qquad\qquad\;\; 9 \quad\;\; 27 \quad 75 \quad$ ← Signs changed
$\qquad\qquad\qquad\;\; 3 \quad\;\; 9 \quad\;\; 25 \quad\;\; 80$

Read the quotient from the bottom row.

$3x^2 + 9x + 25 + \dfrac{80}{x - 3}$

The first three numbers in the bottom row are used to obtain a polynomial of degree 1 less than the degree of the dividend. The last number gives the remainder.

This shortcut procedure is called **synthetic division**. It is used only when dividing a polynomial by a polynomial of the form $x - k$.

EXAMPLE 1 Using Synthetic Division

Use synthetic division to divide $5x^2 + 16x + 15$ by $x + 2$.

As mentioned above, synthetic division can be used only when dividing by a polynomial of the form $x - k$. To get $x + 2$ in this form, write it as

$$x + 2 = x - (-2),$$

where $k = -2$. Now write the coefficients of $5x^2 + 16x + 15$, placing -2 to the left.

$x + 2$ leads to -2 → $-2 \overline{)\, 5 \quad 16 \quad 15\,}$ Coefficients

Bring down the 5, and multiply: $-2 \cdot 5 = -10$.

$-2 \overline{)\, 5 \quad\;\; 16 \quad 15\,}$
$\qquad\quad -10$
$\qquad\; 5$

Add 16 and -10, getting 6. Multiply 6 and -2 to get -12.

$-2 \overline{)\, 5 \quad\;\; 16 \quad\;\; 15\,}$
$\qquad\quad -10 \quad -12$
$\qquad\; 5 \quad\;\; 6$

Add 15 and -12, getting 3.

$-2 \overline{)\, 5 \quad\;\; 16 \quad\;\; 15\,}$
$\qquad\quad -10 \quad -12$
$\qquad\; 5 \quad\;\; 6 \quad\;\; 3$

Read the result from the bottom row.

$$\dfrac{5x^2 + 16x + 15}{x + 2} = 5x + 6 + \dfrac{3}{x + 2} \quad\blacksquare$$

WORK PROBLEM 1 AT THE SIDE.

3.6 SYNTHETIC DIVISION

EXAMPLE 2 Using Synthetic Division with Missing Terms
Use synthetic division to find
$$(-4x^5 + x^4 + 6x^3 + 2x^2 + 50) \div (x - 2).$$

Use the steps given above, inserting a 0 for the missing x term.

```
                        ↓ Insert 0 for missing term.
   2) -4    1     6     2     0    50
            -8   -14   -16   -28   -56
       -4   -7   -8    -14   -28   -6
```

Read the result from the bottom row.

$$\frac{-4x^5 + x^4 + 6x^3 + 2x^2 + 50}{x - 2}$$
$$= -4x^4 - 7x^3 - 8x^2 - 14x - 28 + \frac{-6}{x - 2} \blacksquare$$

WORK PROBLEM 2 AT THE SIDE.

2 Synthetic division can be used to evaluate polynomials. For example, in Example 2, synthetic division was used to divide $-4x^5 + x^4 + 6x^3 + 2x^2 + 50$ by $x - 2$. The remainder in this division was -6. If x is replaced with 2,

$$-4x^5 + x^4 + 6x^3 + 2x^2 + 50$$
$$= -4 \cdot \mathbf{2}^5 + \mathbf{2}^4 + 6 \cdot \mathbf{2}^3 + 2 \cdot \mathbf{2}^2 + 50$$
$$= -4 \cdot 32 + 16 + 6 \cdot 8 + 2 \cdot 4 + 50$$
$$= -128 + 16 + 48 + 8 + 50$$
$$= -6,$$

the same number as the remainder; that is, dividing by $x - 2$ produced a remainder equal to the result when x is replaced with 2. This always happens, as the following **remainder theorem** states.

REMAINDER THEOREM

If the polynomial $P(x)$ is divided by $x - k$, the remainder equals $P(k)$.

This result is proved in more advanced courses.

EXAMPLE 3 Using the Remainder Theorem
Let $P(x) = 2x^3 - 5x^2 - 3x + 11$. Find $P(-2)$.

By the remainder theorem, we can find $P(-2)$ by dividing $P(x)$ by $x - (-2) = x + 2$.

```
Value of x →  -2) 2   -5   -3    11
                      -4   18   -30
                  2   -9   15   -19   ← Remainder
```

By this result, $P(-2) = -19$. $\blacksquare$

WORK PROBLEM 3 AT THE SIDE.

---

2. Divide, using synthetic division.

(a) $\dfrac{3a^3 - 2a + 21}{a + 2}$

(b) $(-4x^4 + 3x^3 + 18x + 2) \div (x - 2)$

(c) $\dfrac{v^4 - 1}{v + 1}$

3. Let $P(x) = x^3 - 5x^2 + 7x - 3$. Use synthetic division to find each of the following.

(a) $P(1)$ (Divide by $x - 1$.)

(b) $P(-2)$

(c) $P(3)$

ANSWERS
2. (a) $3a^2 - 6a + 10 + \dfrac{1}{a + 2}$
   (b) $-4x^3 - 5x^2 - 10x - 2 + \dfrac{-2}{x - 2}$
   (c) $v^3 - v^2 + v - 1$
3. (a) 0 (b) -45 (c) 0

CHAPTER 3 POLYNOMIALS

4. Use synthetic division to decide whether or not 2 is a solution for each of the following.

(a) $3r^3 - 11r^2 + 17r - 14 = 0$

(b) $x^4 - 3x^3 + x^2 - x - 2 = 0$

(c) $y^4 - 3y^2 + 4y - 12 = 0$

(d) $4k^5 - 7k^4 - 11k^2 + 2k + 6 = 0$

3 The remainder theorem can also be used to show that a given number is a solution of an equation.

EXAMPLE 4 Using the Remainder Theorem

Show that -5 is a solution of the equation

$$2x^4 + 12x^3 + 6x^2 - 5x + 75 = 0.$$

One way to show that -5 is a solution is by substituting -5 for x in the equation. However, an easier way is to use synthetic division and the remainder theorem given above.

```
Proposed ⟶ -5)2    12     6    -5    75
solution          -10   -10    20   -75
                ─────────────────────────
                  2     2    -4    15     0  ← Remainder = 0
```

Since the remainder is 0, the value of the polynomial on the left side of the equation when $x = -5$ is 0, and the number -5 is a solution of the equation. ∎

WORK PROBLEM 4 AT THE SIDE.

ANSWERS
4. (a) yes (b) no (c) yes (d) no

3.6 EXERCISES

Use synthetic division in each of the following. See Examples 1 and 2.

1. $\dfrac{x^2 + 6x - 7}{x - 1}$

2. $\dfrac{x^2 - 2x - 15}{x + 3}$

3. $\dfrac{3m^2 + 7m - 40}{m + 5}$

4. $\dfrac{5k^2 - 13k - 6}{k - 3}$

5. $\dfrac{3a^2 + 10a + 11}{a + 2}$

6. $\dfrac{4y^2 - 5y - 20}{y - 3}$

7. $(p^2 - 3p + 5) \div (p + 1)$

8. $(z^2 + 4z - 6) \div (z - 5)$

9. $\dfrac{4a^3 - 3a^2 + 2a + 1}{a - 1}$

10. $\dfrac{5p^3 - 6p^2 + 3p + 14}{p + 1}$

11. $(6x^5 - 2x^3 - 4x^2 + 3x - 2) \div (x - 2)$

12. $(y^5 - 4y^4 + 3y^2 - 5y + 4) \div (y - 3)$

13. $(-4r^6 - 3r^5 - 3r^4 + 5r^3 - 6r^2 + 3r + 3) \div (r - 1)$

14. $(2t^6 - 3t^5 + 2t^4 - 5t^3 + 6t^2 - 3t - 2) \div (t - 2)$

15. $(-3y^5 + 2y^4 - 5y^3 - 6y^2 - 1) \div (y + 2)$

16. $(m^6 + 2m^4 - 5m + 11) \div (m - 2)$

17. $\dfrac{y^3 - 1}{y + 1}$

18. $\dfrac{z^4 + 81}{z + 3}$

CHAPTER 3 POLYNOMIALS

Use the remainder theorem to find P(k). See Example 3.

19. $P(x) = x^3 - 4x^2 + 8x - 1; \quad k = 2$

20. $P(y) = 2y^3 + y^2 - y + 5; \quad k = -1$

21. $P(r) = -r^3 + 8r^2 - 3r - 2; \quad k = -4$

22. $P(z) = z^3 + 5z^2 - 7z + 1; \quad k = 3$

23. $P(y) = 2y^3 - 4y^2 + 5y - 33; \quad k = 3$

24. $P(x) = x^3 - 3x^2 + 4x - 4; \quad k = 2$

25. $P(x) = x^4 - 3x^3 + 5x^2 - 2x + 5; \quad k = -2$

26. $P(t) = -t^4 + t^3 - 5t^2 + 3t - 4; \quad k = -1$

Use synthetic division to decide whether the given number is a solution for each of the following equations. See Example 4.

27. $x^3 - 2x^2 - 2x + 12 = 0; \quad x = -2$

28. $x^5 + 2x^4 - 3x^3 + 8x = 0; \quad x = -2$

29. $m^3 - 3m^2 - 2m + 16 = 0; \quad m = -2$

30. $r^4 - r^3 - 6r^2 + 5r + 10 = 0; \quad r = -2$

31. $3a^3 + 2a^2 - 2a + 11 = 0; \quad a = -2$

32. $3z^3 + 10z^2 + 3z - 9 = 0; \quad z = -2$

33. $2x^3 - x^2 - 13x + 24 = 0; \quad x = -3$

34. $5p^3 + 22p^2 + p - 28 = 0; \quad p = -4$

35. $7z^3 - z^2 + 5z - 3 = 0; \quad z = 3$

36. $2r^3 + 4r^2 - r - 5 = 0; \quad r = -1$

SKILL SHARPENERS

Use the distributive property to rewrite each expression. See Section 1.5.

37. $9 \cdot 6 + 9 \cdot r$

38. $8 \cdot y - 8 \cdot 5$

39. $7(2x) - 7(3y)$

40. $4(8p) + 4(9y)$

41. $x(x + 1) + y(x + 1)$

42. $p(2p - 3) + 5(2p - 3)$

3.7 GREATEST COMMON FACTORS; FACTORING BY GROUPING

Earlier, we defined integer factors as two numbers whose product is a third number. In this chapter, we will extend this concept to the factoring of polynomials.

Integers are often written in **prime factored form,** that is, as a product of prime numbers. A **prime number** is a positive integer greater than 1 that can be divided without remainder only by itself and 1. The first few primes are 2, 3, 5, 7, 11, and 13. Each number in the factored form is a factor of the integer.

Writing a polynomial as the product of two or more simpler polynomials is called **factoring** the polynomial. For example, the product of $3x$ and $5x - 2$ is $15x^2 - 6x$, and $15x^2 - 6x$ can be factored as the product $3x(5x - 2)$.

$$3x(5x - 2) = 15x^2 - 6x \quad \text{Multiplication}$$
$$15x^2 - 6x = 3x(5x - 2) \quad \text{Factoring}$$

Notice that both multiplication and factoring are examples of the distributive property, used in opposite directions.

1 The first step in factoring a polynomial is to find the *greatest common factor* for the terms of the polynomial. The **greatest common factor** is the largest term that is a factor of all the terms in the polynomial. For example, the greatest common factor for $8x + 12$ is 4, since 4 is the largest number that is a factor of both $8x$ and 12. Using the distributive property,

$$8x + 12 = \mathbf{4}(2x) + \mathbf{4}(3) = \mathbf{4}(2x + 3).$$

As a check, multiply 4 and $2x + 3$. The result should be $8x + 12$. This process is called **factoring out the greatest common factor.**

EXAMPLE 1 Factoring Out the Greatest Common Factor
Factor out the greatest common factor.

(a) $9z - 18$

Since 9 is the greatest common factor,
$$9z - 18 = \mathbf{9} \cdot z - \mathbf{9} \cdot 2 = \mathbf{9}(z - 2).$$

(b) $56m + 35p = 7(8m + 5p)$

(c) $2y + 5$ There is no common factor other than 1.

(d) $12 + 24z = 12 \cdot 1 + 12 \cdot 2z$
$\qquad\qquad\quad = 12(1 + 2z)$ 12 is the greatest common factor. ∎

CAUTION   When factoring, it is very common to forget the 1. Be careful to include it as needed. Do not forget that any answer can be checked by multiplication.

WORK PROBLEM 1 AT THE SIDE.

OBJECTIVES

1 Factor out the greatest common factor.

2 Factor by grouping.

1. Factor out the greatest common factor.

   (a) $7k + 28$

   (b) $32m + 24$

   (c) $8a - 9$

   (d) $5z + 5$

ANSWERS
1. (a) $7(k + 4)$ (b) $8(4m + 3)$
   (c) cannot be factored   (d) $5(z + 1)$

CHAPTER 3 POLYNOMIALS

2. Factor out the greatest common factor.

   (a) $16y^4 + 8y^3$

   (b) $14p^2 - 9p^3 + 6p^4$

   (c) $15z^2 + 45z^5 - 60z^6$

   (d) $100m^5 - 50m^4 + 25m^3$

3. Factor out the greatest common factor.

   (a) $8p^3q^2 - 6p^4q$

   (b) $12y^5x^2 + 8y^3x^3$

   (c) $5m^4x^3 + 15m^5x^6 - 20m^4x^6$

ANSWERS
2. (a) $8y^3(2y + 1)$
   (b) $p^2(14 - 9p + 6p^2)$
   (c) $15z^2(1 + 3z^3 - 4z^4)$
   (d) $25m^3(4m^2 - 2m + 1)$
3. (a) $2p^3q(4q - 3p)$
   (b) $4y^3x^2(3y^2 + 2x)$
   (c) $5m^4x^3(1 + 3mx^3 - 4x^3)$

EXAMPLE 2 Factoring Out the Greatest Common Factor
Factor out the greatest common factor.

(a) $9x^2 + 12x^3$

The numerical part of the common factor is 3, the largest number that is a factor of both 9 and 12. For the variable portions, x^2 and x^3, use x to the lowest degree; here the lowest degree is 2. The greatest common factor is $3x^2$.

$$9x^2 + 12x^3 = \mathbf{3x^2}(3) + \mathbf{3x^2}(4x) = 3x^2(3 + 4x)$$

(b) $32p^4 - 24p^3 + 40p^5$

The lowest degree is 3. The greatest common factor is thus $8p^3$.

$$32p^4 - 24p^3 + 40p^5 = \mathbf{8p^3}(4p) + \mathbf{8p^3}(-3) + \mathbf{8p^3}(5p^2)$$
$$= 8p^3(4p - 3 + 5p^2)$$

(c) $3k^4 - 15k^7 + 24k^9 = 3k^4(1 - 5k^3 + 8k^5)$ ∎

WORK PROBLEM 2 AT THE SIDE.

EXAMPLE 3 Factoring Out the Greatest Common Factor
Factor out the greatest common factor.

(a) $24m^3n^2 - 18m^2n + 6m^4n^3$

The numerical part of the greatest common factor is 6. Find the variable part by writing each variable with its lowest degree. Here, 2 is the lowest exponent that appears on m, while 1 is the lowest exponent on n. Finally, $6m^2n$ is the greatest common factor.

$$24m^3n^2 - 18m^2n + 6m^4n^3$$
$$= (6m^2n)(4mn) + (6m^2n)(-3) + (6m^2n)(m^2n^2)$$
$$= 6m^2n(4mn - 3 + m^2n^2)$$

(b) $25x^2y^3 + 30xy^5 - 15x^4y^7 = 5xy^3(5x + 6y^2 - 3x^3y^4)$ ∎

WORK PROBLEM 3 AT THE SIDE.

EXAMPLE 4 Factoring Out a Binomial Factor
Factor out the greatest common factor.

(a) $(x - 5)(x + 6) + (x - 5)(2x + 5)$

The greatest common factor here is $x - 5$.

$$\mathbf{(x - 5)}(x + 6) + \mathbf{(x - 5)}(2x + 5)$$
$$= \mathbf{(x - 5)}[(x + 6) + (2x + 5)]$$
$$= (x - 5)(x + 6 + 2x + 5)$$
$$= (x - 5)(3x + 11)$$

(b) $z^2(m + n) + x^2(m + n) = (m + n)(z^2 + x^2)$

182

3.7 GREATEST COMMON FACTORS; FACTORING BY GROUPING

(c) $p(r + 2s) - q^2(r + 2s) = (r + 2s)(p - q^2)$

(d) $(p - 5)(p + 2) - (p - 5)(3p + 4)$
$= (p - 5)[(p + 2) - (3p + 4)]$
$= (p - 5)(p + 2 - 3p - 4)$
$= (p - 5)(-2p - 2)$
$= (p - 5)(-2)(p + 1)$ or $-2(p - 5)(p + 1)$ ■

WORK PROBLEM 4 AT THE SIDE.

EXAMPLE 5 Factoring Out a Negative Factor
Factor out the greatest common factor from

$$-a^3 + 3a^2 - 5a.$$

There are two ways to factor this polynomial, both of which are correct. We could use a as common factor, giving

$$-a^3 + 3a^2 - 5a = \boldsymbol{a}(-a^2) + \boldsymbol{a}(3a) + \boldsymbol{a}(-5)$$
$$= a(-a^2 + 3a - 5).$$

Alternatively, $-a$ could be used as common factor.

$$-a^3 + 3a^2 - 5a = \boldsymbol{-a}(a^2) + (\boldsymbol{-a})(-3a) + (\boldsymbol{-a})(5)$$
$$= -a(a^2 - 3a + 5)$$

Sometimes in a particular problem there will be a reason to prefer one of these forms over the other, but both are correct. The answer section in this book will give only one of these forms, usually the one where the common factor has a positive coefficient, but either is correct. ■

WORK PROBLEM 5 AT THE SIDE.

2 Sometimes the terms of a polynomial have a greatest common factor of 1, but it still may be possible to write the polynomial as a product of factors. The idea is to look for factors common to *some* of the terms, rearrange the terms accordingly, and then factor. This process is called **factoring by grouping.**

For example, to factor the polynomial

$$ax - ay + bx - by$$

group the terms as follows.

Terms with common factors
$$(ax - ay) + (bx - by)$$

Next, factor $ax - ay$ as $a(x - y)$, and factor $bx - by$ as $b(x - y)$, to get

$$ax - ay + bx - by = \boldsymbol{a}(x - y) + \boldsymbol{b}(x - y).$$

On the right, the common factor is $x - y$. The final factorization is

$$ax - ay + bx - by = (x - y)(a + b).$$

---

4. Factor out the greatest common factor.

(a) $(a + 2)(a - 3) + (a + 2)(a + 6)$

(b) $(y - 1)(y + 3) - (y - 1)(y + 4)$

(c) $k^2(a + 5b) + m^2(a + 5b)$

(d) $r^2(y + 6) + r^2(y + 3)$

5. Factor each polynomial in two ways.

(a) $-k^2 + 3k$

(b) $-6r^2 + 5r$

ANSWERS
4. (a) $(a + 2)(2a + 3)$
 (b) $(y - 1)(-1)$, or $-y + 1$
 (c) $(a + 5b)(k^2 + m^2)$
 (d) $r^2(2y + 9)$
5. (a) $k(-k + 3)$ or $-k(k - 3)$
 (b) $r(-6r + 5)$ or $-r(6r - 5)$

183

CHAPTER 3 POLYNOMIALS

6. Factor.

(a) $mn + 2n + 3m + 6$

(b) $yx - zx + 4y - 4z$

(c) $kn - kp + mn - mp$

7. Factor.

(a) $xy + 2y - 4x - 8$

(b) $10p^2 + 15p - 12p - 18$

(c) $3a^2 - 15a - a + 5$

ANSWERS
6. (a) $(m + 2)(n + 3)$
   (b) $(y - z)(x + 4)$
   (c) $(k + m)(n - p)$
7. (a) $(x + 2)(y - 4)$
   (b) $(2p + 3)(5p - 6)$
   (c) $(a - 5)(3a - 1)$

FACTORING BY GROUPING

Step 1 Collect the terms into groups so that each group has a common factor.

Step 2 Factor out the common factor in each group.

Step 3 If each group now has a common factor, factor it out. If not, try a different grouping.

EXAMPLE 6 Factoring by Grouping

Factor $p^2q^2 - 2q^2 + 5p^2 - 10$.

Group the terms as follows.

$$(p^2q^2 - 2q^2) + (5p^2 - 10)$$

Factoring the common factor from each part gives

$$q^2(p^2 - 2) + 5(p^2 - 2) = (p^2 - 2)(q^2 + 5). \ \blacksquare$$

CAUTION It is a common error to stop at the step

$$q^2(p^2 - 2) + 5(p^2 - 2).$$

This expression is *not in factored form* because it is a *sum* of two terms: $q^2(p^2 - 2)$ and $5(p^2 - 2)$.

■ **WORK PROBLEM 6 AT THE SIDE.**

EXAMPLE 7 Factoring by Grouping

Factor $3x - 3y - ax + ay$.

Grouping terms as we did above and factoring gives

$$(3x - 3y) + (-ax + ay) = 3(x - y) + a(-x + y).$$

There is no simple common factor here. However, if we factor out $-a$ instead of a from the second group, we get

$$3(x - y) - a(x - y).$$

(Be careful with the signs.) Now factor out $x - y$ to get

$$(x - y)(3 - a). \ \blacksquare$$

EXAMPLE 8 Factoring by Grouping

Factor $6x^2 - 4x - 15x + 10$.

Work as above. Note that we must factor -5 rather than 5 from the second group in order to get a common factor of $3x - 2$.

$$6x^2 - 4x - 15x + 10 = 2x(3x - 2) - 5(3x - 2)$$
$$= (3x - 2)(2x - 5) \ \blacksquare$$

■ **WORK PROBLEM 7 AT THE SIDE.**

3.7 EXERCISES

Factor out the greatest common factor. See Examples 1–4.

1. $15k + 30$
2. $7m - 21$
3. $6p^2 + 4p$
4. $12a^3 + 9a^2$

5. $4mx - 5mx^2$
6. $2zb - 3z^2b$
7. $8m^2 + 6m$
8. $2p^2 + 4p$

9. $12x^2 - 3x$
10. $14b^2 + 7b$
11. $144m^4 + 16m^5 - 32m^3$

12. $39a^5 - 26a^7 + 52a^8$
13. $15a^3 + 12a^2 - 3a$
14. $20m^4 + 30m^2 + 50m^3$

15. $14a^3b^2 + 7a^2b - 21a^5b^3$
16. $12km^3 - 24k^3m^2 + 36k^2m^4$

17. $-15m^3p^3 - 6m^3p^4 - 9mp^3$
18. $-25x^3y^2 - 20x^4y^3 + 15x^5y^4$

19. $(m - 9)(m + 1) + (m - 9)(m + 2)$
20. $(a + 5)(a - 6) + (a + 5)(a - 1)$

21. $(3k - 7)(k + 2) + (3k - 7)(k + 5)$
22. $(5m - 11)(2m + 5) + (5m - 11)(m + 5)$

23. $m^5(r + s) + m^5(t + u)$
24. $z^3(k + m) + z^3(p + q)$

25. $4(3 - x)^2 - (3 - x)^3 + 3(3 - x)$
26. $2(t - s) + 4(t - s)^2 - (t - s)^3$

27. $5(m + p)^3 - 10(m + p)^2 - 15(m + p)^4$
28. $-9a^2(p + q) - 3a^3(p + q)^2 + 6a(p + q)^3$

Factor each of the following polynomials twice. First use a common factor with a positive coefficient, and then use a common factor with a negative coefficient. See Example 5.

29. $36y^2 - 72y$

30. $42z^3 - 56z^4$

31. $-2x^5 + 6x^3 + 4x^2$

32. $-5a^3 + 10a^4 - 15a^5$

33. $-32a^4m^5 - 16a^2m^3 - 64a^5m^6$

34. $-144z^{11}n^5 + 16z^3n^{11} - 32z^4n^7$

Factor by grouping. See Examples 6–8.

35. $ax + 2bx + ay + 2by$

36. $2x + 2y + 7ax + 7ay$

37. $2b + 2c + ab + ac$

38. $3am + 3ap + 2bm + 2bp$

39. $p^2 + pq - 3py - 3yq$

40. $r^2 + 6rs - 3rt - 18st$

41. $x^2 - 3x + 2x - 6$

42. $y^2 - 8y + 4y - 32$

43. $3r^2 - 2r + 15r - 10$

44. $2a^2 - 6a + 7a - 21$

45. $-16p^2 - 6pq + 8pq + 3q^2$

46. $-8r^2 + 6rs + 12rs - 9s^2$

47. $-3a^3 - 3ab^2 + 2a^2b + 2b^3$

48. $-16m^3 + 4m^2p^2 - 4mp + p^3$

49. $1 - a + ab - b$

50. $2ab^2 - 8b^2 + a - 4$

51. $8 - 6y^3 - 12y + 9y^4$

52. $x^3y^2 + x^3 - 3y^2 - 3$

Factor out the variable that is raised to the smaller exponent.

53. $3m^{-5} + m^{-3}$

54. $k^{-2} + 2k^{-4}$

55. $3p^{-3} + 2p^{-2}$

SKILL SHARPENERS
Use the FOIL method to find each product. See Section 3.4.

56. $(m - 1)(m + 2)$

57. $(p + 5)(p - 7)$

38. $(8r + 5)(2r - 3)$

59. $(7y - 2)(3y + 5)$

60. $(9z - 7a)(2z + 5a)$

61. $(8p + 3q)(8p - 3q)$

3.8 FACTORING TRINOMIALS

The product of $x + 3$ and $x - 5$ is

$$(x + 3)(x - 5) = x^2 - 5x + 3x - 15$$
$$= x^2 - 2x - 15.$$

By this result, the **factored form** of $x^2 - 2x - 15$ is $(x + 3)(x - 5)$.

$$\text{Factored form} \rightarrow (x + 3)(x - 5) \xrightarrow{\text{Multiplication}} x^2 - 2x - 15 \leftarrow \text{Product}$$
$$\xleftarrow{\text{Factoring}}$$

OBJECTIVES

1. Factor trinomials in which the coefficient of the squared term is 1.
2. Factor trinomials in which the coefficient of the squared term is not 1.
3. Use an alternative method to factor trinomials.
4. Factor by substitution.

1 Trinomials will be factored in this section, beginning with those having 1 as the coefficient of the squared term. To see how to factor these trinomials, let us start with an example: $x^2 - 2x - 15$. As shown below, the x^2 term came from multiplying x and x, and -15 was obtained by multiplying 3 and -5.

Product of x and x is x^2.

$$(x + 3)(x - 5) = x^2 - 2x - 15$$

Product of 3 and -5 is -15.

The $-2x$ in $x^2 - 2x - 15$ was found by multiplying the outside terms, and then the inside terms, and adding.

Outside terms: $x(-5) = -5x$
$(x + 3)(x - 5)$ Add to get $-2x$.
Inside terms: $3 \cdot x = 3x$

Based on this example, we can factor a trinomial with 1 as the coefficient of the squared term by using the steps below.

FACTORING $x^2 + bx + c$

Step 1 Find all pairs of numbers whose product is the third term of the trinomial.

Step 2 Choose the pair whose sum is the coefficient of the middle term.

Step 3 If there are no such numbers, the polynomial cannot be factored.

In Step 3, a polynomial that cannot be factored is called **prime.**

CHAPTER 3 POLYNOMIALS

1. Find all pairs of numbers whose product is 12. Find the sum of each pair of numbers.

   $12 \cdot (\_\_)$; $\quad 12 + \_\_ = \_\_$
   $6 \cdot (\_\_)$; $\quad 6 + \_\_ = \_\_$
   $3 \cdot (\_\_)$; $\quad 3 + \_\_ = \_\_$
   $-12 \cdot (\_\_)$; $\quad -12 + \_\_ = \_\_$
   $-6 \cdot (\_\_)$; $\quad -6 + \_\_ = \_\_$
   $-3 \cdot (\_\_)$; $\quad -3 + \_\_ = \_\_$

2. Factor.

   (a) $p^2 + 6p + 5$

   (b) $a^2 + 9a + 20$

3. Factor.

   (a) $k^2 - k - 6$

   (b) $b^2 - 7b + 10$

   (c) $r^2 - 3r - 4$

   (d) $y^2 - 8y + 6$

4. Factor.

   (a) $m^2 + 2mn - 8n^2$

   (b) $p^2 - 5pq + 6q^2$

   (c) $z^2 - 7zx + 9x^2$

ANSWERS
1. 1, 1, 13; 2, 2, 8; 4, 4, 7;
   $-1, -1, -13; -2, -2, -8;$
   $-4, -4, -7$
2. (a) $(p + 1)(p + 5)$
   (b) $(a + 5)(a + 4)$
3. (a) $(k - 3)(k + 2)$
   (b) $(b - 5)(b - 2)$
   (c) $(r - 4)(r + 1)$
   (d) prime
4. (a) $(m - 2n)(m + 4n)$
   (b) $(p - 3q)(p - 2q)$
   (c) prime

EXAMPLE 1 Factoring a Trinomial $x^2 + bx + c$
Factor $x^2 + 8x + 12$.

We need to find the correct two numbers to place in the blanks.

$$x^2 + 8x + 12 = (x + \_\_\_\_)(x + \_\_\_\_)$$

■ **WORK PROBLEM 1 AT THE SIDE.**

As Problem 1 shows, the numbers 6 and 2 have a product of 12 and a sum of 8. Therefore,

$$x^2 + 8x + 12 = (x + \mathbf{6})(x + \mathbf{2}).$$

Because of the commutative property, it would be equally correct to write $(x + 2)(x + 6)$. ■

■ **WORK PROBLEM 2 AT THE SIDE.**

EXAMPLE 2 Factoring a Trinomial $x^2 + bx + c$
Factor $y^2 + 2y - 35$.

Step 1

Find pairs of numbers whose product is -35.

$-35(1)$
$35(-1)$
$\mathbf{7(-5)}$
$5(-7)$

Step 2

Write sums of the pairs.

$-35 + 1 = -34$
$35 + (-1) = 34$
$\mathbf{7 + (-5) = 2}$ ← Coefficient of middle term
$5 + (-7) = -2$

Use 7 and -5.

$$y^2 + 2y - 35 = (y \mathbf{+ 7})(y \mathbf{- 5})$$

To check, find the product of $y + 7$ and $y - 5$. ■

EXAMPLE 3 Recognizing a Prime Polynomial
Factor $m^2 + 6m + 7$.

Look for two numbers whose product is 7 and whose sum is 6. Only two pairs of numbers, 7 and 1 and -7 and -1, give a product of 7. Neither of these pairs has a sum of 6, so $m^2 + 6m + 7$ cannot be factored and is prime. ■

■ **WORK PROBLEM 3 AT THE SIDE.**

EXAMPLE 4 Factoring a Trinomial in Two Variables
Factor $p^2 + 6ap - 16a^2$.

Look for two expressions whose sum is $6a$ and whose product is $-16a^2$. The quantities $8a$ and $-2a$ have the necessary sum and product, so

$$p^2 + 6ap - 16a^2 = (p \mathbf{+ 8a})(p \mathbf{- 2a}). \blacksquare$$

■ **WORK PROBLEM 4 AT THE SIDE.**

3.8 FACTORING TRINOMIALS

EXAMPLE 5 Factoring a Trinomial with a Common Factor
Factor $16y^3 - 32y^2 - 48y$.

Start by factoring out the greatest common factor, $16y$.

$$16y^3 - 32y^2 - 48y = \mathbf{16y}(y^2 - 2y - 3)$$

To factor $y^2 - 2y - 3$, look for two integers whose sum is -2 and whose product is -3. The necessary integers are -3 and 1, with

$$16y^3 - 32y^2 - 48y = 16y(y - 3)(y + 1). \quad \blacksquare$$

CAUTION In factoring, always look for a common factor first. Do not forget to write the common factor as part of the answer.

2 A generalization of the method shown above can be used to factor a trinomial of the form $ax^2 + bx + c$, where the coefficient of the squared term is not equal to 1. In the first step we find all pairs of factors whose product is ac. In Step 2, we choose the pair whose sum is b. To see how this method works, let us factor $3x^2 + 7x + 2$. First, identify the values of a, b, and c.

$$\begin{array}{ccc} \mathbf{a}x^2 + & \mathbf{b}x + & \mathbf{c} \\ \downarrow & \downarrow & \downarrow \\ \mathbf{3}x^2 + & \mathbf{7}x + & \mathbf{2} \end{array}$$
$$a = 3, \quad b = 7, \quad c = 2$$

The product ac is $3 \cdot 2 = 6$, so we need integers having a product of 6 and a sum of 7 (since the coefficient of the middle term is 7). By inspection, the necessary integers are 1 and 6. Write $7x$ as $1x + 6x$, or $x + 6x$, giving

$$3x^2 + \boxed{7x} + 2 = 3x^2 + \underbrace{\boxed{x + 6x}}_{x + 6x = 7x} + 2.$$

Now factor by grouping.

$$= x(3x + 1) + 2(3x + 1)$$
$$3x^2 + 7x + 2 = (3x + 1)(x + 2)$$

As before, check this result by multiplying $3x + 1$ and $x + 2$.

EXAMPLE 6 Factoring a Trinomial $ax^2 + bx + c$, $a \neq 1$
Factor $12r^2 - 5r - 2$.

Since $a = 12$, $b = -5$, and $c = -2$, the product ac is -24. Two integers whose product is -24 and whose sum is -5 are -8 and 3, with

$$12r^2 \boxed{- 5r} - 2 = 12r^2 \boxed{+ 3r - 8r} - 2$$

where $-5r$ is written as $3r - 8r$. Factor by grouping.

$$= 3r(4r + 1) - 2(4r + 1)$$
$$= (4r + 1)(3r - 2) \quad \blacksquare$$

WORK PROBLEM 5 AT THE SIDE.

3 An alternate approach, the method of trying repeated combinations, is especially helpful when the product ac is large. This method will be shown by using the same polynomials as above.

---

5. Factor.

(a) $2m^2 + 7m + 6$

(b) $3y^2 - 11y - 4$

(c) $6k^2 - 19k + 10$

(d) $10r^2 - 3r - 18$

ANSWERS
5. (a) $(m + 2)(2m + 3)$
 (b) $(y - 4)(3y + 1)$
 (c) $(2k - 5)(3k - 2)$
 (d) $(5r + 6)(2r - 3)$

EXAMPLE 7 Factoring a Trinomial $ax^2 + bx + c$, $a \neq 1$

Factor each of the following.

(a) $3x^2 + 7x + 2$

To factor this polynomial, we need to find the correct numbers to put in the blanks.

$$3x^2 + 7x + 2 = (\_\_\_x + \_\_\_)(\_\_\_x + \_\_\_)$$

Plus signs were used since all the signs in the polynomial are plus. The first two expressions have a product of $3x^2$, so they must be $3x$ and x.

$$3x^2 + 7x + 2 = (3x + \_\_\_)(x + \_\_\_)$$

The product of the two last terms must be 2, so that the numbers must be 2 and 1. We have a choice—we could use the 2 with the $3x$ or with the x. Only one of these choices can give the correct middle term. Try each.

$(3x + 2)(x + 1)$ → $3x + 2x = 5x$ Wrong middle term

$(3x + 1)(x + 2)$ → $6x + x = 7x$ Correct middle term

Therefore, $3x^2 + 7x + 2 = (3x + 1)(x + 2)$.

(b) $12r^2 - 5r - 2$

For factors of 12 we could choose 4 and 3, 6 and 2, or 12 and 1. Let us try 4 and 3. If these do not work, we will make another choice.

$$12r^2 - 5r - 2 = (4r \_\_\_)(3r \_\_\_)$$

We do not know what signs to use yet. The factors of -2 are -2 and 1, or 2 and -1. Try both possibilities.

$(4r - 2)(3r + 1)$ → $4r + (-6r) = -2r$ Wrong middle term

$(4r - 1)(3r + 2)$ → $8r + (-3r) = 5r$ Wrong middle term

The middle term on the right is $5r$, instead of the $-5r$ we need. To get $-5r$, it is only necessary to exchange the middle signs.

$(4r + 1)(3r - 2)$ → $-8r + 3r = -5r$ Correct middle term

Thus, $12r^2 - 5r - 2 = (4r + 1)(3r - 2)$. ∎

The alternate method of factoring a trinomial with the coefficient of the squared term not equal to 1 is summarized on the next page.

3.8 FACTORING TRINOMIALS

FACTORING A TRINOMIAL

Step 1 Write all pairs of factors of the coefficient of the squared term.

Step 2 Write all pairs of factors of the last term.

Step 3 Use various combinations of these factors until the necessary middle term is found.

Step 4 If the necessary combination does not exist, the polynomial is prime.

WORK PROBLEM 6 AT THE SIDE.

EXAMPLE 8 Factoring a Trinomial in Two Variables

Factor $18m^2 - 19mx - 12x^2$.

There is no common factor (except 1). Go through the steps to factor the trinomial. There are many possible factors of both 18 and -12. As a general rule, the middle-sized factors should be tried first. Let's try 6 and 3 for 18 and -3 and 4 for -12.

$$(6m - 3x)(3m + 4x)$$

with $24mx$ and $-9mx$

$$24mx + (-9mx) = 15mx$$

Wrong middle term

(Actually, this could not have been correct, since the terms of $6m$ and $3x$ have a common factor of 3, while the terms of the given polynomial do not.)

$$(6m + 4x)(3m - 3x)$$

Common factor of 2   Common factor of 3

Both of these have a common factor and cannot be correct.

We did not get very far with 6 and 3 as factors of 18; so we will try 9 and 2 instead, and try -4 and 3 as factors of -12.

$$(9m + 3x)(2m - 4x)$$
Common factors

$$(9m - 4x)(2m + 3x)$$
with $27mx$ and $-8mx$

$$27mx + (-8mx) = 19mx$$

The result on the right differs from the correct middle term only in the sign, so we exchange the middle signs on the two factors to get

$$18m^2 - 19mx - 12x^2 = (9m + 4x)(2m - 3x).$$ ∎

CAUTION As shown in Example 8, if the terms of a polynomial have no common factor (except 1), then none of the terms of its factors have a common factor.

WORK PROBLEM 7 AT THE SIDE.

6. Factor.

(a) $2p^2 + 5p - 12$

(b) $6k^2 - k - 2$

(c) $8m^2 + 18m - 5$

(d) $6r^2 + 13r - 5$

7. Factor.

(a) $7p^2 + 15pq + 2q^2$

(b) $6m^2 + 7mn - 5n^2$

(c) $12z^2 - 5zy - 2y^2$

(d) $10a^2 - 13ab - 3b^2$

ANSWERS
6. (a) $(p + 4)(2p - 3)$
   (b) $(3k - 2)(2k + 1)$
   (c) $(4m - 1)(2m + 5)$
   (d) $(2r + 5)(3r - 1)$
7. (a) $(7p + q)(p + 2q)$
   (b) $(3m + 5n)(2m - n)$
   (c) $(3z - 2y)(4z + y)$
   (d) $(5a + b)(2a - 3b)$

8. Factor.

(a) $2m^3 - 4m^2 - 6m$

(b) $12r^4 + 6r^3 - 90r^2$

(c) $30y^5 - 55y^4 - 50y^3$

(d) $8a^3 + 14a^2 - 15a$

9. Factor.

(a) $y^4 + y^2 - 6$

(b) $2p^4 + 7p^2 - 15$

(c) $6r^4 - 13r^2 + 5$

10. Factor.

(a) $6(a - 1)^2 + (a - 1) - 2$

(b) $8(z + 5)^2 - 2(z + 5) - 3$

(c) $15(m - 4)^2 - 11(m - 4) + 2$

EXAMPLE 9 Factoring a Trinomial with Terms That Have a Common Factor

Factor $16y^3 + 24y^2 - 16y$.

The terms of this polynomial have a greatest common factor of $8y$. Factor this out first.

$$16y^3 + 24y^2 - 16y = 8y(2y^2 + 3y - 2)$$

Factor $2y^2 + 3y - 2$ by either method given above.

$$16y^3 + 24y^2 - 16y = 8y(2y - 1)(y + 2)$$

Do not forget to write the common factor in front. ■

WORK PROBLEM 8 AT THE SIDE.

EXAMPLE 10 Factoring a Trinomial $ax^4 + bx^2 + c$

Factor $6y^4 + 7y^2 - 20$.

We know that $y^4 = (y^2)^2$, so factors of $6y^4$ might be $6y^2$ and y^2 or $3y^2$ and $2y^2$. Let us try $3y^2$ and $2y^2$ with -4 and 5 as factors of -20. This gives

$$(3y^2 - 4)(2y^2 + 5).$$

Check the middle term: $(3y^2)(5) + (-4)(2y^2) = 15y^2 - 8y^2 = 7y^2$, as required. Thus,

$$6y^4 + 7y^2 - 20 = (3y^2 - 4)(2y^2 + 5).\ ■$$

WORK PROBLEM 9 AT THE SIDE.

4 Sometimes a more complicated polynomial can be factored by making a substitution of one variable for another. The next example shows this **method of substitution.**

EXAMPLE 11 Factoring a Trinomial Using Substitution

Factor $2(x + 3)^2 + 5(x + 3) - 12$.

Substitute y for $x + 3$, so that the polynomial becomes

$$2y^2 + 5y - 12$$

which factors as

$$2y^2 + 5y - 12 = (2y - 3)(y + 4).$$

Now replace y with $x + 3$ to get

$$2(x + 3)^2 + 5(x + 3) - 12$$
$$= [2(x + 3) - 3][(x + 3) + 4]$$
$$= (2x + 6 - 3)(x + 7)$$
$$= (2x + 3)(x + 7).\ ■$$

WORK PROBLEM 10 AT THE SIDE.

ANSWERS
8. (a) $2m(m + 1)(m - 3)$
   (b) $6r^2(r + 3)(2r - 5)$
   (c) $5y^3(2y - 5)(3y + 2)$
   (d) $a(4a - 3)(2a + 5)$
9. (a) $(y^2 - 2)(y^2 + 3)$
   (b) $(2p^2 - 3)(p^2 + 5)$
   (c) $(3r^2 - 5)(2r^2 - 1)$
10. (a) $(2a - 3)(3a - 1)$
    (b) $(4z + 17)(2z + 11)$
    (c) $(3m - 13)(5m - 22)$

3.8 EXERCISES

Factor each of the following trinomials. See Examples 1–9.

1. $m^2 + 5m + 6$
2. $p^2 + 6p + 8$
3. $a^2 - a - 12$

4. $y^2 + 3y - 28$
5. $r^2 - r - 20$
6. $m^2 - m - 30$

7. $a^2 - 6a - 16$
8. $k^2 - kn - 6n^2$
9. $a^2 + 3ab - 18b^2$

10. $y^2 - 3yx - 10x^2$
11. $p^2 - 2pq - 15q^2$
12. $a^2b^2 - 7ab + 12$

13. $y^2w^2 + 4yw - 21$
14. $3y^2 + 14y + 8$
15. $8y^2 + 13y - 6$

16. $6x^2 + 13x + 6$
17. $18x^2 - 3x - 10$
18. $12m^2 - 8m - 15$

19. $35p^2 - 4p - 15$
20. $6m^2 - 17m - 14$
21. $12a^2 + 8ab - 15b^2$

22. $3m^2 + 7mk + 2k^2$
23. $4k^2 - 12ak + 9a^2$
24. $18a^2 - 3ab - 28b^2$

25. $35x^2 - 41xy - 24y^2$
26. $6x^2 - 5xy - 39y^2$
27. $6k^2p^2 + 13kp + 6$

28. $15z^2x^2 - 22zx - 5$
29. $12m^2 + 14m - 40$
30. $36t^2 + 30t - 50$

31. $18a^2 - 15a - 18$
32. $100r^2 - 90r + 20$
33. $6a^3 + 12a^2 - 90a$

34. $3m^4 + 6m^3 - 72m^2$
35. $13y^3 + 39y^2 - 52y$
36. $4p^3 + 24p^2 - 64p$

37. $2x^3y^3 - 48x^2y^4 + 288xy^5$
38. $6m^3n^2 - 24m^2n^3 - 30mn^4$

Factor each of the following trinomials. See Example 10.

39. $3x^4 - 14x^2 - 5$
40. $3p^4 - 8p^2 - 3$
41. $z^4 - 7z^2 - 30$

42. $k^4 + k^2 - 12$
43. $6x^4 + 5x^2 - 25$
44. $6a^4 - 11a^2 - 10$

45. $12p^4 + 28p^2r - 5r^2$
46. $2y^4 + xy^2 - 6x^2$

Factor each of the following trinomials. See Example 11.

47. $6(p + 3)^2 + 13(p + 3) + 5$
48. $10(m - 5)^2 - 9(m - 5) - 9$

49. $6(z + k)^2 - 7(z + k) - 5$
50. $3(r + m)^2 - 10(r + m) - 25$

51. $a^2(a + b)^2 - ab(a + b)^2 - 6b^2(a + b)^2$
52. $m^2(m - p) + mp(m - p) - 2p^2(m - p)$

53. $p^2(p + q) + 4pq(p + q) + 3q^2(p + q)$
54. $r^2(r - t) - 5rt(r - t) - 6t^2(r - t)$

SKILL SHARPENERS
Find each product. See Section 3.4.

55. $(2m - 5)(2m + 5)$
56. $(3p + 2q)(3p - 2q)$
57. $(x + 6y)^2$

58. $(2z + 5a)^2$
59. $(y - 2)(y^2 + 2y + 4)$
60. $(5a + 3)(25a^2 - 15a + 9)$

3.9 SPECIAL FACTORING

Certain types of factoring occur so often that they deserve special study.

1 As discussed earlier in this chapter, the product of the sum and difference of two terms is

$$(x + y)(x - y) = x^2 - y^2.$$

This result leads to the **difference of two squares,** a formula that is useful in factoring.

DIFFERENCE OF TWO SQUARES

$$x^2 - y^2 = (x + y)(x - y)$$

EXAMPLE 1 Factoring the Difference of Squares

(a) $16m^2 - 49p^2 = (\mathbf{4m})^2 - (\mathbf{7p})^2$
$= (4m + 7p)(4m - 7p)$

(b) $81k^2 - (a + 2)^2 = (9k)^2 - (a + 2)^2$
$= (9k + a + 2)(9k - [a + 2])$
$= (9k + a + 2)(9k - a - 2)$ ∎

CAUTION Assuming no greatest common factor except 1, it is not possible to factor (with real numbers) a *sum* of two squares such as $x^2 + 25$. In particular, $x^2 + y^2 \neq (x + y)^2$, as shown next.

WORK PROBLEM 1 AT THE SIDE.

2 Two other useful patterns come from the rules for squaring binomials.

PERFECT SQUARE TRINOMIALS

$$x^2 + 2xy + y^2 = (x + y)^2$$
$$x^2 - 2xy + y^2 = (x - y)^2$$

The trinomial $x^2 + 2xy + y^2$ is the square of $x + y$. For this reason $x^2 + 2xy + y^2$ is called a **perfect square trinomial.** In these patterns, both the first and last terms of the trinomial must be perfect squares. In the factored form, twice the product of the first and last terms must give the middle term of the trinomial. It is important to understand these patterns in terms of words, since they occur with many different symbols (other than x and y).

OBJECTIVES

1 Factor the difference of two squares.

2 Factor a perfect square trinomial.

3 Factor the difference of two cubes.

4 Factor the sum of two cubes.

1. Factor.

(a) $9a^2 - 16b^2$

(b) $64p^2 - 81q^2$

(c) $(m + 3)^2 - 49z^2$

(d) $144y^2 - (z + 1)^2$

ANSWERS
1. (a) $(3a - 4b)(3a + 4b)$
  (b) $(8p - 9q)(8p + 9q)$
  (c) $(m + 3 + 7z)(m + 3 - 7z)$
  (d) $(12y + z + 1)(12y - z - 1)$

3.9 SPECIAL FACTORING

CHAPTER 3 POLYNOMIALS

2. Factor.

   (a) $49z^2 - 14zk + k^2$

   (b) $9a^2 + 48ab + 64b^2$

   (c) $(k + m)^2 - 12(k + m) + 36$

   (d) $x^2 - 2x + 1 - y^2$

ANSWERS
2. (a) $(7z - k)^2$ (b) $(3a + 8b)^2$
   (c) $[(k + m) - 6]^2$ or $(k + m - 6)^2$
   (d) $(x - 1 + y)(x - 1 - y)$

EXAMPLE 2 Factoring a Perfect Square Trinomial
Factor.

(a) $144p^2 - 120p + 25$

Here $144p^2 = (12p)^2$ and $25 = 5^2$. The sign on the middle term is $-$, so if $144p^2 - 120p + 25$ is a perfect square trinomial, it will have to be
$$(12p - 5)^2.$$
To see if this is correct, take twice the product of these two terms:
$$2(12p)(-5) = -120p$$
which is the middle term of the given trinomial. Thus,
$$144p^2 - 120p + 25 = (12p - 5)^2.$$

(b) $4m^2 + 28mn + 49n^2 = (2m + 7n)^2$

Check the middle term: $2(2m)(7n) = 28mn$, which is correct.

(c) $(r + 5)^2 + 6(r + 5) + 9 = [(r + 5) + 3]^2$ Since $2(3)(r + 5)$
$$= (r + 8)^2 \qquad\qquad\qquad\qquad = 6(r + 5)$$

(d) $m^2 - 8m + 16 - p^2$

Since m^2 and 16 are squares and $2(m)(-4) = -8m$, the first three terms are a perfect square trinomial; group them together and factor the trinomial.
$$(m^2 - 8m + 16) - p^2 = (m - 4)^2 - p^2$$
The result is a difference of squares. Factor again to get
$$(m - 4)^2 - p^2 = (m - 4 + p)(m - 4 - p). \blacksquare$$

WORK PROBLEM 2 AT THE SIDE.

3 The **difference of two cubes,** $x^3 - y^3$, can be factored as follows.

DIFFERENCE OF TWO CUBES

$$x^3 - y^3 = (x - y)(x^2 + xy + y^2)$$

To check this, find the product of $x - y$ and $x^2 + xy + y^2$, as follows.

$$\begin{array}{r} x^2 + xy + y^2 \\ x - y \\ \hline -x^2y - xy^2 - y^3 \\ x^3 + x^2y + xy^2 \\ \hline x^3 \qquad\qquad\quad - y^3 \end{array}$$

This result shows that
$$x^3 - y^3 = (x - y)(x^2 + xy + y^2).$$

3.9 SPECIAL FACTORING

EXAMPLE 3 Factoring the Difference of Cubes
Factor.

(a) $m^3 - 8 = \mathbf{m^3 - 2^3} = (m - 2)(m^2 + 2m + 2^2)$
$= (m - 2)(m^2 + 2m + 4)$

(b) $27x^3 - 8y^3 = (\mathbf{3x})^3 - (\mathbf{2y})^3$
$= (3x - 2y)[(3x)^2 + (3x)(2y) + (2y)^2]$
$= (3x - 2y)(9x^2 + 6xy + 4y^2)$

(c) $1000k^3 - 27n^3 = (10k)^3 - (3n)^3$
$= (10k - 3n)[(10k)^2 + (10k)(3n) + (3n)^2]$
$= (10k - 3n)(100k^2 + 30kn + 9n^2)$ ■

WORK PROBLEM 3 AT THE SIDE.

4 While an expression of the form $x^2 + y^2$ (a sum of two squares) usually cannot be factored with real numbers, the **sum of two cubes** can be factored with a pattern that is very similar to that for the difference of two cubes.

SUM OF TWO CUBES

$$x^3 + y^3 = (x + y)(x^2 - xy + y^2)$$

To verify this result, find the product of $x + y$ and $x^2 - xy + y^2$ or divide $x^3 + y^3$ by $x + y$. The quotient will be $x^2 - xy + y^2$, so that $(x + y)(x^2 - xy + y^2) = x^3 + y^3$.

EXAMPLE 4 Factoring the Sum of Cubes
Factor.

(a) $r^3 + 27 = \mathbf{r^3 + 3^3} = (r + 3)(r^2 - 3r + 3^2)$
$= (r + 3)(r^2 - 3r + 9)$

(b) $27z^3 + 125 = (\mathbf{3z})^3 + \mathbf{5}^3 = (3z + 5)[(3z)^2 - (3z)(5) + 5^2]$
$= (3z + 5)(9z^2 - 15z + 25)$

(c) $125t^3 + 216s^6 = (5t)^3 + (6s^2)^3$
$= (5t + 6s^2)[(5t)^2 - (5t)(6s^2) + (6s^2)^2]$
$= (5t + 6s^2)(25t^2 - 30ts^2 + 36s^4)$ ■

WORK PROBLEM 4 AT THE SIDE.

CAUTION A common error is to think that $x^2 + xy + y^2 = (x + y)^2$. However, $(x + y)^2 = x^2 + 2xy + y^2$. Also, $x^2 - 2xy + y^2 = (x - y)^2$. In general, the expressions $x^2 + xy + y^2$ and $x^2 - xy + y^2$ cannot be factored.

---

3. Factor.

(a) $x^3 - 1000$

(b) $8k^3 - y^3$

(c) $27m^3 - 64$

4. Factor.

(a) $x^3 + 1000$

(b) $8p^3 + 125$

(c) $27m^3 + 125n^3$

ANSWERS
3. (a) $(x - 10)(x^2 + 10x + 100)$
 (b) $(2k - y)(4k^2 + 2ky + y^2)$
 (c) $(3m - 4)(9m^2 + 12m + 16)$
4. (a) $(x + 10)(x^2 - 10x + 100)$
 (b) $(2p + 5)(4p^2 - 10p + 25)$
 (c) $(3m + 5n)(9m^2 - 15mn + 25n^2)$

The special types of factoring discussed in this section are summarized below and should be memorized.

Difference of two squares	$x^2 - y^2 = (x + y)(x - y)$
Perfect square trinomial	$x^2 + 2xy + y^2 = (x + y)^2$ $x^2 - 2xy + y^2 = (x - y)^2$
Difference of two cubes	$x^3 - y^3 = (x - y)(x^2 + xy + y^2)$
Sum of two cubes	$x^3 + y^3 = (x + y)(x^2 - xy + y^2)$

3.9 EXERCISES

Factor each of the following polynomials. See Examples 1–4.

1. $x^2 - 25$
2. $1 - p^2$
3. $36m^2 - 1$
4. $8x^2 - 98$
5. $16y^2 - 81q^2$
6. $9m^2 - 100r^2$
7. $27x^4 - 3y^4$
8. $4a^4 - 16b^4$
9. $(x + y)^2 - 16$
10. $(a + b)^2 - 100$
11. $25 - (r + 3s)^2$
12. $81 - (2k + z)^2$
13. $(a + b)^2 - (a - b)^2$
14. $(c - d)^2 - (c + d)^2$
15. $x^2 + 4x + 4$
16. $y^2 + 6y + 9$
17. $9r^2 - 6rs + s^2$
18. $4a^2 - 20ab + 25b^2$
19. $x^2 - 4x + 4 - w^2$
20. $4y^2 - 4y + 1 - z^2$
21. $9m^2 - 12m + 4 - n^2$
22. $4p^2 - 20p + 25 - q^2$
23. $a^2 - b^2 + 2b - 1$
24. $k^2 - m^2 + 6m - 9$
25. $25x^2y^2 - 20xy + 4$
26. $9k^2q^2 + 24kq + 16$
27. $72m^2 - 120mp + 50p^2$
28. $100y^2 - 100yz + 25z^2$
29. $(a + b)^2 + 2(a + b) + 1$
30. $(x + y)^2 + 6(x + y) + 9$

31. $(m - p)^2 + 4(m - p) + 4$
32. $(w - r)^2 + 8(w - r) + 16$
33. $8a^3 + 1$

34. $125a^3 - 1$
35. $27x^3 - 64y^3$
36. $8a^3 + 125m^3$

37. $250m^3 - 2p^3$
38. $3 + 3000a^3b^3$
39. $64y^6 + 1$

40. $(p - 5)^3 + 125$
41. $(x + y)^3 - 27$

42. $64 - (a - b)^3$
43. $(r + 1)^3 - 1$

44. $m^6 - 8$
45. $m^3 + (m + 3)^3$

46. $a^3 - (a - 4)^3$
47. $(p + q)^3 - (p - q)^3$

SKILL SHARPENERS

Simplify. Write answers with positive exponents. See Section 3.1.

48. $\dfrac{2^{-4} \cdot 2^{-3}}{2^{-2}}$
49. $\dfrac{3^6 \cdot 3^{-1}}{3^3}$
50. $\dfrac{p^5 p^{-2}}{p^{-1} p^4}$ $(p \neq 0)$
51. $\dfrac{2k^{-1}k}{k^2 k^{-2}}$ $(k \neq 0)$

Solve each equation. See Section 2.1.

52. $3x + 1 = 13$
53. $5r - 2 = 8$
54. $-2q + 3 = 7$
55. $-2z + 8 = 9$

SUMMARY OF FACTORING METHODS

A polynomial is completely factored when the polynomial is in the form described below.

> 1. The polynomial is written as a product of prime polynomials with integer coefficients.
> 2. None of the polynomial factors can be factored further, except that a monomial factor need not be factored completely.

For example, $9x^2(x + 2)$ is the factored form of $9x^3 + 18x^2$. Because of the remark in the second rule above, it is not necessary to factor $9x^2$ as $3 \cdot 3 \cdot x \cdot x$. The order of the factors does not matter.

The steps to follow in factoring a polynomial are listed below.

> *Step 1* Factor out any common factor. See Section 3.7.
> *Step 2a* If the polynomial is a binomial, check to see if it is the difference of two squares, the difference of two cubes, or the sum of two cubes. See Section 3.9.
> *Step 2b* If the polynomial is a trinomial, check to see if it is a perfect square trinomial. If it is not, factor as in Section 3.8.
> *Step 2c* If the polynomial has four terms, try to factor by grouping. See Section 3.7.

Factor each of the following polynomials.

1. $100a^2 - 9b^2$
2. $10r^2 + 13r - 3$
3. $18p^5 - 24p^3 + 12p^6$

4. $15x^2 - 20x$
5. $x^2 + 2x - 35$
6. $9 - a^2 + 2ab - b^2$

7. $49z^2 - 16$
8. $225p^2 + 256$
9. $x^3 - 1000$

10. $6b^2 - 17b - 3$
11. $k^2 - 6k - 16$
12. $18m^3n + 3m^2n^2 - 6mn^3$

13. $6t^2 + 19tu - 77u^2$

14. $2p^2 + 11pq + 15q^2$

15. $40p^2 - 32p$

16. $9m^2 - 45m + 18m^3$

17. $4k^2 + 28kr + 49r^2$

18. $54m^3 - 2000$

19. $mn - 2n + 5m - 10$

20. $2a^2 - 7a - 4$

21. $9m^2 - 30mn + 25n^2 - p^2$

22. $x^3 + 3x^2 - 9x - 27$

23. $56k^3 - 875$

24. $9r^2 + 100$

25. $16z^3x^2 - 32z^2x$

26. $8p^3 - 125$

27. $m^2(m - 2) - 4(m - 2)$

28. $6k^2 - k - 1$

29. $27m^2 + 144mn + 192n^2$

30. $x^4 - 625$

name date hour

SUMMARY OF FACTORING METHODS

31. $125m^6 + 216$ **32.** $ab + 6b + ac + 6c$ **33.** $2m^2 - mn - 15n^2$

34. $p^3 + 64$ **35.** $4y^2 - 8y$ **36.** $6a^4 - 11a^2 - 10$

37. $14z^2 - 3zk - 2k^2$ **38.** $12z^3 - 6z^2 + 18z$ **39.** $256b^2 - 400c^2$

40. $z^2 - zp - 20p^2$ **41.** $1000z^3 + 512$ **42.** $64m^2 - 25n^2$

43. $10r^2 + 23rs - 5s^2$ **44.** $12k^2 - 17kq - 5q^2$ **45.** $32x^2 + 16x^3 - 24x^5$

46. $48k^4 - 243$ **47.** $14x^2 - 25xq - 25q^2$ **48.** $5p^2 - 10p$

203

49. $y^2 + 3y - 10$

50. $b^2 - 7ba - 18a^2$

51. $2a^3 + 6a^2 - 4a$

52. $12m^2rx + 4mnrx + 40n^2rx$

53. $18p^2 + 53pr - 35r^2$

54. $21a^2 - 5ab - 4b^2$

55. $(x - 2y)^2 - 4$

56. $(3m - n)^2 - 25$

57. $(5r + 2s)^2 - 6(5r + 2s) + 9$

58. $(p + 8q)^2 - 10(p + 8q) + 25$

59. $z^4 - 9z^2 + 20$

60. $21m^4 - 32m^2 - 5$

3.10 SOLVING EQUATIONS BY FACTORING

Up to now in Chapter 3 we have worked with polynomials, which are algebraic *expressions*. Now we discuss polynomial *equations*, so we will be using the addition and multiplication properties of equality from Chapter 2.

1 In this section we show how factoring is used to solve equations. Some equations that cannot be solved by other methods can be solved by factoring. This process depends on a special property of the number 0, called the **zero-factor property.**

OBJECTIVES

1 Learn the zero-factor property.

2 Use the zero-factor property to solve equations.

3 Solve word problems that need the zero-factor property.

ZERO-FACTOR PROPERTY

If two numbers have a product of 0, then at least one of the numbers in the product must be 0. That is, if $ab = 0$, then $a = 0$ or $b = 0$.

To prove the zero-factor property, first assume $a \neq 0$. (If a does equal 0, then the property is proved already.) If $a \neq 0$, then $\frac{1}{a}$ exists, and we can multiply both sides of $ab = 0$ by $\frac{1}{a}$ to get

$$\frac{1}{a} \cdot ab = \frac{1}{a} \cdot 0 \qquad \text{Multiply by } \frac{1}{a}.$$
$$b = 0.$$

Thus, if $a \neq 0$, then $b = 0$, and the result is proved.

2 The next examples show how to use the zero-factor property to solve equations.

EXAMPLE 1 Using the Zero-Factor Property to Solve an Equation
Solve the equation $(x + 6)(2x - 3) = 0$.

Here the product of $x + 6$ and $2x - 3$ is 0. By the zero-factor property, this can be true only if $x + 6$ equals 0 or if $2x - 3$ equals 0:

$$x + 6 = 0 \qquad \text{or} \qquad 2x - 3 = 0.$$

Solve $x + 6 = 0$ by subtracting 6 from both sides of the equation. Solve $2x - 3 = 0$ by adding 3 on both sides and then dividing by 2.

$$\begin{array}{rcl}
x + 6 = 0 & \text{or} & 2x - 3 = 0 \\
x + 6 - 6 = 0 - 6 & & 2x - 3 + 3 = 0 + 3 \\
x = -6 & & 2x = 3 \\
& & x = \frac{3}{2}
\end{array}$$

This equation has two solutions that should be checked by substitution in the original equation.

If $x = -6$,
then $(x + 6)(2x - 3) = 0$.
$(-6 + 6)[2(-6) - 3] = 0$
$0(-15) = 0$ True

If $x = \frac{3}{2}$, then $(x + 6)(2x - 3) = 0$.
$\left(\frac{3}{2} + 6\right)\left(2 \cdot \frac{3}{2} - 3\right) = 0$
$\frac{15}{2}(0) = 0$ True

Both solutions check; the solution set is $\left\{-6, \frac{3}{2}\right\}$. ■

CHAPTER 3 POLYNOMIALS

1. Solve.

   (a) $(3k + 5)(k + 1) = 0$

   (b) $(4r - 7)(r - 3) = 0$

   (c) $(8a + 3)(2a + 1) = 0$

   (d) $(3r + 11)(5r - 2) = 0$

■ **WORK PROBLEM 1 AT THE SIDE.**

Since the product $(x + 6)(2x - 3)$ equals $2x^2 + 9x - 18$, the equation in Example 1 has a squared term and is an example of a *quadratic equation*.

An equation that can be written in the form

$$ax^2 + bx + c = 0$$

$(a \neq 0)$, is a **quadratic equation.**

Quadratic equations are discussed in more detail in Chapter 6.
The steps involved in solving an equation by factoring are summarized below.

SOLVING AN EQUATION BY FACTORING

Step 1 Rewrite the equation if necessary so that it is in the form $ax^2 + bx + c = 0$.

Step 2 Factor the trinomial.

Step 3 Set each factor equal to 0, using the zero-factor property.

Step 4 Solve each equation in Step 3.

Step 5 Check each solution in the original equation.

EXAMPLE 2 Solving a Quadratic Equation by Factoring
Solve the equation $2p^2 + 3p = 2$.

Rewrite the equation in the form $ap^2 + bp + c = 0$ by subtracting 2 from both sides.

$$2p^2 + 3p = 2$$
$$2p^2 + 3p - 2 = 0 \quad \text{Subtract 2.}$$

Factor on the left.

$$(2p - 1)(p + 2) = 0$$

Set each factor equal to 0, and then solve each of the two resulting equations.

$$2p - 1 = 0 \quad \text{or} \quad p + 2 = 0$$
$$2p = 1 \qquad\qquad p = -2$$
$$p = \frac{1}{2}$$

ANSWERS

1. (a) $\left\{-\dfrac{5}{3}, -1\right\}$ (b) $\left\{\dfrac{7}{4}, 3\right\}$

   (c) $\left\{-\dfrac{3}{8}, -\dfrac{1}{2}\right\}$ (d) $\left\{-\dfrac{11}{3}, \dfrac{2}{5}\right\}$

3.10 SOLVING EQUATIONS BY FACTORING

Check these answers by substituting them into the original equation.

If $p = \dfrac{1}{2}$, then $2p^2 + 3p = 2$.

$$2\left(\dfrac{1}{2}\right)^2 + 3\left(\dfrac{1}{2}\right) = 2$$

$$2\left(\dfrac{1}{4}\right) + \dfrac{3}{2} = 2$$

$$\dfrac{1}{2} + \dfrac{3}{2} = 2$$

$$2 = 2$$
True

If $p = -2$, then $2p^2 + 3p = 2$.

$$2(-2)^2 + 3(-2) = 2$$
$$2(4) - 6 = 2$$
$$8 - 6 = 2$$
$$2 = 2$$
True

Since both solutions check, the solution set is $\left\{\dfrac{1}{2}, -2\right\}$. ■

WORK PROBLEM 2 AT THE SIDE.

EXAMPLE 3 Solving a Quadratic Equation with a Missing Term
Solve $5z^2 - 25z = 0$.

This quadratic equation has a missing term. Comparing it with the general form $ax^2 + bx + c = 0$ shows that $c = 0$. The zero-factor property still can be used, however, since $5z^2 - 25z$ can be factored as

$$5z^2 - 25 = 5z(z - 5).$$

Now use the zero-factor property to solve the equation.

$$5z^2 - 25z = 0$$
$$5z(z - 5) = 0$$
$$\mathbf{5z = 0} \quad \text{or} \quad \mathbf{z - 5 = 0}$$
$$z = 0 \quad \text{or} \quad z = 5$$

The solution set is $\{0, 5\}$, as can be verified by substituting back in the original equation. ■

CAUTION It is important to remember that the zero-factor property works only for a product equal to *zero*. If $ab = 0$, then $a = 0$ or $b = 0$. However, if $ab = 6$, for example, we do not know that $a = 6$ or $b = 6$; it is very likely that *neither $a = 6$ nor $b = 6$.*

WORK PROBLEM 3 AT THE SIDE.

2. Solve.

(a) $3r^2 - r = 4$

(b) $4k^2 + 4k = 3$

(c) $15m^2 + 7m = 2$

(d) $8z^2 + 22z = 21$

3. Solve.

(a) $y^2 = 9y$

(b) $p^2 = -12p$

(c) $k^2 - 16 = 0$

(d) $9r^2 - 64 = 0$

ANSWERS

2. (a) $\left\{\dfrac{4}{3}, -1\right\}$ (b) $\left\{-\dfrac{3}{2}, \dfrac{1}{2}\right\}$
(c) $\left\{\dfrac{1}{5}, -\dfrac{2}{3}\right\}$ (d) $\left\{-\dfrac{7}{2}, \dfrac{3}{4}\right\}$

3. (a) $\{0, 9\}$ (b) $\{0, -12\}$
(c) $\{4, -4\}$ (d) $\left\{\dfrac{8}{3}, -\dfrac{8}{3}\right\}$

CHAPTER 3 POLYNOMIALS

4. Solve.

(a) $(a + 6)(a - 2) = 2 + a - 10$

(b) $(3k - 2)(k + 1) = -2(3k + 2)$

EXAMPLE 4 Solving an Equation Requiring Rewriting

Solve $(2q + 1)(q + 1) = 2(1 - q) + 6$.

To get the equation in the form $aq^2 + bq + c = 0$, first multiply out both sides.

$$(2q + 1)(q + 1) = 2(1 - q) + 6$$
$$2q^2 + 3q + 1 = 2 - 2q + 6 \quad \text{FOIL; distributive property}$$
$$2q^2 + 3q + 1 = 8 - 2q \quad \text{Combine terms.}$$
$$2q^2 + 5q - 7 = 0 \quad \text{Add } -8 + 2q \text{ to both sides.}$$
$$(2q + 7)(q - 1) = 0 \quad \text{Factor.}$$
$$2q + 7 = 0 \quad \text{or} \quad q - 1 = 0 \quad \text{Set each factor equal to 0.}$$
$$q = -\frac{7}{2} \quad \text{or} \quad q = 1 \quad \text{Solve for } q.$$

Check that the solution set is $\left\{-\frac{7}{2}, 1\right\}$. ■

■ **WORK PROBLEM 4 AT THE SIDE.**

5. (a) If the square of a number is added to the number, the result is 12. Find the number.

3 The zero-factor property is also used to solve word problems, as the next example shows.

EXAMPLE 5 Using a Quadratic Equation in an Application

A lot has the shape of a parallelogram. The longer sides are each 8 meters longer than the distance between them. The area of the lot is 48 square meters. Find the length of the longer sides and the distance between them.

Sketch a parallelogram as shown in Figure 1. (See the list of formulas from geometry in Appendix B.) Label your sketch as follows.

Let x be the distance between the longer sides; then $x + 8$ is the length of each longer side.

FIGURE 1

The area of a parallelogram is given by $A = bh$, where b is the length of the longer side and h is the height, the distance between the longer sides. Here, $b = x + 8$ and $h = x$.

(b) The length of a room is 2 meters less than three times the width. The area of the room is 96 square meters. Find the width of the room.

$$A = bh$$
$$48 = (x + 8)x \quad \text{Let } A = 48, b = x + 8, h = x.$$
$$48 = x^2 + 8x \quad \text{Distributive property}$$
$$0 = x^2 + 8x - 48 \quad \text{Subtract 48 on both sides.}$$
$$0 = (x + 12)(x - 4) \quad \text{Factor.}$$
$$x + 12 = 0 \quad \text{or} \quad x - 4 = 0 \quad \text{Let each factor} = 0.$$
$$x = -12 \quad \text{or} \quad x = 4$$

A parallelogram cannot have a negative height, so reject $x = -12$ as a solution. The only solution is $x = 4$; the distance between the longer sides is 4 meters. The length of the longer sides is $4 + 8 = 12$ meters. ■

ANSWERS

4. (a) $\{1, -4\}$ (b) $\left\{-\frac{1}{3}, -2\right\}$

5. (a) 3 or -4 (b) 6 meters

■ **WORK PROBLEM 5 AT THE SIDE.**

3.10 EXERCISES

Find all solutions by using the zero-factor property. See Example 1.

1. $(x - 2)(x - 3) = 0$

2. $(m + 4)(m + 5) = 0$

3. $(3m - 8)(5m + 6) = 0$

4. $(7y + 4)(3y - 5) = 0$

5. $(r - 4)(2r + 5) = 0$

6. $(3x - 5)(2x + 7) = 0$

Find all solutions by factoring. See Examples 2–4.

7. $x^2 - x - 12 = 0$

8. $m^2 + 4m - 5 = 0$

9. $y^2 + 9y + 14 = 0$

10. $p^2 + 12p + 35 = 0$

11. $2m^2 + 5m = 3$

12. $4a^2 - 7a = 2$

13. $15r^2 + 7r = 2$

14. $12x^2 + 4x = 1$

15. $4a^2 + 9a + 2 = 0$

16. $12p^2 - 11p + 2 = 0$

17. $3n^2 + n - 4 = 0$

18. $2y^2 - 12 - 4y = y^2 - 3y$

19. $3p^2 + 9p + 30 = 2p^2 - 2p$

20. $(5y + 1)(y + 3) = -2(5y + 1)$

21. $(3x + 1)(x - 3) = 2 + 3(x + 5)$

22. $6m^2 - 36m = 0$

23. $-3m^2 + 27m = 0$

24. $-3m^2 + 27 = 0$

25. $-2a^2 + 8 = 0$

26. $4p^2 - 16 = 0$

27. $9x^2 - 81 = 0$

Find each solution by writing a quadratic equation and solving it. See Example 5.

28. The sum of two numbers is 10. The sum of their squares is 68. Find the numbers. (*Hint:* Let $x =$ one number; then $10 - x =$ the other number.)

29. Two numbers have a sum of 12. The sum of the squares of the numbers is 90. Find the numbers.

30. If two integers are added, the result is 15. If the squares of the integers are added, the result is 125. Find the integers.

31. Find two consecutive integers whose product is 132.

32. A sign is to have the shape of a triangle with a height 3 meters greater than the length of the base. How long should the base be if the area is to be 14 square meters?

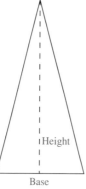

33. The frame of a wall in a new building forms a right triangle with an area of 112 square feet. The base of the triangle is 2 feet longer than the height. Find the height of the triangle.

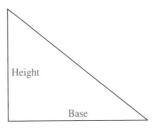

34. A table top has an area of 54 square feet. Its length is 3 feet more than its width. Find the width of the table top.

35. A lot with an area of 140 square meters has the shape of a rectangle, whose length is 4 meters more than the width. Find the width of the lot.

Use the zero-factor property to solve each equation. (Hint: Look for a factoring pattern.)

36. $2(m + 3)^2 = 5(m + 3) - 2$

37. $(x - 1)^2 - (2x - 5)^2 = 0$

SKILL SHARPENERS
Simplify. See Section 3.1.

38. $\dfrac{6m^5}{2m^2}$

39. $\dfrac{9y^7}{6y^3}$

40. $\dfrac{-100a^5b^2}{75a^6b^5}$

41. $\dfrac{-32r^3s^5}{48r^6s^7}$

Write each fraction with the indicated denominator.

42. $\dfrac{3}{4}, \dfrac{}{24}$

43. $\dfrac{5}{9}, \dfrac{}{63}$

44. $\dfrac{11}{15}, \dfrac{}{75}$

45. $\dfrac{7}{5}, \dfrac{}{30}$

CHAPTER 3 SUMMARY

KEY TERMS

3.2 **scientific notation** — In scientific notation, a number is written as the product of a number between 1 and 10 (or -1 and -10) and some integer power of 10.

3.3 **term** — A term is the product of a number and one or more variables raised to a power.

coefficient — A coefficient is a factor in a term (usually used for the numerical factor).

polynomial — A polynomial is a finite sum of terms with only whole number exponents on the variables and no variable denominators.

descending powers — A polynomial is written in descending powers if the exponents on the variables in the terms decrease from left to right.

trinomial — A trinomial is a polynomial with exactly three terms.

binomial — A binomial is a polynomial with exactly two terms.

monomial — A monomial is a polynomial with exactly one term.

degree of a term — The degree of a term with one variable is the exponent on that variable.

degree of a polynomial — The degree of a polynomial is the highest degree of any of the terms in the polynomial.

like terms — Like terms have the same variables to the same power.

combining terms — Using the distributive property to replace two or more like terms with one term is called combining terms.

3.6 **synthetic division** — Synthetic division is a shortcut procedure for dividing a polynomial by a polynomial of the form $x - k$.

3.7 **prime number** — A prime number is a positive integer greater than 1 that has only itself and 1 as factors.

prime factored form — An integer that has been factored as a product of prime numbers is in prime factored form.

greatest common factor — The product of the largest numerical factor and the variable factor of lowest degree of every term in a polynomial is the greatest common factor of the terms of the polynomial.

3.10 **quadratic equation** — A quadratic equation is one that can be written in the form $ax^2 + bx + c = 0$ ($a \neq 0$).

NEW SYMBOLS

P(x) — polynomial in x (read "P of x")

CHAPTER 3 POLYNOMIALS

QUICK REVIEW

Section	Concepts	Examples
3.1 Integer Exponents	**Definitions and Rules for Exponents** Product Rule: $a^m \cdot a^n = a^{m+n}$ Quotient Rule: $\dfrac{a^m}{a^n} = a^{m-n}$ Negative Exponent: $a^{-n} = \dfrac{1}{a^n}$ $\dfrac{a^{-n}}{b^{-m}} = \dfrac{b^m}{a^n}$ Zero Exponent: $a^0 = 1$	$3^4 \cdot 3^2 = 3^6$ $\dfrac{2^5}{2^3} = 2^2$ $5^{-2} = \dfrac{1}{5^2}$ $\dfrac{5^{-3}}{4^{-6}} = \dfrac{4^6}{5^3}$ $27^0 = 1, \quad (-5)^0 = 1$
3.2 Further Properties of Exponents	**Power Rules** $\quad (a^m)^n = a^{mn}$ $(ab)^m = a^m b^m$ $\left(\dfrac{a}{b}\right)^n = \dfrac{a^n}{b^n}$ $\left(\dfrac{a}{b}\right)^{-n} = \left(\dfrac{b}{a}\right)^n$	$(6^3)^4 = 6^{12}$ $(5p)^4 = 5^4 p^4$ $\left(\dfrac{2}{3}\right)^5 = \dfrac{2^5}{3^5}$ $\left(\dfrac{4}{7}\right)^{-2} = \left(\dfrac{7}{4}\right)^2$
3.3 Addition and Subtraction of Polynomials	Add or subtract polynomials by combining like terms.	$(x^2 - 2x + 3) + (2x^2 - 8)$ $= 3x^2 - 2x - 5$ $(5x^4 + 3x^2) - (7x^4 + x^2 - x)$ $= -2x^4 + 2x^2 + x$
3.4 Multiplication of Polynomials	Use the commutative and associative properties and the rules for exponents to multiply two polynomials. $(x + y)(x - y) = x^2 - y^2$ $(x + y)^2 = x^2 + 2xy + y^2$ $(x - y)^2 = x^2 - 2xy + y^2$	$(x^3 + 3x)(4x^2 - 5x + 2)$ $= 4x^5 + 12x^3 - 5x^4 - 15x^2$ $\quad + 2x^3 + 6x$ $= 4x^5 - 5x^4 + 14x^3$ $\quad - 15x^2 + 6x$ $(3m + 8)(3m - 8) = 9m^2 - 64$ $(5a + 3b)^2 = 25a^2 + 30ab + 9b^2$ $(2k - 1)^2 = 4k^2 - 4k + 1$

CHAPTER 3 SUMMARY

Section	Concepts	Examples
3.5 Division of Polynomials	To divide a polynomial by a monomial, divide each term of the polynomial by the monomial and write each quotient in lowest terms. Divide any polynomial by another polynomial using a long division process similar to that used for numbers.	$$\frac{5x^4 - 15x^2 + 25x}{5x} = x^3 - 3x + 5$$ $$\begin{array}{r} x^3 + 2x^2 - x \\ x-2{\overline{\smash{\big)}\,x^4 + 0x^3 - 5x^2 + 2x}} \\ \underline{x^4 - 2x^3} \\ 2x^3 - 5x^2 \\ \underline{2x^3 - 4x^2} \\ -x^2 + 2x \\ \underline{-x^2 + 2x} \\ 0 \end{array}$$
3.6 Synthetic Division	**Remainder Theorem** If $P(x)$ is divided by $x - k$, the remainder is $P(k)$.	Find $P(3)$, given $P(x) = 2x^2 - 5x + 6$. $$\begin{array}{r|rrr} 3 & 2 & -5 & 6 \\ & & 6 & 3 \\ \hline & 2 & 1 & 9 \end{array} \leftarrow P(3) = 9$$
3.7 Greatest Common Factors; Factoring by Grouping	**Factoring out the Greatest Common Factor** The product of the largest numerical factor and the variable of lowest degree of every term in a polynomial is the greatest common factor of the terms of the polynomial. **Factoring by Grouping** *Step 1* Collect the terms into groups so that the terms in each group have a common factor. *Step 2* Factor out the common factor of the terms in each group. *Step 3* If the groups now have a common factor, factor it out. If not, try a different grouping.	Factor $4x^2y - 50xy^2 = 2^2x^2y - 2 \cdot 5^2xy^2$. The greatest common factor is $2xy$. $$4x^2y - 50xy^2 = 2xy(2x - 25y)$$ $5a - 5b - ax + bx$ $= (5a - 5b) + (-ax + bx)$ $= 5(a - b) - x(a - b)$ $= (5 - x)(a - b)$

213

CHAPTER 3 POLYNOMIALS

Section	Concepts	Examples
3.8 Factoring Trinomials	For a trinomial in the form $ax^2 + bx + c$ or $ax^2 + bxy + cy^2$:	Factor $15x^2 + 14x - 8$.
	Step 1 Write all pairs of factors of the coefficient of the first term.	Factors of 15 are 5 and 3, 1 and 15.
	Step 2 Write all pairs of factors of the last term.	Factors of -8 are -4 and 2, 4 and -2, -1 and 8, 8 and -1.
	Step 3 Use various combinations of these factors until the necessary middle term is found.	$15x^2 + 14x - 8 = (5x - 2)(3x + 4)$
	Step 4 If the necessary combination does not exist, the polynomial is prime.	
3.9 Special Factoring	**Difference of Two Squares** $x^2 - y^2 = (x + y)(x - y)$	$4m^2 - 25n^2 = (2m)^2 - (5n)^2$ $= (2m + 5n)(2m - 5n)$
	Perfect Square Trinomials $x^2 + 2xy + y^2 = (x + y)^2$ $x^2 - 2xy + y^2 = (x - y)^2$	$9y^2 + 6y + 1 = (3y + 1)^2$ $16p^2 - 56p + 49 = (4p - 7)^2$
	Difference of Two Cubes $x^3 - y^3 = (x - y)(x^2 + xy + y^2)$	$8 - 27a^3$ $= (2 - 3a)(4 + 6a + 9a^2)$
	Sum of Two Cubes $x^3 + y^3 = (x + y)(x^2 - xy + y^2)$	$64z^3 + 1$ $= (4z + 1)(16z^2 - 4z + 1)$
3.10 Solving Equations by Factoring	*Step 1* Rewrite the equation if necessary so that one side is 0.	Solve $2x^2 + 5x = 3$. $2x^2 + 5x - 3 = 0$
	Step 2 Factor the polynomial.	$(2x - 1)(x + 3) = 0$
	Step 3 Set each factor equal to 0.	$2x - 1 = 0$ or $x + 3 = 0$
	Step 4 Solve each equation.	$2x = 1 \qquad x = -3$ $x = \dfrac{1}{2}$
	Step 5 Check each solution.	A check verifies that the solution set is $\left\{\dfrac{1}{2}, -3\right\}$.

CHAPTER 3 REVIEW EXERCISES

[3.1] *Identify the exponent and the base. Do not evaluate.*

1. 9^3
2. $(-5)^4$
3. -8^2
4. $-5z^{-2}$

Evaluate.

5. 3^5
6. $\left(\dfrac{1}{2}\right)^4$
7. $(-4)^3$
8. $\dfrac{3}{(-7)^{-2}}$
9. $\left(\dfrac{3}{4}\right)^{-3}$
10. $\left(\dfrac{4}{5}\right)^{-2}$

Simplify. Write answers with only positive exponents. Assume that all variables represent nonzero real numbers.

11. $9^{15} \cdot 9^{-7}$
12. $\dfrac{2^4}{2^{-2}}$
13. $y^9 y^3 y$
14. $9(4p^3)(6p^{-7})$
15. $\dfrac{5^{-4}}{5^{-7}}$
16. $\dfrac{m^{-2} m^{-5}}{m^{-8} m^{-4}}$

[3.2] *Write in scientific notation.*

17. 3450
18. .000000076
19. .13

Write without scientific notation.

20. 2.1×10^5
21. 3.8×10^{-3}
22. 3.78×10^{-1}

Simplify. Write answers with only positive exponents. Assume that all variables represent nonzero real numbers.

23. $(3^{-4})^2$
24. 16^0
25. -9^0
26. $(z^{-3})^3 z^{-6}$

27. $(5m^{-3})^2 (m^4)^{-3}$
28. $\dfrac{(3r)^2 r^4}{r^{-2} r^{-3}}$
29. $\dfrac{5z^{-3}}{z^{-1}}$
30. $\left(\dfrac{6m^{-4}}{m^{-9}}\right)^{-1} \cdot \dfrac{m^{-2}}{16}$

31. $\left(\dfrac{3r^5}{5r^{-3}}\right)^{-2} \left(\dfrac{9r^{-1}}{2r^{-5}}\right)^3$
32. $\left(\dfrac{a^{-2} b^{-1}}{3a^2}\right)^{-2} \left(\dfrac{b^{-2} \cdot 3a^4}{2b^{-3}}\right)^{-2}$

CHAPTER 3 POLYNOMIALS

Use scientific notation to compute.

33. $\dfrac{6 \times 10^4}{3 \times 10^8}$

34. $\dfrac{5 \times 10^{-2}}{1 \times 10^{-5}}$

35. $\dfrac{.000000016}{.0004}$

36. $\dfrac{.0009 \times 12{,}000}{400}$

[3.3] *Give the coefficient of each term.*

37. $14p^5$

38. $-z$

39. 29

40. $104p^3r^5$

For each polynomial **(a)** *write in descending powers of the variable,* **(b)** *identify as* monomial, binomial, trinomial, *or* none of these, *and* **(c)** *give the degree of the polynomial.*

41. $9k + 11k^3 - 3k^2$ **(a)** **(b)** **(c)**

42. $14m^6 - 9m^7$ **(a)** **(b)** **(c)**

43. $-5y^4 + 3y^3 + 7y^2 - 2y$ **(a)** **(b)** **(c)**

For each polynomial, find **(a)** $P(-1)$ *and* **(b)** $P(3)$.

44. $P(x) = -5x + 11$ **(a)** **(b)**

45. $P(x) = -3x^2 + 7x - 1$ **(a)** **(b)**

Add or subtract as indicated.

46. Add.
$-5x^2 + 7x - 3$
$\underline{6x^2 + 4x + 9}$

47. Subtract.
$8y - 9$
$\underline{-4y + 2}$

48. Subtract.
$-4y^3 + 7y - 6$
$\underline{ 2y^2 - 9y + 8}$

49. $(-5 + 11w) + (6 + 5w)$

50. $(2k - 1) - (3k - 2)$

51. $(5a^3 - 9a^2 + 11a) - (-a^3 + 6a^2 + 4a)$

52. $(2y^2 + 7y + 4) + (3y^2 - 9y + 2)$

CHAPTER 3 REVIEW EXERCISES

[3.4] *Find each product.*

53. $-2a(4a^2 + 11)$

54. $(2y - 3)(y + 4)$

55. $(8y - 7)(3y + 5)$

56. $(2p^2 + 5p - 3)(p + 4)$

57. $(5r + 3s)(2r - 3s)$

58. $(3k^2 + 5k)(4k^2 - 6)$

59. $(2z^3 - 4z^2 + 6z - 1)(3z + 2)$

60. $(3r^2 + 5)(3r^2 - 5)$

61. $\left(z + \dfrac{2}{3}\right)\left(z - \dfrac{2}{3}\right)$

62. $(y + 2)^2$

63. $(3p - 7)^2$

[3.5] *Divide.*

64. $\dfrac{18r - 32}{12}$

65. $\dfrac{5m^3 - 11m^2 + 10m}{5m}$

66. $\dfrac{8x^2 - 23x + 2}{x - 3}$

67. $(15p^2 + 11p - 17) \div (3p - 2)$

68. $(k^3 - 4k^2 - 15k + 18) \div (k + 3)$

69. $\dfrac{5m^4 + 15m^3 - 33m^2 - 10m + 32}{5m^2 - 3}$

70. $\dfrac{2p^3 + 9p^2 + 27}{2p - 3}$

71. $\dfrac{12y^4 + 7y^2 - 2y + 1}{3y^2 + 1}$

CHAPTER 3 POLYNOMIALS

[3.6] *Use synthetic division for each of the following.*

72. $\dfrac{3p^2 - p - 2}{p - 1}$

73. $\dfrac{10k^2 - 3k - 15}{k + 2}$

74. $(2k^3 - 5k^2 + 12) \div (k - 3)$

75. $(3z^4 + 14z^3 - 22z^2 + 14z + 18) \div (z + 6)$

76. $(9a^5 + 35a^4 + 19a^2 + 18a + 15) \div (a + 4)$

77. $\dfrac{y^5 - 1}{y - 1}$

Use synthetic division to decide whether or not -5 is a solution for each of the following equations.

78. $2w^3 + 8w^2 - 14w - 20 = 0$

79. $-3q^4 + 2q^3 + 5q^2 - 9q + 1 = 0$

Use synthetic division to evaluate P(k) for the given value of k.

80. $P(x) = 3x^3 - 5x^2 + 4x - 1; \quad k = -1$

81. $P(z) = z^4 - 2z^3 - 9z - 5; \quad k = 3$

[3.7–3.8] *Factor completely.*

82. $11k + 12k^2$

83. $15y^2 + 20y$

84. $12m^2 - 6m$

85. $12a^2b + 18ab^3 - 24a^3b^2$

86. $9mn^3 - 72m^2n^2 + 54m^3n$

87. $(x - 4)(3x + 2) - (x - 4)(5x + 3)$

88. $(a + 2)(a + 1) + (a + 2)(a - 4)$

89. $4x + 4y + mx + my$

90. $x^2 + xy + 5x + 5y$

CHAPTER 3 REVIEW EXERCISES

91. $x^2 - 11x + 24$

92. $a^2 + 16a - 36$

93. $4p^2 + 3p - 1$

94. $6m^2 + 7m - 3$

95. $12p^2 + p - 20$

96. $18r^2 - 3r - 10$

97. $20a^2 - 13ab + 2b^2$

98. $9x^2 + 13xy - 3y^2$

99. $30a + am - am^2$

100. $2r^2z - rz - 3z$

101. $r^4 + r^2 - 6$

102. $2k^4 - 5k^2 - 3$

[3.9]

103. $9p^2 - 49$

104. $16z^2 - 121$

105. $144(r + 1)^2 - 81(s + 1)^2$

106. $p^2 + 14p + 49$

107. $9k^2 - 12k + 4$

108. $25z^2 - 30zm + 9m^2$

109. $8 - a^3$

110. $2r^3 + 54$

111. $375x^3 - 3$

[3.10] *Solve each equation.*

112. $(5x - 2)(x + 1) = 0$

113. $p^2 - 5p + 6 = 0$

114. $q^2 + 2q = 8$

115. $6z^2 = 5z + 50$

116. $6r^2 + 7r = 3$

117. $8k^2 + 14k + 3 = 0$

118. If the square of a number is added to five times the number, the result is 36. Find the number.

119. If the square of a number is added to 8, the result is 6 times the number. Find the number.

120. If three times the square of a number is added to five times the number, the result is 100. Find the number.

121. A rectangular room has a width which is 5 meters less than its length. The area of the room is 50 square meters. Find the length of the room.

122. A triangular flower bed has an area of 54 square feet. The height is 3 feet greater than the length of the base. Find the length of the base.

123. The length of a rectangular picture frame is 2 inches longer than its width. The area enclosed by the frame is 48 square inches. What is the width?

MIXED REVIEW EXERCISES
Perform the indicated operations, then simplify. Write answers with only positive exponents. Assume all variables represent nonzero real numbers.

124. $(4x + 1)(2x - 3)$

125. 5^{-3}

126. $(y^6)^{-5}(2y^{-3})^{-4}$

127. $(-5 + 11w) + (6 + 5w) + (-15 - 8w^2)$

128. $\dfrac{m^2 - 8m + 15}{m - 5}$

129. $\dfrac{(-z^{-2})^3}{5(z^{-3})^{-1}}$

130. $-(-3)^2$

131. $\dfrac{(5z^2x^3)^2}{(-10zx^{-3})^{-2}}$

132. $(2k - 1) - (3k^2 - 2k + 6)$

133. $7p^5(3p^4 + p^3 + 2p^2)$

134. $\dfrac{20y^3x^3 + 15y^4x + 25yx^4}{10yx^2}$

Factor completely.

135. $10m^2 - m - 3$

136. $12z - 72z^2$

137. $64 - p^3$

138. $k^2 + kq - 6q^2$

139. $16c^3 + 28c^2$

140. $16d^2 + 24d + 9$

CHAPTER 3 TEST

Evaluate.

1. $\left(\dfrac{3}{2}\right)^{-3}$

2. $-4^2 + (-4)^2$

Simplify. Write answers with only positive exponents. Assume all variables are nonzero.

3. $8^{-2} \cdot 8^5 \cdot 8^{-6}$

4. $(3a^{-2}b^3)^{-3}$

5. $\dfrac{3^{-2}r^{-3}}{5(r^2)^{-3}}$

Perform the indicated operations.

6. $(3k^3 - 5k^2 + 8k - 2) - (4k^3 + 11k + 7) + (2k^2 - 5k)$

7. $(8x - 7)(x + 3)$

8. $(3m - 2)(4m^2 - 6m + 5)$

9. $(3x - 2y)(3x + 2y)$

10. $(2p + q)^2$

Divide.

11. $\dfrac{8z^3 - 16z^2 + 24z}{8z^2}$

12. $\dfrac{6y^4 - 3y^3 + 5y^2 + 6y - 9}{2y + 1}$

13. Use synthetic division to decide whether 4 is a solution of $x^4 - 8x^3 + 21x^2 - 14x - 24 = 0$.

CHAPTER 3 POLYNOMIALS

Factor each polynomial.

14. _____ 14. $12f^2 - 12f$

15. _____ 15. $(g - 1)(2g + 3) + (g - 1)(g - 1)$

16. _____ 16. $ax + ay + bx + by$

17. _____ 17. $2p^2 - 5pq + 3q^2$

18. _____ 18. $9m^2 - 6mn + 4n^2$

19. _____ 19. $x^3 + 1000$

20. _____ 20. $100d^2 - 9m^4$

21. _____ 21. $18k^4 + 9k^2 - 20$

Solve each equation.

22. _____ 22. $2x^2 + 11x + 15 = 0$

23. _____ 23. $8r^2 = 14r - 3$

24. _____ 24. $5m(m - 1) = 2(1 - m)$

25. _____ 25. The length of a rectangle is 2 meters greater than the width. The area of the rectangle is 63 square meters. Find the length and width of the rectangle.

4 RATIONAL EXPRESSIONS

4.1 BASICS OF RATIONAL EXPRESSIONS

1 In arithmetic, a rational number is the quotient of two integers, with the denominator not 0. In algebra, this idea is generalized: A **rational expression** (or **algebraic fraction**) is the quotient of two polynomials, again with the denominator not 0. For example,

$$\frac{m+4}{m-2}, \quad \frac{8x^2 - 2x + 5}{4x^2 + 5x}, \quad \text{and} \quad x^5 \left(\text{or } \frac{x^5}{1}\right)$$

are all rational expressions.

OBJECTIVES

1. Define rational expressions.
2. Find the numbers that make a rational expression undefined.
3. Write rational expressions in lowest terms.

2 Any number can be used as a replacement for the variable in a rational expression, except for values that make the denominator 0. With a zero denominator, the rational expression is undefined. For example, the number 5 cannot be used as a replacement for the x in

$$\frac{2}{x-5}$$

since 5 would make the denominator equal 0.

EXAMPLE 1 Determining When Rational Expressions Are Undefined

Find all numbers that make the rational expression undefined.

(a) $\dfrac{3}{7k - 14}$

The only values that cannot be used are those that make the denominator 0. Find these values by setting the denominator equal to 0 and solving the resulting equation.

$$\begin{aligned} 7k - 14 &= 0 \\ 7k &= 14 \quad \text{Add 14.} \\ k &= 2 \quad \text{Divide by 7.} \end{aligned}$$

The number 2 cannot be used as a replacement for k; 2 makes the rational expression undefined.

CHAPTER 4 RATIONAL EXPRESSIONS

1. Find all numbers that make the rational expression undefined.

(a) $\dfrac{m + 4}{m - 6}$

(b) $\dfrac{8y}{5}$

(c) $\dfrac{r + 6}{r^2 - r - 6}$

(d) $\dfrac{k + 2}{k^2 + 1}$

ANSWERS
1. (a) 6
 (b) Any real number can replace y.
 (c) 3, -2
 (d) Any real number can replace k.

(b) $\dfrac{3 + p}{p^2 - 4p + 3}$

Set the denominator equal to 0.
$$p^2 - 4p + 3 = 0$$
Factor to get
$$(p - 3)(p - 1) = 0.$$
Set each factor equal to 0.
$$p - 3 = 0 \quad \text{or} \quad p - 1 = 0$$
$$p = 3 \quad \text{or} \quad p = 1$$

Both 3 and 1 make the rational expression undefined.

(c) $\dfrac{8x + 2}{3}$

The denominator can never be 0, so any real number can replace x in the rational expression. ■

■ **WORK PROBLEM 1 AT THE SIDE.**

3 In arithmetic, we write the fraction $\dfrac{15}{20}$ in lowest terms by dividing numerator and denominator by 5 to get $\dfrac{3}{4}$. Rational expressions can be written in lowest terms in a similar manner, with the *fundamental principle of rational numbers*.

FUNDAMENTAL PRINCIPLE OF RATIONAL NUMBERS

If $\dfrac{a}{b}$ is a rational number, and if c is any nonzero real number, then
$$\dfrac{a}{b} = \dfrac{ac}{bc}.$$

(The numerator and denominator of a rational number may either be multiplied or divided by the same nonzero number without changing the value of the rational number.)

Since a rational expression is a quotient of two polynomials, and since the value of a polynomial is a real number for all values of its variables, any statement that applies to rational numbers will also apply to rational expressions. In particular, the fundamental principle may be used to write rational expressions in lowest terms. We do this as follows.

WRITING IN LOWEST TERMS

Step 1 Factor both the numerator and denominator to find their greatest common factor.

Step 2 Divide both numerator and denominator by their greatest common factor.

4.1 BASICS OF RATIONAL EXPRESSIONS

EXAMPLE 2 Writing Rational Expressions in Lowest Terms

Write each rational expression in lowest terms.

(a) $\dfrac{8k}{16}$

The greatest common factor of the numerator and denominator is 8.

$$\dfrac{8k}{16} = \dfrac{k \cdot \mathbf{8}}{2 \cdot \mathbf{8}} = \dfrac{k}{2} \cdot \dfrac{\mathbf{8}}{\mathbf{8}} = \dfrac{k}{2} \cdot \mathbf{1} = \dfrac{k}{2}$$

Notice that the identity property for multiplication is used here to write the rational expression in lowest terms. In practice, we usually just divide the numerator and denominator by their greatest common factor.

(b) $\dfrac{12x^3y^2}{6x^4y} = \dfrac{2y \cdot \mathbf{6x^3y}}{x \cdot \mathbf{6x^3y}} = \dfrac{2y}{x}$

Divide by the greatest common factor, $6x^3y$. (Here 3 is the lowest exponent on x, and 1 the lowest exponent on y.)

(c) $\dfrac{a^2 - a - 6}{a^2 + 5a + 6}$

Start by factoring the numerator and denominator.

$$\dfrac{a^2 - a - 6}{a^2 + 5a + 6} = \dfrac{(a - 3)(a + 2)}{(a + 3)(a + 2)}$$

Divide the numerator and denominator by $a + 2$ to get

$$\dfrac{a^2 - a - 6}{a^2 + 5a + 6} = \dfrac{a - 3}{a + 3}. \quad \blacksquare$$

CAUTION One of the most common errors in algebra involves incorrect use of the fundamental principle of rational numbers. Only common *factors* may be divided. For example, it would be incorrect to try to "divide" the a^2 terms in the numerator and denominator in Example 2(c). It is essential to *factor* before writing in lowest terms.

WORK PROBLEM 2 AT THE SIDE.

In the rational expression

$$\dfrac{a^2 - a - 6}{a^2 + 5a + 6}, \quad \text{or} \quad \dfrac{(a - 3)(a + 2)}{(a + 3)(a + 2)},$$

a can take on any value at all except -3 or -2. In the rational expression

$$\dfrac{a - 3}{a + 3}$$

a cannot equal -3. Because of this,

$$\dfrac{a^2 - a - 6}{a^2 + 5a + 6} = \dfrac{a - 3}{a + 3}$$

for all values of a except -3 or -2. From now on we shall write such statements of equality with the understanding that they apply only for those real numbers that make no denominator equal 0.

---

2. Write each rational expression in lowest terms.

(a) $\dfrac{18m^2}{9m^5}$

(b) $\dfrac{6y^2z^2}{12yz^3}$

(c) $\dfrac{y^2 + 2y - 3}{y^2 - 3y + 2}$

(d) $\dfrac{3y + 9}{y^2 - 9}$

ANSWERS
2. (a) $\dfrac{2}{m^3}$ (b) $\dfrac{y}{2z}$ (c) $\dfrac{y + 3}{y - 2}$
(d) $\dfrac{3}{y - 3}$

CHAPTER 4 RATIONAL EXPRESSIONS

3. Write each rational expression in lowest terms.

(a) $\dfrac{y - 2}{2 - y}$

(b) $\dfrac{3k - 12}{4 - k}$

(c) $\dfrac{8 - b}{8 + b}$

(d) $\dfrac{p - 2}{4 - p^2}$

ANSWERS
3. (a) -1 (b) -3
   (c) already in lowest terms
   (d) $\dfrac{-1}{2 + p}$

EXAMPLE 3 Writing Rational Expressions in Lowest Terms

Write each rational expression in lowest terms.

(a) $\dfrac{y^2 - 4}{2y + 4} = \dfrac{(y + 2)(y - 2)}{2(y + 2)} = \dfrac{y - 2}{2}$

(b) $\dfrac{m - 3}{3 - m}$

In this rational expression, the numerator and denominator are exactly opposite. The given expression can be written in lowest terms by writing the denominator as $3 - m = -1(m - 3)$, giving

$$\dfrac{m - 3}{\boxed{3 - m}} = \dfrac{m - 3}{\boxed{-1(m - 3)}} = \dfrac{1}{-1} = -1.$$

The numerator could have been rewritten instead to get the same result.

(c) $\dfrac{r^2 - 16}{4 - r} = \dfrac{(r + 4)(r - 4)}{4 - r}$

$= \dfrac{(r + 4)(r - 4)}{-1(r - 4)}$ Write $4 - r$ as $-1(r - 4)$.

$= \dfrac{r + 4}{-1}$ Lowest terms

$= -(r + 4)$ or $-r - 4$ ∎

Working as in parts (b) and (c) of Example 3, the quotient

$$\dfrac{a}{-a} \quad (a \neq 0)$$

can be simplified as $\dfrac{a}{-a} = \dfrac{a}{-1(a)} = \dfrac{1}{-1} = -1.$

The following generalization applies.

The quotient of two quantities that differ only in sign is -1.

Based on this result, $\dfrac{q - 7}{7 - q} = -1$ and $\dfrac{-5a + 2b}{5a - 2b} = -1$, but $\dfrac{r - 2}{r + 2}$ cannot be simplified further.

■ WORK PROBLEM 3 AT THE SIDE.

CAUTION Rational expressions often can be written in lowest terms in *seemingly* different ways. For example,

$$\dfrac{y - 3}{-5} \quad \text{and} \quad \dfrac{-y + 3}{5}$$

look different, but the second quotient is obtained by multiplying the first by -1 in both the numerator and denominator. If your answer does not exactly match the one given in the text, check to see if it is equivalent to the one given.

4.1 EXERCISES

Find all real numbers that make the rational expression undefined. See Example 1.

1. $\dfrac{m}{m-4}$
2. $\dfrac{k}{k+9}$
3. $\dfrac{2p+5}{3p-7}$
4. $\dfrac{9r+2}{5r+11}$
5. $\dfrac{6+z}{z}$

6. $\dfrac{2-b}{b}$
7. $\dfrac{5p+9}{4}$
8. $\dfrac{2z-6}{5}$
9. $\dfrac{m+5}{m^2-m-12}$
10. $\dfrac{r+2}{r^2+8r+15}$

11. $\dfrac{x^2+5}{2x^2+3x-2}$
12. $\dfrac{a^2+3a-1}{3a^2-10a-8}$
13. $\dfrac{8z-2}{z^2+16}$
14. $\dfrac{-4p+11}{p^2+25}$

Write each rational expression in lowest terms. See Example 2.

15. $\dfrac{18m^5n^3}{12m^4n^5}$
16. $\dfrac{-8y^6z^5}{6y^7z^3}$
17. $\dfrac{64p^4q^6}{16p^3q^7}$
18. $\dfrac{27a^5b^2}{81a^8b^6}$

19. $\dfrac{(x+2)(x-5)}{(x-1)(x+2)}$
20. $\dfrac{(r+6)(r-2)}{(r-2)(r+3)}$
21. $\dfrac{p(p-4)}{3p(p-4)}$
22. $\dfrac{12m(m-9)}{m(m-9)}$

23. $\dfrac{x+6}{x+2}$
24. $\dfrac{m-9}{m-3}$
25. $\dfrac{2x-2}{3x-3}$
26. $\dfrac{5m+10}{6m+12}$

27. $\dfrac{4z^2+2z}{8z+4}$
28. $\dfrac{5m+5}{m^2-1}$
29. $\dfrac{2b^2-2}{4b-4}$
30. $\dfrac{k^2+2k}{6k+12}$

31. $\dfrac{2y-y^2}{4y-y^2}$
32. $\dfrac{6m^2+m}{4m^2+m}$
33. $\dfrac{a^2-4}{a^2+4a+4}$
34. $\dfrac{4p^2-20p}{p^2-4p-5}$

35. $\dfrac{r^2 - r - 20}{r^2 + r - 30}$

36. $\dfrac{y^2 - 2y - 15}{y^2 + 7y + 12}$

37. $\dfrac{c^2 + cd - 30d^2}{c^2 - 6cd + 5d^2}$

38. $\dfrac{s^2 - 3st - 18t^2}{s^2 - 2st - 24t^2}$

39. $\dfrac{z^3 + x^3}{z^2 - x^2}$

40. $\dfrac{p^3 - q^3}{p^2 - q^2}$

41. $\dfrac{xy - yw + xz - zw}{xy + yw + xz + zw}$

42. $\dfrac{ac + ad - bc - bd}{ab + ac - b^2 - bc}$

Write each rational expression in lowest terms. See Example 3.

43. $\dfrac{6 - m}{m - 6}$

44. $\dfrac{x - 2}{2 - x}$

45. $\dfrac{x - y}{y^2 - x^2}$

46. $\dfrac{b - a}{a^2 - b^2}$

47. $\dfrac{(3 - m)(4 - n)}{(m - 3)(4 + n)}$

48. $\dfrac{(y - 4)(4 - y)}{(4 - y)(4 + y)}$

49. $\dfrac{2k - 4}{6 - 3k}$

50. $\dfrac{7m - 21}{6 - 2m}$

51. $\dfrac{x^2 - x}{1 - x}$

52. $\dfrac{r^2 - 16r}{16 - r}$

SKILL SHARPENERS
Multiply or divide as indicated. See Section 1.3.

53. $-\dfrac{3}{8} \cdot \dfrac{16}{9}$

54. $-\dfrac{15}{11} \cdot \left(-\dfrac{22}{5}\right)$

55. $-3 \div \left(-\dfrac{6}{5}\right)$

56. $-10 \div \left(-\dfrac{5}{4}\right)$

4.2 MULTIPLICATION AND DIVISION OF RATIONAL EXPRESSIONS

1 Multiplication of rational expressions follows the same procedure as multiplication of rational numbers:

MULTIPLICATION OF RATIONAL NUMBERS

If a, b, c, and d are integers, with $b \neq 0$ and $d \neq 0$,

$$\frac{a}{b} \cdot \frac{c}{d} = \frac{ac}{bd}.$$

The same idea is used to multiply two rational expressions:

MULTIPLYING RATIONAL EXPRESSIONS

Step 1 Factor all numerators and denominators as completely as possible.

Step 2 Divide both numerator and denominator by any factors that both have in common.

Step 3 Multiply remaining factors in the numerator, and multiply remaining factors in the denominator.

Step 4 Be certain that the product is in lowest terms.

EXAMPLE 1 Multiplying Rational Expressions
Multiply.

(a) $\dfrac{3x^2}{5} \cdot \dfrac{10}{x^3} = \dfrac{3x^2 \cdot 5 \cdot 2}{5 \cdot x^2 \cdot x} = \dfrac{6}{x}$

(b) $\dfrac{5p - 5}{p} \cdot \dfrac{3p^2}{10p - 10}$

Factor where possible.

$$\frac{5p - 5}{p} \cdot \frac{3p^2}{10p - 10} = \frac{5(p - 1)}{p} \cdot \frac{3p^2}{5 \cdot 2(p - 1)}$$
$$= \frac{5 \cdot 3 \cdot p \cdot p(p - 1)}{5 \cdot 2 \cdot p(p - 1)} = \frac{3p}{2}$$

(c) $\dfrac{k^2 + 2k - 15}{k^2 - 4k + 3} \cdot \dfrac{k^2 - k}{k^2 + k - 20}$

$= \dfrac{(k + 5)(k - 3)}{(k - 3)(k - 1)} \cdot \dfrac{k(k - 1)}{(k + 5)(k - 4)} = \dfrac{k}{k - 4}$ ■

WORK PROBLEM 1 AT THE SIDE.

2 Recall that the reciprocal of a nonzero real number a is the real number $\dfrac{1}{a}$.

To find the *reciprocal* of a nonzero rational expression, invert the rational expression.

OBJECTIVES

1 Multiply rational expressions.

2 Find reciprocals for rational expressions.

3 Divide rational expressions.

1. Multiply.

(a) $\dfrac{12r}{7} \cdot \dfrac{14}{r^2}$

(b) $\dfrac{6z^2}{5} \cdot \dfrac{4z}{3}$

(c) $\dfrac{2r + 4}{5r} \cdot \dfrac{3r}{5r + 10}$

(d) $\dfrac{m^2 - 16}{m + 2} \cdot \dfrac{1}{m + 4}$

(e) $\dfrac{c^2 + 2c}{c^2 - 4} \cdot \dfrac{c^2 - 4c + 4}{c^2 - c}$

ANSWERS
1. (a) $\dfrac{24}{r}$ (b) $\dfrac{8z^3}{5}$ (c) $\dfrac{6}{25}$
(d) $\dfrac{m - 4}{m + 2}$ (e) $\dfrac{c - 2}{c - 1}$

CHAPTER 4 RATIONAL EXPRESSIONS

2. Find the reciprocal.

(a) $\dfrac{-3}{r}$

(b) $\dfrac{7}{y+8}$

(c) $\dfrac{a^2 + 7a}{2a - 1}$

3. Divide.

(a) $\dfrac{16k^2}{5} \div \dfrac{3k}{10}$

(b) $\dfrac{5p + 2}{6} \div \dfrac{15p + 6}{5}$

(c) $\dfrac{q^2 + 2q}{5 + q} \div \dfrac{4 - q^2}{3q - 6}$

(d) $\dfrac{y^2 - 2y - 3}{y^2 + 4y + 4} \div \dfrac{y^2 - 1}{y^2 + y - 2}$

EXAMPLE 2 Finding the Reciprocal of a Rational Expression

Find the reciprocal of each rational expression.

Rational Expression	Reciprocal
$\dfrac{5}{k}$	$\dfrac{k}{5}$
$\dfrac{m^2 - 9m}{2}$	$\dfrac{2}{m^2 - 9m}$
$\dfrac{0}{4}$	Undefined—no reciprocal for 0

■ **WORK PROBLEM 2 AT THE SIDE.**

3 Division of rational expressions follows the rule for division of rational numbers:

DIVISION OF RATIONAL NUMBERS

For rational numbers $\dfrac{a}{b}$ and $\dfrac{c}{d}$, with $\dfrac{c}{d} \neq 0$, $\quad \dfrac{a}{b} \div \dfrac{c}{d} = \dfrac{a}{b} \cdot \dfrac{d}{c}$.

This result leads to the following rule.

DIVIDING RATIONAL EXPRESSIONS

To divide two rational expressions, *multiply* the first by the reciprocal of the second (the divisor).

EXAMPLE 3 Dividing Rational Expressions

Divide.

(a) $\dfrac{2z}{9} \div \dfrac{5z^2}{18} = \dfrac{2z}{9} \cdot \dfrac{18}{5z^2} = \dfrac{4}{5z}$ Multiply by the reciprocal of the divisor.

(b) $\dfrac{8k - 16}{3k} \div \dfrac{6 - 3k}{4k^2} = \dfrac{8k - 16}{3k} \cdot \dfrac{4k^2}{6 - 3k}$ Multiply by the reciprocal of the divisor.

$= \dfrac{8(k - 2)}{3k} \cdot \dfrac{4k^2}{3(2 - k)} = -\dfrac{32k}{9}$ Factor.

The negative sign appears in the quotient because the factors $k - 2$ in the numerator and $2 - k$ in the denominator are negatives of each other.

(c) $\dfrac{m^2 - 1}{25m^2 - 9} \cdot \dfrac{5m^2 + 17m - 12}{3m^2 + 7m - 20} \div \dfrac{5m^2 + 2m - 3}{15m^2 - 34m + 15}$

$= \dfrac{m^2 - 1}{25m^2 - 9} \cdot \dfrac{5m^2 + 17m - 12}{3m^2 + 7m - 20} \cdot \dfrac{15m^2 - 34m + 15}{5m^2 + 2m - 3}$

$= \dfrac{(m + 1)(m - 1)}{(5m + 3)(5m - 3)} \cdot \dfrac{(5m - 3)(m + 4)}{(m + 4)(3m - 5)} \cdot \dfrac{(3m - 5)(5m - 3)}{(5m - 3)(m + 1)}$

$= \dfrac{m - 1}{5m + 3}$ ■

■ **WORK PROBLEM 3 AT THE SIDE.**

ANSWERS

2. (a) $\dfrac{-r}{3}$ (b) $\dfrac{y + 8}{7}$ (c) $\dfrac{2a - 1}{a^2 + 7a}$

3. (a) $\dfrac{32k}{3}$ (b) $\dfrac{5}{18}$ (c) $-\dfrac{3q}{5 + q}$

(d) $\dfrac{y - 3}{y + 2}$

4.2 EXERCISES

Multiply or divide as indicated. Write all answers in lowest terms. See Examples 1 and 3.

1. $\dfrac{m^3}{2} \cdot \dfrac{4m}{m^4}$

2. $\dfrac{3y^4}{5y} \cdot \dfrac{8y^2}{9}$

3. $\dfrac{6a^4}{a^3} \div \dfrac{12a^2}{a^5}$

4. $\dfrac{11p^3}{p^2} \div \dfrac{22p^4}{p}$

5. $\dfrac{6y^5x^6}{y^3x^4} \cdot \dfrac{y^4x^2}{3y^5x^7}$

6. $\dfrac{6p^2q^3}{4p^3q} \cdot \dfrac{18p^2q^3}{12p^3q^2}$

7. $\dfrac{25a^2b}{60a^3b^2} \div \dfrac{5a^4b^2}{16a^2b}$

8. $\dfrac{8s^4t^2}{5t^6} \div \dfrac{s^3t^2}{10s^2t^4}$

9. $\dfrac{2r}{8r + 4} \cdot \dfrac{14r + 7}{3}$

10. $\dfrac{6a - 10}{5a} \cdot \dfrac{3}{9a - 15}$

11. $\dfrac{7k + 7}{2} \div \dfrac{4k + 4}{5}$

12. $\dfrac{8y - 16}{5} \div \dfrac{3y - 6}{10}$

13. $\dfrac{(m + 3)^2}{m} \cdot \dfrac{m^2}{m^2 - 9}$

14. $\dfrac{6r}{(r + 1)^2} \cdot \dfrac{r^2 - 1}{r^2}$

15. $\dfrac{3k}{(2k - 3)^2} \div \dfrac{8k^2}{4k^2 - 9}$

16. $\dfrac{9y^2 - 25}{5y^2} \div \dfrac{(3y + 5)^2}{y}$

17. $\dfrac{z^2 - 1}{2z} \cdot \dfrac{1}{1 - z}$

18. $\dfrac{m^2 - 16}{5m} \cdot \dfrac{2}{4 - m}$

19. $\dfrac{p^2 - 36}{p + 1} \div \dfrac{6 - p}{p}$

20. $\dfrac{y^2 - 4}{3y^2} \div \dfrac{2 - y}{8y}$

21. $\dfrac{6r - 5s}{3r + 2s} \cdot \dfrac{6r + 4s}{5s - 6r}$

22. $\dfrac{9y - 12x}{y + x} \div \dfrac{4x - 3y}{x + y}$

23. $\dfrac{a^2 - 16}{a^2 + a - 12} \cdot \dfrac{a^2 + 5a + 6}{a^2 - 2a - 8}$

24. $\dfrac{m^2 - 3m - 10}{m^2 + 3m + 2} \cdot \dfrac{m^2 - 2m - 3}{m^2 + 2m - 15}$

25. $\dfrac{6a^2 + 5ab - 6b^2}{12a^2 - 11ab + 2b^2} \cdot \dfrac{8a^2 - 14ab + 3b^2}{4a^2 - 12ab + 9b^2}$

26. $\dfrac{8p^2 - 6pq - 9q^2}{6p^2 - 5pq - 6q^2} \div \dfrac{4p^2 + 11pq + 6q^2}{9p^2 - 4q^2}$

27. $\dfrac{15x^2 - xy - 2y^2}{15x^2 + 11xy + 2y^2} \div \dfrac{15x^2 + 4xy - 4y^2}{15x^2 + xy - 2y^2}$

28. $\dfrac{18w^2 + 3wx - 10x^2}{6w^2 + 11wx - 10x^2} \cdot \dfrac{6w^2 + 19wx + 10x^2}{3w^2 + 11wx + 6x^2}$

29. $\dfrac{6k^2 + kr - 2r^2}{6k^2 - 5kr + r^2} \div \dfrac{3k^2 + 17kr + 10r^2}{6k^2 + 13kr - 5r^2}$

30. $\dfrac{16m^2 - 9n^2}{16m^2 - 24mn + 9n^2} \div \dfrac{16m^2 + 24mn + 9n^2}{16m^2 - 16mn + 3n^2}$

31. $\dfrac{6k^2 - 13k - 5}{k^2 + 7k} \cdot \dfrac{k^3 + 6k^2 - 7k}{2k - 5} \div \dfrac{3k^2 - 8k - 3}{k^2 - 5k + 6}$

32. $\dfrac{2r^3 + 3r^2 - 2r}{3r - 15} \cdot \dfrac{r^2 - 3r - 10}{2r^3 - r^2} \div \dfrac{3r^2 + 12r + 12}{5r^2 - 10r}$

33. $\dfrac{a^2(2a + b) + 6a(2a + b) + 5(2a + b)}{3a^2(2a + b) - 2a(2a + b) + (2a + b)} \div \dfrac{a + 1}{a - 1}$

34. $\dfrac{2x^2(x - 3z) - 5x(x - 3z) + 2(x - 3z)}{4x^2(x - 3z) - 11x(x - 3z) + 6(x - 3z)} \div \dfrac{4x - 3}{4x + 1}$

SKILL SHARPENERS
Find the least common denominator for each pair of fractions.

35. $\dfrac{3}{5}, \dfrac{3}{10}$

36. $\dfrac{5}{3}, \dfrac{7}{9}$

37. $\dfrac{5}{12}, \dfrac{1}{18}$

38. $\dfrac{2}{15}, \dfrac{17}{25}$

4.3 ADDITION AND SUBTRACTION OF RATIONAL EXPRESSIONS

Fractions with the same denominators are called **like fractions**. For example, $\frac{2}{3}$ and $\frac{4}{3}$ are like fractions. Fractions with different denominators, such as $\frac{3}{4}$ and $\frac{5}{6}$, are **unlike fractions**. To add or subtract like fractions, we use the following rule.

OBJECTIVES

1. Add and subtract rational expressions with the same denominators.
2. Find the least common denominator.
3. Add and subtract rational expressions with different denominators.

ADDITION AND SUBTRACTION OF RATIONAL NUMBERS

If $\frac{a}{b}$ and $\frac{c}{b}$ are rational numbers, then

$$\frac{a}{b} + \frac{c}{b} = \frac{a+c}{b}$$

and

$$\frac{a}{b} - \frac{c}{b} = \frac{a-c}{b}.$$

The following steps are used in adding or subtracting rational expressions.

ADDING OR SUBTRACTING RATIONAL EXPRESSIONS

Step 1(a) If the denominators are the same, add or subtract the numerators. Place the result over the common denominator.

Step 1(b) If the denominators are different, first find the least common denominator. Write all rational expressions with this least common denominator, and then add or subtract the numerators. Place the result over the common denominator.

Step 2 Write all answers in lowest terms.

1 The first example shows how to add and subtract like fractions.

EXAMPLE 1 Adding and Subtracting Like Fractions

(a) $\frac{7}{3} + \frac{1}{3} = \frac{7+1}{3} = \frac{8}{3}$

Since the denominators of these rational numbers are the same, just add the numerators. Place the result over the common denominator.

(b) $\frac{5}{m} - \frac{7}{m} = \frac{5-7}{m} = \frac{-2}{m} = -\frac{2}{m}$

The rule for subtraction is similar to that for addition: since the denominators are the same, subtract the numerators.

(c) $\frac{7}{2r^2} - \frac{11}{2r^2} = \frac{7-11}{2r^2} = \frac{-4}{2r^2} = -\frac{2}{r^2}$ Lowest terms

CHAPTER 4 RATIONAL EXPRESSIONS

1. Add or subtract.

(a) $\dfrac{3m}{8} + \dfrac{5n}{8}$

(b) $\dfrac{7}{3a} + \dfrac{10}{3a}$

(c) $\dfrac{2}{y^2} - \dfrac{5}{y^2}$

(d) $\dfrac{a}{a+b} + \dfrac{b}{a+b}$

(e) $\dfrac{2y-1}{y^2+y-2} - \dfrac{y}{y^2+y-2}$

(d) $\dfrac{m}{m^2-p^2} + \dfrac{p}{m^2-p^2} = \dfrac{m+p}{m^2-p^2}$

$= \dfrac{m+p}{(m+p)(m-p)}$ Factor the denominator.

$= \dfrac{1}{m-p}$ Lowest terms

(e) $\dfrac{4}{x^2+2x-8} + \dfrac{x}{x^2+2x-8} = \dfrac{4+x}{x^2+2x-8}$

$= \dfrac{4+x}{(x-2)(x+4)} = \dfrac{1}{x-2}$ ∎

WORK PROBLEM 1 AT THE SIDE.

2 The rational expressions in Example 1 all had the same denominators. If rational expressions to be added or subtracted have different denominators, it is necessary to first find their **least common denominator**, an expression divisible by the denominator of each of the rational expressions. We find the least common denominator for a group of rational expressions as shown below.

FINDING THE LEAST COMMON DENOMINATOR

Step 1   Factor each denominator.

Step 2   The least common denominator is the product of all the different factors from each denominator, with each factor raised to the *highest* power that occurs in any denominator.

EXAMPLE 2 Finding the Least Common Denominator

Find the least common denominator for each pair of rational expressions.

(a) $\dfrac{6}{5xy^2}, \dfrac{3}{2x^3y}$

Each denominator is already factored, so the least common denominator is obtained as follows.

$5xy^2 \qquad 2x^3y$ Original denominators

Least common denominator $= 5 \cdot 2 \cdot x^3 \cdot y^2$ ← Highest exponent on x is 3. Highest exponent on y is 2.

$= 10x^3y^2$

(b) $\dfrac{9}{p^2-16}, \dfrac{2}{p+4}$

Factor each denominator to get $\dfrac{9}{(p+4)(p-4)}, \dfrac{2}{p+4}$.

Exponents of 1 are understood on both $(p+4)^1(p-4)^1$ and $(p+4)^1$. The highest exponent on $p+4$ is 1, as is the highest exponent on $p-4$. The least common denominator is

$(p+4)^1(p-4)^1 = (p+4)(p-4).$

ANSWERS

1. (a) $\dfrac{3m+5n}{8}$ (b) $\dfrac{17}{3a}$ (c) $-\dfrac{3}{y^2}$
(d) 1   (e) $\dfrac{1}{y+2}$

4.3 ADDITION AND SUBTRACTION OF RATIONAL EXPRESSIONS

(c) $\dfrac{8}{k-3}, \dfrac{7}{k}$

The least common denominator, an expression divisible by both $k-3$ and k, is

$$k(k-3).$$

It is usually best to leave a least common denominator in factored form.

(d) $\dfrac{y}{y^2 - 2y - 8}, \dfrac{3y+1}{y^2 + 3y + 2}$

Factor the denominators to get

$$\dfrac{y}{(y+2)(y-4)}, \dfrac{3y+1}{(y+2)(y+1)}.$$

The highest exponent of each of the factors $y+2$, $y-4$, and $y+1$ is 1. The least common denominator is

$$(y+2)(y-4)(y+1). \blacksquare$$

WORK PROBLEM 2 AT THE SIDE.

3 As mentioned earlier, if the rational expressions to be added or subtracted have different denominators, it is necessary to rewrite the rational expressions with their least common denominator. This is done by multiplying both the numerator and the denominator of each rational expression by the factor required to get the least common denominator. This procedure is valid because each rational expression is being multiplied by a form of 1, the identity element for multiplication. The next example shows how this is done.

EXAMPLE 3 Adding and Subtracting Unlike Fractions
Add or subtract as indicated.

(a) $\dfrac{7}{15} + \dfrac{5}{12}$

The least common denominator for 15 and 12 is 60. Multiply $\dfrac{7}{15}$ by $\dfrac{4}{4}$ and multiply $\dfrac{5}{12}$ by $\dfrac{5}{5}$ so that each fraction has denominator 60, and then add the numerators.

$$\dfrac{7}{15} + \dfrac{5}{12} = \dfrac{7 \cdot 4}{15 \cdot 4} + \dfrac{5 \cdot 5}{12 \cdot 5} \quad \text{Get a common denominator.}$$

$$= \dfrac{28}{60} + \dfrac{25}{60}$$

$$= \dfrac{28 + 25}{60} \quad \text{Add numerators.}$$

$$= \dfrac{53}{60}$$

2. Find the least common denominator for each pair of rational expressions.

(a) $\dfrac{9}{5k^3 s}, \dfrac{7}{10ks^4}$

(b) $\dfrac{8}{z}, \dfrac{8}{z+6}$

(c) $\dfrac{3}{m^2 - 16}, \dfrac{-1}{m^2 + 8m + 16}$

(d) $\dfrac{y-4}{2y^2 - 3y - 2}, \dfrac{2y+7}{2y^2 + 3y + 1}$

ANSWERS
2. (a) $10k^3 s^4$ (b) $z(z+6)$
(c) $(m+4)^2(m-4)$
(d) $(y-2)(2y+1)(y+1)$

CHAPTER 4 RATIONAL EXPRESSIONS

3. Add or subtract.

(a) $\dfrac{6}{m} + \dfrac{1}{4m}$

(b) $\dfrac{8}{3k} - \dfrac{2}{9k}$

(c) $\dfrac{2}{y} - \dfrac{1}{y+4}$

(b) $\dfrac{5}{2p} + \dfrac{3}{8p}$

The least common denominator for $2p$ and $8p$ is $8p$. To write the first rational expression with a denominator of $8p$, multiply by $\dfrac{4}{4}$.

$$\dfrac{5}{2p} + \dfrac{3}{8p} = \dfrac{5 \cdot 4}{2p \cdot 4} + \dfrac{3}{8p}$$
$$= \dfrac{20}{8p} + \dfrac{3}{8p}$$
$$= \dfrac{20 + 3}{8p} = \dfrac{23}{8p}$$

(c) $\dfrac{6}{r} - \dfrac{5}{r-3}$

The least common denominator is $r(r-3)$. Rewrite each rational expression with this denominator.

$$\dfrac{6}{r} - \dfrac{5}{r-3} = \dfrac{6(r-3)}{r(r-3)} - \dfrac{5r}{r(r-3)}$$
$$= \dfrac{6(r-3) - 5r}{r(r-3)} \quad \text{Subtract numerators.}$$
$$= \dfrac{6r - 18 - 5r}{r(r-3)} \quad \text{Distributive property}$$
$$= \dfrac{r - 18}{r(r-3)} \quad \text{Combine terms in numerator.} \blacksquare$$

■ **WORK PROBLEM 3 AT THE SIDE.**

CAUTION One of the most common sign errors in algebra occurs when a rational expression with two or more terms in its numerator is being subtracted. Remember that in this situation, the subtraction sign must be distributed to *every* term in the numerator of the fraction that follows.

The next example illustrates this situation.

EXAMPLE 4 Subtracting a Rational Expression with a Binomial Numerator

Subtract.

(a) $\dfrac{7x}{3x+1} - \dfrac{x-2}{3x+1}$

The denominator is the same for both rational expressions. The subtraction sign must be applied to *both* terms in the numerator of the second rational expression. Notice the careful use of parentheses here.

ANSWERS

3. (a) $\dfrac{25}{4m}$ (b) $\dfrac{22}{9k}$ (c) $\dfrac{y+8}{y(y+4)}$

236

4.3 ADDITION AND SUBTRACTION OF RATIONAL EXPRESSIONS

$$\frac{7x}{3x+1} - \frac{x-2}{3x+1} = \frac{7x - (x-2)}{3x+1}$$ Write as a single rational expression.

$$= \frac{7x - x + 2}{3x+1}$$ Subtract: $-(x-2) = -x+2$.

$$= \frac{6x+2}{3x+1}$$ Combine terms in numerator.

$$= \frac{2(3x+1)}{3x+1}$$ Factor numerator.

$$= 2$$ Reduce to lowest terms.

(b) $\dfrac{1}{q-1} - \dfrac{1}{q+1}$

$$= \frac{1(q+1)}{(q-1)(q+1)} - \frac{1(q-1)}{(q+1)(q-1)}$$ Get a common denominator.

$$= \frac{(q+1) - (q-1)}{(q-1)(q+1)}$$ Subtract.

$$= \frac{q+1-q+1}{(q-1)(q+1)}$$ Distributive property

$$= \frac{2}{(q-1)(q+1)}$$ Combine terms. ■

WORK PROBLEM 4 AT THE SIDE.

In some problems, rational expressions to be added or subtracted have denominators that are negatives of each other. The next example illustrates how to proceed in such a problem.

EXAMPLE 5 Adding Rational Expressions with Denominators That Are Negatives

Add.

$$\frac{y}{y-2} + \frac{8}{2-y}$$

To get a common denominator of $y - 2$, multiply the second expression by -1 in both the numerator and the denominator.

$$\frac{y}{y-2} + \frac{8}{2-y} = \frac{y}{y-2} + \frac{8(-1)}{(2-y)(-1)}$$

$$= \frac{y}{y-2} + \frac{-8}{y-2}$$

$$= \frac{y-8}{y-2}$$ ■

WORK PROBLEM 5 AT THE SIDE.

The next example illustrates addition and subtraction involving more than two rational expressions.

4. Subtract.

(a) $\dfrac{5x+7}{2x+7} - \dfrac{-x-14}{2x+7}$

(b) $\dfrac{2}{r-2} - \dfrac{r+3}{r-1}$

5. Add or subtract as indicated.

(a) $\dfrac{8}{x-4} + \dfrac{2}{4-x}$

(b) $\dfrac{9}{2x-9} - \dfrac{4}{9-2x}$

ANSWERS
4. (a) 3
   (b) $\dfrac{-r^2 + r + 4}{(r-2)(r-1)}$
5. (a) $\dfrac{6}{x-4}$ or $\dfrac{-6}{4-x}$
   (b) $\dfrac{13}{2x-9}$ or $\dfrac{-13}{9-2x}$

CHAPTER 4 RATIONAL EXPRESSIONS

6. Add and subtract as indicated.

$$\frac{4}{x-5} + \frac{-2}{x} - \frac{10}{x^2 - 5x}$$

7. Add or subtract.

(a) $\dfrac{2}{m-3} - \dfrac{1}{m+3}$

(b) $\dfrac{2x}{x^2 + x - 2} + \dfrac{3x}{x^2 + 2x - 3}$

(c) $\dfrac{-a}{a^2 + 3a - 4} - \dfrac{4a}{a^2 + 7a + 12}$

ANSWERS

6. $\dfrac{2}{x-5}$

7. (a) $\dfrac{m+9}{(m-3)(m+3)}$

(b) $\dfrac{5x^2 + 12x}{(x-1)(x+2)(x+3)}$

(c) $\dfrac{-5a^2 + a}{(a+4)(a-1)(a+3)}$

EXAMPLE 6 Adding and Subtracting Three Rational Expressions

Add and subtract as indicated.

$$\frac{3}{x-2} + \frac{5}{x} - \frac{6}{x^2 - 2x}$$

The denominator of the third rational expression is $x(x - 2)$, which is the least common denominator for the three rational expressions.

$$\frac{3}{x-2} + \frac{5}{x} - \frac{6}{x^2 - 2x}$$

$= \dfrac{3x}{x(x-2)} + \dfrac{5(x-2)}{x(x-2)} - \dfrac{6}{x(x-2)}$ Get a common denominator.

$= \dfrac{3x + 5(x-2) - 6}{x(x-2)}$ Add and subtract numerators.

$= \dfrac{3x + 5x - 10 - 6}{x(x-2)}$ Distributive property

$= \dfrac{8x - 16}{x(x-2)}$ Combine terms in numerator.

$= \dfrac{8(x-2)}{x(x-2)}$ Factor numerator.

$= \dfrac{8}{x}$ Lowest terms ■

WORK PROBLEM 6 AT THE SIDE.

EXAMPLE 7 Subtracting Rational Expressions

Subtract.

$$\frac{m+4}{m^2 - 2m - 3} - \frac{2m-3}{m^2 - 5m + 6}$$

Factor each denominator.

$$\frac{m+4}{m^2 - 2m - 3} - \frac{2m-3}{m^2 - 5m + 6}$$

$= \dfrac{m+4}{(m-3)(m+1)} - \dfrac{2m-3}{(m-3)(m-2)}$

The least common denominator is $(m - 3)(m + 1)(m - 2)$.

$= \dfrac{(m+4)(m-2)}{(m-3)(m+1)(m-2)} - \dfrac{(2m-3)(m+1)}{(m-3)(m-2)(m+1)}$ Get a common denominator.

$= \dfrac{(m+4)(m-2) - (2m-3)(m+1)}{(m-3)(m+1)(m-2)}$ Subtract.

$= \dfrac{m^2 + 2m - 8 - (2m^2 - m - 3)}{(m-3)(m+1)(m-2)}$ Multiply in numerator.

$= \dfrac{m^2 + 2m - 8 - 2m^2 + m + 3}{(m-3)(m+1)(m-2)}$ Distributive property

$= \dfrac{-m^2 + 3m - 5}{(m-3)(m+1)(m-2)}$ Combine terms in numerator. ■

WORK PROBLEM 7 AT THE SIDE.

4.3 EXERCISES

Identify the least common denominator for each group of rational expressions. See Example 2.

1. $\dfrac{3}{8}, \dfrac{5}{2x}$

2. $\dfrac{1}{5}, \dfrac{7}{10m}$

3. $\dfrac{3}{k}, \dfrac{9}{5k^2}$

4. $\dfrac{-1}{r}, \dfrac{5}{9r^2}$

5. $\dfrac{3}{7a}, \dfrac{2}{5a}$

6. $\dfrac{12}{11p}, \dfrac{3}{2p}$

7. $\dfrac{3}{mn^2}, \dfrac{-4}{m^2n}$

8. $\dfrac{-2}{p^2q^3}, \dfrac{1}{p^3q}$

9. $\dfrac{3}{z}, \dfrac{2}{z-1}$

10. $\dfrac{8}{m-3}, \dfrac{1}{m}$

11. $\dfrac{7}{2a+6}, \dfrac{4}{a+3}$

12. $\dfrac{1}{3p-9}, \dfrac{2}{p-3}$

13. $\dfrac{5}{8m+16}, \dfrac{3}{5m+10}$

14. $\dfrac{2}{3r+9}, \dfrac{8}{5r+15}$

15. $\dfrac{2}{a^2-16}, \dfrac{1}{(a-4)^2}$

16. $\dfrac{5}{m^2-25}, \dfrac{1}{(m+5)^2}$

17. $\dfrac{9}{x+y}, \dfrac{2}{x-y}$

18. $\dfrac{4}{p-q}, \dfrac{1}{p+q}$

19. $\dfrac{2}{r}, \dfrac{3}{r-1}, \dfrac{4}{r+1}$

20. $\dfrac{6}{y}, \dfrac{-5}{2+y}, \dfrac{1}{2-y}$

21. $\dfrac{3p+2}{p^2-3p-4}, \dfrac{2}{p^2+p}$

22. $\dfrac{6z+9}{z^2-8z+12}, \dfrac{3}{z^2-6z}$

23. $\dfrac{5m+3}{m^2+3m-10}, \dfrac{2m-1}{2m^2+7m-15}$

24. $\dfrac{8x-1}{3x^2+x-2}, \dfrac{-x+2}{x^2-3x-4}$

Add or subtract as indicated. See Examples 1, 3–7.

25. $\dfrac{2}{p} + \dfrac{5}{p}$

26. $\dfrac{8}{m} + \dfrac{4}{m}$

27. $\dfrac{6}{5k} - \dfrac{2}{5k}$

28. $\dfrac{11}{9y} - \dfrac{7}{9y}$

29. $\dfrac{1}{m+1} + \dfrac{m}{m+1}$

30. $\dfrac{k}{k-2} - \dfrac{2}{k-2}$

31. $\dfrac{6r}{r+2} + \dfrac{12}{r+2}$

32. $\dfrac{5m}{m-2} - \dfrac{10}{m-2}$

33. $\dfrac{a^2}{a+b} - \dfrac{b^2}{a+b}$

34. $\dfrac{y}{y^2-x^2} + \dfrac{x}{y^2-x^2}$

35. $\dfrac{5}{p^2+3p-10} + \dfrac{p}{p^2+3p-10}$

36. $\dfrac{2k}{2k^2+k-1} - \dfrac{1}{2k^2+k-1}$

37. $\dfrac{9}{r} + \dfrac{5}{2r}$

38. $\dfrac{7}{3y} + \dfrac{1}{y}$

39. $\dfrac{4}{3z} + \dfrac{1}{z^2}$

40. $\dfrac{1}{2y} + \dfrac{3}{y}$

41. $\dfrac{3}{4x} + \dfrac{5}{3x}$

42. $\dfrac{3}{7p} - \dfrac{1}{3p}$

43. $\dfrac{5}{9k} - \dfrac{3}{2k}$

44. $\dfrac{4}{3y} + \dfrac{1}{y^2}$

45. $\dfrac{1}{m-1} + \dfrac{1}{m}$

46. $\dfrac{2}{p-3} - \dfrac{1}{p}$

4.3 EXERCISES

47. $\dfrac{5}{t+2} - \dfrac{3}{t-2}$

48. $\dfrac{3}{a-5} + \dfrac{2}{a+5}$

49. $\dfrac{5}{m-4} + \dfrac{2}{4-m}$

50. $\dfrac{1}{y-2} + \dfrac{3}{2-y}$

51. $\dfrac{y}{x-y} - \dfrac{x}{y-x}$

52. $\dfrac{p}{p-q} - \dfrac{q}{q-p}$

53. $\dfrac{7}{3a+9} - \dfrac{5}{4a+12}$

54. $\dfrac{-11}{5r-25} + \dfrac{4}{3r-15}$

55. $\dfrac{3}{x^2-5x+6} - \dfrac{2}{x^2-x-2}$

56. $\dfrac{2}{m^2-4m+4} + \dfrac{3}{m^2+m-6}$

57. $\dfrac{4x}{x-1} - \dfrac{2}{x+1} - \dfrac{4}{x^2-1}$

58. $\dfrac{-x}{x-3} + \dfrac{4}{x+3} - \dfrac{18}{x^2-9}$

59. $\dfrac{15}{y^2+3y} + \dfrac{5}{y+3} + \dfrac{2}{y}$

60. $\dfrac{7}{p-2} - \dfrac{6}{p^2-2p} - \dfrac{3}{p}$

61. $\dfrac{2w+3}{w-2} + \dfrac{7}{w} + \dfrac{14}{w^2-2w}$

62. $\dfrac{4y+2}{y+1} - \dfrac{8}{y} - \dfrac{2}{y^2+y}$

63. $\dfrac{5r}{r^2-r-2} - \dfrac{2}{r^2-2r-3}$

64. $\dfrac{7t}{4t^2-7t-2} - \dfrac{3}{t^2-4}$

65. $\dfrac{5x}{x^2+xy-2y^2} - \dfrac{3x}{x^2+5xy-6y^2}$

66. $\dfrac{6x}{6x^2+5xy-4y^2} - \dfrac{2y}{9x^2-16y^2}$

67. $\dfrac{3y}{y^2+yz-2z^2} + \dfrac{4y-1}{y^2-z^2}$

68. $\dfrac{r+s}{3r^2+2rs-s^2} - \dfrac{s-r}{6r^2-5rs+s^2}$

SKILL SHARPENERS
Simplify. See Section 1.4.

69. $\dfrac{\frac{2}{3}+\frac{1}{6}}{\frac{5}{9}-\frac{1}{3}}$

70. $\dfrac{\frac{1}{4}+\frac{3}{8}}{\frac{3}{2}-\frac{7}{8}}$

71. $\dfrac{3+\frac{5}{4}}{2-\frac{1}{4}}$

72. $\dfrac{2-\frac{4}{3}}{\frac{3}{8}-1}$

4.4 COMPLEX FRACTIONS

A **complex fraction** is a fraction that has a fraction in the numerator, the denominator, or both. Examples of complex fractions include

$$\frac{1 + \frac{1}{x}}{2}, \quad \frac{\frac{4}{y}}{6 - \frac{3}{y}}, \quad \text{and} \quad \frac{\frac{m^2 - 9}{m + 1}}{\frac{m + 3}{m^2 - 1}}.$$

1 There are two different methods for simplifying complex fractions.

SIMPLIFYING COMPLEX FRACTIONS

Method 1 Simplify the numerator and denominator separately, as much as possible. Then multiply the numerator by the reciprocal of the denominator.

EXAMPLE 1 Simplifying Complex Fractions Using Method 1
Use Method 1 to simplify each complex fraction.

(a) $\dfrac{\frac{x + 1}{x}}{\frac{x - 1}{2x}}$ Both the numerator and denominator are already simplified, so multiply the numerator by the reciprocal of the denominator.

$$\frac{\frac{x + 1}{x}}{\frac{x - 1}{2x}} = \frac{x + 1}{x} \div \frac{x - 1}{2x}$$

$$= \frac{x + 1}{x} \cdot \frac{2x}{x - 1} \qquad \text{Reciprocal of } \frac{x - 1}{2x}$$

$$= \frac{2(x + 1)}{x - 1}$$

(b) $\dfrac{2 + \frac{1}{y}}{3 - \frac{2}{y}} = \dfrac{\frac{2y}{y} + \frac{1}{y}}{\frac{3y}{y} - \frac{2}{y}} = \dfrac{\frac{2y + 1}{y}}{\frac{3y - 2}{y}}$ Simplify numerator and denominator.

$$= \frac{2y + 1}{y} \cdot \frac{y}{3y - 2} \qquad \text{Reciprocal of } \frac{3y - 2}{y}$$

$$= \frac{2y + 1}{3y - 2} \quad \blacksquare$$

WORK PROBLEM 1 AT THE SIDE.

2 Now we discuss the second method for simplifying complex fractions.

SIMPLIFYING COMPLEX FRACTIONS

Method 2 Multiply the numerator and denominator of the complex fraction by the least common denominator of all fractions appearing in the numerator or denominator of the complex fraction.

OBJECTIVES

1 Simplify complex fractions by simplifying numerator and denominator.

2 Simplify complex fractions by multiplying by a common denominator.

3 Simplify rational expressions written with negative exponents.

1. Use Method 1 to simplify each complex fraction.

(a) $\dfrac{\frac{a + 2}{5a}}{\frac{a - 3}{7a}}$

(b) $\dfrac{\frac{6y - 12}{y}}{\frac{4y - 8}{y}}$

(c) $\dfrac{2 + \frac{1}{k}}{2 - \frac{1}{k}}$

(d) $\dfrac{\frac{r^2 - 4}{4}}{1 + \frac{2}{r}}$

ANSWERS
1. (a) $\dfrac{7(a + 2)}{5(a - 3)}$ (b) $\dfrac{3}{2}$
   (c) $\dfrac{2k + 1}{2k - 1}$ (d) $\dfrac{r(r - 2)}{4}$

4.4 COMPLEX FRACTIONS   **243**

CHAPTER 4 RATIONAL EXPRESSIONS

2. Use Method 2 to simplify each complex fraction.

(a) $\dfrac{\dfrac{5}{y} + 6}{\dfrac{8}{3y} - 1}$

(b) $\dfrac{\dfrac{2p - 5}{8p}}{\dfrac{7p + 1}{4}}$

(c) $\dfrac{\dfrac{1}{y} + \dfrac{1}{y - 1}}{\dfrac{1}{y} - \dfrac{2}{y - 1}}$

3. Simplify each of the following.

(a) $a^{-1} + b^{-1}$

(b) $\dfrac{k^{-1}}{k^{-1} + 1}$

ANSWERS

2. (a) $\dfrac{15 + 18y}{8 - 3y}$ (b) $\dfrac{2p - 5}{2p(7p + 1)}$

(c) $\dfrac{2y - 1}{-y - 1}$ or $\dfrac{1 - 2y}{y + 1}$

3. (a) $\dfrac{b + a}{ab}$ (b) $\dfrac{1}{1 + k}$

EXAMPLE 2 Simplifying a Complex Fraction Using Method 2

Use Method 2 to simplify the following complex fraction.

$$\dfrac{2 + \dfrac{1}{y}}{3 - \dfrac{2}{y}}$$

For Method 2, multiply the numerator and denominator by the least common denominator of all the fractions appearing in either part of the complex fraction. Here the least common denominator is y.

$$\dfrac{2 + \dfrac{1}{y}}{3 - \dfrac{2}{y}} = \dfrac{\left(2 + \dfrac{1}{y}\right) \cdot y}{\left(3 - \dfrac{2}{y}\right) \cdot y} \quad \text{Multiply by } y.$$

$$= \dfrac{2 \cdot y + \dfrac{1}{y} \cdot y}{3 \cdot y - \dfrac{2}{y} \cdot y} \quad \text{Use the distributive property.}$$

$$= \dfrac{2y + 1}{3y - 2}$$

Compare this method of solution with that used for this same complex fraction in Example 1(b). ∎

■ **WORK PROBLEM 2 AT THE SIDE.**

3 Rational expressions and complex fractions often involve negative exponents, as in the following example.

EXAMPLE 3 Simplifying a Rational Expression with Negative Exponents

Simplify the following rational expression.

$$\dfrac{m^{-1} + p^{-2}}{m^{-2} - p^{-1}}$$

Begin by using laws of exponents to write the expression without negative exponents.

$$\dfrac{m^{-1} + p^{-2}}{m^{-2} - p^{-1}} = \dfrac{\dfrac{1}{m} + \dfrac{1}{p^2}}{\dfrac{1}{m^2} - \dfrac{1}{p}}$$

We use Method 2 to simplify this complex fraction, multiplying each term in the numerator and denominator by the least common denominator of all terms in the complex fraction, $m^2 p^2$.

$$\dfrac{\dfrac{1}{m} + \dfrac{1}{p^2}}{\dfrac{1}{m^2} - \dfrac{1}{p}} = \dfrac{m^2 p^2 \cdot \dfrac{1}{m} + m^2 p^2 \cdot \dfrac{1}{p^2}}{m^2 p^2 \cdot \dfrac{1}{m^2} - m^2 p^2 \cdot \dfrac{1}{p}} = \dfrac{mp^2 + m^2}{p^2 - m^2 p} \quad ∎$$

■ **WORK PROBLEM 3 AT THE SIDE.**

4.4 EXERCISES

Use either method to simplify. See Examples 1–3.

1. $\dfrac{\frac{3}{x}}{\frac{6}{x-1}}$

2. $\dfrac{\frac{2}{m}}{\frac{8}{m+4}}$

3. $\dfrac{\frac{4}{p+5}}{\frac{3}{2p}}$

4. $\dfrac{\frac{k+1}{2}}{\frac{3k-1}{4}}$

5. $\dfrac{\frac{m-1}{4m}}{\frac{m+1}{m}}$

6. $\dfrac{\frac{r-s}{3}}{\frac{r+s}{6}}$

7. $\dfrac{\frac{4z^2 x}{9}}{\frac{12xz^3}{5}}$

8. $\dfrac{\frac{9y^3 x^4}{16}}{\frac{3y^2 x^3}{8}}$

9. $\dfrac{\frac{x-3y}{5x}}{\frac{8x-24y}{10}}$

10. $\dfrac{\frac{p+2q}{q^2}}{\frac{p^2-4q^2}{2q}}$

11. $\dfrac{\frac{25a^2-b^2}{4a}}{\frac{5a+b}{7a}}$

12. $\dfrac{1+\frac{1}{y}}{1-\frac{1}{y}}$

13. $\dfrac{\frac{2}{m}-1}{1+\frac{2}{m}}$

14. $\dfrac{\frac{1}{k}+\frac{1}{r}}{\frac{1}{k}-\frac{1}{r}}$

15. $\dfrac{\dfrac{2}{q} - \dfrac{3}{r}}{2 + \dfrac{1}{qr}}$

16. $\dfrac{\dfrac{x}{y} - \dfrac{y}{x}}{\dfrac{x}{y} + 1}$

17. $\dfrac{\dfrac{a}{b} + 2}{\dfrac{a}{b} - \dfrac{4b}{a}}$

18. $\dfrac{\dfrac{1}{r} - \dfrac{4}{s}}{\dfrac{s^2 - 16r^2}{rs}}$

19. $\dfrac{\dfrac{2}{x} - \dfrac{3}{y}}{\dfrac{4y^2 - 9x^2}{2x}}$

20. $x^{-2} - y^{-2}$

21. $a^{-2} + b^{-2}$

22. $\dfrac{x^{-2} - y^{-2}}{x^{-1} - y^{-1}}$

23. $\dfrac{x^{-1} + y^{-1}}{x^{-2} - y^{-2}}$

24. $(a^{-1} + b^{-1})^{-1}$

25. $(p^{-1} - q^{-1})^{-1}$

26. $(2k)^{-1} + 3m^{-1}$

SKILL SHARPENERS
Solve each equation. See Sections 2.1 and 3.10.

27. $4 - 3(2 + p) = 8p - 35$

28. $1 - 5(r + 3) = 7r - 2$

29. $6r^2 + r = 2$

30. $10z^2 + 17z = -3$

4.5 EQUATIONS WITH RATIONAL EXPRESSIONS

OBJECTIVES

1. Solve equations with rational expressions.
2. Know when potential solutions must be checked.

1 The easiest way to solve most equations involving rational expressions is to multiply all the terms in the equation by the least common denominator, as the next examples show.

EXAMPLE 1 Solving an Equation with Rational Expressions

Solve $\dfrac{2x}{5} - \dfrac{x}{3} = 2$.

The least common denominator for $\dfrac{2x}{5}$ and $\dfrac{x}{3}$ is 15, so multiply both sides of the equation by 15.

$$15\left(\dfrac{2x}{5} - \dfrac{x}{3}\right) = 15(2)$$

$$15\left(\dfrac{2x}{5}\right) - 15\left(\dfrac{x}{3}\right) = 15(2) \quad \text{Distributive property}$$

$$6x - 5x = 30 \quad \text{Simplify.}$$

$$x = 30 \quad \text{Combine terms.}$$

Substitute 30 for x in the original equation to check this solution.

$$\dfrac{2 \cdot 30}{5} - \dfrac{30}{3} = 2 \quad \text{Let } x = 30.$$

$$\dfrac{60}{5} - 10 = 2$$

$$12 - 10 = 2$$

$$2 = 2 \quad \text{True}$$

This check shows that the solution set is $\{30\}$. ■

WORK PROBLEM 1 AT THE SIDE.

1. Solve.

(a) $\dfrac{r}{5} + \dfrac{r}{2} = 7$

(b) $\dfrac{2k}{3} - \dfrac{k}{6} = -3$

(c) $\dfrac{3p}{2} - \dfrac{p}{4} = 1$

(d) $\dfrac{7k}{12} - \dfrac{5k}{8} = \dfrac{1}{4}$

EXAMPLE 2 Solving an Equation with Rational Expressions

Solve $\dfrac{2}{y} - \dfrac{3}{2} = \dfrac{7}{2y}$.

Multiply both sides by the least common denominator, $2y$.

$$2y\left(\dfrac{2}{y} - \dfrac{3}{2}\right) = 2y\left(\dfrac{7}{2y}\right)$$

$$2y\left(\dfrac{2}{y}\right) - 2y\left(\dfrac{3}{2}\right) = 2y\left(\dfrac{7}{2y}\right) \quad \text{Distributive property}$$

$$4 - 3y = 7$$

$$-3y = 3 \quad \text{Subtract 4.}$$

$$y = -1 \quad \text{Divide by } -3.$$

To see if -1 is a solution for the equation, replace y with -1 in the original equation:

$$\dfrac{2}{-1} - \dfrac{3}{2} = \dfrac{7}{2(-1)} \quad \text{Let } y = -1$$

$$-\dfrac{4}{2} - \dfrac{3}{2} = -\dfrac{7}{2}. \quad \text{True}$$

The solution set is $\{-1\}$. ■

WORK PROBLEM 2 AT THE SIDE.

2. Solve.

(a) $\dfrac{3}{p} - \dfrac{7}{10} = \dfrac{8}{5p}$

(b) $\dfrac{-3}{20} + \dfrac{2}{m} = \dfrac{5}{4m}$

(c) $\dfrac{3x}{7} + \dfrac{1}{2} = \dfrac{5x}{14}$

ANSWERS

1. (a) $\{10\}$ (b) $\{-6\}$ (c) $\left\{\dfrac{4}{5}\right\}$ (d) $\{-6\}$

2. (a) $\{2\}$ (b) $\{5\}$ (c) $\{-7\}$

CHAPTER 4 RATIONAL EXPRESSIONS

3. Solve.

(a) $\dfrac{y^2}{y-3} = \dfrac{9}{y-3}$

(b) $\dfrac{3}{a+1} = \dfrac{1}{a-1} - \dfrac{2}{a^2-1}$

4. Solve.

(a) $\dfrac{10}{m^2} - \dfrac{3}{m} = 1$

(b) $2 = \dfrac{3}{y} + \dfrac{2}{y^2}$

(c) $\dfrac{x}{x-3} + \dfrac{2}{x+3} = \dfrac{-12}{x^2-9}$

ANSWERS
3. (a) $\{-3\}$ (b) $\emptyset$
4. (a) $\{2, -5\}$ (b) $\left\{-\dfrac{1}{2}, 2\right\}$ (c) $\{-2\}$

---

2 To solve the equation in Example 2, we multiplied both sides by $2y$. Be careful, though, when multiplying by a variable expression.

> **NOTE** When both sides of an equation are multiplied by an expression containing a variable, it is possible that the resulting "solutions" are not actually solutions of the given equation.

EXAMPLE 3 Solving an Equation with Rational Expressions

Solve $\dfrac{2}{m-3} - \dfrac{3}{m+3} = \dfrac{12}{m^2-9}$.

Multiply both sides by the least common denominator, which is $(m+3)(m-3)$.

$$(m+3)(m-3) \cdot \dfrac{2}{m-3} - (m+3)(m-3) \cdot \dfrac{3}{m+3}$$
$$= (m+3)(m-3) \cdot \dfrac{12}{m^2-9}$$

$$2(m+3) - 3(m-3) = 12$$
$$2m + 6 - 3m + 9 = 12 \quad \text{Distributive property}$$
$$-m + 15 = 12 \quad \text{Combine terms.}$$
$$-m = -3$$
$$m = 3$$

Since both sides were multiplied by a term containing a variable, we must check the potential solution.

$$\dfrac{2}{3-3} - \dfrac{3}{3+3} = \dfrac{12}{3^2-9} \quad \text{Let } m = 3.$$

$$\dfrac{2}{0} - \dfrac{3}{6} = \dfrac{12}{0}$$

Division by 0 is not possible; the given equation has no solution and the solution set is $\emptyset$. ∎

WORK PROBLEM 3 AT THE SIDE.

EXAMPLE 4 Solving an Equation That Leads to a Quadratic Equation

Solve $\dfrac{2}{3x+1} = \dfrac{1}{x} - \dfrac{6x}{3x+1}$.

Multiply both sides by $x(3x+1)$. The resulting equation is

$$2x = (3x+1) - 6x^2.$$

Now solve. Since the equation is quadratic, get 0 on the right side.

$$6x^2 - 3x + 2x - 1 = 0$$
$$6x^2 - x - 1 = 0$$
$$(3x+1)(2x-1) = 0$$
$$3x + 1 = 0 \quad \text{or} \quad 2x - 1 = 0 \quad \text{Zero-factor property}$$
$$x = -\dfrac{1}{3} \quad \text{or} \quad x = \dfrac{1}{2}$$

Using $-\dfrac{1}{3}$ in the original equation causes the denominator $3x + 1$ to equal 0, so it is not a solution. The solution set is $\left\{\dfrac{1}{2}\right\}$. ∎

WORK PROBLEM 4 AT THE SIDE.

4.5 EXERCISES

Solve each of the following equations. Be sure to check the potential solutions when multiplying both sides of the equation by a variable expression. See Examples 1–4.

1. $\dfrac{p}{4} - \dfrac{p}{8} = 2$

2. $\dfrac{m}{6} + \dfrac{m}{3} = 6$

3. $\dfrac{r+6}{3} = \dfrac{r+8}{5}$

4. $\dfrac{2k+1}{5} = \dfrac{k+1}{4}$

5. $\dfrac{3y-6}{2} = \dfrac{5y+1}{6}$

6. $\dfrac{8-p}{3} = \dfrac{6+5p}{2}$

7. $\dfrac{1}{x} = \dfrac{3}{x} - 2$

8. $\dfrac{-3}{y} = 2 + \dfrac{1}{y}$

9. $\dfrac{1}{a} + \dfrac{2}{3a} = \dfrac{1}{3}$

10. $\dfrac{2}{m} + \dfrac{3}{2m} = \dfrac{7}{6}$

11. $\dfrac{3}{4p} - \dfrac{2}{p} = \dfrac{5}{12}$

12. $\dfrac{5}{2y} - \dfrac{2}{y} = -\dfrac{1}{12}$

13. $\dfrac{1}{r-1} + \dfrac{2}{3r-3} = -\dfrac{5}{12}$

14. $\dfrac{4}{z+2} - \dfrac{1}{3z+6} = \dfrac{11}{9}$

CHAPTER 4   RATIONAL EXPRESSIONS

15. $\dfrac{3}{4a-8} - \dfrac{2}{3a-6} = \dfrac{1}{36}$

16. $\dfrac{1}{x-2} - \dfrac{1}{6} = \dfrac{2}{3x-6}$

17. $\dfrac{5}{6q+14} - \dfrac{2}{3q+7} = \dfrac{1}{56}$

18. $\dfrac{7}{2y+1} + \dfrac{3}{4y+2} = \dfrac{17}{2}$

19. $\dfrac{2m}{3m+3} - \dfrac{m+2}{6m+6} - \dfrac{m-6}{8m+8} = \dfrac{5}{12}$

20. $\dfrac{3p}{2p+10} + \dfrac{p}{3p+15} - \dfrac{p+3}{4p+20} = -\dfrac{7}{12}$

21. $\dfrac{2}{x-1} + \dfrac{3}{x+1} = \dfrac{9}{x^2-1}$

22. $\dfrac{h-1}{h^2-4} - \dfrac{2}{h+2} = \dfrac{4}{h-2}$

23. $\dfrac{2}{y-5} + \dfrac{1}{y+5} = \dfrac{11}{y^2-25}$

24. $\dfrac{3}{m+2} - \dfrac{1}{m-2} = \dfrac{2}{m^2-4}$

25. $y + 2 = \dfrac{24}{y}$

26. $m + 8 + \dfrac{15}{m} = 0$

4.5 EXERCISES

27. $\dfrac{x}{x-4} + \dfrac{2}{x} = \dfrac{16}{x^2 - 4x}$

28. $\dfrac{x+7}{x-2} - \dfrac{3}{x} = \dfrac{6}{x^2 - 2x}$

29. $\dfrac{a+3}{a} - \dfrac{a+4}{a+5} = \dfrac{15}{a^2 + 5a}$

30. $\dfrac{4t}{9t+18} + \dfrac{t+2}{3t+6} = \dfrac{7}{9}$

31. $\dfrac{a+12}{a^2-16} - \dfrac{3}{a-4} = \dfrac{1}{a+4}$

32. $\dfrac{g-8}{g-3} - \dfrac{g+8}{g+3} = \dfrac{g}{g^2 - 9}$

33. $\dfrac{5}{x-4} - \dfrac{3}{x-1} = \dfrac{x+11}{x^2 - 5x + 4}$

34. $\dfrac{7}{z-5} - \dfrac{6}{z+3} = \dfrac{48}{z^2 - 2z - 15}$

35. $\dfrac{7}{p-5} = \dfrac{p^2 - 10}{p^2 - p - 20} - 1$

36. $\dfrac{22}{2a^2 - 9a - 5} - \dfrac{3}{2a+1} = \dfrac{2}{a-5}$

CHAPTER 4 RATIONAL EXPRESSIONS

37. $\dfrac{1}{x+2} - \dfrac{5}{x^2+9x+14} = \dfrac{-3}{x+7}$

38. $\dfrac{4}{q+5} + \dfrac{1}{q+3} = \dfrac{2}{q^2+8q+15}$

39. $\dfrac{2y-29}{y^2+7y-8} = \dfrac{5}{y+8} - \dfrac{2}{y-1}$

40. $\dfrac{3m-1}{m^2+5m-14} = \dfrac{1}{m-2} - \dfrac{2}{m+7}$

41. $\dfrac{-4x}{6x-3} + \dfrac{2}{6x-3} = \dfrac{9}{x}$

42. $\dfrac{2q}{6-3q} - \dfrac{4}{6-3q} = \dfrac{5}{q}$

SKILL SHARPENERS
Solve the formula for the specified variable. See Section 2.2.

43. $d = rt$ for r

44. $I = prt$ for t

45. $A = \dfrac{1}{2}bh$ for h

46. $P = 2L + 2W$ for W

SUMMARY OF OPERATIONS INVOLVING RATIONAL EXPRESSIONS

A common student error is to confuse an *equation*, such as $\frac{x}{2} + \frac{x}{3} = -5$, with an addition problem, such as $\frac{x}{2} + \frac{x}{3}$. Equations are solved to get a numerical answer, while addition problems result in simplified expressions, as shown below.

Solve: $\frac{x}{2} + \frac{x}{3} = -5$.

Multiply both sides by 6.

$$6\left(\frac{x}{2} + \frac{x}{3}\right) = 6(-5)$$
$$3x + 2x = -30$$
$$5x = -30$$
$$x = -6$$

Check that the solution is -6.

Add: $\frac{x}{2} + \frac{x}{3}$.

The least common denominator is 6.

$$\frac{x}{2} + \frac{x}{3} = \frac{x \cdot 3}{2 \cdot 3} + \frac{x \cdot 2}{3 \cdot 2}$$
$$= \frac{3x}{6} + \frac{2x}{6}$$
$$= \frac{3x + 2x}{6}$$
$$= \frac{5x}{6}$$

In the following exercises, either perform the indicated operation or solve the given equation, as appropriate.

1. $\dfrac{a}{2} - \dfrac{a}{5} = 3$

2. $\dfrac{12z^2 x^3}{8zx^4} \cdot \dfrac{3x^5}{7x}$

3. $\dfrac{-2y}{(y-2)^2} \cdot \dfrac{y^2 - 4}{5}$

4. $\dfrac{5}{m} - \dfrac{10}{3m}$

5. $\dfrac{\frac{1}{x} - \frac{1}{y}}{\frac{1}{x} + \frac{1}{y}}$

6. $\dfrac{3}{z} + \dfrac{5}{7z} = \dfrac{52}{7}$

7. $\dfrac{r-5}{3} - \dfrac{r-2}{5} = -\dfrac{1}{3}$

8. $\dfrac{8}{5r} + \dfrac{1}{3r}$

9. $\dfrac{1}{x} - \dfrac{1}{2x} = 1$

10. $\dfrac{\frac{5}{m} - \frac{2}{m+1}}{\frac{3}{m+1} + \frac{1}{m}}$

11. $\dfrac{2}{q+2} - \dfrac{3}{4q+8}$

12. $\dfrac{1}{r-2} + \dfrac{1}{r+1}$

13. $\dfrac{3p(p-2)}{p+5} \div \dfrac{p^2 - 4}{4p + 20}$

14. $\dfrac{9}{2-x} \cdot \dfrac{4x-8}{5}$

15. $\dfrac{m-4}{3} - \dfrac{m+1}{2} = -\dfrac{11}{6}$

16. $\dfrac{a^2 + a - 2}{a^2 - 4} \cdot \dfrac{a^2 - 2a}{a^2 + 3a - 4}$

17. $\dfrac{2q^2 + 5q - 3}{q^2 + q - 6} \div \dfrac{10q^2 - 5q}{3q^3 - 6q^2}$

18. $\dfrac{5}{a^2 - 2a} + \dfrac{3}{a^2 - 4}$

19. $\dfrac{3}{k + 1} + \dfrac{2}{5k + 5} = \dfrac{17}{15}$

20. $\dfrac{r^2 - 1}{r^2 + 3r + 2} \div \dfrac{r^2 - 2r + 1}{r^2 + 2r - 3}$

21. $\dfrac{\dfrac{3}{k} - \dfrac{5}{m}}{\dfrac{9m^2 - 25k^2}{km^2}}$

22. $\dfrac{4y - 20}{y^2 - 3y - 10} \cdot \dfrac{2y^2 - 3y - 14}{16y^3 + 8y^2}$

23. $\dfrac{p^2 + 5p + 6}{p^2 - 3p + 2} \cdot \dfrac{p^2 + 3p - 10}{p^2 + 8p + 15}$

24. $\dfrac{4}{r^2 - 2r - 8} - \dfrac{1}{2r^2 + 5r + 2}$

25. $\dfrac{8}{3p + 9} - \dfrac{2}{5p + 15} = \dfrac{8}{15}$

26. $\dfrac{4y^2 + 11y - 3}{6y^2 - 5y - 6} \div \dfrac{4y^2 - 13y + 3}{2y^2 - 9y + 9}$

27. $\dfrac{2}{m^2 + m - 2} + \dfrac{3}{m^2 + 5m + 6}$

28. $\dfrac{\dfrac{2}{r} + \dfrac{5}{s}}{\dfrac{4s^2 - 25r^2}{3rs}}$

29. $\dfrac{2}{q + 1} - \dfrac{3}{q^2 - q - 2} = \dfrac{3}{q - 2}$

30. $\dfrac{3}{r + 1} + 5 = \dfrac{-3r}{3r + 3}$

31. $\dfrac{r}{r^2 + 3rp + 2p^2} - \dfrac{1}{r^2 + 5rp + 6p^2}$

32. $\dfrac{7}{2p^2 - 8p} + \dfrac{3}{p^2 - 16}$

33. $\dfrac{2x^2 - 11xy + 15y^2}{2x^2 + 5xy - 3y^2} \cdot \dfrac{8x^2 - 2xy - y^2}{8x^2 - 18xy - 5y^2}$

34. $\dfrac{-2}{a^2 + 2a - 3} + \dfrac{5}{3a - 3} = \dfrac{4}{3a + 9}$

4.6 FORMULAS WITH RATIONAL EXPRESSIONS

Many common formulas involve rational expressions. Methods of working with these formulas are shown in this section.

1 The first example shows how to find the value of an unknown variable in a formula.

EXAMPLE 1 Finding a Value Using a Formula

In physics, the focal length, f, of a lens is given by the formula

$$\frac{1}{f} = \frac{1}{p} + \frac{1}{q}$$

where p is the distance from the object to the lens and q is the distance from the lens to the image. Find q if $p = 20$ centimeters and $f = 10$ centimeters.

Replace f with 10 and p with 20.

$$\frac{1}{f} = \frac{1}{p} + \frac{1}{q}$$

$$\frac{1}{10} = \frac{1}{20} + \frac{1}{q} \qquad \text{Let } f = 10 \text{ and } p = 20.$$

Multiply both sides by the least common denominator, $20q$.

$$20q \cdot \frac{1}{10} = 20q \cdot \frac{1}{20} + 20q \cdot \frac{1}{q}$$

$$2q = q + 20$$

$$q = 20$$

The distance from the lens to the image is 20 centimeters. ∎

WORK PROBLEM 1 AT THE SIDE.

2 The next example shows how to solve a formula for a particular variable.

EXAMPLE 2 Solving a Formula for a Particular Variable

Solve $\frac{1}{f} = \frac{1}{p} + \frac{1}{q}$ for p.

To solve the formula for p, begin by multiplying both sides by the least common denominator, fpq. Get p alone on one side of the equals sign by subtracting fp on both sides, so that both terms with p are on one side.

$$fpq \cdot \frac{1}{f} = fpq\left(\frac{1}{p} + \frac{1}{q}\right)$$

$$pq = fq + fp \qquad \text{Distributive property}$$

$$pq - fp = fq \qquad \text{Subtract } fp.$$

$$p(q - f) = fq \qquad \text{Distributive property}$$

$$p = \frac{fq}{q - f} \qquad \text{Divide by } q - f.$$

The last step requires that $q \neq f$. ∎

WORK PROBLEM 2 AT THE SIDE.

OBJECTIVES

1 Find the value of an unknown variable.

2 Solve a formula for a given variable.

---

1. Use the formula given in Example 1 to answer each part.

(a) Find p if $f = 15$ and $q = 25$.

(b) Find f if $p = 6$ and $q = 9$.

(c) Find q if $f = 12$ and $p = 16$.

2. (a) Solve $\frac{1}{f} = \frac{1}{p} + \frac{1}{q}$ for q.

(b) Solve $\frac{3}{p} + \frac{3}{q} = \frac{5}{r}$ for q.

(c) Solve $\frac{8}{7x} - \frac{9}{5y} = \frac{2}{z}$ for y.

ANSWERS

1. (a) $\frac{75}{2}$ (b) $\frac{18}{5}$ (c) 48

2. (a) $q = \frac{fp}{p - f}$ (b) $q = \frac{3rp}{5p - 3r}$

   (c) $y = \frac{63xz}{40z - 70x}$

3. **(a)** Solve for R.

$$A = \frac{Rr}{R + r}$$

(b) Solve for r.

$$I = \frac{nE}{R + nr}$$

ANSWERS

3. (a) $R = \dfrac{-Ar}{A - r}$ or $R = \dfrac{Ar}{r - A}$

   (b) $R = \dfrac{nE - IR}{In}$

EXAMPLE 3 Solving a Formula for a Particular Variable

Solve $I = \dfrac{nE}{R + nr}$ for n.

First, multiply both sides by $R + nr$.

$$(R + nr)I = (R + nr)\frac{nE}{R + nr}$$

$$RI + nrI = nE$$

Get the terms with n (the specified variable) together on one side of the equals sign. To do this, subtract nrI from both sides.

$$RI = nE - nrI$$

$$RI = n(E - rI) \qquad \text{Distributive property}$$

Finally, divide both sides by $E - rI$.

$$\frac{RI}{E - rI} = n \quad \blacksquare$$

WORK PROBLEM 3 AT THE SIDE.

CAUTION Refer to the steps in Examples 2 and 3 that use the distributive property. This is a step that often gives students difficulty. Remember that the variable for which you are solving *must* be a factor on one side of the equation so that in the last step, both sides are divided by the remaining factor there. The *distributive property* allows us to perform this factorization.

4.6 EXERCISES

Solve. See Example 1.

1. Ohm's law in electricity says that
$$I = \frac{E}{R}.$$
Find R if $I = 20$ and $E = 8$.

2. A law from physics says that
$$m = \frac{F}{a}.$$
Find F if $m = 30$ and $a = 9$.

3. A formula from anthropology says that
$$c = \frac{100b}{L}.$$
Find L if $c = 80$ and $b = 5$.

4. The gravitational force between two masses is given by
$$F = \frac{GMm}{d^2}.$$
Find M if $F = 10$, $G = 6.67 \times 10^{-11}$, $m = 1$, and $d = 3 \times 10^{-6}$.

5. In work with electric circuits, the formula
$$\frac{1}{a} = \frac{1}{b} + \frac{1}{c}$$
occurs. Find b if $a = 8$ and $c = 12$.

6. A gas law in chemistry says that
$$\frac{PV}{T} = \frac{pv}{t}.$$
Suppose that $T = 300$, $t = 350$, $V = 9$, $P = 50$, and $v = 8$. Find p.

Solve each formula for the specified variable. See Examples 2 and 3.

7. $F = \dfrac{GMm}{d^2}$ for M (physics)

8. $F = \dfrac{GMm}{d^2}$ for G (physics)

9. $\dfrac{1}{a} = \dfrac{1}{b} + \dfrac{1}{c}$ for b (electricity)

10. $\dfrac{1}{a} = \dfrac{1}{b} + \dfrac{1}{c}$ for a (electricity)

11. $\dfrac{PV}{T} = \dfrac{pv}{t}$ for T (chemistry)

12. $\dfrac{PV}{T} = \dfrac{pv}{t}$ for v (chemistry)

13. $a = \dfrac{V - v}{t}$ for V (physics)

14. $A = P + Prt$ for P (finance)

15. $S = \dfrac{n}{2}(a + \ell)d$ for n (mathematics)

16. $A = \dfrac{1}{2}h(B + b)$ for b (mathematics)

17. $t = a + (n - 1)d$ for n (mathematics)

18. $S = \dfrac{n}{2}[2a + (n - 1)d]$ for d (mathematics)

19. $\dfrac{E}{e} = \dfrac{R + r}{r}$ for r (engineering)

20. $\dfrac{E}{e} = \dfrac{R + r}{r}$ for e (engineering)

SKILL SHARPENERS

The skill sharpeners for this section are a combination of the work with word problems from Section 2.3 and the work on solving equations from Section 4.5. Write an equation for each of the following, and then solve to find the unknown number.

21. The reciprocal of a number is added to 3, giving 4.

22. When 2 is subtracted from the reciprocal of a number, the result is 1.

23. The reciprocal of 2, plus the reciprocal of 3, gives the reciprocal of what number?

24. When the reciprocals of both 5 and 4 are added together, the result is the reciprocal of what number?

4.7 APPLICATIONS

In this section, we will look at several applications that produce equations with rational expressions.

1 The first examples show word problems about numbers. These problems are included mainly to give practice in setting up problems.

EXAMPLE 1 Solving a Problem About Numbers

The numerator of $\frac{6}{7}$ is multiplied by a number, and the same number is added to the denominator. What number should be used to make the resulting fraction equal to $\frac{9}{5}$?

Let x = the unknown number.

From the statement of the problem, the numerator of $\frac{6}{7}$, which is 6, should be multiplied by x. Also, x is to be added to the denominator, 7. The result should equal $\frac{9}{5}$. This gives the equation

$$\frac{6x}{7+x} = \frac{9}{5}.$$

The least common denominator is $5(7 + x)$. Multiply both sides of the equation by $5(7 + x)$.

$$5(7+x) \cdot \frac{6x}{7+x} = 5(7+x) \cdot \frac{9}{5}$$

$$30x = 9(7+x)$$
$$30x = 63 + 9x \quad \text{Distributive property}$$
$$21x = 63 \quad \text{Subtract } 9x.$$
$$x = 3 \quad \text{Divide by 21.}$$

The required number is 3. Check this result in the words of the original problem. ■

WORK PROBLEM 1 AT THE SIDE.

For the next example, recall that the reciprocal of a nonzero number x is $\frac{1}{x}$.

EXAMPLE 2 Solving a Problem About Reciprocals

The difference between the reciprocal of a number and the reciprocal of twice the number is $\frac{1}{7}$. Find the number.

Let x = the number;

$2x$ = twice the number.

The reciprocals of the two numbers are $\frac{1}{x}$ and $\frac{1}{2x}$. According to the problem, the difference between these is $\frac{1}{7}$, so the equation is

$$\frac{1}{x} - \frac{1}{2x} = \frac{1}{7}.$$

OBJECTIVES

1 Solve word problems about numbers.

2 Solve word problems about work.

3 Solve word problems about distance, rate, and time.

1. **(a)** What number added to the numerator and subtracted from the denominator of $\frac{1}{9}$ makes the resulting fraction equal to $\frac{3}{2}$?

(b) The numerator of $\frac{3}{5}$ is multiplied by a number. The same number is subtracted from the denominator. The resulting fraction equals 2. Find the number.

ANSWERS
1. (a) 5   (b) 2

CHAPTER 4 RATIONAL EXPRESSIONS

2. The sum of the reciprocal of a number and the reciprocal of three times the number is $\frac{4}{9}$. Find the number.

Solve this equation by multiplying both sides by the least common denominator, $14x$.

$$14x\left(\frac{1}{x}\right) - 14x\left(\frac{1}{2x}\right) = 14x\left(\frac{1}{7}\right)$$

$$14 - 7 = 2x$$

$$7 = 2x \qquad \text{Combine terms.}$$

$$x = \frac{7}{2} \qquad \text{Divide by 2.}$$

To check, use the words of the problem. The number is $\frac{7}{2}$, so twice the number is $2\left(\frac{7}{2}\right) = 7$. The reciprocal of the number is $\frac{2}{7}$, and the reciprocal of twice the number is $\frac{1}{7}$. Since $\frac{2}{7} - \frac{1}{7} = \frac{1}{7}$, the words of the problem are satisfied. The number is $\frac{7}{2}$. ∎

WORK PROBLEM 2 AT THE SIDE.

2 The next example shows a problem involving different amounts of time to do a job.

EXAMPLE 3 Solving a Problem About Work
Frank can clean all the trash from a strip of beach in 7 hours, while Mary takes only 5 hours. How long will it take them if they work together?

Let x = the number of hours it will take the two working together;

$\frac{1}{x}$ = the fraction of the job done in one hour by the two working together.

In one hour working alone, Frank will do $\frac{1}{7}$ of the job, while in one hour alone, Mary will do $\frac{1}{5}$ of the job. The total part of the job done by both people in one hour $\left(\frac{1}{x}\right)$ is equal to the sum of the parts of the job done by Frank and Mary in one hour $\left(\frac{1}{7} + \frac{1}{5}\right)$. That is,

Part done by Frank ⟶ $\frac{1}{7} + \frac{1}{5} = \frac{1}{x}$. ⟵ Part done by both working together
                                                ↑ Part done by Mary

3. (a) One worker can paint a room in 6 hours, while another takes 8 hours. How long will it take the two workers if they work together on the room?

(b) Stan needs 45 minutes to do the dishes, while Bobbie can do them in 30 minutes. How long will it take them if they work together?

Here $35x$ is the least common denominator.

$$35x \cdot \frac{1}{7} + 35x \cdot \frac{1}{5} = 35x \cdot \frac{1}{x}$$

$$5x + 7x = 35$$

$$12x = 35$$

$$x = \frac{35}{12}$$

Working together, Frank and Mary can do the entire job in $\frac{35}{12}$ hours, or 2 hours and 55 minutes. Check this result in the original problem. ∎

WORK PROBLEM 3 AT THE SIDE.

ANSWERS
2. 3
3. (a) $\frac{24}{7}$ hours (b) 18 minutes

3

The final examples use the distance formula, $d = rt$.

EXAMPLE 4 Solving a Problem About Distance, Rate, and Time

Tom's boat can go 10 miles against the current in a small river in the same time that it goes 15 miles with the current. If the speed of the current is 3 miles per hour, find the speed of the boat in still water.

Use the distance formula:

$$\text{distance} = \text{rate} \times \text{time}, \quad \text{or} \quad d = rt.$$

Let $\quad x =$ the speed of the boat in still water;
$x - 3 =$ the speed of the boat against the current;
$x + 3 =$ the speed of the boat with the current.

Since the time is the same going against the current as with the current, find time in terms of distance and rate (speed) for each situation.

Start with $d = rt$ and divide both sides by r to get

$$t = \frac{d}{r}.$$

Going against the current, the distance is 10 miles and the rate is $x - 3$, giving

$$t = \frac{d}{r} = \frac{10}{x - 3}.$$

Going with the current, the distance is 15 miles and the rate is $x + 3$, so that

$$t = \frac{d}{r} = \frac{15}{x + 3}.$$

This information is summarized in the following chart.

Direction	Distance	Rate	Time
Against current	10	$x - 3$	$\frac{10}{x-3}$
With current	15	$x + 3$	$\frac{15}{x+3}$

Times are equal.

Since the times are equal, $\frac{10}{x-3} = \frac{15}{x+3}$. The least common denominator is $(x - 3)(x + 3)$.

Multiplying on both sides by the least common denominator gives

$$10(x + 3) = 15(x - 3)$$
$$10x + 30 = 15x - 45 \quad \text{Distributive property}$$
$$30 = 5x - 45 \quad \text{Subtract } 10x.$$
$$75 = 5x \quad \text{Add 45.}$$
$$15 = x. \quad \text{Divide by 5.}$$

The speed of the boat in still water is 15 miles per hour. ■

WORK PROBLEM 4 AT THE SIDE.

---

4. A plane goes 100 miles against the wind in the same time that it takes to travel 120 miles with the wind. The wind speed is 20 miles per hour.

(a) Complete this table.

Direction	d	r	t
Against wind	100	$x - 20$	
With wind	120	$x + 20$	

(b) Find the speed of the plane in still air.

ANSWERS

4. (a) $\frac{100}{x - 20}$; $\frac{120}{x + 20}$
  (b) 220 miles per hour

5. A truck driver delivered a load to Richmond from Washington, D.C., averaging 50 miles per hour for the trip. He picked up another load in Richmond and delivered it to New York City, averaging 55 miles per hour on this part of the trip. The distance from Richmond to New York is 120 miles more than the distance from Washington to Richmond. If the entire trip required a driving time of 6 hours, find the distance from Washington to Richmond.

EXAMPLE 5 Solving a Problem About Distance, Rate, and Time

Claire drives 300 miles east from Indianapolis, mostly on the freeway. She usually averages 55 miles per hour, but an accident slows her speed through Columbus to 15 miles per hour. If her trip took 6 hours, how many miles did she drive at reduced speed?

Let x = the number of miles through Columbus;

$300 - x$ = the number of miles at 55 miles per hour.

The number of hours at 55 miles per hour is

$$t = \frac{d}{r} = \frac{300 - x}{55}$$

while the time at reduced speed is

$$t = \frac{x}{15}.$$

This information is summarized in the following chart.

	Distance	Rate	Time
In Columbus	x	15	$\frac{x}{15}$
On highway	$300 - x$	55	$\frac{300 - x}{55}$

Total time is 6 hours.

The total time is 6 hours, so

$$\frac{300 - x}{55} + \frac{x}{15} = 6.$$

Multiply both sides by the least common denominator, 165.

$$165 \cdot \frac{300 - x}{55} + 165 \cdot \frac{x}{15} = 165 \cdot 6$$

$$3(300 - x) + 11x = 990$$

$$900 - 3x + 11x = 990 \qquad \text{Distributive property}$$

$$8x = 90 \qquad \text{Subtract 900.}$$

$$x = \frac{45}{4} = 11\frac{1}{4}$$

Claire drove $11\frac{1}{4}$ miles at reduced speed through Columbus. ∎

WORK PROBLEM 5 AT THE SIDE.

ANSWER
5. 100 miles

4.7 EXERCISES

Work each of the following problems. See Examples 1 and 2.

1. What number must be added to the denominator of $\frac{12}{13}$ to make the result equal to 4?

2. What number must be added to both the numerator and denominator of $\frac{19}{17}$ to make the result equal to $\frac{13}{12}$?

3. Half the reciprocal of a number is added to $\frac{1}{4}$, giving a sum of $\frac{5}{12}$. Find the number.

4. One-third of the reciprocal of a number is added to $\frac{1}{2}$, giving a sum of 1. Find the number.

5. If the reciprocal of a number is added to the reciprocal of one more than the number, the result is $-\frac{19}{90}$. Find the number.

6. When the reciprocal of five times a number is subtracted from 2, the result is the reciprocal of the number. Find the number.

7. If the reciprocal of a number is subtracted from twice the reciprocal of the number, the result is 15. Find the number.

8. If the reciprocal of twice a number is added to twice the reciprocal of the number, the result is 3. Find the number.

Solve each of the following problems. See Example 3.

9. A.J. can do a job in 6 hours, but his wife, Audrey, can do the same job in 5 hours. How long will it take them if they work together?

10. Carroll and Elaine want to clean up an office that they share. Carroll can do the job alone in 5 hours, while Elaine can do it alone in 2 hours. How long will it take them if they work together?

11. Mike can paint a room in 6 hours working alone. If Dee helps him, the job takes 4 hours. How long would it take Dee to do the job if she worked alone?

12. Bernard and Carolyn Schmulen are refinishing a table. Working alone, Bernard could do the job in 7 hours. If the two work together, the job takes 5 hours. How long will it take Carolyn to refinish the table if she works alone?

13. If a vat of acid can be filled by an inlet pipe in 10 hours, and emptied by an outlet pipe in 20 hours, how long will it take to fill the vat if both pipes are open?

14. An ink factory has a vat to hold blue ink. An inlet pipe can fill the vat in 9 hours, while an outlet pipe can empty it in 12 hours. How long will it take to fill the vat if both the outlet and the inlet pipes are open?

15. Machine A can complete a certain job in 2 hours. To speed up the work, Machine B, which could complete the job alone in 3 hours, is brought in to help. How long will it take the two machines to complete the job working together?

16. A hot-water tap can fill a tub in 12 minutes, and a cold-water tap can fill it in 10 minutes. How long would it take to fill the tub with both taps open?

4.7 EXERCISES

17. Suppose Arlene and Mort can clean their entire house in 7 hours, while their toddler, Mimi, just by being around, can completely mess it up in only 2 hours. If Arlene and Mort clean the house while Mimi is at her grandma's, and then start cleaning up after Mimi the minute she gets home, how long does it take from the time Mimi gets home until the whole place is a shambles?

18. An inlet pipe can fill a barrel of wine in 6 hours, while an outlet pipe can empty it in 8 hours. Through an error both pipes are left on. How long will it take for the barrel to fill?

Work each of the following problems. See Examples 4 and 5.

19. Alfredo's boat goes 12 miles per hour. Find the speed of the current in the river if he can go 3 miles upstream in the same time that he takes to go 5 miles downstream.

	Distance	Rate	Time
Downstream	5	$12 + x$	
Upstream	3	$12 - x$	

20. Imelda averages 12 miles per hour riding her bike to school. Averaging 36 miles per hour by car takes her $\frac{1}{2}$ hour less time. How far does she travel to school?

	Distance	Rate	Time
Bike	x	12	
Car	x	36	

21. A canal has a current of 2 miles per hour. Find the speed of Byron Hopkins' boat in still water if it goes 22 miles downstream on the canal in the same time as it takes to go 16 miles upstream.

22. Aurelio's boat can go 20 miles per hour in still water. How far downstream can Aurelio go if a river has a current of 5 miles per hour and he must go down and back in 2 hours?

265

23. An airplane traveled from New York to a secret destination. The trip there was at 200 miles per hour, and the trip back was at 300 miles per hour. The total traveling time was 2 hours. How far was it to the secret destination?

24. Lucinda can fly her plane 200 miles against the wind in the same time it takes her to fly 300 miles with the wind. The wind blows at 30 miles per hour. Find the speed of her plane in still air.

25. Joe Tredeau averages 30 miles per hour when he drives on the old highway to his favorite fishing hole, while he averages 50 miles per hour when most of his route is on the interstate. If both routes are the same length, and he saves 2 hours by traveling on the interstate, how far away is the fishing hole?

26. On the first part of a trip to Carmel traveling on the freeway, Marge averaged 60 miles per hour. On the rest of the trip, which was 10 miles longer than the first part, she averaged 50 miles per hour. Find the total distance to Carmel if the first part of the trip took $\frac{1}{2}$ hour less time than the second part.

SKILL SHARPENERS
Evaluate each expression. See Section 1.4.

27. $\sqrt{4}$

28. $\sqrt{9}$

29. $\sqrt[3]{8}$

30. $\sqrt[3]{27}$

31. $\sqrt[5]{32}$

32. $\sqrt[4]{16}$

Historical Reflections

Euclid's *Elements*

Euclid's famous text *Elements* was translated into English by Billingsley in 1570. Primarily a geometry text, it also contains some number theory and elementary algebra (which is primarily geometric in nature). It consists of thirteen books and some 465 propositions. It is a classic example of deductive reasoning: Euclid begins with ten statements that are assumed to be true without proof (five "axioms" and five "postulates") and deduces his propositions from them. They include many of the topics that are studied in elementary mathematics today: the Euclidean algorithm for finding the greatest common factor of two numbers, the idea of the infinitude of the prime numbers, and the Fundamental Theorem of Arithmetic, for example.

No copy of *Elements* exists that dates back to the time of Euclid (circa 300 B.C.), and most current translations are based on a revision of the work prepared by Theon of Alexandria. Although *Elements* was only one of several works by Euclid, it is, by far, the most important. The most influential mathematics text ever written, it ranks second to the Bible as the most published book in history.

Art: From "The Elements" by Euclid

CHAPTER 4 SUMMARY

KEY TERMS

4.1 rational expression   A rational expression (or algebraic fraction) is the quotient of two polynomials with denominator not 0.

4.3 like fractions   Like fractions are fractions with the same denominators.

    unlike fractions   Unlike fractions are fractions with different denominators.

    common denominator   A common denominator of several denominators is an expression that is divisible by each of these denominators.

4.4 complex fraction   A fraction that has a fraction in the numerator, the denominator, or both, is called a complex fraction.

QUICK REVIEW

Section	Concepts	Examples
4.1 Basics of Rational Expressions	**Writing in Lowest Terms** *Step 1* Factor both numerator and denominator to find their greatest common factor. *Step 2* Divide both numerator and denominator by their greatest common factor.	Write in lowest terms. $$\frac{2x+8}{x^2-16} = \frac{2(x+4)}{(x-4)(x+4)}$$ $$= \frac{2}{x-4}$$
4.2 Multiplication and Division of Rational Expressions	**Multiplying Rational Expressions** *Step 1* Factor all numerators and denominators as completely as possible. *Step 2* Divide both numerator and denominator by any factors that both have in common. *Step 3* Multiply remaining factors in the numerator, and multiply remaining factors in the denominator. *Step 4* Be certain that the product is in lowest terms.	Multiply. $$\frac{x^2+2x+1}{x^2-1} \cdot \frac{5}{3x+3}$$ $$= \frac{(x+1)^2}{(x-1)(x+1)} \cdot \frac{5}{3(x+1)}$$ $$= \frac{5}{3(x-1)}$$
	Dividing Rational Expressions To divide two rational expressions, *multiply* the first by the reciprocal of the second (the divisor).	Divide. $$\frac{2x+5}{x-3} \div \frac{2x^2+3x-5}{x^2-9}$$ $$= \frac{2x+5}{x-3} \cdot \frac{(x+3)(x-3)}{(2x+5)(x-1)}$$ $$= \frac{x+3}{x-1}$$

CHAPTER 4 SUMMARY

Section	Concepts	Examples
4.3 Addition and Subtraction of Rational Expressions	**Adding or Subtracting Rational Expressions** *Step 1(a)* If the denominators are the same, add or subtract the numerators. Place the result over the common denominator. *Step 1(b)* If the denominators are different, first find the least common denominator. Write all rational expressions with this least common denominator, and then add or subtract the numerators. Place the result over the common denominator. *Step 2* Write all answers in lowest terms.	Subtract. $$\frac{1}{x+6} - \frac{3}{x+2}$$ $$= \frac{x+2}{(x+6)(x+2)} - \frac{3(x+6)}{(x+6)(x+2)}$$ $$= \frac{x+2-3x-18}{(x+6)(x+2)}$$ $$= \frac{-2x-16}{(x+6)(x+2)}$$
4.4 Complex Fractions	**Simplifying Complex Fractions** *Method 1* Simplify the numerator and denominator separately, as much as possible. Then multiply the numerator by the reciprocal of the denominator. *Method 2* Multiply the numerator and denominator of the complex fraction by the least common denominator of all fractions appearing in the numerator or denominator of the complex fraction.	Simplify the complex fraction. *Method 1* $$\frac{\frac{1}{2}+\frac{1}{3}}{\frac{1}{4}-\frac{1}{2}} = \frac{\frac{3}{6}+\frac{2}{6}}{\frac{1}{4}-\frac{2}{4}}$$ $$= \frac{\frac{5}{6}}{\frac{-1}{4}} = \frac{5}{6} \cdot \frac{4}{-1}$$ $$= \frac{20}{-6} = -\frac{10}{3}$$ *Method 2* $$\frac{\frac{1}{2}+\frac{1}{3}}{\frac{1}{4}-\frac{1}{2}} = \frac{12\left(\frac{1}{2}\right)+12\left(\frac{1}{3}\right)}{12\left(\frac{1}{4}\right)-12\left(\frac{1}{2}\right)}$$ $$= \frac{6+4}{3-6} = \frac{10}{-3} = -\frac{10}{3}$$

Section	Concepts	Examples
4.5 Equations with Rational Expressions	To solve an equation involving rational expressions, multiply all the terms in the equation by the least common denominator. Each potential solution must be checked to make sure that no denominator in the original equation is 0.	Solve for x: $$\frac{1}{x} + x = \frac{26}{5}.$$ $5 + 5x^2 = 26x$ Multiply by $5x$. $5x^2 - 26x + 5 = 0$ $(5x - 1)(x - 5) = 0$ $x = \frac{1}{5}$ or $x = 5$ Both check. The solution set is $\left\{\frac{1}{5}, 5\right\}$.
4.6 Formulas with Rational Expressions	To solve a formula for a particular variable, get that variable alone on one side by following the method described in Section 4.5.	Solve for L. $$c = \frac{100b}{L}$$ $cL = 100b$ Multiply by L. $L = \dfrac{100b}{c}$ Divide by c.
4.7 Applications	If a word problem translates into an equation with rational expressions, solve the equation using the method described in Section 4.5.	If the same number is added to the numerator and subtracted from the denominator of $\frac{5}{7}$, the result is 1. Find the number. Let x represent the number. The equation is $$\frac{5 + x}{7 - x} = 1.$$ Multiply both sides by $7 - x$. $5 + x = 7 - x$ $2x = 2$ $x = 1$ The number is 1. To check, add 1 to 5 to get 6, and subtract 1 from 7 to get 6. Since $\frac{6}{6} = 1$, the answer is correct.

CHAPTER 4 REVIEW EXERCISES

[4.1] *Find all numbers that make the rational expression undefined.*

1. $\dfrac{1}{p + 11}$

2. $\dfrac{-7}{9k + 18}$

3. $\dfrac{5r - 1}{r^2 - 7r + 10}$

Write in lowest terms.

4. $\dfrac{25m^4 n^3}{10m^5 n}$

5. $\dfrac{12x^2 + x}{24x + 2}$

6. $\dfrac{y^2 + 3y - 10}{y^2 - 5y + 6}$

7. $\dfrac{25m^2 - n^2}{25m^2 - 10mn + n^2}$

8. $\dfrac{p^3 + q^3}{p + q}$

9. $\dfrac{r - 2}{4 - r^2}$

[4.2] *Multiply or divide. Write all answers in lowest terms.*

10. $\dfrac{y^5}{6} \cdot \dfrac{9}{y^4}$

11. $\dfrac{25p^3 q^2}{8p^4 q} \div \dfrac{15pq^2}{16p^5}$

12. $\dfrac{3m + 12}{8} \div \dfrac{5m + 20}{4}$

13. $\dfrac{(2y + 3)^2}{5y} \cdot \dfrac{15y^3}{4y^2 - 9}$

14. $\dfrac{w^2 - 16}{w} \cdot \dfrac{3}{4 - w}$

15. $\dfrac{y^2 + 2y}{y^2 + y - 2} \div \dfrac{y - 5}{y^2 + 4y - 5}$

16. $\dfrac{z^2 - z - 6}{z - 6} \cdot \dfrac{z^2 - 6z}{z^2 + 2z - 15}$

17. $\dfrac{2p^2 - 5p - 12}{5p^2 - 18p - 8} \cdot \dfrac{25p^2 - 4}{30p - 12}$

18. $\dfrac{m^3 - n^3}{m^2 - n^2} \div \dfrac{m^2 + mn + n^2}{m + n}$

[4.3] *Find the least common denominator for each group of rational expressions.*

19. $\dfrac{5}{12}, \dfrac{8}{15x}$

20. $\dfrac{5a}{32b^3}, \dfrac{-1}{24b^5}$

21. $\dfrac{r}{p}, \dfrac{9 - r}{p + 6}$

22. $\dfrac{7}{9r^2}, \dfrac{5r - 3}{3r + 1}$

23. $\dfrac{4}{6k + 3}, \dfrac{7k^2 + 2k + 1}{10k + 5}, \dfrac{-11}{18k + 9}$

24. $\dfrac{x - 9}{6x^2 + 13x - 5}, \dfrac{x + 5}{9x^2 + 9x - 4}$

Add or subtract as indicated.

25. $\dfrac{7}{3k} - \dfrac{16}{3k}$

26. $\dfrac{2}{y} - \dfrac{3}{8y}$

27. $\dfrac{8}{z} - \dfrac{3}{2z^2}$

28. $\dfrac{3}{t-2} - \dfrac{5}{2-t}$

29. $\dfrac{2}{y+1} + \dfrac{6}{y-1}$

30. $\dfrac{6}{5a+10} + \dfrac{7}{6a+12}$

31. $\dfrac{2z}{2z^2 - zy - y^2} - \dfrac{3z}{3z^2 - 5zy + 2y^2}$

32. $\dfrac{3r}{10r^2 - 3rs - s^2} + \dfrac{2r}{2r^2 + rs - s^2}$

[4.4] Simplify each complex fraction.

33. $\dfrac{\dfrac{6}{p}}{\dfrac{8}{p+5}}$

34. $\dfrac{\dfrac{4m^5 n^6}{mn}}{\dfrac{8m^7 n^3}{m^4 n^2}}$

35. $\dfrac{\dfrac{r+2s}{10}}{\dfrac{8r+16s}{5}}$

36. $\dfrac{\dfrac{3}{x} - 5}{6 + \dfrac{1}{x}}$

37. $\dfrac{\dfrac{3}{p} - \dfrac{2}{q}}{\dfrac{9q^2 - 4p^2}{qp}}$

38. $\dfrac{\dfrac{a}{5} - \dfrac{1}{a}}{\dfrac{3}{a} - \dfrac{1}{4}}$

[4.5] Solve each equation.

39. $\dfrac{-1}{3h} - \dfrac{1}{h} = -\dfrac{4}{3}$

40. $\dfrac{1}{m+4} - \dfrac{3}{2m+8} = -\dfrac{1}{2}$

41. $\dfrac{5y}{y+1} - \dfrac{y}{3y+3} = \dfrac{-56}{6y+6}$

42. $\dfrac{2}{m-1} + \dfrac{1}{m+1} = \dfrac{4m+1}{m^2-1}$

43. $\dfrac{k^2}{k-3} = \dfrac{9}{k-3}$

CHAPTER 4 REVIEW EXERCISES

44. $\dfrac{3+p}{p^2-5p+4} - \dfrac{2}{p^2-4p} = \dfrac{1}{p}$

[4.6]

45. According to a law from physics, $\dfrac{1}{A} = \dfrac{1}{B} + \dfrac{1}{C}$. Find A if $B = 30$ and $C = 10$.

46. Using the law given in Exercise 45, find B if $A = 10$ and $C = 15$.

Solve each formula for the specified variable.

47. $F = \dfrac{GMm}{d^2}$ for m

48. $\dfrac{VP}{T} = \dfrac{vp}{t}$ for t

49. $S = \dfrac{n}{2}(a + \ell)$ for ℓ

50. $t = a + (n-1)d$ for d

[4.7]

51. When twice the reciprocal of three times a number is subtracted from 3, the result is 7 times the reciprocal of three times the number. Find the number.

52. Jane Brandsma and Mark Firmin need to sort a pile of bottles at the recycling center. Working alone, Jane could do the entire job in 9 hours, while Mark could do the entire job in 6 hours. How long will it take them if they work together?

53. A sink can be filled by a cold-water tap in 8 minutes, and filled by the hot-water tap in 12 minutes. How long would it take to fill the sink with both taps open?

54. A bathtub can be filled in 20 minutes. The drain can empty it in 30 minutes. How long will it take to fill the bathtub if both the water faucet and the drain are open?

CHAPTER 4 RATIONAL EXPRESSIONS

55. A river has a current of 4 kilometers per hour. Find the speed of Lynn McTernan's boat in still water if it goes 40 kilometers downstream in the same time that it takes to go 24 kilometers upstream.

56. A bus can travel 80 miles in the same time that a train goes 96 miles. The speed of the train is 10 miles per hour faster than the speed of the bus. Find both speeds.

MIXED REVIEW EXERCISES
Perform the indicated operations.

57. $\dfrac{2}{z} - \dfrac{5}{3z^2}$

58. $\dfrac{6p^2 + 17p - 3}{p^2 - 9} \cdot \dfrac{p^2 - 6p + 9}{216p^3 - 1}$

59. $\dfrac{\dfrac{y}{2} - \dfrac{3}{y}}{1 + \dfrac{y+1}{y}}$

60. $\dfrac{9y^2 + 46y + 5}{3y^2 - 2y - 1} \cdot \dfrac{y^3 + 5y^2 - 6y}{y^2 + 11y + 30}$

61. $\dfrac{4x + 16}{25} \div \dfrac{2x + 8}{5}$

62. $\dfrac{2}{x - 3} - \dfrac{9}{3 - x}$

63. $\dfrac{5}{6x + 12} + \dfrac{7}{2x + 4}$

64. $\dfrac{4x}{x^2 - xy - 2y^2} - \dfrac{6y}{x^2 + 4xy + 3y^2}$

Solve.

65. $A = \dfrac{Rr}{R + r}$ for r

66. $1 - \dfrac{5}{x} + \dfrac{4}{x^2} = 0$

67. $\dfrac{2z - 2}{z^2 + 3z + 2} + \dfrac{3}{z + 1} = \dfrac{4}{z + 2}$

68. The hot water tap can fill a tub in 20 minutes. The cold water tap takes 15 minutes to fill the tub. How long would it take to fill the tub with both taps open?

CHAPTER 4 TEST

1. Find all numbers that make the expression $\dfrac{m-5}{2m^2+5m-3}$ undefined.

2. Write $\dfrac{3p^2+5p}{3p^2-7p-20}$ in lowest terms.

Multiply or divide.

3. $\dfrac{5k^5}{9k^2} \div \dfrac{10k^6}{8k}$

4. $\dfrac{x^2-7x+12}{x^2+2x-15} \cdot \dfrac{x^2-25}{x^2-16}$

5. $\dfrac{y^2+y-12}{y^3+9y^2+20y} \div \dfrac{y^2-9}{y^3+3y^2}$

Find the least common denominator for each group of rational expressions.

6. $\dfrac{5}{9z}$, $-\dfrac{17}{12z^2}$, $\dfrac{1}{6z^3}$

7. $\dfrac{m}{m^2+m-6}$, $\dfrac{17m^2}{m^2+3m}$

Add or subtract.

8. $\dfrac{4}{k} + \dfrac{1}{5k}$

9. $\dfrac{1}{a+b} + \dfrac{3}{a-b}$

10. $\dfrac{4}{t^2-16} - \dfrac{5}{2t+8}$

Simplify each complex fraction.

11. $\dfrac{\dfrac{6}{m-2}}{\dfrac{8}{3m-6}}$

12. $\dfrac{\dfrac{1}{s}-\dfrac{1}{t}}{\dfrac{s}{t}-\dfrac{t}{s}}$

1. _____

2. _____

3. _____

4. _____

5. _____

6. _____

7. _____

8. _____

9. _____

10. _____

11. _____

12. _____

CHAPTER 4 RATIONAL EXPRESSIONS

Solve each of the following.

13. $1 - \dfrac{3}{x} = \dfrac{1}{2}$

14. $\dfrac{2}{m-3} - \dfrac{3}{m+3} = \dfrac{12}{m^2 - 9}$

15. $\dfrac{1}{a} - \dfrac{1}{a-2} = \dfrac{4}{3a}$

Solve for the specified variable.

16. $\dfrac{2}{z} - \dfrac{5}{m} = \dfrac{1}{y}$ for m

17. $A = \dfrac{1}{2}h(B + b)$ for B

Solve each word problem.

18. What number must be added to the numerator and the denominator of $\dfrac{9}{4}$ to make the result equal 3?

19. Wayne can do a job in 9 hours, while Susan can do the same job in 5 hours. How long would it take them to do the job if they worked together?

20. The current of a river runs at 3 miles per hour. Vera's boat can go 36 miles downstream in the same time that it takes to go 24 miles upstream. Find the speed of the boat in still water.

ROOTS AND RADICALS

5.1 RADICALS

1 Recall from Chapter 1 that the *square root* of a number a is a number b that can be squared to give a. That is, b is a square root of a if $b^2 = a$. For example,

4 is a square root of 16 since $4^2 = 16$;

-9 is a square root of 81 since $(-9)^2 = 81$.

EXAMPLE 1 Finding Square Roots of a Number
Find the square roots of each number.

(a) 36

Since $6^2 = 36$, and $(-6)^2 = 36$, both 6 and -6 are square roots of 36.

(b) 100

Both 10 and -10 are square roots of 100.

(c) -4

There is no real number that can be squared to give -4, so -4 has no real number square root. (However, -4 does have an *imaginary* number square root, as explained at the end of this chapter.) ■

WORK PROBLEM 1 AT THE SIDE.

A square root can be written with a **radical sign,** $\sqrt{}$. For example,

$\sqrt{64} = 8,$ since $8^2 = 64;$

$\sqrt{144} = 12,$ $\sqrt{9} = 3,$ $\sqrt{0} = 0,$ and so on.

While 0 has only one square root ($\sqrt{0} = 0$), a given positive number has two square roots, one positive and one negative.

The symbol $\sqrt{a}$ is used only for the *nonnegative* square root of a; the negative square root of a is written $-\sqrt{a}$.

As an abbreviation, the two square roots of the positive number a are sometimes written together as $\pm\sqrt{a}$, with the sign $\pm$ read "plus or minus."

OBJECTIVES

1 Find square roots of numbers.

2 Find higher roots of numbers.

3 Find principal roots.

4 Use a calculator with a square root key.

1. Find all square roots.

(a) 64

(b) 169

(c) 441

(d) -9

ANSWERS
1. (a) 8, -8 (b) 13, -13
(c) 21, -21 (d) no real number square roots

5.1 RADICALS 277

CHAPTER 5 ROOTS AND RADICALS

2. Simplify.

   (a) $\sqrt{49}$

   (b) $-\sqrt{16}$

   (c) $-\sqrt{\dfrac{36}{25}}$

   (d) $\sqrt{(-6)^2}$

   (e) $\sqrt{10^2}$

   (f) $\sqrt{r^2}$

3. Simplify.

   (a) $\sqrt[3]{8}$

   (b) $\sqrt[3]{1000}$

   (c) $\sqrt[3]{-1}$

   (d) $\sqrt[4]{81}$

   (e) $\sqrt[4]{-1}$

   (f) $\sqrt[6]{64}$

ANSWERS

2. (a) 7 (b) -4 (c) $-\dfrac{6}{5}$ (d) 6
   (e) 10 (f) $|r|$
3. (a) 2 (b) 10 (c) -1 (d) 3
   (e) not a real number (f) 2

A square root of a^2 (where $a \neq 0$) is a number that can be squared to give a^2. This number is either a or $-a$. Since one root is negative and one is positive, and since the symbol $\sqrt{a^2}$ represents the *nonnegative* square root, this root is written with absolute value bars as $|a|$.

EXAMPLE 2 Simplifying Square Roots

Find each square root.

(a) $\sqrt{7^2} = |7| = 7$

(b) $\sqrt{(-7)^2} = |-7| = 7$

(c) $\sqrt{k^2} = |k|$

(d) $\sqrt{(-k)^2} = |-k| = |k|$ ■

■ WORK PROBLEM 2 AT THE SIDE.

Radical signs also can be used with higher roots, such as **cube roots,** **fourth roots,** and so on. By definition,

$$\sqrt[n]{a} = b \quad \text{means} \quad b^n = a.$$

n is called the **index,** a is called the **radicand,** and the entire expression $\sqrt[n]{a}$ is called a **radical.**

EXAMPLE 3 Simplifying Higher Roots

Simplify.

(a) $\sqrt[3]{27} = 3$, since $3^3 = 27$.

(b) $\sqrt[3]{125} = 5$, since $5^3 = 125$.

(c) $\sqrt[3]{-216} = -6$, since $(-6)^3 = -216$.

(d) $\sqrt[4]{16} = 2$, since $2^4 = 16$.

(e) $\sqrt[4]{-16}$ is not a real number.

(f) $\sqrt[5]{32} = 2$, since $2^5 = 32$. ■

■ WORK PROBLEM 3 AT THE SIDE.

Both 2 and -2 are fourth roots of 16; however, the symbol $\sqrt[4]{16}$ is used for the *positive* root. The positive root is called the **principal root.**

$\sqrt[n]{a}$ represents the *principal* nth root of a, and $-\sqrt[n]{a}$ is the negative nth root of a.
If a is positive, then $\sqrt[n]{a}$ is positive.
If a is negative, then
  $\sqrt[n]{a}$ is negative when n is odd,
  $\sqrt[n]{a}$ is not a real number when n is even.

5.1 RADICALS

EXAMPLE 4 Finding Principal Roots
Find each root.

(a) $\sqrt{100} = 10$

Here the radicand 100 is positive. There are two square roots, 10 and -10, but 10 is the principal root.

(b) $\sqrt[4]{81} = 3$

(c) $\sqrt[6]{-8}$

The index is even and the radicand is negative, so the principal root is not a real number.

(d) $\sqrt[3]{-8} = -2$ since $(-2)^3 = -8$. ∎

WORK PROBLEM 4 AT THE SIDE.

As shown above, $\sqrt{a^2} = |a|$. The fourth root of a^4, the sixth root of a^6, and other even nth roots of a^n also must be written as absolute values. For example, $\sqrt[4]{a^4} = |a|$ and $\sqrt[6]{a^6} = |a|$. On the other hand, the cube (or third) root of a positive number is positive and the cube root of a negative number is negative, so $\sqrt[3]{a^3} = a$ whether a is positive or negative. All odd roots behave like cube roots. These examples suggest the following rule.

If n is an **even** positive integer, then $\sqrt[n]{a^n} = |a|$,
and if n is an **odd** positive integer, then $\sqrt[n]{a^n} = a$.

EXAMPLE 5 Simplifying Roots with Absolute Value
Simplify each root.

(a) $\sqrt[6]{(-3)^6} = |-3| = 3$

(b) $\sqrt[5]{(-4)^5} = -4$

(c) $-\sqrt[4]{(-9)^4} = -|-9| = -9$

(d) $\sqrt{\dfrac{9}{16}} = \dfrac{3}{4}$

(e) $-\sqrt[4]{m^4} = -|m|$

(f) $\sqrt[3]{a^{12}} = a^4$ [since $(a^4)^3 = a^{12}$]

No absolute value bars are needed here, since a^4 is nonnegative for any real number value of a. Also, there is only *one* cube root for each number.

(g) $\sqrt[4]{x^{12}} = |x^3|$

Absolute value bars are used to guarantee that the result is not negative (since x^3 can be either positive or negative, depending on x). If desired, $|x^3|$ can be written as $x^2 \cdot |x|$. ∎

WORK PROBLEM 5 AT THE SIDE.

---

4. Find each root.

(a) $\sqrt{4}$

(b) $\sqrt[3]{27}$

(c) $\sqrt[3]{-27}$

(d) $\sqrt[4]{625}$

(e) $\sqrt[5]{-32}$

5. Simplify.

(a) $\sqrt[6]{64}$

(b) $-\sqrt[4]{16}$

(c) $\sqrt[3]{\dfrac{216}{125}}$

(d) $\sqrt[5]{-243}$

(e) $\sqrt[6]{(-p)^6}$

(f) $\sqrt[6]{y^{24}}$

ANSWERS
4. (a) 2 (b) 3 (c) -3 (d) 5
   (e) -2
5. (a) 2 (b) -2 (c) $\dfrac{6}{5}$ (d) -3
   (e) $|p|$ (f) y^4

CHAPTER 5 ROOTS AND RADICALS

6. Find a decimal approximation for each of the following.

(a) $\sqrt{10}$

(b) $\sqrt{51}$

(c) $-\sqrt{99}$

(d) $\sqrt{950}$

(e) $-\sqrt{670}$

4 Not all square roots are rational numbers. For example, there is no rational number whose square is 15, so $\sqrt{15}$ is not a rational number. (Recall from Chapter 1 that $\sqrt{15}$ is an *irrational* number.) We can find an approximation to $\sqrt{15}$ by using a calculator.

To use a calculator to approximate $\sqrt{15}$, enter the number 15 and press the key marked $\sqrt{\ }$. The display will then read 3.8729833. (There may be fewer or more decimal places, depending upon the model of calculator used.) In this book we will show approximations correct to three decimal places. Therefore,

$$\sqrt{15} \approx 3.873$$

where $\approx$ means "is approximately equal to."

EXAMPLE 6 Finding Approximations for Square Roots

Find a decimal approximation for each of the following, using a calculator.

(a) $\sqrt{39} \approx 6.245$

(b) $\sqrt{83} \approx 9.110$

(c) $-\sqrt{72} \approx -8.485$

To find the negative square root, first find the positive square root and then take its negative.

(d) $\sqrt{770} \approx 27.749$

(e) $-\sqrt{420} \approx -20.494$ ■

WORK PROBLEM 6 AT THE SIDE.

Many calculators can be used to find higher roots. For instructions, consult the owner's manual.

To help in problems with principal roots, the following table shows some of the most commonly used powers.

n Number	n^3 Cube	n^4 Fourth power	n^5 Fifth power	n^6 Sixth power
2	8	16	32	64
3	27	81	243	729
4	64	256	1024	4096
5	125	625	3125	15,625
6	216	1296	7776	46,656
7	343	2401	16,807	117,649
8	512	4096	32,768	262,144
9	729	6561	59,049	531,441
10	1000	10,000	100,000	1,000,000
11	1331	14,641	161,051	1,771,561
12	1728	20,736	248,832	2,985,984

ANSWERS
6. (a) 3.162 (b) 7.141
   (c) −9.950 (d) 30.822
   (e) −25.884

5.1 EXERCISES

Find all square roots of the following numbers. Use a calculator as needed. Round to the nearest thousandth in Exercises 11 and 12. See Example 1.

1. 25

2. 49

3. 361

4. 400

5. 625

6. 1444

7. 2209

8. 4624

9. 6241

10. −3481

11. 758.31

12. 589.62

Find each square root that is a real number. See Example 2. Use a calculator as needed.

13. $\sqrt{4}$

14. $\sqrt{64}$

15. $\sqrt{256}$

16. $\sqrt{784}$

17. $-\sqrt{2025}$

18. $-\sqrt{-36}$

19. $-\sqrt{-225}$

20. $\sqrt{y^2}$

21. $\sqrt{x^4}$

Find a decimal approximation for each of the following by using a calculator. Round to the nearest thousandth. See Example 6.

22. $\sqrt{7}$

23. $\sqrt{11}$

24. $-\sqrt{19}$

25. $-\sqrt{56}$

26. $-\sqrt{82}$

27. $-\sqrt{91}$

28. $\sqrt{150}$

29. $\sqrt{280}$

30. $-\sqrt{510}$

Find each root. Use the table at the end of this section as necessary. See Examples 3–5.

31. $\sqrt{81}$

32. $-\sqrt{100}$

33. $\sqrt[3]{64}$

34. $\sqrt[3]{125}$

35. $\sqrt[3]{-216}$

36. $\sqrt[3]{-343}$

37. $-\sqrt[3]{512}$

38. $-\sqrt[3]{1000}$

39. $\sqrt[4]{81}$

40. $\sqrt[4]{256}$

41. $-\sqrt[4]{625}$

42. $-\sqrt[4]{1296}$

CHAPTER 5 ROOTS AND RADICALS

43. $\sqrt[4]{-2401}$

44. $\sqrt[4]{-4096}$

45. $-\sqrt[4]{6561}$

46. $-\sqrt[4]{10{,}000}$

47. $\sqrt[5]{32}$

48. $\sqrt[5]{243}$

49. $-\sqrt[5]{-1024}$

50. $-\sqrt[5]{3125}$

51. $-\sqrt[6]{64}$

52. $\sqrt[6]{-729}$

53. $\sqrt[6]{4096}$

54. $\sqrt[5]{-q^{15}}$

55. $\sqrt[7]{-z^{21}}$

56. $\sqrt[4]{y^{16}}$

57. $\sqrt[3]{m^{15}}$

Work each problem.

58. *Heron's formula* gives a method of finding the area of a triangle if the lengths of its sides are known. Suppose that a, b, and c are the lengths of the sides. Let s denote one-half of the perimeter of the triangle (called the *semiperimeter*); that is,

$$s = \frac{1}{2}(a + b + c).$$

Then the area of the triangle is

$$K = \sqrt{s(s-a)(s-b)(s-c)}.$$

Find the area of a triangle with sides measuring 4, 3, and 5 meters.

59. In accident reconstruction, the formula

$$S = 5.5\sqrt{DF}$$

is used to determine the speed S in miles per hour of a vehicle that leaves skid marks of D feet. F is a constant that represents the friction of the road surface. Let $F = 8$. Find the speed of a vehicle that left skid marks of
(a) 50 feet;
(b) 10 feet;
(c) 20 feet.
Round answers to the nearest whole number.

SKILL SHARPENERS

Use laws of exponents to simplify the following. See Sections 3.1 and 3.2.

60. $x^2 \cdot x^5$

61. $x^{-3} \cdot x^6$

62. $\dfrac{y^5}{y^2}$

63. $\dfrac{m^{12}}{m^{10}}$

64. $(t^4)^5$

65. $(s^3)^5$

5.2 RATIONAL EXPONENTS

1 In this section the definitions and rules for exponents are extended to include rational exponents as well as integer exponents. If past rules for exponents are still to be valid, then the product $(3^{1/2})^2 = 3^{1/2} \cdot 3^{1/2}$ should be found by adding exponents.

$$(3^{1/2})^2 = 3^{1/2} \cdot 3^{1/2}$$
$$= 3^{1/2 + 1/2}$$
$$= 3^1 = 3$$

In Section 5.1, by definition, $(\sqrt{3})^2 = \sqrt{3} \cdot \sqrt{3} = 3$. Then, since both $(3^{1/2})^2$ and $(\sqrt{3})^2$ are equal to 3,

$$3^{1/2} = \sqrt{3}.$$

Generalizing from this example, $a^{1/n}$ is defined as the principal nth root of a.

$a^{1/n} = \sqrt[n]{a}$ whenever $\sqrt[n]{a}$ is real.

EXAMPLE 1 Evaluating Exponentials of the Form $a^{1/n}$

(a) $64^{1/3} = \sqrt[3]{64} = 4$

(b) $100^{1/2} = \sqrt{100} = 10$

(c) $-256^{1/4} = -\sqrt[4]{256} = -4$

(d) $m^{1/5} = \sqrt[5]{m}$ ∎

WORK PROBLEM 1 AT THE SIDE.

2 What about the symbol $8^{2/3}$? For past rules of exponents to be valid,

$$8^{2/3} = 8^{(1/3)2} = (8^{1/3})^2.$$

Since $8^{1/3} = \sqrt[3]{8}$,

$$8^{2/3} = (\sqrt[3]{8})^2 = 2^2 = 4.$$

Generalizing, $a^{m/n}$ is defined as follows.

If m and n are positive integers with m/n in lowest terms, then

$$a^{m/n} = (a^{1/n})^m$$

provided that $a^{1/n}$ is a real number. If $a^{1/n}$ is not a real number, then $a^{m/n}$ is not real.

OBJECTIVES

1 Define $a^{1/n}$.

2 Define $a^{m/n}$.

3 Use the rules for exponents with rational exponents.

4 Convert radicals to numbers with rational exponents.

1. Simplify.

(a) $8^{1/3}$

(b) $9^{1/2}$

(c) $64^{1/6}$

(d) $1000^{1/3}$

(e) $-16^{1/4}$

(f) $a^{1/5}$

ANSWERS
1. (a) 2 (b) 3 (c) 2
   (d) 10 (e) -2
   (f) $\sqrt[5]{a}$

CHAPTER 5 ROOTS AND RADICALS

2. Simplify.

(a) $64^{2/3}$

(b) $100^{3/2}$

(c) $-16^{3/4}$

3. Simplify.

(a) $8^{-2/3}$

(b) $36^{-3/2}$

(c) $32^{-4/5}$

(d) $\left(\dfrac{4}{9}\right)^{-5/2}$

ANSWERS
2. (a) 16 (b) 1000 (c) -8
3. (a) $\dfrac{1}{4}$ (b) $\dfrac{1}{216}$ (c) $\dfrac{1}{16}$ (d) $\dfrac{243}{32}$

EXAMPLE 2 Evaluating Exponentials of the Form $a^{m/n}$

(a) $36^{3/2} = (36^{1/2})^3 = 6^3 = 216$

(b) $125^{2/3} = (125^{1/3})^2 = 5^2 = 25$

(c) $-4^{5/2} = -(4^{5/2}) = -(4^{1/2})^5 = -(2)^5 = -32$

(d) $(-27)^{2/3} = [(-27)^{1/3}]^2 = (-3)^2 = 9$

Notice how the $-$ sign is used in (c) and in (d). In (c), we first evaluate the exponential and then find its negative. In (d), the $-$ sign is part of the base, -27.

(e) $(-100)^{3/2}$ is not a real number, since $(-100)^{1/2}$ is not a real number. ∎

■ WORK PROBLEM 2 AT THE SIDE.

EXAMPLE 3 Evaluating Exponentials with Negative Rational Exponents

Simplify each of the following.

(a) $16^{-3/4}$

By the definition of a negative exponent.

$$16^{-3/4} = \dfrac{1}{16^{3/4}}.$$

Since $16^{3/4} = (\sqrt[4]{16})^3 = 2^3 = 8$,

$$16^{-3/4} = \dfrac{1}{16^{3/4}} = \dfrac{1}{8}.$$

(b) $25^{-3/2} = \dfrac{1}{25^{3/2}} = \dfrac{1}{(\sqrt{25})^3} = \dfrac{1}{5^3} = \dfrac{1}{125}$

(c) $\left(\dfrac{8}{27}\right)^{-2/3} = \dfrac{1}{\left(\dfrac{8}{27}\right)^{2/3}} = \dfrac{1}{\left(\dfrac{2}{3}\right)^2} = \dfrac{1}{\dfrac{4}{9}} = \dfrac{9}{4}$ ∎

■ WORK PROBLEM 3 AT THE SIDE.

The exponential expression $a^{m/n}$ was defined as $(a^{1/n})^m$. An alternative definition can be obtained by using the power rule for exponents in another way. If all indicated roots are real,

$$a^{m/n} = a^{m(1/n)}$$
$$= (a^m)^{1/n},$$

so that $a^{m/n} = (a^m)^{1/n}.$

With this result, $a^{m/n}$ can be defined in either of two ways.

If all indicated roots are real, then

$$a^{m/n} = (a^{1/n})^m = (a^m)^{1/n}.$$

5.2 RATIONAL EXPONENTS

An expression such as $27^{2/3}$ can now be evaluated in two ways:

$$27^{2/3} = (27^{1/3})^2 = 3^2 = 9$$
$$27^{2/3} = (27^2)^{1/3} = 729^{1/3} = 9.$$

In most cases, it is easier to use $(a^{1/n})^m$.

3 With the definition of a rational exponent given above, all the rules for exponents are still valid. They are repeated here for reference.

DEFINITIONS AND RULES FOR EXPONENTS

If a and b are real numbers, all powers are real, and m and n are rational numbers:

Product rule	$a^m \cdot a^n = a^{m+n}$	
Quotient rule	$\dfrac{a^m}{a^n} = a^{m-n}$	$(a \neq 0)$
Power rules	$(a^m)^n = a^{mn}$	
	$(ab)^m = a^m b^m$	
	$\left(\dfrac{a}{b}\right)^m = \dfrac{a^m}{b^m}$	$(b \neq 0)$
Zero exponent	$a^0 = 1$	$(a \neq 0)$
Negative exponent	$a^{-n} = \dfrac{1}{a^n}$	$(a \neq 0).$

The next example shows applications of these rules.

EXAMPLE 4 Applying Exponent Rules to Rational Exponents

(a) $3^{3/2} \cdot 3^{1/2} = 3^{3/2 + 1/2} = 3^2 = 9$

(b) $\dfrac{5^{2/3}}{5^{5/3}} = 5^{2/3 - 5/3} = 5^{-1} = \dfrac{1}{5}$

(c) $6^{5/8} \cdot 6^{-3/8} = 6^{5/8 + (-3/8)} = 6^{1/4}$

(d) $(3^{4/3})^5 = 3^{(4/3) \cdot 5} = 3^{20/3}$

(e) $\dfrac{(y^{2/3})^4}{y} = \dfrac{y^{8/3}}{y^1} = y^{(8/3) - 1} = y^{5/3}$ $(y \neq 0)$ ∎

CAUTION Errors often occur in exercises like the ones in Example 4 because students try to convert the expressions to radical form. Remember that the *rules of exponents* apply here.

WORK PROBLEM 4 AT THE SIDE.

---

4. Simplify.

(a) $11^{3/4} \cdot 11^{5/4}$

(b) $\dfrac{7^{3/4}}{7^{7/4}}$

(c) $\dfrac{9^{2/3}}{9^{-1/3}}$

(d) $(x^{3/2})^4$, $x > 0$

ANSWERS
4. (a) 11^2 or 121
   (b) $\dfrac{1}{7}$ (c) 9   (d) x^6

5. Simplify. Assume all variables represent positive real numbers.

(a) $\sqrt{y^{10}}$

(b) $\sqrt[6]{27y^9}$

(c) $\sqrt[5]{y} \cdot \sqrt[10]{y^7}$

(d) $\dfrac{\sqrt[3]{m^2}}{\sqrt[5]{m}}$

4 Using the definition of rational exponents, many problems involving radicals can be simplified by converting the radicals to numbers with rational exponents. After simplifying, the answer can be converted back to radical form.

In the next example, we assume that all variables represent positive real numbers. We make this assumption in order to eliminate the need to use absolute value symbols when the root index is even.

EXAMPLE 5 Converting Between Radicals and Rational Exponents

Simplify. Assume all variables represent positive real numbers.

(a) $\sqrt{m^4} = m^{4/2} = m^2$

(b) $\sqrt[4]{25y^2} = \sqrt[4]{5^2 y^2}$
$= (5^2 y^2)^{1/4} = 5^{1/2} y^{1/2} = (5y)^{1/2} = \sqrt{5y}$

(c) $\sqrt[3]{w} \cdot \sqrt[4]{w} = w^{1/3} \cdot w^{1/4}$
$= w^{1/3 + 1/4} = w^{7/12} = \sqrt[12]{w^7}$

(d) $\sqrt[8]{k^5} \cdot \sqrt{k} \cdot \sqrt[4]{k^3} = k^{5/8} \cdot k^{1/2} \cdot k^{3/4}$
$= k^{15/8} = k^{1 + 7/8} = k \cdot k^{7/8} = k\sqrt[8]{k^7}$

(e) $\dfrac{\sqrt{x^3}}{\sqrt[3]{x^2}} = \dfrac{x^{3/2}}{x^{2/3}} = x^{3/2 - 2/3}$
$= x^{5/6} = \sqrt[6]{x^5}$ ∎

WORK PROBLEM 5 AT THE SIDE.

ANSWERS
5. (a) y^5 (b) $y\sqrt{3y}$ (c) $\sqrt[10]{y^9}$
(d) $\sqrt[15]{m^7}$

5.2 EXERCISES

Simplify. Use a calculator or the table of powers at the end of Section 5.1 as necessary. See Examples 1–3.

1. $121^{1/2}$

2. $169^{1/2}$

3. $512^{1/3}$

4. $729^{1/3}$

5. $256^{1/4}$

6. $-4096^{1/4}$

7. $-6561^{1/4}$

8. $\left(\dfrac{1}{256}\right)^{1/4}$

9. $\left(\dfrac{1}{32}\right)^{1/5}$

10. $\left(\dfrac{4}{9}\right)^{1/2}$

11. $\left(\dfrac{16}{25}\right)^{1/2}$

12. $\left(-\dfrac{64}{27}\right)^{1/3}$

13. $\left(\dfrac{8}{125}\right)^{1/3}$

14. $8^{2/3}$

15. $100^{3/2}$

16. $32^{2/5}$

17. $32^{6/5}$

18. $(-125)^{2/3}$

19. $-144^{3/2}$

20. $-49^{3/2}$

21. $(-1728)^{2/3}$

22. $\left(\dfrac{1}{9}\right)^{3/2}$

23. $\left(\dfrac{16}{81}\right)^{3/4}$

24. $-\left(\dfrac{81}{625}\right)^{3/4}$

25. $-\left(\dfrac{9}{4}\right)^{5/2}$

26. $\left(\dfrac{27}{64}\right)^{-1/3}$

27. $\left(\dfrac{16}{625}\right)^{-1/4}$

28. $\left(\dfrac{25}{36}\right)^{-3/2}$

Use the rules of exponents to simplify each of the following. Write all answers with positive exponents. Assume that all variables represent positive real numbers. See Example 4.

29. $2^{1/2} \cdot 2^{3/2}$

30. $5^{4/3} \cdot 5^{2/3}$

31. $9^{3/5} \cdot 9^{7/5}$

32. $7^{3/4} \cdot 7^{9/4}$

33. $\dfrac{81^{5/4}}{81^{3/4}}$

34. $\dfrac{125^{2/3}}{125^{1/3}}$

35. $x^{5/3} \cdot x^{-2/3}$

36. $m^{-8/9} \cdot m^{17/9}$

37. $\dfrac{k^{2/3} k^{-1}}{k^{1/3}}$

38. $\dfrac{z^{5/4} z^{-2}}{z^{3/4}}$

39. $\dfrac{(n^{2/3})^2}{n^2}$

40. $\dfrac{(k^{5/4})^2}{k}$

41. $(8p^9 q^6)^{2/3}$

42. $(25m^8 r^{10})^{3/2}$

43. $\dfrac{(r^{2/3})^2}{(r^2)^{5/3}}$

44. $\dfrac{(k^{5/4})^2}{(k^3)^{4/3}}$

CHAPTER 5 ROOTS AND RADICALS

45. $\left(\dfrac{z^{10}}{x^{12}}\right)^{1/4}$

46. $\left(\dfrac{r^5}{s^7}\right)^{2/5}$

47. $\dfrac{(m^2h)^{1/2}}{m^{3/4}h^{-1/4}}$

48. $\dfrac{(a^{-3}b^2)^{1/6}}{(a^2b^5)^{-1/4}}$

49. $\dfrac{p^{1/5}p^{7/10}p^{1/2}}{(p^3)^{-1/5}}$

50. $\dfrac{z^{1/3}z^{-2/3}z^{1/6}}{(z^{-1/6})^3}$

51. $\left(\dfrac{m^{-2/3}}{a^{-3/4}}\right)^4 (m^{-3/8}a^{1/4})^{-2}$

52. $\left(\dfrac{b^{-3/2}}{c^{-5/3}}\right)^2 (b^{-1/4}c^{-1/3})^{-1}$

Simplify each of the following by first converting to rational exponents. Assume that all variables represent positive real numbers. In Exercises 57–64, convert answers back to radical form. See Example 5.

53. $\sqrt{2^{10}}$

54. $\sqrt{5^8}$

55. $\sqrt[3]{6^{12}}$

56. $\sqrt[4]{7^{12}}$

57. $-\sqrt[3]{11^5}$

58. $-\sqrt[5]{23^8}$

59. $\sqrt[4]{x^7}$

60. $\sqrt[5]{y^{12}}$

61. $\sqrt[7]{a^9} \cdot \sqrt{a}$

62. $\sqrt[4]{p^9} \cdot \sqrt[3]{p}$

63. $\dfrac{\sqrt[3]{x^5}}{\sqrt{x^3}}$

64. $\dfrac{\sqrt[3]{k^7}}{\sqrt{k^5}}$

Work each problem.

65. The length L of an animal in centimeters is related to its surface area S by the equation
$$L = \left(\dfrac{S}{a}\right)^{1/2}$$
where a is a constant that depends on the type of animal. Find the length of an animal with a surface area of 1000 square centimeters if $a = \dfrac{2}{5}$.

66. The threshold weight T for a person is the weight above which the risk of death increases greatly. One researcher found that the threshold weight in pounds for men aged 40–49 is related to height in inches by the equation
$$h = 12.3T^{1/3}.$$
What height corresponds to a threshold of 216 pounds for a man in this age group?

SKILL SHARPENERS

Simplify each of the following. See Sections 3.1 and 3.2.

67. $p^8 \cdot p^{-2} \cdot p^5$

68. $y^3 \cdot y^{-7} \cdot y^{-9}$

69. $(6x^2)^{-1}(2x^3)^{-2}$

70. $(5m^{-1})^2(3m^{-2})^{-1}$

71. $(8r^2s)^{-1}(2r^{-1}s^{-2})^{-1}$

72. $(3p^{-1}q^{-2})^{-1}(2pq^{-1})^{-1}$

5.3 SIMPLIFYING RADICALS

1 The product of $\sqrt{36}$ and $\sqrt{4}$ is
$$\sqrt{36} \cdot \sqrt{4} = 6 \cdot 2 = 12.$$
Multiplying 36 and 4 and then taking the square root gives
$$\sqrt{36} \cdot \sqrt{4} = \sqrt{36 \cdot 4} = \sqrt{144} = 12,$$
the same answer. This result is an example of the **product rule for radicals.**

PRODUCT RULE FOR RADICALS

If a and b are real numbers, not both negative, all roots are real, and n is a natural number,
$$\sqrt[n]{a} \cdot \sqrt[n]{b} = \sqrt[n]{ab}.$$
(The product of two radicals is the radical of the product.)

OBJECTIVES

1 Use the product rule for radicals.

2 Use the quotient rule for radicals.

3 Simplify radicals.

4 Write rational exponents in lowest terms.

The product rule can be justified using rational exponents. Since $\sqrt[n]{a} = a^{1/n}$ and $\sqrt[n]{b} = b^{1/n}$,
$$\sqrt[n]{a} \cdot \sqrt[n]{b} = a^{1/n} \cdot b^{1/n} = (ab)^{1/n} = \sqrt[n]{ab}.$$

Examples in which a or b is negative or both are negative are given in Section 5.7.

EXAMPLE 1 Using the Product Rule
Multiply. Assume that all variables represent positive real numbers.

(a) $\sqrt{5} \cdot \sqrt{7} = \sqrt{5 \cdot 7} = \sqrt{35}$

(b) $\sqrt{2} \cdot \sqrt{19} = \sqrt{2 \cdot 19} = \sqrt{38}$

(c) $\sqrt{7} \cdot \sqrt{11xyz} = \sqrt{77xyz}$

(d) $\sqrt{\dfrac{7}{y}} \cdot \sqrt{\dfrac{3}{p}} = \sqrt{\dfrac{21}{yp}}$ ∎

1. Multiply. Assume all variables represent positive real numbers.

   (a) $\sqrt{5} \cdot \sqrt{13}$

   (b) $\sqrt{15} \cdot \sqrt{2}$

   (c) $\sqrt{10y} \cdot \sqrt{3k}$

   (d) $\sqrt{\dfrac{5}{a}} \cdot \sqrt{\dfrac{11}{z}}$

WORK PROBLEM 1 AT THE SIDE.

EXAMPLE 2 Using the Product Rule
Multiply. Assume that all variables represent positive real numbers.

(a) $\sqrt[3]{3} \cdot \sqrt[3]{12} = \sqrt[3]{3 \cdot 12} = \sqrt[3]{36}$

(b) $\sqrt[4]{8y} \cdot \sqrt[4]{3r^2} = \sqrt[4]{24yr^2}$

(c) $\sqrt[6]{10m^4} \cdot \sqrt[6]{5m} = \sqrt[6]{50m^5}$

(d) $\sqrt[4]{5} \cdot \sqrt[5]{2}$ cannot be simplified by the product rule, since the two indexes (4 and 5) are different. ∎

WORK PROBLEM 2 AT THE SIDE.

2. Multiply. Assume all variables represent positive real numbers.

   (a) $\sqrt[3]{2} \cdot \sqrt[3]{7}$

   (b) $\sqrt[6]{8r^2} \cdot \sqrt[6]{2r^3}$

   (c) $\sqrt[5]{9y^2x} \cdot \sqrt[5]{8xy^2}$

   (d) $\sqrt{7} \cdot \sqrt[3]{5}$

ANSWERS
1. (a) $\sqrt{65}$ (b) $\sqrt{30}$
   (c) $\sqrt{30yk}$ (d) $\sqrt{\dfrac{55}{az}}$
2. (a) $\sqrt[3]{14}$ (b) $\sqrt[6]{16r^5}$
   (c) $\sqrt[5]{72y^4x^2}$
   (d) cannot be simplified by the product rule

CHAPTER 5 ROOTS AND RADICALS

3. Simplify. Assume that all variables represent positive real numbers.

(a) $\sqrt{\dfrac{100}{81}}$

(b) $\sqrt{\dfrac{11}{25}}$

(c) $\sqrt{\dfrac{k}{64}}$

(d) $\sqrt{\dfrac{y^8}{16}}$

(e) $\sqrt[3]{\dfrac{18}{125}}$

(f) $\sqrt[3]{\dfrac{x^2}{r^{12}}}$

ANSWERS

3. (a) $\dfrac{10}{9}$ (b) $\dfrac{\sqrt{11}}{5}$ (c) $\dfrac{\sqrt{k}}{8}$

(d) $\dfrac{y^4}{4}$ (e) $\dfrac{\sqrt[3]{18}}{5}$

(f) $\dfrac{\sqrt[3]{x^2}}{r^4}$

---

2 The **quotient rule for radicals** is similar to the product rule.

QUOTIENT RULE FOR RADICALS

If a and b are real numbers, not both negative, all roots are real, if $b \neq 0$, and if n is a natural number, then

$$\sqrt[n]{\dfrac{a}{b}} = \dfrac{\sqrt[n]{a}}{\sqrt[n]{b}}.$$

(The radical of a quotient is the quotient of the radicals.)

The quotient rule can be justified in a manner similar to the one used for the product rule.

EXAMPLE 3 Using the Quotient Rule
Simplify. Assume that all variables represent positive real numbers.

(a) $\sqrt{\dfrac{16}{25}} = \dfrac{\sqrt{16}}{\sqrt{25}} = \dfrac{4}{5}$

(b) $\sqrt{\dfrac{7}{36}} = \dfrac{\sqrt{7}}{\sqrt{36}} = \dfrac{\sqrt{7}}{6}$

(c) $\sqrt[3]{-\dfrac{8}{125}} = \sqrt[3]{\dfrac{-8}{125}} = \dfrac{\sqrt[3]{-8}}{\sqrt[3]{125}} = \dfrac{-2}{5} = -\dfrac{2}{5}$

(d) $\sqrt[3]{\dfrac{7}{216}} = \dfrac{\sqrt[3]{7}}{\sqrt[3]{216}} = \dfrac{\sqrt[3]{7}}{6}$

(e) $\sqrt{\dfrac{m^4}{25}} = \dfrac{\sqrt{m^4}}{\sqrt{25}} = \dfrac{m^2}{5}$ ∎

WORK PROBLEM 3 AT THE SIDE.

3 One of the main uses of the product and quotient rules is in simplifying radicals. A radical is **simplified** if the following four conditions are met.

SIMPLIFIED RADICAL

1. The radicand has no factor raised to a power greater than or equal to the index.

2. The radicand has no fractions.

3. No denominator contains a radical.

4. Exponents in the radicand and the index of the radical have no common factors (except 1).

5.3 SIMPLIFYING RADICALS

EXAMPLE 4 Simplifying Radicals Involving Numbers
Simplify each radical.

(a) $\sqrt{24}$

Check to see if 24 is divisible by one of the perfect squares 4, 9, ..., and choose the largest perfect square (square of a natural number) that divides into 24. The largest such number is 4. Write 24 as the product of 4 and 6, and then use the product rule.

$$\sqrt{24} = \sqrt{4 \cdot 6} = \sqrt{4} \cdot \sqrt{6} = 2\sqrt{6}$$

(b) $\sqrt{108}$

The number 108 is divisible by the perfect square 36. If this is not obvious, try factoring 108 into its prime factors.

$$\sqrt{108} = \sqrt{2^2 \cdot 3^3}$$
$$= \sqrt{2^2 \cdot 3^2 \cdot 3}$$
$$= 2 \cdot 3\sqrt{3} \quad \text{Product rule}$$
$$= 6\sqrt{3}$$

(c) $\sqrt{500} = \sqrt{100 \cdot 5} = \sqrt{100} \cdot \sqrt{5} = 10\sqrt{5}$

(d) $\sqrt{10}$

No perfect square (other than 1) divides into 10, so $\sqrt{10}$ cannot be simplified further.

(e) $\sqrt[3]{16}$

Look for the largest perfect *cube* that divides into 16. The number 8 satisfies this condition, so write 16 as 8 · 2.

$$\sqrt[3]{16} = \sqrt[3]{8 \cdot 2} = \sqrt[3]{8} \cdot \sqrt[3]{2} = 2\sqrt[3]{2}$$

(f) $\sqrt[4]{162} = \sqrt[4]{81 \cdot 2}$ 81 is a perfect 4th power.
$$= \sqrt[4]{81} \cdot \sqrt[4]{2}$$
$$= 3\sqrt[4]{2} \quad \blacksquare$$

WORK PROBLEM 4 AT THE SIDE.

EXAMPLE 5 Simplifying Radicals Involving Variables
Simplify. Assume that all variables represent positive real numbers.

(a) $\sqrt{16m^3} = \sqrt{16m^2 \cdot m}$
$$= \sqrt{16m^2} \cdot \sqrt{m}$$
$$= 4m\sqrt{m}$$

No absolute value bars are needed around the *m* because of the assumption that all the variables represent *positive* real numbers.

(b) $\sqrt{200k^7q^8} = \sqrt{10^2 \cdot 2 \cdot (k^3)^2 \cdot k \cdot (q^4)^2}$
$$= 10k^3q^4\sqrt{2k}$$

---

4. Simplify.

(a) $\sqrt{8}$

(b) $\sqrt{32}$

(c) $\sqrt{45}$

(d) $\sqrt{300}$

(e) $\sqrt{35}$

(f) $\sqrt[3]{54}$

(g) $\sqrt[4]{243}$

ANSWERS
4. (a) $2\sqrt{2}$ (b) $4\sqrt{2}$ (c) $3\sqrt{5}$
    (d) $10\sqrt{3}$ (e) cannot be further simplified   (f) $3\sqrt[3]{2}$ (g) $3\sqrt[4]{3}$

CHAPTER 5 ROOTS AND RADICALS

5. Simplify. Assume all variables represent positive real numbers.

(a) $\sqrt{25p^7}$

(b) $\sqrt{72y^3 x}$

(c) $\sqrt[3]{y^7 x^5 z^6}$

(d) $\sqrt[4]{64a^8 b^6}$

(c) $\sqrt{75p^6 q^9} = \sqrt{(25p^6 q^8)(3q)} = \sqrt{25p^6 q^8} \cdot \sqrt{3q} = 5p^3 q^4 \sqrt{3q}$

(d) $\sqrt[3]{8x^4 y^5} = \sqrt[3]{(8x^3 y^3)(xy^2)} = \sqrt[3]{8x^3 y^3} \cdot \sqrt[3]{xy^2} = 2xy\sqrt[3]{xy^2}$

(e) $\sqrt[4]{32y^9} = \sqrt[4]{(16y^8)(2y)} = \sqrt[4]{16y^8} \cdot \sqrt[4]{2y} = 2y^2 \sqrt[4]{2y}$ ■

■ WORK PROBLEM 5 AT THE SIDE.

4 The conditions for a simplifed radical given earlier state that an exponent in the radicand and the index of the radical should have no common factor (except 1). The next example shows how to simplify radicals with such common factors.

EXAMPLE 6 Simplifying Radicals Using Smaller Indexes
Simplify. Assume that all variables represent positive real numbers.

(a) $\sqrt[9]{5^6}$

Write this radical using rational exponents and then write the exponent in lowest terms. Express the answer as a radical.

$$\sqrt[9]{5^6} = 5^{6/9} = 5^{2/3} = \sqrt[3]{5^2} \text{ or } \sqrt[3]{25}$$

(b) $\sqrt[4]{p^2} = p^{2/4} = p^{1/2} = \sqrt{p}$ (Recall the assumption $p > 0$.) ■

These examples suggest the following rule.

If m is an integer, n is a positive integer, k is a positive integer, and a is a positive real number,

$$\sqrt[kn]{a^{km}} = \sqrt[n]{a^m}.$$

---

6. Simplify. Assume all variables represent positive real numbers.

(a) $\sqrt[12]{2^3}$

(b) $\sqrt[6]{t^2}$

■ WORK PROBLEM 6 AT THE SIDE.

The final example shows how to simplify the product of two radicals having different indexes.

EXAMPLE 7 Multiplying Radicals with Different Indexes
Simplify $\sqrt{7} \cdot \sqrt[3]{2}$.

Since the indexes, 2 and 3, have a least common index of 6, use rational exponents to write each radical as a sixth root.

$$\sqrt{7} = 7^{1/2} = 7^{3/6} = \sqrt[6]{7^3} = \sqrt[6]{343}$$

$$\sqrt[3]{2} = 2^{1/3} = 2^{2/6} = \sqrt[6]{2^2} = \sqrt[6]{4}$$

$$\sqrt{7} \cdot \sqrt[3]{2} = \sqrt[6]{343} \cdot \sqrt[6]{4} = \sqrt[6]{1372} \quad \text{Product rule} \quad ■$$

7. Simplify $\sqrt{5} \cdot \sqrt[3]{4}$.

■ WORK PROBLEM 7 AT THE SIDE.

ANSWERS
5. (a) $5p^3\sqrt{p}$ (b) $6y\sqrt{2yx}$
   (c) $y^2 xz^2 \sqrt[3]{yx^2}$
   (d) $2a^2 b\sqrt[4]{4b^2}$
6. (a) $\sqrt[4]{2}$ (b) $\sqrt[3]{t}$
7. $\sqrt[6]{2000}$

5.3 EXERCISES

Multiply. See Examples 1 and 2.

1. $\sqrt{7} \cdot \sqrt{11}$
2. $\sqrt{2} \cdot \sqrt{5}$
3. $\sqrt{14} \cdot \sqrt{3}$
4. $\sqrt{15} \cdot \sqrt{10}$

5. $\sqrt[3]{2} \cdot \sqrt[3]{7}$
6. $\sqrt[3]{5} \cdot \sqrt[3]{9}$
7. $\sqrt[4]{8} \cdot \sqrt[4]{3}$
8. $\sqrt[4]{9} \cdot \sqrt[4]{5}$

Simplify each radical. Assume that all variables represent positive real numbers. See Example 3.

9. $\sqrt{\dfrac{16}{25}}$
10. $\sqrt{\dfrac{49}{64}}$
11. $\sqrt{\dfrac{7}{25}}$
12. $\sqrt{\dfrac{11}{49}}$

13. $\sqrt{\dfrac{r}{100}}$
14. $\sqrt{\dfrac{p}{4}}$
15. $\sqrt{\dfrac{m^8}{25}}$
16. $\sqrt{\dfrac{z^4}{81}}$

17. $\sqrt[3]{-\dfrac{27}{64}}$
18. $\sqrt[3]{\dfrac{125}{216}}$
19. $\sqrt[3]{\dfrac{k^2}{125}}$
20. $\sqrt[3]{\dfrac{p}{216}}$

Write each of the following in simplified form. See Example 4.

21. $\sqrt{18}$
22. $\sqrt{12}$
23. $\sqrt{72}$
24. $\sqrt{288}$

25. $-\sqrt{48}$
26. $-\sqrt{32}$
27. $-\sqrt{24}$
28. $-\sqrt{28}$

29. $\sqrt{60}$
30. $\sqrt{80}$
31. $\sqrt{150}$
32. $\sqrt{300}$

CHAPTER 5 ROOTS AND RADICALS

33. $\sqrt[3]{-16}$ 34. $\sqrt[3]{-250}$ 35. $\sqrt[3]{24}$ 36. $\sqrt[3]{128}$

37. $\sqrt[3]{375}$ 38. $\sqrt[3]{40}$ 39. $\sqrt[4]{32}$ 40. $\sqrt[4]{162}$

41. $-\sqrt[4]{1250}$ 42. $-\sqrt[4]{512}$ 43. $\sqrt[5]{128}$ 44. $\sqrt[5]{486}$

Write each of the following in simplified form. Assume that all variables represent positive real numbers. See Example 5.

45. $\sqrt{25k^2}$ 46. $\sqrt{18m^2}$ 47. $\sqrt[3]{\dfrac{32}{125}}$ 48. $\sqrt[3]{\dfrac{81}{1000}}$

49. $\sqrt{100y^{10}}$ 50. $\sqrt{256z^6}$ 51. $-\sqrt[3]{8k^9}$ 52. $-\sqrt[3]{27y^{15}}$

53. $-\sqrt{144m^{10}z^2}$ 54. $-\sqrt{4k^2z^{18}}$ 55. $-\sqrt[3]{-125m^9b^{18}c^{24}}$ 56. $\sqrt[3]{-216y^{12}x^3z^{18}}$

57. $\sqrt[4]{\dfrac{1}{16}m^{12}x^{16}}$ 58. $\sqrt[4]{\dfrac{81}{256}k^4m^8}$ 59. $\sqrt{75y^3}$ 60. $\sqrt{200z^3}$

61. $\sqrt{1000m^9}$ 62. $\sqrt{800k^{15}}$ 63. $\sqrt{7x^5y^6}$ 64. $\sqrt{12k^9p^{12}}$

65. $\sqrt[3]{8z^9r^{12}}$

66. $\sqrt[3]{125k^{15}n^9}$

67. $\sqrt[3]{-16z^4t^{11}}$

68. $\sqrt[3]{-81w^7y^8}$

69. $\sqrt[4]{16a^8b^{12}}$

70. $\sqrt[4]{81z^{16}y^{20}}$

71. $\sqrt[4]{32k^5m^{10}}$

72. $\sqrt[4]{162r^{15}s^{10}}$

73. $\sqrt{\dfrac{m^9}{16}}$

74. $\sqrt{\dfrac{y^{15}}{100}}$

75. $\sqrt[3]{\dfrac{y^{10}}{27}}$

76. $\sqrt[3]{\dfrac{r^{26}}{125}}$

Simplify each of the following. Assume that all variables represent positive real numbers. See Example 6.

77. $\sqrt[4]{12^2}$

78. $\sqrt[6]{21^3}$

79. $\sqrt[10]{m^{15}}$

80. $\sqrt[8]{y^6}$

Write as radicals with the same index and simplify. See Example 7.

81. $\sqrt[3]{5} \cdot \sqrt{3}$

82. $\sqrt[3]{3} \cdot \sqrt[4]{6}$

83. $\sqrt[3]{2} \cdot \sqrt[5]{3}$

84. $\sqrt[4]{8} \cdot \sqrt[5]{3}$

CHAPTER 5 ROOTS AND RADICALS

Work each problem. Give answers as simplified radicals.

85. The illumination I produced by a light source is related to the distance d from the light source by the equation

$$d = \sqrt{\frac{k}{I}}$$

where k is a constant. If $k = 700$, how far from the light source will the illumination be 12 footcandles?

86. The length of the diagonal of a box is $d = \sqrt{L^2 + W^2 + H^2}$, where L, W, and H are the length, width, and height of the box. Find the length of the diagonal of a box that is $9\frac{1}{5}$ inches long, $8\frac{2}{5}$ inches wide, and 2 inches deep.

SKILL SHARPENERS
Combine terms. See Section 3.3.

87. $8x^2 - 4x^2 + 2x^2$

88. $3p^3 - 5p^3 - 8p^2$

89. $9q + 2q - 5q^2 - q^2$

90. $7z^5 - 2z^3 + 8z^5 - z^3$

91. $16a^5 - 9a^2 + 4a$

92. $12x^4 - 5x^3 + 2x^2$

5.4 ADDITION AND SUBTRACTION OF RADICAL EXPRESSIONS

OBJECTIVES

1. Define a radical expression.
2. Simplify sums and differences of radical expressions by using the distributive property.

1 A **radical expression** is an algebraic expression that contains radicals. For example,

$$\sqrt[4]{3} + \sqrt{6}, \quad \sqrt{2} - 1, \quad \text{and} \quad \sqrt{8} - \sqrt{2}$$

are radical expressions. The examples in the preceding section required simplifying radical expressions involving only multiplication and division. Now we will simplify radical expressions that involve addition and subtraction.

2 An expression such as $4\sqrt{2} + 3\sqrt{2}$ can be simplified by using the distributive property.

$$4\sqrt{2} + 3\sqrt{2} = (4 + 3)\sqrt{2} = 7\sqrt{2}$$

As another example, $2\sqrt{3} - 5\sqrt{3} = (2 - 5)\sqrt{3} = -3\sqrt{3}$. This is much like simplifying $2x + 3x$ to $5x$ or $5y - 8y$ to $-3y$.

Only radical expressions with the same index and the same radicand may be combined. Expressions such as $5\sqrt{3} + 2\sqrt{2}$ and $3\sqrt{3} + 2\sqrt[3]{3}$ cannot be simplified.

EXAMPLE 1 Adding and Subtracting Radicals

Add or subtract the following radical expressions.

(a) $3\sqrt{24} + \sqrt{54}$

Begin by simplifying each radical; then use the distributive property.

$$3\sqrt{24} + \sqrt{54} = 3\sqrt{4} \cdot \sqrt{6} + \sqrt{9} \cdot \sqrt{6}$$
$$= 3 \cdot 2\sqrt{6} + 3\sqrt{6}$$
$$= 6\sqrt{6} + 3\sqrt{6}$$
$$= 9\sqrt{6}$$

(b) $-3\sqrt{8} + 4\sqrt{18} = -3\sqrt{4}\sqrt{2} + 4\sqrt{9}\sqrt{2}$
$$= -3 \cdot 2\sqrt{2} + 4 \cdot 3\sqrt{2}$$
$$= -6\sqrt{2} + 12\sqrt{2} = 6\sqrt{2}$$

(c) $2\sqrt{20x} - \sqrt{45x} = 2\sqrt{4}\sqrt{5x} - \sqrt{9}\sqrt{5x}$
$$= 2 \cdot 2\sqrt{5x} - 3\sqrt{5x}$$
$$= 4\sqrt{5x} - 3\sqrt{5x} = \sqrt{5x} \quad \text{(if } x \geq 0\text{)}$$

(d) $2\sqrt{125} - 3\sqrt{180} + 2\sqrt{500}$
$$= 2\sqrt{25} \cdot \sqrt{5} - 3\sqrt{36} \cdot \sqrt{5} + 2\sqrt{100} \cdot \sqrt{5}$$
$$= 2 \cdot 5\sqrt{5} - 3 \cdot 6\sqrt{5} + 2 \cdot 10\sqrt{5}$$
$$= 10\sqrt{5} - 18\sqrt{5} + 20\sqrt{5} = 12\sqrt{5}$$

(e) $2\sqrt{3} - 4\sqrt{5}$

Here the radicals differ and are already simplified, so $2\sqrt{3} - 4\sqrt{5}$ cannot be simplified further. ■

WORK PROBLEM 1 AT THE SIDE.

1. Simplify.

(a) $3\sqrt{5} + 7\sqrt{5}$

(b) $2\sqrt{11} - \sqrt{11} + 3\sqrt{44}$

(c) $5\sqrt{12y} + 6\sqrt{75y}, \quad y \geq 0$

(d) $3\sqrt{8} - 6\sqrt{50} + 2\sqrt{200}$

(e) $9\sqrt{5} - 4\sqrt{10}$

ANSWERS
1. (a) $10\sqrt{5}$ (b) $7\sqrt{11}$ (c) $40\sqrt{3y}$
   (d) $-4\sqrt{2}$ (e) cannot be further simplified

CHAPTER 5 ROOTS AND RADICALS

2. Simplify each radical expression. Assume that all variables represent positive real numbers.

   (a) $7\sqrt[3]{81} + 3\sqrt[3]{24}$

   (b) $6\sqrt[3]{54} + 5\sqrt[3]{16}$

   (c) $-2\sqrt[4]{32} - 7\sqrt[4]{162}$

   (d) $\sqrt[3]{p^4q^7} - \sqrt[3]{64pq}$

EXAMPLE 2 Adding and Subtracting Radicals with Higher Indexes

Add or subtract the following radical expressions. Assume that all variables represent positive real numbers.

(a) $2\sqrt[3]{16} - 5\sqrt[3]{54} = 2\sqrt[3]{8 \cdot 2} - 5\sqrt[3]{27 \cdot 2}$
$= 2\sqrt[3]{8} \cdot \sqrt[3]{2} - 5\sqrt[3]{27} \cdot \sqrt[3]{2}$
$= 2 \cdot 2 \cdot \sqrt[3]{2} - 5 \cdot 3 \cdot \sqrt[3]{2}$
$= 4\sqrt[3]{2} - 15\sqrt[3]{2}$
$= -11\sqrt[3]{2}$

(b) $2\sqrt[3]{x^2y} + \sqrt[3]{8x^5y^4} = 2\sqrt[3]{x^2y} + \sqrt[3]{(8x^3y^3)x^2y}$
$= 2\sqrt[3]{x^2y} + 2xy\sqrt[3]{x^2y}$
$= (2 + 2xy)\sqrt[3]{x^2y}$ Distributive property ■

▶ **WORK PROBLEM 2 AT THE SIDE.**

CAUTION It is a common error to forget to write the index when working with cube roots, fourth roots, and so on. Be careful not to forget to write these indexes.

ANSWERS
2. (a) $27\sqrt[3]{3}$ (b) $28\sqrt[3]{2}$
   (c) $-25\sqrt[4]{2}$ (d) $(pq^2 - 4)\sqrt[3]{pq}$

5.4 EXERCISES

Simplify. Assume that all variables represent positive real numbers. See Examples 1 and 2.

1. $\sqrt{36} + \sqrt{100}$

2. $\sqrt{25} + \sqrt{81}$

3. $2\sqrt{48} + 3\sqrt{75}$

4. $3\sqrt{32} - 2\sqrt{8}$

5. $4\sqrt{12} - 7\sqrt{27}$

6. $3\sqrt{80} + 5\sqrt{45}$

7. $6\sqrt{18} + \sqrt{32} - 2\sqrt{50}$

8. $5\sqrt{8} - 3\sqrt{72} + 3\sqrt{50}$

9. $2\sqrt{63} - 2\sqrt{28} + 3\sqrt{7}$

10. $6\sqrt{27} - 2\sqrt{48} + \sqrt{75}$

11. $2\sqrt{5} - 3\sqrt{20} - 4\sqrt{45}$

12. $5\sqrt{54} + 2\sqrt{24} - 2\sqrt{96}$

13. $2\sqrt{40} + 6\sqrt{90} - 3\sqrt{160}$

14. $5\sqrt{28} - 3\sqrt{63} + 2\sqrt{112}$

15. $3\sqrt{2x} - \sqrt{8x} - \sqrt{72x}$

16. $4\sqrt{18k} - \sqrt{72k} + 4\sqrt{50k}$

17. $3\sqrt{72m^2} + 2\sqrt{32m^2} - 3\sqrt{18m^2}$

18. $9\sqrt{27p^2} - 4\sqrt{108p^2} - 2\sqrt{48p^2}$

19. $\sqrt[3]{54} - 2\sqrt[3]{16}$

20. $5\sqrt[3]{81} - 4\sqrt[3]{24}$

21. $2\sqrt[3]{27x} + 2\sqrt[3]{8x}$

22. $6\sqrt[3]{128m} - 3\sqrt[3]{16m}$

23. $5\sqrt[4]{32} + 3\sqrt[4]{162}$

24. $2\sqrt[4]{512} - 4\sqrt[4]{32}$

25. $\sqrt[3]{x^2y} - \sqrt[3]{8x^2y}$

26. $3\sqrt[3]{x^2y^2} - 2\sqrt[3]{64x^2y^2}$

CHAPTER 5   ROOTS AND RADICALS

27. $3x\sqrt[3]{xy^2} - 2\sqrt[3]{8x^4y^2}$ **28.** $6q^2\sqrt[3]{5q} - 2q\sqrt[3]{40q^4}$ **29.** $5\sqrt[4]{32} + 3\sqrt[4]{162}$ **30.** $2\sqrt[4]{512} - 4\sqrt[4]{32}$

31. $2\sqrt[4]{32a^3} + 5\sqrt[4]{2a^3}$ **32.** $-\sqrt[4]{16r} + 5\sqrt[4]{r}$ **33.** $3\sqrt[4]{x^5y} - 2x\sqrt[4]{xy}$ **34.** $2\sqrt[4]{m^9p^6} - 3m^2p\sqrt[4]{mp^2}$

Work the following problems. Give answers as simplified radicals.

35. If the lengths of the sides of a triangle are $2\sqrt{45}$, $\sqrt{75}$, and $3\sqrt{20}$ centimeters, find the perimeter.

36. A rectangular yard has a length of $\sqrt{192}$ meters and a width of $\sqrt{48}$ meters. What is the perimeter?

37. Find the perimeter of a lot with sides measuring $3\sqrt{18}$, $2\sqrt{32}$, $4\sqrt{50}$, and $5\sqrt{12}$ yards.

38. What is the perimeter of a triangle with sides measuring $3\sqrt{54}$, $4\sqrt{24}$, and $\sqrt{80}$ meters?

39. Find decimal approximations for $\sqrt{3}$ and $\sqrt{12}$. Do the approximations suggest that $\sqrt{12} = 2\sqrt{3}$?

40. Find decimal approximations for $2\sqrt{7}$ and $\dfrac{14}{\sqrt{7}}$. Do the approximations suggest that $\dfrac{14}{\sqrt{7}} = 2\sqrt{7}$?

SKILL SHARPENERS
Multiply. See Section 5.3.

41. $\sqrt{5} \cdot \sqrt{5}$ **42.** $\sqrt{7} \cdot \sqrt{7}$ **43.** $\sqrt[3]{4} \cdot \sqrt[3]{2}$ **44.** $\sqrt[4]{2} \cdot \sqrt[4]{8}$

Simplify. See Section 3.4.

45. $-6p(2p^2 - 1)$ **46.** $8r(6r^2 - 5)$ **47.** $(3a - 2b)(5a + 7b)$ **48.** $(8p - 9q)(8p + 9q)$

5.5 MULTIPLICATION AND DIVISION OF RADICAL EXPRESSIONS

OBJECTIVES
1. Multiply binomial expressions involving radicals.
2. Rationalize denominators involving a single square root.
3. Rationalize denominators in roots of fractions.
4. Rationalize denominators involving cube roots.
5. Rationalize denominators with binomials involving radicals.
6. Write radical expressions in lowest terms.

1 We can multiply binomial expressions involving radicals by using the FOIL (First, Outside, Inside, Last) method. For example, the product of the binomials $\sqrt{5} + 3$ and $\sqrt{6} + 1$ is found as follows.

$$(\sqrt{5} + 3)(\sqrt{6} + 1) = \overbrace{\sqrt{5} \cdot \sqrt{6}}^{\text{First}} + \overbrace{\sqrt{5} \cdot 1}^{\text{Outside}} + \overbrace{3 \cdot \sqrt{6}}^{\text{Inside}} + \overbrace{3 \cdot 1}^{\text{Last}}$$
$$= \sqrt{30} + \sqrt{5} + 3\sqrt{6} + 3$$

This result cannot be simplified further.

EXAMPLE 1 Multiplying Binomials Involving Radical Expressions
Multiply.

(a) $(7 - \sqrt{3})(\sqrt{5} + \sqrt{2}) = 7\sqrt{5} + 7\sqrt{2} - \sqrt{3} \cdot \sqrt{5} - \sqrt{3} \cdot \sqrt{2}$
$\qquad = 7\sqrt{5} + 7\sqrt{2} - \sqrt{15} - \sqrt{6}$

(b) $(\sqrt{10} + \sqrt{3})(\sqrt{10} - \sqrt{3})$
$\qquad = \sqrt{10}\sqrt{10} - \sqrt{10}\sqrt{3} + \sqrt{10}\sqrt{3} - \sqrt{3}\sqrt{3}$
$\qquad = 10 - 3$
$\qquad = 7$

Notice that this is an example of the kind of product that results in the difference of two squares:

$$(a + b)(a - b) = a^2 - b^2.$$

Here, $a = \sqrt{10}$ and $b = \sqrt{3}$.

(c) $(\sqrt{7} - 3)^2 = (\sqrt{7} - 3)(\sqrt{7} - 3)$
$\qquad = \sqrt{7} \cdot \sqrt{7} - 3\sqrt{7} - 3\sqrt{7} + 3 \cdot 3$
$\qquad = 7 - 6\sqrt{7} + 9$
$\qquad = 16 - 6\sqrt{7}$

(d) $(5 - \sqrt[3]{3})(5 + \sqrt[3]{3}) = 5 \cdot 5 + 5\sqrt[3]{3} - 5\sqrt[3]{3} - \sqrt[3]{3} \cdot \sqrt[3]{3}$
$\qquad = 25 - \sqrt[3]{3^2}$
$\qquad = 25 - \sqrt[3]{9}$ ■

WORK PROBLEM 1 AT THE SIDE.

2 Before the advent of hand-held calculators, in order to find a decimal approximation for a radical expression such as $\frac{1}{\sqrt{2}}$, it was advantageous to rewrite the expression so that no radical appeared in the denominator. This made it easier to perform the division. While this is no longer necessary because of the availability of calculators, there are other reasons for rationalizing, so it is still worthwhile to learn how to do it. A radical expression is not considered to be in simplified form if the denominator contains a radical.

The process of removing radicals from the denominator so that the denominator contains only rational quantities is called **rationalizing the denominator**. The next few examples show how this process works.

1. Multiply.

    (a) $(2 + \sqrt{3})(1 + \sqrt{5})$

    (b) $(\sqrt{6} + \sqrt{2})(\sqrt{5} - \sqrt{7})$

    (c) $(4 + \sqrt{3})(4 - \sqrt{3})$

    (d) $(4 + \sqrt[3]{7})(4 - \sqrt[3]{7})$

ANSWERS
1. (a) $2 + 2\sqrt{5} + \sqrt{3} + \sqrt{15}$
   (b) $\sqrt{30} - \sqrt{42} + \sqrt{10} - \sqrt{14}$
   (c) 13   (d) $16 - \sqrt[3]{49}$

CHAPTER 5 ROOTS AND RADICALS

2. Rationalize the denominators.

(a) $\dfrac{8}{\sqrt{3}}$

(b) $\dfrac{9}{\sqrt{6}}$

(c) $\dfrac{\sqrt{3}}{\sqrt{7}}$

(d) $\dfrac{\sqrt{10}}{\sqrt{3}}$

3. Rationalize the denominators.

(a) $\dfrac{7}{\sqrt{8}}$

(b) $\dfrac{3}{\sqrt{48}}$

(c) $\dfrac{9}{\sqrt{20}}$

(d) $\dfrac{-16}{\sqrt{32}}$

ANSWERS

2. (a) $\dfrac{8\sqrt{3}}{3}$ (b) $\dfrac{3\sqrt{6}}{2}$
(c) $\dfrac{\sqrt{21}}{7}$ (d) $\dfrac{\sqrt{30}}{3}$

3. (a) $\dfrac{7\sqrt{2}}{4}$ (b) $\dfrac{\sqrt{3}}{4}$
(c) $\dfrac{9\sqrt{5}}{10}$ (d) $-2\sqrt{2}$

EXAMPLE 2 Rationalizing Denominators with Square Roots

Rationalize each denominator.

(a) $\dfrac{3}{\sqrt{7}}$

Multiply numerator and denominator by $\sqrt{7}$.

$$\dfrac{3}{\sqrt{7}} = \dfrac{3 \cdot \sqrt{7}}{\sqrt{7} \cdot \sqrt{7}}$$

Since $\sqrt{7} \cdot \sqrt{7} = \sqrt{7 \cdot 7} = 7$,

$$\dfrac{3}{\sqrt{7}} = \dfrac{3\sqrt{7}}{7}.$$

The radical is now removed from the denominator.

(b) $\dfrac{5\sqrt{2}}{\sqrt{5}}$

Multiply the numerator and denominator by $\sqrt{5}$.

$$\dfrac{5\sqrt{2}}{\sqrt{5}} = \dfrac{5\sqrt{2} \cdot \sqrt{5}}{\sqrt{5} \cdot \sqrt{5}} = \dfrac{5\sqrt{10}}{5} = \sqrt{10} \quad \blacksquare$$

WORK PROBLEM 2 AT THE SIDE.

EXAMPLE 3 Rationalizing Denominators with Square Roots

Rationalize each denominator.

(a) $\dfrac{6}{\sqrt{12}}$

Less work is involved if the radical in the denominator is simplified first.

$$\dfrac{6}{\sqrt{12}} = \dfrac{6}{\sqrt{4 \cdot 3}} = \dfrac{6}{2\sqrt{3}} = \dfrac{3}{\sqrt{3}}$$

Now rationalize the denominator by multiplying numerator and denominator by $\sqrt{3}$.

$$\dfrac{3 \cdot \sqrt{3}}{\sqrt{3} \cdot \sqrt{3}} = \dfrac{3\sqrt{3}}{3} = \sqrt{3}$$

(b) $\dfrac{9}{\sqrt{72}}$

First simplify $\sqrt{72}$ in the denominator.

$$\dfrac{9}{\sqrt{72}} = \dfrac{9}{\sqrt{36 \cdot 2}} = \dfrac{9}{6\sqrt{2}} = \dfrac{3}{2\sqrt{2}}$$

Now multiply the numerator and denominator by $\sqrt{2}$.

$$\dfrac{3}{2\sqrt{2}} = \dfrac{3 \cdot \sqrt{2}}{2\sqrt{2} \cdot \sqrt{2}} = \dfrac{3\sqrt{2}}{2 \cdot 2} = \dfrac{3\sqrt{2}}{4} \quad \blacksquare$$

WORK PROBLEM 3 AT THE SIDE.

5.5 MULTIPLICATION AND DIVISION OF RADICAL EXPRESSIONS

3 The next example shows how to rationalize denominators in expressions involving quotients under a radical sign.

EXAMPLE 4 Rationalizing Denominators in Roots of Fractions

Simplify each of the following.

(a) $\sqrt{\dfrac{18}{125}}$

$\sqrt{\dfrac{18}{125}} = \dfrac{\sqrt{18}}{\sqrt{125}}$ Quotient rule

$= \dfrac{\sqrt{9 \cdot 2}}{\sqrt{25 \cdot 5}}$ Factor.

$= \dfrac{3\sqrt{2}}{5\sqrt{5}}$ Product rule

$= \dfrac{3\sqrt{2} \cdot \sqrt{5}}{5\sqrt{5} \cdot \sqrt{5}}$ Multiply by $\sqrt{5}$ in numerator and denominator.

$= \dfrac{3\sqrt{10}}{5 \cdot 5}$ Product rule

$= \dfrac{3\sqrt{10}}{25}$

(b) $\sqrt{\dfrac{50m^4}{p^5}}, \quad p > 0$

$\sqrt{\dfrac{50m^4}{p^5}} = \dfrac{\sqrt{50m^4}}{\sqrt{p^5}}$ Quotient rule

$= \dfrac{5m^2\sqrt{2}}{p^2\sqrt{p}}$ Product rule

$= \dfrac{5m^2\sqrt{2} \cdot \sqrt{p}}{p^2\sqrt{p} \cdot \sqrt{p}}$ Multiply by $\sqrt{p}$ in numerator and denominator.

$= \dfrac{5m^2\sqrt{2p}}{p^2 \cdot p}$ Product rule

$= \dfrac{5m^2\sqrt{2p}}{p^3}$ ∎

WORK PROBLEM 4 AT THE SIDE.

4. Simplify. Assume all variables represent positive real numbers.

(a) $\sqrt{\dfrac{8}{45}}$

(b) $\sqrt{\dfrac{25}{m}}$

(c) $\sqrt{\dfrac{72}{y}}$

(d) $\sqrt{\dfrac{200k^6}{y^7}}$

ANSWERS

4. (a) $\dfrac{2\sqrt{10}}{15}$ (b) $\dfrac{5\sqrt{m}}{m}$

(c) $\dfrac{6\sqrt{2y}}{y}$ (d) $\dfrac{10k^3\sqrt{2y}}{y^4}$

5. Simplify.

(a) $\sqrt[3]{\dfrac{1}{4}}$

(b) $\sqrt[3]{\dfrac{125}{36}}$

(c) $\sqrt[3]{\dfrac{15}{32}}$

(d) $\sqrt[3]{\dfrac{m^{12}}{n}}, \quad n \neq 0$

ANSWERS

5. (a) $\dfrac{\sqrt[3]{2}}{2}$ (b) $\dfrac{5\sqrt[3]{6}}{6}$
   (c) $\dfrac{\sqrt[3]{30}}{4}$ (d) $\dfrac{m^4\sqrt[3]{n^2}}{n}$

---

4 The next example shows how denominators are rationalized in expressions with higher roots, such as cube roots.

EXAMPLE 5 Rationalizing a Denominator Involving a Cube Root

Simplify $\sqrt[3]{\dfrac{27}{16}}$.

Use the quotient rule.

$$\sqrt[3]{\dfrac{27}{16}} = \dfrac{\sqrt[3]{27}}{\sqrt[3]{16}} = \dfrac{3}{\sqrt[3]{16}}$$

Simplify the denominator: $\sqrt[3]{16} = \sqrt[3]{8 \cdot 2} = 2\sqrt[3]{2}$. To get a rational denominator, multiply numerator and denominator by a number that will produce a perfect cube under the radical in the denominator. Since

$$\sqrt[3]{2} \cdot \sqrt[3]{2^2} = \sqrt[3]{2^3} = 2,$$

multiply numerator and denominator by $\sqrt[3]{2^2}$ or $\sqrt[3]{4}$.

$$\sqrt[3]{\dfrac{27}{16}} = \dfrac{3}{2\sqrt[3]{2}} = \dfrac{3 \cdot \sqrt[3]{4}}{2\sqrt[3]{2} \cdot \sqrt[3]{4}} = \dfrac{3\sqrt[3]{4}}{2\sqrt[3]{8}} = \dfrac{3\sqrt[3]{4}}{2 \cdot 2} = \dfrac{3\sqrt[3]{4}}{4} \quad \blacksquare$$

CAUTION It is easy to make mistakes in problems like the one in Example 5. A typical error is to multiply numerator and denominator by $\sqrt[3]{2}$, forgetting that

$$\sqrt[3]{2} \cdot \sqrt[3]{2} \neq 2.$$

You must think of a number that, when multiplied by 2, gives a perfect *cube* under the radical. As shown in the example, 4 satisfies this condition.

WORK PROBLEM 5 AT THE SIDE.

5 Recall the special product

$$(a + b)(a - b) = a^2 - b^2.$$

In order to rationalize a denominator that contains a binomial expression (one that contains exactly two terms) involving radicals, such as

$$\dfrac{3}{1 + \sqrt{2}},$$

we must use conjugates. The **conjugate** of $1 + \sqrt{2}$ is $1 - \sqrt{2}$. In general, $a + b$ and $a - b$ are conjugates.

Whenever a radical expression has a sum or difference with square root radicals in the denominator, rationalize by multiplying both the numerator and the denominator by the conjugate of the denominator.

5.5 MULTIPLICATION AND DIVISION OF RADICAL EXPRESSIONS

For the expression $\frac{3}{1+\sqrt{2}}$, rationalize the denominator by multiplying both the numerator and denominator by the conjugate of the denominator.

$$\frac{3}{1+\sqrt{2}} = \frac{3(1-\sqrt{2})}{(1+\sqrt{2})(1-\sqrt{2})}$$

According to the special product mentioned earlier, $(1+\sqrt{2})(1-\sqrt{2}) = 1^2 - (\sqrt{2})^2 = 1 - 2 = -1$. Placing -1 in the denominator gives

$$= \frac{3(1-\sqrt{2})}{-1}$$

which simplifies to $-3(1-\sqrt{2})$, or $-3 + 3\sqrt{2}$.

EXAMPLE 6 Rationalizing a Binomial Denominator

Rationalize the denominator of $\frac{5}{4-\sqrt{3}}$.

Multiply numerator and denominator by $4 + \sqrt{3}$.

$$\frac{5}{4-\sqrt{3}} = \frac{5(4+\sqrt{3})}{(4-\sqrt{3})(4+\sqrt{3})}$$

$$= \frac{5(4+\sqrt{3})}{16-3}$$

$$= \frac{5(4+\sqrt{3})}{13}$$

Notice that the numerator is left in factored form. Doing this makes it easier to determine whether the expression can be reduced to lowest terms. ■

WORK PROBLEM 6 AT THE SIDE.

EXAMPLE 7 Rationalizing Binomial Denominators

Rationalize each denominator.

(a) $\frac{\sqrt{2}-\sqrt{3}}{\sqrt{5}+\sqrt{3}} = \frac{(\sqrt{2}-\sqrt{3})(\sqrt{5}-\sqrt{3})}{(\sqrt{5}+\sqrt{3})(\sqrt{5}-\sqrt{3})}$

$$= \frac{\sqrt{10}-\sqrt{6}-\sqrt{15}+3}{5-3}$$

$$= \frac{\sqrt{10}-\sqrt{6}-\sqrt{15}+3}{2}$$

(b) $\frac{3}{\sqrt{5m}-\sqrt{p}} = \frac{3(\sqrt{5m}+\sqrt{p})}{(\sqrt{5m}-\sqrt{p})(\sqrt{5m}+\sqrt{p})}$

$$= \frac{3(\sqrt{5m}+\sqrt{p})}{5m-p} \quad (5m \neq p, m > 0, p > 0) \quad ■$$

6. Rationalize the denominators.

(a) $\frac{7}{3-\sqrt{2}}$

(b) $\frac{-4}{\sqrt{5}+2}$

(c) $\frac{3}{2-\sqrt{7}}$

(d) $\frac{18}{9+\sqrt{18}}$

ANSWERS

6. (a) $3 + \sqrt{2}$ (b) $-4(\sqrt{5}-2)$
   (c) $-(2+\sqrt{7})$ (d) $\frac{2(9-3\sqrt{2})}{7}$

CHAPTER 5 ROOTS AND RADICALS

7. Rationalize the denominators.

(a) $\dfrac{15}{\sqrt{7} + \sqrt{2}}$

(b) $\dfrac{-6}{\sqrt{11} - \sqrt{7}}$

(c) $\dfrac{7}{\sqrt{2} + \sqrt{13}}$

(d) $\dfrac{\sqrt{3} + \sqrt{5}}{\sqrt{2} - \sqrt{7}}$

(e) $\dfrac{2}{\sqrt{k} + \sqrt{z}}$, $k \neq z$, $k > 0$, $z > 0$

8. Write in lowest terms.

(a) $\dfrac{15 - 5\sqrt{3}}{5}$

(b) $\dfrac{24 - 36\sqrt{7}}{16}$

ANSWERS
7. (a) $3(\sqrt{7} - \sqrt{2})$
 (b) $\dfrac{-3(\sqrt{11} + \sqrt{7})}{2}$
 (c) $\dfrac{-7(\sqrt{2} - \sqrt{13})}{11}$
 (d) $\dfrac{-(\sqrt{6} + \sqrt{21} + \sqrt{10} + \sqrt{35})}{5}$
 (e) $\dfrac{2(\sqrt{k} - \sqrt{z})}{k - z}$
8. (a) $3 - \sqrt{3}$ (b) $\dfrac{6 - 9\sqrt{7}}{4}$

WORK PROBLEM 7 AT THE SIDE.

6 The final example shows how to write radical expressions in lowest terms.

EXAMPLE 8 Writing Radical Expressions in Lowest Terms
Write in lowest terms.

(a) $\dfrac{6 + 2\sqrt{5}}{4}$

Factor the numerator, then write in lowest terms.

$$\dfrac{6 + 2\sqrt{5}}{4} = \dfrac{2(3 + \sqrt{5})}{4}$$
$$= \dfrac{3 + \sqrt{5}}{2}$$

Here is an alternative method to write this expression in lowest terms.

$$\dfrac{6 + 2\sqrt{5}}{4} = \dfrac{6}{4} + \dfrac{2\sqrt{5}}{4}$$
$$= \dfrac{3}{2} + \dfrac{\sqrt{5}}{2}$$
$$= \dfrac{3 + \sqrt{5}}{2}$$

(b) $\dfrac{5y - \sqrt{8y^2}}{6y} = \dfrac{5y - 2y\sqrt{2}}{6y}$
$$= \dfrac{y(5 - 2\sqrt{2})}{6y}$$
$$= \dfrac{5 - 2\sqrt{2}}{6} \quad (y > 0) \quad \blacksquare$$

CAUTION Refer to part (a) in Example 8. A common error occurs when students try to write in lowest terms *before* factoring. Be sure to factor before attempting to write in lowest terms.

WORK PROBLEM 8 AT THE SIDE.

5.5 EXERCISES

Multiply, then simplify the products. Assume that all variables represent positive real numbers. See Example 1.

1. $\sqrt{3}(\sqrt{12} + 2)$

2. $\sqrt{5}(\sqrt{15} - \sqrt{5})$

3. $\sqrt{2}(\sqrt{18} + \sqrt{3})$

4. $(\sqrt{3} + 2)(\sqrt{3} - 2)$

5. $(\sqrt{5} - 1)(\sqrt{5} + 1)$

6. $(\sqrt{2} - \sqrt{3})(\sqrt{2} + \sqrt{3})$

7. $(\sqrt{7} + \sqrt{3})(\sqrt{7} - \sqrt{3})$

8. $(\sqrt{8} - \sqrt{2})(\sqrt{8} + \sqrt{2})$

9. $(\sqrt{20} - \sqrt{5})(\sqrt{20} + \sqrt{5})$

10. $(\sqrt{11} + \sqrt{2})(\sqrt{11} - \sqrt{2})$

11. $(\sqrt{7} - \sqrt{5})(\sqrt{7} + \sqrt{5})$

12. $(\sqrt{3} + 1)(\sqrt{5} - 1)$

13. $(\sqrt{2} + 3)(\sqrt{3} - 2)$

14. $(3\sqrt{x} - \sqrt{5})(2\sqrt{x} + 1)$

15. $(4\sqrt{p} + \sqrt{7})(\sqrt{p} - 9)$

16. $(3\sqrt{r} - \sqrt{s})(3\sqrt{r} + \sqrt{s})$

17. $(\sqrt{k} + 4\sqrt{m})(\sqrt{k} - 4\sqrt{m})$

18. $(5\sqrt{z} + 1)^2$

19. $(6\sqrt{a} - 5)^2$

20. $(2\sqrt{p} - 3)^2$

21. $(4\sqrt{t} + 1)^2$

22. $(\sqrt[3]{2y} - 5)(4\sqrt[3]{2y} + 1)$

23. $(\sqrt[3]{9z} - 2)(5\sqrt[3]{9z} + 7)$

CHAPTER 5 ROOTS AND RADICALS

Rationalize the denominators. Assume that all variables represent positive real numbers. See Examples 2 and 3.

24. $\dfrac{5}{\sqrt{5}}$ 25. $\dfrac{12}{\sqrt{3}}$ 26. $\dfrac{50}{\sqrt{15}}$ 27. $\dfrac{\sqrt{7}}{\sqrt{5}}$ 28. $\dfrac{\sqrt{5}}{\sqrt{11}}$

29. $\dfrac{9\sqrt{3}}{\sqrt{5}}$ 30. $\dfrac{12\sqrt{7}}{\sqrt{2}}$ 31. $\dfrac{3\sqrt{2}}{\sqrt{11}}$ 32. $\dfrac{2\sqrt{15}}{\sqrt{2}}$ 33. $\dfrac{3}{\sqrt{8}}$

34. $\dfrac{5}{\sqrt{18}}$ 35. $\dfrac{9}{\sqrt{20}}$ 36. $\dfrac{7}{\sqrt{27}}$ 37. $\dfrac{-6\sqrt{5}}{\sqrt{12}}$ 38. $\dfrac{-4\sqrt{3}}{\sqrt{32}}$

39. $\dfrac{8\sqrt{3}}{\sqrt{k}}$ 40. $\dfrac{6\sqrt{7}}{\sqrt{r}}$ 41. $\dfrac{5\sqrt{2m}}{\sqrt{y^3}}$ 42. $\dfrac{2\sqrt{5r}}{\sqrt{m^3}}$

Simplify. Assume all variables represent positive real numbers. See Example 4.

43. $\sqrt{\dfrac{9}{2}}$ 44. $\sqrt{\dfrac{10}{3}}$ 45. $-\sqrt{\dfrac{18}{7}}$ 46. $\sqrt{\dfrac{7}{8}}$ 47. $\sqrt{\dfrac{3}{50}}$

48. $\sqrt{\dfrac{16}{x}}$ 49. $\sqrt{\dfrac{25}{y}}$ 50. $\sqrt{\dfrac{18}{m}}$ 51. $\sqrt{\dfrac{72}{r}}$ 52. $-\sqrt{\dfrac{48k^2}{z}}$

53. $-\sqrt{\dfrac{75m^3}{n}}$ 54. $\sqrt{\dfrac{32p^4}{q^3}}$ 55. $\sqrt{\dfrac{72x^8}{y^3}}$ 56. $\sqrt{\dfrac{288a^5}{b^7}}$ 57. $\sqrt{\dfrac{1000r^7}{s^5}}$

5.5 EXERCISES

Rationalize the denominators. Assume that all variables represent nonzero real numbers. See Example 5.

58. $\sqrt[3]{\dfrac{3}{4}}$

59. $\sqrt[3]{\dfrac{1}{25}}$

60. $\sqrt[3]{\dfrac{9}{32}}$

61. $\sqrt[3]{\dfrac{10}{9}}$

62. $\sqrt[3]{\dfrac{x^6}{y}}$

63. $\sqrt[3]{\dfrac{m^9}{n}}$

64. $\sqrt[3]{\dfrac{r^{15}}{s^8}}$

65. $\sqrt[3]{\dfrac{p^{12}}{q^{11}}}$

Rationalize the denominators. Assume that all variables represent positive real numbers and no denominators are 0. See Examples 6 and 7.

66. $\dfrac{3}{4 + \sqrt{5}}$

67. $\dfrac{4}{3 - \sqrt{7}}$

68. $\dfrac{2}{\sqrt{5} + 2}$

69. $\dfrac{3}{\sqrt{3} - 1}$

70. $\dfrac{2}{\sqrt{2} + \sqrt{5}}$

71. $\dfrac{-5}{\sqrt{3} - \sqrt{7}}$

72. $\dfrac{\sqrt{8}}{3 - \sqrt{2}}$

73. $\dfrac{\sqrt{27}}{2 + \sqrt{3}}$

74. $\dfrac{\sqrt{2} - 1}{\sqrt{2} + \sqrt{3}}$

75. $\dfrac{\sqrt{3} + 1}{\sqrt{5} + \sqrt{3}}$

76. $\dfrac{2 - \sqrt{3}}{\sqrt{6} - \sqrt{5}}$

77. $\dfrac{5 + \sqrt{6}}{\sqrt{3} - \sqrt{2}}$

78. $\dfrac{3\sqrt{x}}{\sqrt{x} - 2\sqrt{y}}$

79. $\dfrac{5\sqrt{k}}{2\sqrt{k} + \sqrt{q}}$

80. $\dfrac{\sqrt{m} - \sqrt{3r}}{\sqrt{m} + \sqrt{3r}}$

81. $\dfrac{3\sqrt{2p} + \sqrt{5s}}{\sqrt{2p} - 3\sqrt{5s}}$

Write in lowest terms. Assume that all variables represent positive real numbers. See Example 8.

82. $\dfrac{15 - 10\sqrt{6}}{10}$

83. $\dfrac{9 - 6\sqrt{5}}{12}$

84. $\dfrac{3 - 3\sqrt{5}}{3}$

85. $\dfrac{-5 + 5\sqrt{2}}{5}$

86. $\dfrac{16 - 4\sqrt{8}}{12}$

87. $\dfrac{12 - 9\sqrt{72}}{18}$

88. $\dfrac{6p - \sqrt{24p^3}}{3p}$

89. $\dfrac{11y - \sqrt{242y^5}}{22y}$

SKILL SHARPENERS

Solve each equation. See Sections 2.1 and 3.10.

90. $8x - 7 = 9$

91. $-2r + 3 = 5$

92. $6m^2 = 7m + 3$

93. $15a^2 + 2 = 11a$

Give the square of each expression. Assume each radicand is positive.

94. $\sqrt{x - 12}$

95. $\sqrt{5 - 3x^2}$

96. $\sqrt{x^2 - 2x - 3}$

5.6 EQUATIONS WITH RADICALS

1 The equation $x = 1$ has only one solution. Its solution set is $\{1\}$. If we square both sides of this equation, another equation is obtained: $x^2 = 1$. This new equation has two solutions: -1 and 1. Notice that the solution of the original equation is also a solution of the squared equation. However, the squared equation has another solution, -1, that is *not* a solution of the original equation.

When solving equations with radicals, we will use this idea of raising both sides to a power. It is an application of the *power rule*.

POWER RULE

If both sides of an equation are raised to the same power, all solutions of the original equation are also solutions of the new equation.

Read the power rule carefully—it does *not* say that all solutions to the new equation are solutions to the original equation. They may or may not be.

When the power rule is used to solve an equation, every solution to the new equation must be checked in the original equation.

Solutions that do not satisfy the original equation are called **extraneous**; they must be discarded.

2 The first example shows how to use the power rule in solving an equation.

EXAMPLE 1 Using the Power Rule
Solve $\sqrt{3x + 4} = 8$.

Use the power rule and square both sides to get

$$(\sqrt{3x + 4})^2 = 8^2$$
$$3x + 4 = 64$$
$$3x = 60$$
$$x = 20.$$

Check the proposed solution in the original equation.

$$\sqrt{3x + 4} = 8$$
$$\sqrt{3 \cdot 20 + 4} = 8 \quad \text{Let } x = 20.$$
$$\sqrt{64} = 8$$
$$8 = 8 \quad \text{True}$$

Since 20 satisfies the *original* equation, the solution set is $\{20\}$. ∎

WORK PROBLEM 1 AT THE SIDE.

OBJECTIVES

1 Learn the power rule.

2 Solve radical equations such as $\sqrt{3x + 4} = 8$.

3 Solve radical equations such as $\sqrt{4 - x} = x + 2$.

4 Solve radical equations such as $\sqrt{5m + 6} + \sqrt{3m + 4} = 2$.

5 Solve radical equations with roots higher than 2, such as $\sqrt[4]{m + 5} = \sqrt[4]{2m - 6}$.

1. Solve.

(a) $\sqrt{r} = 3$

(b) $\sqrt{5x + 1} = 4$

(c) $\sqrt{12m + 9} = 3$

ANSWERS
1. (a) $\{9\}$ (b) $\{3\}$ (c) $\{0\}$

CHAPTER 5 ROOTS AND RADICALS

2. Solve.

(a) $\sqrt{k} + 4 = -3$

(b) $\sqrt{x - 9} - 3 = 0$

(c) $\sqrt{p + 2} + 5 = 0$

The solution of the equation in Example 1 can be generalized to give a method for solving equations with radicals.

SOLVING EQUATIONS WITH RADICALS

Step 1 Make sure that one radical is alone on one side of the equals sign.

Step 2 Raise each side of the equation to a power that is the same as the index of the radical.

Step 3 Solve the resulting equation; if it still contains a radical, repeat Steps 1 and 2.

Step 4 It is essential that all potential solutions be checked in the *original* equation.

EXAMPLE 2 Using the Power Rule

Solve $\sqrt{5q - 1} + 3 = 0$.

To get the radical alone on one side, subtract 3 from both sides.

$$\sqrt{5q - 1} = -3$$

Now square both sides.

$$(\sqrt{5q - 1})^2 = (-3)^2$$
$$5q - 1 = 9$$
$$5q = 10$$
$$q = 2$$

The potential solution, 2, must be checked by substituting it in the original equation.

$$\sqrt{5q - 1} + 3 = 0$$
$$\sqrt{5 \cdot 2 - 1} + 3 = 0 \quad \text{Let } q = 2.$$
$$3 + 3 = 0 \quad \text{False}$$

This false result shows that 2 is *not* a solution of the original equation; it is extraneous. The solution set is ∅. ∎

NOTE We could have determined after the first step that the equation in Example 2 had no solution. The equation $\sqrt{5q - 1} = -3$ has no solution because the expression on the left cannot be negative.

■ **WORK PROBLEM 2 AT THE SIDE.**

3 The next examples involve finding the square of a binomial. Recall from Chapter 3 that $(x + y)^2 = x^2 + 2xy + y^2$.

ANSWERS
2. (a) ∅ (b) {18} (c) ∅

5.6 EQUATIONS WITH RADICALS

EXAMPLE 3 Using the Power Rule; Squaring a Binomial
Solve $\sqrt{4 - x} = x + 2$.

Square both sides; the square of $x + 2$ is $(x + 2)^2 = x^2 + 4x + 4$.

$$(\sqrt{4 - x})^2 = (x + 2)^2$$
$$4 - x = x^2 + 4x + 4$$

↑ Twice the product of 2 and x

$$0 = x^2 + 5x \quad \text{Subtract 4 and add } x.$$
$$0 = x(x + 5) \quad \text{Factor.}$$
$$x = 0 \quad \text{or} \quad x + 5 = 0 \quad \text{Zero-factor property}$$
$$x = -5.$$

Check each potential solution in the original equation.

If $x = 0$,
$\sqrt{4 - x} = x + 2$
$\sqrt{4 - 0} = 0 + 2$
$\sqrt{4} = 2$
$2 = 2.$ True

If $x = -5$,
$\sqrt{4 - x} = x + 2$
$\sqrt{4 - (-5)} = -5 + 2$
$\sqrt{9} = -3$
$3 = -3.$ False

The solution set is $\{0\}$. The other potential solution, -5, is extraneous. ∎

CAUTION Errors are often made when a radical equation involves squaring a binomial, as in Example 3. The middle term is often omitted. Remember that $(x + 2)^2 \neq x^2 + 4$ because $(x + 2)^2 = x^2 + 4x + 4$.

WORK PROBLEM 3 AT THE SIDE.

EXAMPLE 4 Using the Power Rule; Squaring a Binomial
Solve $\sqrt{m^2 - 4m + 9} = m - 1$.

Square both sides. The square of the binomial $m - 1$ is $(m - 1)^2 = m^2 - 2(m)(1) + 1^2$.

$$(\sqrt{m^2 - 4m + 9})^2 = (m - 1)^2$$
$$m^2 - 4m + 9 = m^2 - 2m + 1$$

↑——— Twice the product of m and -1

Subtract m^2 and 1 from both sides, and then add $4m$ to both sides to get

$$8 = 2m$$
$$4 = m.$$

Check this potential solution in the original equation.

$$\sqrt{m^2 - 4m + 9} = m - 1$$
$$\sqrt{4^2 - 4 \cdot 4 + 9} = 4 - 1 \quad \text{Let } m = 4.$$
$$3 = 3 \quad \text{True}$$

The solution set of the given equation is $\{4\}$. ∎

WORK PROBLEM 4 AT THE SIDE.

3. Solve each equation.

(a) $\sqrt{3z - 5} = z - 1$

(b) $\sqrt{4q + 5} = 2q - 5$

4. Solve each equation.

(a) $\sqrt{1 - 2p + p^2} = p + 1$

(b) $\sqrt{4a^2 + 2a - 3} = 2a + 7$

ANSWERS
3. (a) $\{2, 3\}$ (b) $\{5\}$
4. (a) $\{0\}$ (b) $\{-2\}$

CHAPTER 5 ROOTS AND RADICALS

5. Solve each equation.

(a) $\sqrt{p+1} - \sqrt{p-4} = 1$

(b) $\sqrt{2y+3} + \sqrt{y+1} = 1$

ANSWERS
5. (a) $\{8\}$ (b) $\{-1\}$

---

4 The next example shows an equation in which both sides must be squared twice.

EXAMPLE 5 Using the Power Rule; Squaring Twice

Solve $\sqrt{5m+6} + \sqrt{3m+4} = 2$.

Start by getting one radical alone on one side of the equals sign. Do this by subtracting $\sqrt{3m+4}$ from both sides.

$$\sqrt{5m+6} = 2 - \sqrt{3m+4}$$

Now square both sides.

$$(\sqrt{5m+6})^2 = (2 - \sqrt{3m+4})^2$$
$$5m + 6 = 4 \;\mathbf{-\; 4\sqrt{3m+4}}\; + (3m+4)$$

Twice the product of 2 and $-\sqrt{3m+4}$

This equation still contains a radical, so it will be necessary to square both sides again. Before doing this, combine terms on the right.

$$5m + 6 = 8 + 3m - 4\sqrt{3m+4}$$
$$2m - 2 = -4\sqrt{3m+4} \qquad \text{Subtract 8 and } 3m.$$
$$m - 1 = -2\sqrt{3m+4} \qquad \text{Divide by 2.}$$

Now square both sides again.

$$(m-1)^2 = (-2\sqrt{3m+4})^2$$
$$m^2 - 2m + 1 = (-2)^2(\sqrt{3m+4})^2 \qquad (ab)^2 = a^2b^2$$
$$m^2 - 2m + 1 = 4(3m+4)$$
$$m^2 - 2m + 1 = 12m + 16 \qquad \text{Distributive property}$$

This equation is quadratic and may be solved with the zero-factor property. Start by getting 0 on one side of the equation; then factor.

$$m^2 - 14m - 15 = 0$$
$$(m - 15)(m + 1) = 0$$

By the zero-factor property,

$$m - 15 = 0 \quad \text{or} \quad m + 1 = 0$$
$$m = 15 \quad \text{or} \quad m = -1.$$

Check each of these potential solutions in the original equation. Only -1 works, so the equation has solution set $\{-1\}$. The other potential solution, 15, is extraneous. ∎

WORK PROBLEM 5 AT THE SIDE.

5 The power rule also works for powers higher than 2, as the next example shows.

5.6 EQUATIONS WITH RADICALS

EXAMPLE 6 Using the Power Rule; Higher Powers

Solve $\sqrt[4]{m+5} = \sqrt[4]{2m-6}$.

Raise both sides to the fourth power.

$$(\sqrt[4]{m+5})^4 = (\sqrt[4]{2m-6})^4$$
$$m + 5 = 2m - 6$$
$$11 = m$$

Check this result in the original equation.

$$\sqrt[4]{m+5} = \sqrt[4]{2m-6}$$
$$\sqrt[4]{11+5} = \sqrt[4]{2 \cdot 11 - 6} \quad \text{Let } m = 11.$$
$$\sqrt[4]{16} = \sqrt[4]{16} \quad \text{True}$$

The solution set is {11}. ∎

WORK PROBLEM 6 AT THE SIDE.

6. Solve each equation.

(a) $\sqrt[3]{p^2 + 3p + 12} = \sqrt[3]{p^2}$

(b) $\sqrt[4]{2m - 2} = 2$

(c) $\sqrt[4]{2k + 5} + 1 = 0$

ANSWERS
6. (a) {−4} (b) {9} (c) ∅

Historical Reflections

WOMEN IN MATHEMATICS:
Ada Augusta, Countess of Lovelace (1815–1852)

Ada Augusta, the daughter of British poet Lord Byron, has been described as the world's first computer programmer. She was one of the few people who understood what Charles Babbage was trying to accomplish with his calculating machines (forerunners of today's computers). She provided some corrections to his work, and on her own, she devised the concept of repetition of instructions known today as a "loop" or "subroutine" in computer programming.

Lady Lovelace met an untimely death at age 36 from cancer of the uterus, and her programming ideas were not implemented for another century. In 1987 the U.S. Department of Defense solidified its insistence that contractors designing programs to govern military aircraft and missile systems use a language called *Ada*, named for Lady Lovelace and first introduced in 1980. As a result of this standardization move, it was expected that the demand for Ada programmers would quadruple by the 1990s.

Art: From *Faster than Thought: A Symposium on Digital Computing Machines*. B.V. Browden, Pitman Publishing Corp. © 1953. Used with permission.

5.6 EXERCISES

Solve each equation. See Examples 1 and 2.

1. $\sqrt{k} = 2$
2. $\sqrt{p} = 5$
3. $\sqrt{r-2} = 3$

4. $\sqrt{y+1} = 7$
5. $\sqrt{6k-1} = 1$
6. $\sqrt{7n-3} = 5$

7. $\sqrt{a} + 5 = 0$
8. $\sqrt{y} + 2 = 0$
9. $\sqrt{w} - 3 = 0$

10. $\sqrt{z} - 8 = 0$
11. $\sqrt{3k+1} - 4 = 0$
12. $\sqrt{5z+1} - 11 = 0$

13. $4 - \sqrt{x-2} = 0$
14. $9 - \sqrt{4k+1} = 0$
15. $\sqrt{r+1} = \sqrt{2r-3}$

16. $\sqrt{2z+3} = \sqrt{3z-5}$
17. $\sqrt{9a-4} = \sqrt{8a+1}$
18. $\sqrt{4p-2} = \sqrt{3p+5}$

19. $\sqrt{5k-2} - \sqrt{4k+3} = 0$
20. $\sqrt{8p-4} - \sqrt{7p+2} = 0$
21. $2\sqrt{k} = \sqrt{3k+4}$

CHAPTER 5 ROOTS AND RADICALS

Solve each equation. See Examples 3 and 4.

22. $\sqrt{3y + 10} = y + 2$

23. $\sqrt{6q - 5} = 2q - 5$

24. $\sqrt{4r + 13} = r + 4$

25. $\sqrt{7z + 50} = z + 8$

26. $\sqrt{k^2 + 9k + 3} = -k$

27. $\sqrt{p^2 - 15p + 15} = p - 5$

28. $\sqrt{r^2 + 12r - 4} + 4 - r = 0$

29. $\sqrt{m^2 + 3m + 12} - m - 2 = 0$

Solve each equation. See Example 5.

30. $\sqrt{k + 2} - \sqrt{k - 3} = 1$

31. $\sqrt{r + 6} - \sqrt{r - 2} = 2$

32. $\sqrt{x + 2} + \sqrt{x - 1} = 3$

33. $\sqrt{y + 4} + \sqrt{y - 4} = 4$

34. $\sqrt{3p + 4} - \sqrt{2p - 4} = 2$

35. $\sqrt{5y + 6} - \sqrt{y + 3} = 3$

5.6 EXERCISES

Solve each equation. See Example 6.

36. $\sqrt[3]{a^2 + 5a + 1} = \sqrt[3]{a^2 + 4a}$

37. $\sqrt[3]{r^2 + 2r + 8} = \sqrt[3]{r^2}$

38. $\sqrt[3]{2m - 1} = \sqrt[3]{m + 13}$

39. $\sqrt[3]{2k - 11} - \sqrt[3]{5k + 1} = 0$

40. $\sqrt[4]{a + 8} = \sqrt[4]{2a}$

41. $\sqrt[4]{z + 11} = \sqrt[4]{2z + 6}$

42. $\sqrt[3]{x - 8} + 2 = 0$

43. $\sqrt[3]{r + 1} + 1 = 0$

For the following equations, rewrite the expressions with rational exponents as radical expressions, and then solve. Be sure to check for extraneous solutions.

44. $(5p - 6)^{1/2} - (3p - 6)^{1/2} = 2$

45. $(3y + 7)^{1/2} - (y + 2)^{1/2} = 1$

46. $(2p - 1)^{2/3} = p^{1/3}$

47. $(m^2 - 2m)^{1/3} - m^{1/3} = 0$

319

CHAPTER 5 ROOTS AND RADICALS

Work each problem. Give answers as decimals to the nearest tenth.

48. Carpenters stabilize wall frames with a diagonal brace as shown in the figure. The length of the brace is given by $L = \sqrt{H^2 + W^2}$. If the bottom of the brace is attached 9 feet from the corner and the brace is 12 feet long, how far up the corner post should it be nailed?

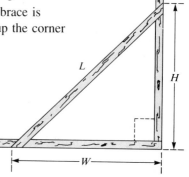

49. The period of a pendulum in seconds depends on its length L in feet, and is given by

$$P = 2\pi \sqrt{\frac{L}{32}}.$$

How long is a pendulum with a period of 2 seconds? Use 3.14 for π.

SKILL SHARPENERS
Perform the indicated operations. See Sections 3.3 and 3.4.

50. $(5 + 9x) + (-4 - 8x)$

51. $(12 + 7y) - (-3 + 2y)$

52. $(x + 3)(2x - 5)$

Simplify each radical expression. Rationalize the denominator. See Section 5.5.

53. $\dfrac{2}{4 + \sqrt{3}}$

54. $\dfrac{-7}{5 - \sqrt{2}}$

55. $\dfrac{\sqrt{2} + \sqrt{7}}{\sqrt{5} + \sqrt{3}}$

5.7 COMPLEX NUMBERS

As discussed in Chapter 1, the set of real numbers includes as subsets many other number sets (the rational numbers, integers, and natural numbers, for example). In this section a new set of numbers is introduced—one that includes the set of real numbers as a subset.

1 The equation $x^2 + 1 = 0$ has no real number solutions, since any solution must be a number whose square is -1. In the set of real numbers all squares are nonnegative numbers, because multiplication is defined in such a way that the product of two positive numbers or two negative numbers is always positive. To provide a solution for the equation $x^2 + 1 = 0$, a new number, i, is defined so that

$$i^2 = -1.$$

That is, i is a number whose square is -1. This definition of i makes it possible to define any square root of a negative number as follows.

OBJECTIVES

1. Simplify numbers of the form $\sqrt{-b}$, where $b > 0$.
2. Recognize a complex number.
3. Add and subtract complex numbers.
4. Find products of complex numbers.
5. Find quotients of complex numbers.
6. Find powers of i.

For any positive number b,

$$\sqrt{-b} = i\sqrt{b}.$$

EXAMPLE 1 Simplifying Square Roots of Negative Numbers
Write each number as a product of a real number and i.

(a) $\sqrt{-100} = i\sqrt{100} = 10i$

(b) $\sqrt{-2} = i\sqrt{2}$ ■

CAUTION It is easy to mistake $\sqrt{2}i$ for $\sqrt{2i}$, with the i under the radical. For this reason, it is common to write $\sqrt{2}i$ as $i\sqrt{2}$.

WORK PROBLEM 1 AT THE SIDE.

When finding a product such as $\sqrt{-4} \cdot \sqrt{-9}$, the product rule for radicals cannot be used, since that rule applies only when no more than one radicand is negative. For this reason, change $\sqrt{-b}$ to the form $i\sqrt{b}$ before performing any multiplications or divisions. For example,

$$\sqrt{-4} \cdot \sqrt{-9} = i\sqrt{4} \cdot i\sqrt{9}$$
$$= i \cdot 2 \cdot i \cdot 3$$
$$= 6i^2.$$

Since $i^2 = -1$, $6i^2 = 6(-1) = -6$. An *incorrect* use of the product rule for radicals would give a *wrong* answer.

$$\sqrt{-4} \cdot \sqrt{-9} = \sqrt{(-4)(-9)}$$
$$= \sqrt{36}$$
$$= 6. \qquad \text{Incorrect}$$

1. Write each as a product of a real number and i.

   (a) $\sqrt{-16}$

   (b) $\sqrt{-25}$

   (c) $-\sqrt{-81}$

   (d) $\sqrt{-7}$

   (e) $\sqrt{-15}$

ANSWERS
1. (a) $4i$ (b) $5i$ (c) $-9i$
   (d) $i\sqrt{7}$ (e) $i\sqrt{15}$

CHAPTER 5 ROOTS AND RADICALS

2. Multiply.

(a) $\sqrt{-7} \cdot \sqrt{-7}$

(b) $\sqrt{-5} \cdot \sqrt{-10}$

(c) $\sqrt{-6} \cdot \sqrt{-3}$

(d) $\sqrt{-15} \cdot \sqrt{2}$

(e) $\sqrt{-5} \cdot \sqrt{7}$

3. Divide.

(a) $\dfrac{\sqrt{-32}}{\sqrt{-2}}$

(b) $\dfrac{\sqrt{-27}}{\sqrt{-3}}$

(c) $\dfrac{\sqrt{-40}}{\sqrt{10}}$

ANSWERS
2. (a) -7 (b) $-5\sqrt{2}$ (c) $-3\sqrt{2}$
   (d) $i\sqrt{30}$ (e) $i\sqrt{35}$
3. (a) 4 (b) 3 (c) $2i$

EXAMPLE 2 Multiplying Square Roots of Negative Numbers
Multiply.

(a) $\sqrt{-3} \cdot \sqrt{-7} = i\sqrt{3} \cdot i\sqrt{7}$
$= i^2 \sqrt{3 \cdot 7}$
$= (-1)\sqrt{21}$
$= -\sqrt{21}$

(b) $\sqrt{-2} \cdot \sqrt{-8} = i\sqrt{2} \cdot i\sqrt{8}$
$= i^2 \sqrt{2 \cdot 8}$
$= (-1)\sqrt{16}$
$= (-1)4 = -4$

(c) $\sqrt{-5} \cdot \sqrt{6} = i\sqrt{5} \cdot \sqrt{6} = i\sqrt{30}$ ■

■ **WORK PROBLEM 2 AT THE SIDE.**

The methods used to find the products in Example 2 also apply to quotients, as the next example shows.

EXAMPLE 3 Dividing Square Roots of Negative Numbers
Divide.

(a) $\dfrac{\sqrt{-75}}{\sqrt{-3}} = \dfrac{i\sqrt{75}}{i\sqrt{3}} = \sqrt{\dfrac{75}{3}} = \sqrt{25} = 5$

(b) $\dfrac{\sqrt{-32}}{\sqrt{8}} = \dfrac{i\sqrt{32}}{\sqrt{8}} = i\sqrt{\dfrac{32}{8}} = i\sqrt{4} = 2i$ ■

■ **WORK PROBLEM 3 AT THE SIDE.**

2 With the new number i and the real numbers, a new set of numbers can be formed that includes the real numbers as a subset. The *complex numbers* are defined as follows.

If a and b are real numbers, then

$$a + bi$$

is called a **complex number.**

In the complex number $a + bi$, the number a is called the **real part** and b is called the **imaginary part.** When $b = 0$, $a + bi$ is a real number, so the real numbers are a subset of the complex numbers. Complex numbers with $b \neq 0$ are called **imaginary numbers.** In spite of their name, imaginary numbers are very useful in applications, particularly in work with electricity.

The relationships among the various sets of numbers discussed in this book are shown in Figure 1.

5.7 COMPLEX NUMBERS

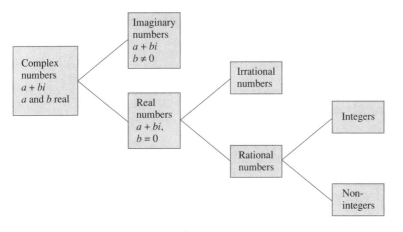

FIGURE 1

3 The commutative, associative, and distributive properties for real numbers are also valid for complex numbers. To add complex numbers, add their real parts and add their imaginary parts.

ADDING COMPLEX NUMBERS

The **sum** of the complex numbers $a + bi$ and $c + di$ is

$$(a + bi) + (c + di) = (a + c) + (b + d)i.$$

EXAMPLE 4 Adding Complex Numbers
Add.

(a) $(2 + 3i) + (6 + 4i)$
$= (2 + 6) + (3 + 4)i$ Commutative and associative properties
$= 8 + 7i$

(b) $5 + (9 - 3i) = (5 + 9) - 3i$
$= 14 - 3i$ ∎

WORK PROBLEM 4 AT THE SIDE.

Since the sum of the complex numbers $8 - 2i$ and $-8 + 2i$ is $0 + 0i$, or just 0, these two numbers are additive inverses of each other. The additive inverse of the complex number $a + bi$ is $-(a + bi) = -a - bi$. Additive inverses are used to subtract complex numbers.

SUBTRACTING COMPLEX NUMBERS

To *subtract* two complex numbers, *add* the first complex number and the additive inverse of the second.

4. Add.

(a) $(4 + 6i) + (-3 + 5i)$

(b) $(-1 + 8i) + (9 - 3i)$

(c) $(6 - 3i) + (8 + 0i)$

(d) $3 + (5 + 2i)$

ANSWERS
4. (a) $1 + 11i$ (b) $8 + 5i$
    (c) $14 - 3i$ (d) $8 + 2i$

CHAPTER 5 ROOTS AND RADICALS

5. Subtract.

(a) $(7 + 3i) - (4 + 2i)$

(b) $(-1 + 2i) - (4 + i)$

(c) $(-6 - i) - (-5 - 4i)$

(d) $8 - (3 - 2i)$

6. Multiply.

(a) $(3 + 5i)(4 - i)$

(b) $(1 - 5i)(3 + 7i)$

(c) $(6 - 4i)(2 + 4i)$

(d) $(3 - 2i)(3 + 2i)$

(e) $(5 + 4i)(5 - 4i)$

ANSWERS
5. (a) $3 + i$ (b) $-5 + i$
   (c) $-1 + 3i$ (d) $5 + 2i$
6. (a) $17 + 17i$ (b) $38 - 8i$
   (c) $28 + 16i$ (d) 13 (e) 41

EXAMPLE 5 Subtracting Complex Numbers
Subtract.

(a) $(6 + 5i) - (3 + 2i) = (6 + 5i)$ **+ (−3 − 2i)**   Additive inverse
$= 3 + 3i$

(b) $(7 - 3i) - (8 - 6i) = (7 - 3i) + (-8 + 6i)$
$= -1 + 3i$

(c) $(-9 + 4i) - (-9 + 8i) = (-9 + 4i) + (9 - 8i)$
$= -4i$ ∎

WORK PROBLEM 5 AT THE SIDE.

4 Complex numbers of the form $a + bi$ have the same form as a binomial, so the product of two complex numbers can be found by using the FOIL method for multiplying binomials. (Recall that FOIL stands for *First-Outside-Inside-Last*.) The next example shows how to apply this method.

EXAMPLE 6 Multiplying Complex Numbers
(a) Multiply $3 + 5i$ and $4 - 2i$.

$$(3 + 5i)(4 - 2i) = \underbrace{3(4)}_{\text{First}} + \underbrace{3(-2i)}_{\text{Outside}} + \underbrace{5i(4)}_{\text{Inside}} + \underbrace{5i(-2i)}_{\text{Last}}.$$

Now simplify. (Remember that $i^2 = -1$.)

$= 12 - 6i + 20i - 10i^2$
$= 12 + 14i - 10(-1)$ Let $i^2 = -1$.
$= 12 + 14i + 10$
$= 22 + 14i$

(b) $(2 + 3i)(1 - 5i) = 2(1) + 2(-5i) + 3i(1) + 3i(-5i)$
$= 2 - 10i + 3i - 15i^2$
$= 2 - 7i - 15(-1)$
$= 2 - 7i + 15$
$= 17 - 7i$ ∎

WORK PROBLEM 6 AT THE SIDE.

The two complex numbers $a + bi$ and $a - bi$ are called **conjugates** of each other. The product of a complex number and its conjugate is always a real number, as shown here.

$(a + bi)(a - bi) = a \cdot a - abi + abi - b^2i^2$
$= a^2 - b^2(-1)$
$\boxed{(a + bi)(a - bi) = a^2 + b^2}$

For example, $(3 + 7i)(3 - 7i) = 3^2 + 7^2 = 9 + 49 = 58$.

5.7 COMPLEX NUMBERS

5 Conjugates are used to find the quotient of two complex numbers, as shown in the next example.

EXAMPLE 7 Dividing Complex Numbers
Find the quotients.

(a) $\dfrac{4 - 3i}{5 + 2i}$

Multiply the numerator and denominator by the conjugate of the denominator. The conjugate of $5 + 2i$ is $5 - 2i$.

$$\frac{4 - 3i}{5 + 2i} = \frac{(4 - 3i)(5 - 2i)}{(5 + 2i)(5 - 2i)}$$

$$= \frac{20 - 8i - 15i + 6i^2}{5^2 + 2^2}$$

$$= \frac{14 - 23i}{29} \text{ or } \frac{14}{29} - \frac{23}{29}i$$

(b) $\dfrac{1 + i}{i}$

The conjugate of i is $-i$. Multiply the numerator and denominator by $-i$.

$$\frac{1 + i}{i} = \frac{(1 + i)(-i)}{i(-i)}$$

$$= \frac{-i - i^2}{-i^2}$$

$$= \frac{-i - (-1)}{-(-1)}$$

$$= \frac{-i + 1}{1} = 1 - i \quad \blacksquare$$

WORK PROBLEM 7 AT THE SIDE.

6 The fact that i^2 is equal to -1 can be used to find higher powers of i, as shown below.

$$i^3 = i \cdot i^2 = i(-1) = -i$$
$$i^4 = i^2 \cdot i^2 = (-1)(-1) = 1$$
$$i^5 = i \cdot i^4 = i \cdot 1 = i$$
$$i^6 = i^2 \cdot i^4 = (-1) \cdot 1 = -1$$
$$i^7 = i^3 \cdot i^4 = (-i) \cdot 1 = -i$$
$$i^8 = i^4 \cdot i^4 = 1 \cdot 1 = 1$$

A few powers of i are listed here.

POWERS OF i

$i^1 = i$	$i^5 = i$	$i^9 = i$	$i^{13} = i$
$i^2 = -1$	$i^6 = -1$	$i^{10} = -1$	$i^{14} = -1$
$i^3 = -i$	$i^7 = -i$	$i^{11} = -i$	$i^{15} = -i$
$i^4 = 1$	$i^8 = 1$	$i^{12} = 1$	$i^{16} = 1$

7. Find the quotients.

(a) $\dfrac{2 + i}{3 - i}$

(b) $\dfrac{6 + 2i}{4 - 3i}$

(c) $\dfrac{5}{3 - 2i}$

(d) $\dfrac{5 - i}{i}$

ANSWERS
7. (a) $\dfrac{1 + i}{2}$ (b) $\dfrac{18 + 26i}{25}$
   (c) $\dfrac{15 + 10i}{13}$ (d) $-1 - 5i$

CHAPTER 5 ROOTS AND RADICALS

8. Find each power of i.

(a) i^7

(b) i^{21}

(c) i^{36}

(d) i^{50}

(e) i^{-9}

As these examples show, the powers of i rotate through the four numbers i, -1, $-i$, and 1. Larger powers of i can be simplified by using the fact that $i^4 = 1$. For example, $i^{75} = (i^4)^{18} \cdot i^3 = 1^{18} \cdot i^3 = 1 \cdot i^3 = i^3 = -i$. This example suggests a quick method for simplifying large powers of i.

SIMPLIFYING LARGE POWERS OF i

Step 1 Divide the exponent by 4.

Step 2 Observe the remainder obtained in Step 1. The large power of i is the same as i raised to the power determined by this remainder. Refer to the chart above to complete the simplification. (If the remainder is 0, the power simplifies to $i^0 = 1$.)

EXAMPLE 8 Simplifying Powers of i

Find each power of i.

(a) $i^{12} = (i^4)^3 = 1^3 = 1$

(b) $i^{39} = i^{36} \cdot i^3$
$= (i^4)^9 \cdot i^3$
$= 1^9 \cdot (-i)$
$= -i$

(c) $i^{-2} = \dfrac{1}{i^2} = \dfrac{1}{-1} = -1$

(d) $i^{-1} = \dfrac{1}{i}$

To simplify this quotient, multiply numerator and denominator by the conjugate of i, which is $-i$.

$$\dfrac{1}{i} = \dfrac{1(-i)}{i(-i)}$$
$$= \dfrac{-i}{-i^2}$$
$$= \dfrac{-i}{-(-1)}$$
$$= \dfrac{-i}{1}$$
$$= -i$$

WORK PROBLEM 8 AT THE SIDE.

ANSWERS

8. (a) $-i$ (b) i (c) 1 (d) -1
   (e) $-i$

5.7 EXERCISES

Simplify. See Examples 1–3.

1. $\sqrt{-144}$

2. $\sqrt{-196}$

3. $-\sqrt{-225}$

4. $-\sqrt{-400}$

5. $\sqrt{-3}$

6. $\sqrt{-19}$

7. $\sqrt{-75}$

8. $\sqrt{-125}$

9. $\sqrt{-500}$

10. $4\sqrt{-12} - 3\sqrt{-3}$

11. $6\sqrt{-18} + 2\sqrt{-32}$

12. $2\sqrt{-45} + 4\sqrt{-80}$

13. $-3\sqrt{-28} + 2\sqrt{-63}$

14. $\sqrt{-5} \cdot \sqrt{-5}$

15. $\sqrt{-3} \cdot \sqrt{-3}$

16. $\sqrt{-9} \cdot \sqrt{-36}$

17. $\sqrt{-4} \cdot \sqrt{-81}$

18. $\sqrt{-16} \cdot \sqrt{-100}$

19. $\dfrac{\sqrt{-200}}{\sqrt{-100}}$

20. $\dfrac{\sqrt{-50}}{\sqrt{-2}}$

21. $\dfrac{\sqrt{-54}}{\sqrt{6}}$

22. $\dfrac{\sqrt{-90}}{\sqrt{10}}$

23. $\dfrac{\sqrt{-288}}{\sqrt{-8}}$

24. $\dfrac{\sqrt{-48}\,\sqrt{-3}}{\sqrt{-2}}$

CHAPTER 5 ROOTS AND RADICALS

Add or subtract as indicated. Write your answers in the form ***a + bi***. *See Examples 4 and 5.*

25. $(6 + 2i) + (4 + 3i)$

26. $(1 + i) + (2 - i)$

27. $(3 - 2i) + (-4 + 5i)$

28. $(7 + 15i) + (-11 - 14i)$

29. $(5 - i) + (-5 + i)$

30. $(-2 + 6i) + (2 - 6i)$

31. $(6 - 3i) - (-2 - 2i)$

32. $(6 - 3i) - (2 + 2i)$

33. $(4 + i) + (-3 - 2i)$

34. $(4 + i) - (3 + 2i)$

35. $(-3 - 4i) - (-1 - i)$

36. $(-2 - 3i) - (5 + 5i)$

37. $(-4 + 11i) + (-2 - 4i) + (-3i)$

38. $(4 + i) + (2 + 5i) + (3 + 2i)$

39. $[(7 + 3i) - (4 - 2i)] + (3 - i)$

40. $[(3 + 2i) + (-4 - i)] - (2 + 5i)$

Multiply. See Example 6.

41. $(2i)(5i)$

42. $(8i)(2i)$

43. $(4i)(7i)$

44. $(-3i)(-2i)$

45. $(-6i)(-5i)$

46. $2i(4 - 3i)$

5.7 EXERCISES

47. $6i(1 - i)$

48. $(3 + 2i)(-1 + 2i)$

49. $(7 + 3i)(-5 + i)$

50. $(4 + 3i)(6 + 5i)$

51. $(-1 + 3i)(1 + 2i)$

52. $(2 + 2i)^2$

53. $(-1 + 3i)^2$

54. $(5 + 6i)(5 - 6i)$

55. $(2 + 3i)(2 - 3i)$

56. $(4 + 2i)(4 - 2i)$

57. $(1 + i)(1 - i)$

58. $(5 - 2i)(5 + 2i)$

Find each quotient. See Example 7.

59. $\dfrac{2}{1 + i}$

60. $\dfrac{-5}{2 - i}$

61. $\dfrac{i}{2 + i}$

62. $\dfrac{-i}{3 - 2i}$

63. $\dfrac{2 - 2i}{1 + i}$

64. $\dfrac{2 - 3i}{2 + 3i}$

65. $\dfrac{1 - i}{1 + i}$

66. $\dfrac{-1 + 5i}{3 + 2i}$

Find each power of i. See Example 8.

67. i^6

68. i^9

69. i^{36}

70. i^{17}

71. i^{26}

72. i^{35}

73. i^{-4}

74. i^{-9}

CHAPTER 5 ROOTS AND RADICALS

Perform the indicated operations. Give answers in the form a + bi.

75. $\dfrac{3}{2-i} + \dfrac{5}{1+i}$

76. $\dfrac{2}{3+4i} + \dfrac{4}{1-i}$

Ohm's law for the current I in a circuit with voltage E, resistance R, capacitance reactance X_c, and inductive reactance X_L is

$$I = \dfrac{E}{R + (X_L - X_c)i}.$$

77. Find I if $E = 2 + 3i$, $R = 5$, $X_L = 4$, and $X_c = 3$.

78. Using the law given for Exercise 77, find E if $I = 1 - i$, $R = 2$, $X_L = 3$, and $X_c = 1$.

Work each problem.

79. Show that $1 - 5i$ is a solution of $x^2 - 2x + 26 = 0$.

80. Show that $3 - 2i$ is a solution of $p^2 - 6p + 13 = 0$.

SKILL SHARPENERS

Give all square roots of each number. See Section 5.1.

81. 25

82. 81

83. 45

Solve each equation. See Section 3.10.

84. $x^2 - 2x - 24 = 0$

85. $x^2 + 3x - 40 = 0$

86. $2x^2 - 5x = 7$

CHAPTER 5 SUMMARY

KEY TERMS

5.1 **square root** — A square root of a number a is a number b that can be squared to give a.

cube root, fourth root — These are defined in a manner similar to square roots (see above).

principal root — For a positive number a and even value of n, the principal nth root of a is the positive nth root of a.

radicand, index — In the expression $\sqrt[n]{a}$, a is the radicand and n is the index.

5.4 **radical expression** — A radical expression is an algebraic expression that contains radicals.

5.5 **rationalizing the denominator** — The process of removing radicals from the denominator so that the denominator contains only rational quantities is called rationalizing the denominator.

5.6 **extraneous solution** — An extraneous solution of a radical equation is a solution of $x = a^2$ that is not a solution of $\sqrt{x} = a$.

5.7 **complex number** — A complex number is a number that can be written in the form $a + bi$, where a and b are real numbers.

real part — The real part of $a + bi$ is a.

imaginary part — The imaginary part of $a + bi$ is b.

imaginary number — A complex number $a + bi$ with $b \neq 0$ is called an imaginary number.

conjugate — The conjugate of $a + bi$ is $a - bi$.

NEW SYMBOLS

$\sqrt{}$	radical sign
$\pm$	plus or minus
$\sqrt[n]{a}$	principal nth root of a
$\approx$	is approximately equal to
i	a number whose square is -1
$a^{1/n}$	a to the power $\frac{1}{n}$
$a^{m/n}$	a to the power $\frac{m}{n}$

QUICK REVIEW

Section	Concepts	Examples
5.1 Radicals	$\sqrt[n]{a} = b$ means $b^n = a$. $\sqrt[n]{a}$ is the principal nth root of a. $\sqrt[n]{a^n} = \lvert a \rvert$ if n is even. $\sqrt[n]{a^n} = a$ if n is odd.	The two square roots of 64 are $\sqrt{64} = 8$ and $-\sqrt{64} = -8$. Of these, 8 is the principal square root of 64. $\sqrt[3]{-27} = -3 \qquad \sqrt[4]{(-2)^4} = \lvert -2 \rvert = 2$

CHAPTER 5 ROOTS AND RADICALS

Section	Concepts	Examples
5.2 Rational Exponents	$a^{1/n} = \sqrt[n]{a}$ whenever $\sqrt[n]{a}$ exists. If m and n are positive integers with m/n in lowest terms, then $$a^{m/n} = (a^{1/n})^m$$ provided that $a^{1/n}$ is a real number. All of the usual rules for exponents are valid for rational exponents.	$25^{1/2} = \sqrt{25} = 5$ $(-64)^{1/3} = \sqrt[3]{-64} = -4$ $8^{5/3} = (8^{1/3})^5 = 2^5 = 32$ $5^{-1/2} \cdot 5^{1/4} = 5^{-1/2 + 1/4} = 5^{-1/4}$ $(y^{2/5})^{10} = y^4$ $\dfrac{x^{-1/3}}{x^{-1/2}} = x^{-1/3 - (-1/2)}$ $\qquad = x^{-1/3 + 1/2} = x^{1/6} \quad (x > 0)$
5.3 Simplifying Radicals	**Product and Quotient Rules for Radicals** If a and b are real numbers, not both negative, all roots are real, and n is a natural number, $$\sqrt[n]{a} \cdot \sqrt[n]{b} = \sqrt[n]{ab}$$ and $$\sqrt[n]{\dfrac{a}{b}} = \dfrac{\sqrt[n]{a}}{\sqrt[n]{b}} \quad (b \neq 0).$$ **Simplified Radical** 1. The radicand has no factor raised to a power greater than or equal to the index. 2. The radicand has no fractions. 3. No denominator contains a radical. 4. Exponents in the radicand and the index of the radical have no common factors (except 1).	$\sqrt{3} \cdot \sqrt{7} = \sqrt{21}$ $\sqrt[5]{x^3 y} \cdot \sqrt[5]{xy^2} = \sqrt[5]{x^4 y^3}$ $\dfrac{\sqrt{x^5}}{\sqrt{x^4}} = \sqrt{\dfrac{x^5}{x^4}} = \sqrt{x} \quad (x > 0)$ $\sqrt{18} = \sqrt{9 \cdot 2} = 3\sqrt{2}$ $\sqrt[3]{54 x^5 y^3} = \sqrt[3]{27 x^3 y^3 \cdot 2x^2} = 3xy\sqrt[3]{2x^2}$ $\sqrt{\dfrac{7}{4}} = \dfrac{\sqrt{7}}{2}$ $\sqrt[9]{x^3} = x^{3/9} = x^{1/3}$ or $\sqrt[3]{x}$
5.4 Addition and Subtraction of Radical Expressions	Only radical expressions with the same index and the same radicand may be combined.	$3\sqrt{17} + 2\sqrt{17} - 8\sqrt{17}$ $\quad = (3 + 2 - 8)\sqrt{17}$ $\quad = -3\sqrt{17}$ $\sqrt[3]{2} - \sqrt[3]{250} = \sqrt[3]{2} - 5\sqrt[3]{2}$ $\qquad\qquad\qquad = -4\sqrt[3]{2}$ $\left.\begin{array}{l}\sqrt{15} + \sqrt{30} \\ \sqrt{3} + \sqrt[3]{9}\end{array}\right\}$ cannot be further simplified.

CHAPTER 5 SUMMARY

Section	Concepts	Examples
5.5 Multiplication and Division of Radical Expressions	Radical expressions may often be multiplied by using the FOIL method. Special products from Section 3.4 may apply.	$(\sqrt{2} + \sqrt{7})(\sqrt{3} - \sqrt{6})$ $= \sqrt{6} - \sqrt{12} + \sqrt{21} - \sqrt{42}$ $= \sqrt{6} - 2\sqrt{3} + \sqrt{21} - \sqrt{42}$ $(\sqrt{5} - \sqrt{10})(\sqrt{5} + \sqrt{10})$ $= 5 - 10 = -5$ $(\sqrt{3} - \sqrt{2})^2$ $= 3 - 2\sqrt{3} \cdot \sqrt{2} + 2$ $= 5 - 2\sqrt{6}$
	Rationalize the denominator by multiplying both the numerator and denominator by the same expression.	$\dfrac{\sqrt{7}}{\sqrt{5}} = \dfrac{\sqrt{7}}{\sqrt{5}} \cdot \dfrac{\sqrt{5}}{\sqrt{5}} = \dfrac{\sqrt{35}}{5}$ $\dfrac{\sqrt[3]{2}}{\sqrt[3]{4}} = \dfrac{\sqrt[3]{2}}{\sqrt[3]{4}} \cdot \dfrac{\sqrt[3]{2}}{\sqrt[3]{2}} = \dfrac{\sqrt[3]{4}}{\sqrt[3]{8}} = \dfrac{\sqrt[3]{4}}{2}$ $\dfrac{4}{\sqrt{5}-\sqrt{2}} = \dfrac{4}{\sqrt{5}-\sqrt{2}} \cdot \dfrac{\sqrt{5}+\sqrt{2}}{\sqrt{5}+\sqrt{2}}$ $= \dfrac{4(\sqrt{5}+\sqrt{2})}{5-2}$ $= \dfrac{4(\sqrt{5}+\sqrt{2})}{3}$
	To write a quotient involving radicals, such as $$\dfrac{5 + 15\sqrt{6}}{10}$$ in lowest terms, factor the numerator and denominator, and then divide both by the greatest common factor.	$\dfrac{5 + 15\sqrt{6}}{10} = \dfrac{5(1 + 3\sqrt{6})}{5 \cdot 2}$ $= \dfrac{1 + 3\sqrt{6}}{2}$
5.6 Equations with Radicals	**Solving Equations with Radicals** *Step 1* Make sure that one radical is alone on one side of the equals sign. *Step 2* Raise each side of the equation to a power that is the same as the index of the radical. *Step 3* Solve the resulting equation; if it still contains a radical, repeat Steps 1 and 2. *Step 4* It is essential that all potential solutions be checked in the *original* equation. Potential solutions that do not check are *extraneous*; they are not part of the solution set.	Solve $\sqrt{2x + 3} - x = 0$. $\sqrt{2x + 3} = x$ $2x + 3 = x^2$ $x^2 - 2x - 3 = 0$ $3(x - 3)(x + 1) = 0$ $x - 3 = 0$ or $x + 1 = 0$ $x = 3$ $\qquad x = -1$ A check shows that 3 is a solution, but -1 is extraneous. The solution set is $\{3\}$.

CHAPTER 5 ROOTS AND RADICALS

Section	Concepts	Examples
5.7 Complex Numbers	$i^2 = -1$ For any positive number b, $$\sqrt{-b} = i\sqrt{b}.$$ To perform a multiplication such as $\sqrt{-3} \cdot \sqrt{-27}$, first change each factor to the form $i\sqrt{b}$, then multiply. The same procedure applies to quotients such as $$\frac{\sqrt{-18}}{\sqrt{-2}}.$$ **Adding and Subtracting Complex Numbers** $(a + bi) + (c + di)$ $= (a + c) + (b + d)i$ $(a + bi) - (c + di)$ $= (a - c) + (b - d)i$ **Multiplying and Dividing Complex Numbers** Multiply complex numbers by using the FOIL method. Divide complex numbers by multiplying the numerator and the denominator by the conjugate of the denominator.	$\sqrt{-3} \cdot \sqrt{-27} = i\sqrt{3} \cdot i\sqrt{27}$ $\phantom{\sqrt{-3} \cdot \sqrt{-27}} = i^2\sqrt{81}$ $\phantom{\sqrt{-3} \cdot \sqrt{-27}} = -1 \cdot 9 = -9$ $\dfrac{\sqrt{-18}}{\sqrt{-2}} = \dfrac{i\sqrt{18}}{i\sqrt{2}} = \dfrac{\sqrt{18}}{\sqrt{2}} = \sqrt{9} = 3$ $(5 + 3i) + (8 - 7i) = 13 - 4i$ $(5 + 3i) - (8 - 7i)$ $= (5 + 3i) + (-8 + 7i)$ $= -3 + 10i$ $(2 + i)(5 - 3i) = 10 - 6i + 5i - 3i^2$ $ = 10 - i - 3(-1)$ $ = 10 - i + 3$ $ = 13 + i$ $\dfrac{2}{3 + i} = \dfrac{2}{3 + i} \cdot \dfrac{3 - i}{3 - i}$ $\phantom{\dfrac{2}{3 + i}} = \dfrac{2(3 - i)}{9 - i^2}$ $\phantom{\dfrac{2}{3 + i}} = \dfrac{2(3 - i)}{10}$ $\phantom{\dfrac{2}{3 + i}} = \dfrac{3 - i}{5}$

CHAPTER 5 REVIEW EXERCISES

[5.1] *Find each of the following real number roots. Use a calculator if necessary.*

1. $\sqrt{1764}$
2. $-\sqrt{289}$
3. $-\sqrt{-841}$
4. $\sqrt[3]{216}$

5. $\sqrt[3]{-125}$
6. $-\sqrt[3]{27z^{12}}$
7. $\sqrt[5]{-32}$
8. $\sqrt[5]{32r^{10}s^{20}}$

Find decimal approximations for each of the following. Round to the nearest thousandth.

9. $\sqrt{40}$
10. $\sqrt{77}$
11. $\sqrt{310}$

[5.2] *Simplify. Assume that all variables represent positive real numbers.*

12. $16^{5/4}$
13. $-8^{2/3}$
14. $-\left(\dfrac{36}{25}\right)^{3/2}$

15. $\left(-\dfrac{1}{8}\right)^{-5/3}$
16. $\left(\dfrac{81}{10,000}\right)^{-3/4}$
17. $5^{1/4} \cdot 5^{7/4}$

18. $\dfrac{96^{2/3}}{96^{-1/3}}$
19. $\dfrac{(p^{5/4})^2}{p^{-3/2}}$
20. $\dfrac{k^{2/3}k^{-1/2}k^{3/4}}{2(k^2)^{-1/4}}$

Simplify by first converting to rational exponents. Convert all answers back to radical form in Exercises 22–24. Assume all variables represent positive real numbers.

21. $\sqrt{3^{18}}$
22. $\sqrt{7^9}$
23. $\sqrt[3]{m^5} \cdot \sqrt[3]{m^8}$
24. $\sqrt[4]{k^2} \cdot \sqrt[4]{k^7}$

[5.3] *Simplify each of the following. Assume that all variables represent positive real numbers.*

25. $\sqrt{6} \cdot \sqrt{13}$
26. $\sqrt{5} \cdot \sqrt{r}$
27. $\sqrt[3]{6} \cdot \sqrt[3]{5}$
28. $\sqrt[4]{7} \cdot \sqrt[4]{3}$

29. $\sqrt{20}$
30. $\sqrt{75}$
31. $\sqrt{125}$
32. $\sqrt[3]{108}$

33. $\sqrt{100y^7}$

34. $\sqrt[3]{64p^8q^5}$

35. $\sqrt{\dfrac{49}{81}}$

36. $\sqrt{\dfrac{y^3}{144}}$

37. $\sqrt[3]{\dfrac{m^{15}}{27}}$

38. $\sqrt[3]{\dfrac{r^2}{8}}$

39. $\sqrt[3]{108a^8b^5}$

40. $\sqrt[3]{632r^8t^4}$

[5.4] *Perform the indicated operations. Assume that all variables represent positive real numbers.*

41. $2\sqrt{8} - 3\sqrt{50}$

42. $8\sqrt{80} - 3\sqrt{45}$

43. $-\sqrt{27y} + 2\sqrt{75y}$

44. $2\sqrt{54m^3} + 5\sqrt{96m^3}$

45. $3\sqrt[3]{54} + 5\sqrt[3]{16}$

46. $-6\sqrt[4]{32} + \sqrt[4]{512}$

[5.5] *Multiply, then simplify the products.*

47. $(\sqrt{3} + 1)(\sqrt{3} - 2)$

48. $(\sqrt{7} + \sqrt{5})(\sqrt{7} - \sqrt{5})$

49. $(3\sqrt{2} + 1)(2\sqrt{2} - 3)$

50. $(\sqrt{11} + 3\sqrt{5})(\sqrt{11} + 5\sqrt{5})$

51. $(\sqrt{13} - \sqrt{2})^2$

52. $(\sqrt{5} - \sqrt{7})^2$

Rationalize the denominators. Assume that all variables represent positive real numbers.

53. $\dfrac{2}{\sqrt{7}}$

54. $\dfrac{\sqrt{6}}{\sqrt{5}}$

55. $\dfrac{-6\sqrt{3}}{\sqrt{2}}$

56. $\dfrac{3\sqrt{7p}}{\sqrt{y}}$

57. $\sqrt{\dfrac{11}{8}}$

58. $-\sqrt[3]{\dfrac{9}{25}}$

59. $\sqrt[3]{\dfrac{108m^3}{n^5}}$

60. $\dfrac{1}{2 - \sqrt{7}}$

61. $\dfrac{-5}{\sqrt{6} - \sqrt{3}}$

CHAPTER 5 REVIEW EXERCISES

[5.6] *Solve each equation.*

62. $\sqrt{8y + 9} = 5$

63. $\sqrt{2z - 3} - 3 = 0$

64. $\sqrt{3m + 1} = -1$

65. $\sqrt{7z + 1} = z + 1$

66. $3\sqrt{m} = \sqrt{10m - 9}$

67. $\sqrt{p^2 + 3p + 7} = p + 2$

68. $\sqrt{a + 2} - \sqrt{a - 3} = 1$

69. $\sqrt[3]{5m - 1} = \sqrt[3]{3m - 2}$

70. $\sqrt[4]{b + 6} = \sqrt[4]{2b}$

[5.7] *Simplify.*

71. $\sqrt{-25}$

72. $\sqrt{-200}$

73. $2\sqrt{-50} - 6\sqrt{-18}$

Perform the indicated operations. Write each imaginary number answer in the form a + bi.

74. $(-2 + 5i) + (-8 - 7i)$

75. $(5 + 4i) - (-9 - 3i)$

76. $\sqrt{-5} \cdot \sqrt{-7}$

77. $\sqrt{-25} \cdot \sqrt{-81}$

78. $\dfrac{\sqrt{-72}}{\sqrt{-8}}$

79. $(2 + 3i)(1 - i)$

80. $(5 + 4i)(2 - 8i)$

81. $\dfrac{3 - i}{2 + i}$

82. $\dfrac{6 - 5i}{2 + 3i}$

Find each power of i.

83. i^{11}

84. i^{52}

85. i^{-13}

CHAPTER 5 ROOTS AND RADICALS

MIXED REVIEW EXERCISES
Simplify. Assume all variables represent positive real numbers.

86. $-\sqrt[4]{256}$

87. $-\sqrt{169a^2b^4}$

88. $1000^{-2/3}$

89. $\dfrac{y^{-1/3} \cdot y^{5/6}}{y}$

90. $\dfrac{z^{-1/4}x^{1/2}}{z^{1/2}x^{-1/4}}$

91. $\sqrt[4]{k^{24}}$

92. $\sqrt[3]{54z^9t^8}$

93. $-5\sqrt{18} + 12\sqrt{72}$

94. $8\sqrt[3]{x^3y^2} - 2x\sqrt[3]{y^2}$

95. $(\sqrt{5} - \sqrt{3})(\sqrt{7} + \sqrt{3})$

96. $\dfrac{-1}{\sqrt{12}}$

97. $\sqrt[3]{\dfrac{12}{25}}$

98. $\dfrac{2\sqrt{z}}{\sqrt{z} - 2}$

99. $\sqrt{-49}$

100. $(4 - 9i) + (-1 + 2i)$

101. $\dfrac{\sqrt{50}}{\sqrt{-2}}$

102. i^{-1000}

103. $\dfrac{5 - 2i}{3 + i}$

Solve each equation.

104. $\sqrt{x + 4} = x - 2$

105. $\sqrt[3]{2x - 9} = \sqrt[3]{5x + 3}$

CHAPTER 5 TEST

Find the following. Use a calculator if necessary.

1. All square roots of 225

2. $-\sqrt{1681}$

3. $\sqrt[3]{-64}$

Simplify each of the following. Assume that all variables represent positive real numbers.

4. $27^{-4/3}$

5. $3^{1/2} \cdot 3^{1/4}$

6. $\dfrac{x^{-3/4}}{x^{-2/3}}$

Simplify. Assume that all variables represent positive real numbers.

7. $\sqrt{108x^5}$

8. $\sqrt[4]{16a^5b^{11}}$

9. $\sqrt[3]{3} \cdot \sqrt[3]{18}$

10. $8\sqrt{20} + 3\sqrt{80} - 2\sqrt{500}$

11. $(2\sqrt{5} - 3)(\sqrt{5} + 6)$

1. _____

2. _____

3. _____

4. _____

5. _____

6. _____

7. _____

8. _____

9. _____

10. _____

11. _____

12. $(3\sqrt{2} - 5\sqrt{3})^2$

13. $\dfrac{-9}{\sqrt{40}}$

14. $\dfrac{3}{\sqrt[3]{4}}$

15. $\dfrac{-8}{\sqrt{7} - \sqrt{5}}$

Solve each equation.

16. $\sqrt{y^2 - 5y + 3} = 4 + y$

17. $\sqrt{y + 5} + \sqrt{2y + 9} = 2$

Perform the indicated operations.

18. $(-6 + 4i) - (8 - 7i) + 10i$

19. $(1 + 5i)(3 + i)$

20. $\dfrac{1 + 2i}{3 - i}$

QUADRATIC EQUATIONS AND INEQUALITIES

6.1 COMPLETING THE SQUARE

In Section 3.10, we solved quadratic equations by using the zero-factor property. In this section, we introduce additional methods for solving quadratic equations.

Recall that a *quadratic equation* is defined as follows.

An equation that can be written in the form

$$ax^2 + bx + c = 0,$$

where a, b, and c are real numbers, with $a \neq 0$, is a **quadratic equation.**

OBJECTIVES

1 Extend the zero-factor property to complex number factors.

2 Learn the square root property.

3 Solve quadratic equations of the form $(ax - k)^2 = n$ by using the square root property.

4 Solve quadratic equations by completing the square.

For example, $m^2 - 4m + 5 = 0$ and $3q^2 = 4q - 8$ are quadratic equations.

1 The zero-factor property stated in Chapter 3 was restricted to real number factors. Now that property can be extended to complex number factors.

ZERO-FACTOR PROPERTY

If a and b are complex numbers and if $ab = 0$, then either $a = 0$ or $b = 0$, or both.

For example, the equation $x^2 + 4 = 0$ can now be solved with the zero-factor property as follows.

$$x^2 + 4 = 0$$
$$x^2 - (-4) = 0 \qquad \text{Write as a difference.}$$
$$(x + \sqrt{-4})(x - \sqrt{-4}) = 0 \qquad \text{Factor.}$$
$$x + \sqrt{-4} = 0 \quad \text{or} \quad x - \sqrt{-4} = 0 \qquad \text{Zero-factor property}$$
$$x = -\sqrt{-4} \qquad\qquad x = \sqrt{-4}$$
$$x = -2i \qquad\qquad x = 2i \qquad \sqrt{-4} = 2i$$

1. Solve.

   (a) $m^2 = 64$

   (b) $p^2 = 7$

2 Although factoring is the simplest way to solve quadratic equations, not every quadratic equation can be solved easily by factoring. In this section and the next, other methods of solving quadratic equations are developed based on the following property.

SQUARE ROOT PROPERTY

If a and b are complex numbers and if $a^2 = b$, then $a = \sqrt{b}$ or $a = -\sqrt{b}$.

To see why this property works, write $a^2 = b$ as $a^2 - b = 0$. Then factor to get $(a - \sqrt{b})(a + \sqrt{b}) = 0$. Set each factor equal to zero to get $a = \sqrt{b}$ or $a = -\sqrt{b}$. For example, by this property, $x^2 = 16$ implies that $x = 4$ or $x = -4$.

CAUTION Do not forget that if $b \neq 0$, using the square root property always produces *two* square roots, one positive and one negative.

EXAMPLE 1 Using the Square Root Property
Solve $r^2 = 5$.

From the square root property

$$r = \sqrt{5} \quad \text{or} \quad r = -\sqrt{5}$$

and the solution set is $\{\sqrt{5}, -\sqrt{5}\}$. ■

WORK PROBLEM 1 AT THE SIDE.

Recall from Chapter 5 that solutions such as those in Example 1 are sometimes abbreviated with the symbol $\pm$ (read "plus or minus"); with this symbol the solutions in Example 1 would be written $\pm\sqrt{5}$.

3 The square root property can be used to solve more complicated equations, such as

$$(x - 5)^2 = 36,$$

by substituting $(x - 5)^2$ for a^2 and 36 for b in the square root property to get

$$x - 5 = 6 \quad \text{or} \quad x - 5 = -6$$
$$x = 11 \qquad\qquad x = -1.$$

Check that both 11 and -1 satisfy the given equation, so that the solution set is $\{11, -1\}$.

ANSWERS
1. (a) $\{-8, 8\}$ (b) $\{-\sqrt{7}, \sqrt{7}\}$

6.1 COMPLETING THE SQUARE

EXAMPLE 2 Using the Square Root Property
Solve $(2a - 3)^2 = 18$.

By the square root property,

$$2a - 3 = \sqrt{18} \quad \text{or} \quad 2a - 3 = -\sqrt{18}$$

from which

$$2a = 3 + \sqrt{18} \qquad 2a = 3 - \sqrt{18}$$
$$a = \frac{3 + \sqrt{18}}{2} \qquad a = \frac{3 - \sqrt{18}}{2}.$$

Since $\sqrt{18} = \sqrt{9 \cdot 2} = 3\sqrt{2}$, the solution set can be written

$$\left\{ \frac{3 + 3\sqrt{2}}{2}, \frac{3 - 3\sqrt{2}}{2} \right\}. \quad \blacksquare$$

WORK PROBLEM 2 AT THE SIDE.

The next example shows an equation with imaginary number solutions.

EXAMPLE 3 Using the Square Root Property
Solve $(b + 2)^2 = -16$.

By the square root property,

$$b + 2 = \sqrt{-16} \quad \text{or} \quad b + 2 = -\sqrt{-16}.$$

Since $\sqrt{-16} = 4i$,

$$b + 2 = 4i \qquad \text{or} \qquad b + 2 = -4i$$
$$b = -2 + 4i \qquad \qquad b = -2 - 4i.$$

The solution set is $\{-2 + 4i, -2 - 4i\}$. $\blacksquare$

WORK PROBLEM 3 AT THE SIDE.

4 The square root property can be used to solve any quadratic equation by writing it in the form $(x + k)^2 = n$. That is, the left side of the equation must be rewritten as a perfect square trinomial [that can be factored as $(x + k)^2$] and the right side must be a constant. Rewriting the equation in this form is called **completing the square.**

For example,

$$m^2 + 8m + 10 = 0$$

is a quadratic equation that cannot be solved easily by factoring. To get a perfect square trinomial on the left side of the equation $m^2 + 8m + 10 = 0$, first subtract 10 from both sides:

$$m^2 + 8m = -10.$$

The left side should be a perfect square, say $(m + k)^2$. Since $(m + k)^2 = m^2 + 2mk + k^2$, comparing $m^2 + 8m$ with $m^2 + 2mk$ shows that

$$2mk = 8m$$
$$k = 4.$$

2. Solve.

(a) $(a - 3)^2 = 25$

(b) $(y + 9)^2 = 81$

(c) $(3k + 1)^2 = 2$

(d) $(2r - 3)^2 = 8$

3. Solve.

(a) $r^2 = -81$

(b) $(k + 5)^2 = -100$

(c) $(2k - 5)^2 + 9 = 0$

ANSWERS
2. (a) $\{8, -2\}$ (b) $\{0, -18\}$
(c) $\left\{ \dfrac{-1 + \sqrt{2}}{3}, \dfrac{-1 - \sqrt{2}}{3} \right\}$
(d) $\left\{ \dfrac{3 + 2\sqrt{2}}{2}, \dfrac{3 - 2\sqrt{2}}{2} \right\}$
3. (a) $\{9i, -9i\}$
(b) $\{-5 + 10i, -5 - 10i\}$
(c) $\left\{ \dfrac{5 + 3i}{2}, \dfrac{5 - 3i}{2} \right\}$

CHAPTER 6 QUADRATIC EQUATIONS AND INEQUALITIES

4. Determine the number that will complete each perfect square trinomial.

(a) $x^2 + 6x +$ _____

(b) $m^2 - 14m +$ _____

(c) $r^2 + 3r +$ _____

ANSWERS

4. (a) 9 (b) 49 (c) $\dfrac{9}{4}$

If $k = 4$, then $(m + k)^2$ becomes $(m + 4)^2$, or $m^2 + 8m + 16$. To get the necessary $+16$, add 16 on both sides so that

$$m^2 + 8m = -10$$

becomes

$$m^2 + 8m + 16 = -10 + 16.$$

Factor the left side and add on the right to get

$$(m + 4)^2 = 6.$$

This equation can be solved with the square root property:

$$m + 4 = \sqrt{6} \quad \text{or} \quad m + 4 = -\sqrt{6},$$

leading to the solution set $\{-4 + \sqrt{6}, -4 - \sqrt{6}\}$.

Based on the work of this example, an equation of the form $x^2 + px = q$ can be converted into an equation of the form $(x + k)^2 = n$ by adding the square of half the coefficient of x to both sides of the equation.

■ **WORK PROBLEM 4 AT THE SIDE.**

In summary, to find the solutions of $ax^2 + bx + c = 0$ ($a \neq 0$) by completing the square, proceed as follows.

COMPLETING THE SQUARE

Step 1 If $a \neq 1$, divide both sides by a.

Step 2 Rewrite the equation so that both terms containing variables are on one side of the equals sign and the constant is on the other side.

Step 3 Take half the coefficient of x and square it.

Step 4 Add the square to both sides.

Step 5 One side should now be a perfect square trinomial. Write it as the square of a binomial.

Step 6 Use the square root property to complete the solution.

EXAMPLE 4 Solving a Quadratic Equation by Completing the Square

Solve $k^2 + 5k - 1 = 0$ by completing the square.

Follow the steps listed above. Since $a = 1$, Step 1 is not needed here. Begin by adding 1 to both sides.

$$k^2 + 5k = 1 \qquad \text{Step 2}$$

Take half of 5 (the coefficient of the first-degree term) and square the result.

$$\frac{1}{2} \cdot 5 = \mathbf{\frac{5}{2}} \quad \text{and} \quad \left(\mathbf{\frac{5}{2}}\right)^2 = \frac{25}{4} \qquad \text{Step 3}$$

Add the square to each side of the equation to get

$$k^2 + 5k + \left(\frac{5}{2}\right)^2 = 1 + \frac{25}{4}.$$ Step 4

Write the left side as a square and add on the right. Then use the square root property.

$$\left(k + \frac{5}{2}\right)^2 = \frac{29}{4}$$ Step 5

$$k + \frac{5}{2} = \sqrt{\frac{29}{4}} \quad \text{or} \quad k + \frac{5}{2} = -\sqrt{\frac{29}{4}}$$ Step 6

Simplify: $\sqrt{\frac{29}{4}} = \frac{\sqrt{29}}{\sqrt{4}} = \frac{\sqrt{29}}{2}$.

$$k + \frac{5}{2} = \frac{\sqrt{29}}{2} \quad \text{or} \quad k + \frac{5}{2} = \frac{-\sqrt{29}}{2}$$

$$k = -\frac{5}{2} + \frac{\sqrt{29}}{2} \qquad k = -\frac{5}{2} - \frac{\sqrt{29}}{2}$$

Combine terms on the right to get the two solutions,

$$k = \frac{-5 + \sqrt{29}}{2} \quad \text{or} \quad k = \frac{-5 - \sqrt{29}}{2}.$$

The solution set is $\left\{\dfrac{-5 + \sqrt{29}}{2}, \dfrac{-5 - \sqrt{29}}{2}\right\}$. ■

WORK PROBLEM 5 AT THE SIDE.

EXAMPLE 5 Solving a Quadratic Equation by Completing the Square

Solve $2a^2 - 4a - 5 = 0$.

Go through the steps for completing the square. First divide both sides of the equation by 2 to make the coefficient of the second-degree term equal to 1, getting

$$a^2 - 2a - \frac{5}{2} = 0.$$ Step 1

$$a^2 - 2a = \frac{5}{2}$$ Step 2

$$\frac{1}{2}(-2) = -1 \text{ and } (-1)^2 = 1.$$ Step 3

$$a^2 - 2a + 1 = \frac{5}{2} + 1$$ Step 4

$$(a - 1)^2 = \frac{7}{2}$$ Step 5

5. Solve.

(a) $y^2 + 6y + 4 = 0$

(b) $x^2 - 2x + 10 = 0$

ANSWERS
5. (a) $\{-3 + \sqrt{5}, -3 - \sqrt{5}\}$
 (b) $\{1 + 3i, 1 - 3i\}$

CHAPTER 6 QUADRATIC EQUATIONS AND INEQUALITIES

6. Solve by completing the square.

(a) $3a^2 + 6a - 2 = 0$

(b) $2r^2 - 4r + 1 = 0$

(c) $5t^2 - 15t + 12 = 0$

$$a - 1 = \sqrt{\frac{7}{2}} \quad \text{or} \quad a - 1 = -\sqrt{\frac{7}{2}} \quad \text{Step 6}$$

$$a = 1 + \sqrt{\frac{7}{2}} \quad\quad a = 1 - \sqrt{\frac{7}{2}}.$$

Since $\sqrt{\dfrac{7}{2}} = \dfrac{\sqrt{14}}{2}$,

$$a = 1 + \frac{\sqrt{14}}{2} \quad \text{or} \quad a = 1 - \frac{\sqrt{14}}{2}.$$

Add the two terms in each solution as follows:

$$1 + \frac{\sqrt{14}}{2} = \frac{2}{2} + \frac{\sqrt{14}}{2} = \frac{2 + \sqrt{14}}{2}$$

$$1 - \frac{\sqrt{14}}{2} = \frac{2}{2} - \frac{\sqrt{14}}{2} = \frac{2 - \sqrt{14}}{2}$$

The solution set is

$$\left\{ \frac{2 + \sqrt{14}}{2}, \frac{2 - \sqrt{14}}{2} \right\}. \quad \blacksquare$$

■ **WORK PROBLEM 6 AT THE SIDE.**

ANSWERS

6. (a) $\left\{ \dfrac{-3 + \sqrt{15}}{3}, \dfrac{-3 - \sqrt{15}}{3} \right\}$

(b) $\left\{ \dfrac{2 + \sqrt{2}}{2}, \dfrac{2 - \sqrt{2}}{2} \right\}$

(c) $\left\{ \dfrac{15 + i\sqrt{15}}{10}, \dfrac{15 - i\sqrt{15}}{10} \right\}$

6.1 EXERCISES

Use the square root property to solve each equation. (All the solutions for these equations are real numbers.) See Examples 1 and 2.

1. $b^2 = 49$
2. $n^2 = 100$
3. $t^2 = 13$

4. $y^2 = 11$
5. $k^2 = 12$
6. $w^2 = 18$

7. $(p + 6)^2 = 9$
8. $(q - 2)^2 = 25$
9. $(3a - 1)^2 = 3$

10. $(2x + 4)^2 = 5$
11. $(4p + 1)^2 = 12$
12. $(5k - 2)^2 = 24$

Find the imaginary number solutions of the following equations. See Example 3.

13. $m^2 = -72$
14. $r^2 = -24$
15. $(x - 5)^2 = -1$

16. $(z + 2)^2 = -2$
17. $(2m - 1)^2 = -3$
18. $(3k + 2)^2 = -5$

Solve each equation by completing the square. (All the solutions for these equations are real numbers.) See Examples 4 and 5.

19. $x^2 - 2x - 15 = 0$
20. $m^2 - 4m = 32$
21. $2y^2 + y = 15$

22. $2z^2 - 7z = 5$
23. $2m^2 + 5m - 2 = 0$
24. $3a^2 + 2a = 2$

25. $x^2 - 2x - 1 = 0$

26. $p^2 - 4p + 1 = 0$

27. $m^2 = -6m - 7$

Solve each equation by completing the square. (Some of these equations have imaginary number solutions.) See Examples 4 and 5.

28. $y^2 - \dfrac{4}{3}y = -\dfrac{1}{9}$

29. $x^2 - x = \dfrac{1}{2}$

30. $r^2 + \dfrac{4}{3}r + \dfrac{4}{3} = 0$

31. $9a^2 - 24a = -13$

32. $x^2 - x - 1 = 0$

33. $m^2 + 6m + 10 = 0$

34. $x^2 + 4x + 13 = 0$

35. $25m^2 - 20m - 1 = 0$

36. $3p^2 + 3 = 8p$

SKILL SHARPENERS

Evaluate $\sqrt{b^2 - 4ac}$ for the following values of a, b, and c. See Section 1.4.

37. $a = 3, b = 1, c = -1$

38. $a = 4, b = 11, c = -3$

39. $a = 18, b = -25, c = -3$

40. $a = 6, b = 7, c = 2$

41. $a = 7, b = 5, c = -5$

42. $a = 3, b = 9, c = -4$

6.2 THE QUADRATIC FORMULA

The examples in the previous section showed that any quadratic equation can be solved by completing the square; however, completing the square is often tedious and time consuming. Later in this section, we will complete the square on the general quadratic equation to develop a formula that can be used to solve any quadratic equation. For now, we state the formula and show how it is used.

OBJECTIVES

1. Solve quadratic equations by using the quadratic formula.
2. Solve word problems by using the quadratic formula.
3. Use the discriminant to determine the number and type of solutions.
4. Use the discriminant to decide whether a quadratic equation can be factored.

QUADRATIC FORMULA

The solutions of $ax^2 + bx + c = 0$ ($a \neq 0$) are

$$\frac{-b \pm \sqrt{b^2 - 4ac}}{2a}.$$

1 To use the quadratic formula, first write the given equation in the form $ax^2 + bx + c = 0$; then identify the values of a, b, and c and substitute them into the quadratic formula, as shown in the next examples.

WORK PROBLEM 1 AT THE SIDE.

EXAMPLE 1 Using the Quadratic Formula
Solve $6x^2 - 5x - 4 = 0$.

First, identify the letters a, b, and c of the general quadratic equation, $ax^2 + bx + c = 0$. Here a, the coefficient of the second-degree term, is 6, while b, the coefficient of the first-degree term, is -5, and the constant c is -4. Substitute these values into the quadratic formula.

$$x = \frac{-b \pm \sqrt{b^2 - 4ac}}{2a}.$$

$$x = \frac{-(-5) \pm \sqrt{(-5)^2 - 4(6)(-4)}}{2(6)} \quad a = 6, b = -5, c = -4$$

$$= \frac{5 \pm \sqrt{25 + 96}}{12}$$

$$= \frac{5 \pm \sqrt{121}}{12}$$

$$x = \frac{5 \pm 11}{12}$$

This last statement leads to two solutions, one from $+$ and one from $-$, giving

$$x = \frac{5 + 11}{12} = \frac{16}{12} = \frac{4}{3} \quad \text{or} \quad x = \frac{5 - 11}{12} = \frac{-6}{12} = -\frac{1}{2}.$$

Check these solutions by substituting each one in the original equation. The solution set is $\left\{\frac{4}{3}, -\frac{1}{2}\right\}$. ■

1. Identify the letters a, b, and c. (*Hint:* Get the polynomial equal to 0 first.) *Do not solve.*

(a) $m^2 - 9m + 12 = 0$

(b) $-3q^2 + 9q - 4 = 0$

(c) $3y^2 = 6y + 2$

(d) $3m^2 - 4 = 8m$

ANSWERS
1. (a) 1; −9; 12 (b) −3; 9; −4
   (c) 3; −6; −2 (d) 3; −8; −4

CHAPTER 6 QUADRATIC EQUATIONS AND INEQUALITIES

2. Solve by using the quadratic formula.

(a) $6x^2 + 7x - 5 = 0$

(b) $4y^2 - 11y - 3 = 0$

3. Solve by using the quadratic formula.

(a) $2k^2 + 19 = 14k$

(b) $z^2 = 4z - 5$

CAUTION Notice in the quadratic formula that the square root is added to or subtracted from the value of $-b$ *before* dividing by $2a$.

■ **WORK PROBLEM 2 AT THE SIDE.**

EXAMPLE 2 Using the Quadratic Formula
Solve $4r^2 = 8r - 1$.

Rewrite the equation as $4r^2 - 8r + 1 = 0$, and identify $a = 4$, $b = -8$, and $c = 1$. Now use the quadratic formula.

$$r = \frac{-b \pm \sqrt{b^2 - 4ac}}{2a}$$

$$r = \frac{-(-8) \pm \sqrt{(-8)^2 - 4(4)(1)}}{2(4)} \quad a = 4, b = -8, c = 1$$

$$= \frac{8 \pm \sqrt{64 - 16}}{8}$$

$$= \frac{8 \pm \sqrt{48}}{8} = \frac{8 \pm 4\sqrt{3}}{8}$$

$$r = \frac{4(2 \pm \sqrt{3})}{8} = \frac{2 \pm \sqrt{3}}{2}$$

The solution set for $4r^2 = 8r - 1$ is $\left\{\frac{2 + \sqrt{3}}{2}, \frac{2 - \sqrt{3}}{2}\right\}$. ■

The solutions to the equation in the next example are imaginary numbers.

EXAMPLE 3 Using the Quadratic Formula
Solve $9q^2 + 5 = 6q$.

After writing the equation in the required form, $9q^2 - 6q + 5 = 0$, identify $a = 9$, $b = -6$, and $c = 5$. Then use the quadratic formula.

$$q = \frac{-(-6) \pm \sqrt{(-6)^2 - 4(9)(5)}}{2(9)}$$

$$= \frac{6 \pm \sqrt{-144}}{18} = \frac{6 \pm 12i}{18} \quad i = \sqrt{-1}$$

$$= \frac{6(1 \pm 2i)}{18} = \frac{1 \pm 2i}{3}$$

The solution set is $\left\{\frac{1 + 2i}{3}, \frac{1 - 2i}{3}\right\}$. ■

■ **WORK PROBLEM 3 AT THE SIDE.**

ANSWERS
2. (a) $\left\{\frac{1}{2}, -\frac{5}{3}\right\}$ (b) $\left\{3, -\frac{1}{4}\right\}$

3. (a) $\left\{\frac{7 + \sqrt{11}}{2}, \frac{7 - \sqrt{11}}{2}\right\}$

(b) $\{2 + i, 2 - i\}$

2 The next example shows how the quadratic formula is used to solve word problems.

EXAMPLE 4 Solving a Problem About Motion in a Straight Line

If a rock is dropped from a 144-foot building, its position (in feet above the ground) is given by $s = -16t^2 + 112t + 144$, where t is time in seconds after it was dropped. When does it hit the ground?

When the rock hits the ground, its distance above the ground is zero. Find t when s is zero by solving the equation

$$0 = -16t^2 + 112t + 144. \quad \text{Let } s = 0.$$
$$0 = t^2 - 7t - 9 \quad \text{Divide both sides by } -16.$$
$$t = \frac{7 \pm \sqrt{49 + 36}}{2} \quad \text{Quadratic formula}$$
$$t = \frac{7 \pm \sqrt{85}}{2}$$

A calculator gives $\sqrt{85} \approx 9.2$, so

$$t \approx \frac{7 \pm 9.2}{2},$$

giving the solutions $t \approx 8.1$ or $t \approx -1.1$. Discard the negative solution. The rock will hit the ground about 8.1 seconds after it is dropped. ∎

WORK PROBLEM 4 AT THE SIDE.

As mentioned earlier, the quadratic formula is developed by solving the general quadratic equation $ax^2 + bx + c = 0$ ($a \neq 0$) by completing the square. We show the steps here. First divide both sides by a (using the fact that $a \neq 0$).

$$x^2 + \frac{b}{a}x + \frac{c}{a} = 0 \quad \text{Step 1}$$

Now subtract $\frac{c}{a}$ from each side.

$$x^2 + \frac{b}{a}x = -\frac{c}{a} \quad \text{Step 2}$$

Take half of $\frac{b}{a}$ and square it.

$$\frac{1}{2}\left(\frac{b}{a}\right) = \frac{b}{2a} \quad \text{Step 3}$$
$$\left(\frac{b}{2a}\right)^2 = \frac{b^2}{4a^2}$$

Add the square to both sides.

$$x^2 + \frac{b}{a}x + \frac{b^2}{4a^2} = -\frac{c}{a} + \frac{b^2}{4a^2} \quad \textbf{(1)} \quad \text{Step 4}$$

4. A ball is thrown vertically upward from the ground. Its distance in feet from the ground at t seconds is $s = -16t^2 + 64t$. At what times will the ball be 32 feet from the ground? (*Hint:* There are two answers.)

ANSWERS
4. At .59 second and at 3.4 seconds

The left side is now a perfect square.

$$x^2 + \frac{b}{a}x + \frac{b^2}{4a^2} = \left(x + \frac{b}{2a}\right)^2$$

Using this result and rearranging the right side, equation (1) can be written as

$$\left(x + \frac{b}{2a}\right)^2 = \frac{b^2}{4a^2} + \frac{-c}{a}. \quad (2) \quad \text{Step 5}$$

Now simplify the right side.

$$\frac{b^2}{4a^2} + \frac{-c}{a} = \frac{b^2}{4a^2} + \frac{-4ac}{4a^2}$$
$$= \frac{b^2 - 4ac}{4a^2}$$

Finally, equation (2) becomes

$$\left(x + \frac{b}{2a}\right)^2 = \frac{b^2 - 4ac}{4a^2}.$$

By the square root property,

$$x + \frac{b}{2a} = \sqrt{\frac{b^2 - 4ac}{4a^2}} \quad \text{or} \quad x + \frac{b}{2a} = -\sqrt{\frac{b^2 - 4ac}{4a^2}}. \quad \text{Step 6}$$

Since

$$\sqrt{\frac{b^2 - 4ac}{4a^2}} = \frac{\sqrt{b^2 - 4ac}}{\sqrt{4a^2}} = \frac{\sqrt{b^2 - 4ac}}{2a},$$

the result above can be expressed as

$$x + \frac{b}{2a} = \frac{\sqrt{b^2 - 4ac}}{2a} \quad \text{or} \quad x + \frac{b}{2a} = \frac{-\sqrt{b^2 - 4ac}}{2a}$$

$$x = -\frac{b}{2a} + \frac{\sqrt{b^2 - 4ac}}{2a} \quad \text{or} \quad x = -\frac{b}{2a} - \frac{\sqrt{b^2 - 4ac}}{2a}$$

$$x = \frac{-b + \sqrt{b^2 - 4ac}}{2a} \quad \text{or} \quad x = \frac{-b - \sqrt{b^2 - 4ac}}{2a}.$$

This solution is the quadratic formula $x = \dfrac{-b \pm \sqrt{b^2 - 4ac}}{2a}$.

3 The quadratic formula gives the solutions of the quadratic equation $ax^2 + bx + c = 0$ as

$$x = \frac{-b \pm \sqrt{b^2 - 4ac}}{2a}.$$

6.2 THE QUADRATIC FORMULA

If a, b, and c are integers, the type of solutions of a quadratic equation (that is, rational, irrational, or imaginary) is determined by the quantity under the square root sign, $b^2 - 4ac$. Because it distinguishes among the three types of solutions, the quantity $b^2 - 4ac$ is called the **discriminant.** By calculating the discriminant before solving a quadratic equation, we can predict whether the solutions will be rational numbers, irrational numbers, or imaginary numbers. This can be useful in an applied problem, for example, where irrational or imaginary number solutions are not acceptable.

DISCRIMINANT

The discriminant of $ax^2 + bx + c = 0$ is given by $b^2 - 4ac$. If a, b, and c are integers, then the type of solution is determined as follows.

Discriminant	Solutions
Positive, and the square of an integer	Two different rational solutions
Positive, but not the square of an integer	Two different irrational solutions
Zero	One rational solution
Negative	Two different imaginary solutions

EXAMPLE 5 Using the Discriminant

Given the equation $6x^2 - x - 15 = 0$, find the discriminant and determine whether the solutions of the equation will be rational, irrational, or imaginary.

The discriminant is found by evaluating $b^2 - 4ac$. In this example, $a = 6$, $b = -1$, $c = -15$, and the discriminant is

$$b^2 - 4ac = (-1)^2 - 4(6)(-15) = 1 + 360 = 361.$$

A calculator or square root table shows that $361 = 19^2$. Because the discriminant is a perfect square and a, b, and c are integers, the solutions to the given equation will be two different rational numbers. ∎

WORK PROBLEM 5 AT THE SIDE.

EXAMPLE 6 Using the Discriminant

Predict the number and type of solutions for each of the following equations.

(a) $4y^2 + 9 = 12y$

Rewrite the equation as $4y^2 - 12y + 9 = 0$ to find $a = 4$, $b = -12$, and $c = 9$. The discriminant is

$$b^2 - 4ac = (-12)^2 - 4(4)(9) = 144 - 144 = 0.$$

Since the discriminant is 0, the quantity under the radical in the quadratic formula is 0, and there is only one rational solution.

---

5. Find the discriminant and decide whether it is a perfect square.

(a) $6m^2 - 13m - 28 = 0$

(b) $4y^2 + 2y + 1 = 0$

(c) $15k^2 + 11k = 14$

ANSWERS
5. (a) 841; yes (b) −12; no
   (c) 961; yes

CHAPTER 6 QUADRATIC EQUATIONS AND INEQUALITIES

6. Predict the number and type of solutions for each equation.

(a) $2x^2 + 3x = 4$

(b) $2x^2 + 3x + 4 = 0$

(c) $9m^2 - 7m - 23 = 0$

(d) $7w^2 + 6w + 12 = 0$

7. Use the discriminant to decide whether each trinomial can be factored; then factor it, if possible.

(a) $2y^2 + 13y - 7$

(b) $3r^2 + 22r - 16$

(c) $6z^2 - 11z + 18$

ANSWERS
6. (a) two; irrational (b) two; imaginary
   (c) two; irrational (d) two; imaginary
7. (a) $(2y - 1)(y + 7)$
   (b) $(r + 8)(3r - 2)$
   (c) cannot be factored

(b) $3m^2 - 4m = 5$

Rewrite the equation as $3m^2 - 4m - 5 = 0$. Then $a = 3$, $b = -4$, $c = -5$, and the discriminant is

$$b^2 - 4ac = (-4)^2 - 4(3)(-5) = 16 + 60 = 76.$$

Since 76 is not a perfect square, $\sqrt{76}$ is irrational, and since a, b, and c are integers, the given equation will have two different irrational solutions, one from using $\sqrt{76}$ and one from using $-\sqrt{76}$.

(c) $4x^2 + x + 1 = 0$

Here $a = 4$, $b = 1$, $c = 1$, and the discriminant is

$$b^2 - 4ac = 1^2 - 4(4)(1) = 1 - 16 = -15.$$

Since the discriminant is negative, the equation $4x^2 + x + 1 = 0$ will have two imaginary number solutions. ■

■ **WORK PROBLEM 6 AT THE SIDE.**

4 It can be shown that a quadratic trinomial can be factored with rational coefficients only if the corresponding quadratic equation has rational solutions. Thus, the discriminant can be used to decide whether or not a given trinomial is factorable.

EXAMPLE 7 Deciding Whether a Trinomial Is Factorable
Decide whether or not the following trinomials can be factored.

(a) $24x^2 + 7x - 5$

To decide whether the solutions of $24x^2 + 7x - 5 = 0$ are rational numbers, evaluate the discriminant.

$$b^2 - 4ac = 7^2 - 4(24)(-5) = 49 + 480 = 529 = 23^2$$

Since 529 is a perfect square, the solutions are rational and the trinomial can be factored. In fact, it factors as

$$24x^2 + 7x - 5 = (3x - 1)(8x + 5).$$

(b) $11m^2 - 9m + 12$

The discriminant is $b^2 - 4ac = (-9)^2 - 4(11)(12) = -447$. This number is negative, so the corresponding quadratic equation has imaginary number solutions and therefore the trinomial cannot be factored. ■

■ **WORK PROBLEM 7 AT THE SIDE.**

6.2 EXERCISES

Use the quadratic formula to find the real number solutions of each of the following equations. In Exercises 9–12, multiply on the left side first. See Examples 1 and 2.

1. $m^2 + 8m + 15 = 0$

2. $4x^2 - 8x + 1 = 0$

3. $2x^2 + 4x + 1 = 0$

4. $2k^2 + 3k - 1 = 0$

5. $m^2 + 18 = 10m$

6. $2y^2 = 2y + 1$

7. $9r^2 + 6r = 1$

8. $2p^2 - 4p = 5$

9. $4x(x + 1) = 1$

10. $4r(r - 1) = 19$

11. $(g + 2)(g - 3) = 1$

12. $(y - 5)(y + 2) = 6$

Use the quadratic formula to find the imaginary number solutions of the following equations. See Example 3.

13. $3x^2 + 4x + 2 = 0$

14. $2k^2 + 3k = -2$

15. $m^2 - 6m + 14 = 0$

16. $p^2 + 4p + 11 = 0$

17. $4z^2 = 4z - 7$

18. $9p^2 + 7 = 6p$

19. $1 + \dfrac{1}{m^2} = -\dfrac{1}{m}$

20. $1 = \dfrac{2}{y} - \dfrac{2}{y^2}$

21. $3 - \dfrac{1}{w} + \dfrac{4}{w^2} = 0$

Solve each equation. Round each solution to the nearest thousandth.

22. $3x^2 + 2x = 2$

23. $3r^2 = -2 + 26r$

24. $y = \dfrac{5 - y}{3(y + 1)}$

Solve the following problems. Give answers to the nearest tenth in Exercises 25 and 26. See Example 4.

25. An object is moving in a straight line so that its distance in feet from a fixed point is given by $d = t^2 + 5t + 2$, where t is time in seconds. When is the object 10 feet from its starting point?

26. The straight-line distance (in feet) traveled by a particle from a starting point is given by $d = 3t^2 + 4t + 2$, where t is in seconds. At what time is the particle 12 feet from the starting point?

27. The demand for a certain product is given by $D = -2p^2 + 4p + 65$, where p is the price in dollars. Find the price when the demand is for 50 items.

28. The profit from selling a certain product is $P = 10{,}000 + 320x - 20x^2$, where x is the amount in dollars spent on advertising. Find the amount to the nearest dollar that should be spent on advertising to realize a profit of $9000.

name date hour

6.2 EXERCISES

Use the discriminant to determine whether the solutions to each of the following equations are:

(a) *two distinct rational numbers;* **(b)** *exactly one rational number;*
(c) *two distinct irrational numbers;* **(d)** *two distinct imaginary numbers.*

Do not solve. In Exercises 41–43, multiply both sides by the least common denominator so that the coefficients will be integers. See Examples 5 and 6.

29. $x^2 + 4x + 1 = 0$

30. $2y^2 - y + 1 = 0$

31. $4x^2 - 4x = 3$

32. $9y^2 - 12y = 1$

33. $4y^2 - y - 10 = 0$

34. $25m^2 - 70m + 49 = 0$

35. $49p^2 - 42p + 5 = 0$

36. $3m^2 - 5m = 2$

37. $9y^2 - 30y + 45 = 0$

38. $9z^2 - 30z + 21 = 0$

39. $4p^2 + 28p + 49 = 0$

40. $9m^2 + 30m + 41 = 0$

41. $\dfrac{w^2}{2} + 5w - 1 = 0$

42. $\dfrac{3}{2}k^2 + \dfrac{1}{2}k + 4 = 0$

43. $\dfrac{z^2}{2} = \dfrac{3}{10}z + 2$

Use the discriminant to tell which polynomials can be factored. If a polynomial can be factored, factor it. See Example 7.

44. $6k^2 + 7k - 5$

45. $8r^2 - 2r - 21$

46. $24z^2 - 14z - 5$

47. $8k^2 + 38k - 33$

48. $9r^2 - 11r + 8$

49. $15m^2 - 17m - 12$

Solve each equation. Square both sides of the equation first.

50. $\sqrt{2x} = \sqrt{5x + 2}$

51. $p = \sqrt{p + 3}$

52. $\sqrt{3t} = \sqrt{t + 1}$

53. $\sqrt{y + 1} = -1 + \sqrt{2y}$

SKILL SHARPENERS

Rewrite each expression making the indicated substitutions, then factor the trinomial. See Section 3.8.

54. $(m + 1)^2 + 4(m + 1) - 5; \quad m + 1 = a$

55. $4(3k + 2)^2 - 7(3k + 2) - 2; \quad 3k + 2 = r$

56. $3(2x - 1)^2 + 4(2x - 1) + 1; \quad 2x - 1 = m$

57. $2(r + 3)^2 - 5(r + 3) + 2; \quad r + 3 = t$

6.3 EQUATIONS QUADRATIC IN FORM

Four methods have now been introduced for solving quadratic equations written in the form $ax^2 + bx + c = 0$. The chart below gives some advantages and disadvantages of each method.

METHODS FOR SOLVING QUADRATIC EQUATIONS

Method	Advantages	Disadvantages
Factoring	Usually the fastest method	Not all polynomials are factorable; some factorable polynomials are hard to factor
Completing the square	None for solving equations (the procedure is useful in other areas of mathematics)	Requires more steps than other methods
Quadratic formula	Can always be used	More difficult than factoring because of the square root
Square root method	Simplest method for solving equations of the form $(ax - k)^2 = n$	Few equations are given in this form

OBJECTIVES

1. Write an equation with fractions in quadratic form.
2. Use quadratic equations to solve word problems with fractions.
3. Write an equation with radicals in quadratic form.
4. Use substitution to write an equation in quadratic form.

1 A variety of nonquadratic equations can be written in the form of a quadratic equation and solved by using these methods. For example, some equations with fractions lead to quadratic equations. As you solve the equations in this section try to decide which is the best method for each equation.

EXAMPLE 1 Writing an Equation in Quadratic Form

Solve $\dfrac{1}{x} + \dfrac{1}{x - 1} = \dfrac{7}{12}$.

To clear the equation of fractions, multiply each term by the common denominator, $12x(x - 1)$.

$$\mathbf{12x(x - 1)}\dfrac{1}{x} + \mathbf{12x(x - 1)}\dfrac{1}{x - 1} = \mathbf{12x(x - 1)}\dfrac{7}{12}$$

$$12(x - 1) + 12x = 7x(x - 1)$$
$$12x - 12 + 12x = 7x^2 - 7x \quad \text{Distributive property}$$
$$24x - 12 = 7x^2 - 7x \quad \text{Combine terms.}$$

A quadratic equation must be in the form $ax^2 + bx + c = 0$ before it can be solved. Combine terms and arrange them so that one side of the equation is zero.

$$0 = 7x^2 - 31x + 12 \quad \text{Subtract } 24x \text{ and add 12.}$$
$$0 = (7x - 3)(x - 4) \quad \text{Factor.}$$

CHAPTER 6 QUADRATIC EQUATIONS AND INEQUALITIES

1. Solve.

(a) $\dfrac{5}{m} + \dfrac{12}{m^2} = 2$

(b) $\dfrac{2}{x} + \dfrac{1}{x-2} = \dfrac{5}{3}$

(c) $\dfrac{4}{m-1} + 9 = -\dfrac{7}{m}$

ANSWERS
1. (a) $\left\{-\dfrac{3}{2}, 4\right\}$ (b) $\left\{\dfrac{4}{5}, 3\right\}$
(c) $\left\{\dfrac{7}{9}, -1\right\}$

Setting each factor equal to 0 and solving the two linear equations gives the solutions $\dfrac{3}{7}$ and 4. Check by substituting these solutions in the original equation. The solution set is $\left\{\dfrac{3}{7}, 4\right\}$. ∎

■ **WORK PROBLEM 1 AT THE SIDE.**

2 Distance-rate-time applications and problems about work often lead to equations with fractions, such as those in the next examples.

EXAMPLE 2 Solving a Motion Problem
A riverboat for tourists averages 12 miles per hour in still water. It takes the boat 1 hour, 4 minutes to go 6 miles upstream and return. Find the speed of the current. See Figure 1.

FIGURE 1

For a problem about rate (or speed), use the distance formula, $d = rt$.

Let $x =$ the speed of the current;
       $12 - x =$ the rate upstream;
       $12 + x =$ the rate downstream.

The rate upstream is the difference of the speed of the boat in still water and the speed of the current, or $12 - x$. The speed downstream is, in the same way, $12 + x$. To find the time, rewrite the formula $d = rt$ as

$$t = \dfrac{d}{r}.$$

This information was used to complete the following chart.

	d	r	t
Upstream	6	$12 - x$	$\dfrac{6}{12-x}$
Downstream	6	$12 + x$	$\dfrac{6}{12+x}$

← Times in hours

The total time, 1 hour and 4 minutes, can be written as

$$1 + \dfrac{4}{60} = 1 + \dfrac{1}{15} = \dfrac{16}{15} \text{ hours.}$$

Since the time upstream plus the time downstream equals $\frac{16}{15}$ hours,

$$\frac{6}{12-x} + \frac{6}{12+x} = \frac{16}{15}.$$

Now multiply both sides of the equation by the common denominator $15(12-x)(12+x)$ and solve the resulting quadratic equation.

$$15(12+x)6 + 15(12-x)6 = 16(12-x)(12+x)$$
$$90(12+x) + 90(12-x) = 16(144-x^2)$$
$$1080 + 90x + 1080 - 90x = 2304 - 16x^2 \quad \text{Distributive property}$$
$$2160 = 2304 - 16x^2 \quad \text{Combine terms.}$$
$$16x^2 = 144$$
$$x^2 = 9$$

Solve $x^2 = 9$ by using the square root property to get the two solutions

$$x = 3 \quad \text{or} \quad x = -3.$$

The speed of the current cannot be -3, so the solution is $x = 3$ miles per hour. ∎

WORK PROBLEM 2 AT THE SIDE.

CAUTION As shown in Example 2, when a quadratic equation is used to solve a word problem, usually only *one* answer satisfies the application. It is *always necessary* to check each answer in the words of the stated problem.

EXAMPLE 3 Solving a Work Problem

It takes two carpet layers 4 hours to carpet a room. If each worked alone, one of them could do the job in one hour less time than the other. How long would it take the slower one to complete the job alone?

Let x represent the number of hours for the slower carpet layer to complete the job alone. Then the faster carpet layer could do the entire job in $x - 1$ hours. Together, they do the job in 4 hours. In one hour the slower person can do $\frac{1}{x}$ part of the job, the faster person can do $\frac{1}{x-1}$ part of the job, and together they can do $\frac{1}{4}$ part of the job. Thus,

$$\frac{1}{x} + \frac{1}{x-1} = \frac{1}{4}.$$

Multiply both sides by the common denominator, $4x(x-1)$.

$$4(x-1) + 4x = x(x-1)$$
$$4x - 4 + 4x = x^2 - x \quad \text{Distributive property}$$

Simplify and get 0 on one side.

$$0 = x^2 - 9x + 4$$

Now use the quadratic formula.

$$x = \frac{9 \pm \sqrt{81-16}}{2} = \frac{9 \pm \sqrt{65}}{2} \quad a = 1, b = -9, c = 4$$

2. (a) In 4 hours Kerrie can go 15 miles upriver and come back. The speed of the current is 5 miles per hour. Complete this chart.

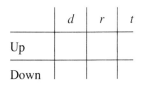

(b) Find the speed of the boat from part (a) in still water.

(c) In $1\frac{3}{4}$ hours Khe rows his boat 5 miles upriver and comes back. The speed of the current is 3 miles per hour. How fast does Khe row?

ANSWERS

2. **(a)** row 1: 15; $x - 5$; $\frac{15}{x-5}$; row 2: 15; $x + 5$; $\frac{15}{x+5}$ **(b)** 10 miles per hour **(c)** 7 miles per hour

CHAPTER 6 QUADRATIC EQUATIONS AND INEQUALITIES

3. Solve each word problem.

(a) Carlos can do a job in 2 hours less time than Jaime. If they can finish the job together in 7 hours, how long would it take each of them working alone? Give answers to the nearest tenth.

(b) Mary can do a job in 3 hours less than Jack. Working together, they can do the job in 12 hours. How long would it take each of them working alone? Give answers to the nearest tenth.

4. Solve. Check each answer.

(a) $x = \sqrt{7x - 10}$

(b) $m = \sqrt{\dfrac{5 - 9m}{2}}$

(c) $2x = \sqrt{x + 1}$

ANSWERS
3. (a) Jaime, 15.1 hours; Carlos, 13.1 hours   (b) Mary, 22.6 hours; Jack, 25.6 hours
4. (a) {2, 5}   (b) $\left\{\dfrac{1}{2}\right\}$ (c) {1}

From a calculator, $\sqrt{65} \approx 8.062$, so

$$x \approx \frac{9 \pm 8.062}{2}.$$

Using the + sign gives $x \approx 8.5$, while the − sign leads to $x \approx .5$. (Here we rounded to the nearest tenth.) Only the solution 8.5 makes sense in the original problem. (Why?) Thus, the slower carpet layer can do the job in about 8.5 hours and the faster in about $8.5 - 1 = 7.5$ hours. ■

■ **WORK PROBLEM 3 AT THE SIDE.**

3 In Section 5.6 we saw that some equations with radicals lead to quadratic equations.

EXAMPLE 4 Writing an Equation in Quadratic Form
Solve each equation.

(a) $k = \sqrt{6k - 8}$

This equation is not quadratic. However, squaring both sides of the equation gives $k^2 = 6k - 8$, which is a quadratic equation that can be solved by factoring.

$$k^2 = 6k - 8$$
$$k^2 - 6k + 8 = 0$$
$$(k - 4)(k - 2) = 0$$
$$k = 4 \quad \text{or} \quad k = 2 \quad \text{Potential solutions}$$

Check both of these numbers in the original (and *not* the squared) equation to be sure they are solutions.

If $k = \mathbf{4}$,   If $k = \mathbf{2}$,
$\mathbf{4} = \sqrt{6(\mathbf{4}) - 8}$ $\mathbf{2} = \sqrt{6(\mathbf{2}) - 8}$
$4 = \sqrt{16}$ $2 = \sqrt{4}$
$4 = 4.$ True $2 = 2.$ True

Both numbers check, so the solution set is {2, 4}.

(b) $x + \sqrt{x} = 6$

$\sqrt{x} = 6 - x$ Isolate the radical on one side.
$x = 36 - 12x + x^2$ Square both sides.
$0 = x^2 - 13x + 36$ Get 0 on one side.
$0 = (x - 4)(x - 9)$ Factor.
$x - 4 = 0 \quad \text{or} \quad x - 9 = 0$ Set each factor equal to 0.
$x = 4$ $x = 9$

Check both potential solutions.

If $x = 4$,   If $x = 9$,
$4 + \sqrt{4} = 6.$ True $9 + \sqrt{9} = 6.$ False

The solution set is {4}. ■

■ **WORK PROBLEM 4 AT THE SIDE.**

6.3 EQUATIONS QUADRATIC IN FORM

4 An equation that can be written in the form $au^2 + bu + c = 0$, for $a \neq 0$ and u an algebraic expression, is called **quadratic in form.**

EXAMPLE 5 Solving an Equation That Is Quadratic in Form
Solve each of the following.

(a) $x^4 - 13x^2 + 36 = 0$.

Since $x^4 = (x^2)^2$, this equation can be written as $(x^2)^2 - 13(x^2) + 36 = 0$, so it is in quadratic form with $u = x^2$ and can be solved by factoring.

$$x^4 - 13x^2 + 36 = 0$$
$$(x^2 - 4)(x^2 - 9) = 0 \quad \text{Factor.}$$
$$x^2 - 4 = 0 \quad \text{or} \quad x^2 - 9 = 0 \quad \text{Set each factor equal to 0.}$$
$$x^2 = 4 \quad \text{or} \quad x^2 = 9 \quad \text{Solve for } x^2.$$
$$x = \pm 2 \quad \text{or} \quad x = \pm 3 \quad \text{Square root property}$$

The equation $x^4 - 13x^2 + 36 = 0$, a fourth-degree equation, has four solutions. The solution set is $\{2, -2, 3, -3\}$, as can be verified by substituting into the equation.

(b) $4x^4 + 1 = 5x^2$

Use the fact that $x^4 = (x^2)^2$ again.

$$4x^4 + 1 = 5x^2$$
$$4x^4 - 5x^2 + 1 = 0 \quad \text{Get 0 on one side.}$$
$$(4x^2 - 1)(x^2 - 1) = 0 \quad \text{Factor.}$$
$$4x^2 - 1 = 0 \quad \text{or} \quad x^2 - 1 = 0 \quad \text{Set each factor equal to 0.}$$
$$x^2 = \frac{1}{4} \quad \text{or} \quad x^2 = 1 \quad \text{Solve for } x^2.$$
$$x = \pm \frac{1}{2} \quad \text{or} \quad x = \pm 1 \quad \text{Square root property}$$

The solution set is $\left\{\frac{1}{2}, -\frac{1}{2}, 1, -1\right\}$. ■

WORK PROBLEM 5 AT THE SIDE.

EXAMPLE 6 Solving an Equation That Is Quadratic in Form
Solve each equation.

(a) $2(4m - 3)^2 + 7(4m - 3) + 5 = 0$

Because of the repeated quantity $4m - 3$, this equation is quadratic in form with $u = 4m - 3$. (Any letter, except m, could be used instead of u.) Write

$$2(\mathbf{4m - 3})^2 + 7(\mathbf{4m - 3}) + 5 = 0$$

as
$$2u^2 + 7u + 5 = 0. \quad \text{Let } 4m - 3 = u.$$
$$(2u + 5)(u + 1) = 0 \quad \text{Factor.}$$

The solutions to $2u^2 + 7u + 5 = 0$ are

$$u = -\frac{5}{2} \quad \text{or} \quad u = -1.$$

5. Solve.

(a) $m^4 - 10m^2 + 9 = 0$

(b) $9k^4 - 37k^2 + 4 = 0$

(c) $y^4 + 25 = 26y^2$

ANSWERS
5. (a) $\{1, -1, 3, -3\}$
(b) $\left\{\frac{1}{3}, -\frac{1}{3}, 2, -2\right\}$
(c) $\{5, -5, 1, -1\}$

CHAPTER 6 QUADRATIC EQUATIONS AND INEQUALITIES

6. Solve.

(a) $5(r + 3)^2 + 9(r + 3) = 2$

(b) $3(2p + 1)^2 - 2(2p + 1) = 8$

(c) $4m^{2/3} = 3m^{1/3} + 1$

ANSWERS

6. (a) $\left\{-\dfrac{14}{5}, -5\right\}$ (b) $\left\{-\dfrac{7}{6}, \dfrac{1}{2}\right\}$

(c) $\left\{-\dfrac{1}{64}, 1\right\}$

To find m, substitute $4m - 3$ for u.

$$4m - 3 = -\dfrac{5}{2} \quad \text{or} \quad 4m - 3 = -1$$

$$4m = \dfrac{1}{2} \quad \text{or} \quad 4m = 2$$

$$m = \dfrac{1}{8} \quad \text{or} \quad m = \dfrac{1}{2}$$

The solution set of the original equation is $\left\{\dfrac{1}{8}, \dfrac{1}{2}\right\}$.

(b) $2a^{2/3} - 11a^{1/3} + 12 = 0$

Let $a^{1/3} = u$; then $a^{2/3} = u^2$. Substitute into the given equation.

$$2u^2 - 11u + 12 = 0 \qquad \text{Let } a^{1/3} = u,\ a^{2/3} = u^2.$$
$$(2u - 3)(u - 4) = 0 \qquad \text{Factor.}$$

$$2u - 3 = 0 \quad \text{or} \quad u - 4 = 0$$

$$u = \dfrac{3}{2} \quad \text{or} \quad u = 4$$

$$a^{1/3} = \dfrac{3}{2} \quad \text{or} \quad a^{1/3} = 4 \qquad u = a^{1/3}$$

$$a = \left(\dfrac{3}{2}\right)^3 = \dfrac{27}{8} \quad \text{or} \quad a = 4^3 = 64 \qquad \text{Cube both sides.}$$

(Recall that equations with radicals were solved in Section 5.6 by raising both sides to the same power.) Check that the solution set is $\left\{\dfrac{27}{8}, 64\right\}$. ∎

■ **WORK PROBLEM 6 AT THE SIDE.**

6.3 EXERCISES

Solve by first clearing each equation of fractions. See Example 1.

1. $3b - 5 = \dfrac{2}{b}$

2. $7 + \dfrac{5}{y} = -2y$

3. $1 + \dfrac{3}{x} - \dfrac{28}{x^2} = 0$

4. $4 + \dfrac{7}{k} = \dfrac{2}{k^2}$

5. $3 = \dfrac{1}{y} + \dfrac{2}{y^2}$

6. $\dfrac{1}{2} + \dfrac{2}{x} = \dfrac{5}{2x^2}$

7. $1 + \dfrac{2}{y} - \dfrac{1}{y^2} = 0$

8. $\dfrac{3}{2x} - \dfrac{1}{2(x+2)} = 1$

9. $1 - \dfrac{2}{m} - \dfrac{1}{2m^2} = 0$

Find all solutions by first squaring. Be sure to check all answers. See Example 4.

10. $4y = \sqrt{1 - 6y}$

11. $2a = -\sqrt{-11a - 6}$

12. $x + 4 = \sqrt{-2x}$

13. $p = 8 + 2\sqrt{p}$

14. $m = \sqrt{\dfrac{6 - 13m}{5}}$

15. $x = \sqrt{\dfrac{-7x - 5}{2}}$

16. $x - 5 = \sqrt{5x - 29}$

17. $3k + 1 = -\sqrt{3k + 3}$

CHAPTER 6 QUADRATIC EQUATIONS AND INEQUALITIES

Find all solutions to the following equations. See Examples 5–6.

18. $x^4 - 2x^2 + 1 = 0$

19. $t^4 - 29t^2 + 100 = 0$

20. $z^4 - 37z^2 + 36 = 0$

21. $9m^4 - 25m^2 + 16 = 0$

22. $(2 + x)^2 + 5(2 + x) + 6 = 0$

23. $(m - 3)^2 + (m - 3) = 20$

24. $2(3k - 1)^2 + 5(3k - 1) + 2 = 0$

25. $3(2p + 2)^2 - 7(2p + 2) - 6 = 0$

26. $1 + \dfrac{6}{(r + 5)^2} = \dfrac{7}{r + 5}$

27. $3 - \dfrac{2}{a + 4} = \dfrac{8}{(a + 4)^2}$

28. $2 - (y - 1)^{-1} = 6(y - 1)^{-2}$

29. $3 = 2(p - 1)^{-1} + (p - 1)^{-2}$

Solve each problem by writing an equation with fractions and solving it. See Examples 2 and 3.

30. Karen flew her plane for 6 hours at a constant speed. She traveled 810 miles with the wind, then turned around and traveled 720 miles against the wind. The wind speed was a constant 15 miles per hour. Find the speed of the plane.

31. The distance from Jackson to Lodi is about 40 miles, as is the distance from Lodi to Manteca. Rico drove from Jackson to Lodi during the rush hour, stopped in Lodi for a root beer, and then drove on to Manteca at 10 miles per hour faster. Driving time for the entire trip was 88 minutes. Find his speed from Jackson to Lodi.

32. On a windy day Yoshiaki found that he could go 16 miles downstream and then 4 miles back upstream at top speed in a total of 48 minutes. What was the top speed of Yoshiaki's boat if the current was 15 miles per hour?

33. Albuquerque and Amarillo are 300 miles apart. Steve rides his Honda 20 miles per hour faster than Paula rides her Yamaha. Find Steve's average speed if he travels from Albuquerque to Amarillo in $1\frac{1}{4}$ hours less time than Paula.

34. Two pipes together can fill a large tank in 2 hours. One of the pipes, used alone, takes 3 hours longer than the other to fill the tank. How long would each pipe take to fill the tank alone?

35. A washing machine can be filled in 6 minutes if both the hot and cold water taps are fully opened. To fill the washer with hot water alone takes 9 minutes longer than filling with cold water alone. How long does it take to fill the tank with cold water?

CHAPTER 6 QUADRATIC EQUATIONS AND INEQUALITIES

Use substitution to solve the following equations. See Example 6.

36. $q^{2/3} - 3q^{1/3} + 2 = 0$

37. $z^{2/3} = 4z^{1/3} - 3$

38. $(x^2 + x)^2 = 8(x^2 + x) - 12$

39. $2(1 + \sqrt{x})^2 = 11(1 + \sqrt{x}) - 12$

SKILL SHARPENERS

Solve each equation for the specified variable. See Section 2.2.

40. $P = 2L + 2W$ for L

41. $A = \frac{1}{2}bh$ for h

42. $F = \frac{9}{5}C + 32$ for C

43. $S = 2\pi rh + 2\pi r^2$ for h

44. $\frac{m^2}{2} + \frac{p^2}{4} = 1$ for m^2

45. $\frac{r^2 - 6}{3} = q^2$ for r^2

6.4 FORMULAS AND APPLICATIONS

1 Many useful formulas have a second-degree term. The methods presented earlier in this chapter can be used to solve a formula for a variable that is squared.

The formula in Example 1 is the Pythagorean formula of geometry.

PYTHAGOREAN FORMULA

If c is the length of the longest side of a right triangle and a and b are the lengths of the shorter sides, then

$$c^2 = a^2 + b^2.$$

See the figure.

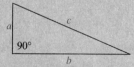

The longest side is the **hypotenuse** and the two shorter sides are the **legs** of the triangle.

OBJECTIVES

1. Solve second-degree formulas for a specified variable.
2. Solve word problems about motion along a straight line.
3. Solve word problems using the Pythagorean formula.
4. Solve word problems using formulas for area.
5. Solve word problems about work.

EXAMPLE 1 Solving for a Squared Variable
Solve the Pythagorean formula $c^2 = a^2 + b^2$ for b.

Think of c^2 and a^2 as constants. Solve for b by first getting b^2 alone on one side of the equation. Begin by subtracting a^2 from both sides.

$$c^2 = a^2 + b^2$$
$$c^2 - a^2 = b^2$$

Now use the square root property.

$$b = \sqrt{c^2 - a^2} \quad \text{or} \quad b = -\sqrt{c^2 - a^2}$$

Since b represents the side of a triangle, b must be positive. Because of this,

$$b = \sqrt{c^2 - a^2}$$

is the only solution. The solution cannot be simplified further. ■

WORK PROBLEM 1 AT THE SIDE.

EXAMPLE 2 Solving for a Squared Variable
Solve $s = 2t^2 + kt$ for t.

Since the equation has terms with t^2 and t, first put it in the quadratic form $ax^2 + bx + c = 0$ with t as the variable instead of x.

$$s = 2t^2 + kt$$
$$0 = 2t^2 + kt - s$$

Now use the quadratic formula with $a = 2$, $b = k$, and $c = -s$.

$$t = \frac{-k \pm \sqrt{k^2 - 4(2)(-s)}}{2(2)}$$
$$= \frac{-k \pm \sqrt{k^2 + 8s}}{4}$$

The solutions are $t = \dfrac{-k + \sqrt{k^2 + 8s}}{4}$ and $t = \dfrac{-k - \sqrt{k^2 + 8s}}{4}$. ■

WORK PROBLEM 2 AT THE SIDE.

---

1. Solve for x. Assume the variables represent the lengths of the sides of a triangle.

   (a) $3k^2 = x^2 - 5$

   (b) $5y^2 = z^2 + 9x^2$

2. Solve for t.

   (a) $2t^2 - 5t + k = 0$

   (b) $5st^2 - 7st + 1 = 0$

ANSWERS

1. (a) $x = \sqrt{3k^2 + 5}$ (b) $x = \dfrac{\sqrt{5y^2 - z^2}}{3}$

2. (a) $t = \dfrac{5 + \sqrt{25 - 8k}}{4}$, $t = \dfrac{5 - \sqrt{25 - 8k}}{4}$

   (b) $t = \dfrac{7s + \sqrt{49s^2 - 20s}}{10s}$, $t = \dfrac{7s - \sqrt{49s^2 - 20s}}{10s}$

CHAPTER 6 QUADRATIC EQUATIONS AND INEQUALITIES

3. (a) How long will it take the projectile in Example 3 to move 63 feet?

CAUTION The following examples show that it is important to check all proposed solutions of word problems against the information of the original problem. Numbers that are valid solutions of the equation may not satisfy the physical conditions of the problem.

The next example shows how quadratic equations are used to solve problems about motion in a straight line.

EXAMPLE 3 Solving a Straight-Line Motion Problem
The position of a projectile moving in a straight line is given by

$$s = 2t^2 - 5t,$$

where s is the distance in feet from a starting point and t is the time in seconds the projectile has been in motion. How many seconds will it take for the object to move 12 feet?

We must find t when $s = 12$. First substitute 12 for s in the original equation, and then solve for t.

$$12 = 2t^2 - 5t \quad \text{Let } s = 12.$$
$$0 = 2t^2 - 5t - 12 \quad \text{Get 0 on one side.}$$
$$0 = (2t + 3)(t - 4) \quad \text{Factor.}$$
$$t = -\frac{3}{2} \quad \text{or} \quad t = 4$$

(b) How long (to the nearest tenth of a second) will it take a projectile to move 10 feet if its position equation is $s = t^2 - 6t$?

Here, the negative answer is not of interest. (In some problems of this type it might be.) It will take 4 seconds for the projectile to move 12 feet. ■

WORK PROBLEM 3 AT THE SIDE.

3 The Pythagorean formula is used again in the solution of the next example.

EXAMPLE 4 Using the Pythagorean Formula
Two cars left an intersection at the same time, one heading due north, the other due west. Some time later, they were exactly 100 miles apart. The car headed north had gone 20 miles farther than the car headed west. How far had each car traveled?

Let x be the distance traveled by the car headed west. Then $x + 20$ is the distance traveled by the car headed north. These distances are shown in Figure 2. The cars are 100 miles apart, so the hypotenuse of the right triangle equals 100 and the two legs are equal to x and $x + 20$.

By the Pythagorean formula,

$$c^2 = a^2 + b^2$$
$$100^2 = x^2 + (x + 20)^2$$
$$10{,}000 = x^2 + x^2 + 40x + 400$$
$$0 = 2x^2 + 40x - 9600$$
$$0 = 2(x^2 + 20x - 4800).$$

Factor out the common factor.

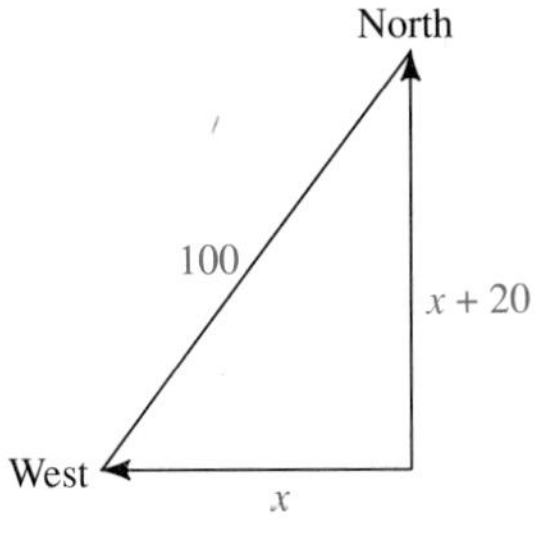

FIGURE 2

ANSWERS
3. (a) 7 seconds (b) 7.4 seconds

6.4 FORMULAS AND APPLICATIONS

Divide both sides by 2 to get $x^2 + 20x - 4800 = 0$ with $a = 1$, $b = 20$, and $c = -4800$. Use the quadratic formula to find x.

$$x = \frac{-20 \pm \sqrt{400 - 4(-4800)}}{2} \quad a = 1, b = 20, c = -4800$$

$$= \frac{-20 \pm \sqrt{19{,}600}}{2}$$

From a calculator, $\sqrt{19{,}600} = 140$, so

$$x = \frac{-20 \pm 140}{2}.$$

The solutions are $x = 60$ or $x = -80$. Discard the negative solution, so that 60 and $60 + 20 = 80$ are the required distances in miles. ■

WORK PROBLEM 4 AT THE SIDE.

4 Formulas for area may also result in quadratic equations, as the next example shows.

EXAMPLE 5 Solving an Area Problem

A reflecting pool in a park is 20 feet wide and 30 feet long. The park gardener wants to plant a strip of grass of uniform width around the edge of the pool. She has enough seed to cover 336 square feet. How wide will the strip be?

FIGURE 3

The pool is shown in Figure 3. If x represents the unknown width of the grass strip, the width of the large rectangle is given by $20 + 2x$ (the width of the pool plus two grass strips), and the length is given by $30 + 2x$. The area of the large rectangle is given by the product of its length and width, $(20 + 2x)(30 + 2x)$. The area of the pool is $20 \cdot 30 = 600$ square feet. The area of the large rectangle, minus the area of the pool, should equal the area of the grass strip. Since the area of the grass strip is to be 336 square feet, the equation is

Area of rectangle | Area of pool | Area of strip
$(20 + 2x)(30 + 2x) - 600 = 336.$
$600 + 100x + 4x^2 - 600 = 336$ Multiply.
$4x^2 + 100x - 336 = 0$ Collect terms.
$x^2 + 25x - 84 = 0$ Divide by 4.
$(x + 28)(x - 3) = 0$ Factor.
$x = -28$ or $x = 3$

4. (a) A 13-foot ladder is leaning against a house. The distance from the bottom of the ladder to the house is 7 feet less than the distance from the top of the ladder to the ground. How far is the bottom of the ladder from the house?

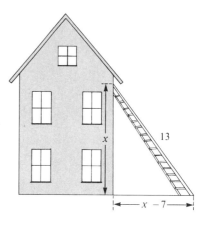

(b) A girl is flying a kite that is 30 feet farther above her hand than its horizontal distance from her. The string from her hand to the kite is 150 feet long. How high is the kite?

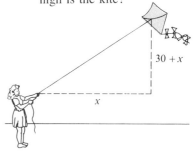

ANSWERS
4. (a) 5 feet   (b) 120 feet

5. (a) Suppose the pool in Example 5 is 20 feet by 40 feet and there is enough seed to cover 700 square feet. How wide should the grass strip be?

(b) A rectangle has a length 2 meters more than its width. If one meter is cut from the length and one meter is added to the width, the resulting figure is a square with an area of 121 square meters. Find the dimensions of the original rectangle.

6. (a) If the new worker in Example 6 takes 1 hour longer than the experienced worker to clean the building, and together they complete the job in 5 hours, how long would the experienced worker require to complete the job working alone?

(b) Working together, two people can cut a large lawn in 2 hours. One person can do the job alone in 1 hour more than the other. How long, to the nearest tenth of an hour, would it take the faster person to do the job?

ANSWERS
5. (a) 5 feet (b) 10 meters by 12 meters
6. (a) 9.5 hours (b) 3.6 hours

The width of a grass strip cannot be -28 feet, so 3 feet is the desired width of the strip. ■

WORK PROBLEM 5 AT THE SIDE.

5 As shown earlier, word problems about work may result in quadratic equations.

EXAMPLE 6 Solving a Work Problem

A janitorial service provides two people to clean an office building. Working together, the two can clean the building in 5 hours. One person is new to the job and would take 2 hours longer than the other person to clean the building working alone. How long would it take the experienced worker to clean the building working alone?

Let $\quad x =$ time in hours for the experienced worker to clean the building;

$x + 2 =$ time in hours for the new worker to clean the building;

$5 =$ time in hours for the two workers to clean the building together.

Then, in one hour,

$\dfrac{1}{x} =$ fraction of the job done by the experienced worker;

$\dfrac{1}{x+2} =$ fraction of the job done by the new worker;

$\dfrac{1}{5} =$ fraction of the job done by the two together.

From the words of the problem,

$\dfrac{1}{x} + \dfrac{1}{x+2} = \dfrac{1}{5}.$

$5(x + 2) + 5x = x(x + 2)$ Multiply by the common denominator.

$5x + 10 + 5x = x^2 + 2x$ Distributive property

$0 = x^2 - 8x - 10$ Collect terms.

$x = \dfrac{8 \pm \sqrt{64 + 40}}{2}$ Quadratic formula

$= \dfrac{8 \pm \sqrt{104}}{2}$

$\approx \dfrac{8 \pm 10.20}{2}$ From a calculator

$x = 9.1$ Discard negative solution.

To the nearest tenth, the experienced worker requires 9.1 hours to clean the building working alone. ■

WORK PROBLEM 6 AT THE SIDE.

6.4 EXERCISES

Solve each equation for the indicated variable. (Leave ± in the answers.) See Examples 1 and 2.

1. $c^2 = a^2 + b^2$ for a

2. $d = kt^2$ for t

3. $s = kwd^2$ for d

4. $I = \dfrac{ks}{d^2}$ for d

5. $R = \dfrac{k}{d^2}$ for d

6. $F = \dfrac{kA}{v^2}$ for v

7. $L = \dfrac{kd^4}{h^2}$ for h

8. $F = \dfrac{kwv^2}{r}$ for v

9. $V = \dfrac{1}{3}\pi r^2 h$ for r

10. $V = \pi(r^2 + R^2)h$ for r

11. $At^2 + Bt = -C$ for t

12. $S = 2\pi rh + \pi r^2$ for r

13. $D = \sqrt{kh}$ for h

14. $F = \dfrac{k}{\sqrt{d}}$ for d

15. $p = \sqrt{\dfrac{k\ell}{g}}$ for ℓ

Solve the following word problems using methods for solving quadratic equations. See Example 3.

16. (a) An object is projected directly upward from the ground. After t seconds its distance above the ground is $s = 144t - 16t^2$ feet. After how many seconds will it be 128 feet above the ground? (*Hint:* Look for a common factor before solving the equation.)

    (b) When does the object in part (a) strike the ground?

17. The formula $D = 13t^2 - 100t$ gives the distance a car going approximately 68 miles per hour will skid in t seconds. Find the time it would take for the car to skid 200 feet.

    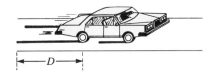

18. The formula in Exercise 17 becomes $D = 13t^2 - 73t$ for a car going 50 miles per hour. Find the time for this car to skid 248 feet.

Solve the following problems. See Example 4.

19. If the hypotenuse of a right triangle is 1 foot longer than the longer leg, and the shorter leg is 7 feet long, find the length of the longer leg.

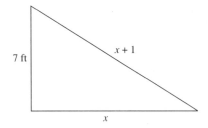

20. The hypotenuse of a right triangle is 1 meter longer than twice the length of the shorter leg, and the longer leg is 9 meters shorter than 3 times the length of the shorter leg. Find the three sides of the triangle.

21. At a point 30 meters from the base of a tower, the distance to the top of the tower is 2 meters more than twice the height of the tower. Find the height of the tower.

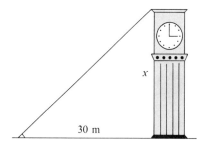

22. The longer leg of a right triangle is 2 meters more than twice as long as the shorter leg, while the hypotenuse is 1 meter longer than the longer leg. Find the three sides.

Solve the following problems. See Example 5.

23. A rectangular piece of sheet metal has a length that is 4 inches less than twice the width. A square piece 2 inches on a side is cut from each corner. The sides are then turned up to form an uncovered box of volume 256 cubic inches. Find the dimensions of the original piece of metal.

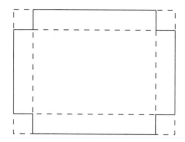

24. A square has an area of 64 square centimeters. If the same amount is removed from one dimension and added to the other, the resulting rectangle has an area 9 square centimeters less. Find the dimensions of the rectangle.

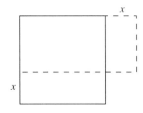

25. A couple wants to buy a rug for a room that is 20 feet long and 15 feet wide. They want to leave an even strip of flooring uncovered around the edges of the room. How wide a strip will they have if they buy a rug with an area of 234 square feet?

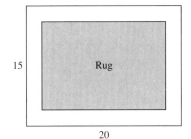

26. A club swimming pool is 30 feet wide and 40 feet long. The club members want an exposed aggregate border in a strip of uniform width around the pool. They have enough material for 296 square feet. How wide can the strip be?

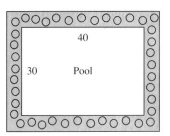

Use the quadratic formula to solve the following problems. Give answers to the nearest tenth. See Example 6.

27. Michael and Jessie are distributing brochures for a political campaign. Together they can complete the job in 3 hours. If Michael could do the job alone in one hour more than Jessie, how long would it take Jessie working alone?

28. Carmen and Juan can clean the house together in 2 hours. Working alone, it takes Carmen $\frac{1}{2}$ hour longer than Juan to do the job. How long would it take Juan alone?

29. It takes a drain 5 minutes more to empty a sink than it takes the faucet to fill the sink. If both are open the sink will fill in 30 minutes. How long would it take to fill the sink with the drain closed?

30. Two pipes can fill a tank in 4 hours when used together. Alone, one can fill the tank in $\frac{1}{2}$ hour less time than the other. How long will it take each pipe to fill the tank alone?

Use a quadratic equation to solve the following problems.

 31. A bicyclist traveled 20 miles at a speed that was 20 miles per hour slower than that of a motorist, arriving $\frac{5}{6}$ of an hour after the motorist. How fast did the motorist travel?

32. Janet and Russ are runners. Janet runs 1 mile per hour faster than Russ. When both ran in a 5-mile race, Russ's time was $\frac{5}{6}$ of an hour longer than Janet's. Find Janet's rate.

33. A small plane flew 600 miles against a 20-mile-per-hour wind, then returned with the wind. If the return trip was 2.5 hours shorter, find the speed of the plane in still air.

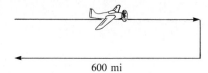

34. A boat traveled 36 miles upstream against a 10-mile-per-hour current and then returned downstream. The entire trip took 4.8 hours. What was the speed of the boat in still water?

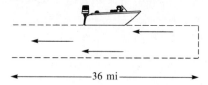

35. The formula $A = P(1 + r)^2$ gives the amount A in dollars that P dollars will grow to in 2 years at an interest rate r. What interest rate will cause $1,000 to grow to $1210 in 2 years?

36. In Exercise 35, if $2000 grows to $2205 in 2 years, what is the interest rate?

37. In one area the demand for compact discs is $\frac{700}{P}$ per day, where P is the price in dollars per disc. The supply is $5P - 1$ per day. At what price does supply equal demand?

38. A certain bakery has found that the daily demand for bran muffins is $\frac{3200}{p}$, where p is the price of a muffin in cents. The daily supply is $3p - 200$. Find the price at which supply and demand are equal.

SKILL SHARPENERS

Graph the following intervals on the number line. See Sections 1.2 and 2.6.

39. $[2, 4]$

40. $(-3, 0)$

41. $(-1, 5]$

42. $(-\infty, 1] \cup [4, \infty)$

43. $(-\infty, -2) \cup (1, \infty)$

44. $(-\infty, 0) \cup \left[\frac{3}{2}, \infty\right)$

6.5 NONLINEAR INEQUALITIES

1 We have discussed methods of solving linear inequalities (earlier) and methods of solving quadratic equations (in this chapter). Now this work can be extended to include solving *quadratic inequalities*.

A **quadratic inequality** can be written in the form
$$ax^2 + bx + c < 0 \quad \text{or} \quad ax^2 + bx + c > 0$$
where a, b, and c are real numbers, with $a \neq 0$.

As before, $<$ and $>$ can be replaced with $\leq$ and $\geq$ as necessary.

2 A method for solving quadratic inequalities is shown in the next example.

EXAMPLE 1 Solving Quadratic Inequalities
Solve $x^2 - x - 12 > 0$.

First solve the quadratic equation $x^2 - x - 12 = 0$ by factoring.
$$(x - 4)(x + 3) = 0$$
$$x - 4 = 0 \quad \text{or} \quad x + 3 = 0$$
$$x = 4 \quad \text{or} \quad x = -3$$

The numbers 4 and -3 divide the number line into the three regions shown in Figure 4. (Be careful to put the smaller number on the left.)

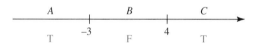

FIGURE 4

If one number in a region satisfies the inequality, then all the numbers in that region will satisfy the inequality. Choose any number from Region A in Figure 4 (any number less than -3). Substitute this number for x in the inequality $x^2 - x - 12 > 0$. If the result is *true*, then all numbers in Region A satisfy the original inequality.

Let us choose -5 from Region A. Substitute -5 into $x^2 - x - 12 > 0$, getting
$$(-5)^2 - (-5) - 12 > 0$$
$$25 + 5 - 12 > 0$$
$$18 > 0. \quad \text{True}$$

Since -5 from Region A satisfies the inequality, all numbers from Region A are solutions.

Try 0 from Region B. If $x = 0$, then
$$0^2 - 0 - 12 > 0$$
$$-12 > 0. \quad \text{False}$$

The numbers in Region B are *not* solutions. ■

WORK PROBLEM 1 AT THE SIDE.

OBJECTIVES

1 Recognize a quadratic inequality.

2 Solve quadratic inequalities.

3 Solve polynomial inequalities of degree 3 or more.

4 Solve fractional inequalities.

5 Understand special cases of quadratic inequalities.

---

1. Does the number 5 from Region C satisfy $x^2 - x - 12 > 0$?

ANSWERS
1. yes

CHAPTER 6 QUADRATIC EQUATIONS AND INEQUALITIES

2. Solve. Graph each solution.

(a) $x^2 + x - 6 > 0$

(b) $2p^2 + 3p \geq 2$

(c) $3m^2 - 13m - 10 \leq 0$

Since the number 5 satisfies the inequality, the numbers in Region C are also solutions to the inequality.

Based on these results (shown by the colored letters in Figure 4), the solution set includes the numbers in Regions A and C, as shown on the graph in Figure 5. The solution set is written in interval notation as

$$(-\infty, -3) \cup (4, \infty). \blacksquare$$

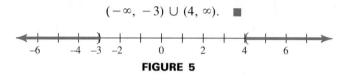

FIGURE 5

In summary, a quadratic inequality is solved by following these steps.

SOLVING A QUADRATIC INEQUALITY

Step 1 Write the inequality as an equation and solve the equation.

Step 2 Place the numbers found in Step 1 on a number line. These numbers divide the number line into regions.

Step 3 Substitute a number from each region into the inequality to determine the intervals that make the inequality true. If one number in a region satisfies the inequality, then all the numbers in that region will satisfy the inequality. The numbers in those intervals that make the inequality true are in the solution set.

Step 4 The numbers found in Step 1 are included in the solution set if the symbol is $\leq$ or $\geq$; they are not included if it is $<$ or $>$.

■ **WORK PROBLEM 2 AT THE SIDE.**

3 Higher-degree polynomial inequalities that are factorable can be solved in the same way as quadratic inequalities.

EXAMPLE 2 Solving a Third-Degree Polynomial Inequality

Solve $(x - 1)(x + 2)(x - 4) \leq 0$.

This is a *cubic* (third-degree) inequality rather than a quadratic inequality, but it can be solved by the method shown above and by extending the zero-factor property to more than two factors. Begin by setting the factored polynomial *equal* to 0 and solving the equation.

$$(x - 1)(x + 2)(x - 4) = 0$$

$x - 1 = 0$ or $x + 2 = 0$ or $x - 4 = 0$

$x = 1$ or $x = -2$ or $x = 4$

Locate the numbers -2, 1, and 4 on a number line as in Figure 6 to determine the regions A, B, C, and D. The numbers -2, 1, and 4 are in the solution set because of the "or equal to" part of the inequality symbol.

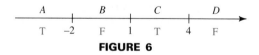

FIGURE 6

ANSWERS
2. (a) $(-\infty, -3) \cup (2, \infty)$

(b) $(-\infty, -2] \cup \left[\frac{1}{2}, \infty\right)$

(c) $\left[-\frac{2}{3}, 5\right]$

378

Substitute a number from each region into the original inequality to determine which regions satisfy the inequality. These results are shown below the number line in Figure 6. For example, in Region A, using $x = -3$ gives
$$(-3 - 1)(-3 + 2)(-3 - 4) \le 0$$
$$(-4)(-1)(-7) \le 0$$
$$-28 \le 0. \quad \text{True}$$

The numbers in Region A are in the solution set. Verify that the numbers in Region C are also in the solution set, which is written
$$(-\infty, -2] \cup [1, 4].$$

The solution set is graphed in Figure 7. ■

FIGURE 7

WORK PROBLEM 3 AT THE SIDE.

4 Inequalities involving fractions are solved in a similar manner, by going through the following steps.

SOLVING INEQUALITIES WITH FRACTIONS

Step 1 Write the inequality as an equation and solve the equation.

Step 2 Set the denominator equal to zero and solve that equation.

Step 3 Use the solutions from Steps 1 and 2 to divide a number line into regions.

Step 4 Test a number from each region by substitution in the inequality to determine the intervals that satisfy the inequality.

Step 5 Be sure to exclude any values that make the denominator equal to zero.

CAUTION Don't forget Step 2. Any number that makes the denominator zero *must* separate two regions on the number line.

EXAMPLE 3 Solving an Inequality with a Fraction

Solve the inequality $\dfrac{-1}{p - 3} > 1$.

Step 1 Write the corresponding equation and solve it.
$$\frac{-1}{p - 3} = 1$$
$$-1 = p - 3 \quad \text{Multiply by the common denominator.}$$
$$2 = p$$

6.5 NONLINEAR INEQUALITIES

3. Solve. Graph each solution.

(a) $(y - 3)(y + 2)(y + 1) > 0$

(b) $(k - 5)(k + 1)(k - 3) \le 0$

ANSWERS
3. (a) $(-2, -1) \cup (3, \infty)$

(b) $(\infty, -1] \cup [3, 5]$

379

4. Solve. Graph each solution.

(a) $\dfrac{2}{x-4} < 3$

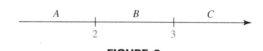

(b) $\dfrac{5}{y+1} > 4$

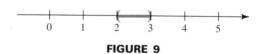

ANSWERS

4. (a) $(-\infty, 4) \cup \left(\dfrac{14}{3}, \infty\right)$

(b) $\left(-1, \dfrac{1}{4}\right)$

Step 2 Find the number that makes the denominator 0.

$$p - 3 = 0$$
$$p = 3$$

Step 3 These two numbers, 2 and 3, divide a number line into three regions. (See Figure 8.)

```
      A       B       C
  ————+———————+————→
       2       3
```

FIGURE 8

Step 4 Testing one number from each region in the given inequality shows that the solution set is the interval (2, 3).

Step 5 This interval does not include any value that might make the denominator of the original inequality equal to zero. A graph of this solution set is given in Figure 9. ∎

```
  ——+——+——(——)——+——+——→
    0  1  2  3  4  5
```

FIGURE 9

WORK PROBLEM 4 AT THE SIDE.

EXAMPLE 4 Solving an Inequality with a Fraction

Solve $\dfrac{m-2}{m+2} \leq 2$.

Write the corresponding equation and solve it. (Step 1)

$$\dfrac{m-2}{m+2} = 2$$

$$m - 2 = 2(m + 2) \quad \text{Multiply by the common denominator.}$$

$$m - 2 = 2m + 4 \quad \text{Distributive property}$$

$$-6 = m$$

Set the denominator equal to zero and solve the equation. (Step 2)

$$m + 2 = 0$$
$$m = -2$$

The numbers -6 and -2 determine three regions (Step 3). Test one number from each region (Step 4) to see that the solution set is the interval

$$(-\infty, -6] \cup (-2, \infty).$$

The number -6 satisfies the equality in $\leq$, but -2 cannot be used as a solution since it makes the denominator equal to zero (Step 5). The graph of the solution set is shown in Figure 10. ∎

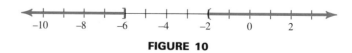

FIGURE 10

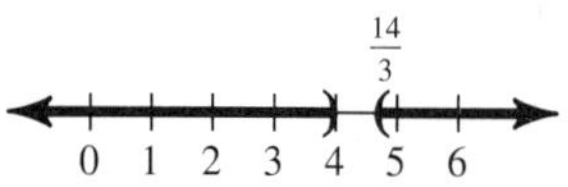

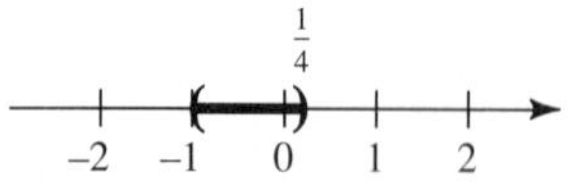

6.5 NONLINEAR INEQUALITIES

5. Solve. Graph each solution.

(a) $\dfrac{1}{a+3} \geq 2$

(b) $\dfrac{k+2}{k-1} \leq 5$

WORK PROBLEM 5 AT THE SIDE.

5 Special cases of quadratic inequalities may occur, such as those discussed in the next example.

EXAMPLE 5 Solving Special Cases

Solve $(2y - 3)^2 > -1$.

Since $(2y - 3)^2$ is never negative, it is always greater than -1. Thus, the solution is the set of all real numbers. In the same way, there is no solution for $(2y - 3)^2 < -1$ and the solution set is $\emptyset$. ∎

WORK PROBLEM 6 AT THE SIDE.

6. Solve.

(a) $(3k - 2)^2 > -2$

(b) $(5z + 3)^2 < -3$

ANSWERS

5. (a) $\left(-3, -\dfrac{5}{2}\right]$

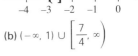

(b) $(-\infty, 1) \cup \left[\dfrac{7}{4}, \infty\right)$

6. (a) {all real numbers} (b) $\emptyset$

Historical Reflections

WOMEN IN MATHEMATICS:
Amalie ("Emmy") Noether (1882–1935)

Emmy Noether was an outstanding mathematician in the field of abstract algebra. She studied and worked in Germany at a time when it was very difficult for a woman to do so. At the University of Erlangen in 1900, she was one of only two women. Although she could attend classes, professors could and did deny her the right to take exams for their courses. Not until 1904 was she allowed to register officially. She completed her doctorate four years later.

In 1916 she went to Göttingen to work, but it was not until three years later that she become a *Privatdozent*, a member of the lowest rank in the faculty. In 1922 she was made an unofficial professor (or assistant). She received no pay for this post, although she was given a small stipend to lecture in algebra. She left Germany in 1933 and accepted a position at Bryn Mawr College in Pennsylvania. She died two years later at the age of fifty-three.

Art: Courtesy Gottfried Noether

6.5 EXERCISES

Solve the following inequalities. Graph each solution. See Example 1. (Hint: In Exercises 14 and 15, use the quadratic formula.)

1. $(2x + 1)(x - 5) > 0$

2. $(m + 6)(m - 2) \geq 0$

3. $(r + 4)(r - 6) \leq 0$

4. $(y + 5)(y - 6) < 0$

5. $x^2 - 4x + 3 < 0$

6. $m^2 - 3m - 10 > 0$

7. $10a^2 + 9a \geq 9$

8. $4x^2 + 5 \leq 21x$

9. $3r^2 + 10r \leq 8$

10. $9p^2 + 3p \geq 2$

11. $2y^2 + y < 15$

12. $6x^2 + x > 1$

13. $4y^2 + 7y + 3 < 0$

14. $y^2 - 6y + 6 \geq 0$

15. $3k^2 - 6k + 2 \leq 0$

Solve the following inequalities. Graph each solution. See Example 2.

16. $(p - 1)(p - 2)(p - 3) < 0$

17. $(2r + 1)(3r - 2)(4r + 5) < 0$

18. $(a - 4)(2a + 3)(3a - 1) \leq 0$

19. $(z + 2)(4z - 3)(2z + 5) \geq 0$

CHAPTER 6 QUADRATIC EQUATIONS AND INEQUALITIES

20. $(2k + 1)(3k - 1)(k + 4) > 0$

21. $(3m + 5)(2m + 1)(m - 3) < 0$

Solve the following inequalities. Graph each solution. See Examples 3 and 4.

22. $\dfrac{8}{x - 2} \geq 2$

23. $\dfrac{20}{y - 1} \leq 1$

24. $\dfrac{-6}{p - 5} \geq 10$

25. $\dfrac{3}{m - 1} \leq 1$

26. $\dfrac{-2}{k + 1} > 5$

27. $\dfrac{3}{2t - 1} < 2$

28. $\dfrac{1 - p}{4 - p} > 0$

29. $\dfrac{x + 1}{x - 5} < 0$

30. $\dfrac{a}{a + 2} \geq 2$

31. $\dfrac{m}{m + 5} \leq 1$

32. $\dfrac{t}{2t - 3} < -1$

33. $\dfrac{y}{3y - 2} > -2$

34. $\dfrac{4k}{2k - 1} < k$

35. $\dfrac{r}{r + 2} \geq 2r$

36. $\dfrac{x^2 - 4}{3x} \geq 1$

SKILL SHARPENERS

Suppose that $3x - 2y = 12$. Find y for the following values of x. See Section 1.4.

37. 0

38. 4

39. -2

40. -3

CHAPTER 6 SUMMARY

KEY TERMS

6.2 **quadratic formula** — The quadratic formula is a formula for solving quadratic equations.

discriminant — The discriminant is the quantity under the radical in the quadratic formula.

6.3 **quadratic in form** — An equation that can be written as a quadratic equation is called quadratic in form.

6.4 **Pythagorean formula** — The Pythagorean formula states that in a right triangle the square of the length of the hypotenuse equals the sum of the squares of the lengths of the legs.

hypotenuse — The hypotenuse is the longest side in a right triangle.

leg — The two shorter sides of a right triangle are called the legs.

6.5 **quadratic inequality** — A quadratic inequality is an inequality that can be written in the form $ax^2 + bx + c < 0$ or $ax^2 + bx + c > 0$, or with $\leq$ or $\geq$.

QUICK REVIEW

Section	Concepts	Examples
6.1 Completing the Square	**Square Root Property** If a and b are complex numbers and if $a^2 = b$, then $\quad a = \sqrt{b}$ or $a = -\sqrt{b}$.	Solve $(x - 1)^2 = 8$. $\quad x - 1 = \pm\sqrt{8} = \pm 2\sqrt{2}$ $\quad x = 1 \pm 2\sqrt{2}$
	Completing the Square *Step 1* If $a \neq 1$, divide both sides by a.	Given: $2x^2 - 4x - 18 = 0$. $\quad x^2 - 2x - 9 = 0$
	Step 2 Rewrite the equation so that both terms containing variables are on one side of the equals sign and the constant is on the other side.	$x^2 - 2x = 9$
	Step 3 Take half the coefficient of x and square it.	$\left[\dfrac{1}{2}(-2)\right]^2 = (-1)^2 = 1$
	Step 4 Add the square to both sides.	$x^2 - 2x + 1 = 9 + 1$
	Step 5 One side should now be a perfect square. Write it as the square of a binomial.	$(x - 1)^2 = 10$
	Step 6 Use the square root property to complete the solution.	$x - 1 = \pm\sqrt{10}$ $x = 1 \pm \sqrt{10}$

CHAPTER 6 QUADRATIC EQUATIONS AND INEQUALITIES

Section	Concepts	Examples
6.2 The Quadratic Formula	**Quadratic Formula** The solutions of $ax^2 + bx + c = 0$ ($a \neq 0$) are $$x = \frac{-b \pm \sqrt{b^2 - 4ac}}{2a}.$$	Solve $3x^2 + 5x + 2 = 0$. $$x = \frac{-5 \pm \sqrt{5^2 - 4(3)(2)}}{2(3)}$$ $x = -1$ or $x = -\dfrac{2}{3}$
6.5 Nonlinear Inequalities	**Solving a Quadratic Inequality** *Step 1* Write the inequality as an equation and solve. *Step 2* Place the numbers found in Step 1 on a number line. These numbers divide the line into regions. *Step 3* Substitute a number from each region into the inequality to determine the intervals that belong in the solution set—those intervals containing numbers that make the inequality true. **Solving Inequalities With Fractions** *Step 1* Write the inequality as an equation and solve the equation. *Step 2* Set the denominator equal to zero and solve that equation. *Step 3* Use the solutions from Steps 1 and 2 to divide a number line into regions. *Step 4* Test a number from each region in the inequality to determine the regions that satisfy the inequality. *Step 5* Exclude any values that make the denominator zero.	Given: $2x^2 + 5x + 2 < 0$. $$2x^2 + 5x + 2 = 0$$ $$x = -\frac{1}{2}, x = -2$$ $x = -3$ makes it false; $x = -1$ makes it true; $x = 0$ makes it false. Solution: $\left(-2, -\dfrac{1}{2}\right)$ Given: $\dfrac{x}{x+2} > 4$. $$\frac{x}{x+2} = 4; \quad x = -\frac{8}{3}$$ $$x + 2 = 0$$ $$x = -2$$ -4 makes it false; $-\dfrac{7}{3}$ makes it true; 0 makes it false. The solution is $\left(-\dfrac{8}{3}, -2\right)$, since -2 makes the denominator 0 and $-\dfrac{8}{3}$ gives a false sentence.

CHAPTER 6 REVIEW EXERCISES

[6.1] *Solve each equation by using the square root property. Complete the square first, if necessary.*

1. $t^2 = 16$
2. $p^2 = 7$
3. $(2x + 5)^2 = 49$

*4. $(3k - 2)^2 = -16$
5. $x^2 - 4x = 15$
6. $2m^2 - 3m + 1 = 0$

[6.2] *Solve each equation by the quadratic formula.*

7. $2y^2 - y - 21 = 0$
8. $k^2 + 5k - 7 = 0$
9. $2S^2 + 3S = 10$

10. $9p^2 + 49 = 42p$
*11. $3p^2 = 4p - 2$
12. $m(2m - 7) = 3(m^2 + 1)$

*13. $k^4 = 2k^2 + 3$
*14. $3p^4 - p^2 - 2 = 0$
15. $\sqrt{2}p^2 - 3p + \sqrt{2} = 0$

16. A paint-mixing machine has two inlet pipes. One takes 1 hour longer than the other to fill the tank. Together they fill the tank in 3 hours. How long would it take each of them alone to fill the tank? Round your answer to the nearest tenth.

17. A new machine processes a batch of checks in one hour less time than an old one. How long would it take the old machine to process a batch of checks that the two machines together process in 2 hours? Round your answer to the nearest tenth.

*Exercises identified with asterisks have imaginary number solutions.

CHAPTER 6 QUADRATIC EQUATIONS AND INEQUALITIES

Use the discriminant to predict whether the solutions to the following equations are

(a) *two distinct rational numbers;*   **(b)** *exactly one rational number;*
(c) *two distinct irrational numbers;*   **(d)** *two distinct imaginary numbers.*

18. $a^2 + 5a + 1 = 0$

19. $4c^2 + 4c = 3$

20. $4x^2 = 6x - 9$

21. $y^2 + 2y + 5 = 0$

22. $9z^2 - 30z + 25 = 0$

23. $11m^2 - 7m - 9 = 0$

Use the discriminant to tell which polynomials can be factored. If a polynomial can be factored, factor it.

24. $24x^2 + 74x + 45$

25. $36x^2 + 69x - 35$

[6.3] *Solve each equation.*

26. $2k - 1 = \dfrac{15}{k}$

27. $4 = \dfrac{1}{t} + \dfrac{3}{t^2}$

28. $-r^4 = 16 - 10r^2$

29. $8(2a + 3)^2 + 2(2a + 3) - 1 = 0$

30. $\sqrt{3x} = \sqrt{-4x - 1}$

31. $r = \sqrt{8r + 35}$

32. $\sqrt{t - 1} = 1 - \sqrt{2t}$

33. Ann Bezzone drove 8 kilometers to pick up her friend Lenore, and then drove 11 kilometers to a shopping center at a speed 15 kilometers per hour faster. If Ann's total travel time was 24 minutes, what was her speed on the trip to pick up Lenore?

34. Bill Poole can write a report for his boss in 2 hours less time than Linda Youngman can. Working together, they can do the report in $\dfrac{24}{7}$ hours. How long would it take each person working alone to write the report?

CHAPTER 6 REVIEW EXERCISES

[6.4] *Solve each formula for the indicated variable. (Give answers with $\pm$.)*

35. $S = \dfrac{Id^2}{k}$ for d

36. $k = \dfrac{rF}{wv^2}$ for v

37. $S = 2\pi rh + 2\pi r^2$ for r

38. $mt^2 = 3mt + 5$ for t

Find the time it takes for an object to move 16 feet in a straight line, given the following equations for the distance s (in feet) it moves in t seconds. Give your answers to the nearest tenth of a second.

39. $s = t^2 + 6t + 13$

40. $s = 2t^2 - 4t + 7$

Solve each of the following problems.

41. If the hypotenuse of a right triangle is 25 meters in length and the length of one leg is 8 meters, find the length of the other leg.

42. The hypotenuse of a right triangle is 9 feet shorter than twice the length of the longer leg. The shorter leg is 3 feet shorter than the longer leg. Find the lengths of the three sides of the triangle.

43. The length of a rectangle is 3 inches longer than 5 times the width. The area is 26 square inches. Find the dimensions of the rectangle.

$5w + 3$
w

44. Masami wants to buy a mat for a photograph that measures 14 inches by 20 inches. She wants to have an even border around the picture when it is mounted on the mat. If the area of the mat she chooses is 352 square inches, how wide will the border be?

45. Janet Tilden can clean all the snow from a strip of driveway in 4 hours less time than it takes Laurie Golson. Working together, the job takes them $\frac{8}{3}$ hours. How long would it take each person working alone?

46. Ben Whitney can work through a stack of invoices in 1 hour less time than Arnold Parker can. Working together, they take $\frac{3}{2}$ hours. How long would it take each person working alone?

47. Two cars traveled from Elmhurst to Oakville, a distance of 100 miles. One car traveled 10 miles per hour faster than the other and arrived $\frac{5}{6}$ of an hour sooner. Find the speed of the faster car.

48. An excursion boat traveled 20 miles upriver and then traveled back. If the current was 10 miles per hour and the entire trip took 2 hours, find the speed of the boat in still water.

49. The manager of a fast food outlet has determined that the demand for frozen yogurt is $\frac{25}{p}$ units per day, where p is the price (in dollars) per unit. The supply is $70p + 15$ units per day. Find the price at which supply and demand are equal.

50. Use the formula $A = P(1 + r)^2$ to find the interest rate at which a deposit of $10,000 will increase to $11,664 in 2 years.

51. The distance of a particle from a starting point after t seconds is given by $s = t^2 - 2t - 8$. At what positive time t is the particle 5 units from the starting point? Round the answer to the nearest tenth.

CHAPTER 6 REVIEW EXERCISES

■ **52.** The expression $s = 16t^2 + 15t + 20$ gives the distance in feet an object thrown off a building has fallen in t seconds. Find the time t when the object has fallen 25 feet. Round the answer to the nearest tenth.

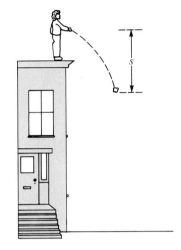

[6.5] *Solve the following inequalities. Graph each solution.*

53. $(2p + 3)(p - 1) > 0$

54. $k^2 + k < 12$

55. $2m^2 \leq 5m + 3$

56. $\dfrac{(y + 1)(y - 1)}{2y - 1} < 0$

MIXED REVIEW EXERCISES
Solve.

57. $\sqrt{2y + 12} = 1 + \sqrt{y + 7}$

58. $V = r^2 + R^2 h$ for R

*59. $3p^2 - 6p + 4 = 0$

391

60. $(b^2 - 2b)^2 = 11(b^2 - 2b) - 24$

61. $(r - 1)(2r + 3)(r + 7) \geq 0$

62. Two pipes together can fill a large tank in 2 hours. One of the pipes, used alone, takes 3 hours longer than the other to fill the tank. How long would each pipe take to fill the tank alone?

63. Phong paddled his canoe 10 miles upstream, then paddled back. If the speed of the current was 3 miles per hour and the total trip took $3\frac{1}{2}$ hours, what was Phong's speed?

64. $(m - 4)^2 = 6$

65. $p = \sqrt{\dfrac{xy}{3}}$ for y

66. $5y^2 = 8y - 3$

67. $6 + \dfrac{19}{m} = -\dfrac{15}{m^2}$

68. $\dfrac{2}{x + 5} \geq 5$

69. $4y^2 + 8y + 3 = 0$

CHAPTER 6 TEST

Solve by the square root property.

1. $r^2 = 8$

2. $(3m + 2)^2 = 4$

Solve by completing the square.

3. $z^2 - 2z = 1$

Solve by the quadratic formula.

4. $2m^2 - 3m + 1 = 0$

*5. $3r^2 + 4r + 5 = 0$

6. $2x = \sqrt{\dfrac{5x + 2}{3}}$

7. Mario and Luis are cutting firewood. Luis can cut a cord of firewood in 2 hours less time than Mario. Together they can do the job in 5 hours. How long will it take each of them if they work alone? Give your answers to the nearest tenth.

Use the discriminant to predict the number and type of solutions. Do not solve.

8. $2p^2 + 5 = 4p$

9. $2y^2 - 4y - 3 = 0$

10. $m^2 - 4m = -3$

1. _____
2. _____
3. _____
4. _____
5. _____
6. _____
7. _____
8. _____
9. _____
10. _____

*Exercises identified with asterisks have imaginary number solutions.

Solve by any method.

11. $\dfrac{3}{p^2} = 1 + \dfrac{1}{p}$

12. $2m^4 + 20 = 13m^2$

13. $12 = (2p - 1)^2 + (2p - 1)$

14. On a 60-mile bicycle trip, Jenny took one hour less time than Bill. If Jenny's speed was 5 miles per hour faster than Bill's, find Bill's speed.

Solve each equation for the indicated variable. (Give answers with $\pm$.)

15. $S = 4\pi r^2$ for r

16. $r = kp(1 - p)$ for p

17. Ken has a pool 24 feet long and 20 feet wide. He wants to plant grass in a strip of uniform width around the pool. He has enough grass seed for 192 square feet. How wide will the strip be?

18. A lot is in the shape of a right triangle. The shorter leg measures 50 meters. The hypotenuse is 110 meters shorter than twice the length of the longer leg. How long is the longer leg?

Solve. Graph each solution set.

19. $2x^2 + 7x < 15$

20. $\dfrac{4}{p - 1} \geq 1$

7 THE STRAIGHT LINE

7.1 THE RECTANGULAR COORDINATE SYSTEM

An **ordered pair** is a pair of numbers written within parentheses in which the order of the numbers matters. By this definition, the ordered pairs (2, 5) and (5, 2) are different. The two numbers are called **components** of the ordered pair. It is customary for x to represent the first component and y the second component. Graph an ordered pair by using two number lines that intersect at right angles at the zero points, as shown in Figure 1. The common zero point is called the **origin.** The horizontal line, the *x*-axis, represents the first number in an ordered pair, and the vertical line, the *y*-axis, represents the second. The *x*-axis and the *y*-axis make up a **rectangular coordinate system.**

OBJECTIVES

1 Plot ordered pairs.

2 Find ordered pairs that satisfy a given equation.

3 Graph lines.

4 Find *x*- and *y*-intercepts.

5 Recognize equations of vertical or horizontal lines.

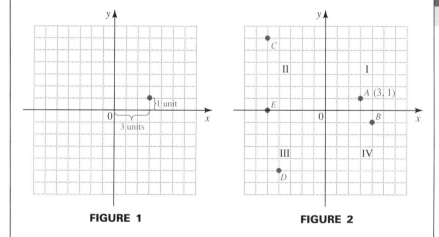

FIGURE 1 **FIGURE 2**

1 To locate, or **plot,** the point on the graph that corresponds to the ordered pair (3, 1), go three units from zero to the right along the *x*-axis, and then go one unit up parallel to the *y*-axis. The point corresponding to the ordered pair (3, 1) is labeled A in Figure 2. The point (4, −1) is labeled B, (−5, 6) is labeled C, and (−4, −5) is labeled D. Point E corresponds to (−5, 0). The phrase "the point corresponding to the ordered pair (2, 1)" is often abbreviated as "the point (2, 1)." The numbers in the ordered pairs are called the **coordinates** of the corresponding point.

1. Plot the following points.

   (a) $(-4, 2)$ (b) $(3, -2)$

   (c) $(-5, -6)$ (d) $(4, 6)$

   (e) $(-3, 0)$ (f) $(0, -5)$

2. Complete the following ordered pairs for $3x - 4y = 12$:

   $(0, \ )$, $(\ , 0)$, $(\ , -2)$, $(-4, \ )$, $(-6, \ )$.

The four regions of the graph, shown in Figure 2, are called **quadrants I, II, III,** and **IV,** reading counterclockwise from the upper right quadrant. The points on the x-axis and y-axis do not belong to any quadrant. For example, point E in Figure 2 belongs to no quadrant.

■ **WORK PROBLEM 1 AT THE SIDE.**

2 The solutions of an equation with two variables can be written as ordered pairs. Find the ordered pairs that satisfy the equation by selecting any number for one of the variables, substituting it into the equation for that variable, and then solving for the other variable. For example, suppose $x = 0$ in the equation $2x + 3y = 6$. Then, by substitution,

becomes
$$2x + 3y = 6$$
$$2(0) + 3y = 6 \quad \text{Let } x = 0.$$
$$0 + 3y = 6$$
$$3y = 6$$
$$y = 2,$$

giving the ordered pair $(0, 2)$. Some other ordered pairs satisfying $2x + 3y = 6$ are $(6, -2)$ and $(3, 0)$.

EXAMPLE 1 Completing Ordered Pairs

Complete the following ordered pairs for $2x + 3y = 6$.

(a) $(-3, \ )$

Let $x = -3$. Substitute into the equation.
$$2(-3) + 3y = 6 \quad \text{Let } x = -3.$$
$$-6 + 3y = 6$$
$$3y = 12$$
$$y = 4$$

The ordered pair is $(-3, 4)$.

(b) $(\ , -4)$

Replace y with -4.
$$2x + 3y = 6$$
$$2x + 3(-4) = 6 \quad \text{Let } y = -4.$$
$$2x - 12 = 6$$
$$2x = 18$$
$$x = 9$$

The ordered pair is $(9, -4)$. ■

■ **WORK PROBLEM 2 AT THE SIDE.**

ANSWERS

1.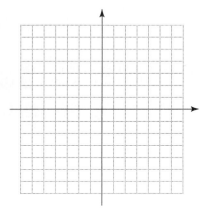

2. $(0, -3)$, $(4, 0)$, $\left(\dfrac{4}{3}, -2\right)$, $(-4, -6)$, $\left(-6, -\dfrac{15}{2}\right)$

3 The **graph** of an equation is the set of points that correspond to all the ordered pairs that satisfy the equation. It gives a "picture" of the equation. Since most equations with two variables have an infinite set of ordered pairs, their graphs include an infinite number of points. To graph an equation, we plot a number of ordered pairs that satisfy the equation until we have enough points to suggest the shape of the graph. For example, to graph $2x + 3y = 6$, first graph all the ordered pairs mentioned earlier, as shown in Figure 3(a). The resulting points appear to lie on a straight line. If all the ordered pairs that satisfy the equation $2x + 3y = 6$ were graphed, they would form a straight line. In fact, the graph of any first-degree equation in two variables is always a straight line. The graph of $2x + 3y = 6$ is the line shown in Figure 3(b).

3. Graph $3x - 4y = 12$. Use the points from Problem 2.

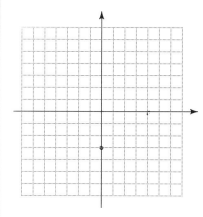

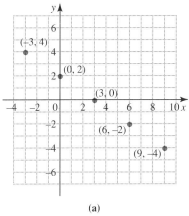 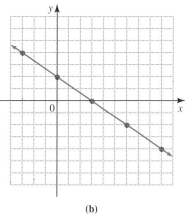

(a)   (b)

FIGURE 3

WORK PROBLEM 3 AT THE SIDE.

4 Since first-degree equations with two variables have straight-line graphs, they are called **linear equations in two variables.** (We discussed linear equations in one variable in Chapter 2.)

An equation that can be put in the **standard form**

$$ax + by = c$$

is a linear equation.

A straight line is determined if any two different points on the line are known, so finding two different points is enough to graph the line, but it is wise to find a third point as a check. Two points that are useful for graphing are the x- and y-intercepts. The **x-intercept** is the point (if any) where the line crosses the x-axis; likewise, the **y-intercept** is the point (if any) where the line crosses the y-axis. Look at Figure 3(b). The y-value of the point where the line crosses the x-axis is 0. Similarly, the x-value of the point where the line crosses the y-axis is 0. This suggests a method for finding the x- and y-intercepts.

ANSWER
3.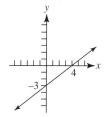

In the equation of a line, let $y = 0$ to find the x-intercept; let $x = 0$ to find the y-intercept.

EXAMPLE 2 Finding Intercepts

Find the x- and y-intercepts of $4x - y = -3$ and graph the equation.

To find the x-intercept, let $y = 0$.

$$4x - \mathbf{0} = -3 \quad \text{Let } y = 0.$$
$$4x = -3$$
$$x = -\frac{3}{4}$$

The x-intercept is $\left(-\frac{3}{4}, 0\right)$.

For the y-intercept, let $x = 0$.

$$4(\mathbf{0}) - y = -3 \quad \text{Let } x = 0.$$
$$-y = -3$$
$$y = 3$$

The y-intercept is $(0, 3)$.

Use the two intercepts to draw the graph, shown in Figure 4. ∎

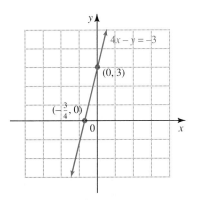

FIGURE 4

CAUTION If the graph goes through the origin, the x- and y-intercepts will be the same point, $(0, 0)$. In this case an additional point *must* be used to graph the equation.

5 The next example shows that a graph can fail to have an x-intercept, which is why we added the phrase "if any" when discussing intercepts.

EXAMPLE 3 Graphing a Horizontal Line
Graph $y = 2$.

Writing $y = 2$ as $0x + 1y = 2$ shows that any value of x, including $x = 0$, gives $y = 2$, making the y-intercept $(0, 2)$. Every ordered pair that satisfies this equation has a y-coordinate of 2. Since y is always 2, there is no value of x corresponding to $y = 0$ and the graph has no x-intercept. The graph, shown in Figure 5, is a horizontal line. ■

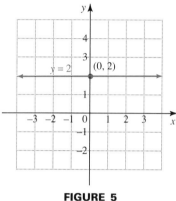

FIGURE 5

It is also possible for a graph to have no y-intercept, as in the next example.

EXAMPLE 4 Graphing a Vertical Line
Graph $x = -1$.

The form $1x + 0y = -1$ shows that *every* value of y leads to $x = -1$, so no value of y makes x equal to 0. The only way a straight line can have no y-intercept is to be vertical, as in Figure 6. Notice that every point on the line has $x = -1$, while all real numbers are used for the y-values. ■

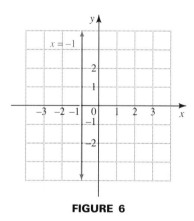

FIGURE 6

WORK PROBLEM 4 AT THE SIDE.

---

4. Find the intercepts and graph each line.

(a) $2x - y = 4$

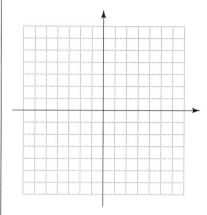

(b) $x + 2 = 0$

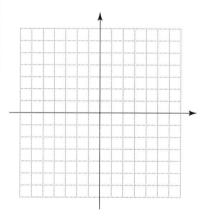

ANSWERS
4. (a) *x*-intercept is (2, 0); *y*-intercept is (0, −4).
   (b) no *y*-intercept; *x*-intercept is (−2, 0)

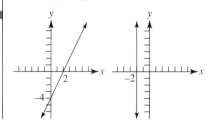

Historical Reflections

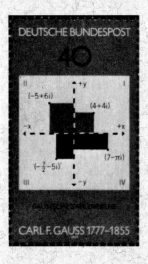

Gauss (1777–1855) and the Complex Numbers

The stamp shown here honors the many contributions made by the German mathematician Carl F. Gauss. Gauss is known as the "Prince of Mathematicians."

He showed genius at an early age. One day while he was in elementary school, his class was asked to find the sum of the first 100 counting numbers: $1 + 2 + 3 \ldots + 98 + 99 + 100$. Young Carl immediately wrote down the answer without showing any work. He explained that he noticed that $1 + 100 = 101$, $2 + 99 = 101$, $3 + 98 = 101$, and so on. Since there are 50 pairs of numbers that each add up to 101, the sum must be $50(101) = 5050$. According to E. T. Bell, Gauss' teacher then bought Gauss the best arithmetic text available, admitting, "He is beyond me . . . I can teach him nothing more."

In 1831 Gauss was able to show that numbers of the form $a + bi$ can be represented as points (as the stamp indicates) in the *complex plane*. He shares this contribution with Robert Argand, a bookkeeper in Paris who wrote an essay on the geometry of the complex numbers in 1806. It went unnoticed at the time.

name   date   hour

7.1 EXERCISES

Name the quadrant in which each point is located.

1. (1, 5)   **2.** (−2, 4)   **3.** (−3, −2)   **4.** (5, −1)   **5.** (2, −3)

6. (−7, −4)   **7.** (−1, 4)   **8.** (0, 4)   **9.** (2, 0)   **10.** (5, 2)

Locate the following points on the rectangular coordinate system.

11. (2, 3)   **12.** (−1, 2)

13. (−3, −2)   **14.** (1, −4)

15. (0, 5)   **16.** (−2, −4)

17. (−2, 4)   **18.** (3, 0)

19. (−2, 0)   **20.** (3, −3)

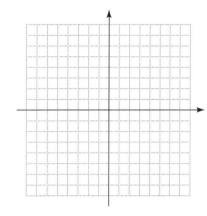

In each exercise, complete the given ordered pairs for the equation, and then graph the equation. See Example 1.

21. $2x + y = 5$
(0,), (, 0), (1,),
(, 1)

22. $3x - 4y = 24$
(0,), (, 0), $\left(6, \right)$,
(, −3)

23. $x - y = 4$
(0,), (, 0), (2,),
(, −1)

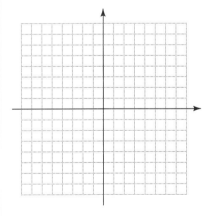   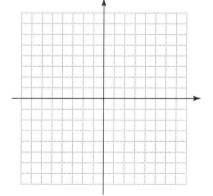   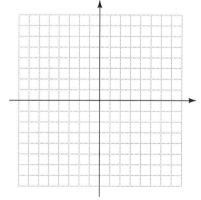

24. $x + 3y = 12$
(0,), (, 0), (3,),
(, 6)
(*Hint:* Let each grid square represent 2 units.)

25. $4x + 5y = 20$
(0,), (, 0), $\left(3, \quad\right)$,
$\left(\quad, 2\right)$

26. $2x - 5y = 12$
$\left(0, \quad\right)$, (, 0),
(, −2), $\left(-2, \quad\right)$

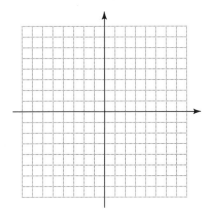
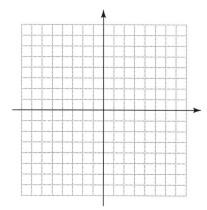
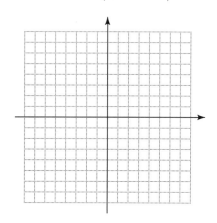

27. $3x + 2y = 8$
(0,), $\left(\quad, 0\right)$, (2,),
(, −2)

28. $5x + y = 12$
(0,), $\left(\quad, 0\right)$, (, −3),
(2,)

29. $4x + 3y = 7$
$\left(0, \quad\right)$, $\left(\quad, 0\right)$, (1,),
(, −3)

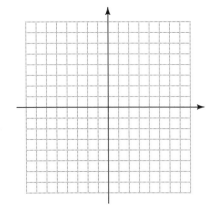
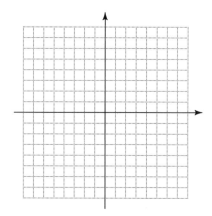
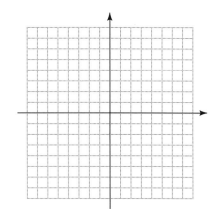

7.1 EXERCISES

For each equation, find the x-intercept and the y-intercept and graph. See Examples 2–4.

30. $3x + 2y = 12$

31. $2x + 5y = 10$

32. $5x + 6y = 10$

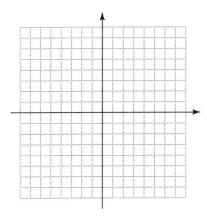
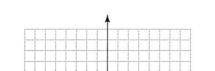
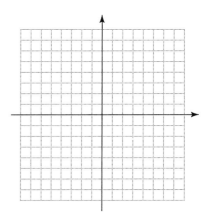

33. $3y + x = 6$

34. $y + x = 0$

35. $7x - 3y = 8$

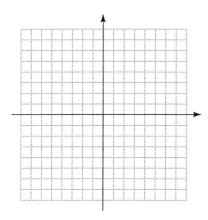
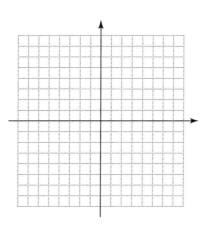
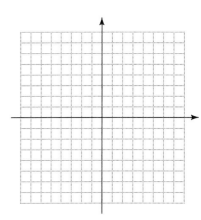

36. $4x = 8 + 5y$

37. $x + 2y = 5$

38. $y = -\dfrac{1}{2}$

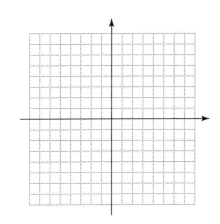
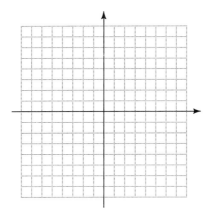
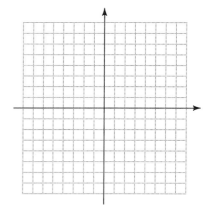

39. Recently the U.S. population has been growing according to the equation

$$y = 1.7x + 230$$

where y gives the population (in millions) in year x, measured from year 1980. For example, in 1980 $x = 0$ and $y = 1.7(0) + 230 = 230$. This means that the population was about 230 million in 1980. To find the population in 1985, let $x = 5$, and so on. Find the population in each of the following years.

(a) 1982 (b) 1985 (c) 1990

(d) In what year will the population reach 315 million?

(e) Graph the equation.

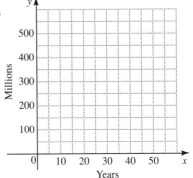

40. It is estimated that y, the number of bicycles (in millions) sold in the United States in year x, where x represents the number of years since 1990, is given by

$$y = 1.71x + 2.98.$$

That is, $x = 0$ represents 1990, $x = 1$ represents 1991, and so on. Find the number of bicycles sold in the following years.

(a) 1990 (b) 1991 (c) 1992

(d) In what year will about 10 million bicycles be sold?

(e) Graph the equation.

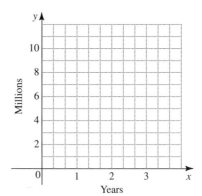

SKILL SHARPENERS
Find each quotient. See Section 1.3.

41. $\dfrac{5 - 3}{6 - 2}$

42. $\dfrac{-4 - 2}{5 - 7}$

43. $\dfrac{-3 - (-5)}{4 - (-1)}$

44. $\dfrac{6 - (-2)}{-3 - (-1)}$

45. $\dfrac{-2 - (-3)}{-1 - 4}$

46. $\dfrac{-8 - 2}{6 - (-3)}$

7.2 THE SLOPE OF A LINE

Two different points determine a line. A line can also be determined if we know a point on the line and some measure of the "steepness" of the line. One way to get a measure of the steepness, or slope, of a line is to compare the vertical change in the line (the *rise*) to the horizontal change (the *run*) while moving along the line from one fixed point to another.

OBJECTIVES

1. Find the slope of a line, given two points on the line.
2. Find the slope of a line, given an equation of the line.
3. Graph a line, using its slope and a point on the line.
4. Use slopes to determine whether two lines are parallel, perpendicular, or neither.

1 The notation x_1 (read "x-sub-one"), x_2, y_1, y_2, and so on represents specific *x*-values or *y*-values. Suppose (x_1, y_1) and (x_2, y_2) are two different points on a line. Then, as we move along the line from (x_1, y_1) to (x_2, y_2), the *y*-value changes from y_1 to y_2, an amount equal to $y_2 - y_1$. As *y* changes from y_1 to y_2, the value of *x* changes from x_1 to x_2 by the amount $x_2 - x_1$. See Figure 7.

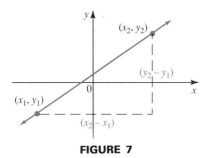

FIGURE 7

The ratio of the change in *y* to the change in *x* is called the **slope** of the line, with the letter *m* used for slope.

SLOPE

If $x_1 \neq x_2$, the slope of the line through the distinct points (x_1, y_1) and (x_2, y_2) is

$$m = \frac{\text{change in } y}{\text{change in } x} = \frac{y_2 - y_1}{x_2 - x_1}.$$

The larger the absolute value of *m*, the steeper the corresponding line will be, since a larger value of *m* indicates a greater change in *y* as compared to the change in *x*.

EXAMPLE 1 Using the Definition of Slope
Find the slope of the line through the points $(2, -1)$ and $(-5, 3)$.

Let $(2, -1) = (x_1, y_1)$ and $(-5, 3) = (x_2, y_2)$; then

$$m = \frac{y_2 - y_1}{x_2 - x_1} = \frac{3 - (-1)}{-5 - 2} = \frac{4}{-7} = -\frac{4}{7}.$$

On the other hand, if the pairs are reversed, so that $(2, -1) = (x_2, y_2)$ and $(-5, 3) = (x_1, y_1)$, the slope is

$$m = \frac{-1 - 3}{2 - (-5)} = \frac{-4}{7} = -\frac{4}{7},$$

the same answer. ∎

CHAPTER 7 THE STRAIGHT LINE

1. Find the slope of the line through each pair of points.

(a) $(-2, 7), (4, -3)$

Example 1 suggests that the slope is the same no matter which point is considered first. Also, using similar triangles from geometry, it can be shown that the slope is the same no matter which two different points on the line are chosen.

CAUTION In calculating the slope, be careful to subtract the *y*-values and the *x*-values in the *same* order.

Correct	Incorrect
$\dfrac{y_2 - y_1}{x_2 - x_1}$ or $\dfrac{y_1 - y_2}{x_1 - x_2}$	$\dfrac{y_2 - y_1}{x_1 - x_2}$ or $\dfrac{y_1 - y_2}{x_2 - x_1}$

■ **WORK PROBLEM 1 AT THE SIDE.**

(b) $(5, 1), (2, 8)$

2 When an equation of a line is given, the slope can be found from the definition of slope by first finding two different points on the line.

EXAMPLE 2 Finding the Slope of a Line
Find the slope of the line $4x - y = 8$.

The intercepts can be used as the two different points needed to find the slope. Replace y with 0 to find that the *x*-intercept is $(2, 0)$; the *y*-intercept is $(0, -8)$. The slope is

$$m = \frac{-8 - 0}{0 - 2} = \frac{-8}{-2} = 4. \blacksquare$$

EXAMPLE 3 Finding the Slope of a Line
Find the slope of the following lines.

(a) $x = -3$

(c) $(-6, 9), (3, -5)$

By inspection, $(-3, 5)$ and $(-3, -4)$ are two points that satisfy the equation $x = -3$. Use these two points to find the slope.

$$m = \frac{-4 - 5}{-3 - (-3)} = \frac{-9}{0}$$

Since division by zero is undefined, the slope is undefined. As shown in Section 7.1, the graph of an equation such as $x = -3$ is a vertical line. See Figure 8(a).

(d) $(8, -2), (3, -2)$

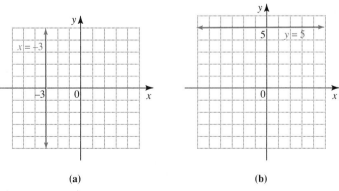

(a)   (b)

FIGURE 8

ANSWERS
1. (a) $-\dfrac{5}{3}$ (b) $-\dfrac{7}{3}$ (c) $-\dfrac{14}{9}$ (d) 0

7.2 THE SLOPE OF A LINE

(b) $y = 5$

Find the slope by selecting two different points on the line, such as $(3, 5)$ and $(-1, 5)$, and using the definition of slope.

$$m = \frac{5 - 5}{3 - (-1)} = \frac{0}{4} = 0$$

The graph of $y = 5$ is the horizontal line in Figure 8(b). ∎

WORK PROBLEM 2 AT THE SIDE.

Example 3 suggests the following generalization.

The slope of a vertical line is undefined; the slope of a horizontal line is 0.

---

3 The following example shows how to graph a straight line by using the slope and one point on the line.

EXAMPLE 4 Using the Slope and a Point to Graph a Line

Graph the line that has slope $\frac{2}{3}$ and goes through the point $(-1, 4)$.

First locate the point $(-1, 4)$ on a graph (see Figure 9). Then, from the definition of slope,

$$m = \frac{\text{change in } y}{\text{change in } x} = \frac{2}{3},$$

move 2 units *up* in the y direction and then 3 units to the *right* in the x direction to locate another point on the graph, P. The line through $(-1, 4)$ and P is the required graph. An additional point can be found by moving 2 units *down* and 3 units to the *left*, since

$$\frac{2}{3} = \frac{-2}{-3}. \quad \blacksquare$$

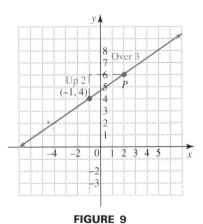

FIGURE 9

2. Find the slope of each line.

(a) $2x + y = 6$

(b) $3x - 4y = 12$

(c) $8x - 9y = 2$

(d) $x = -6$

(e) $y + 5 = 0$

ANSWERS

2. (a) -2 (b) $\frac{3}{4}$ (c) $\frac{8}{9}$
   (d) undefined (e) 0

3. Graph the following lines.

  (a) Through $(1, -3)$; $m = -\dfrac{3}{4}$

  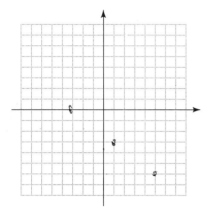

  (b) Through $(-1, -4)$; $m = \dfrac{2}{3}$

  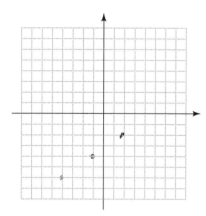

ANSWERS
3. (a)   (b)

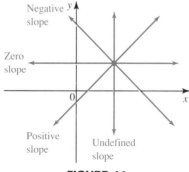

EXAMPLE 5 Using the Slope and a Point to Graph a Line

Graph the line through $(2, 3)$ with slope $-\dfrac{4}{3}$.

Start by locating the point $(2, 3)$ on the graph. Use the definition of slope to find a second point on the line.

$$\text{Slope} = \frac{\text{change in } y}{\text{change in } x} = \frac{-4}{3}$$

Move *down* 4 units from $(2, 3)$ and then 3 units to the *right*. Draw a line through this second point and $(2, 3)$, as in Figure 10. ■

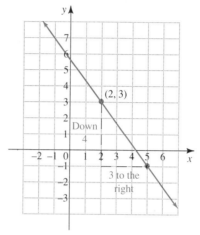

FIGURE 10

▬ WORK PROBLEM 3 AT THE SIDE.

In Problem 3 at the side, the slope for part (b) is *positive*. As shown in the answer, the graph for this line goes up from left to right. The line in part (a) has a *negative* slope, and the graph goes down from left to right. This suggests the following conclusion.

A positive slope indicates that the line goes up from left to right; a negative slope indicates that the line goes down from left to right.

Figure 11 shows lines of positive, zero, negative, and undefined slopes.

FIGURE 11

7.2 THE SLOPE OF A LINE

4 The slopes of a pair of parallel or perpendicular lines are related in a special way. The slope of a line measures the steepness of a line. Since parallel lines have equal steepness, their slopes must be equal; also, lines with the same slope are parallel.

SLOPES OF PARALLEL LINES

Two nonvertical lines with the same slope are parallel; two nonvertical parallel lines have the same slope.

EXAMPLE 6 Determining Parallel Lines

Are the lines L_1, through $(-2, 1)$ and $(4, 5)$ and L_2, through $(3, 0)$ and $(0, -2)$, parallel?

The slope of L_1 is

$$m_1 = \frac{5 - 1}{4 - (-2)} = \frac{4}{6} = \frac{2}{3}.$$

The slope of L_2 is

$$m_2 = \frac{-2 - 0}{0 - 3} = \frac{-2}{-3} = \frac{2}{3}.$$

Since the slopes are equal, the lines are parallel. ■

If two lines (neither one vertical) are perpendicular (cross at right angles) their slopes must have opposite signs. In fact, it can be shown that the slopes of perpendicular lines have a product of -1. This means the two slopes are negative reciprocals of each other. For example, if the slopes of two lines are $\frac{3}{4}$ and $-\frac{4}{3}$, then the lines are perpendicular because $\left(\frac{3}{4}\right)\left(-\frac{4}{3}\right) = -1.$

SLOPES OF PERPENDICULAR LINES

If neither is vertical, perpendicular lines have slopes that are negative reciprocals, so that their product is -1. Also, lines with slopes that are negative reciprocals are perpendicular.

EXAMPLE 7 Determining Perpendicular Lines

Are the lines with equations $2y = 3x - 6$ and $2x + 3y = -6$ perpendicular?

Find the slope of each line by first finding two points on the line. The points $(0, -3)$ and $(2, 0)$ are on the first line. The slope is

$$m_1 = \frac{0 - (-3)}{2 - 0} = \frac{3}{2}.$$

CHAPTER 7 THE STRAIGHT LINE

4. Write *parallel, perpendicular,* or *neither* for each pair of lines.

(a) The line through $(-1, 2)$ and $(3, 5)$ and the line through $(4, 7)$ and $(8, 10)$

(b) The line through $(5, -9)$ and $(3, 7)$ and the line through $(0, 2)$ and $(8, 3)$

(c) $3x + 5y = 6$ and $5x - 3y = 2$

(d) $2x - y = 4$ and $2x + y = 6$

The second line goes through $(-3, 0)$ and $(0, -2)$ and has slope

$$m_2 = \frac{-2 - 0}{0 - (-3)} = \boxed{-\frac{2}{3}}.$$

Since the product of the slopes is $\left(\frac{3}{2}\right)\left(-\frac{2}{3}\right) = -1$, the lines are perpendicular. ■

■ **WORK PROBLEM 4 AT THE SIDE.**

ANSWERS
4. (a) parallel   (b) perpendicular
   (c) perpendicular   (d) neither

7.2 EXERCISES

Graph the line through each pair of points. Find the slope in each case. See Examples 1 and 3.

1. $(2, -3)$ and $(1, 5)$

2. $(4, -1)$ and $(-2, -6)$

3. $(6, 3)$ and $(5, 4)$

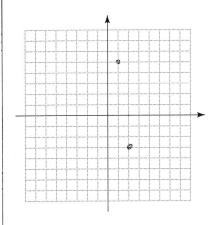
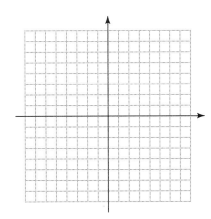
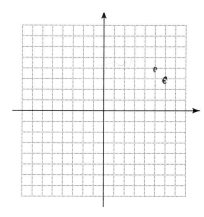

4. $(3, 3)$ and $(-5, -6)$

5. $(2, 5)$ and $(-4, 5)$

6. $(-5, 3)$ and $(-1, 3)$

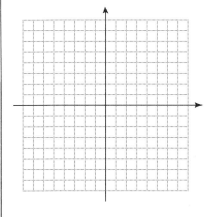
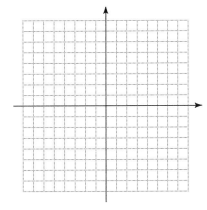
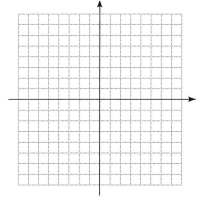

CHAPTER 7 THE STRAIGHT LINE

Find the slope of each of the following lines and sketch the graph. See Examples 1 and 2.

7. $2x + 4y = 5$

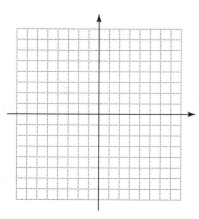

8. $3x - 6y = 12$

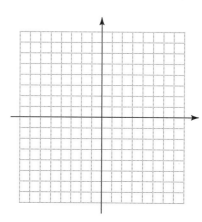

9. $-x + y = 5$

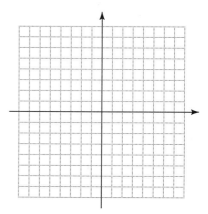

10. $x + y = 1$

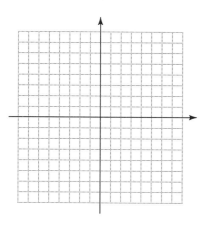

11. $6x - 5y = 30$

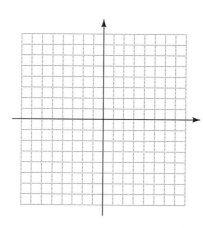

12. $4x + 3y = 12$

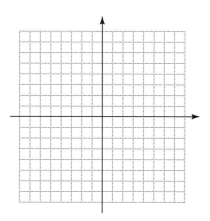

13. $2x - 5y = 10$

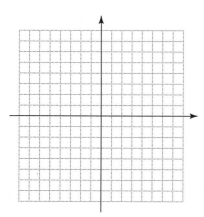

14. $y = 2x$

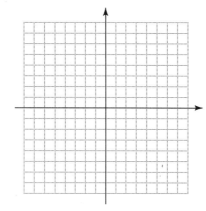

15. $3x + y = 0$

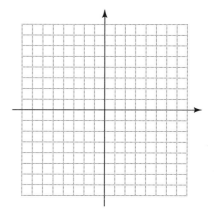

7.2 EXERCISES

Use the method shown in Examples 4 and 5 to graph each of the following lines.

16. Through $(-3, 2)$; $m = \dfrac{1}{2}$

17. Through $(0, 1)$; $m = -\dfrac{2}{3}$

18. Through $(-2, -1)$; $m = \dfrac{5}{4}$

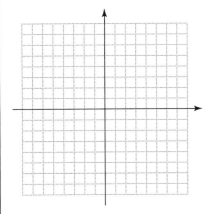
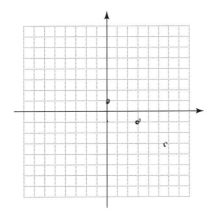
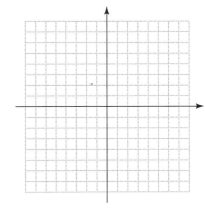

19. Through $(1, 1)$; $m = 5$

20. Through $(-1, -4)$; $m = -2$

21. Through $(-1, -2)$; $m = -\dfrac{3}{2}$

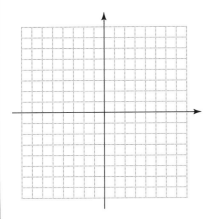
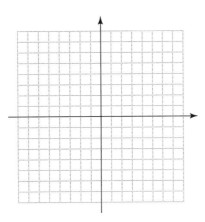
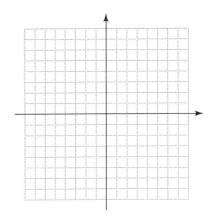

Decide which pairs of lines are parallel. See Example 6.

22. $3x = y$ and $2y - 6x = 5$

23. $2x + 5y = -8$ and $6 + 2x = 5y$
$6 = 5y - 2x$

24. $4x + y = 0$ and $5x - 8 = 2y$

25. $x = 6$ and $6 - x = 8$

26. The line through (4, 6) and (−8, 7) and the line through (7, 4) and (−5, 5)

27. The line through (9, 15) and (−7, 12) and the line through (−4, 8) and (−20, 5)

Decide which pairs of lines are perpendicular. See Example 7.

28. $4x - 3y = 8$ and $4y + 3x = 12$

29. $2x = y + 3$ and $2y + x = 3$

30. $4x - 3y = 5$ and $3x - 4y = 2$

31. $5x - y = 7$ and $5x = 3 + y$

32. $2x + y = 1$ and $x - y = 2$

33. $2y - x = 3$ and $y + 2x = 1$

Solve the following problems using your knowledge of slopes of parallel and perpendicular lines.

34. Show that (−13, −9), (−11, −1), (4, 6), and (2, −2), are the vertices of a parallelogram. (*Hint:* A parallelogram is a four-sided figure with opposite sides parallel.)

35. Is the figure with vertices at (−11, −5), (−2, −19), (12, −10), and (3, 4) a parallelogram? Is it a rectangle? (*Hint:* A rectangle is a parallelogram with two sides perpendicular.)

SKILL SHARPENERS
Solve each equation for y. See Section 2.2.

36. $2x + 3y = 8$

37. $5x - y = 12$

38. $-3x = 4y - 7$

Write without fractions and combine terms if possible. See Section 2.2.

39. $y - 7 = -\dfrac{2}{3}(x - 3)$

40. $y - (-2) = \dfrac{3}{2}(x - 5)$

41. $y - (-1) = \dfrac{5}{3}[x - (-4)]$

7.3 LINEAR EQUATIONS IN TWO VARIABLES

1 To find the slope of the line $4x - y = 8$ in the last section, we first found two points on the line and then used the definition of slope to find that the slope is 4. For an alternative method of finding the slope of this line, solve $4x - y = 8$ for y, getting

$$4x - y = 8$$
$$-y = -4x + 8 \quad \text{Subtract } 4x.$$
$$y = 4x - 8. \quad \text{Multiply by } -1.$$

When the equation is solved for y, the coefficient of x is the slope, 4. Writing the equation in this form also shows that the y-intercept is $(0, -8)$. This suggests the following generalization.

SLOPE-INTERCEPT FORM

The equation of a line with slope m and y-intercept $(0, b)$ is written in **slope-intercept form** as

$$y = mx + b.$$
$$\uparrow \uparrow$$
slope y-intercept is $(0, b)$.

OBJECTIVES

1. Find the slope and y-intercept of a line, given its equation.
2. Write the equation of a line, given its slope and y-intercept.
3. Write the equation of a line, given its slope and a point on the line.
4. Write the equation of a line, given two points on the line.
5. Write the equation of a line parallel or perpendicular to a given line.

EXAMPLE 1 Writing a Linear Equation in Slope-Intercept Form
Write $3y + 2x = 9$ in slope-intercept form and find the slope and y-intercept.

Solve for y to put the equation in slope-intercept form.

$$3y + 2x = 9$$
$$3y = -2x + 9 \quad \text{Subtract } 2x.$$
$$y = -\frac{2}{3}x + 3 \quad \text{Divide by 3.}$$

The slope-intercept form gives $-\frac{2}{3}$ for the slope, with y-intercept $(0, 3)$. ∎

WORK PROBLEM 1 AT THE SIDE.

2 In Example 1, we started with an equation of a line and produced the slope and the y-intercept. It is also possible to start with the slope and y-intercept and end up with an equation of the line, as in the next example.

EXAMPLE 2 Finding the Equation of a Line
Find an equation of the line with slope $-\frac{4}{5}$ and y-intercept $(0, -2)$.

Here $m = -\frac{4}{5}$ and $b = -2$. Substitute these values into the slope-intercept form.

$$y = mx + b = -\frac{4}{5}x - 2$$

Clear of fractions by multiplying both sides by 5 to get

$$5y = -4x - 10, \quad \text{or} \quad 4x + 5y = -10. \quad \blacksquare$$

1. Find the slope and the y-intercept of each line.

   (a) $x + y = 2$

   (b) $3x - y = 7$

   (c) $2x - 5y = 1$

   (d) $8x + 3y = 1$

ANSWERS
1. (a) -1, $(0, 2)$ (b) 3, $(0, -7)$
   (c) $\frac{2}{5}$, $\left(0, -\frac{1}{5}\right)$ (d) $-\frac{8}{3}$, $\left(0, \frac{1}{3}\right)$

CHAPTER 7 THE STRAIGHT LINE

2. Write an equation in standard form for each line with the given slope and y-intercept.

(a) Slope 2; y-intercept $(0, -3)$

(b) Slope $-\frac{7}{8}$; y-intercept $\left(0, \frac{1}{2}\right)$

(c) Slope $-\frac{2}{3}$; y-intercept $(0, 0)$

(d) Slope 0; y-intercept $(0, 3)$

It is convenient to have an agreement on the form in which a linear equation should be written. In Section 7.1, we defined *standard form* for a linear equation as

$$ax + by = c.$$

In addition, from now on, let us agree that a, b, and c will be integers, with $a \geq 0$. For example, the final equation found in Example 2, $4x + 5y = -10$, is written in standard form.

CAUTION The definition of "standard form" is not standard from one text to another. Any linear equation can be written in many different (all equally correct) forms. For example, the equation $2x + 3y = 8$ can be written as $2x = 8 - 3y$, $3y = 8 - 2x$, $x + \frac{3}{2}y = 4$, $4x + 6y = 16$, and so on. In addition to writing it in the form $ax + by = c$ (with $a \geq 0$), let us agree that the form $2x + 3y = 8$ is preferred over any multiples of both sides, such as $4x + 6y = 16$.

■ **WORK PROBLEM 2 AT THE SIDE.**

3 The slope-intercept form can be used to write an equation of a line from its slope and y-intercept. Often, however, the slope is known, along with some point on the line other than the intercept. To find the equation of such a line, suppose that (x_1, y_1) is a point on a line having slope m. If (x, y) is any other point on the line, then by the definition of slope,

$$m = \frac{y - y_1}{x - x_1}.$$

Multiplying both sides by $x - x_1$ gives the *point-slope form* of the equation, which shows the given point and the slope.

POINT-SLOPE FORM

The point-slope form of the equation of a line is

$$y - y_1 = m(x - x_1).$$

where m is the Slope and (x_1, y_1) is the Given point.

EXAMPLE 3 Finding the Equation of a Line

Find an equation of the line with slope $\frac{1}{3}$ that goes through the point $(-2, 5)$.

Use the point-slope form of the equation of a line, with $(x_1, y_1) = (-2, 5)$ and $m = \frac{1}{3}$.

ANSWERS
2. (a) $2x - y = 3$ (b) $7x + 8y = 4$
   (c) $2x + 3y = 0$ (d) $y = 3$

7.3 LINEAR EQUATIONS IN TWO VARIABLES

$y - y_1 = m(x - x_1)$

$y - \mathbf{5} = \dfrac{\mathbf{1}}{\mathbf{3}}[x - (\mathbf{-2})]$ $y_1 = 5, m = \dfrac{1}{3}, x_1 = -2$

$y - 5 = \dfrac{1}{3}(x + 2)$

$3y - 15 = x + 2$ Multiply by 3.

$x - 3y = -17$ Standard form ∎

WORK PROBLEM 3 AT THE SIDE.

Notice that the point-slope form does not apply to a vertical line, since the slope of a vertical line is undefined. A vertical line through the point (k, y), where k is a constant, has equation $x = k$.

A horizontal line has slope 0. From the point-slope form, the equation of a horizontal line through the point (x, k), where k is a constant, is

$y - y_1 = m(x - x_1)$

$y - k = 0(x - x)$ Let $m = 0$, $x_1 = x$, and $y_1 = k$.

$y - k = 0$

$y = k$.

In summary, horizontal and vertical lines have special equations as follows.

If k is a constant, the vertical line through (k, y) has equation $x = k$, and the horizontal line through (x, k) has equation $y = k$.

---

4 When two points on a line are known, the definition of slope can be used to find the slope of the line. Then the point-slope form can be used to write an equation for the line.

EXAMPLE 4 Finding the Equation of a Line

Find an equation of the line through the points $(-4, 3)$ and $(5, -7)$.

First find the slope by using the definition.

$$m = \dfrac{-7 - 3}{5 - (-4)} = -\dfrac{10}{9}$$

Either $(-4, 3)$ or $(5, -7)$ can be used as (x_1, y_1) in the point-slope form of the equation of a line. If $(-4, 3)$ is used, then $-4 = x_1$ and $3 = y_1$.

$y - y_1 = m(x - x_1)$

$y - \mathbf{3} = -\dfrac{\mathbf{10}}{\mathbf{9}}[x - (\mathbf{-4})]$ $y_1 = 3, m = -\dfrac{10}{9}, x_1 = -4$

$9y - 27 = -10x - 40$ Multiply by 9.

$10x + 9y = -13$ Standard form

---

3. Write equations of the following lines in standard form.

(a) Through $(-2, 7)$; $m = 3$

(b) Through $(1, 3)$; $m = -\dfrac{5}{4}$

(c) Through $(3, -4)$; $m = \dfrac{2}{5}$

(d) Through $(8, -2)$; $m = 0$

ANSWERS
3. (a) $3x - y = -13$
   (b) $5x + 4y = 17$
   (c) $2x - 5y = 26$
   (d) $y = -2$

CHAPTER 7 THE STRAIGHT LINE

4. Write equations in standard form of the following lines.

   (a) Through $(-1, 2)$ and $(5, 7)$

   (b) Through $(8, 1)$ and $(3, 9)$

   (c) Through $(-2, 6)$ and $(1, 4)$

   (d) Through $(0, 5)$ and $(-4, 3)$

5. Write equations in standard form of lines satisfying the given conditions.

   (a) Through $(5, 7)$; parallel to $2x - 5y = 15$

   (b) Through $(-3, 5)$; parallel to $4x + 3y = 7$

   (c) Through $(-9, 0)$; parallel to $x - 6y = 1$

ANSWERS
4. (a) $5x - 6y = -17$
   (b) $8x + 5y = 69$
   (c) $2x + 3y = 14$
   (d) $x - 2y = -10$
5. (a) $2x - 5y = -25$
   (b) $4x + 3y = 3$
   (c) $x - 6y = -9$

On the other hand, if $(5, -7)$ were used, the equation would be

$$y - (-7) = -\frac{10}{9}(x - 5) \quad y_1 = -7, m = -\frac{10}{9}, x_1 = 5$$

$$9y + 63 = -10x + 50 \quad \text{Multiply by 9.}$$

$$10x + 9y = -13, \quad \text{Standard form}$$

the same equation. Either way, the line through $(-4, 3)$ and $(5, -7)$ has equation $10x + 9y = -13$. ■

■ **WORK PROBLEM 4 AT THE SIDE.**

5 In the last section it was shown that parallel lines have the same slope and perpendicular lines have slopes that are negative reciprocals. These results are used in the next two examples.

EXAMPLE 5 Finding the Equation of a Parallel Line
Find an equation of the line going through the point $(-2, -3)$ and parallel to the line $2x + 3y = 6$.
   The slope of a line parallel to the line $2x + 3y = 6$ can be found by solving for y.

$$2x + 3y = 6$$

$$3y = -2x + 6 \quad \text{Subtract 2x on both sides.}$$

$$y = -\frac{2}{3}x + 2 \quad \text{Divide both sides by 3.}$$

The slope is given by the coefficient of x, so $m = -\frac{2}{3}$. This means that the line through $(-2, -3)$ and parallel to $2x + 3y = 6$ has slope $-\frac{2}{3}$. Use the point-slope form, with $(x_1, y_1) = (-2, -3)$ and $m = -\frac{2}{3}$ to write the equation.

$$y - y_1 = m(x - x_1)$$

$$y - (-3) = -\frac{2}{3}[x - (-2)] \quad y_1 = -3, m = -\frac{2}{3}, x_1 = -2$$

$$y + 3 = -\frac{2}{3}(x + 2)$$

$$3(y + 3) = -2(x + 2) \quad \text{Multiply by 3.}$$

$$3y + 9 = -2x - 4 \quad \text{Distributive property}$$

$$2x + 3y = -13 \quad \text{Standard form} \quad ■$$

■ **WORK PROBLEM 5 AT THE SIDE.**

EXAMPLE 6 Finding the Equation of a Perpendicular Line
Find an equation of the line perpendicular to $2x + 5y = 8$ and going through $(2, 3)$.
   First, find the slope of the line $2x + 5y = 8$ by solving for y.

$$5y = -2x + 8$$

$$y = -\frac{2}{5}x + \frac{8}{5}$$

7.3 LINEAR EQUATIONS IN TWO VARIABLES

The slope of this line is $-\frac{2}{5}$. Since the negative reciprocal of $-\frac{2}{5}$ is $\frac{5}{2}$, a line perpendicular to $2x + 5y = 8$ would have slope $\frac{5}{2}$. We want an equation of a line with slope $\frac{5}{2}$ and going through $(2, 3)$. Use the point-slope form.

$$y - 3 = \frac{5}{2}(x - 2) \qquad y_1 = 3, m = \frac{5}{2}, x_1 = 2$$
$$2(y - 3) = 5(x - 2) \qquad \text{Multiply by 2.}$$
$$2y - 6 = 5x - 10 \qquad \text{Distributive property}$$
$$5x - 2y = 4 \qquad \text{Standard form} \quad \blacksquare$$

WORK PROBLEM 6 AT THE SIDE.

A summary of the forms of linear equations follows.

$ax + by = c$ (a, b, and c integers, neither a nor b equal to 0)	**Standard form** Slope is $-\frac{a}{b}$. x-intercept is $\left(\frac{c}{a}, 0\right)$. y-intercept is $\left(0, \frac{c}{b}\right)$.
$x = k$	**Vertical line** Slope is undefined. x-intercept is $(k, 0)$.
$y = k$	**Horizontal line** Slope is 0. y-intercept is $(0, k)$.
$y = mx + b$	**Slope-intercept form** Slope is m. y-intercept is $(0, b)$.
$y - y_1 = m(x - x_1)$	**Point-slope form** Slope is m. Line passes through (x_1, y_1).

6. Write equations in standard form of lines satisfying the given conditions.

   (a) Through $(1, 6)$; perpendicular to $x + y = 9$

   (b) Through $(-8, 3)$; perpendicular to $2x - 3y = 10$

   (c) Through $(3, -2)$; perpendicular to $4x - 5y = 1$

ANSWERS
6. (a) $x - y = -5$
   (b) $3x + 2y = -18$
   (c) $5x + 4y = 7$

Historical Reflections

Raymond Smullyan, Puzzler Extraordinaire

Raymond Smullyan is one of today's foremost writers of logic puzzles. One of his books of puzzles is titled *The Lady or the Tiger?* after the classic Frank Stockton short story. In the original story, a prisoner must make a choice between two doors: behind one door is a beautiful lady and behind the other is a hungry tiger.

Smullyan proposes the following twist. Suppose that each door has a sign, and the man knows that only one sign is true.

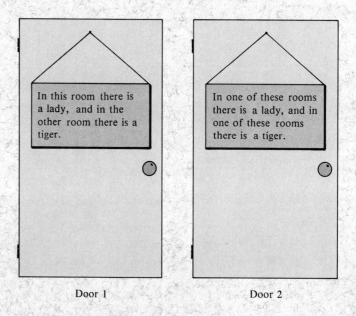

Door 1: In this room there is a lady, and in the other room there is a tiger.

Door 2: In one of these rooms there is a lady, and in one of these rooms there is a tiger.

With this information, the man is able to choose the correct door. Which door is it?

Answer: The lady is behind Door 2. Suppose that the sign on Door 1 is true. Then the sign on Door 2 is also true, but this is impossible. So the sign on Door 2 must be true, making the sign on Door 1 false. Since the sign on Door 1 is false, and it says that the lady is behind Door 1, the lady must be behind Door 2.

Art: (Photograph) Bill Ray

7.3 EXERCISES

Write the following in slope-intercept form, then give the slope of the line and the y-intercept.
See Example 1.

| | Equation | Slope | y-intercept |

1. $x + y = 8$

2. $x - y = 2$

3. $5x + 2y = 10$

4. $6x - 5y = 18$

5. $2x - 3y = 5$

6. $4x + 3y = 10$

7. $-5x - 3y = 4$

8. $-2x - 7y = 15$

9. $8x + 11y = 9$

10. $4x + 13y = 19$

Write equations in slope-intercept form of lines satisfying the following conditions. See Example 2.

11. $m = 5; b = 4$

12. $m = -2; b = 1$

13. $m = -\dfrac{2}{3}; b = \dfrac{1}{2}$

14. $m = -\dfrac{5}{8}; b = \dfrac{1}{4}$

15. Slope $\dfrac{2}{5}$; y-intercept $(0, -1)$

16. Slope $-\dfrac{3}{4}$; y-intercept $(0, 2)$

17. Slope 0; y-intercept $(0, 4)$

18. Slope 0; y-intercept $(0, -3)$

19. Slope -1.573; y-intercept $(0, 4.209)$

CHAPTER 7 THE STRAIGHT LINE

Write equations in standard form of lines satisfying the following conditions. See Example 3.

20. Through $(-2, 5)$; $m = -\dfrac{3}{4}$

21. Through $(4, -3)$; $m = -\dfrac{5}{6}$

22. Through $(1, 5)$; $m = -2$

23. Through $(-2, 3)$; $m = 1$

24. Through $(7, 4)$; $m = \dfrac{1}{2}$

25. Through $(1, -2)$; $m = \dfrac{1}{4}$

26. Through $(-3, 2)$; horizontal

27. Through $(1, 5)$; vertical

28. x-intercept $(3, 0)$; $m = 4$

29. x-intercept $(-2, 0)$; $m = -5$

Write equations in standard form for the following lines. (Hint: What kind of line has undefined slope?)

30. Through $(2, 8)$; undefined slope

31. Through $(-4, 1)$; undefined slope

32. Through $(-7, 1)$; vertical

33. Through $(3, -9)$; vertical

name　　　　　　　　　　　　　　　　　　　　　　　　date　　　　　　　　　　　　hour

7.3 EXERCISES

Write equations in standard form of the lines passing through the following pairs of points. See Example 4.

34. (3, 4) and (2, 6)　　　　**35.** (5, −2) and (−3, 1)　　　　**36.** (6, 1) and (−2, 5)

37. $\left(-\dfrac{2}{5}, \dfrac{2}{5}\right)$ and $\left(\dfrac{4}{3}, \dfrac{2}{3}\right)$　　　　**38.** $\left(\dfrac{3}{4}, \dfrac{8}{3}\right)$ and $\left(\dfrac{2}{5}, \dfrac{2}{3}\right)$　　　　**39.** (2, 5) and (1, 5)

40. (−2, 2) and (4, 2)　　　　**41.** (1, $\sqrt{5}$) and (3, 2$\sqrt{5}$)　　　　**42.** (−4, $\sqrt{2}$) and (5, −$\sqrt{2}$)

Write equations in standard form of the lines satisfying the following conditions. See Examples 5 and 6.

43. Through (−7, 3); parallel to $3x - y = 8$　　　　**44.** Through (4, 7); parallel to $2x + 5y = 10$

45. Through (−2, −2); parallel to $-x + 2y = 3$　　　　**46.** Through (−1, 3); perpendicular to $3x + 2y = 6$

47. Through (8, 5); perpendicular to $2x - y = 4$　　　　**48.** Through (2, −7); perpendicular to $5x + 2y = 7$

49. Through (−2, 7); parallel to $y = 4$　　　　**50.** Through (8, 4); parallel to $x - 2 = 0$

CHAPTER 7 THE STRAIGHT LINE

Many real-world situations can be described approximately by a straight-line graph. One way to find the equation of such a line is to use two typical data points from the graph and the point-slope form of the equation of a line. Assume the following problems have straight-line graphs, and use the given information to find an equation of the line.

51. A weekly magazine had 28 pages of advertising one week that produced revenue of $9700. Another week, $18,500 was produced by 34 pages of advertising. Let y be the revenue from x pages of advertising.

52. The owner of a variety store found that in 1980, year 0, profits were $28,000, while in 1987, year 7, profits had increased to $42,000. Let y be the profit in year x.

SKILL SHARPENERS

Solve each inequality. See Section 2.5.

53. $2x + 5 < 6$

54. $-x + 10 > 8$

55. $6 - 3x \leq 12$

56. $4x - 2 \geq 8$

57. $-5x - 5 > 3$

58. $12 - 8x < 12$

7.4 LINEAR INEQUALITIES IN TWO VARIABLES

1 Linear inequalities with one variable were graphed on the number line in Chapter 2. In this section linear inequalities in two variables are graphed on a rectangular coordinate system.

OBJECTIVES

1 Graph linear inequalities.

2 Graph the intersection of two linear inequalities.

3 Graph the union of two linear inequalities.

4 Graph first-degree absolute value inequalities.

An inequality that can be written as

$$ax + by < c \quad \text{or} \quad ax + by > c,$$

where a, b, and c are real numbers and a and b are not both 0, is a **linear inequality in two variables.**

Also, $\leq$ and $\geq$ may replace $<$ and $>$ in the definition.

A line divides the plane into three regions: the line itself and the two half-planes on either side of the line. Recall that the graphs of linear inequalities in one variable are intervals on the number line that sometimes include an end-point. The graphs of linear inequalities are *regions* in the real number plane and may include a *boundary line*. The **boundary line** for the inequality $ax + by < c$ or $ax + by > c$ is the graph of the *equation* $ax + by = c$. To graph a linear inequality, we go through the following steps.

GRAPHING A LINEAR INEQUALITY

Step 1 Draw the graph of the straight line that is the boundary. Make the line solid if the inequality involves $\leq$ or $\geq$; make the line dashed if the inequality involves $<$ or $>$.

Step 2 Choose any point not on the line as a test point.

Step 3 Shade the region that includes the test point if the test point satisfies the original inequality; otherwise, shade the region on the other side of the boundary line.

EXAMPLE 1 Graphing a Linear Inequality

Graph the solutions of $3x + 2y \geq 6$.

Graph the linear inequality $3x + 2y \geq 6$ by first graphing the straight line $3x + 2y = 6$. The graph of this line, the boundary of the graph of the inequality, is shown in Figure 12. The graph of $3x + 2y \geq 6$ includes

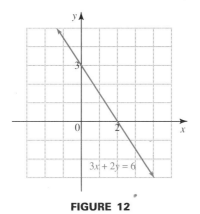

FIGURE 12

CHAPTER 7 THE STRAIGHT LINE

1. Graph the solutions of each inequality.

(a) $x + y \leq 4$

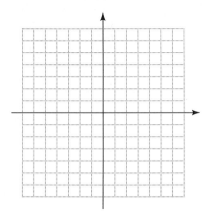

(b) $3x + y \geq 6$

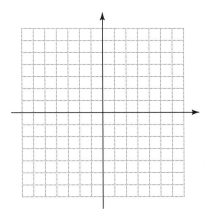

ANSWERS
1. (a)

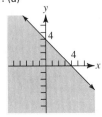

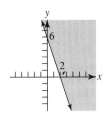

the points of the line $3x + 2y = 6$, and either the points *above* the line $3x + 2y = 6$ or the points *below* that line. To decide which side belongs to the graph, first select any point not on the line $3x + 2y = 6$. The origin, $(0, 0)$, is often a good choice. Substitute the values from the test point $(0, 0)$ for x and y in the inequality $3x + 2y > 6$:

$$3(\mathbf{0}) + 2(\mathbf{0}) > 6$$
$$0 > 6. \qquad \text{False}$$

Since the result is false, $(0, 0)$ does not satisfy the inequality, and the solutions include all points on the other side of the line. This region is shaded in Figure 13. ■

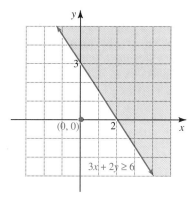

FIGURE 13

▨ **WORK PROBLEM 1 AT THE SIDE.**

EXAMPLE 2 Graphing an Inequality

Graph the solutions of $x - 3y > 4$.

First graph the boundary line, $x - 3y = 4$. The graph is shown in Figure 14. The points of the boundary line do not belong to the inequality $x - 3y > 4$ (since the inequality symbol is $>$ and not $\geq$). For this reason, the line is dashed. To decide which side of the line is the graph of the solutions, choose any point that is not on the line, say $(1, 2)$. Substitute 1 for x and 2 for y in the original inequality.

$$\mathbf{1} - 3(\mathbf{2}) > 4$$
$$-5 > 4 \qquad \text{False}$$

Because of this false result, the solutions lie on the side of the boundary line that does *not* contain the test point $(1, 2)$. The solutions, graphed in Figure 14, include only those points in the shaded half-plane (not those on the line). ■

7.4 LINEAR INEQUALITIES IN TWO VARIABLES

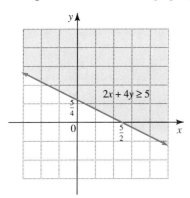

FIGURE 14

WORK PROBLEM 2 AT THE SIDE.

2 The graph of the intersection of the solutions of two or more inequalities is that region of the plane in which all of the points satisfy all of the inequalities.

EXAMPLE 3 Graphing the Intersection of Two Inequalities
Graph the intersection of the solutions of $2x + 4y \geq 5$ and $x \geq 1$.

To begin, graph each of the two inequalities $2x + 4y \geq 5$ and $x \geq 1$ separately. The graph of $2x + 4y \geq 5$ is shown in Figure 15(a), and the graph of $x \geq 1$ is shown in Figure 15(b). The graph of the intersection is the region common to both graphs, as shown in Figure 15(c). ∎

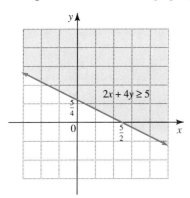
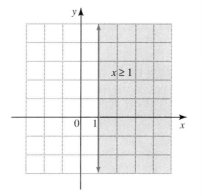
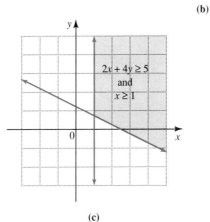

FIGURE 15

2. Graph the solutions of each inequality.

(a) $x - y > 2$

(b) $3x + 4y < 12$

ANSWERS
2. (a)
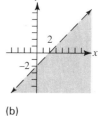

(b)
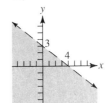

427

CHAPTER 7 THE STRAIGHT LINE

3. Graph the intersection of the solutions of the inequalities $x - y \leq 4$ and $x \geq -2$.

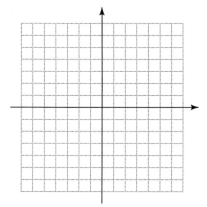

4. Graph the union of the solutions of $7x - 3y < 21$ and $x > 2$.

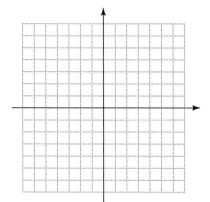

ANSWERS

3.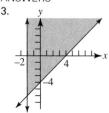

4.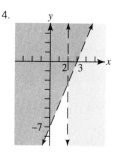

■ WORK PROBLEM 3 AT THE SIDE.

3 The graph of the *union* of the solutions of two inequalities includes all points that satisfy either inequality.

EXAMPLE 4 Graphing the Union of Two Inequalities

Graph the union of the solutions of $2x + 4y \geq 5$ and $x \geq 1$.

The graphs of the two inequalities are shown in Figures 15(a) and 15(b). The graph of the union is shown in Figure 16. ■

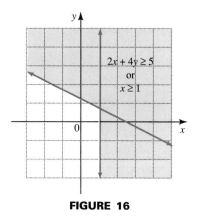

FIGURE 16

■ WORK PROBLEM 4 AT THE SIDE.

4 An absolute value inequality should first be written without absolute value bars as shown in Chapter 2. Then the methods given in the previous examples can be used to graph the inequality.

EXAMPLE 5 Graphing an Absolute Value Inequality

Graph the solutions of $|x| \leq 4$.

Rewrite $|x| \leq 4$ as $-4 \leq x \leq 4$. The topic of this section is linear inequalities in *two* variables, so the boundary lines are the vertical lines $x = -4$ and $x = 4$. Since points from the region between these lines satisfy the inequality, that region is shaded. See Figure 17. ■

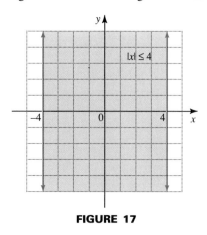

FIGURE 17

EXAMPLE 6 Graphing an Absolute Value Inequality

Graph the solutions of $|y + 2| > 3$.

As shown in Section 2.7, the equation of the boundary, $|y + 2| = 3$, can be rewritten as

$$y + 2 = 3 \quad \text{or} \quad y + 2 = -3$$
$$y = 1 \quad \text{or} \quad y = -5.$$

From this, $y = 1$ and $y = -5$ are boundary lines. Checking points from each of the three regions determined by the horizontal boundary lines gives the graph shown in Figure 18. ∎

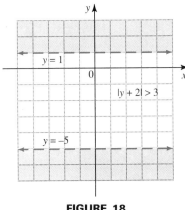

FIGURE 18

WORK PROBLEM 5 AT THE SIDE.

5. Graph the solutions of each absolute value inequality.

(a) $|x| \geq 4$

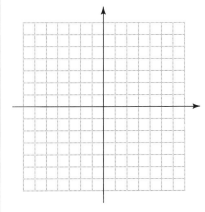

(b) $|x - 2| < 1$

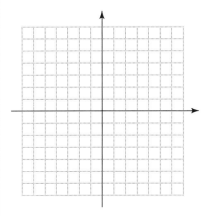

ANSWERS
5. (a)

(b)

Historical Reflections

The Computation of Pi

Because π is an irrational number, its decimal representation will never terminate and will never repeat. Despite this fact, the computation of the digits of π continues. Researchers at Columbia University have determined one billion decimal places of π; the first few are 3.14159.

The computation of π has fascinated mathematicians and laymen for centuries. In the nineteenth century the British mathematician William Shanks spent many years of his life calculating π to 707 decimal places. It turned out that only the first 527 were correct. The advent of the computer greatly revolutionized the quest to calculate π. In fact, the accuracy of computers and computer programs is sometimes tested by performing the computation of π.

Despite the fact that, in 1767, J.H. Lambert proved that π is irrational (and thus its decimal will never terminate and never repeat), the 1897 Indiana state legislature considered a bill that would have *legislated* the value of π. In one part of the bill, the value was stated to be 4, and in another part, 3.2. Amazingly, the bill passed the House, but the Senate postponed action on the bill indefinitely!

The following expressions are some that may be used to compute π to more and more decimal places.

$$\frac{\pi}{2} = \frac{2 \cdot 2 \cdot 4 \cdot 4 \cdot 6 \cdot 6 \cdot 8 \cdots}{1 \cdot 3 \cdot 3 \cdot 5 \cdot 5 \cdot 7 \cdot 7 \cdots}$$

$$\frac{\pi}{4} = 1 - \frac{1}{3} + \frac{1}{5} - \frac{1}{7} + \cdots$$

$$\frac{2}{\pi} = \frac{\sqrt{2}}{2} \cdot \frac{\sqrt{2+\sqrt{2}}}{2} \cdot \frac{\sqrt{2+\sqrt{2+\sqrt{2}}}}{2} \cdots$$

The fascinating history of π has been chronicled by Petr Beckman in the book *A History of Pi*.

7.4 EXERCISES

Graph each linear inequality. See Examples 1 and 2.

1. $x + y \leq 2$

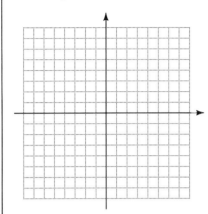

2. $-x \leq 3 - y$

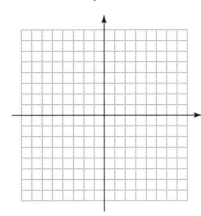

3. $4x - y \leq 5$

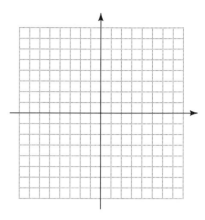

4. $3x + y \geq 6$

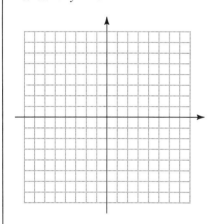

5. $x + 3y \geq -2$

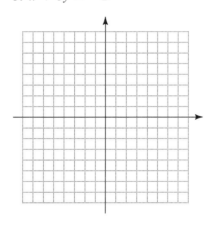

6. $4x + 6y \leq -3$

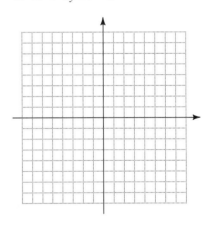

7. $5x + 3y > 15$

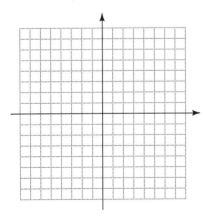

8. $x + y > 0$

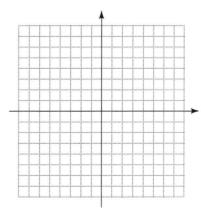

9. $y < x$

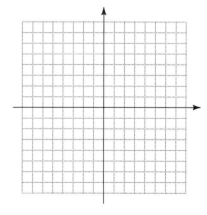

10. $x < 1$

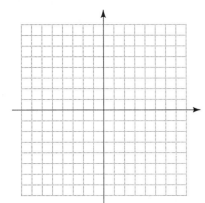

11. $y \geq -3$

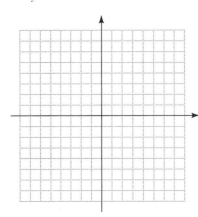

12. $y < 2$

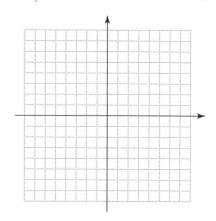

Graph the intersections or unions of the solutions of the following pairs of linear inequalities. Recall from Section 2.6 that "and" means intersection and "or" means union. See Examples 3 and 4.

13. $x + y \leq 1$ and $x \geq 0$

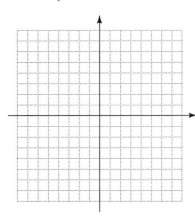

14. $3x - 4y \leq 6$ and $y \geq 1$

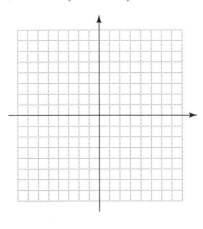

15. $2x - y \geq 1$ and $x \geq -4$

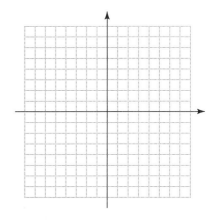

16. $x + 3y > 6$ and $x < 3$

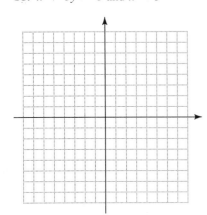

17. $-x - y < 5$ and $y < -2$

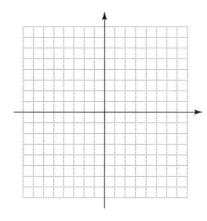

18. $6x - 4y < 8$ and $y > 1$

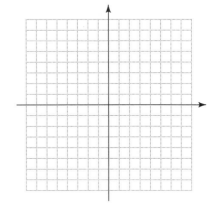

7.4 EXERCISES

19. $x - y > 1$ or $x + y < 4$

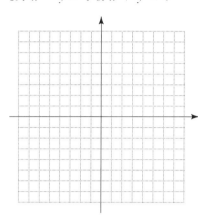

20. $3x + 2y < 6$ or $x - 2y > 2$

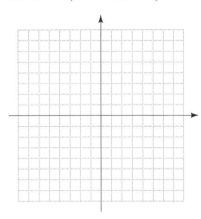

21. $x - 2 > y$ or $x < 1$

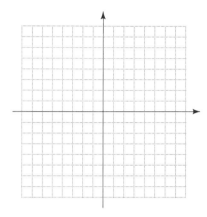

Graph each of the following. See Examples 5 and 6.

22. $|x| > 1$

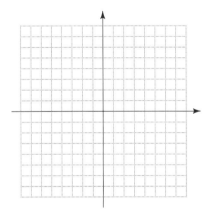

23. $|x| < 3$

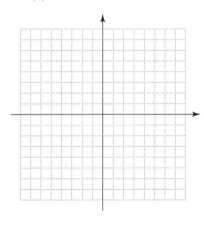

24. $|y| < 5$

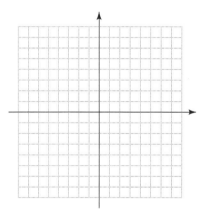

25. $|y| > 4$

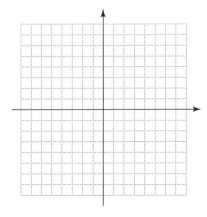

26. $|x| \geq 2$ and $y \leq 1$

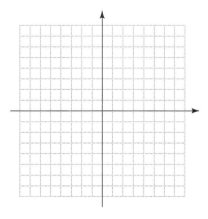

27. $|y| \leq 2$ and $x \geq 1$

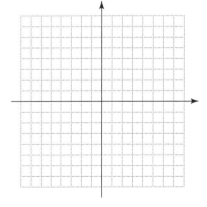

433

28. $|x + 1| < 2$

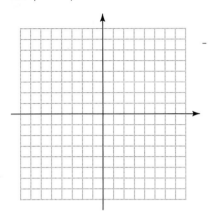

29. $|y - 2| > 3$

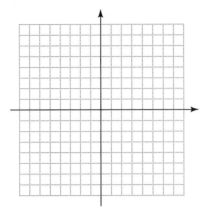

30. $|y - 3| < 2$

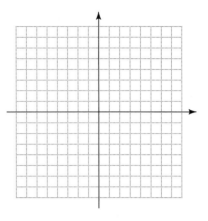

31. $|x + 3| > 5$

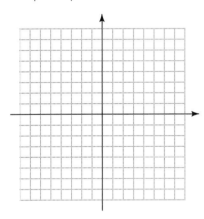

32. $|x| \geq y$ and $y \geq 0$

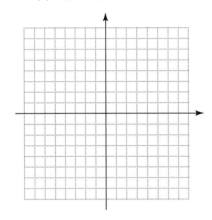

33. $|y - 1| < x$ and $x \leq 0$

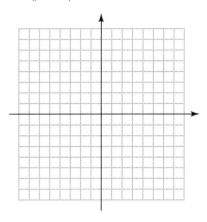

SKILL SHARPENERS
Work each problem. See Section 2.2.

34. If rate, time, and simple interest for an investment are known, write a formula for the principal.

35. Given the volume, length, and height of a box, write a formula for the width.

36. The area of a trapezoid is 20 square meters. One base is 8 meters and the height is 4 meters. What is the length of the other base?

37. What amount of principal must be invested at 10% per year for 6 years to earn $84 in interest?

7.5 FUNCTIONS

It is often useful to describe one quantity in terms of another. For example, the growth of a plant is related to the amount of light it receives, the demand for a product is related to the price of the product, the cost of a trip is related to the distance traveled, and so on. To represent these corresponding quantities, it is helpful to use ordered pairs.

For example, we can indicate the relationship between the demand for a product and its price by writing ordered pairs in which the first number represents the price and the second number represents the demand. The ordered pair (5, 1000) then could indicate a demand for 1000 items when the price of the item is $5. Since the demand depends on the price charged, we place the price first and the demand second. The ordered pair is an abbreviation for the sentence "If the price is 5 (dollars), then the demand is for 1000 (items)." Similarly, the ordered pairs (3, 5000) and (10, 250) show that a price of $3 produces a demand for 5000 items, and a price of $10 produces a demand for 250 items.

In this example, the demand depends on the price of the item. For this reason, demand is called the *dependent variable*, and price the *independent variable*. Generalizing, if the value of the variable y depends on the variable x, then y is the **dependent variable** and x the **independent variable.**

1 Since related quantities can be written with ordered pairs, a **relation** can be defined as a set of ordered pairs. A special kind of relation, called a *function*, is very important in mathematics and its applications.

OBJECTIVES

1 Identify a function.

2 Find the domain and range of a function.

3 Use the vertical line test.

4 Use $f(x)$ notation.

5 Write a linear equation as a linear function.

A **function** is a relation in which, for each value of the first component of the ordered pairs, there is exactly one value of the second component.

EXAMPLE 1 Identifying a Function
The following relations are functions.

$$F = \{(1, 2), (-2, 5), (3, -1)\}$$
$$G = \{(-2, 1)(-1, 0), (0, 1), (1, 2), (2, 2)\}$$

In set G the last two ordered pairs have the same y-value. This does not violate the definition of a function since each first component (x-value) has only one second component (y-value). ■

EXAMPLE 2 Identifying a Function
The following relations are not functions.

$$H = \{(-4, 1), (-2, 1), (-2, 0)\}$$
$$J = \{(3, 1), (0, 2), (0, 3), (4, 0)\}$$

Set H is not a function because the last two pairs have the same x-value, but different y-values. Since set J includes the pairs (0, 2) and (0, 3), it is not a function. ■

1. Which are functions?

   (a) {(1, 2), (2, 4), (3, 3), (4, 2)}

   (b) {(0, 3), (−1, 2), (−1, 3)}

   (c)

   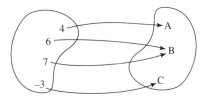

   (d)

   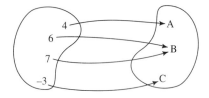

2. Give the domain and range of each function.

   (a) {(1, 2), (2, 4), (3, 3), (4, 2)}

   (b)

   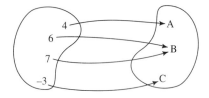

ANSWERS
1. (a) and (d) are functions.
2. (a) domain: {1, 2, 3, 4}
       range: {2, 3, 4}
   (b) domain: {−3, 4, 6, 7}
       range: {A, B, C}

EXAMPLE 3 Recognizing a Function Expressed as a Mapping

A function can also be expressed as a correspondence or *mapping* from one set to another. The mapping in Figure 19 is a function that assigns to a state its population (in millions) expected by the year 2000. ■

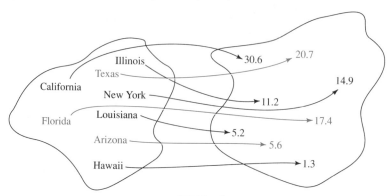

FIGURE 19

WORK PROBLEM 1 AT THE SIDE.

The set of all first components (*x*-values) of the ordered pairs of a relation is called the **domain** of the relation, and the set of all second components (*y*-values) is called the **range.** For example, the domain of function F in Example 1 is $\{1, -2, 3\}$; the range is $\{2, 5, -1\}$. Also, the domain of function G is $\{-2, -1, 0, 1, 2\}$ and the range is $\{0, 1, 2\}$. Domains and ranges can also be defined in terms of independent and dependent variables.

In a relation, the set of all values of the independent variable (*x*) is the **domain;** the set of all values of the dependent variable (*y*) is the **range**.

EXAMPLE 4 Finding Domains and Ranges

Give the domain and range of each function.

(a) {(3, −1), (4, 2), (0, 5)}

The domain, the set of *x*-values, is $\{3, 4, 0\}$; the range is the set of *y*-values, $\{-1, 2, 5\}$.

(b) The function in Figure 19

The domain is {Illinois, Texas, California, New York, Florida, Louisiana, Arizona, Hawaii} and the range is {1.3, 5.2, 5.6, 11.2, 14.9, 17.4, 20.7, 30.6}. ■

WORK PROBLEM 2 AT THE SIDE.

Although some functions (such as the ones in Example 4) have a finite number of ordered pairs, the more useful ones generally include an infinite number of ordered pairs. A rule, usually an equation, is used to tell how to find the ordered pairs.

EXAMPLE 5 Identifying Finite and Infinite Functions
(a) $F = \{(-1, 4), (0, 2), (1, 5), (2, -3)\}$ is an example of a function with a finite number of ordered pairs.

(b) $y = 2x + 3$ defines a function with an infinite number of ordered pairs. For any value of x, a corresponding value of y can be found by substituting that value of x in the equation and solving for y. ∎

The domain of a function defined by an equation such as $y = 2x + 3$ is assumed to be all real numbers that can be used for x in the equation. Since any real number can be used as a replacement for x in $y = 2x + 3$, the domain of this function is the set of real numbers. As another example, the function defined by $y = \dfrac{1}{x}$ has all real numbers except 0 as a domain, since the denominator cannot be 0.

EXAMPLE 6 Identifying a Function from an Equation or Inequality
For each of the following expressions, decide whether it defines a function and give the domain.

(a) $y = \sqrt{2x - 1}$

Here, for any choice of x in the domain, there is exactly one corresponding value for y (the radical is a nonnegative number), so this equation defines a function. Since the radicand cannot be negative,

$$2x - 1 \geq 0$$
$$2x \geq 1$$
$$x \geq \frac{1}{2}.$$

The domain is $\left[\dfrac{1}{2}, \infty\right)$.

(b) $y^2 = x$

The ordered pairs $(16, 4)$ and $(16, -4)$ both satisfy this equation. Since one value of x, 16, corresponds to two values of y, 4 and -4, this equation does not define a function. Solving $y^2 = x$ for y gives $y = \sqrt{x}$ or $y = -\sqrt{x}$, which shows that two values of y correspond to each positive value of x. Because x is equal to the square of y, the values of x must always be nonnegative. The domain is $[0, \infty)$.

(c) $y \leq x - 1$

By definition, y is a function of x if a value of x leads to exactly one value of y. In this example, a particular value of x, say 1, corresponds to many values of y. The ordered pairs $(1, 0), (1, -1), (1, -2), (1, -3)$, and so on, all satisfy the inequality. For this reason, this expression does not define a function. Any number can be used for x, so the domain is the set of real numbers. This can be expressed as $(-\infty, \infty)$.

3. Decide whether or not each expression is a function and give the domain.

   (a) $y = 6x + 12$

   (b) $y \leq 4x$

   (c) $y = -\sqrt{3x - 2}$

   (d) $y^2 = 25x$

4. Which of the following graphs represent functions?

   (a)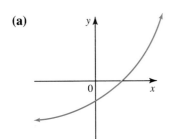

   (b)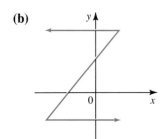

   (c)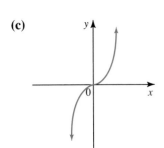

ANSWERS
3. (a) yes; $(-\infty, \infty)$
   (b) no; $(-\infty, \infty)$
   (c) yes; $\left[\dfrac{2}{3}, \infty\right)$
   (d) no; $[0, \infty)$
4. (a) and (c) are the graphs of functions.

(d) $y = \dfrac{5}{x^2 - 1}$

Given any value of x, we find y by squaring x, subtracting 1, then dividing the result into 5. This process produces exactly one value of y for each x-value, so this equation defines a function. The domain includes all real numbers except those that make the denominator zero. We find these numbers by setting the denominator equal to zero and solving for x.

$$x^2 - 1 = 0$$
$$x^2 = 1$$
$$x = 1 \quad \text{or} \quad x = -1 \quad \text{Square root property}$$

Thus, the domain includes all real numbers except 1 and -1. In interval notation this is written as

$$(-\infty, -1) \cup (-1, 1) \cup (1, \infty).$$

WORK PROBLEM 3 AT THE SIDE.

3 The **graph of a relation** is the graph of its ordered pairs. The graph gives a picture of the relation. In a function each value of x leads to only one value of y, so that any vertical line drawn through the graph of a function would cut the graph in at most one point. This is the **vertical line test for a function.**

If a vertical line cuts the graph of a relation in more than one point, then the relation is not a function.

For example, the graph shown in Figure 20(a) is not the graph of a function, since a vertical line cuts the graph in more than one point. The graph of Figure 20(b) does represent a function. Any vertical line will cross the graph at most once.

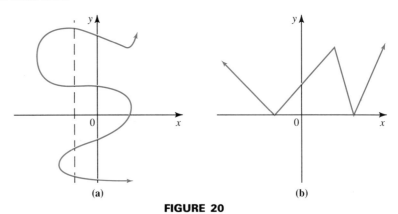

FIGURE 20

WORK PROBLEM 4 AT THE SIDE.

7.5 FUNCTIONS

4 To say that *y* is a function of *x* means that for each value of *x* from the domain of the function, there is exactly one value of *y*. To emphasize that *y is a function of x*, or that *y* depends on *x*, it is common to write

$$y = f(x),$$

with *f(x)* read "*f* of *x*." (In this special notation, the parentheses do not indicate multiplication.) The letter *f* stands for *function*. For example, if $y = 9x - 5$, we emphasize that *y* is a function of *x* by writing $y = 9x - 5$ as

$$f(x) = 9x - 5.$$

This **functional notation** can be used to simplify certain statements. For example, if $y = 9x - 5$, then replacing *x* with 2 gives

$$y = 9 \cdot 2 - 5$$
$$= 18 - 5$$
$$= 13.$$

The statement "if $x = 2$, then $y = 13$" is abbreviated with functional notation as

$$f(\mathbf{2}) = 13.$$

Also, $f(0) = 9 \cdot 0 - 5 = -5$, and $f(-3) = -32$.

CAUTION The symbol *f(x) does not* indicate "*f* times *x*," but represents the *y*-value for the indicated *x*-value. As shown above, *f*(2) is the *y*-value that corresponds to the *x*-value 2.

WORK PROBLEM 5 AT THE SIDE.

Before functional notation can be used to write the equation that defines a function, the equation must first be solved for *y*. Then *y* can be replaced by *f(x)*.

EXAMPLE 7 Writing Equations Using Functional Notation

Rewrite each equation using functional notation; then find $f(-2)$ and $f(a)$.

(a) $y = x^2 + 1$

This equation is already solved for *y*. Since $y = f(x)$,

$$f(x) = x^2 + 1.$$

To find $f(-2)$, let $x = -2$:

$$f(\mathbf{-2}) = (\mathbf{-2})^2 + 1$$
$$= 4 + 1$$
$$= 5.$$

Find $f(a)$ by letting $x = a$: $f(\mathbf{a}) = \mathbf{a}^2 + 1$.

5. Find $f(-3)$ and $f(6)$.

(a) $f(x) = 6x - 2$

(b) $f(x) = \dfrac{-3x + 5}{2}$

(c) $f(x) = \dfrac{1}{6}x^2 - 1$

ANSWERS

5. (a) -20; 34   (b) 7; $-\dfrac{13}{2}$ (c) $\dfrac{1}{2}$; 5

6. Find $f(x)$.

   (a) $y = 2x^2 - 4$

   (b) $3x + 4y = 8$

   (c) $x^2 - 4y = 3$

(b) $x - 4y = 5$

First solve $x - 4y = 5$ for y.

$$x - 4y = 5$$
$$x - 5 = 4y$$
$$y = \frac{x-5}{4} \quad \text{so} \quad f(x) = \frac{x-5}{4}$$

Now find $f(-2)$ and $f(a)$.

$$f(\mathbf{-2}) = \frac{\mathbf{-2} - 5}{4} = \frac{-7}{4} = -\frac{7}{4}$$

and

$$f(\mathbf{a}) = \frac{\mathbf{a} - 5}{4} \quad \blacksquare$$

■ **WORK PROBLEM 6 AT THE SIDE.**

5 By the vertical line test, linear equations (except for the type where $x = k$) define functions, since their graphs are straight lines.

LINEAR FUNCTION

A function with an equation of the form

$$f(x) = mx + b$$

is a **linear function**.

As shown in Section 7.3, this form of the equation is the slope-intercept form, where m is the slope, $(0, b)$ is the y-intercept, and $f(x)$ is just another name for y.

ANSWERS
6. (a) $f(x) = 2x^2 - 4$
   (b) $f(x) = \dfrac{8 - 3x}{4}$
   (c) $f(x) = \dfrac{x^2 - 3}{4}$

7.5 EXERCISES

Give the domain and range of each relation. Identify any functions. See Examples 1–4.

1. {(5, 1), (3, 2), (4, 9), (7, 6)}

2. {(8, 0), (5, 4), (9, 3), (3, 8)}

3. {(2, 4), (0, 2), (2, 5)}

4. {(9, −2), (−3, 5), (9, 2)}

5. {(−3, 1), (4, 1), (−2, 7)}

6. {(1, 3), (4, 7), (0, 6), (7, 2)}

7.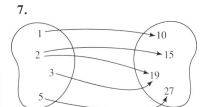

8.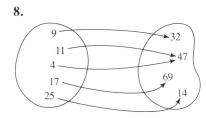

9.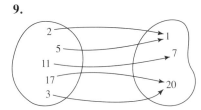

Decide whether each of the following defines a function and give the domain. Identify any that are linear functions. See Examples 5 and 6.

10. $x = y^2$

11. $y = x^2$

12. $x + y = 1$

13. $3x = 5 - 2y$

CHAPTER 7 THE STRAIGHT LINE

14. $x = y^3$

15. $y = x^3$

16. $x < y$

17. $x \neq y$

18. $y \geq 3x$

19. $x + y < 2$

20. $xy = 1$

21. $x = \dfrac{1}{y}$

22. $|x| \leq y$

23. $|x| \leq 1 + y$

24. $x = \sqrt{y}$

25. $y = \sqrt{x}$

Let f be the function defined by $x + 2y = 6$. *Find each of the following. See Example 7. In Exercise 32, first find* $f(1)$.

26. $f(0)$

27. $f(-2)$

28. $f(3)$

29. $f(5)$

30. $f(a)$

31. $f(-p)$

32. $f[f(1)]$

33. $f[f(-3)]$

7.5 EXERCISES

Let $f(x) = -9x + 12$ and $g(x) = -x^2 + 4x - 1$. Find each of the following. See Example 7.

34. $f(0)$ **35.** $f(-3)$ **36.** $f(-8)$

37. $g\left(\dfrac{1}{2}\right)$ **38.** $g(10)$ **39.** $g(-2)$

40. $g(-4)$ **41.** $f[g(0)]$ **42.** $g[f(1)]$

Identify any of the following that represent functions. Use the vertical line test.

43.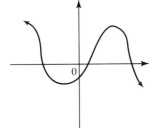

44.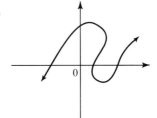

45.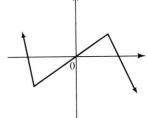

In Exercises 46–51, suppose a taxi driver charges $.50 per mile. Let $R(x)$ be the cost for a ride of x miles. Find the following.

46. $R(3)$ **47.** $R\left(4\dfrac{1}{2}\right)$

48. $R\left(2\dfrac{3}{5}\right)$ **49.** $R(5)$

50. Graph $R(x)$.

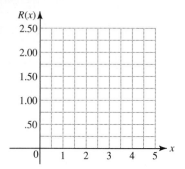

51. Does $R(x)$ represent a function?

SKILL SHARPENERS
Solve each equation. See Section 2.1.

52. $-2(y - 1) + 3y = 8$

53. $-5(2m + 3) - m = 10$

54. $2\left(\dfrac{3x - 1}{2}\right) - 2x = -4$

55. $3\left(\dfrac{5p + 4}{3}\right) + p = 6$

56. $-4\left(\dfrac{t - 3}{3}\right) + 2t = 8$

57. $-2\left(\dfrac{3n + 1}{5}\right) + n = -4$

7.6 VARIATION

1 The circumference of a circle is given by the formula $C = 2\pi r$, where r is the radius of the circle. As the formula shows, the circumference is always a constant multiple of the radius (C is always found by multiplying r by the constant 2π). Because of this, the circumference is said to *vary directly* as the radius. Whenever the ratio of two related variables is constant, they are said to *vary directly*.

OBJECTIVES

1. Write an equation expressing direct variation.
2. Find the constant of variation and solve direct variation problems.
3. Solve inverse variation problems.
4. Solve joint variation problems.
5. Solve combined variation problems.

DIRECT VARIATION

y varies directly as x if there exists some constant k such that

$$y = kx.$$

Also, y is said to be **proportional** to x. The number k is called the **constant of variation**.

The direct variation equation defines a linear function. In applications, functions are often defined by variation equations. For example, if Tom earns $8 per hour, his wages vary directly as, or are proportional to, the number of hours he works. If y represents his total wages and x the number of hours he has worked, then

$$y = 8x.$$

Here k, the constant of variation, is 8.

2 The following example shows how to find the value of the constant k.

EXAMPLE 1 Finding the Constant of Variation

Suppose y varies directly as z, and $y = 50$ when $z = 100$. Find k and the equation relating y and z.

Since y varies directly as z,

$$y = kz,$$

for some constant k. We know that $y = 50$ when $z = 100$. Substituting these values into the equation $y = kz$ gives

$$y = kz$$
$$50 = k \cdot 100. \quad \text{Let } y = 50 \text{ and } z = 100.$$

Now solve for k.

$$k = \frac{50}{100} = \frac{1}{2}$$

The variables y and z are related by the equation

$$y = \frac{1}{2}z.$$

WORK PROBLEMS 1 AND 2 AT THE SIDE.

1. In Example 1, find y for each of the following values of z.

   (a) $z = 80$

   (b) $z = 6$

   (c) $z = -50$

2. Assume m varies directly as n. Find each equation relating m and n.

   (a) $m = 7$ when $n = 35$

   (b) $m = 30$ when $n = 20$

   (c) $m = -8$ when $n = 5$

ANSWERS
1. (a) 40 (b) 3 (c) -25
2. (a) $m = \frac{1}{5}n$ (b) $m = \frac{3}{2}n$
   (c) $m = -\frac{8}{5}n$

CHAPTER 7 THE STRAIGHT LINE

3. It costs $52 to use 800 kwh of electricity. How much would the following kilowatt-hours cost?

(a) 1000

(b) 650

(c) 900

EXAMPLE 2 Solving a Direct Variation Problem

Power consumption is measured in kilowatt-hours (kwh). The charge to customers varies directly as the number of hours of consumption. If it costs $76.50 to use 850 kwh, how much will 1000 kwh cost?

If c represents the cost and h is the number of kilowatt-hours, then $c = kh$ for some constant k. Since 850 kwh cost $76.50, let $c = 76.50$ and $h = 850$ in the equation $c = kh$ to find k.

$$76.50 = k(850)$$
$$k = \frac{76.50}{850}$$
$$k = .09$$

Thus, $k = .09$. For 1000 kwh,

$$c = (.09)(1000)$$
$$c = 90.$$

It will cost $90 to use 1000 kilowatts of power. ∎

■ WORK PROBLEM 3 AT THE SIDE.

The equation $C = 2\pi r$ mentioned in the introduction to this section shows the functional relationship between the circumference C of a circle and its radius r. Similarly, in Example 2, cost c is a function of the number of kilowatt-hours h that are used.

The direct variation equation $y = kx$ defines a linear function. However, other kinds of variations are expressed as nonlinear functions. Since the methods of solving all variation problems are basically the same, we include them all in this section.

Often, one variable is directly proportional to a *power* of another variable; by definition, **y varies directly as the *n*th power of *x*** if there exists a real number k such that

$$y = kx^n.$$

These variation equations define *polynomial functions* when kx^n has degree greater than one.

EXAMPLE 3 Solving a Direct Variation Problem

The distance a body falls from rest varies directly as the square of the time it falls (here we disregard air resistance). If an object falls 64 feet in 2 seconds, how far will it fall in 8 seconds?

If d represents the distance the object falls, and t the time it takes to fall, then d is a function of t, and

$$d = kt^2$$

for some constant k. To find the value of k, use the fact that the object falls 64 feet in 2 seconds.

$$d = kt^2$$
$$64 = k(2)^2 \quad\quad d = 64 \text{ and } t = 2$$
$$k = 16$$

ANSWERS
3. (a) $65 (b) $42.25 (c) $58.50

7.6 VARIATION

Using 16 for k, the variation equation becomes

$$d = 16t^2.$$

Now let $t = 8$ to find the number of feet the object will fall in 8 seconds.

$$d = 16(8)^2 \qquad t = 8$$
$$= 1024$$

The object will fall 1024 feet in 8 seconds. ∎

WORK PROBLEM 4 AT THE SIDE.

3 Whenever the product of two variables is constant, the variables are said to *vary inversely*.

INVERSE VARIATION

y varies inversely as x if there exists a real number k such that

$$y = \frac{k}{x}.$$

Also, **y varies inversely as the nth power of x** if there exists a real number k such that

$$y = \frac{k}{x^n}.$$

Notice that the inverse variation equation also defines a function. Since x is in the denominator, these functions are called *rational functions*.

EXAMPLE 4 Solving an Inverse Variation Problem

The weight of an object above the earth varies inversely as the square of its distance from the center of the earth. A space vehicle in an elliptical orbit has a maximum distance from the center of the earth (apogee) of 6700 miles. Its minimum distance from the center of the earth (perigee) is 4090 miles. See Figure 21. If an astronaut in the vehicle weighs 57 pounds at its apogee, what does the astronaut weigh at its perigee?

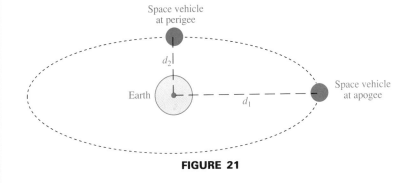

FIGURE 21

4. Suppose y varies directly as the cube of x, and $y = 24$ when $x = 2$. Find y, given the following.

(a) $x = 1$

(b) $x = 4$

(c) $x = 7$

(d) $x = -5$

ANSWERS
4. (a) 3 (b) 192 (c) 1029
   (d) −375

447

CHAPTER 7 THE STRAIGHT LINE

5. Suppose p varies inversely as the cube of q and $p = 100$ when $q = 3$. Find p, given the following.

(a) $q = 1$

(b) $q = 5$

(c) $q = -6$

If w is the weight and d is the distance from the center of the earth, then

$$w = \frac{k}{d^2}$$

for some constant k. At the apogee the astronaut weighs 57 pounds and the distance from the center of the earth is 6700 miles. Use these values to find k.

$$57 = \frac{k}{(6700)^2} \qquad \text{Let } w = 57 \text{ and } d = 6700.$$
$$k = 57(6700)^2$$

Then the weight at the perigee with $d = 4090$ miles is

$$w = \frac{57(6700)^2}{(4090)^2} \approx 153 \text{ pounds.} \quad \blacksquare$$

■ **WORK PROBLEM 5 AT THE SIDE.**

4 It is common for the value of one variable to depend on the values of several others. For example, if the value of one variable varies directly as the product of the values of several other variables (perhaps raised to powers), the first variable is said to **vary jointly** as the others. The next example illustrates joint variation.

EXAMPLE 5 Solving a Joint Variation Problem
The strength of a rectangular beam varies jointly as its width and the square of its depth. If the strength of a beam 2 inches wide and 10 inches deep is 1000 pounds per square inch, what is the strength of a beam 4 inches wide and 8 inches deep?

If S represents the strength, w the width, and d the depth, then,

$$S = kwd^2$$

for some constant k. Since $S = 1000$ if $w = 2$ and $d = 10$,

$$1000 = k(2)(10)^2. \qquad S = 1000, w = 2, \text{ and } d = 10$$

Solving this equation for k gives

$$1000 = k \cdot 2 \cdot 100$$
$$1000 = 200k,$$

or
$$k = 5,$$
so that
$$S = 5wd^2.$$

Now find S when $w = 4$ and $d = 8$.

$$S = 5(4)(8)^2 \qquad w = 4 \text{ and } d = 8$$
$$= 1280$$

The strength of the beam is 1280 pounds per square inch. ■

ANSWERS
5. (a) 2700 (b) $\frac{108}{5}$ (c) $-\frac{25}{2}$

7.6 VARIATION

5 There are many combinations of direct and inverse variation. The final example shows a typical one.

EXAMPLE 6 Solving a Combined Variation Problem

The maximum load that a cylindrical column with a circular cross section can hold varies directly as the fourth power of the diameter of the cross section and inversely as the square of the height. A 9-meter column 1 meter in diameter will support 8 metric tons. How many metric tons can be supported by a column 12 meters high and $\frac{2}{3}$ meter in diameter?

Let L represent the load, d the diameter, and h the height. Then

$$L = \frac{kd^4}{h^2}.$$

← Load varies directly as the 4th power of the diameter.
← Load varies inversely as the square of the height.

Now find k. Let $h = 9$, $d = 1$, and $L = 8$.

$$8 = \frac{k(1)^4}{9^2} \qquad h = 9, d = 1, L = 8$$

Solve for k:

$$8 = \frac{k}{81}$$

$$k = 648.$$

Substitute 648 for k in the first equation.

$$L = \frac{648d^4}{h^2}$$

Now find L when $h = 12$ and $d = \frac{2}{3}$ by substituting the values into the last equation.

$$L = \frac{648\left(\frac{2}{3}\right)^4}{12^2} \qquad h = 12, d = \frac{2}{3}$$

$$L = \frac{648\left(\frac{16}{81}\right)}{144}$$

$$= 648 \cdot \frac{16}{81} \cdot \frac{1}{144} = \frac{8}{9}$$

The maximum load is about $\frac{8}{9}$ metric ton. ■

WORK PROBLEM 6 AT THE SIDE.

---

6. Suppose z varies jointly as x and y^2 and inversely as w. Also, $z = \frac{3}{8}$ when $x = 2$, $y = 3$, and $w = 12$. Find z, given the following.

(a) $x = 4$; $y = 1$; $w = 6$

(b) $x = 7$; $y = 2$; $w = 10$

(c) $x = -4$; $y = 9$; $w = 3$

ANSWERS
6. (a) $\frac{1}{6}$ (b) $\frac{7}{10}$ (c) -27

Historical Reflections

"The Man Who Loved Numbers": Srinivasa Ramanujan (1887–1920)

The Indian mathematician Srinivasa Ramanujan developed many ideas in the branch of mathematics known as number theory. His friend and collaborator on occasion was G. H. Hardy, also a number theorist and professor at Cambridge University in England.

Ramanujan introduced himself to Hardy in 1913 by a letter in which he stated without proof a number of complicated formulas. Hardy at first thought it was a crank letter, but he soon realized that the formulas had to have been the work of a genius. Hardy arranged for Ramanujan to receive a modest stipend, enough to live fairly well at Cambridge.

A story has been told about Ramanujan that illustrates his genius. Hardy once mentioned to Ramanujan that he had just taken a taxicab with a rather dull number: 1729. Ramanujan countered by saying that this number isn't dull at all: it is the smallest natural number that can be expressed as the sum of two cubes in two different ways:

$$1^3 + 12^3 = 1729 \quad \text{and} \quad 9^3 + 10^3 = 1729.$$

Ramanujan was the subject of a 1988 installment of the *Nova* series aired on the Public Broadcasting System. It was aptly titled "The Man Who Loved Numbers."

7.6 EXERCISES

Solve the following. See Examples 1–5.

1. a varies directly as the square of b, and $a = 4$ when $b = 3$. Find a when b is 2.

2. h varies directly as the square of m, and $h = 15$ when $m = 5$. Find h when $m = 7$.

3. z varies inversely as w, and $z = 5$ when $w = \frac{1}{2}$. Find z when $w = 8$.

4. t varies inversely as s, and $t = 3$ when $s = 5$. Find t when $s = 3$.

5. p varies jointly as q and r^2, and $p = 100$ when $q = 2$ and $r = 3$. Find p when $q = 5$ and $r = 2$.

6. f varies jointly as g^2 and h, and $f = 25$ when $g = 4$ and $h = 2$. Find f when $g = 3$ and $h = 6$.

Work the following word problems. See Examples 2–6.

7. The resistance in ohms of a platinum wire temperature sensor varies directly as the temperature in *degrees Kelvin* (°K). If the resistance is 646 ohms at a temperature of 190°K, find the resistance at a temperature of 250°K.

8. For a body falling freely from rest (disregarding air resistance), the distance the body falls varies directly as the square of the time. If an object is dropped from the top of a tower 576 feet high and hits the ground in 6 seconds, how far did it fall in the first 4 seconds?

9. The frequency of a vibrating string varies inversely as its length. That is, a longer string vibrates fewer times in a second than a shorter string. Suppose a piano string 2 feet long vibrates 250 cycles per second. What frequency would a string 1.5 feet long have?

10. The current in a simple electrical circuit is inversely proportional to the resistance. If the current is 20 amperes (an *ampere* is a unit for measuring current) when the resistance is 5 ohms, find the current when the resistance is 8 ohms.

11. The force with which the earth attracts an object above the earth's surface varies inversely with the square of the distance from the center of the earth. If an object 4000 miles from the center of the earth is attracted with a force of 160 pounds, find the force of attraction on an object 4500 miles from the center of the earth.

12. Hooke's law states that the force required to compress a spring is proportional to the change in length of the spring. If a force of 20 newtons* is required to compress a certain spring 2 centimeters, how much force is required to compress the spring from 20 centimeters to 10 centimeters?

13. Two electrons repel each other with a force inversely proportional to the square of the distance between them. When the electrons are 5×10^{-10} meter apart, they repel one another with a force of 100 units. How far apart are they when the force is 196 units?

14. The illumination produced by a light source varies inversely as the square of the distance from the source. If the illumination produced 4 meters from a light source is 50 footcandles, find the illumination produced 9 meters from the same source.

15. The collision impact of an automobile varies jointly as its mass and the square of its speed. Suppose a 2000-pound car traveling at 55 miles per hour has a collision impact of 6.1. What is the collision impact of the same car at 65 miles per hour?

16. The volume of a gas varies inversely as the pressure and directly as the temperature. If a certain gas occupies a volume of 1.3 liters at 300°K and a pressure of 18 kilograms per square centimeter, find the volume at 340°K and a pressure of 24 kilograms per square centimeter.

SKILL SHARPENERS
Complete the indicated ordered pairs for the given equations. See Section 7.1.

17. $3x + y = 4$

    (0,   ), (5,   ), (−2,   )

18. $-5x + 3y = 15$

    (0,   ), (3,   ), (−3,   )

19. $y = x^2$

    (−1,   ), (0,   ), (1,   )

20. $x = y^2$

    (, 3), (, 2), (, −2), (, −3)

21. $x = |y|$

    (, 3), (, 4), (, −4), (, −3)

22. $y = \sqrt{x}$

    (0,   ), (1,   ), (4,   )

---

*A newton is a unit of measure of force used in physics.

Historical Reflections

A Case of Mathematical Plagiarism: Girolamo Cardano (1501–1576) and Nicolo Tartaglia (ca. 1499–1557)

Methods of solving linear and quadratic equations have been known since the time of the Babylonians. Nevertheless, for centuries mathematicians wrestled with the problem of finding a formula that could be used to solve cubic (third-degree) equations. A story that unfolded during the sixteenth century in Italy matches any plot that today's soap opera writers could concoct. The story is sprinkled with minor characters whose names are del Ferro, Fior, and Ferrari, but the two main characters are Girolamo Cardano and Nicolo Tartaglia. As was the custom in those days, mathematicians often kept secret their methods and participated in contests. Tartaglia had developed a method of solving a cubic equation of the form $x^3 + mx = n$ and had used it to his advantage in disposing of an opponent in one of these contests. Cardano pleaded that Tartaglia give him his method, and swore secrecy. Now Cardano was not the kindest of men; he supposedly cut off the ears of one of his sons. True to form, Cardano published Tartaglia's method in his 1545 work *Ars Magna*, despite the vow of secrecy.

The formula for finding one real solution of the equation given above is

$$x = \sqrt[3]{\left(\frac{n}{2}\right) + \sqrt{\left(\frac{n}{2}\right)^2 + \left(\frac{m}{3}\right)^3}} - \sqrt[3]{-\left(\frac{n}{2}\right) + \sqrt{\left(\frac{n}{2}\right)^2 + \left(\frac{m}{3}\right)^3}}.$$

It is certainly more complicated than the quadratic formula, and is limited to solving cubic equations that lack a quadratic term. Cubic equations with a quadratic term can be solved in a roundabout way using this formula. This method is discussed in texts dealing with the theory of equations.

Try solving for one solution of the equation

$$x^3 + 9x = 26$$

using this formula. (The solution given by the formula is 2.)

Art: Historical Pictures Service, Chicago

CHAPTER 7 SUMMARY

KEY TERMS

7.1 **origin** — When two number lines intersect at a right angle, the origin is the common zero point.

x-axis — The horizontal number line in a rectangular coordinate system is called the x-axis.

y-axis — The vertical number line in a rectangular coordinate system is called the y-axis.

rectangular coordinate system — Two number lines that intersect at a right angle at their zero points form a rectangular coordinate system.

plot — To plot an ordered pair is to locate it on a rectangular coordinate system.

coordinates — The numbers in an ordered pair are called the coordinates of the corresponding point.

quadrant — A quadrant is one of the four regions in the plane determined by a rectangular coordinate system.

linear equation in two variables — A first-degree equation with two variables is a linear equation in two variables.

standard form — A linear equation is in standard form when written as $ax + by = c$, with $a \geq 0$, and a, b, and c integers.

x-intercept — The point where a line crosses the x-axis is the x-intercept.

y-intercept — The point where a line crosses the y-axis is the y-intercept.

7.2 **slope** — The ratio of the change in y compared to the change in x along a line is the slope of the line.

7.4 **boundary line** — In the graph of a linear inequality, the boundary line separates the region that satisfies the inequality from the region that does not satisfy the inequality.

half-plane — Either of the two regions determined by a boundary line is called a half-plane.

7.5 **dependent variable** — If the quantity y depends on x, then y is called the dependent variable in an equation relating x and y.

independent variable — If y depends on x, then x is the independent variable in an equation relating x and y.

relation — A relation is any set of ordered pairs.

function — A function is a set of ordered pairs in which each value of the first component, x, corresponds to exactly one value of the second component, y.

domain — The domain of a relation is the set of first components (x-values) of the ordered pairs of a relation.

range — The range of a relation is the set of second components (y-values) of the ordered pairs of a relation.

graph of a relation — The graph of a relation is the graph of the ordered pairs of the relation.

vertical line test — The vertical line test says that if a vertical line cuts the graph of a relation in more than one point, then the relation is not a function.

CHAPTER 7 SUMMARY

NEW SYMBOLS

(a, b)	ordered pair
x_1	a specific value of the variable x (read "x sub one")
m	slope
$f(x)$	function of x (read "f of x")

QUICK REVIEW

Section	Concepts	Examples
7.1 The Rectangular Coordinate System	**Finding Intercepts** To find the x-intercept, let $y = 0$. To find the y-intercept, let $x = 0$.	$2x + 3y = 12$: x-intercept is $(6, 0)$ y-intercept is $(0, 4)$
7.2 The Slope of a Line	$m = \dfrac{\text{change in } y}{\text{change in } x} = \dfrac{y_2 - y_1}{x_2 - x_1}$	$2x + 3y = 12$: $m = \dfrac{4 - 0}{0 - 6} = -\dfrac{2}{3}$
	A vertical line has undefined slope.	$x = 3$: undefined slope
	A horizontal line has 0 slope.	$y = -5$: $m = 0$
	Parallel lines have equal slopes.	$y = 2x + 3$ $4x - 2y = 6$ $m = 2$ $m = 2$ Lines are parallel.
	The slopes of perpendicular lines have a product of -1.	$y = 3x - 1$ $x + 3y = 4$ $m = 3$ $m = -\dfrac{1}{3}$ Lines are perpendicular.
7.3 Linear Equations in Two Variables	**Standard Form** $ax + by = c$	$2x - 5y = 8$
	Vertical Line $x = k$	$x = -1$
	Horizontal Line $y = k$	$y = 4$
	Slope-Intercept Form $y = mx + b$	$y = 2x + 3$ $m = 2$, y-intercept is $(0, 3)$
	Point-Slope Form $y - y_1 = m(x - x_1)$	$y - 3 = 4(x - 5)$ $(5, 3)$ is on the line, $m = 4$

CHAPTER 7 THE STRAIGHT LINE

Section	Concepts	Examples
7.4 Linear Inequalities in Two Variables	**Graphing a Linear Inequality** *Step 1* Draw the graph of the line that is the boundary. Make the line solid if the inequality involves $\leq$ or $\geq$; make the line dashed if the inequality involves $<$ or $>$. *Step 2* Choose any point not on the line as a test point. *Step 3* Shade the region that includes the test point if the test point satisfies the original inequality; otherwise, shade the region on the other side of the boundary line.	$2x - 3y \leq 5$ Draw the graph of $2x - 3y = 5$. Use a solid line because of $\leq$. Choose $(1, 2)$. $$2(1) - 3(2) = 2 - 6 \leq 5$$ Shade the side of the line that includes $(1, 2)$.
7.5 Functions	To write the equation that defines a function in functional notation, solve the equation for y. Then replace y with $f(x)$.	Given: $2x + 3y = 12$. $$3y = -2x + 12$$ $$y = -\frac{2}{3}x + 4$$ $$f(x) = -\frac{2}{3}x + 4$$
7.6 Variation	If there is some constant k such that: $y = kx^n$, then y varies directly as x^n; $y = \dfrac{k}{x^n}$, then y varies inversely as x^n.	Area of a circle varies directly as the square of the radius. $$A = kr^2$$ Pressure varies inversely as volume. $$p = \frac{k}{V}$$

CHAPTER 7 REVIEW EXERCISES

[7.1] *Complete the given ordered pairs for each equation, and then graph the equation.*

1. $3x + 2y = 10$;
   (0,), (, 0), (2,), (, −2)

2. $x - y = 8$;
   (2,), (, −3), (3,), (, −2)

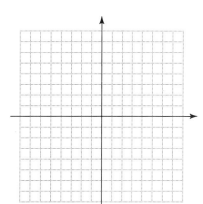

Find the x- and y-intercepts and graph each of the following equations.

3. $4x - 3y = 12$

4. $5x + 7y = 28$

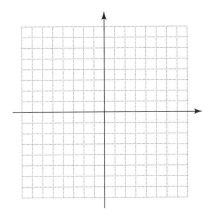
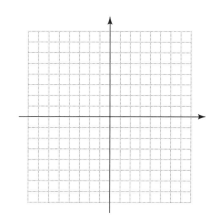

[7.2] *Find the slope for each line in Exercises 5–11.*

5. Through $(-1, 2)$ and $(4, -5)$

6. Through $(0, 3)$ and $(-2, 4)$

7. $y = 2x + 3$

8. $3x - 4y = 5$

9. $x = 5$

10. Parallel to $3y = 2x + 5$

11. Perpendicular to $3x - y = 4$

12. Use the slope formula to show that the figure with vertices (corners) at $(-2, -1)$, $(-2, 3)$, $(4, 3)$, and $(4, -1)$ is a rectangle. (*Hint:* A rectangle has opposite sides parallel and adjacent sides perpendicular.)

[7.3] *Write an equation for each line. In Exercises 16–22, write the equations in standard form.*

13. Slope $\dfrac{3}{5}$; y-intercept $(0, 2)$

14. Slope $-\dfrac{1}{3}$; y-intercept $(0, -1)$

15. Slope 0; y-intercept $(0, -2)$

16. Slope $-\dfrac{4}{3}$; through $(2, 7)$

17. Slope 3; through $(-1, 4)$

18. Vertical; through $(2, 5)$

19. Through $(2, -5)$ and $(1, 4)$

20. Through $(-3, -1)$ and $(2, 6)$

21. Parallel to $4x - y = 3$ and through $(7, -1)$

22. Perpendicular to $2x - 5y = 7$ and through $(4, 3)$

CHAPTER 7 REVIEW EXERCISES

[7.4] *Graph the solution of each inequality.*

23. $3x - 2y \leq 5$

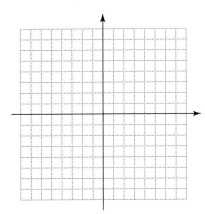

24. $5x + y > 6$

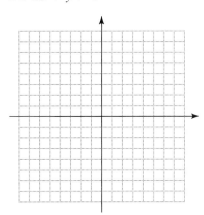

25. $y \leq 2$

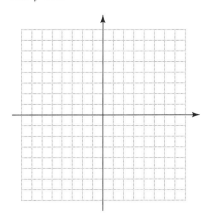

26. $-3 \leq x < 5$

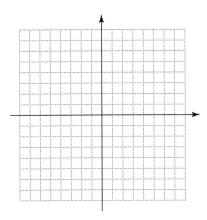

27. $2x + y \leq 1$ and $x > 2y$

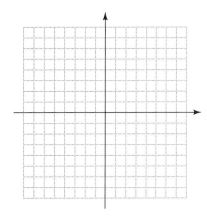

28. $x - 2y \leq 4$ or $x + y < 3$

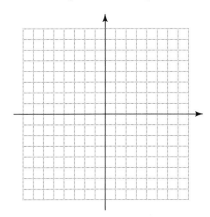

29. $|y - 3| < 4$

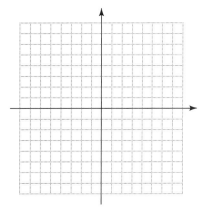

[7.5] *Give the domain and range of each relation. Identify any functions.*

30. {(1, 5), (2, 7), (3, 9), (4, 11)}

31. {(−4, 2), (−4, −2), (1, 5), (1, −5)}

32.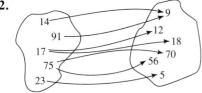

33.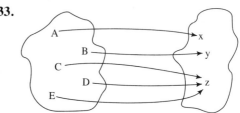

Give the domain for each of the following and identify any functions.

34. $3y = 2x - 1$

35. $2x = y^2 + 1$

36. $y - 1 = x^2$

37. $|y| = x$

38. $y \leq 2x + 3$

39. $y = \dfrac{2}{x - 1}$

Given $f(x) = x^2 + 2x - 1$, find the following.

40. $f(-1)$

41. $f(0)$

42. $f(3)$

43. $f\left(-\dfrac{1}{2}\right)$

44. $f(k)$

45. $f[f(0)]$

CHAPTER 7 REVIEW EXERCISES

Identify any graphs of functions.

46.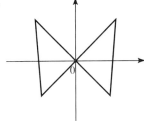

47.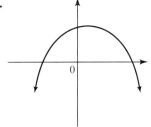

48.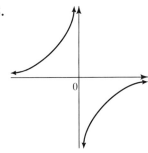

[7.6] *Solve the following problems.*

49. m varies directly as p^2 and inversely as q, and $m = 32$ when $p = 8$ and $q = 10$. Find m when $p = 12$ and $q = 15$.

50. x varies jointly as y and z and inversely as $\sqrt{w}$. If $x = 12$, when $y = 3$, $z = 8$, and $w = 36$, find x when $y = 5$, $z = 4$, and $w = 25$.

51. For the subject in a photograph to appear in the same perspective in the photograph as in real life, the viewing distance must be properly related to the amount of enlargement. For a particular camera, the viewing distance varies directly as the amount of enlargement. A picture taken with this camera that is enlarged 5 times should be viewed from a distance of 250 millimeters. Suppose a print 8.6 times the size of the negative is made. From what distance should it be viewed?

52. The distance that a person can see to the horizon from a point above the surface of the earth varies directly as the square root of the height (disregarding mountains, smog, and haze). If a person 144 meters above the surface can see for 18 kilometers to the horizon, how far can a person see to the horizon from a point 1600 meters high?

461

53. The frequency of vibration, f, of a guitar string varies directly as the square root of the tension, t, and inversely as the length, L, of the string. If the frequency is 20 when the tension is 9 (in appropriate units) and the length is 30 inches, find f when the tension is doubled and the length remains the same.

54. The force needed to keep a car from skidding on a curve varies inversely as the radius of the curve and jointly as the weight of the car and the square of the speed. If 3000 pounds of force keep a 2000-pound car from skidding as it travels at 30 miles per hour on a curve of radius 500 feet, what force would keep the same car from skidding on a curve of radius 750 feet at 50 miles per hour?

55. The cost of a dam is proportional to the square of the height and to the length of the crest (distance across the top). If a dam with a height of 100 meters and a length of 600 meters costs 1.5 million dollars, how much would it cost to build a dam 80 meters high with a crest of 500 meters?

56. A meteorite approaching the earth has velocity inversely proportional to the square root of its distance from the center of the earth. If the velocity is 5 kilometers per second when the distance is 8100 kilometers from the center of the earth, find the velocity at a distance of 6400 kilometers.

name date hour

CHAPTER 7 TEST

1. Find the slope of the line through (6, 4) and (−1, 2).

1. _____

For each line, find the slope and the x- and y-intercepts.

2. $3x - 2y = 20$

2. _____

3. $y = 3$

3. _____

4. $x = 3$

4. _____

Write the equation of each line in standard form.

5. Through (−3, 4); horizontal

5. _____

6. Through (4, −1); $m = -3$

6. _____

7. Through (−7, 2); parallel to $3x + 5y = 12$

7. _____

Graph each of the following.

8. $4x + 3y = 16$

8.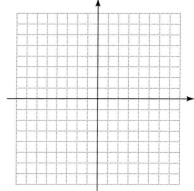

9. $y + 3 = 0$

9.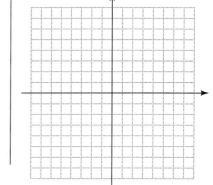

10.

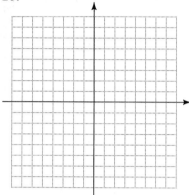

11.

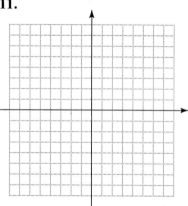

12. _____

13. _____

14. _____

15. _____

16. _____

10. $3x + 4y \leq 12$

11. $2x > y - 4$

Give the domain. Identify any functions.

12. $x = y^2 - 1$

13. $x = \sqrt{y - 1}$

For $f(x) = -x^2 + 4x - 7$, find the following.

14. $f(-2)$

15. $f\left(\dfrac{1}{m}\right)$

16. The force of the wind blowing on a vertical surface varies jointly as the area of the surface and the square of the velocity. If a wind blowing at 40 miles per hour exerts a force of 50 pounds on a surface of $\dfrac{1}{2}$ square foot, how much force will a wind of 80 miles per hour place on a surface of 2 square feet?

8 SYSTEMS OF LINEAR EQUATIONS

8.1 LINEAR SYSTEMS OF EQUATIONS IN TWO VARIABLES

Methods shown in this chapter help to identify numbers that make two or more equations true at the same time. Such a set of equations is called a **system of equations.**

1 Recall from Chapter 7 that the graph of a first-degree equation of the form $ax + by = c$ is a straight line. For this reason, such an equation is called a linear equation. Two or more linear equations form a **linear system.** The solution set of a linear system of equations contains all ordered pairs that satisfy all the equations of the system at the same time.

EXAMPLE 1 Deciding Whether an Ordered Pair Is a Solution
Is the given ordered pair a solution of the system?

(a) $x + y = 6 \quad (4, 2)$
$\quad 4x - y = 14$

Replace x with 4 and y with 2 in each equation of the system.

$$x + y = 6 \qquad 4x - y = 14$$
$$4 + 2 = 6 \qquad 4(4) - 2 = 14$$
$$\text{True} \qquad\qquad \text{True}$$

Since (4, 2) makes both equations true, (4, 2) is a solution of the system.

(b) $3x + 2y = 11 \quad (-1, 7)$
$\quad x + 5y = 36$

$$3x + 2y = 11 \qquad x + 5y = 36$$
$$3(-1) + 2(7) = 11 \qquad -1 + 5(7) = 36$$
$$-3 + 14 = 11 \qquad -1 + 35 = 36$$
$$\text{True} \qquad\qquad \text{False}$$

The ordered pair $(-1, 7)$ is not a solution of the system, since it does not make *both* equations true. ■

WORK PROBLEM 1 AT THE SIDE.

OBJECTIVES

1 Decide whether an ordered pair is a solution of a linear system.

2 Solve linear systems by graphing.

3 Solve linear systems with two equations and two unknowns by the elimination method.

4 Solve linear systems with two equations and two unknowns by the substitution method.

1. Are the ordered pairs solutions of the given systems?

(a) $2x + y = -6 \quad (-4, 2)$
$\quad x + 3y = 2$

(b) $7x - 3y = 25 \quad (4, 1)$
$\quad 8x + 5y = 37$

(c) $9x - y = -4 \quad (-1, 5)$
$\quad 4x + 3y = 11$

ANSWERS
1. (a) yes (b) yes (c) no

CHAPTER 8 SYSTEMS OF LINEAR EQUATIONS

2. Solve each system by graphing.

(a) $x - y = 3$
$2x - y = 4$

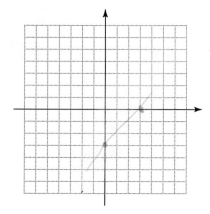

(b) $2x + y = -5$
$-x + 3y = 6$

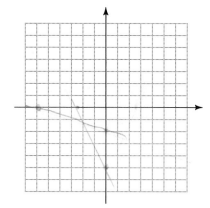

2 The solution set of a linear system of equations sometimes can be estimated by graphing the equations of the system on the same axes and then estimating the coordinates of any point of intersection.

EXAMPLE 2 Solving a System by Graphing

Solve the following system by graphing.

$$x + y = 5$$
$$2x - y = 4$$

The graphs of these linear equations are shown in Figure 1. The graph suggests that the point of intersection is the ordered pair (3, 2). Check this by substituting these values for x and y in both of the equations. As the check shows, the solution set of the system is {(3, 2)}. ■

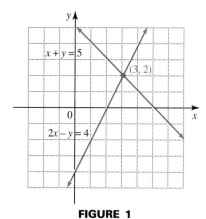

FIGURE 1

■ WORK PROBLEM 2 AT THE SIDE.

Since the graph of a linear equation is a straight line, there are three possibilities for the solution of a system of two linear equations, as shown in Figure 2.

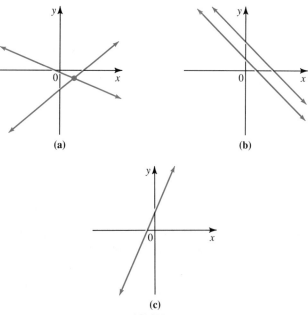

FIGURE 2

ANSWERS
2. (a) {(1, −2)}

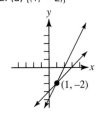

(b) {(−3, 1)}

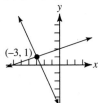

8.1 LINEAR SYSTEMS OF EQUATIONS IN TWO VARIABLES

GRAPHS OF A LINEAR SYSTEM

1. The two graphs intersect in a single point. The coordinates of this point give the only solution of the system. This is the most common case. See Figure 2(a).

2. The graphs are parallel lines. In this case the system is **inconsistent;** that is, there is no solution common to both equations of the system and the solution set is $\emptyset$. See Figure 2(b).

3. The graphs are the same line. In this case the equations are **dependent,** since any solution of one equation of the system is also a solution of the other. The solution set is an infinite set of ordered pairs. See Figure 2(c).

3 It is possible to find the solution of a system of equations by graphing. However, since it can be hard to read exact coordinates from a graph, an algebraic method is usually used to solve a system. One such algebraic method, called the **elimination method** (or the **addition method**), is explained in the following examples.

EXAMPLE 3 Solving a System by Elimination
Solve the system

$$2x + 3y = -6 \quad (1)$$
$$x - 3y = 6. \quad (2)$$

The elimination method involves combining the two equations so that one variable is eliminated. This is done using the following fact.

If $a = b$ and $c = d$, then $a + c = b + d$.

Adding corresponding sides of equations (1) and (2) gives

$$\begin{array}{r}2x + 3y = -6 \\ \underline{x - 3y = 6} \\ 3x = 0,\end{array}$$

and dividing both sides of the equation $3x = 0$ by 3 gives

$$x = 0.$$

To find y, replace x with 0 in either equation (1) or equation (2). Choosing equation (1) gives

$$\begin{aligned}2x + 3y &= -6 \\ 2(0) + 3y &= -6 \quad \text{Let } x = 0. \\ 0 + 3y &= -6 \\ 3y &= -6 \\ y &= -2.\end{aligned}$$

The solution of the system is $x = 0$ and $y = -2$, written as the ordered pair $(0, -2)$. Check this solution by substituting 0 for x and -2 for y in both equations of the original system. The solution set is $\{(0, -2)\}$. ■

WORK PROBLEM 3 AT THE SIDE.

3. Solve each system by elimination. Graph both equations.

(a) $3x - y = -7$
$2x + y = -3$

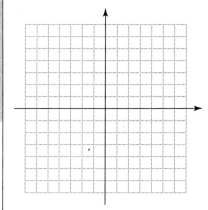

(b) $2x - 3y = 12$
$-2x + y = -4$

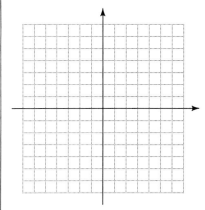

ANSWERS
3. (a) $\{(-2, 1)\}$

(b) $\{(0, -4)\}$

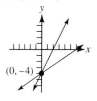

467

CHAPTER 8 SYSTEMS OF LINEAR EQUATIONS

4. Solve each system.

(a) $x + 3y = 8$
$2x - 5y = -17$

(b) $4x - y = 14$
$3x + 4y = 20$

(c) $2x + 3y = 19$
$3x - 7y = -6$

By adding the equations in Example 3, the variable y was eliminated because the coefficients of y were opposites. In many cases the coefficients will *not* be opposites. In these cases it is necessary to transform one or both equations so that the coefficients of one of the variables are opposites. The general method of solving a system by the elimination method is summarized as follows.

SOLVING LINEAR SYSTEMS OF TWO EQUATIONS BY ELIMINATION

Step 1 Write both equations in the form $ax + by = c$.

Step 2 Multiply one or both equations by appropriate numbers so that the sum of the coefficients of either x or y is zero.

Step 3 Add the new equations. The sum should be an equation with just one variable.

Step 4 Solve the equation from Step 3.

Step 5 Substitute the result of Step 4 into either of the given equations and solve for the other variable.

Step 6 Check the solution in both of the given equations.

EXAMPLE 4 Solving a System by Elimination
Solve the system

$$5x - 2y = 4 \quad (3)$$
$$2x + 3y = 13. \quad (4)$$

Both equations are in the form $ax + by = c$. Suppose that we wish to eliminate the variable x. In order to do this, multiply equation (3) by 2 and equation (4) by -5.

$10x - 4y = 8$ 2 times each side of equation (3)
$-10x - 15y = -65$ -5 times each side of equation (4)

Now add.

$$\begin{array}{r} 10x - 4y = 8 \\ -10x - 15y = -65 \\ \hline -19y = -57 \\ y = 3 \end{array}$$

To find x, substitute 3 for y in either equation (3) or (4). Substituting in equation (4) gives

$2x + 3y = 13$
$2x + 3(3) = 13$ Let $y = 3$.
$2x + 9 = 13$
$2x = 4$ Subtract 9.
$x = 2$. Divide by 2.

The solution set of the system is $\{(2, 3)\}$. Check this solution in both equations of the given system. ∎

ANSWERS
4. (a) $\{(-1, 3)\}$ (b) $\{(4, 2)\}$
(c) $\{(5, 3)\}$

WORK PROBLEM 4 AT THE SIDE.

8.1 LINEAR SYSTEMS OF EQUATIONS IN TWO VARIABLES

EXAMPLE 5 Solving a System of Dependent Equations
Solve the system

$$2x - y = 3 \quad (5)$$
$$6x - 3y = 9. \quad (6)$$

Multiply each side of equation (5) by -3, and then add the result to equation (6).

$$\begin{array}{r} -6x + 3y = -9 \\ 6x - 3y = 9 \\ \hline 0 = 0 \end{array}$$

Each side of equation (5) multiplied by -3

True

Adding these equations gave the true statement $0 = 0$. In the original system, equation (6) could be obtained from equation (5) by multiplying both sides of equation (5) by 3. Because of this, equations (5) and (6) are equivalent and have the same line as graph, as shown in Figure 3. The equations are dependent, and the solution set is written as $\{(x, y) | 2x - y = 3\}$. ∎

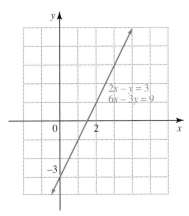

FIGURE 3

WORK PROBLEM 5 AT THE SIDE.

EXAMPLE 6 Solving an Inconsistent System
Solve the system

$$x + 3y = 4 \quad (7)$$
$$-2x - 6y = 3. \quad (8)$$

Multiply both sides of equation (7) by 2, and then add the result to equation (8).

$$\begin{array}{r} 2x + 6y = 8 \\ -2x - 6y = 3 \\ \hline 0 = 11 \end{array}$$

Equation (7) multiplied by 2

False

The result of the addition step here is a false statement, which shows that the system is inconsistent. As shown in Figure 4 on the next page, the graphs of the equations of the system are parallel lines. There are no ordered pairs that satisfy both equations, so there is no solution for the system and the solution set is ∅. ∎

5. Solve the system below. Graph both equations.

$$2x + y = 6$$
$$-8x - 4y = -24$$

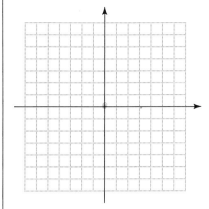

ANSWER
5. $\{(x, y) | 2x + y = 6\}$

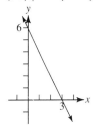

CHAPTER 8 SYSTEMS OF LINEAR EQUATIONS

6. Solve the system below. Graph both equations.

$$4x - 3y = 8$$
$$8x - 6y = 14$$

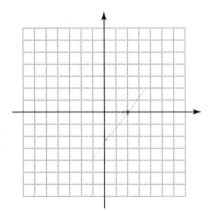

7. Write the equations of Example 5 in slope-intercept form.

8. Write the equations of Example 6 in slope-intercept form.

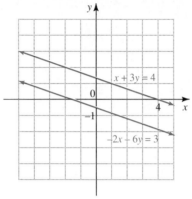

FIGURE 4

The results of Examples 5 and 6 are generalized as follows.

If both variables are eliminated when a system of linear equations is solved,

1. there is no solution if the resulting statement is *false*;
2. there are infinitely many solutions if the resulting statement is *true*.

■ WORK PROBLEM 6 AT THE SIDE.

Slopes and y-intercepts can be used to decide if the graphs of a system of equations are parallel lines or if they coincide. For the system of Example 5, writing each equation in slope-intercept form shows that both lines have a slope of 2 and a y-intercept of $(0, -3)$, so the graphs are the same line.

■ WORK PROBLEM 7 AT THE SIDE.

In Example 6, both equations have a slope of $-\frac{1}{3}$, but the y-intercepts are $\left(0, \frac{4}{3}\right)$ and $\left(0, -\frac{1}{2}\right)$, showing that the graphs are two distinct parallel lines.

■ WORK PROBLEM 8 AT THE SIDE.

4 Linear systems can also be solved by the **substitution method.** The substitution method is most useful in solving linear systems in which one variable has a coefficient of 1. However, as shown in the next chapter, the substitution method is the best choice for solving many *nonlinear* systems.

The method of solving a system by substitution is summarized as follows.

ANSWERS
6. ∅

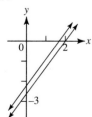

7. Both equations are $y = 2x - 3$.
8. $y = -\frac{1}{3}x + \frac{4}{3}$; $y = -\frac{1}{3}x - \frac{1}{2}$

8.1 LINEAR SYSTEMS OF EQUATIONS IN TWO VARIABLES

SOLVING LINEAR SYSTEMS BY SUBSTITUTION

Step 1 Solve one of the equations for either variable. (If one of the variables has coefficient 1 or -1, choose it, since the substitution method is usually easier this way.)

Step 2 Substitute for that variable in the other equation. The result should be an equation with just one variable.

Step 3 Solve the equation from Step 2.

Step 4 Substitute the result from Step 3 into the equation from Step 1 to find the value of the other variable.

Step 5 Check the solution in both of the given equations.

The next two examples illustrate this method.

EXAMPLE 7 Solving a System by Substitution

Solve the system

$$\frac{x}{2} + \frac{y}{3} = \frac{13}{6} \quad (9)$$

$$4x - y = -1. \quad (10)$$

Write the first equation without fractions by multiplying both sides by the common denominator, 6.

$$6 \cdot \left(\frac{x}{2} + \frac{y}{3}\right) = 6 \cdot \frac{13}{6} \quad \text{Multiply by the common denominator, 6.}$$

$$\mathbf{6} \cdot \frac{x}{2} + \mathbf{6} \cdot \frac{y}{3} = \mathbf{6} \cdot \frac{13}{6} \quad \text{Distributive property}$$

$$3x + 2y = 13 \quad (11)$$

The new system is

$$3x + 2y = 13 \quad (11)$$
$$4x - y = -1. \quad (10)$$

To use the substitution method, first solve one of the equations for either x or y. Since the coefficient of y in equation (10) is -1, it is easiest to solve for y in equation (10).

$$-y = -1 - 4x$$
$$y = 1 + 4x$$

Substitute $1 + 4x$ for y in equation (11) and solve for x.

$$3x + 2(\mathbf{1 + 4x}) = 13 \quad \text{Let } y = 1 + 4x.$$
$$3x + 2 + 8x = 13 \quad \text{Distributive property}$$
$$11x = 11 \quad \text{Combine terms; subtract 2.}$$
$$x = 1 \quad \text{Divide by 11.}$$

CHAPTER 8 SYSTEMS OF LINEAR EQUATIONS

9. Solve by substitution.

(a) $\dfrac{x}{4} + \dfrac{y}{2} = \dfrac{1}{4}$
$2x - 3y = 9$

(b) $\dfrac{x}{5} + \dfrac{2y}{3} = -\dfrac{8}{5}$
$\dfrac{3x}{4} - \dfrac{y}{3} = \dfrac{5}{2}$

10. Solve by substitution.

(a) $7x - 2y = -2$
$y = 3x$

(b) $2x - 3y = 1$
$3x = 2y + 9$

Since $y = 1 + 4x$,
$$y = 1 + 4(\mathbf{1}) = 5. \quad \text{Let } x = 1.$$
Check that the solution set is $\{(1, 5)\}$. ∎

■ **WORK PROBLEM 9 AT THE SIDE.**

EXAMPLE 8 Solving a System by Substitution
Solve the system
$$4x - 3y = 7 \quad (12)$$
$$3x - 2y = 6. \quad (13)$$

If the substitution method is to be used, one equation must be solved for one of the two variables. Let us solve equation (13) for x.
$$3x = 2y + 6$$
$$x = \dfrac{2y + 6}{3}$$

Now substitute $\dfrac{2y + 6}{3}$ for x in equation (12).
$$4x - 3y = 7 \quad (12)$$
$$4\left(\dfrac{\mathbf{2y + 6}}{\mathbf{3}}\right) - 3y = 7$$

Multiply both sides of the equation by 3 to eliminate the fraction.

$4(2y + 6) - 9y = 21$	Multiply by the common denominator, 3.
$8y + 24 - 9y = 21$	Distributive property
$24 - y = 21$	Combine terms.
$-y = -3$	Add -24.
$y = 3$	Divide by -1.

Since $x = \dfrac{2y + 6}{3}$ and $y = 3$,
$$x = \dfrac{2(3) + 6}{3} = \dfrac{6 + 6}{3} = 4.$$

The solution set is $\{(4, 3)\}$. ∎

■ **WORK PROBLEM 10 AT THE SIDE.**

ANSWERS
9. (a) $\{(3, -1)\}$ (b) $\{(2, -3)\}$
10. (a) $\{(-2, -6)\}$ (b) $\{(5, 3)\}$

8.1 EXERCISES

Decide whether the ordered pair is a solution for the given system. See Example 1.

1. $x + y = 5$ $(4, 1)$
   $x - y = 3$

2. $x - y = -1$ $(2, 3)$
   $x + y = 5$

3. $2x - y = 8$ $(5, 2)$
   $3x + 2y = 18$

4. $3x - 5y = -12$ $(-1, 2)$
   $x - y = -7$

Use the elimination method to solve the following systems of linear equations. Identify any pairs of equations that are inconsistent or dependent. See Examples 2–6. In Exercises 5–6, graph the system.

5. $x + y = 10$
   $2x - y = 5$

6. $3x - 2y = 4$
   $3x + y = -2$

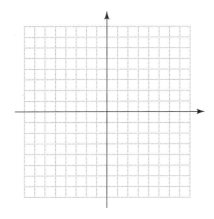
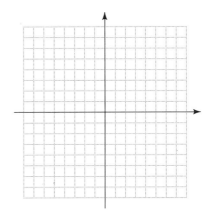

7. $2x - 5y = 10$
   $3x + y = 15$

8. $4x - y = 3$
   $-2x + 3y = 1$

9. $7x + 2y = 3$
   $-14x - 4y = -6$

10. $4x + 2y = 3$
    $-3x - 3y = 0$

11. $-3x + 5y = 2$
    $2x - 3y = 1$

12. $4x - 16y = 4$
    $x - 4y = 1$

13. $5x + 3y = 1$
$-3x - 4y = 6$

14. $6x - 6y = 5$
$x - y = 8$

15. $\dfrac{x}{2} + \dfrac{y}{3} = -\dfrac{1}{3}$
$\dfrac{x}{2} + 2y = -7$

16. $x + 5y = 3$
$\dfrac{2}{3}x + \dfrac{10}{3}y = 2$

17. $2x - 3y = 2$
$\dfrac{x}{3} - \dfrac{y}{2} = -\dfrac{1}{3}$

18. $\dfrac{1}{4}x + \dfrac{5}{4}y = -\dfrac{1}{2}$
$-x - 5y = 2$

Solve these systems of linear equations by the substitution method. See Examples 7 and 8.

19. $4x + y = 9$
$x = 2y$

20. $3x - 4y = -22$
$y = 3x$

21. $2x = y + 6$
$y = 5x$

22. $-3x = -y - 5$
$x = -2y$

23. $x + 4y = 14$
$2x = y + 1$

24. $3x + 5y = 17$
$4x = y - 8$

25. $5x - 4y = 9$
$3 + x = 2y$

26. $6x = y - 9$
$4 + y = -7x$

27. $x = 3y + 5$
$y = \dfrac{2}{3}x$

28. $3x = y - 2$
$x = \dfrac{1}{4}y$

29. $\dfrac{x}{2} + \dfrac{y}{3} = 3$
$y = 3x$

30. $\dfrac{x}{4} - \dfrac{y}{5} = 9$
$y = 5x$

In the following systems, let $p = 1/x$ and $q = 1/y$. Substitute, solve for p and q, and then find x and y.

31. $\dfrac{3}{x} + \dfrac{4}{y} = \dfrac{5}{2}$
$\dfrac{5}{x} - \dfrac{3}{y} = \dfrac{7}{4}$

32. $\dfrac{4}{x} - \dfrac{9}{y} = -1$
$-\dfrac{7}{x} + \dfrac{6}{y} = -\dfrac{3}{2}$

33. $\dfrac{2}{x} - \dfrac{5}{y} = \dfrac{3}{2}$
$\dfrac{4}{x} + \dfrac{1}{y} = \dfrac{4}{5}$

34. $\dfrac{2}{x} + \dfrac{3}{y} = \dfrac{11}{2}$
$-\dfrac{1}{x} + \dfrac{2}{y} = -1$

8.1 EXERCISES

Solve by any method. Here a and b represent nonzero constants.

35. $ax + by = 2$
$-ax + 2by = 1$

36. $2ax - y = 3$
$y = 5ax$

37. $3ax + 2y = 1$
$-ax + y = 2$

38. $ax + by = c$
$bx + ay = c$

Work the following problems that involve graphs of linear systems.

39. The following graph shows the expected annual return on $10,000 for two different investment opportunities.

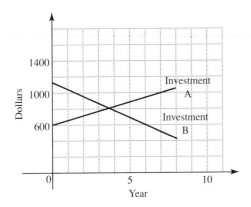

(a) What is the annual return for each investment after seven years?

(b) In what year is the return from the two investments the same? What is the return at that time?

(c) How many years does it take in each case to get an annual return of $1000?

40. Julio Nicolai compared the monthly interest costs for two types of home mortgages: a fixed-rate mortgage and a graduated-payment mortgage. His results are shown in the figure.

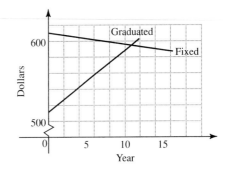

(a) In how many years will the monthly interest costs be equal for the two plans?

(b) What is the monthly interest in the seventh year for each plan?

(c) In what year will the monthly interest cost be $590 for each plan?

41. The following graph shows a company's costs to produce computer parts and the revenue from the sale of computer parts.

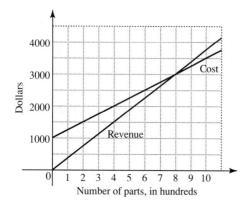

(a) At what production level does the cost equal the revenue? What is the revenue at that point?

(b) Profit is revenue less cost. Estimate the profit on the sale of 400 parts.

(c) The cost to produce 0 items is called the *fixed cost*. Find the fixed cost.

42. The following figure shows graphs that represent the supply and demand for one brand of ice cream at various prices per half-gallon.

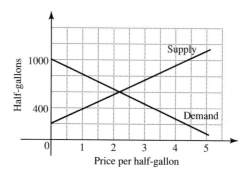

(a) At what price does supply equal demand?

(b) For how many half-gallons does supply equal demand?

(c) What are the supply and demand at a price of $1 per half-gallon?

SKILL SHARPENERS

Multiply both sides of each equation by the given number. See Section 2.1.

43. $2x - 4y + z = 8$; by -3

44. $3x + 6y - 2z = 12$; by 4

45. $-x + 3y - 2z = 18$; by -2

Add. See Section 3.3.

46. $3x + 2y - z = 7$
$5x - 2y + 2z = 10$

47. $x + 5y - 3z = 12$
$2x + 6y + 3z = -15$

48. $-2x - 4y + 5z = 20$
$2x + 3y - 4z = 15$

8.2 LINEAR SYSTEMS OF EQUATIONS IN THREE VARIABLES

A solution of an equation in three variables, such as $2x + 3y - z = 4$, is called an **ordered triple** and written (x, y, z). For example, the ordered triples $(1, 1, 1)$ and $(10, -3, 7)$ are solutions of $2x + 3y - z = 4$, since the numbers in these triples satisfy the equation when used as replacements for x, y, and z, respectively.

In the rest of this chapter, the term *linear equation* is extended to equations of the form $ax + by + cz + \cdots + dw = k$, where not all the coefficients a, b, c, . . ., d equal zero. For example, $2x + 3y - 5z = 7$ and $x - 2y - z + 3u - 2w = 8$ are linear equations, the first with three variables and the second with five variables.

In this section we discuss the solution of a system of linear equations in three variables such as

$$4x + 8y + z = 2$$
$$x + 7y - 3z = -14$$
$$2x - 3y + 2z = 3.$$

Theoretically, a system of this type can be solved by graphing. However, the graph of a linear equation with three variables is a *plane* and not a line. Since the graph of each equation of the system is a plane, which requires three-dimensional graphing, this method is not practical. However, it does serve to illustrate the number of solutions possible for such systems, as Figure 5 shows.

OBJECTIVES

1 Solve linear systems with three equations and three unknowns by the elimination method.

2 Solve linear systems with three equations and three unknowns where some of the equations have missing terms.

3 Solve linear systems with three equations and three unknowns that are inconsistent or that include dependent equations.

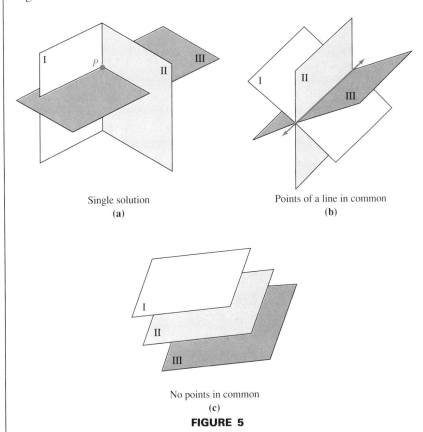

FIGURE 5

GRAPHS OF LINEAR SYSTEMS IN THREE VARIABLES

1. The three planes may meet at a single, common point that is the solution of the system. See Figure 5(a).
2. The three planes may have the points of a line in common so that the set of points that satisfy the equation of the line is the solution of the system. See Figure 5(b).
3. The planes may have no points common to all three, in which case there is no solution for the system. See Figure 5(c).
4. The three planes may coincide, in which case the solution of the system is the set of all points on a plane.

1 Since a graphic solution of a system of three equations in three variables is impractical, these systems are solved with an extension of the elimination method, summarized as follows.

SOLVING LINEAR SYSTEMS IN THREE VARIABLES

Step 1 Use the elimination method to eliminate any variable from any two of the given equations. The result is an equation in two variables.

Step 2 Eliminate the *same* variable from any *other* two equations. The result is an equation in the same two variables as in Step 1.

Step 3 Use the elimination method to eliminate a second variable from the two equations in two variables that result from Steps 1 and 2. The result is an equation in one variable that gives the value of that variable.

Step 4 Substitute the value of the variable found in Step 3 into either of the equations in two variables to find the value of the second variable.

Step 5 Use the values of the two variables from Steps 3 and 4 to find the value of the third variable by substituting into any of the original equations.

EXAMPLE 1 Solving a System in Three Variables
Solve the system

$$4x + 8y + z = 2 \quad (1)$$
$$x + 7y - 3z = -14 \quad (2)$$
$$2x - 3y + 2z = 3. \quad (3)$$

As before, the elimination method involves eliminating a variable from the sum of two equations. To begin, choose equations (1) and (2) and eliminate z by multiplying both sides of equation (1) by 3 and then adding the result to equation (2).

8.2 LINEAR SYSTEMS OF EQUATIONS IN THREE VARIABLES

$$12x + 24y + 3z = 6 \quad \text{3 times both sides of equation (1)}$$
$$\underline{x + 7y - 3z = -14} \quad (2)$$
$$13x + 31y = -8 \quad (4)$$

Equation (4) has only two variables. To get another equation without z, multiply both sides of equation (1) by -2 and add the result to equation (3). It is important at this point to eliminate the *same variable*, z.

$$-8x - 16y - 2z = -4 \quad \text{-2 times both sides of equation (1)}$$
$$\underline{2x - 3y + 2z = 3} \quad (3)$$
$$-6x - 19y = -1 \quad (5)$$

Now solve the system of equations (4) and (5) for x and y.

WORK PROBLEM 1 AT THE SIDE.

As shown by Problem 1 at the side, the solution of the system of equations (4) and (5) is $x = -3$ and $y = 1$. To find z, substitute -3 for x and 1 for y in equation (1). (Any of the three given equations could be used.)

$$4x + 8y + z = 2 \quad (1)$$
$$4(-3) + 8(1) + z = 2$$
$$z = 6$$

The ordered triple $(-3, 1, 6)$ is the only solution of the system. Check that the solution satisfies all three equations of the system so the solution set is $\{(-3, 1, 6)\}$. ∎

WORK PROBLEM 2 AT THE SIDE.

2 When one or more of the equations of a system has a missing term, one elimination step can be omitted.

EXAMPLE 2 Solving a System of Equations with Missing Terms
Solve the system

$$6x - 12y = -5 \quad (6)$$
$$8y + z = 0 \quad (7)$$
$$9x - z = 12. \quad (8)$$

Since equation (8) is missing the variable y, one way to begin the solution is to eliminate y again with equations (6) and (7). Multiply both sides of equation (6) by 2 and both sides of equation (7) by 3, and then add.

$$12x - 24y = -10 \quad \text{2 times both sides of equation (6)}$$
$$\underline{24y + 3z = 0} \quad \text{3 times both sides of equation (7)}$$
$$12x + 3z = -10 \quad (9)$$

Use this result, together with equation (8), to eliminate z. Multiply both sides of equation (8) by 3. This gives

$$27x - 3z = 36 \quad \text{3 times both sides of equation (8)}$$
$$\underline{12x + 3z = -10} \quad (9)$$
$$39x = 26.$$

1. Solve the system of equations for x and y.

$$13x + 31y = -8$$
$$-6x - 19y = -1$$

2. Solve the system.

$$x + y + z = 2$$
$$x - y + 2z = 2$$
$$-x + 2y - z = 1$$

ANSWERS
1. $\{(-3, 1)\}$
2. $\{(-1, 1, 2)\}$

CHAPTER 8 SYSTEMS OF LINEAR EQUATIONS

3. Solve the system.

$$x - y = 6$$
$$2y + 5z = 1$$
$$3x - 4z = 8$$

From this equation, $x = \frac{26}{39} = \frac{2}{3}$. Substitution into equation (8) gives

$$9x - z = 12 \qquad (8)$$
$$9\left(\frac{2}{3}\right) - z = 12 \qquad x = \frac{2}{3}$$
$$6 - z = 12$$
$$z = -6.$$

Substitution of -6 for z in equation (7) gives

$$8y + z = 0 \qquad (7)$$
$$8y - 6 = 0 \qquad z = -6$$
$$8y = 6$$
$$y = \frac{3}{4}.$$

Check in each of the original equations of the system to verify that the solution set of the system is $\left\{\left(\frac{2}{3}, \frac{3}{4}, -6\right)\right\}$. ■

4. Solve each system.

(a) $\quad 3x - 5y + 2z = 1$
$\quad\quad 5x + 8y - z = 4$
$\quad -6x + 10y - 4z = 5$

■ **WORK PROBLEM 3 AT THE SIDE.**

3 Linear systems with three variables may be inconsistent or may include dependent equations. The next two examples illustrate these cases.

EXAMPLE 3 Solving an Inconsistent System with Three Variables

Solve the following system.

$$2x - 4y + 6z = 5 \qquad (10)$$
$$-x + 3y - 2z = -1 \qquad (11)$$
$$x - 2y + 3z = 1 \qquad (12)$$

Eliminate x by adding equations (11) and (12) to get the equation

$$y + z = 0.$$

Now to eliminate x again, multiply both sides of equation (12) by -2 and add the result to equation (10).

$$-2x + 4y - 6z = -2$$
$$\underline{2x - 4y + 6z = 5}$$
$$0 = 3 \quad \text{False}$$

(b) $7x - 9y + 2z = 0$
$\quad\quad\quad y + z = 0$
$\quad 8x \quad\quad - z = 0$

The resulting false statement indicates that equations (10) and (12) have no common solution; the system is inconsistent and the solution set is ∅. The graph of the equations of the system would show two of the planes parallel to one another. ■

■ **WORK PROBLEM 4 AT THE SIDE.**

ANSWERS
3. {(4, −2, 1)}
4. (a) ∅ (b) {(0, 0, 0)}

480

8.2 LINEAR SYSTEMS OF EQUATIONS IN THREE VARIABLES

EXAMPLE 4 Solving a System of Dependent Equations with Three Variables

Solve the system.

$$2x - 3y + 4z = 8 \quad (13)$$

$$-x + \frac{3}{2}y - 2z = -4 \quad (14)$$

$$6x - 9y + 12z = 24 \quad (15)$$

Multiplying both sides of equation (13) by 3 gives equation (15). Multiplying both sides of equation (14) by -6 also results in equation (15). Because of this, the three equations are dependent. All three equations have the same graph. The solution set is written

$$\{(x, y, z) | 2x - 3y + 4z = 8\}. \blacksquare$$

WORK PROBLEM 5 AT THE SIDE.

The method discussed in this section can be extended to solve larger systems. For example, to solve a system of four equations in four variables, eliminate a variable from three pairs of equations to get a system of three equations in three unknowns. Then proceed as shown above.

5. Solve the system.

$$x - y + z = 4$$
$$-3x + 3y - 3z = -12$$
$$2x - 2y + 2z = 8$$

ANSWER
5. $\{(x, y, z) | x - y + z = 4\}$

Historical Reflections

Pierre de Fermat (ca. 1601–1665) and His Last Theorem

Pierre de Fermat was a French government official who did not interest himself in mathematics until he was past the age of 30. He faithfully and accurately carried out the duties of his post, devoting leisure time to the study of mathematics. He was a worthy scholar, best known for his work in number theory. His other major contributions involved certain applications of geometry and the study of probability.

Unfortunately, much of Fermat's best work survived only on loose sheets or jotted, without proof, in the margins of works that he read. In fact, one of his marginal comments continues to haunt mathematicians. As we know from the Pythagorean Theorem, there are infinitely many triples *(a, b, c)* of positive integers that satisfy

$$a^2 + b^2 = c^2.$$

Are there any such triples that satisfy

$$a^n + b^n = c^n$$

for natural numbers *n* greater than 2? Fermat thought not, and indicated in a marginal note that he had a "truly wonderful proof" for this, but that the margin was "too small to contain it." Was Fermat just joking, did he indeed have a proof, or did he have an incorrect proof? We do not know for certain, but as of early 1990, no one has ever proved the theorem that is now known as Fermat's Last Theorem. It is one of the most famous unsolved problems of mathematics.

In March, 1988, national news services carried the story that Yoichi Miyaoka, a Japanese mathematician working in Germany, had finally succeeded in proving Fermat's Last Theorem. A few weeks later, however, it was reported that the proof was flawed. Thus, work continues on this 400-year-old problem.

8.2 EXERCISES

Solve each system of equations. See Example 1.

1. $2x + y + z = 3$
$3x - y + z = -2$
$4x - y + 2z = 0$

2. $x + y + z = 6$
$2x + 3y - z = 7$
$3x - y - z = 6$

3. $3x + 2y + z = 4$
$2x - 3y + 2z = -7$
$x + 4y - z = 10$

4. $-3x + y - z = 8$
$-4x + 2y + 3z = -3$
$2x + 3y - 2z = -1$

5. $2x + 5y + 2z = 9$
$4x - 7y - 3z = 7$
$3x - 8y - 2z = 9$

6. $5x - 2y + 3z = 13$
$4x + 3y + 5z = -10$
$2x + 4y - 2z = -22$

7. $x + y - z = -2$
$2x - y + z = -5$
$x - 2y + 3z = 4$

8. $x + 3y - 6z = 7$
$2x - y + z = 1$
$x + 2y + 2z = -1$

CHAPTER 8 SYSTEMS OF LINEAR EQUATIONS

Solve each system of equations. See Example 2.

9. $2x - 3y + 2z = -1$
   $x + 2y = 14$
   $x - 3z = -5$

10. $2x - y + 3z = 0$
    $x + 2y - z = 5$
    $2y + z = 1$

11. $4x + 2y - 3z = 6$
    $x - 4y + z = -4$
    $-x + 2z = 2$

12. $-x + y = 1$
    $y - z = 2$
    $x + z = -2$

13. $x + y = 1$
    $2x - z = 0$
    $y + 2z = -2$

14. $2x + y = 6$
    $3y - 2z = -4$
    $3x - 5z = -7$

Solve each system of equations. See Examples 1, 3, and 4.

15. $x + 3y + 2z = 4$
    $x - y - z = 3$
    $3x + 9y + 6z = 5$

16. $x - 2y + 4z = -10$
    $-3x + 6y - 12z = 20$
    $2x + 5y + z = 12$

484

17. $\begin{aligned} x - 2y &= 0 \\ 3y + z &= -1 \\ 4x - z &= 11 \end{aligned}$

18. $\begin{aligned} 2x - 5y - z &= 3 \\ 5x + 14y - z &= -11 \\ 7x + 9y - 2z &= -5 \end{aligned}$

19. $\begin{aligned} x + 4y - z &= 3 \\ -2x - 8y + 2z &= -6 \\ 3x + 12y - 3z &= 9 \end{aligned}$

20. $\begin{aligned} 2x + 2y - 6z &= 5 \\ 3x - y + z &= 2 \\ -x - y + 3z &= 4 \end{aligned}$

21. $\begin{aligned} 2x + y - z &= 6 \\ 4x + 2y - 2z &= 12 \\ -x - \tfrac{1}{2}y + \tfrac{1}{2}z &= -3 \end{aligned}$

22. $\begin{aligned} -5x + 5y - 20z &= -40 \\ x - y + 4z &= 8 \\ 3x - 3y + 12z &= 24 \end{aligned}$

23. $\begin{aligned} 2x - 8y + 2z &= -10 \\ -x + 4y - z &= 5 \\ 3x - 12y + 3z &= -15 \end{aligned}$

24. $\begin{aligned} -5x - y + z &= 3 \\ 8x + 2y - z &= -5 \\ 3x + y + 5z &= 3 \end{aligned}$

CHAPTER 8 SYSTEMS OF LINEAR EQUATIONS

Solve each system.

25. $x + y + z - w = 5$
     $2x + y - z + w = 3$
     $x - 2y + 3z + w = 18$
     $-x - y + z + 2w = 8$

26. $3x + y - z + 2w = 9$
     $x + y + 2z - w = 10$
     $x - y - z + 3w = -2$
     $-x + y - z + w = -6$

SKILL SHARPENERS

Solve each of the following problems. See Sections 2.3 and 2.4.

27. To make a 10% acid solution for chemistry class, John wants to mix some 5% solution with 10 centiliters of 20% solution. How much 5% solution should he use?

28. In a 100-mile bicycle race, the winner finished one hour before the person who came in last. The person in last place averaged 20 miles per hour. Find the winner's average speed.

29. The perimeter of a triangle is 163 inches. The shortest side measures five-sixths the length of the longest side, and the medium side measures 7 inches less than the longest side. Find the lengths of the sides of the triangle.

30. The sum of the three angles of a triangle is 180°. The largest angle is twice the measure of the smallest, and the third angle measures 20° less than the largest. Find the measures of the three angles.

31. The sum of three numbers is 16. The largest number is -3 times the smallest, while the middle number is 4 less than the largest. Find the three numbers.

32. Alexis LeBeau has a collection of pennies, dimes, and quarters. The number of dimes is one less than twice the number of pennies. If there are 27 coins in all worth a total of $4.20, how many of each kind of coin is in the collection?

8.3 APPLICATIONS OF LINEAR SYSTEMS

Many word problems involve more than one unknown quantity. Although some problems with two unknowns can be solved using just one variable, many times it is easier to use two variables. To solve a problem with two unknowns, write two equations that relate the unknown quantities. The system formed by the pair of equations then can be solved using the methods of Section 8.1.

OBJECTIVES

1. Solve geometry problems using two variables.
2. Solve money problems using two variables.
3. Solve mixture problems using two variables.
4. Solve problems about distance, rate, and time using two variables.
5. Solve problems with three unknowns using a system of three equations.

1 Problems about the perimeter of a geometric figure often involve two unknowns. The next example shows how to write a system of equations to solve such a problem.

EXAMPLE 1 Solving a Geometry Problem

The length of a rectangular house is to be 6 meters more than its width. Find the length and width of the house if the perimeter must be 48 meters.

Begin by sketching a rectangle to represent the foundation of the house. Let

$$x = \text{the length}$$

and

$$y = \text{the width}.$$

See Figure 6.

FIGURE 6

The length, x, is 6 meters more than the width, y. Therefore,

$$x = 6 + y.$$

The formula for the perimeter of a rectangle is $P = 2L + 2W$. Here $P = 48$, $L = x$, and $W = y$, so

$$48 = 2x + 2y.$$

The length and width can now be found by solving the system

$$x = 6 + y$$
$$48 = 2x + 2y. \blacksquare$$

WORK PROBLEMS 1 AND 2 AT THE SIDE.

1. Solve the system shown in Example 1.

2. (a) The perimeter of a rectangle is 76 inches. If the width were doubled, it would be 13 inches more than the length. Find the width and length.

   (b) A total of 284 votes were cast in the fire district election. Jones received 60 votes less than Gomez. How many votes did each candidate receive?

ANSWERS
1. width = 9; length = 15
2. (a) 17 and 21 inches
   (b) 112 for Jones, 172 for Gomez

CHAPTER 8 SYSTEMS OF LINEAR EQUATIONS

3. (a) Depham bought 4 pounds of peaches and 2 pounds of apricots, paying $5. Later, she bought 7 pounds of peaches and 3 pounds of apricots for $8.25. Find the cost per pound for each fruit.

2 Another type of problem that often leads to a system of equations is one about different amounts of money.

EXAMPLE 2 Solving a Problem About Money

For an art project Kay bought 8 pieces of poster board and 3 marker pens for $6.50. She later needed 2 pieces of poster board and 2 pens. These items cost $3.00. Find the cost of 1 marker pen and 1 sheet of poster board.

Let x represent the cost of a piece of poster board and y represent the cost of a pen. For the first purchase, $8x$ represents the cost of the pieces of poster board and $3y$ the cost of the pens. The total cost was $6.50, so

$$8x + 3y = 6.50.$$

For the second purchase,

$$2x + 2y = 3.00.$$

To solve the system, multiply both sides of the second equation by -4 and add the result to the first equation.

$$\begin{aligned} 8x + 3y &= 6.50 \\ -8x - 8y &= -12.00 \\ \hline -5y &= -5.50 \\ y &= 1.10 \end{aligned}$$

By substituting 1.10 for y in either of the equations, verify that $x = .40$. Kay paid $.40 for a piece of poster board and $1.10 for a pen. ∎

■ **WORK PROBLEM 3 AT THE SIDE.**

3 Many mixture problems can be solved by writing a system of equations.

(b) A cashier has $1260 in tens and twenties, with a total of 98 bills. How many of each type are there?

EXAMPLE 3 Solving a Mixture Problem

How many ounces each of 5% hydrochloric acid and 20% hydrochloric acid must be combined to get 10 ounces of solution that is 12.5% hydrochloric acid?

Let x represent the number of ounces of 5% solution and y represent the number of ounces of 20% solution. A table summarizes the information from the problem.

Kind of solution	Ounces of solution	Ounces of acid
5%	x	$.05x$
20%	y	$.20y$
12.5%	10	$(.125)10$

When the x ounces of 5% solution and the y ounces of 20% solution are combined, the total number of ounces is 10, so that

$$x + y = 10. \qquad (1)$$

ANSWERS
3. (a) $.75 for peaches, $1 for apricots
   (b) 70 tens, 28 twenties

The ounces of acid in the 5% solution (.05x) plus the ounces of acid in the 20% solution (.20y) should equal the total ounces of acid in the mixture, which is (.125)10. That is,

$$.05x + .20y = (.125)10. \quad (2)$$

Eliminate x by first multiplying both sides of equation (2) by 100 to clear it of decimals, and then multiplying both sides of equation (1) by -5.

$5x + 20y = 125$	100 times both sides of equation (2)
$-5x - 5y = -50$	-5 times both sides of equation (1)
$15y = 75$	
$y = 5$	

Since $y = 5$ and $x + y = 10$, x is also 5, and 5 ounces each of the 5% and the 20% solutions are required. ∎

WORK PROBLEM 4 AT THE SIDE.

4 Constant rate applications require the distance formula, $d = rt$, where d is distance, r is rate (or speed), and t is time. These applications often lead to a system of equations, as in the next example.

EXAMPLE 4 Solving a Motion Problem

A car travels 250 kilometers in the same time that a truck travels 225 kilometers. If the speed of the car is 8 kilometers per hour faster than the speed of the truck, find both speeds.

A table is useful to organize the information in problems about distance, rate, and time. Fill in the given information for each vehicle (in this case distance) and use variables for the unknown speeds (rate) as follows.

	d	r	t
Car	250	x	
Truck	225	y	

The table shows nothing about time. Get an expression for time by solving the distance formula, $d = rt$, for t.

$$\frac{d}{r} = t$$

The two times can be written as $\frac{250}{x}$ and $\frac{225}{y}$.

The problem states that the car travels 8 kilometers per hour faster than the truck. Since the two speeds are x and y,

$$x = y + 8.$$

Both vehicles travel for the same time, so

$$\frac{250}{x} = \frac{225}{y}.$$

4. (a) A grocer has some $4 per pound coffee and some $8 per pound coffee which she will mix to make 50 pounds of $5.60 per pound coffee. How many pounds of each should be used?

(b) Some 40% ethyl alcohol solution is to be mixed with some 80% solution to get 200 liters of a 50% mixture. How many liters of each should be used?

ANSWERS
4. (a) 30 pounds of $4; 20 pounds of $8
   (b) 150 liters of 40%; 50 liters of 80%

CHAPTER 8 SYSTEMS OF LINEAR EQUATIONS

5. A train travels 600 miles in the same time that a truck travels 520 miles. Find the speed of each vehicle if the train's average speed is 8 miles per hour faster than the truck's.

This is not a linear equation. However, multiplying both sides by xy gives

$$250y = 225x,$$

which is linear. Now solve the system.

$$x = y + 8 \quad (3)$$
$$250y = 225x \quad (4)$$

The substitution method can be used. Replace x with $y + 8$ in equation (4).

$$250y = 225(y + 8) \quad \text{Let } x = y + 8.$$
$$250y = 225y + 1800 \quad \text{Distributive property}$$
$$25y = 1800$$
$$y = 72$$

Since $x = y + 8$, the value of x is $72 + 8 = 80$. It is important to check the solution in the original problem since one of the equations had variable denominators. Checking verifies that the speeds are 80 kilometers per hour for the car and 72 kilometers per hour for the truck. ∎

■ **WORK PROBLEM 5 AT THE SIDE.**

5 Some applications involve three unknowns. When three variables are used, three equations are necessary to solve the problem. We can then use the methods of Section 8.2 to solve the system. The next two examples illustrate this.

EXAMPLE 5 Solving a Problem About Food Prices

Joe Schwartz bought apples, hamburger, and milk at the grocery store. Apples cost $.70 a pound, hamburger was $1.50 a pound, and milk was $.80 a quart. He bought twice as many pounds of apples as hamburger. The number of quarts of milk was one more than the number of pounds of hamburger. If his total bill was $8.20, how much of each item did he buy?

First choose variables to represent the three unknowns.

Let x = the number of pounds of apples;
 y = the number of pounds of hamburger;
 z = the number of quarts of milk.

Next, use the information in the problem to write three equations. Since Joe bought twice as many pounds of apples as hamburger,

$$x = 2y$$

or

$$x - 2y = 0.$$

The number of quarts of milk amounted to one more than the number of pounds of hamburger, so

$$z = 1 + y$$

or

$$-y + z = 1.$$

Multiplying the cost of each item by the amount of that item and adding gives the total bill.

$$.70x + 1.50y + .80z = 8.20$$

ANSWER
5. The train travels at 60 miles per hour and the truck at 52 miles per hour.

Multiply both sides of this equation by 10 to clear it of decimals.

$$7x + 15y + 8z = 82$$

Use the method shown in the last section to solve the system

$$\begin{aligned} x - 2y &= 0 \\ -y + z &= 1 \\ 7x + 15y + 8z &= 82. \end{aligned}$$

Verify that the solution is (4, 2, 3). Now go back to the statements defining the variables to decide what the numbers of the solution represent. Doing this shows that Joe bought 4 pounds of apples, 2 pounds of hamburger, and 3 quarts of milk. ∎

WORK PROBLEM 6 AT THE SIDE.

Business problems involving production sometimes require the solution of a system of equations. The final example shows how to set up such a system.

EXAMPLE 6 Solving a Business Production Problem

A company produces three color television sets, models X, Y, and Z. Each model X set requires 2 hours of electronics work, 2 hours of assembly time, and 1 hour of finishing time. Each model Y requires 1, 3, and 1 hours of electronics, assembly, and finishing time, respectively. Each model Z requires 3, 2, and 2 hours of the same work, respectively. There are 100 hours available for electronics, 100 hours available for assembly, and 65 hours available for finishing per week. How many of each model should be produced each week if all available time must be used?

Let x = the number of model X produced per week;
  y = the number of model Y produced per week;
  z = the number of model Z produced per week.

A chart is useful for organizing the information in a problem of this type.

	Model X	Model Y	Model Z	Totals
Hours of electronics work	2	1	3	100
Hours of assembly time	2	3	2	100
Hours of finishing time	1	1	2	65

The x model X sets require $2x$ hours of electronics, the y model Y sets require $1y$ (or y) hours of electronics, and the z model Z sets require $3z$ hours of electronics. Since 100 hours are available for electronics,

$$2x + y + 3z = 100.$$

6. **(a)** Joann has $1, $5, and $10 bills. The number of $1 and $5 bills is the same. The number of $10 bills is 4 more than the number of $5 bills. The total value of the money is $184. How many of each type of bill does she have?

(b) A department store has three kinds of perfume: cheap, better, and best. It has 10 more bottles of cheap than better, and 3 fewer bottles of the best than better. Each bottle of cheap costs $8, better costs $15, and best costs $32. The total value of all the perfume is $589. How many bottles of each are there?

ANSWERS
6. (a) 9 each of $1 and $5; 13 of $10
  (b) 21 bottles of cheap, 11 of better, and 8 of best

7. A paper mill makes newsprint, bond, and copy machine paper. Each ton of newsprint requires 3 tons of recycled paper and 1 ton of wood pulp. Each ton of bond requires 2 tons of recycled paper, 4 tons of wood pulp, and 3 tons of rags. A ton of copy machine paper requires 2 tons of recycled paper, 3 tons of wood pulp, and 2 tons of rags. The mill has 4200 tons of recycled paper, 5800 tons of wood pulp, and 3900 tons of rags. How much of each kind of paper can be made from these supplies?

Similarly, from the fact that 100 hours are available for assembly,

$$2x + 3y + 2z = 100,$$

and the fact that 65 hours are available for finishing leads to the equation

$$x + y + 2z = 65.$$

Solve the system

$$2x + y + 3z = 100$$
$$2x + 3y + 2z = 100$$
$$x + y + 2z = 65$$

to find $x = 15$, $y = 10$, and $z = 20$. The company should produce 15 model X, 10 model Y, and 20 model Z sets per week. ■

Notice the advantage of setting up the chart in Example 6. By reading across, we can easily determine the coefficients and the constants in the system.

■ **WORK PROBLEM 7 AT THE SIDE.**

ANSWERS
7. 400 tons of newsprint, 900 tons of bond, and 600 tons of copy machine paper

8.3 EXERCISES

For each word problem in this exercise set, select variables to represent the unknowns, write equations using the variables, and solve the resulting system.

Solve each problem. See Example 1.

1. The perimeter of a triangle is 51 centimeters. Two sides have the same length. The remaining side is 25 centimeters less than twice the length of either of the other sides. Find the lengths of the sides of the triangle.

2. The perimeter of a rectangle is 80 inches. The width is 1 inch more than one-half the length. Find the length and width of the rectangle.

3. The side of a square is 4 centimeters longer than the side of an equilateral triangle. The perimeter of the square is 22 inches more than the perimeter of the triangle. Find the lengths of a side of the square and a side of the triangle.

4. The length of a rectangle is 7 feet more than the width. If the length were decreased by 3 feet and the width were increased by 2 feet, the perimeter would be 44 feet. Find the length and width of the original rectangle.

Solve each problem. See Examples 2 and 6.

5. Kenneth and Peggy are planning to move, and they need some cardboard boxes. They can buy 10 small and 20 large boxes for $65, or 6 small and 10 large boxes for $34. Find the cost of each size of box.

6. At the Chalmette Nut Shop, 6 pounds of peanuts and 12 pounds of cashews cost $60, while 3 pounds of peanuts and 4 pounds of cashews cost $22. Find the cost of each type of nut.

7. A factory makes use of two basic machines, A and B, which turn out two different products, yarn and thread. Each unit of yarn requires 1 hour on machine A and 2 hours on machine B, while each unit of thread requires 1 hour on A and 1 hour on B. Machine A runs 8 hours per day, while machine B runs 14 hours per day. How many units each of yarn and thread should the factory make to keep its machines running at capacity?

8. A company that makes personal computers has found that each standard model requires 4 hours to manufacture electronics and 2 hours for the case. The top-of-the-line model requires 5 hours for the electronics and 1.5 hours for the case. On a particular production run, the company has available 200 hours in the electronics department and 76 hours in the cabinet department. How many of each model can be made?

9. Harry bought 2 kilograms of dark clay and 3 kilograms of light clay, paying $13 for the clay. He later needed 1 kilogram of dark clay and 2 kilograms of light clay, costing $7 altogether. How much did he pay for each type of clay?

10. A biologist wants to grow two types of algae, green and brown. She has 15 kilograms of nutrient X and 26 kilograms of nutrient Y. A vat of green algae needs 2 kilograms of nutrient X and 3 kilograms of nutrient Y, while a vat of brown algae needs 1 kilogram of nutrient X and 2 kilograms of nutrient Y. How many vats of each type of algae should the biologist grow in order to use all the nutrients?

Solve each problem. See Examples 2 and 3.

11. Mabel Johnston bought apples and oranges at DeVille's Grocery. She bought 6 pounds of fruit. Oranges cost $.90 per pound, while apples cost $.70 per pound. If she spent a total of $5.20, how many pounds of each kind of fruit did she buy?

12. Dan Foley has been saving dimes and quarters. He has 50 coins in all. If the total value is $10.70, how many dimes and how many quarters does he have?

13. A teller at South Savings and Loan received a checking account deposit in twenty-dollar bills and fifty-dollar bills. She received a total of 35 bills, and the amount of the deposit was $1600. How many of each denomination were deposited?

14. Tickets to the senior play at Slidell High School cost $2.50 for general admission or $2.00 with a student identification. If 92 people paid to see a performance and $203 was collected, how many of each type of admission were sold?

15. A total of $6000 is invested, part at 8% simple interest and part at 4%. If the annual return from the two investments is the same, how much is invested at each rate?

Rate	Amount	Interest
8%	x	
4%	y	
Total	$6000	

16. An investor must invest a total of $15,000 in two accounts, one paying 8% annual simple interest, and the other 7%. If he wants to earn $1100 annual interest, how much should he invest at each rate?

Rate	Amount	Interest
8%	x	
7%	y	
Total	$15,000	$1100

17. Teresita de Zayas invested $56,000, part at 14% annual simple interest and the rest at 7%. Her annual interest from the two investments was the same as if she had invested the entire amount at 10%. How much did she invest at each rate?

18. Terry Watkins has money invested in two accounts. He has $3000 more invested at 7% annual simple interest than he has at 5%. If he receives a total of $450 annual interest from the two investments, how much does he have invested at each rate?

19. Pure acid is to be added to a 10% acid solution to obtain 9 liters of a 20% acid solution. What amounts of each should be used?

Kind	Amount	Pure acid
100%	x	
10%	y	
20%	9	

20. How many gallons each of 25% alcohol and 35% alcohol should be mixed to get 10 gallons of 32% alcohol?

Kind	Amount	Pure alcohol
25%	x	
35%	y	
32%	10	

21. A truck radiator holds 24 liters of fluid. How much pure antifreeze must be added to a mixture that is 4% antifreeze in order to fill the radiator with a mixture that is 20% antifreeze?

22. A grocer plans to mix candy that sells for $1.20 a pound with candy that sells for $2.40 a pound to get a mixture that he plans to sell for $1.65 a pound. How much of the $1.20 and $2.40 candy should he use if he wants 20 pounds of the mix?

CHAPTER 8 SYSTEMS OF LINEAR EQUATIONS

23. A party mix is made by adding nuts that sell for $2.50 a kilogram to a cereal mixture that sells for $1 a kilogram. How much of each should be added to get 60 kilograms of a mix that will sell for $1.70 a kilogram?

24. A popular fruit drink is made by mixing fruit juice and soda. Such a mixture with 50% juice is to be mixed with another mixture that is 30% juice to get 100 liters of a mixture which is 45% juice. How much of each should be used?

Solve each problem. See Example 4.

25. A freight train and an express train leave towns 390 kilometers apart, traveling toward one another. The express train travels 30 kilometers per hour faster than the freight train. They pass one another 3 hours later. What are their speeds?

26. Two cars start out from the same spot and travel in opposite directions. At the end of 6 hours, they are 690 kilometers apart. If one car travels 15 kilometers per hour faster than the other, what are their speeds?

27. A train travels 150 kilometers in the same time that a plane covers 400 kilometers. If the speed of the plane is 20 kilometers per hour less than 3 times the speed of the train, find both speeds.

28. In his motorboat, Tri travels upstream at top speed to his favorite fishing spot, a distance of 18 miles, in 1 hour. Returning, he finds that the trip downstream, still at top speed, takes only $\frac{3}{4}$ hour. What is the speed of the current? What is the speed of Tri's boat in still water? (*Hint:* Let b represent the speed of the boat in still water and let c represent the speed of the current.)

	d	r	t
Upstream		$b - c$	
Downstream		$b + c$	

Solve each problem involving three unknowns. See Examples 5 and 6.

29. Find three numbers whose sum is 32, if the second is 1 more than the first, and the first is 4 less than the third.

30. Find three numbers whose sum is 8, if the first number is 26 less than the sum of the second and third, and the second number is 11 more than the first.

31. The sum of the measures of the angles of a triangle is 180°. The measure of the largest angle is 12° less than the sum of the measures of the other two. The smallest angle measures 58° less than the largest. Find the measures of the angles.

32. The sum of the measures of the angles of a triangle is 180°. In a certain triangle, the measure of the second angle is 10° more than three times the first. The third angle measure is equal to the sum of the measures of the other two. Find the measures of the three angles.

33. The perimeter of a triangle is 56 inches. The longest side measures 12 inches less than the sum of the other two sides. Three times the shortest side is 26 inches more than the longest side. Find the lengths of the three sides.

34. The perimeter of a triangle is 70 centimeters. The longest side is 6 centimeters less than the sum of the other two sides. Twice the shortest side is 4 centimeters more than the longest side. Find the length of each side of the triangle.

35. Kasey has a collection of nickels, dimes, and quarters. She has 10 more dimes than nickels, and there are 4 less quarters than dimes and nickels together. The total value of the coins is $7.05. How many of each coin does she have?

36. Marin has some fives, tens, and twenties. Altogether she has 27 bills worth $255. There are 6 less fives than tens. Find the number of each type of bill she has.

37. To meet his sales quota, Jesus Barreto must sell 20 new cars. He must sell 4 more sub-compact cars than compact cars, and one more full-size car than compact cars. How many of each type must he sell?

38. A manufacturer supplies three wholesalers, A, B, and C. The output from a day's production is 160 cases of goods. She must send wholesaler A three times as many cases as she sends B, and she must send wholesaler C 80 cases less than she provides A and B together. How many cases should she send to each wholesaler to distribute the entire day's production to them?

39. Three kinds of tickets are available for a Rhonda Rock concert: "up close," "in the middle," and "far out." "Up close" tickets cost $2 more than "in the middle" tickets, while "in the middle" tickets cost $1 more than "far out" tickets. Twice the cost of an "up close" ticket is $1 less than 3 times the cost of a "far out" seat. Find the price of each kind of ticket.

40. A shop manufactures three kinds of bolts, types A, B, and C. Production restrictions require them to make 5 units more type C bolts than the total of the other types and twice as many type B bolts as type A. The shop must produce a total of 245 units of bolts per day. How many units of each type can be made per day?

41. Nancy Dunn has inherited $80,000 from her uncle. She invests part of the money in a video rental firm which produces a return of 7% per year, and divides the rest equally between a tax-free bond at 6% a year and a money market fund at 12% a year. Her annual return on these investments is $6800. How much is invested in each?

42. Jeff's Rental Properties, Inc., borrowed in three loans for major renovations. The company borrowed a total of $75,000. Some of the money was borrowed at 8% interest, and $30,000 more than that amount was borrowed at 10%. The rest was borrowed at 7%. How much was borrowed at each rate if the total annual simple interest was $6950?

43. The owner of a tea shop wants to mix three kinds of tea to make 100 ounces of a mixture that will sell for $.83 an ounce. He uses Orange Pekoe, which sells for $.80 an ounce, Irish Breakfast, for $.85 an ounce, and Earl Grey, for $.95 an ounce. If he wants to use twice as much Orange Pekoe as Irish Breakfast, how much of each kind of tea should he use?

44. The manager of a candy store wants to feature a special Easter candy mixture of jelly beans, small chocolate eggs, and marshmallow chicks. She plans to make 15 pounds of mix to sell at $1 a pound. Jelly beans sell for $.80 a pound, chocolate eggs for $2 a pound, and marshmallow chicks for $1 a pound. She will use twice as many pounds of jelly beans as eggs and chicks combined and fives times as many pounds of jelly beans as chocolate eggs. How many pounds of each candy should she use?

SKILL SHARPENERS

Find the value of each expression. See Section 1.3.

45. $(-4)(-2) + (2)(-3) - (-1)(-6)$

46. $(-3)(1) - 2(4) - (-1)(-5)$

47. $2[(3)(2) - (-2)(1)]$

48. $3[(-2)(2) - (-3)(1)]$

49. $[(5)(-4) + (-1)(-6)] + [(-2)(-4) - (-1)(3)]$

50. $[(-1)(4) + (-4)(-2)] - [(-2)(3) - (-3)(1)]$

8.4 DETERMINANTS

A method of solving linear systems by using determinants will be introduced in Section 8.5. An ordered array of numbers is called a *matrix*. A *square matrix* is one that has the same number of rows as columns. Associated with every square matrix is a real number called the **determinant** of the matrix. A determinant is symbolized by the entries of the matrix placed between two vertical lines, such as

$$\begin{vmatrix} 2 & 3 \\ 7 & 1 \end{vmatrix} \quad \text{or} \quad \begin{vmatrix} 7 & 4 & 3 \\ 0 & 1 & 5 \\ -6 & 0 & -1 \end{vmatrix}.$$

Determinants are named according to the number of rows and columns they contain. For example, the first determinant above is a 2 × 2 (read "two by two") determinant and the second is a 3 × 3 determinant.

1 The value of the 2 × 2 determinant

$$\begin{vmatrix} a & b \\ c & d \end{vmatrix}$$

is defined as follows.

VALUE OF A 2 × 2 DETERMINANT

$$\begin{vmatrix} a & b \\ c & d \end{vmatrix} = ad - bc.$$

EXAMPLE 1 Evaluating a 2 × 2 Determinant
Evaluate the determinant.

$$\begin{vmatrix} -1 & -3 \\ 4 & -2 \end{vmatrix}$$

Here $a = -1$, $b = -3$, $c = 4$, and $d = -2$. Using these values, the determinant equals

$$\begin{vmatrix} -1 & -3 \\ 4 & -2 \end{vmatrix} = (-1)(-2) - (-3)(4)$$
$$= 2 + 12 = 14. \quad \blacksquare$$

WORK PROBLEM 1 AT THE SIDE.

A 3 × 3 determinant can be evaluated in a similar way.

VALUE OF A 3 × 3 DETERMINANT

$$\begin{vmatrix} a_1 & b_1 & c_1 \\ a_2 & b_2 & c_2 \\ a_3 & b_3 & c_3 \end{vmatrix} = (a_1 b_2 c_3 + b_1 c_2 a_3 + c_1 a_2 b_3) - (a_3 b_2 c_1 + b_3 c_2 a_1 + c_3 a_2 b_1)$$

OBJECTIVES

1 Evaluate 2 × 2 determinants.

2 Use expansion by minors about the first column to evaluate 3 × 3 determinants.

3 Use expansion by minors about any row or column to evaluate determinants.

4 Evaluate larger determinants.

---

1. Evaluate each determinant.

(a) $\begin{vmatrix} -4 & 6 \\ 2 & 3 \end{vmatrix}$

(b) $\begin{vmatrix} 3 & -1 \\ 0 & 2 \end{vmatrix}$

(c) $\begin{vmatrix} -2 & 5 \\ 1 & 5 \end{vmatrix}$

ANSWERS
1. (a) −24 (b) 6 (c) −15

CHAPTER 8 SYSTEMS OF LINEAR EQUATIONS

2. Expand by minors about the first column.

(a) $\begin{vmatrix} 2 & 1 & 4 \\ -3 & 0 & 2 \\ -2 & 1 & 5 \end{vmatrix}$

(b) $\begin{vmatrix} 0 & -1 & 0 \\ 2 & 4 & 2 \\ 3 & 1 & 5 \end{vmatrix}$

This rule for evaluating a 3 × 3 determinant is difficult to remember. A method for calculating a 3 × 3 determinant that is easier to use is based on the rule above. Rearranging terms and factoring gives

$$\begin{vmatrix} a_1 & b_1 & c_1 \\ a_2 & b_2 & c_2 \\ a_3 & b_3 & c_3 \end{vmatrix} = a_1(b_2c_3 - b_3c_2) - a_2(b_1c_3 - b_3c_1) + a_3(b_1c_2 - b_2c_1). \quad (1)$$

Each of the quantities in parentheses represents a 2 × 2 determinant which is that part of the 3 × 3 determinant remaining when the row and column of the multiplier are eliminated, as shown below.

$$a_1(b_2c_3 - b_3c_2) \quad \begin{vmatrix} a_1 & b_1 & c_1 \\ a_2 & b_2 & c_2 \\ a_3 & b_3 & c_3 \end{vmatrix}$$

$$a_2(b_1c_3 - b_3c_1) \quad \begin{vmatrix} a_1 & b_1 & c_1 \\ a_2 & b_2 & c_2 \\ a_3 & b_3 & c_3 \end{vmatrix}$$

$$a_3(b_1c_2 - b_2c_1) \quad \begin{vmatrix} a_1 & b_1 & c_1 \\ a_2 & b_2 & c_2 \\ a_3 & b_3 & c_3 \end{vmatrix}$$

These 2 × 2 determinants are called **minors** of the elements in the 3 × 3 determinant. In the determinant above, the minors of a_1, a_2, and a_3 are, respectively,

$$\begin{vmatrix} b_2 & c_2 \\ b_3 & c_3 \end{vmatrix}, \quad \begin{vmatrix} b_1 & c_1 \\ b_3 & c_3 \end{vmatrix}, \quad \begin{vmatrix} b_1 & c_1 \\ b_2 & c_2 \end{vmatrix}.$$

2 A 3 × 3 determinant can be evaluated by multiplying each element in the first column by its minor and combining the products as indicated in equation (1). This is called the **expansion of the determinant by minors** about the first column.

EXAMPLE 2 Evaluating a 3 × 3 Determinant

Evaluate the determinant by expansion by minors about the first column.

$$\begin{vmatrix} 1 & 3 & -2 \\ -1 & -2 & -3 \\ 1 & 1 & 2 \end{vmatrix} = 1 \begin{vmatrix} -2 & -3 \\ 1 & 2 \end{vmatrix} - (-1) \begin{vmatrix} 3 & -2 \\ 1 & 2 \end{vmatrix} + 1 \begin{vmatrix} 3 & -2 \\ -2 & -3 \end{vmatrix}$$

$$= 1[(-2)(2) - (-3)(1)] + 1[(3)(2) - (-2)(1)]$$
$$\quad + 1[(3)(-3) - (-2)(-2)]$$
$$= 1(-1) + 1(8) + 1(-13)$$
$$= -1 + 8 - 13$$
$$= -6 \quad ■$$

WORK PROBLEM 2 AT THE SIDE.

ANSWERS
2. (a) −5 (b) 4

3 To get equation (1) we could have rearranged terms in the definition of the determinant and factored out the three elements of the second or third columns or of any of the three rows. Therefore, expanding by minors about any row or any column results in the same value for a 3 × 3 determinant. To determine the correct signs for the terms of other expansions, the following **array of signs** is helpful.

ARRAY OF SIGNS FOR A 3 × 3 DETERMINANT

$$\begin{array}{ccc} + & - & + \\ - & + & - \\ + & - & + \end{array}$$

The signs alternate for each row and column beginning with a + in the first row, first column position. For example, if the expansion is to be about the second column, the first term would have a minus sign associated with it, the second term a plus sign, and the third term a minus sign.

EXAMPLE 3 Evaluating a 3 × 3 Determinant

Evaluate the determinant of Example 2 by expansion by minors about the second column.

$$\begin{vmatrix} 1 & \mathbf{3} & -2 \\ -1 & \mathbf{-2} & -3 \\ 1 & \mathbf{1} & 2 \end{vmatrix}$$

$$= -\mathbf{3}\begin{vmatrix} -1 & -3 \\ 1 & 2 \end{vmatrix} + \mathbf{(-2)}\begin{vmatrix} 1 & -2 \\ 1 & 2 \end{vmatrix} - \mathbf{1}\begin{vmatrix} 1 & -2 \\ -1 & -3 \end{vmatrix}$$

$$= -3(1) - 2(4) - 1(-5)$$

$$= -3 - 8 + 5$$

$$= -6$$

As expected, the result is the same as in Example 2. ∎

WORK PROBLEM 3 AT THE SIDE.

4 The method of expansion by minors can be extended to evaluate larger determinants such as 4 × 4 or 5 × 5. For a larger determinant, the sign array also is extended. For example, the signs for a 4 × 4 determinant are arranged as follows.

ARRAY OF SIGNS FOR A 4 × 4 DETERMINANT

$$\begin{array}{cccc} + & - & + & - \\ - & + & - & + \\ + & - & + & - \\ - & + & - & + \end{array}$$

3. Evaluate each determinant by expansion by minors about the second column.

(a) $\begin{vmatrix} 2 & 1 & 3 \\ -1 & 0 & 4 \\ 2 & 4 & 3 \end{vmatrix}$

(b) $\begin{vmatrix} 5 & -1 & 2 \\ 0 & 4 & 3 \\ -1 & 2 & 0 \end{vmatrix}$

ANSWERS
3. (a) −33 (b) −19

CHAPTER 8 SYSTEMS OF LINEAR EQUATIONS

4. Evaluate.

$$\begin{vmatrix} 1 & 0 & 2 & 0 \\ 3 & 0 & 0 & 4 \\ 0 & -1 & 1 & 0 \\ 2 & 0 & -1 & 0 \end{vmatrix}$$

EXAMPLE 4 Evaluating a 4×4 Determinant

Evaluate the determinant below.

$$\begin{vmatrix} -1 & -2 & 3 & 2 \\ 0 & 1 & 4 & -2 \\ 3 & -1 & 4 & 0 \\ 2 & 1 & 0 & 3 \end{vmatrix}$$

The work can be reduced by choosing a row or column with zeros, say the fourth row. Expand by minors about the fourth row using the elements of the fourth row and the signs from the fourth row of the sign array, as shown below. The minors are 3×3 determinants.

$$\begin{vmatrix} -1 & -2 & 3 & 2 \\ 0 & 1 & 4 & -2 \\ 3 & -1 & 4 & 0 \\ \mathbf{2} & \mathbf{1} & \mathbf{0} & \mathbf{3} \end{vmatrix} = -\mathbf{2}\begin{vmatrix} -2 & 3 & 2 \\ 1 & 4 & -2 \\ -1 & 4 & 0 \end{vmatrix} + \mathbf{1}\begin{vmatrix} -1 & 3 & 2 \\ 0 & 4 & -2 \\ 3 & 4 & 0 \end{vmatrix}$$

$$-\mathbf{0}\begin{vmatrix} -1 & -2 & 2 \\ 0 & 1 & -2 \\ 3 & -1 & 0 \end{vmatrix} + \mathbf{3}\begin{vmatrix} -1 & -2 & 3 \\ 0 & 1 & 4 \\ 3 & -1 & 4 \end{vmatrix}$$

Now evaluate each 3×3 determinant.

$$= -2(6) + 1(-50) - 0 + 3(-41)$$
$$= -185 \quad \blacksquare$$

WORK PROBLEM 4 AT THE SIDE.

Each of the four 3×3 determinants in Example 4 is evaluated by expansion of three 2×2 minors. Thus, a great deal of work is needed to evaluate a 4×4 or larger determinant. However, such large determinants can be evaluated quickly with the aid of a computer.

ANSWER
4. 20

8.4 EXERCISES

Evaluate the following determinants. See Example 1.

1. $\begin{vmatrix} 2 & 5 \\ -1 & 4 \end{vmatrix}$
2. $\begin{vmatrix} 3 & 6 \\ 2 & -2 \end{vmatrix}$
3. $\begin{vmatrix} 1 & 3 \\ 7 & 2 \end{vmatrix}$
4. $\begin{vmatrix} 5 & -1 \\ 1 & 0 \end{vmatrix}$

5. $\begin{vmatrix} -2 & 2 \\ 2 & -2 \end{vmatrix}$
6. $\begin{vmatrix} 3 & -3 \\ 5 & 1 \end{vmatrix}$
7. $\begin{vmatrix} 0 & 1 \\ 0 & 1 \end{vmatrix}$
8. $\begin{vmatrix} 0 & 1 \\ 1 & 0 \end{vmatrix}$

Evaluate the following determinants by expansion by minors about the first column. See Example 2.

9. $\begin{vmatrix} 1 & 2 & 3 \\ -3 & -2 & -1 \\ 2 & -1 & 1 \end{vmatrix}$
10. $\begin{vmatrix} 2 & -3 & 5 \\ 1 & 2 & -1 \\ 5 & 3 & 1 \end{vmatrix}$
11. $\begin{vmatrix} 1 & 0 & 2 \\ 0 & 2 & 3 \\ 1 & 0 & 5 \end{vmatrix}$

12. $\begin{vmatrix} 2 & -1 & 0 \\ 0 & -1 & 1 \\ 1 & 2 & 3 \end{vmatrix}$
13. $\begin{vmatrix} 1 & 0 & 0 \\ 0 & 1 & 0 \\ 0 & 0 & 1 \end{vmatrix}$
14. $\begin{vmatrix} 0 & 0 & 1 \\ 0 & 1 & 0 \\ 1 & 0 & 0 \end{vmatrix}$

Evaluate the following determinants by expansion about any row or column. (Hint: Choose a row or column with zeros.) See Example 3.

15. $\begin{vmatrix} 2 & 2 & 1 \\ 1 & -1 & -2 \\ 1 & 0 & 2 \end{vmatrix}$
16. $\begin{vmatrix} 3 & -1 & 2 \\ 1 & 5 & -2 \\ 0 & 1 & 0 \end{vmatrix}$
17. $\begin{vmatrix} 2 & 0 & 3 \\ -1 & 0 & 3 \\ 5 & 0 & 7 \end{vmatrix}$

18. $\begin{vmatrix} 1 & 6 & -5 \\ -3 & 2 & 2 \\ 0 & 0 & 0 \end{vmatrix}$
19. $\begin{vmatrix} 2 & 4 & 0 \\ 3 & -5 & 0 \\ 6 & -7 & 0 \end{vmatrix}$
20. $\begin{vmatrix} 0 & 0 & 3 \\ 4 & 0 & -2 \\ 2 & -1 & 3 \end{vmatrix}$

21. $\begin{vmatrix} 1 & 1 & 2 \\ 5 & 5 & 7 \\ 3 & 3 & 1 \end{vmatrix}$
22. $\begin{vmatrix} 2 & 4 & -1 \\ 1 & 0 & 1 \\ 2 & 4 & -1 \end{vmatrix}$
23. $\begin{vmatrix} 3 & 0 & -2 \\ 1 & -4 & 1 \\ 3 & 1 & -2 \end{vmatrix}$

24. $\begin{vmatrix} 1 & 3 & 2 \\ 0 & -1 & -2 \\ 1 & 10 & 20 \end{vmatrix}$
25. $\begin{vmatrix} 1 & 3m & 2 \\ 0 & 2m & 4 \\ 1 & 10m & 20 \end{vmatrix}$
26. $\begin{vmatrix} 0 & 0 & 0 \\ a & 0 & b \\ c & d & e \end{vmatrix}$

Evaluate the following. Expand by minors about the second row. See Example 4.

27. $\begin{vmatrix} 1 & 4 & 2 & 0 \\ 0 & 2 & 0 & -1 \\ 3 & -1 & 2 & 0 \\ 1 & 4 & -1 & 2 \end{vmatrix}$

28. $\begin{vmatrix} 4 & 1 & 0 & 2 \\ 1 & 0 & 0 & -2 \\ 3 & 4 & 1 & -3 \\ -2 & 1 & 1 & -1 \end{vmatrix}$

29. $\begin{vmatrix} 3 & 5 & 1 & 9 \\ 0 & 5 & 2 & 0 \\ 2 & -1 & -1 & -1 \\ -4 & 2 & 2 & 2 \end{vmatrix}$

30. $\begin{vmatrix} 2 & 3 & 2 & 2 \\ 1 & -1 & 0 & 0 \\ 2 & 1 & 1 & 1 \\ 0 & 0 & -3 & -2 \end{vmatrix}$

Solve each equation involving determinants.

31. $\begin{vmatrix} 4 & 3 \\ 12 & x \end{vmatrix} = 8$

32. $\begin{vmatrix} x & x \\ 3 & x \end{vmatrix} = 4$

33. By expanding the determinant, show that the straight line through (x_1, y_1) and (x_2, y_2) has equation
$$\begin{vmatrix} x & y & 1 \\ x_1 & y_1 & 1 \\ x_2 & y_2 & 1 \end{vmatrix} = 0.$$

34. Write the equations in slope-intercept form and expand the determinant to show that the lines $a_1x + b_1y = c_1$ and $a_2x + b_2y = c_2$, when $c_1 \neq c_2$ are parallel if
$$\begin{vmatrix} a_1 & b_1 \\ a_2 & b_2 \end{vmatrix} = 0.$$

SKILL SHARPENERS
Evaluate the following expressions. See Section 1.4.

35. $\dfrac{3(-4) - 5(-6)}{-2(-3) - 4(-6)}$

36. $\dfrac{-9(6) - 4(-3)}{6(-1) - 2(-4)}$

37. $\dfrac{-5(-5) - 2(3)}{-7(-3) - (-4)(-5)}$

38. $\dfrac{4(9) - 3(5)}{3(-1) - 4(-2)}$

8.5 SOLUTION OF LINEAR SYSTEMS OF EQUATIONS BY DETERMINANTS—CRAMER'S RULE

OBJECTIVES

1. Understand the derivation of Cramer's rule.
2. Apply Cramer's rule to a linear system with two equations and two unknowns.
3. Apply Cramer's rule to a linear system with three equations and three unknowns.

1 The elimination method is used in this section to solve the general system of two equations with two variables,

$$a_1 x + b_1 y = c_1 \qquad (1)$$
$$a_2 x + b_2 y = c_2. \qquad (2)$$

The result will be a formula that can be used directly for any system of two equations with two unknowns. To get this general solution, eliminate y and solve for x by first multiplying both sides of equation (1) by b_2 and both sides of equation (2) by $-b_1$. Add these results and solve for x.

$$\begin{aligned} a_1 b_2 x + b_1 b_2 y &= c_1 b_2 \\ -a_2 b_1 x - b_1 b_2 y &= -c_2 b_1 \\ \hline (a_1 b_2 - a_2 b_1) x &= c_1 b_2 - c_2 b_1 \end{aligned}$$

b_2 times both sides of equation (1)
$-b_1$ times both sides of equation (2)

$$x = \frac{c_1 b_2 - c_2 b_1}{a_1 b_2 - a_2 b_1}, \qquad a_1 b_2 - a_2 b_1 \ne 0$$

To solve for y, multiply both sides of equation (1) by $-a_2$ and both sides of equation (2) by a_1 and add.

$$\begin{aligned} -a_1 a_2 x - a_2 b_1 y &= -a_2 c_1 \\ a_1 a_2 x + a_1 b_2 y &= a_1 c_2 \\ \hline (a_1 b_2 - a_2 b_1) y &= a_1 c_2 - a_2 c_1 \end{aligned}$$

$-a_2$ times both sides of equation (1)
a_1 times both sides of equation (2)

$$y = \frac{a_1 c_2 - a_2 c_1}{a_1 b_2 - a_2 b_1}, \qquad a_1 b_2 - a_2 b_1 \ne 0$$

Both numerators and the common denominator of these values for x and y can be written as determinants, since

$$a_1 c_2 - a_2 c_1 = \begin{vmatrix} a_1 & c_1 \\ a_2 & c_2 \end{vmatrix},$$

$$c_1 b_2 - c_2 b_1 = \begin{vmatrix} c_1 & b_1 \\ c_2 & b_2 \end{vmatrix},$$

and

$$a_1 b_2 - a_2 b_1 = \begin{vmatrix} a_1 & b_1 \\ a_2 & b_2 \end{vmatrix}.$$

Using these results, the solutions for x and y become

$$x = \frac{\begin{vmatrix} c_1 & b_1 \\ c_2 & b_2 \end{vmatrix}}{\begin{vmatrix} a_1 & b_1 \\ a_2 & b_2 \end{vmatrix}} \quad \text{and} \quad y = \frac{\begin{vmatrix} a_1 & c_1 \\ a_2 & c_2 \end{vmatrix}}{\begin{vmatrix} a_1 & b_1 \\ a_2 & b_2 \end{vmatrix}}, \qquad \begin{vmatrix} a_1 & b_1 \\ a_2 & b_2 \end{vmatrix} \ne 0.$$

For convenience, denote the three determinants in the solution as

$$\begin{vmatrix} a_1 & b_1 \\ a_2 & b_2 \end{vmatrix} = D, \qquad \begin{vmatrix} c_1 & b_1 \\ c_2 & b_2 \end{vmatrix} = D_x, \qquad \begin{vmatrix} a_1 & c_1 \\ a_2 & c_2 \end{vmatrix} = D_y.$$

The elements of D are the four coefficients of the variables in the given system; the elements of D_x are obtained by replacing the coefficients of x by the respective constants; the elements of D_y are obtained by replacing the coefficients of y by the respective constants.

1. Solve by Cramer's rule.

   (a) $x + y = 5$
   $x - y = 1$

   (b) $2x - 3y = -26$
   $3x + 4y = 12$

   (c) $3x + 7y = -6$
   $4x - 5y = -8$

These results are summarized as **Cramer's rule.**

CRAMER'S RULE FOR 2 × 2 SYSTEMS

Given the system
$$a_1x + b_1y = c_1$$
$$a_2x + b_2y = c_2 \quad \text{with} \quad a_1b_2 - a_2b_1 = D \neq 0,$$

then

$$x = \frac{\begin{vmatrix} c_1 & b_1 \\ c_2 & b_2 \end{vmatrix}}{\begin{vmatrix} a_1 & b_1 \\ a_2 & b_2 \end{vmatrix}} = \frac{D_x}{D} \quad \text{and} \quad y = \frac{\begin{vmatrix} a_1 & c_1 \\ a_2 & c_2 \end{vmatrix}}{\begin{vmatrix} a_1 & b_1 \\ a_2 & b_2 \end{vmatrix}} = \frac{D_y}{D}.$$

2 To use Cramer's rule to solve a system of equations, find the three determinants, D, D_x and D_y and then write the necessary quotients for x and y.

EXAMPLE 1 Using Cramer's Rule for a 2 × 2 System

Use Cramer's rule to solve the system

$$5x + 7y = -1$$
$$6x + 8y = 1.$$

By Cramer's rule, $x = D_x/D$ and $y = D_y/D$. It is a good idea to find D first, since if $D = 0$, Cramer's rule does not apply. If $D \neq 0$, then find D_x and D_y.

$$D = \begin{vmatrix} 5 & 7 \\ 6 & 8 \end{vmatrix} = 5(8) - 6(7) = -2;$$

$$D_x = \begin{vmatrix} -1 & 7 \\ 1 & 8 \end{vmatrix} = (-1)8 - 7(1) = -15;$$

$$D_y = \begin{vmatrix} 5 & -1 \\ 6 & 1 \end{vmatrix} = 5(1) - (-1)6 = 11.$$

From Cramer's rule,

$$x = \frac{D_x}{D} = \frac{-15}{-2} = \frac{15}{2},$$

and

$$y = \frac{D_y}{D} = \frac{11}{-2} = -\frac{11}{2}.$$

The solution set is $\left\{\left(\frac{15}{2}, -\frac{11}{2}\right)\right\}$, as can be verified by checking in the given system. ∎

WORK PROBLEM 1 AT THE SIDE.

ANSWERS
1. (a) {(3, 2)} (b) {(-4, 6)}
   (c) {(-2, 0)}

8.5 SOLUTION OF LINEAR SYSTEMS OF EQUATIONS BY DETERMINANTS—CRAMER'S RULE

3 In a similar manner, Cramer's rule can be applied to systems of three equations with three variables.

2. Find D_y and D_z.

CRAMER'S RULE FOR 3 × 3 SYSTEMS

Given the system
$$a_1x + b_1y + c_1z = d_1$$
$$a_2x + b_2y + c_2z = d_2$$
$$a_3x + b_3y + c_3z = d_3$$

with

$$D_x = \begin{vmatrix} d_1 & b_1 & c_1 \\ d_2 & b_2 & c_2 \\ d_3 & b_3 & c_3 \end{vmatrix}, \quad D_y = \begin{vmatrix} a_1 & d_1 & c_1 \\ a_2 & d_2 & c_2 \\ a_3 & d_3 & c_3 \end{vmatrix},$$

$$D_z = \begin{vmatrix} a_1 & b_1 & d_1 \\ a_2 & b_2 & d_2 \\ a_3 & b_3 & d_3 \end{vmatrix}, \quad D = \begin{vmatrix} a_1 & b_1 & c_1 \\ a_2 & b_2 & c_2 \\ a_3 & b_3 & c_3 \end{vmatrix} \neq 0,$$

then

$$x = \frac{D_x}{D}, \quad y = \frac{D_y}{D}, \quad z = \frac{D_z}{D}.$$

EXAMPLE 2 Using Cramer's Rule for a 3 × 3 System

Use Cramer's rule to solve the system

$$x + y - z + 2 = 0$$
$$2x - y + z + 5 = 0$$
$$x - 2y + 3z - 4 = 0.$$

To use Cramer's rule, first rewrite the system in the form

$$x + y - z = -2$$
$$2x - y + z = -5$$
$$x - 2y + 3z = 4.$$

Expand by minors about row 1 to find D.

$$D = \begin{vmatrix} 1 & 1 & -1 \\ 2 & -1 & 1 \\ 1 & -2 & 3 \end{vmatrix} = 1\begin{vmatrix} -1 & 1 \\ -2 & 3 \end{vmatrix} - 1\begin{vmatrix} 2 & 1 \\ 1 & 3 \end{vmatrix} + (-1)\begin{vmatrix} 2 & -1 \\ 1 & -2 \end{vmatrix}$$

$$= 1(-1) - 1(5) - 1(-3) = -3$$

Expanding D_x by minors about row 1 gives

$$D_x = \begin{vmatrix} -2 & 1 & -1 \\ -5 & -1 & 1 \\ 4 & -2 & 3 \end{vmatrix}$$

$$= -2\begin{vmatrix} -1 & 1 \\ -2 & 3 \end{vmatrix} - 1\begin{vmatrix} -5 & 1 \\ 4 & 3 \end{vmatrix} + (-1)\begin{vmatrix} -5 & -1 \\ 4 & -2 \end{vmatrix}$$

$$= -2(-1) - 1(-19) - 1(14) = 7.$$

WORK PROBLEM 2 AT THE SIDE.

ANSWERS
2. $D_y = -22$ and $D_z = -21$

CHAPTER 8 SYSTEMS OF LINEAR EQUATIONS

3. Solve by Cramer's rule.

   (a) $x + y + z = 2$
   $2x - z = -3$
   $ y + 2z = 4$

Using the results for D and D_x and the results from Problem 2 at the side, apply Cramer's rule to get

$$x = \frac{D_x}{D} = \frac{7}{-3} = -\frac{7}{3}, \quad y = \frac{D_y}{D} = \frac{-22}{-3} = \frac{22}{3},$$

$$z = \frac{D_z}{D} = \frac{-21}{-3} = 7.$$

The solution set is $\left\{\left(-\frac{7}{3}, \frac{22}{3}, 7\right)\right\}$. ∎

■ WORK PROBLEM 3 AT THE SIDE.

(b) $3x - 2y + 4z = 5$
$4x + y + z = 14$
$x - y - z = 1$

EXAMPLE 3 Determining When Cramer's Rule Does Not Apply

Use Cramer's rule to solve the following system.

$$2x - 3y + 4z = 8$$
$$6x - 9y + 12z = 24$$
$$x + 2y - 3z = 5$$

Find D, D_x, D_y, and D_z. Here

$$D = \begin{vmatrix} 2 & -3 & 4 \\ 6 & -9 & 12 \\ 1 & 2 & -3 \end{vmatrix} = 2\begin{vmatrix} -9 & 12 \\ 2 & -3 \end{vmatrix} - 6\begin{vmatrix} -3 & 4 \\ 2 & -3 \end{vmatrix} + 1\begin{vmatrix} -3 & 4 \\ -9 & 12 \end{vmatrix}$$

$$= 2(3) - 6(1) + 1(0)$$

$$= 0.$$

4. Solve by Cramer's rule where applicable.

   (a) $x - y + z = 6$
   $3x + 2y + z = 4$
   $2x - 2y + 2z = 14$

As mentioned previously, Cramer's rule does not apply if $D = 0$. When $D = 0$, the system is inconsistent or contains equations that are dependent. To determine which situation is true, use the elimination method. Multiplying the first equation on both sides by 3 shows that the first two equations have the same solutions, so this system has dependent equations. ∎

■ WORK PROBLEM 4 AT THE SIDE.

Cramer's rule can be extended to 4×4 or larger systems. See a standard college algebra text for details.

(b) $2x - y + z = 0$
$3x + 4y - 2z = 0$
$5x - y + 6z = 0$

ANSWERS
3. (a) $\{(-1, 2, 1)\}$ (b) $\{(3, 2, 0)\}$
4. (a) Cramer's rule does not apply.
   (b) $\{(0, 0, 0)\}$

8.5 EXERCISES

Use Cramer's rule where applicable to solve the following linear systems of two variables. See Example 1.

1. $8x - 4y = 8$
   $x + 3y = 22$

2. $3x + 5y = -5$
   $2x - 3y = -16$

3. $4x + 3y = 0$
   $3x - 4y = 25$

4. $x + 5y = 6$
   $2x + 3y = -2$

5. $3x + 8y = 3$
   $2x - 4y = 2$

6. $6x + 2y = -10$
   $-3x + 5y = 11$

7. $5x - 4y = 11$
   $-3x + 3y = -6$

8. $6x - y = 9$
   $x - 6y = 19$

9. $3x + 5y = -21$
   $-4x - 2y = 14$

Use Cramer's rule where applicable to solve the following linear systems. See Examples 2 and 3.

10. $2x + 3y + 2z = 10$
    $x - y + 2z = 3$
    $x + 2y - 6z = -15$

11. $x - y + 6z = 14$
    $3x + 3y - z = 10$
    $x + 9y + 2z = 16$

12. $2x + 2y + z = 6$
    $4x - y + z = 12$
    $-x + y - 2z = -3$

13. $2x + 2y + z = 1$
    $x + 3y + 2z = 5$
    $x - y - z = 6$

14. $2x - 3y + 4z = 8$
    $6x - 9y + 12z = 24$
    $-4x + 6y - 8z = -16$

15. $x + y - z = 0$
    $2x - 3y + z = 0$
    $-3x + 2y - 3z = 0$

CHAPTER 8 SYSTEMS OF LINEAR EQUATIONS

16.
$x + 2y - z = 0$
$2x - y + z = 0$
$3x + 2y - 2z = 0$

17.
$x + 2y + 3z = 1$
$-x - y + 3z = 2$
$6x - y - z = 2$

18.
$x - 3y = 13$
$2y + z = 5$
$x - z = 7$

19.
$3x + 5z = 0$
$2x + 3y = 1$
$y - 2z = 11$

20.
$x - 2y = -4$
$3x + y = -5$
$2x + z = -1$

21.
$5x + 2z = -20$
$x + 3y = -1$
$z = 1$

22.
$x + 2y - z + w = 8$
$2x - y - w = 12$
$y + 3z = 11$
$x - z = 4$

23.
$3x + 2y - w = 0$
$2x + z + 2w = 5$
$x + 2y - z = -2$
$2x - y + z + w = 2$

SKILL SHARPENERS

For each function, find $f(3)$ and $f(-2)$. See Section 7.5.

24. $f(x) = 3x - 9$

25. $f(x) = -4x + 3$

26. $f(x) = x^2 + 3x - 4$

27. $f(x) = -2x^2 - 5x + 7$

28. $f(x) = \dfrac{1}{x^2 + 1}$

29. $f(x) = \dfrac{-3}{x^3 + 2}$

Historical Reflections

Galileo Galilei (1564–1642)

Galileo died in the year Sir Isaac Newton was born; his work was important in Newton's development of the calculus. He did more than construct theories to explain physical phenomena—he set up experiments to test his ideas. According to legend, Galileo dropped objects of different weights from the tower of Pisa to disprove the Aristotelian view that heavier objects fall faster than lighter objects. He developed a formula for freely falling objects that stated

$$d = 16t^2$$

where d is the distance in feet that an object falls (neglecting air resistance) in a given time t, in seconds, regardless of weight.

During Galileo's lifetime his findings met strong resistance, and despite the fact that he was a devout Catholic, he was condemned by the Church for using telescopes to prove that the earth revolves around the sun. In 1980, the Vatican began to review Galileo's heresy conviction.

Art: Scripta Mathematica, Yeshiva University, New York

CHAPTER 8 SUMMARY

KEY TERMS

8.1 **system of equations** — Two or more equations that are to be solved at the same time form a system of equations.

linear system — A linear system is a system of equations that contains only linear equations.

inconsistent system — A system is inconsistent if it has no solution.

dependent equations — Dependent equations are equations whose graphs are the same line.

elimination (or addition) method — The elimination (or addition) method of solving a system of equations involves the elimination of a variable by adding the two equations.

substitution method — The substitution method of solving a system involves substituting an expression for one variable in terms of another.

8.4 **determinant** — Associated with every square matrix is a real number called its determinant, symbolized by the entries of the matrix between two vertical lines.

8.5 **Cramer's rule** — Cramer's rule is a method of solving a system of equations using determinants.

NEW SYMBOLS

(x, y, z) ordered triple

$\begin{vmatrix} a & b \\ c & d \end{vmatrix}$ 2 × 2 determinant

$\begin{vmatrix} a & b & c \\ d & e & f \\ g & h & i \end{vmatrix}$ 3 × 3 determinant

QUICK REVIEW

Section	Concepts	Examples
8.1 Linear Systems of Equations in Two Variables	**Solving Linear Systems of Two Equations by Elimination** *Step 1* Write both equations in the form $ax + by = c$. *Step 2* Multiply one or both equations by appropriate numbers so that the sum of the coefficients of either x or y is zero. *Step 3* Add the new equations. The sum should be an equation with just one variable. *Step 4* Solve the equation from Step 3. *Step 5* Substitute the result of Step 4 into either of the given equations and solve for the other variable. *Step 6* Check the solution in both of the given equations.	Solve by elimination. $5x + y = 2$ $2x - 3y = 11$ To eliminate y, multiply the top equation by 3, and add. $15x + 3y = 6$ $2x - 3y = 11$ $\overline{17x = 17}$ $x = 1$ Let $x = 1$ in the top equation, and solve for y. $5(1) + y = 2$ $y = -3$ Check to verify that $\{(1, -3)\}$ is the solution set.

CHAPTER 8 SUMMARY

Section	Concepts	Examples
	Solving Linear Systems of Two Equations by Substitution	Solve by substitution. $$4x - y = 7$$ $$3x + 2y = 30$$
	Step 1 Solve one of the equations for either variable.	Solve for y in the top equation. $$y = 4x - 7$$
	Step 2 Substitute for that variable in the other equation. The result should be an equation with just one variable.	Substitute $4x - 7$ for y in the bottom equation, and solve for x. $$3x + 2(4x - 7) = 30$$ $$3x + 8x - 14 = 30$$ $$11x - 14 = 30$$
	Step 3 Solve the equation from Step 2.	$$11x = 44$$ $$x = 4$$
	Step 4 Substitute the result from Step 3 into the equation from Step 1 to find the value of the other variable.	Substitute 4 for x in the equation $y = 4x - 7$ to find that $y = 9$.
	Step 5 Check the solution in both of the given equations.	Check to see that $\{(4, 9)\}$ is the solution set.
8.2 Linear Systems of Equations in Three Variables	**Solving Linear Systems in Three Variables**	Solve the system $$x + 2y - z = 6$$ $$x + y + z = 6$$ $$2x + y - z = 7.$$
	Step 1 Use the elimination method to eliminate any variable from any two of the given equations. The result is an equation in two variables.	Add the first and second equations; z is eliminated and the result is $2x + 3y = 12$.
	Step 2 Eliminate the *same* variable from any *other* two equations. The result is an equation in the same two variables as in Step 1.	Eliminate z again by adding the second and third equations to get $3x + 2y = 13$. Now solve the system $$2x + 3y = 12 \quad (*)$$ $$3x + 2y = 13.$$
	Step 3 Use the elimination method to eliminate a second variable from the two equations in two variables that result from Steps 1 and 2. The result is an equation in one variable that gives the value of that variable.	To eliminate x, multiply the top equation by -3 and the bottom equation by 2. $$-6x - 9y = -36$$ $$\underline{6x + 4y = 26}$$ $$-5y = -10$$ $$y = 2$$

513

CHAPTER 8 SYSTEMS OF LINEAR EQUATIONS

Section	Concepts	Examples
	Solving Linear Systems in Three Variables (continued)	
	Step 4 Substitute the value of the variable found in Step 3 into either of the equations in two variables to find the value of the second variable.	Let $y = 2$ in equation (*). $$2x + 3(2) = 12$$ $$2x + 6 = 12$$ $$2x = 6$$ $$x = 3$$
	Step 5 Use the values of the two variables from Steps 3 and 4 to find the value of the third variable by substituting into any of the original equations.	Let $y = 2$ and $x = 3$ in any of the original equations to find $z = 1$. The solution set is $\{(3, 2, 1)\}$.
8.3 Applications of Linear Systems	To solve a word problem with two (three) unknowns, write two (three) equations that relate the unknowns. Then solve the system.	The perimeter of a rectangle is 18 feet. The length is 3 feet more than twice the width. Find the dimensions of the rectangle. Let x represent the length and y represent the width. From the perimeter formula, one equation is $2x + 2y = 18$. From the problem, another equation is $x = 3 + 2y$. Now solve the system $$2x + 2y = 18$$ $$x = 3 + 2y.$$ The solution of the system is (7, 2). Therefore, the length is 7 feet and the width is 2 feet.

CHAPTER 8 SUMMARY

Section	Concepts	Examples
8.4 Determinants	**Value of a 2 × 2 Determinant** $$\begin{vmatrix} a & b \\ c & d \end{vmatrix} = ad - bc.$$ Determinants larger than 2 × 2 are evaluated by expansion by minors about a column or row.	$\begin{vmatrix} 3 & 4 \\ -2 & 6 \end{vmatrix} = (3)(6) - (4)(-2) = 26$
	Array of Signs for a 3 × 3 Determinant $$\begin{matrix} + & - & + \\ - & + & - \\ + & - & + \end{matrix}$$	Evaluate $\begin{vmatrix} 2 & -3 & -2 \\ -1 & -4 & -3 \\ -1 & 0 & 2 \end{vmatrix}$ by expanding about the second column. $\begin{vmatrix} 2 & -3 & -2 \\ -1 & -4 & -3 \\ -1 & 0 & 2 \end{vmatrix}$ $= 3(-5) + (-4)(2) - (0)(-8)$ $= -15 - 8 + 0$ $= -23$
	Array of Signs for a 4 × 4 Determinant $$\begin{matrix} + & - & + & - \\ - & + & - & + \\ + & - & + & - \\ - & + & - & + \end{matrix}$$	Evaluate $\begin{vmatrix} 2 & -1 & 4 & 1 \\ 0 & -2 & 3 & 7 \\ -4 & 1 & 0 & 2 \\ -2 & -1 & 3 & 5 \end{vmatrix}$ by expanding about the first column. Expanding about the first column gives $2\begin{vmatrix} -2 & 3 & 7 \\ 1 & 0 & 2 \\ -1 & 3 & 5 \end{vmatrix} - 0\begin{vmatrix} -1 & 4 & 1 \\ 1 & 0 & 2 \\ -1 & 3 & 5 \end{vmatrix}$ $+ (-4)\begin{vmatrix} -1 & 4 & 1 \\ -2 & 3 & 7 \\ -1 & 3 & 5 \end{vmatrix} - (-2)\begin{vmatrix} -1 & 4 & 1 \\ -2 & 3 & 7 \\ 1 & 0 & 2 \end{vmatrix}$ Evaluating the 3 × 3 determinants gives $2(12) - 0(-19) + (-4)(15) - (-2)(35)$ $= 34.$

CHAPTER 8 SYSTEMS OF LINEAR EQUATIONS

Section	Concepts	Examples
8.5 Solution of Linear Systems of Equations by Determinants—Cramer's Rule	**Cramer's Rule for 2 × 2 Systems** Given the system $$a_1x + b_1y = c_1$$ $$a_2x + b_2y = c_2$$ with $a_1b_2 - a_2b_1 = D \neq 0$, then $$x = \frac{\begin{vmatrix} c_1 & b_1 \\ c_2 & b_2 \end{vmatrix}}{\begin{vmatrix} a_1 & b_1 \\ a_2 & b_2 \end{vmatrix}} = \frac{D_x}{D}$$ and $$y = \frac{\begin{vmatrix} a_1 & c_1 \\ a_2 & c_2 \end{vmatrix}}{\begin{vmatrix} a_1 & b_1 \\ a_2 & b_2 \end{vmatrix}} = \frac{D_y}{D}.$$	Solve using Cramer's rule. $$x - 2y = -1$$ $$2x + 5y = 16$$ $$x = \frac{\begin{vmatrix} -1 & -2 \\ 16 & 5 \end{vmatrix}}{\begin{vmatrix} 1 & -2 \\ 2 & 5 \end{vmatrix}} = \frac{-5 + 32}{5 + 4}$$ $$= \frac{27}{9} = 3$$ $$y = \frac{\begin{vmatrix} 1 & -1 \\ 2 & 16 \end{vmatrix}}{\begin{vmatrix} 1 & -2 \\ 2 & 5 \end{vmatrix}} = \frac{16 + 2}{5 + 4} = \frac{18}{9} = 2$$ The solution set is $\{(3, 2)\}$.
	Cramer's Rule for 3 × 3 Systems Given the system $$a_1x + b_1y + c_1z = d_1$$ $$a_2x + b_2y + c_2z = d_2$$ $$a_3x + b_3y + c_3z = d_3$$ with $$D_x = \begin{vmatrix} d_1 & b_1 & c_1 \\ d_2 & b_2 & c_2 \\ d_3 & b_3 & c_3 \end{vmatrix},$$ $$D_y = \begin{vmatrix} a_1 & d_1 & c_1 \\ a_2 & d_2 & c_2 \\ a_3 & d_3 & c_3 \end{vmatrix},$$ $$D_z = \begin{vmatrix} a_1 & b_1 & d_1 \\ a_2 & b_2 & d_2 \\ a_3 & b_3 & d_3 \end{vmatrix},$$ $$D = \begin{vmatrix} a_1 & b_1 & c_1 \\ a_2 & b_2 & c_2 \\ a_3 & b_3 & c_3 \end{vmatrix} \neq 0,$$ then $$x = \frac{D_x}{D}, \quad y = \frac{D_y}{D}, \quad z = \frac{D_z}{D}.$$	Solve using Cramer's rule. $$3x + 2y + z = -5$$ $$x - y + 3z = -5$$ $$2x + 3y + z = 0$$ Using the methods of expansion by minors, it can be shown that $D_x = 45$, $D_y = -30$, $D_z = 0$, and $D = -15$. Therefore, $$x = \frac{D_x}{D} = \frac{45}{-15} = -3,$$ $$y = \frac{D_y}{D} = \frac{-30}{-15} = 2,$$ $$z = \frac{D_z}{D} = \frac{0}{-15} = 0.$$ The solution set is $\{(-3, 2, 0)\}$.

CHAPTER 8 REVIEW EXERCISES

[8.1] *Solve the following systems of equations by the elimination method. In Exercises 1–2, graph the system.*

1. $5x - 3y = 19$
   $4x + y = 5$

2. $-x + 4y = 15$
   $2x + y = 6$

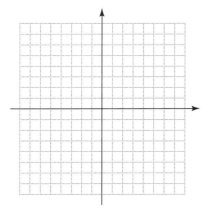
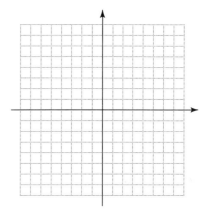

3. $6x + 5y = 4$
   $4x - 2y = -8$

4. $3x + 2y + 6 = 2 - x$
   $x + y + 30 = 1 - 2x + 6y$

Solve the following systems of equations by the substitution method.

5. $-3x - y = 4$
   $x = \dfrac{2}{3}y$

6. $4x = 2y + 3$
   $x = \dfrac{1}{2}y$

7. $5x - 2y = 2$
   $x + 6y = 26$

8. $3x - 2y = 11$
   $4x + 3y = 26$

CHAPTER 8 SYSTEMS OF LINEAR EQUATIONS

[8.2] *Solve the following systems of equations by the elimination method.*

9. $2x + 3y - z = -16$
   $x + 2y + 2z = -3$
   $3x - y - z = 5$

10. $4x - y = 2$
    $3y + z = 9$
    $x + 2z = 7$

11. $-x + 2y + 3z = 5$
    $2x + 3y + z = 12$
    $3x - 6y - 9z = -1$

12. $-3x + y + z = 8$
    $4x + 2y + 3z = 15$
    $-6x + 2y + 2z = 10$

[8.3] *Solve the following word problems.*

13. The length of a rectangle is 3 meters more than the width. The perimeter is 42 meters. Find the length and width of the rectangle.

14. On a 10-day business trip Gale rented a car for $21 a day at weekend rates and $30 a day at weekday rates. If his rental bill was $246, how many days did he rent at each rate?

15. Sweet's Candy Store is offering a special mix for Valentine's Day. Ms. Sweet will mix some $2-a-pound candy with some $1-a-pound candy to get 50 pounds of mix which she will sell at $1.30 a pound. How many pounds of each should she use?

16. A plane flies 560 miles in 1.75 hours traveling with the wind. The return trip later against the same wind takes the plane 2 hours. Find the speed of the plane and the speed of the wind.

17. Sue sells real estate. On three recent sales, she made 10% commission, 6% commission, and 5% commission. Her total commissions on these sales were $8500, and she sold property worth $140,000. If the 5% sale amounted to the sum of the other two, what were the three sales prices?

18. The sum of the measures of the angles of a triangle is 180°. One angle measures 10° less than the sum of the other two. The measure of the middle-sized angle is the average of the other two. Find the measures of the three angles.

19. How many liters each of 8%, 10%, and 20% hydrogen peroxide should be mixed together to get 8 liters of 12.5% solution, if the amount of 20% solution used must be 2 liters less than the amount of 8% solution used?

20. A farmer wishes to satisfy her fertilizer needs with three brands, A, B, and C. She needs to apply a total of 26.4 pounds of nitrogen, 28 pounds of potash, and 26.8 pounds of sulfate of ammonia. Brand A contains 8% nitrogen, 5% potash, and 10% sulfate of ammonia. Brand B contains 6% nitrogen, 10% potash, and 6% sulfate of ammonia. Brand C contains 10% nitrogen, 8% potash, and 5% sulfate of ammonia. How much of each fertilizer should be used?

[8.4] *Evaluate the following determinants.*

21. $\begin{vmatrix} 1 & 2 \\ 5 & 4 \end{vmatrix}$

22. $\begin{vmatrix} 3 & 7 \\ -2 & 1 \end{vmatrix}$

23. $\begin{vmatrix} 10 & 0 \\ -1 & 2 \end{vmatrix}$

24. $\begin{vmatrix} 5 & 4 \\ -1 & 9 \end{vmatrix}$

25. $\begin{vmatrix} 1 & 5 & 2 \\ 0 & 1 & 3 \\ 0 & 6 & -1 \end{vmatrix}$

26. $\begin{vmatrix} 2 & 5 & 0 \\ 1 & -3 & 4 \\ 0 & 2 & -1 \end{vmatrix}$

27. $\begin{vmatrix} 0 & 5 & 3 \\ 7 & 0 & -2 \\ 0 & -1 & 6 \end{vmatrix}$

28. $\begin{vmatrix} 0 & 0 & 0 \\ 0 & 2 & 5 \\ -3 & 7 & 12 \end{vmatrix}$

29. $\begin{vmatrix} 0 & 0 & 2 \\ 2 & 1 & 0 \\ -1 & 0 & 0 \end{vmatrix}$

30. $\begin{vmatrix} a & 0 & 1 \\ 0 & b & 0 \\ 1 & 1 & c \end{vmatrix}$

CHAPTER 8 SYSTEMS OF LINEAR EQUATIONS

[8.5] *Use Cramer's rule to solve the following systems of equations.*

31. $3x - 4y = 5$
    $2x + y = 8$

32. $4x - 3y = 12$
    $2x + 6y = 15$

33. $5x + 6y = 10$
    $4x - 3y = 12$

34. $4x + y + z = 11$
    $x - y - z = 4$
    $y + 2z = 0$

35. $-x + 3y - 4z = 2$
    $2x + 4y + z = 3$
    $3x - z = 9$

36. $5x + y = 10$
    $3x + 2y + z = -3$
    $-y - 2z = -13$

MIXED REVIEW EXERCISES
Solve by any method.

37. $x + 4y = 17$
    $3x - 2y = 9$

38. $2x + 5y - z = 12$
    $x - y + 4z = 10$
    $-8x - 20y + 4z = 29$

39. $x + 2y + 5z = 9$
    $2x - y = 3$
    $3x + 2y + z = 9$

40. $7x - 3y = -12$
    $5x + 2y = 8$

41. $x + 5y - 3z = 0$
    $2x + 6y + z = 0$
    $3x - y + 4z = 0$

42. $-x + 7y = -10$
    $2x + 3y = 3$

43. $\dfrac{2}{3}x + \dfrac{y}{6} = \dfrac{19}{2}$
    $\dfrac{1}{3}x - \dfrac{2}{9}y = 2$

44. $2x - 5y = 8$
    $3x + 4y = 10$

CHAPTER 8 TEST

Solve each system by substitution.

1. $2x - 3y = -8$
   $x = y - 3$

2. $2x + 5y = 1$
   $3x - 2y = -8$

Solve each system by elimination.

3. $2x + 3y = 10$
   $-3x + 2y = 11$

4. $3x + 4y = 8$
   $6x = 7 - 8y$

5. $12x - 5y = 8$
   $3x = \dfrac{5}{4}y + 2$

6. $2x - y + z = 9$
   $3x + y - 2z = 4$
   $x + y - 4z = -6$

7. $4x - 2y = -8$
   $3y - 5z = 14$
   $2x + z = -10$

1. _____
2. _____
3. _____
4. _____
5. _____
6. _____
7. _____

Translate each problem into a system of equations and solve.

8. A chemist needs 6 liters of a 40% alcohol solution. She must mix a 20% solution and a 50% solution. How many liters of each will be required to obtain what she needs?

9. Two cars start from points 420 miles apart and travel toward each other. They meet after 3.5 hours. Find the average speed of each car if one travels 30 miles per hour faster than the other.

10. The perimeter of a triangle is 18 centimeters. The difference between the lengths of the longest side and the shortest side is 2 centimeters. The sum of the lengths of the two shorter sides is 4 centimeters more than the length of the longest side. Find the lengths of the three sides.

Evaluate the following determinants.

11. $\begin{vmatrix} -3 & 2 \\ -1 & 4 \end{vmatrix}$

12. $\begin{vmatrix} 4 & 6 \\ 6 & 9 \end{vmatrix}$

13. $\begin{vmatrix} 2 & -1 & 0 \\ 0 & 3 & 1 \\ -2 & 0 & 1 \end{vmatrix}$

Solve by Cramer's rule.

14. $\begin{aligned} 2x + 6y &= 3 \\ -3x + y &= 8 \end{aligned}$

15. $\begin{aligned} x + y + z &= 4 \\ 2x \phantom{{}+y} - z &= -5 \\ 3y + z &= 9 \end{aligned}$

GRAPHS OF FUNCTIONS AND CONIC SECTIONS

9.1 GRAPHING PARABOLAS

As was shown in Chapter 7, the graphs of first-degree equations are straight lines. In this chapter the graphs of second-degree equations, which are equations with one or more second-degree terms, are discussed. These graphs result from cutting an infinite cone with a plane, as shown in Figure 1. Because of this, the graphs are called **conic sections**.

OBJECTIVES

1. Graph parabolas that are examples of quadratic functions.
2. Find the vertex of a parabola.
3. Predict the shape and direction of the graph of a parabola from the coefficient of x^2.

Circle

Ellipse

Parabola

Hyperbola

FIGURE 1

1 Let us begin by graphing the equation $y = x^2$. First, make a table of ordered pairs satisfying the equation.

x	-2	$-\frac{3}{2}$	-1	$-\frac{1}{2}$	0	$\frac{1}{2}$	1	$\frac{3}{2}$	2
y	4	$\frac{9}{4}$	1	$\frac{1}{4}$	0	$\frac{1}{4}$	1	$\frac{9}{4}$	4

Plot these points and draw a smooth curve through them to get the graph shown in Figure 2. This graph is called a **parabola**. The point (0, 0), with the smallest y-value of any point on the curve, is the **vertex** of this parabola. The vertical line through the vertex is the **axis** of this parabola. The parabola is symmetric about its axis; that is, if the graph were folded along the axis, the two portions of the curve would coincide.

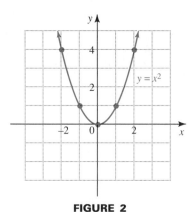

FIGURE 2

Because the graph of $y = x^2$ satisfies the conditions of the graph of a function (Section 7.5), we may write its equation as $f(x) = x^2$. This is the simplest example of a quadratic function.

QUADRATIC FUNCTION

A function that can be written in the form

$$f(x) = ax^2 + bx + c$$

for real numbers a, b, and c, with $a \neq 0$, is a **quadratic function.**

The graph of any quadratic function is a parabola with a vertical axis.

Parabolas have many applications. If an object is thrown upward, then (disregarding air resistance) the path it follows is a parabola. The large disks seen on the sidelines of televised football games, which are used by television crews to pick up the shouted signals of the players on the field, have cross sections that are parabolas. Cross sections of radar dishes and automobile headlights also form parabolas. Additional applications of parabolas are given in Section 9.2.

For the rest of this section, we shall use the function notation $f(x)$ in discussing parabolas.

2 Parabolas need not have their vertices at the origin, as does $f(x) = x^2$. For example, to graph a parabola of the form $f(x) = x^2 + k$, start by selecting the sample values of x that were used to graph $f(x) = x^2$. The corresponding values of $f(x)$ in $f(x) = x^2 + k$ differ by k from those of $f(x) = x^2$. For this reason, the graph of $f(x) = x^2 + k$ is shifted k units vertically compared with that of $f(x) = x^2$.

EXAMPLE 1 Graphing a Parabola with a Vertical Shift
Graph $f(x) = x^2 - 2$.

As we mentioned before, this graph has the same shape as $f(x) = x^2$, but since k here is -2, the graph is shifted 2 units downward, with vertex at

(0, −2). Every function value is 2 less than the corresponding function value of $f(x) = x^2$. Plotting points gives the graph in Figure 3. ■

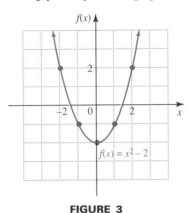

FIGURE 3

The graph of $f(x) = x^2 + k$ is a parabola with the same shape as the graph of $f(x) = x^2$. The parabola is shifted k units upward if $k > 0$, and $|k|$ units downward if $k < 0$. The vertex is $(0, k)$.

WORK PROBLEM 1 AT THE SIDE.

The graph of $f(x) = (x - h)^2$ is also a parabola with the same shape as $f(x) = x^2$. The vertex of the parabola $f(x) = (x - h)^2$ is the lowest point on the parabola. The lowest point occurs here when $f(x)$ is 0. To get $f(x)$ equal to 0, let $x = h$. so the vertex of $f(x) = (x - h)^2$ is at $(h, 0)$. Based on this, the graph of $f(x) = (x - h)^2$ is shifted h units horizontally compared with that of $f(x) = x^2$.

EXAMPLE 2 Graphing a Parabola with a Horizontal Shift
Graph $f(x) = (x - 2)^2$.

When $x = 2$, then $f(x) = 0$, giving the vertex (2, 0). The parabola $f(x) = (x - 2)^2$ has the same shape as $f(x) = x^2$ but is shifted 2 units to the right, as shown in Figure 4. ■

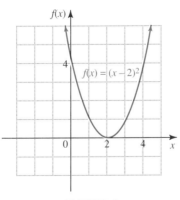

FIGURE 4

The graph of $f(x) = (x - h)^2$ is a parabola with the same shape as the graph of $f(x) = x^2$. The parabola is shifted h units horizontally: h units to the right if $h > 0$, and $|h|$ units to the left if $h < 0$. The vertex is $(h, 0)$.

1. Graph each parabola.

(a) $f(x) = x^2 + 3$

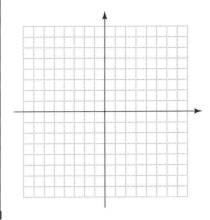

(b) $f(x) = x^2 - 1$

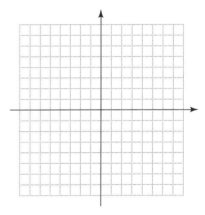

ANSWERS
1. (a)   (b)

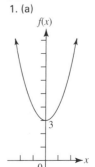
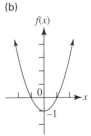

CHAPTER 9 GRAPHS OF FUNCTIONS AND CONIC SECTIONS

2. Graph each parabola.

(a) $f(x) = (x - 3)^2$

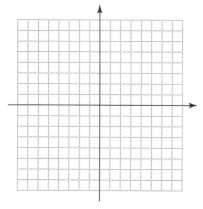

(b) $f(x) = (x + 2)^2$

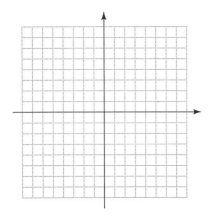

CAUTION Errors frequently occur when horizontal shifts are involved. In order to determine the direction and magnitude of horizontal shifts, find the value that would cause the expression $x - h$ to equal 0. For example, the graph of $f(x) = (x - 5)^2$ would be shifted 5 units to the *right*, because $+5$ would cause $x - 5$ to equal 0. On the other hand, the graph of $f(x) = (x + 4)^2$ would be shifted 4 units to the *left*, because -4 would cause $x + 4$ to equal 0.

■ **WORK PROBLEM 2 AT THE SIDE.**

A parabola can have both a horizontal and a vertical shift, as in Example 3.

EXAMPLE 3 Graphing a Parabola with Horizontal and Vertical Shifts

Graph $f(x) = (x + 3)^2 - 2$.

This graph has the same shape as $f(x) = x^2$, but is shifted 3 units to the left (since $x + 3 = 0$ if $x = -3$), and 2 units downward (because of the -2). As shown in Figure 5, the vertex is at $(-3, -2)$. ■

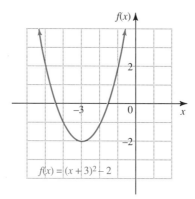

FIGURE 5

EXAMPLE 4 Graphing a Parabola with Horizontal and Vertical Shifts

Graph $f(x) = (x - 1)^2 + 3$.

The graph is shifted one unit to the right and three units up, so the vertex is at $(1, 3)$. See Figure 6. ■

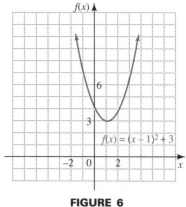

FIGURE 6

ANSWERS
2. (a) 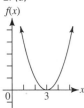 (b)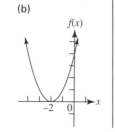

The method of graphing a parabola of the form $f(x) = (x - h)^2 + k$ is summarized as follows.

The graph of $f(x) = (x - h)^2 + k$ is a parabola with the same shape as $f(x) = x^2$ and with vertex at (h, k). The axis is the vertical line $x = h$.

WORK PROBLEM 3 AT THE SIDE.

3 Not all parabolas open upward, and not all parabolas have the same shape as $f(x) = x^2$. The next example shows how to identify parabolas opening downward and having a different shape from that of $f(x) = x^2$.

EXAMPLE 5 Graphing a Parabola That Opens Downward

Graph $f(x) = -\frac{1}{2}x^2$.

This parabola is shown in Figure 7. Some ordered pairs that satisfy the equation are $(0, 0)$, $\left(1, -\frac{1}{2}\right)$ $(2, -2)$, $\left(-1, -\frac{1}{2}\right)$, and $(-2, -2)$. The coefficient $-\frac{1}{2}$ affects the shape of the graph; the $\frac{1}{2}$ makes the parabola wider [since the values of $f(x)$ grow more slowly than they would for $f(x) = x^2$], and the negative sign makes the parabola open downward. The graph is not shifted in any direction; the vertex is still at $(0, 0)$. Here, the vertex has the *largest* function value of any point on the graph. ∎

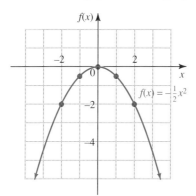

FIGURE 7

Some general principles concerning the graph of $f(x) = a(x - h)^2 + k$ are summarized as follows.

---

1. The graph of the quadratic function
$$f(x) = a(x - h)^2 + k, \quad a \neq 0$$
is a parabola with vertex at (h, k) and the vertical line $x = h$ as axis.
2. The graph opens upward if a is positive and downward if a is negative.
3. The graph is wider than $f(x) = x^2$ if $0 < |a| < 1$. The graph is narrower than $f(x) = x^2$ if $|a| > 1$.

---

3. Graph $f(x) = (x + 2)^2 - 1$. Identify the vertex.

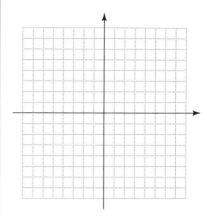

ANSWERS
3. vertex: $(-2, -1)$

CHAPTER 9 GRAPHS OF FUNCTIONS AND CONIC SECTIONS

4. Decide whether each parabola opens upward or downward.

   (a) $f(x) = -\dfrac{2}{3}x^2$

   (b) $f(x) = \dfrac{3}{4}x^2 + 1$

   (c) $f(x) = -2x^2 - 3$

   (d) $f(x) = 3x^2 + 2$

5. Decide whether each parabola in Problem 4 is wider or narrower than $f(x) = x^2$.

6. Graph $f(x) = \dfrac{1}{2}(x - 2)^2 + 1$.

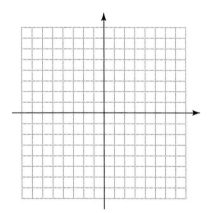

EXAMPLE 6 Using the General Principles to Graph a Parabola

Graph $f(x) = -2(x + 3)^2 + 4$.

The parabola opens downward (because $a < 0$), and is narrower than the graph of $f(x) = x^2$, since the values of $f(x)$ grow more quickly than they would for $f(x) = x^2$. This parabola has vertex at $(-3, 4)$, as shown in Figure 8. To complete the graph, we plotted the ordered pairs $(-4, 2)$ and $(-2, 2)$. ∎

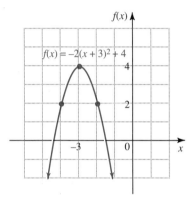

FIGURE 8

■ WORK PROBLEMS 4, 5, AND 6 AT THE SIDE. ■

ANSWERS
4. (a) downward (b) upward
   (c) downward (d) upward
5. (a) wider (b) wider
   (c) narrower (d) narrower
6.
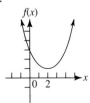

9.1 EXERCISES

Identify the vertex of each parabola. See Examples 1–4.

1. $f(x) = x^2 - 5$

2. $f(x) = x^2 + 7$

3. $f(x) = 5x^2$

4. $f(x) = -8x^2$

5. $f(x) = (x + 4)^2$

6. $f(x) = (x + 2)^2$

7. $f(x) = (x - 9)^2 + 12$

8. $f(x) = (x - 7)^2 - 3$

For each quadratic function, tell whether the graph opens upward or downward and whether the graph is wider, narrower, or the same as $f(x) = x^2$. See Examples 5 and 6.

9. $f(x) = -5x^2$

10. $f(x) = .8x^2$

11. $f(x) = \frac{1}{3}x^2 - 1$

12. $f(x) = -2x^2 + 4$

CHAPTER 9 GRAPHS OF FUNCTIONS AND CONIC SECTIONS

Sketch the graph of each parabola. Plot at least two points in addition to the vertex. See Examples 1–6.

13. $f(x) = 3x^2$

14. $f(x) = -2x^2$

15. $f(x) = -\dfrac{1}{4}x^2$

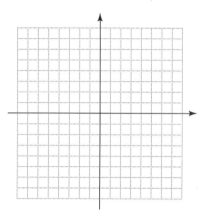
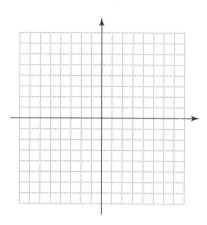
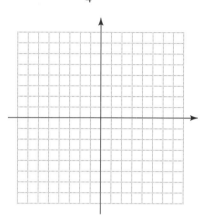

16. $f(x) = \dfrac{1}{3}x^2$

17. $f(x) = x^2 - 1$

18. $f(x) = x^2 + 3$

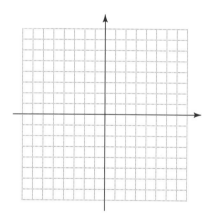
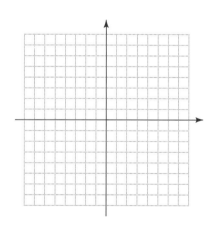
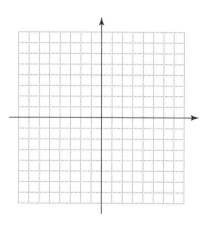

19. $f(x) = -x^2 + 2$

20. $f(x) = -x^2 - 4$

21. $f(x) = 2x^2 - 2$

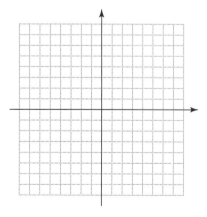
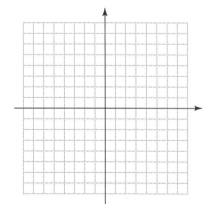
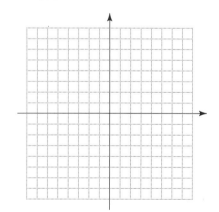

9.1 EXERCISES

22. $f(x) = -3x^2 + 1$

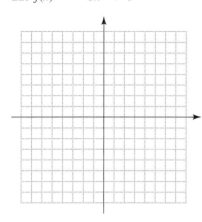

23. $f(x) = (x - 4)^2$

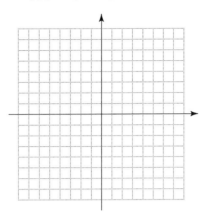

24. $f(x) = 3(x + 1)^2$

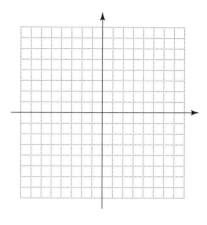

25. $f(x) = -2(x + 1)^2$

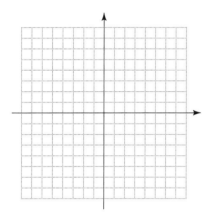

26. $f(x) = (x - 3)^2$

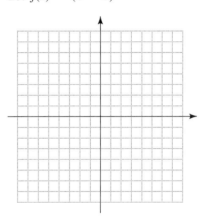

27. $f(x) = (x + 1)^2 - 2$

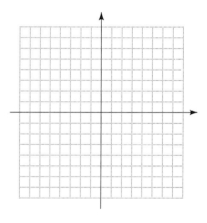

28. $f(x) = (x - 2)^2 + 3$

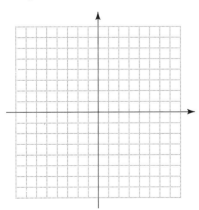

29. $f(x) = 2(x - 1)^2 - 3$

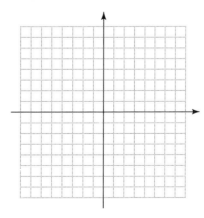

30. $f(x) = -3(x + 4)^2 + 5$

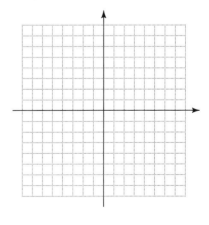

31. $f(x) = -3(x + 2)^2 + 2$

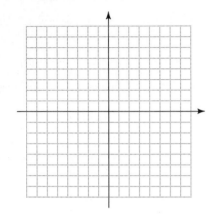

32. $f(x) = -2(x + 1)^2 - 3$

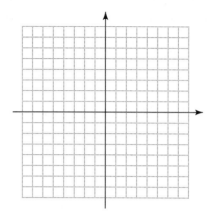

33. $f(x) = \dfrac{2}{3}(x - 1)^2 - 2$

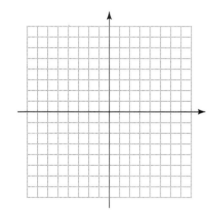

34. $f(x) = \dfrac{5}{4}(x - 2)^2 - 3$

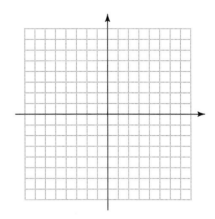

SKILL SHARPENERS

Solve each equation by completing the square. Give only real number answers. See Section 6.1.

35. $x^2 + 6x - 3 = 0$

36. $x^2 + 8x = 4$

37. $2x^2 - 12x = 5$

38. $3x^2 - 12x - 10 = 0$

39. $-x^2 - 3x + 2 = 0$

40. $-2x^2 + 5x - 4 = 0$

9.2 MORE ABOUT PARABOLAS

1 When the equation of a parabola is given in the form $f(x) = ax^2 + bx + c$, it is necessary to locate the vertex in order to sketch an accurate graph. This can be done in two ways. The first is by completing the square, as shown in Examples 1 and 2. The second is by using a formula which may be derived by completing the square.

EXAMPLE 1 Completing the Square to Find the Vertex

Find the vertex of the graph of $f(x) = x^2 - 4x + 5$.

To find the vertex, we need to express $x^2 - 4x + 5$ in the form $(x - h)^2 + k$. This is done by completing the square on $x^2 - 4x$ as in Section 6.1. The process is a little different here because we want to keep $f(x)$ alone on one side of the equation. Instead of adding the appropriate number to both sides, *add and subtract* it on the right. This is equivalent to adding 0.

$$x^2 - 4x + 5 = (x^2 - 4x) + 5$$

Half of -4 is -2; $(-2)^2 = 4$.

$$= (x^2 - 4x + 4 - 4) + 5 \quad \text{Add and subtract 4.}$$
$$= (x^2 - 4x + 4) - 4 + 5 \quad \text{Bring } -4 \text{ outside the parentheses.}$$

Writing $x^2 - 4x + 4$ as $(x - 2)^2$ gives

$$x^2 - 4x + 5 = (x - 2)^2 + 1.$$

Now write the original equation as $f(x) = (x - 2)^2 + 1$. The method of Section 9.1 shows that the vertex of this parabola is (2, 1). ∎

WORK PROBLEM 1 AT THE SIDE.

EXAMPLE 2 Completing the Square to Find the Vertex

Find the vertex of the graph of $f(x) = -3x^2 + 6x - 1$.

We must complete the square on $-3x^2 + 6x$. Because the x^2 term has a coefficient other than 1, factor that coefficient out of the first two terms, and then proceed as in Example 1.

$$-3x^2 + 6x - 1 = -3(x^2 - 2x) - 1$$

Half of -2 is -1, and $(-1)^2 = 1$.

$$= -3(x^2 - 2x + 1 - 1) - 1 \quad \text{Add and subtract 1.}$$
$$= -3(x^2 - 2x + 1) + (-3)(-1) - 1 \quad \text{Distributive property}$$
$$= -3(x^2 - 2x + 1) + 3 - 1$$
$$f(x) = -3(x - 1)^2 + 2$$

The vertex is at (1, 2). ∎

WORK PROBLEM 2 AT THE SIDE.

OBJECTIVES

1 Find the vertex of a vertical parabola.

2 Graph a quadratic function.

3 Use the discriminant to find the number of x-intercepts of a vertical parabola.

4 Use quadratic functions to solve problems involving maximum or minimum value.

5 Graph horizontal parabolas.

---

1. Find the vertex of each parabola by completing the square.

   (a) $f(x) = x^2 - 6x + 7$

   (b) $f(x) = x^2 + 4x - 9$

2. Find the vertex of each parabola.

   (a) $f(x) = 2x^2 - 4x + 1$

   (b) $f(x) = -\dfrac{1}{2}x^2 + 2x - 3$

ANSWERS
1. (a) (3, −2) (b) (−2, −13)
2. (a) (1, −1) (b) (2, −1)

CHAPTER 9 GRAPHS OF FUNCTIONS AND CONIC SECTIONS

3. Use the formula to find the vertex of the graph of each quadratic function.

(a) $f(x) = -2x^2 + 3x - 1$

(b) $f(x) = 4x^2 - x + 5$

A formula for the vertex of the graph of the quadratic function $f(x) = ax^2 + bx + c$ can be found by completing the square for the general form of the equation.

$$f(x) = ax^2 + bx + c \quad (a \neq 0)$$

$$f(x) = a\left(x^2 + \frac{b}{a}x\right) + c \qquad \text{Factor } a \text{ from the first two terms.}$$

$$f(x) = a\left(x^2 + \frac{b}{a}x + \frac{b^2}{4a^2} - \frac{b^2}{4a^2}\right) + c \qquad \text{Add and subtract } \frac{b^2}{4a^2} \text{ within the parentheses.}$$

$$f(x) = a\left(x^2 + \frac{b}{a}x + \frac{b^2}{4a^2}\right) - \frac{b^2}{4a} + c \qquad \text{Distributive property}$$

$$f(x) = a\left(x + \frac{b}{2a}\right)^2 + \frac{4ac - b^2}{4a} \qquad \text{Factor and combine terms.}$$

$$f(x) = a\left[x - \left(\frac{-b}{2a}\right)\right]^2 + \underbrace{\frac{4ac - b^2}{4a}}_{k}$$
$$\underbrace{\phantom{x - \left(\frac{-b}{2a}\right)}}_{h}$$

The final equation shows that the vertex (h, k) can be expressed in terms of a, b, and c. However, it is not necessary to memorize k, since $k = f(h)$.

VERTEX FORMULA

The graph of the quadratic function $f(x) = ax^2 + bx + c$ has its vertex at

$$\left(\frac{-b}{2a}, f\left(\frac{-b}{2a}\right)\right),$$

and the axis of the parabola is the line $x = \frac{-b}{2a}$.

EXAMPLE 3 Using the Formula to Find the Vertex

Use the vertex formula to find the vertex of the graph of the function
$$f(x) = x^2 - x - 6.$$

For this function, $a = 1$, $b = -1$, and $c = -6$. The x-coordinate of the vertex of the parabola is given by

$$\frac{-b}{2a} = \frac{-(-1)}{2(1)} = \frac{1}{2}.$$

The y-coordinate is $f\left(\frac{-b}{2a}\right) = f\left(\frac{1}{2}\right)$.

$$f\left(\frac{1}{2}\right) = \left(\frac{1}{2}\right)^2 - \frac{1}{2} - 6 = \frac{1}{4} - \frac{1}{2} - 6 = -\frac{25}{4}$$

Finally, the vertex is $\left(\frac{1}{2}, -\frac{25}{4}\right)$. ∎

■ WORK PROBLEM 3 AT THE SIDE.

ANSWERS

3. (a) $\left(\frac{3}{4}, \frac{1}{8}\right)$ (b) $\left(\frac{1}{8}, \frac{79}{16}\right)$

9.2 MORE ABOUT PARABOLAS

2 Parabolas were graphed in Section 9.1. A more general approach involving finding intercepts is given here.

GRAPHING A QUADRATIC FUNCTION f(x)

Step 1 Find the y-intercept by evaluating $f(0)$.
Step 2 Find any x-intercepts by solving $f(x) = 0$.
Step 3 Find the vertex either by using the formula or by completing the square.
Step 4 Find and plot additional points as needed, using the symmetry about the axis.
Step 5 Verify that the graph opens upward (if $a > 0$) or opens downward (if $a < 0$).

EXAMPLE 4 Using the Steps for Graphing a Quadratic Function
Graph the quadratic function $f(x) = x^2 - x - 6$.

Begin by finding the y-intercept.

$$f(x) = x^2 - x - 6$$
$$f(0) = 0^2 - 0 - 6 \quad \text{Find } f(0).$$
$$f(0) = -6$$

The y-intercept is $(0, -6)$. Now find any x-intercepts.

$$f(x) = x^2 - x - 6$$
$$0 = x^2 - x - 6 \quad \text{Let } f(x) = 0.$$
$$0 = (x - 3)(x + 2) \quad \text{Factor.}$$
$$x - 3 = 0 \quad \text{or} \quad x + 2 = 0 \quad \text{Set each factor equal to 0 and solve.}$$
$$x = 3 \quad \text{or} \quad x = -2$$

The x-intercepts are $(3, 0)$ and $(-2, 0)$. The vertex, found in Example 3, is $\left(\frac{1}{2}, -\frac{25}{4}\right)$. Plot the points found so far, and plot any additional points as needed. The symmetry of the graph is helpful here. The graph is shown in Figure 9. ■

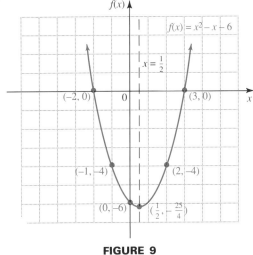

FIGURE 9

WORK PROBLEM 4 AT THE SIDE.

4. Graph the quadratic function

$$f(x) = x^2 - 6x + 5.$$

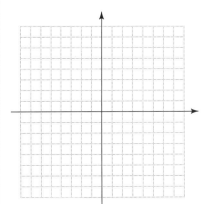

ANSWER
4.
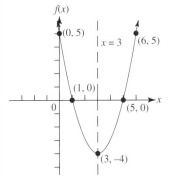

5. Use the discriminant to determine the number of x-intercepts for each graph.

(a) $f(x) = 4x^2 - 20x + 25$

(b) $f(x) = 2x^2 + 3x + 5$

(c) $f(x) = -3x^2 - x + 2$

ANSWERS
5. (a) discriminant is 0; one x-intercept
   (b) discriminant is -31; no x-intercepts
   (c) discriminant is 25; two x-intercepts

3 The graph of a quadratic function may have one x-intercept, two x-intercepts, or no x-intercepts, as shown in Figure 10.

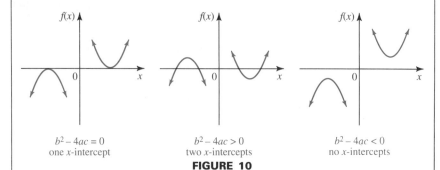

$b^2 - 4ac = 0$ one x-intercept

$b^2 - 4ac > 0$ two x-intercepts

$b^2 - 4ac < 0$ no x-intercepts

FIGURE 10

Recall from Section 6.2 that the value of $b^2 - 4ac$ is called the *discriminant* of the quadratic equation $ax^2 + bx + c = 0$. It can be used to determine the number of real solutions of a quadratic equation. In a similar way, the discriminant of a quadratic *function* can be used to determine the number of x-intercepts of its graph. If the discriminant is positive, the parabola will have two x-intercepts. If the discriminant is 0, there will be only one x-intercept, and it will be the vertex of the parabola. If the discriminant is negative, the graph will have no x-intercepts.

EXAMPLE 5 Using the Discriminant to Determine the Number of x-Intercepts

Determine the number of x-intercepts of the graph of each quadratic function. Use the discriminant.

(a) $f(x) = 2x^2 + 3x - 5$

The discriminant is $b^2 - 4ac$. Here $a = 2$, $b = 3$, and $c = -5$, so
$$b^2 - 4ac = 9 - 4(2)(-5) = 49.$$

Since the discriminant is positive, the parabola has two x-intercepts.

(b) $f(x) = -3x^2 - 1$

In this equation, $a = -3$, $b = 0$, and $c = -1$. The discriminant is
$$b^2 - 4ac = 0 - 4(-3)(-1) = -12.$$

The discriminant is negative and so the graph has no x-intercepts.

(c) $f(x) = 9x^2 + 6x + 1$

Here, $a = 9$, $b = 6$, and $c = 1$. The discriminant is
$$b^2 - 4ac = 36 - 4(9)(1) = 0.$$

The parabola has only one x-intercept (its vertex) since the value of the discriminant is 0. ■

WORK PROBLEM 5 AT THE SIDE.

4 As we have seen, the vertex of a parabola is either the highest or the lowest point on the parabola. The y-value of the vertex gives the maximum or minimum value of y, while the x-value tells where that maximum or minimum occurs. In many practical problems we want to know the largest or smallest value of some quantity. When that quantity can be expressed as a quadratic function $y = ax^2 + bx + c$, as in the next example, the vertex can be used to find the desired value.

EXAMPLE 6 Finding the Maximum Area of a Rectangular Region
A farmer has 120 feet of fencing. He wants to put a fence around a rectangular plot of land next to a river. Find the maximum area he can enclose.

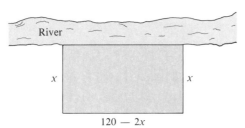

FIGURE 11

Figure 11 shows the field. Let x represent the width of the field. Then, since there are 120 feet of fencing,

$x + x + \text{length} = 120$ Sum of the sides is 120 feet.
$2x + \text{length} = 120$ Combine terms.
$\text{length} = 120 - 2x.$ Subtract $2x$.

The area is given by the product of the length and width, or

$$A = x(120 - 2x) = 120x - 2x^2.$$

To make the area (and thus $120x - 2x^2$) as large as possible, first find the vertex of the parabola $A = 120x - 2x^2$. Do this by completing the square on $120x - 2x^2$.

$A = 120x - 2x^2$
$ = -2x^2 + 120x$
$ = -2(x^2 - 60x)$ Factor out -2.
$ = -2(x^2 - 60x + 900 - 900)$ Add and subtract 900 within parentheses.
$ = -2(x^2 - 60x + 900) + 1800$ $-2(-900) = 1800$
$A = -2(x - 30)^2 + 1800$ Factor $x^2 - 60x + 900$.

The graph is a parabola that opens downward, and its vertex is (30, 1800). The vertex of the graph shows that the maximum area will be 1800 square feet. This area will occur if x, the width of the field, is 30 feet. ∎

WORK PROBLEM 6 AT THE SIDE.

5 If x and y are exchanged in the equation $y = ax^2 + bx + c$, the equation becomes $x = ay^2 + by + c$. Because of the interchange of the roles of x and y, these parabolas are horizontal (with horizontal lines as axes), compared with the vertical ones graphed previously.

EXAMPLE 7 Graphing a Horizontal Parabola
Graph $x = (y - 2)^2 - 3$.

This graph has its vertex at $(-3, 2)$, since the roles of x and y are reversed. It opens to the right, the positive x-direction, and has the same

6. Solve Example 6 if the farmer has only 100 feet of fencing.

ANSWER
6. The field should be 25 feet by 50 feet with a maximum area of 1250 square feet.

CHAPTER 9 GRAPHS OF FUNCTIONS AND CONIC SECTIONS

7. Graph each parabola.

(a) $x = (y + 1)^2 - 4$

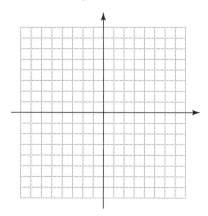

(b) $x = (y - 3)^2 + 2$

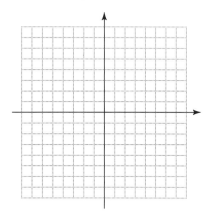

8. Find the vertex of each parabola. Tell whether the graph opens to the right or to the left.

(a) $x = 2y^2 - 6y + 5$

(b) $x = -y^2 + 2y + 5$

ANSWERS

7. (a)   (b)

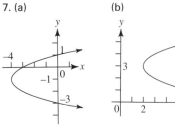

8. (a) $\left(\dfrac{1}{2}, \dfrac{3}{2}\right)$; right   (b) $(6, 1)$; left

shape as $y = x^2$. Plotting a few additional points gives the graph shown in Figure 12. ∎

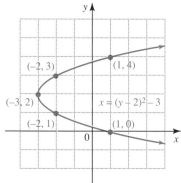

FIGURE 12

WORK PROBLEM 7 AT THE SIDE.

When a quadratic equation is given in the form $x = ay^2 + by + c$, completing the square on y will put the equation into a form in which the vertex can be identified.

EXAMPLE 8 Completing the Square to Graph a Horizontal Parabola

Graph $x = -2y^2 + 4y - 3$.

Complete the square on the right to express the equation in the form $x = a(y - k)^2 + h$.

$$-2y^2 + 4y - 3 = -2(y^2 - 2y) - 3$$
$$= -2(y^2 - 2y + 1 - 1) - 3 \quad \text{Add } 0\ (1 - 1 = 0).$$
$$= -2(y^2 - 2y + 1) + 2 - 3 \quad \text{Distributive property}$$
$$= -2(y - 1)^2 - 1 \quad \text{Factor.}$$

Because of the negative coefficient (-2), the graph opens to the left (the negative x direction) and is narrower than $y = x^2$. As shown in Figure 13, the vertex is $(-1, 1)$. ∎

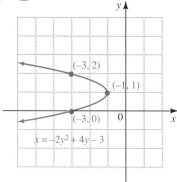

FIGURE 13

WORK PROBLEM 8 AT THE SIDE.

CAUTION Only quadratic equations that are solved for y are examples of functions. The graphs of the equations in Examples 7 and 8 are not graphs of functions. They do not satisfy the conditions of the vertical line test.

9.2 EXERCISES

Find the vertex of each parabola. For each equation, decide whether the graph opens upward, downward, to the left, or to the right, and state whether it is wider, narrower, or the same shape as the graph of $y = x^2$. If it is a vertical parabola, use the discriminant to determine the number of x-intercepts. See Examples 1–3, 5, 7, and 8.

1. $y = x^2 + 8x - 5$

2. $y = x^2 - 2x + 4$

3. $y = -x^2 + 3x + 3$

4. $y = -x^2 + 5x - 1$

5. $y = 2x^2 - 6x + 3$

6. $y = 3x^2 + 12x - 10$

7. $x = 4y^2 + 8y + 2$

8. $x = -3y^2 + 12y - 9$

9. $x = 2y^2 - 8y + 9$

10. $x = 3y^2 + 6y + 1$

CHAPTER 9 GRAPHS OF FUNCTIONS AND CONIC SECTIONS

Graph each parabola by using the techniques described in this section. See Examples 4, 7, and 8.

11. $f(x) = x^2 + 8x + 14$

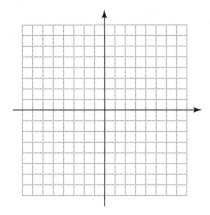

12. $f(x) = x^2 + 10x + 23$

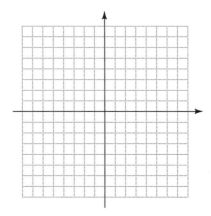

13. $f(x) = 3x^2 - 9x + 8$

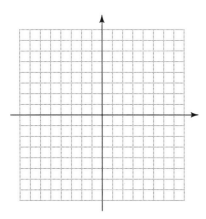

14. $f(x) = 2x^2 + 6x - 1$

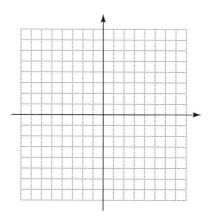

15. $f(x) = -2x^2 + 4x + 5$

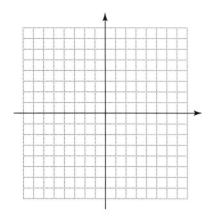

16. $f(x) = -5x^2 - 10x + 2$

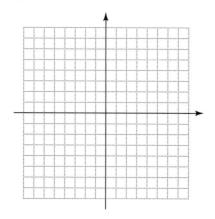

9.2 EXERCISES

17. $x = y^2 - 6y + 4$

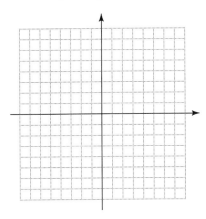

18. $x = y^2 + 2y - 2$

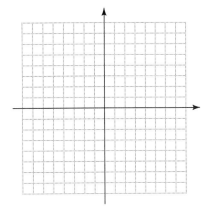

19. $x = -y^2 + 6y - 7$

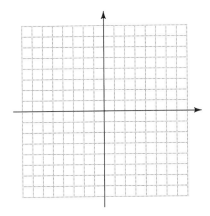

20. $x = -2y^2 - 8y - 1$

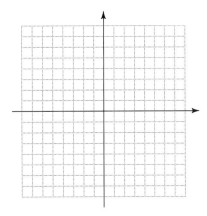

Solve the following word problems. See Example 6.

21. Christina Santiago runs a taco stand. Her past records indicate that the cost of operating the stand is given by

$$c = 2x^2 - 28x + 160$$

where x is the units of tacos sold daily and c is in dollars. Find the number of units of tacos she must sell to produce the lowest cost, and find this lowest cost.

22. Hilary Langlois owns a video store. He has found that the profits of the store are approximately given by

$$p = -x^2 + 16x + 34$$

where p represents profit in dollars, and x is the units of videos rented daily. Find the number of units of videos that he should rent daily to produce the maximum profit. Also find the maximum profit.

CHAPTER 9 GRAPHS OF FUNCTIONS AND CONIC SECTIONS

23. A projectile is fired straight upward so that its distance (in feet) above the ground t seconds after firing is
$$s = -16t^2 + 400t.$$
Find the maximum height it reaches and the number of seconds it takes to reach that height.

24. If an object is thrown upward with an initial velocity of 32 feet per second, then its height after t seconds is given by
$$h = 32t - 16t^2.$$
Find the maximum height attained by the object. Find the number of seconds it takes the object to hit the ground.

25. Of all pairs of numbers whose sum is 80, find the pair with the maximum product. (*Hint:* Let x and $80 - x$ represent the two numbers. Write a quadratic expression for the product.)

26. The length and width of a rectangle have a sum of 52 meters. What width will lead to the maximum area? (*Hint:* Let the width be x and the length be $52 - x$. Write a quadratic expression for the area.)

27. Suppose the price p of a product is related to the demand x for the product by the equation
$$p = 980 - 5x^2$$
where x is measured in hundreds. Find p for each of the following values of x.
(a) 5   (b) 10   (c) 14

28. For a trip to a resort, a charter bus company charges a fare of $48 per person, plus $2 per person for each unsold seat on the bus. If the bus has 42 seats and x represents the number of unsold seats, find the following:
(a) an expression for the total revenue R from the trip (*Hint:* Multiply the total number riding, $42 - x$, by the price per ticket, $48 + 2x$);

(b) the number of unsold seats that produces the maximum revenue;

(c) the maximum revenue.

SKILL SHARPENERS
Express each radical in simplest form. See Section 5.3.

29. $\sqrt{3^2 + 4^2}$

30. $\sqrt{5^2 + 12^2}$

31. $\sqrt{(8 - 2)^2 + (7 + 1)^2}$

32. $\sqrt{(6 + 1)^2 + (3 + 21)^2}$

9.3 THE DISTANCE FORMULA AND THE CIRCLE

OBJECTIVES

1. Find the distance between two points.
2. Find the equation of a circle, given the center and radius.
3. Find the center and radius of a circle, given its equation.
4. Graph square root functions.

1 The second-degree equations discussed in Sections 9.1 and 9.2 were of the form $y = ax^2 + bx + c$ or $x = ay^2 + by + c$ ($a \neq 0$) with just one of the variables squared. Before investigating the graphs of other second-degree equations, in which both variables are squared, we need a formula to find the distance between two points. For example, Figure 14 shows the points $(3, -4)$ and $(-5, 3)$. To find the distance between these points, use the *Pythagorean formula* from geometry, given in Section 6.4. By this formula, the square of the length of the hypotenuse (the longest side), d, of a right triangle is equal to the sum of the squares of the lengths of the two shorter sides, a and b. That is,

$$d^2 = a^2 + b^2.$$

As shown in Figure 14,

$$a = 3 - (-5) = 8 \quad \text{and} \quad b = 3 - (-4) = 7.$$

By the Pythagorean formula,

$$d^2 = 7^2 + 8^2 = 49 + 64 = 113,$$

and

$$d = \sqrt{113}.$$

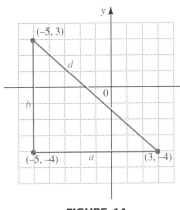

FIGURE 14

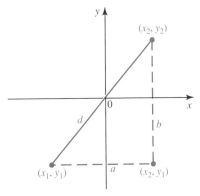

FIGURE 15

Now this result can be generalized. Figure 15 shows the two points (x_1, y_1) and (x_2, y_2). We want to find a general formula for the distance d between these two points. The distance between (x_1, y_1) and (x_2, y_1) is given by

$$a = x_2 - x_1,$$

and the distance between (x_2, y_2) and (x_2, y_1) is given by

$$b = y_2 - y_1.$$

From the Pythagorean formula,

$$d^2 = a^2 + b^2$$
$$= (x_2 - x_1)^2 + (y_2 - y_1)^2.$$

Taking square roots on both sides gives the distance formula.

DISTANCE FORMULA

The distance between (x_1, y_1) and (x_2, y_2) is

$$d = \sqrt{(x_2 - x_1)^2 + (y_2 - y_1)^2}.$$

1. Find the distance between the points in each pair.

   (a) $(2, -1)$ and $(5, 3)$

   (b) $(-3, 2)$ and $(0, -4)$

   (c) $(9, 4)$ and $(2, 4)$

2. Find the equation of the circle with radius 4 and center $(0, 0)$. Sketch its graph.

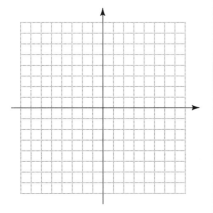

ANSWERS
1. (a) 5 (b) $\sqrt{45}$ or $3\sqrt{5}$ (c) 7
2. $x^2 + y^2 = 16$

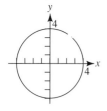

EXAMPLE 1 Using the Distance Formula

Find the distance between $(-3, 5)$ and $(6, 4)$.

Using the distance formula,
$$d = \sqrt{[6 - (-3)]^2 + (4 - 5)^2} = \sqrt{9^2 + (-1)^2} = \sqrt{82}. \blacksquare$$

■ **WORK PROBLEM 1 AT THE SIDE.**

2 A **circle** is the set of all points in a plane that lie a fixed distance from a fixed point. The fixed point is called the **center** and the fixed distance is called the **radius.** The distance formula can be used to find the equation of a circle.

EXAMPLE 2 Finding the Equation of a Circle and Graphing It

Find the equation of the circle with radius 3 and center at $(0, 0)$. Give the graph.

If the point (x, y) is on the circle, the distance from (x, y) to the center $(0, 0)$ is 3. By the distance formula,

$$\sqrt{(x_2 - x_1)^2 + (y_2 - y_1)^2} = d$$
$$\sqrt{(x - 0)^2 + (y - 0)^2} = 3$$
$$x^2 + y^2 = 9. \quad \text{Square both sides.}$$

The equation of this circle is $x^2 + y^2 = 9$. Its graph is given in Figure 16. ■

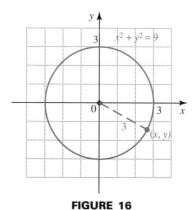

FIGURE 16

■ **WORK PROBLEM 2 AT THE SIDE.**

EXAMPLE 3 Graphing a Circle Given Its Center and Radius

Find an equation for the circle that has its center at $(4, -3)$ and radius 5, and give the graph.

Again use the distance formula with the points (x, y) and $(4, -3)$. The graph of this circle is shown in Figure 17. The equation is

$$\sqrt{(x - 4)^2 + (y + 3)^2} = 5$$
$$(x - 4)^2 + (y + 3)^2 = 25. \blacksquare$$

9.3 THE DISTANCE FORMULA AND THE CIRCLE

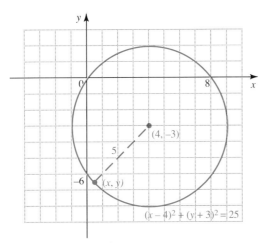

FIGURE 17

The results of Examples 2 and 3 can be generalized to get an equation of a circle with radius r and center at (h, k). If (x, y) is a point on the circle, the distance from the center (h, k) to the point (x, y) is r. Then by the distance formula, with $d = r$, $x_1 = x$, $x_2 = h$, $y_1 = y$, and $y_2 = k$,

$$\sqrt{(x - h)^2 + (y - k)^2} = r.$$

Squaring both sides of the equation gives the following result.

EQUATION OF A CIRCLE

The circle with radius r and center at (h, k) has an equation of the form

$$(x - h)^2 + (y - k)^2 = r^2.$$

WORK PROBLEMS 3 AND 4 AT THE SIDE.

In the equation found in Example 3, multiplying out $(x - 4)^2$ and $(y + 3)^2$ and then combining like terms gives

$$(x - 4)^2 + (y + 3)^2 = 25$$
$$x^2 - 8x + 16 + y^2 + 6y + 9 = 25$$
$$x^2 + y^2 - 8x + 6y = 0.$$

This result suggests that an equation that has both x^2 and y^2 terms with the same coefficients may represent a circle. In many cases it does, and the next example shows how to determine the center and radius of the circle.

EXAMPLE 4 Finding the Center and Radius of a Circle
Find the center and radius of the circle whose equation is

$$x^2 + y^2 + 2x + 4y - 4 = 0.$$

Since the equation has x^2 and y^2 terms with equal coefficients, its graph might be that of a circle. To find the center and radius, complete the square on x and the square on y as follows. Keep only the terms with the variables on the left side and group the x terms and the y terms.

$$(x^2 + 2x \quad) + (y^2 + 4y \quad) = 4$$

---

3. Find the equation of the circle with center at $(3, -2)$ and radius 4. Graph the circle.

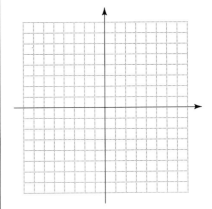

4. Find the center and radius of $(x - 5)^2 + (y + 2)^2 = 9$ and graph the circle.

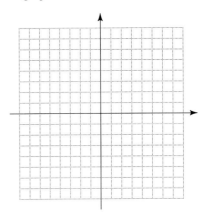

ANSWERS
3. $(x - 3)^2 + (y + 2)^2 = 16$

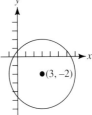

4. center at $(5, -2)$; radius 3

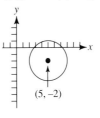

545

CHAPTER 9 GRAPHS OF FUNCTIONS AND CONIC SECTIONS

5. Find the center and radius of each circle.

(a) $x^2 + y^2 - 6x + 8y - 11 = 0$

(b) $x^2 + y^2 + 6x - 4y - 51 = 0$

Add the appropriate constants to complete both squares on the left.

$(x^2 + 2x + 1 - 1) + (y^2 + 4y + 4 - 4) = 4$ Add 0 twice.

$(x^2 + 2x + 1) - 1 + (y^2 + 4y + 4) - 4 = 4$ Associative property

$(x^2 + 2x + 1) + (y^2 + 4y + 4) = 4 + 5$ Add 5 on both sides.

$(x + 1)^2 + (y + 2)^2 = 9$ Factor.

The last equation shows that the center of the circle is $(-1, -2)$ and the radius is 3. ∎

CAUTION If the procedure of Example 4 leads to an equation of the form $(x - h)^2 + (y - k)^2 = 0$, the graph is the single point (h, k). If the constant on the right side is negative, the equation has no graph.

■ **WORK PROBLEM 5 AT THE SIDE.**

4 The final example illustrates the graph of a square root function.

SQUARE ROOT FUNCTION

A function of the form

$$f(x) = \sqrt{u}$$

for an algebraic expression u, with $u \geq 0$, is called a **square root function**.

6. Graph the function $f(x) = \sqrt{36 - x^2}$.

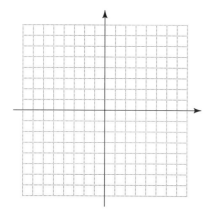

EXAMPLE 5 Graphing a Square Root Function

Graph $f(x) = \sqrt{25 - x^2}$.

Replace $f(x)$ with y and square both sides to get the equation

$$y^2 = 25 - x^2, \quad \text{or} \quad x^2 + y^2 = 25.$$

This is the graph of a circle with center at $(0, 0)$ and radius 5. Since $f(x)$, or y, represents a principal square root in the original equation, $f(x)$ must be nonnegative. This restricts the graph to the upper half of the circle, as shown in Figure 18. Use the graph and the vertical line test to verify that it is indeed a function. ∎

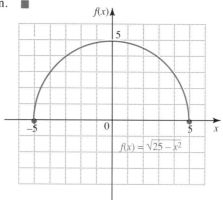

FIGURE 18

■ **WORK PROBLEM 6 AT THE SIDE.**

ANSWERS
5. (a) center at $(3, -4)$; radius 6
   (b) center at $(-3, 2)$; radius 8
6.

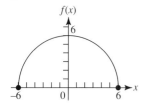

9.3 EXERCISES

Find the distance between each pair of points. See Example 1.

1. $(-2, 1)$ and $(3, -2)$

2. $(3, 4)$ and $(-2, 1)$

3. $(1, -5)$ and $(6, 3)$

4. $(-2, 4)$ and $(3, -2)$

5. $(-8, 7)$ and $(4, -3)$

6. $(3, -2)$ and $(4, 3)$

7. $(a, a - b)$ and $(b, a + b)$

8. $(m + n, n)$ and $(m - n, m)$

Write an equation of each circle with the given center and radius. See Examples 2 and 3.

9. Center $(0, 0)$; radius 8

10. Center $(0, 0)$; radius 7

11. Center $(2, 4)$; radius 5

12. Center $(-1, 5)$; radius 3

13. Center $(0, 3)$; radius $\sqrt{2}$

14. Center $(1, 0)$; radius $\sqrt{3}$

15. Center $(0, -1)$; radius 4

16. Center $(-2, -1)$; radius 1

Find the center and radius of each circle. (Hint: In Exercises 21 and 22, divide both sides by a common factor.) See Example 4.

17. $x^2 + y^2 + 6x - 2y + 6 = 0$

18. $x^2 + y^2 - 10x - 8y - 23 = 0$

19. $x^2 + y^2 + 8x + 4y - 29 = 0$

20. $x^2 + y^2 - 6x - 4y + 9 = 0$

21. $2x^2 + 2y^2 - 4x + 12y + 2 = 0$

22. $3x^2 + 3y^2 - 18x + 30y - 6 = 0$

Graph the following. See Examples 2, 3, and 4.

23. $x^2 + y^2 = 16$

24. $x^2 + y^2 = 9$

25. $2x^2 + 2y^2 = 8$

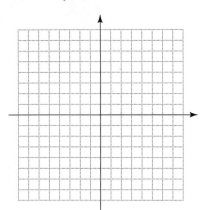
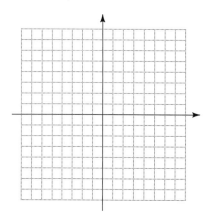
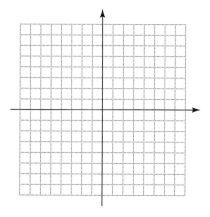

26. $3x^2 + 3y^2 = 15$

27. $y^2 = 144 - x^2$

28. $4x^2 = 16 - 4y^2$

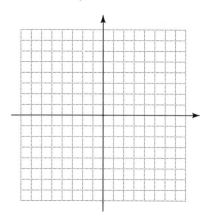
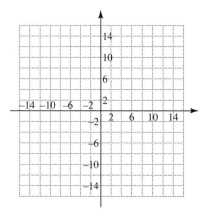
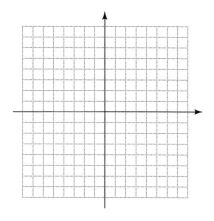

29. $(x - 1)^2 + (y + 3)^2 = 4$

30. $(x + 2)^2 + (y - 4)^2 = 9$

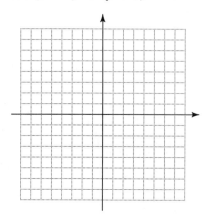
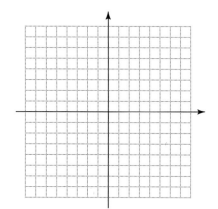

31. $x^2 + 8x + y^2 + 10y + 5 = 0$

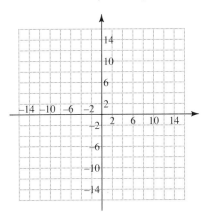

32. $x^2 - 6x + y^2 + 10y + 9 = 0$

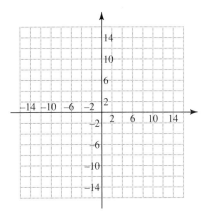

Graph each square root function. See Example 5.

33. $f(x) = \sqrt{4 - x^2}$

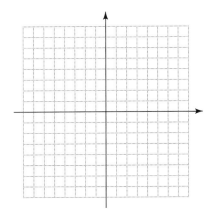

34. $f(x) = \sqrt{16 - x^2}$

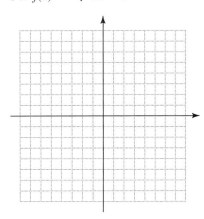

35. $f(x) = -\sqrt{49 - x^2}$

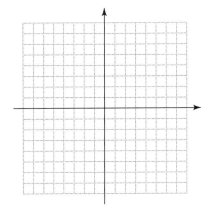

36. $f(x) = -\sqrt{9 - x^2}$

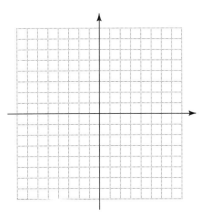

37. $f(x) = \sqrt{4 - x}$

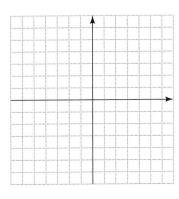

38. $f(x) = -\sqrt{4 - x}$

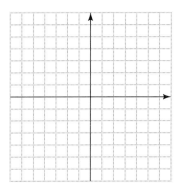

In the following exercise, the distance formula is used to develop the equation of a parabola.

39. A parabola can be defined as the set of all points in a plane equally distant from a given point and a given line not containing the point. See the figure.

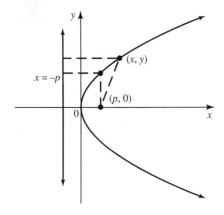

(a) Suppose (x, y) is to be on the parabola. Suppose the line mentioned in the definition is given by $x = -p$. Find the distance between (x, y) and the line. (The distance from a point to a line is the length of the perpendicular from the point to the line.)

(b) If $x = -p$ is the line mentioned in the definition, why should the given point have coordinates $(p, 0)$? (*Hint:* See the figure.)

(c) Find an expression for the distance from (x, y) to $(p, 0)$.

(d) Find an equation for the parabola of the figure. [*Hint:* Use the results of parts (a) and (c) and the fact that (x, y) is equally distant from the point and the line.]

SKILL SHARPENERS
Find the intercepts of the graph of each equation. See Section 7.1.

40. $2x + 5y = 10$

41. $5x - 8y = 3$

42. $\dfrac{x}{4} - \dfrac{y}{3} = 1$

43. $\dfrac{x}{2} + \dfrac{y}{3} = 1$

9.4 THE ELLIPSE AND THE HYPERBOLA

1 An **ellipse** is the set of all points in a plane the sum of whose distances from two fixed points is constant. These fixed points are called **foci** (singular: *focus*). Figure 19 shows an ellipse centered at the origin, with foci at $(c, 0)$ and $(-c, 0)$, x-intercepts $(a, 0)$ and $(-a, 0)$, and y-intercepts $(0, b)$ and $(0, -b)$. From the definition above, it can be shown by the distance formula that an ellipse has the following equation.

OBJECTIVES

1. Recognize the equation of an ellipse.
2. Graph ellipses.
3. Recognize the equation of a hyperbola.
4. Graph hyperbolas by using the asymptotes.
5. Identify conic sections by name from their equations.
6. Graph square root functions.

EQUATION OF AN ELLIPSE

The ellipse whose x-intercepts are $(a, 0)$ and $(-a, 0)$ and whose y-intercepts are $(0, b)$ and $(0, -b)$ has an equation of the form

$$\frac{x^2}{a^2} + \frac{y^2}{b^2} = 1.$$

Note that a circle is a special case of an ellipse, where $a^2 = b^2$.

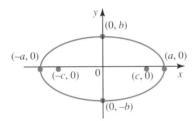

FIGURE 19

The paths of the earth and other planets around the sun are approximately ellipses; the sun is at one focus and a point in space is at the other. The orbits of communication satellites and other space vehicles are elliptical.

2 To graph an ellipse, plot the four intercepts $(a, 0)$, $(-a, 0)$, $(0, b)$, and $(0, -b)$, and sketch an ellipse through the intercepts.

EXAMPLE 1 Graphing an Ellipse

Graph $\dfrac{x^2}{49} + \dfrac{y^2}{36} = 1$.

The x-intercepts of this ellipse are $(7, 0)$ and $(-7, 0)$. The y-intercepts are $(0, 6)$ and $(0, -6)$. Plotting the intercepts and sketching the ellipse through them gives the graph shown in Figure 20. ∎

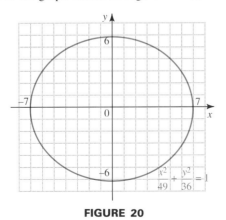

FIGURE 20

1. Graph.

   (a) $\dfrac{x^2}{4} + \dfrac{y^2}{25} = 1$

   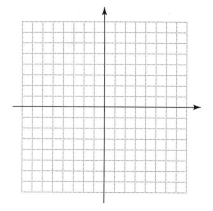

   (b) $\dfrac{x^2}{64} + \dfrac{y^2}{49} = 1$

   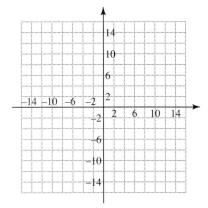

ANSWERS
1. (a) (b)

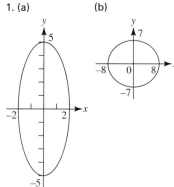

EXAMPLE 2 Graphing an Ellipse

Graph $\dfrac{x^2}{36} + \dfrac{y^2}{121} = 1$.

The x-intercepts for this ellipse are $(6, 0)$ and $(-6, 0)$, and the y-intercepts are $(0, 11)$ and $(0, -11)$. The graph has been sketched in Figure 21. ∎

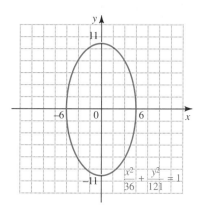

FIGURE 21

■ **WORK PROBLEM 1 AT THE SIDE.**

3 A **hyperbola** is the set of all points in a plane the *difference* of whose distances from two fixed points (called foci) is constant. Figure 22 shows a hyperbola; it can be shown, using this definition and the distance formula, that this hyperbola has equation

$$\dfrac{x^2}{16} - \dfrac{y^2}{12} = 1.$$

The x-intercepts are $(4, 0)$ and $(-4, 0)$. When $x = 0$ the equation becomes

$$-\dfrac{y^2}{12} = 1 \quad \text{or} \quad y^2 = -12.$$

This equation has no real number solutions, so there is no real number value for y corresponding to $x = 0$; that is, there are no y-intercepts.

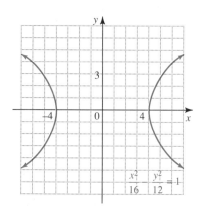

FIGURE 22

Figure 23 shows the graph of the hyperbola

$$\frac{y^2}{25} - \frac{x^2}{9} = 1.$$

Here the y-intercepts are (0, 5) and (0, −5), and there are no x-intercepts.

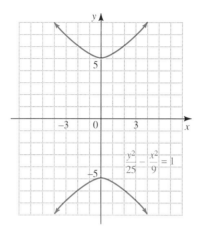

FIGURE 23

This discussion about hyberbolas is summarized as follows.

EQUATIONS OF HYPERBOLAS

A hyperbola with x-intercepts (a, 0) and (−a, 0) has an equation of the form

$$\frac{x^2}{a^2} - \frac{y^2}{b^2} = 1,$$

and a hyperbola with y-intercepts (0, b) and (0, −b) has an equation of the form

$$\frac{y^2}{b^2} - \frac{x^2}{a^2} = 1.$$

4 The two branches of the graph of a hyperbola approach a pair of intersecting straight lines called **asymptotes.** (See Figure 24.) These lines are useful for sketching the graph of the hyperbola. Find the asymptotes as follows.

ASYMPTOTES OF HYPERBOLAS

The extended diagonals of the rectangle with corners at the points (a, b), (−a, b), (−a, −b), and (a, −b) are the **asymptotes** of either of the hyperbolas

$$\frac{x^2}{a^2} - \frac{y^2}{b^2} = 1 \quad \text{or} \quad \frac{y^2}{b^2} - \frac{x^2}{a^2} = 1.$$

This rectangle is called the **fundamental rectangle.**

2. Use asymptotes to graph the following.

(a) $\dfrac{x^2}{4} - \dfrac{y^2}{25} = 1$

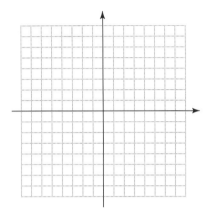

(b) $\dfrac{x^2}{81} - \dfrac{y^2}{64} = 1$

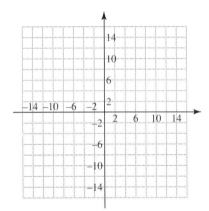

ANSWERS
2. (a) 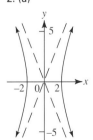 (b)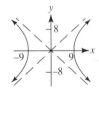

EXAMPLE 3 Graphing a Hyperbola

Graph $\dfrac{x^2}{9} - \dfrac{y^2}{25} = 1$.

Here $a = 3$ and $b = 5$. The x-intercepts are $(3, 0)$ and $(-3, 0)$. The four points $(3, 5)$, $(3, -5)$, $(-3, 5)$, and $(-3, -5)$ are the corners of the fundamental rectangle that determines the asymptotes shown in Figure 24. The hyperbola approaches these lines as x and y get larger and larger in absolute value. ■

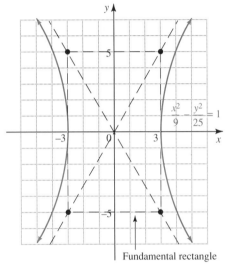

FIGURE 24

In summary, to graph either of the two forms of hyperbolas,

$$\dfrac{x^2}{a^2} - \dfrac{y^2}{b^2} = 1 \quad \text{or} \quad \dfrac{y^2}{b^2} - \dfrac{x^2}{a^2} = 1,$$

follow these steps.

GRAPHING A HYPERBOLA

Step 1 Locate the intercepts: at $(a, 0)$ and $(-a, 0)$ if the x^2 term has a positive coefficient, or at $(0, b)$ and $(0, -b)$ if the y^2 term has a positive coefficient.

Step 2 Locate the corners of a rectangle at (a, b), $(a, -b)$, $(-a, -b)$, and $(-a, b)$.

Step 3 Sketch the asymptotes (the extended diagonals of the rectangle).

Step 4 Sketch each branch of the hyperbola through an intercept and approaching the asymptotes.

■ **WORK PROBLEM 2 AT THE SIDE.**

5 By rewriting a second-degree equation in one of the forms given for ellipses, hyperbolas, circles, or parabolas, we can determine when the graph is one of these figures. A summary of the equations and graphs of the conic sections is given here.

9.4 THE ELLIPSE AND THE HYPERBOLA

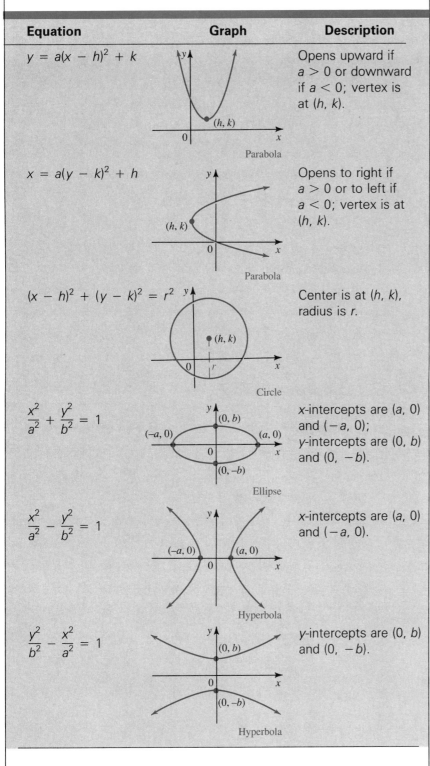

EXAMPLE 4 Identifying the Graph of a Given Equation

Identify the graph of each equation.

(a) $9x^2 = 108 + 12y^2$

Both variables are squared, so the graph is either an ellipse or a hyperbola. (This situation also occurs for a circle, which may be considered a

3. Identify the graph of each equation.

   (a) $3x^2 = 27 - 4y^2$

   (b) $x^2 - y^2 = 4$

   (c) $6x^2 = 100 + 2y^2$

   (d) $3x^2 = 27 - 4y$

   (e) $3x^2 = 27 - 3y^2$

4. Graph $\dfrac{y}{3} = \sqrt{1 - \dfrac{x^2}{4}}$.

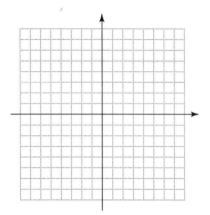

ANSWERS
3. (a) ellipse (b) hyperbola
   (c) hyperbola (d) parabola
   (e) circle
4.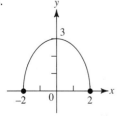

special case of the ellipse.) To see which one it is, rewrite the equation so that the x and y terms are on one side of the equation and 1 is on the other. Divide both sides by 108 so the constant is 1, and rewrite the equation so that both variables are on the same side of the equals sign to get

$$9x^2 = 108 + 12y^2$$

$$\frac{x^2}{12} = 1 + \frac{y^2}{9} \qquad \text{Divide by 108.}$$

$$\frac{x^2}{12} - \frac{y^2}{9} = 1. \qquad \text{Subtract } \frac{y^2}{9}.$$

Because of the minus sign, the graph of this equation is a hyperbola.

(b) $x^2 = y - 3$

Only one of the two variables is squared, x, so this is the vertical parabola $y = x^2 + 3$.

(c) $x^2 = 9 - y^2$

Get the variable terms on the same side of the equation.

$$x^2 = 9 - y^2$$

$$x^2 + y^2 = 9 \qquad \text{Add } y^2.$$

This equation represents a circle with center at the origin and radius 3. ∎

WORK PROBLEM 3 AT THE SIDE.

6 Square root functions were studied in Section 9.3. The next example illustrates another square root function.

EXAMPLE 5 Graphing a Square Root Function

Graph $\dfrac{y}{6} = \sqrt{1 - \dfrac{x^2}{16}}$.

Square both sides to get an equation whose form is known.

$$\frac{y^2}{36} = 1 - \frac{x^2}{16}$$

$$\frac{x^2}{16} + \frac{y^2}{36} = 1 \qquad \text{Add } \frac{x^2}{16}.$$

This is the equation of an ellipse with x-intercepts $(4, 0)$ and $(-4, 0)$ and y-intercepts $(0, 6)$ and $(0, -6)$. Since $\dfrac{y}{6}$ equals a principal square root in the original equation, y must be non-negative, restricting the graph to the upper half of the ellipse, as shown in Figure 25. Verify that this is the graph of a function, using the vertical line test. ∎

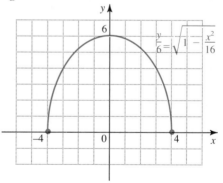

FIGURE 25

WORK PROBLEM 4 AT THE SIDE.

name   date   hour

9.4 EXERCISES

Graph each ellipse. See Examples 1 and 2.

1. $\dfrac{x^2}{4} + \dfrac{y^2}{9} = 1$ **2.** $\dfrac{x^2}{16} + \dfrac{y^2}{25} = 1$ **3.** $\dfrac{x^2}{9} + \dfrac{y^2}{16} = 1$

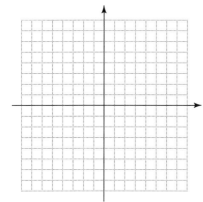 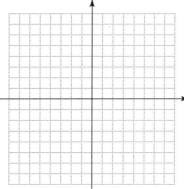 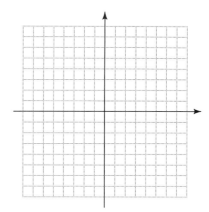

4. $\dfrac{x^2}{36} + \dfrac{y^2}{9} = 1$ **5.** $\dfrac{x^2}{16} + \dfrac{y^2}{4} = 1$ **6.** $\dfrac{x^2}{49} + \dfrac{y^2}{81} = 1$

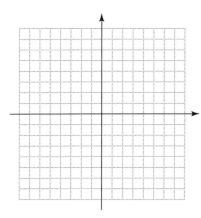 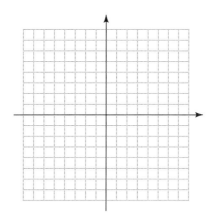 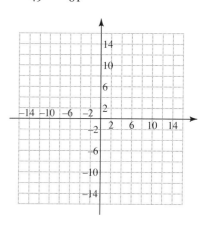

Graph each hyperbola. See Example 3.

7. $\dfrac{x^2}{25} - \dfrac{y^2}{9} = 1$ **8.** $\dfrac{y^2}{16} - \dfrac{x^2}{16} = 1$ **9.** $\dfrac{y^2}{49} - \dfrac{x^2}{36} = 1$

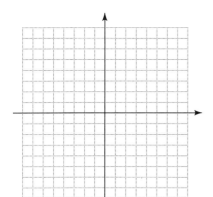 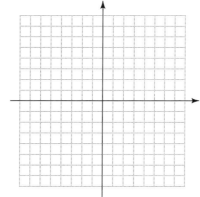 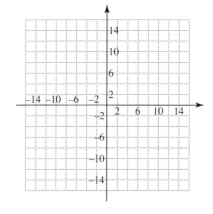

10. $\dfrac{x^2}{144} - \dfrac{y^2}{49} = 1$

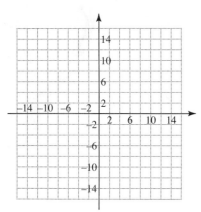

11. $\dfrac{x^2}{64} - \dfrac{y^2}{100} = 1$

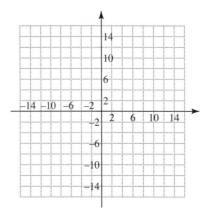

12. $\dfrac{x^2}{4} - \dfrac{y^2}{25} = 1$

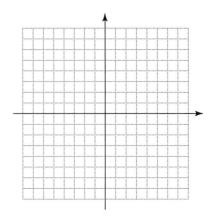

Identify each of the following as a parabola, circle, ellipse, *or* hyperbola. *Sketch the graph. See Example 4.*

13. $x^2 + y^2 = 16$

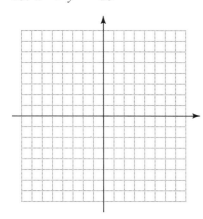

14. $9x^2 = 36 + 4y^2$

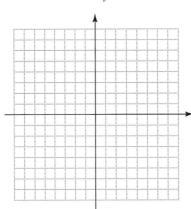

15. $4x^2 = 16 - 4y^2$

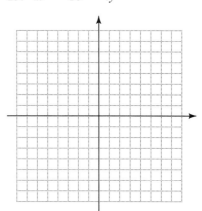

16. $4x^2 + 9y^2 = 36$

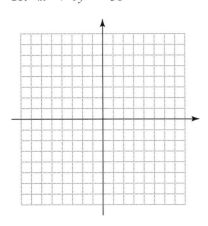

17. $2x^2 - y = 0$

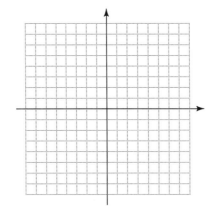

18. $x^2 + 2y^2 = 8$

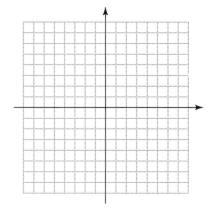

9.4 EXERCISES

19. $y^2 = 144 - x^2$

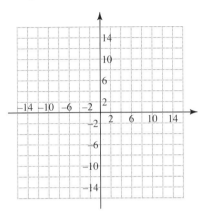

20. $x^2 - y^2 = 36$

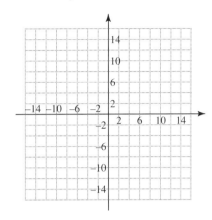

21. $x^2 + 9y^2 = 9$

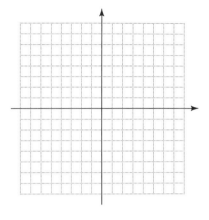

22. $25x^2 + 9y^2 = 225$

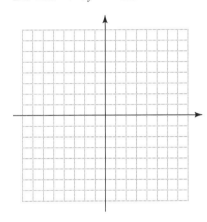

23. $25x^2 = 225 + 9y^2$

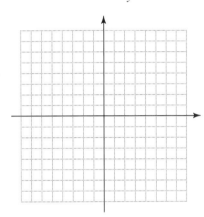

24. $4x^2 - 25y^2 = 100$

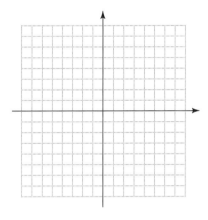

Graph each square root function. See Example 5.

25. $\dfrac{y}{3} = \sqrt{1 + \dfrac{x^2}{9}}$

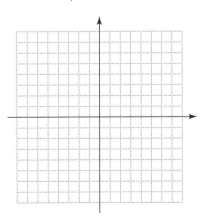

26. $y = \sqrt{\dfrac{x+6}{3}}$

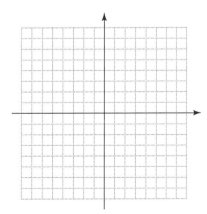

27. $y = -\sqrt{4 - 4x^2}$

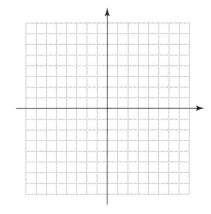

28. A pair of buildings in a sports complex are shaped and positioned like a portion of the branches of the hyperbola $400x^2 - 625y^2 = 250{,}000$ where x and y are in meters.

    (a) How far apart are the buildings at their closest point?

    (b) Find the distance d in the figure.

    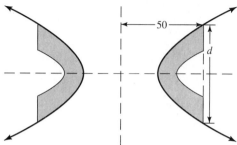

29. An arch has the shape of half an ellipse. The equation of the ellipse is $100x^2 + 324y^2 = 32{,}400$, where x and y are in meters.

    (a) How high is the center of the arch?

    (b) How wide is the arch across the bottom? (See the figure.)

    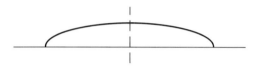

30. The orbit of Venus around the sun (one of the foci) is an ellipse with equation

    $$\frac{x^2}{5013} + \frac{y^2}{4970} = 1$$

    where x and y are measured in millions of miles.

    (a) Find the farthest distance between Venus and the sun.

    (b) Find the smallest distance between Venus and the sun. (*Hint:* See Figure 19 and use the fact that $c^2 = a^2 - b^2$.)

SKILL SHARPENERS

Solve each equation. See Section 6.3.

31. $2x^4 - 5x^2 - 3 = 0$

32. $3x^4 + 26x^2 + 35 = 0$

33. $x^4 - 16 = 0$

34. $x^4 - 81 = 0$

9.5 NONLINEAR SYSTEMS OF EQUATIONS

Any equation that cannot be written in the form $ax + by = c$, for real numbers a, b, and c, is a **nonlinear equation.** In this section, only nonlinear equations whose graphs are conic sections are considered. A **nonlinear system of equations** is a system with at least one nonlinear equation. Nonlinear systems can be solved by the elimination method, the substitution method, or a combination of the two. The following examples illustrate the use of these methods for solving nonlinear systems.

OBJECTIVES

1. Solve a nonlinear system by substitution.
2. Use the elimination method to solve a system with two second-degree equations.
3. Solve a system that requires a combination of methods.

1 The substitution method usually is most useful when one of the equations is linear. The first two examples illustrate this kind of system.

EXAMPLE 1 Using Substitution When One Equation Is Linear
Solve the system

$$x^2 + y^2 = 9 \quad (1)$$
$$2x - y = 3. \quad (2)$$

Solve the linear equation for one of the two variables, then substitute the resulting expression into the nonlinear equation to obtain an equation in one variable. Let us solve equation (2) for y.

$$2x - y = 3 \quad (2)$$
$$y = 2x - 3 \quad (3)$$

Substituting $2x - 3$ for y in equation (1) gives

$$x^2 + (\mathbf{2x - 3})^2 = 9$$
$$x^2 + 4x^2 - 12x + 9 = 9$$
$$5x^2 - 12x = 0.$$

Solve this quadratic equation by factoring.

$$x(5x - 12) = 0$$
$$x = 0 \quad \text{or} \quad x = \frac{12}{5} \qquad \text{Set each factor equal to 0; solve.}$$

Let $x = 0$ in the equation $y = 2x - 3$ to get $y = -3$. If $x = \frac{12}{5}$, then $y = \frac{9}{5}$. The solution set of the system is $\left\{(0, -3), \left(\frac{12}{5}, \frac{9}{5}\right)\right\}$. The graph of the system, shown in Figure 26, confirms the solution. ■

1. Solve each system.

    (a) $x^2 + y^2 = 10$
    $x = y + 2$

    (b) $x^2 - 2y^2 = 8$
    $y + x = 6$

    (c) $4x^2 + 3y^2 = 7$
    $x - y = 0$

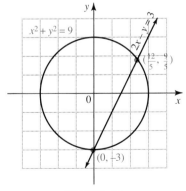

FIGURE 26

WORK PROBLEM 1 AT THE SIDE. ■

ANSWERS
1. (a) $\{(3, 1), (-1, -3)\}$
   (b) $\{(4, 2), (20, -14)\}$
   (c) $\{(1, 1), (-1, -1)\}$

2. Solve each system.

(a) $xy = 8$
$x + y = 6$

(b) $xy = -4$
$2x + y = -2$

(c) $xy + 10 = 0$
$4x + 9y = -2$

EXAMPLE 2 Using Substitution When One Equation Is Linear

Solve the system
$$6x - y = 5 \quad (4)$$
$$xy = 4. \quad (5)$$

Solve either equation for one of the variables and then substitute the result into the other equation. Solving $xy = 4$ for x gives $x = \frac{4}{y}$. Substituting $\frac{4}{y}$ for x in equation (4) gives

$$6\left(\frac{4}{y}\right) - y = 5.$$

To clear fractions, multiply both sides by y, noting the restriction that y cannot be 0. Then solve for y.

$$\frac{24}{y} - y = 5$$
$$24 - y^2 = 5y \quad \text{Multiply by } y.$$
$$0 = y^2 + 5y - 24 \quad \text{Get 0 on one side.}$$

Solve this quadratic equation by factoring.

$$0 = (y - 3)(y + 8)$$
$$y = 3 \quad \text{or} \quad y = -8 \quad \text{Set each factor equal to 0; solve.}$$

Substitute these results into $x = \frac{4}{y}$ to obtain the corresponding values for x.

$$\text{If } y = 3, \text{ then } x = \frac{4}{3}.$$
$$\text{If } y = -8, \text{ then } x = -\frac{1}{2}.$$

The solution set is $\left\{\left(\frac{4}{3}, 3\right), \left(-\frac{1}{2}, -8\right)\right\}$. ∎

■ WORK PROBLEM 2 AT THE SIDE.

2 The elimination method is often useful when both equations are second-degree equations. This method is used in the following example.

EXAMPLE 3 Solving a Nonlinear System by Elimination

Solve the system
$$x^2 + y^2 = 9 \quad (6)$$
$$2x^2 - y^2 = -6. \quad (7)$$

Adding the two equations will eliminate y, leaving an equation that can be solved for x.

$$x^2 + y^2 = 9 \quad (6)$$
$$\underline{2x^2 - y^2 = -6} \quad (7)$$
$$3x^2 = 3$$
$$x^2 = 1$$
$$x = 1 \quad \text{or} \quad x = -1$$

ANSWERS
2. (a) $\{(4, 2), (2, 4)\}$
(b) $\{(-2, 2), (1, -4)\}$
(c) $\left\{(-5, 2), \left(\frac{9}{2}, -\frac{20}{9}\right)\right\}$

Each value of x gives corresponding values for y when substituted into one of the original equations. Using equation (6) gives the following results.

If $x = 1$,
$(\mathbf{1})^2 + y^2 = 9$
$y^2 = 8$
$y = \sqrt{8}$ or $-\sqrt{8}$
$y = 2\sqrt{2}$ or $-2\sqrt{2}$.

If $x = -1$,
$(\mathbf{-1})^2 + y^2 = 9$
$y^2 = 8$
$y = 2\sqrt{2}$ or $-2\sqrt{2}$.

The solution set is $\{(1, 2\sqrt{2}), (1, -2\sqrt{2}), (-1, 2\sqrt{2}), (-1, -2\sqrt{2})\}$. The graph in Figure 27 shows the four points of intersection. ∎

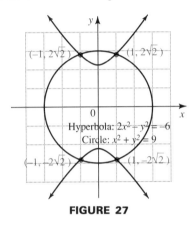

FIGURE 27

WORK PROBLEM 3 AT THE SIDE.

3 The next example shows a system of second-degree equations that can be solved only by a combination of methods.

EXAMPLE 4 Solving a Nonlinear System by a Combination of Methods

Solve the system

$$x^2 + 2xy - y^2 = 7 \quad (8)$$
$$x^2 - y^2 = 3. \quad (9)$$

The elimination method is used here in combination with the substitution method. To begin, eliminate the squared terms by multiplying both sides of equation (9) by -1 and then adding the result to (8).

$$\begin{aligned} x^2 + 2xy - y^2 &= 7 \\ -x^2 + y^2 &= -3 \\ \hline 2xy &= 4 \end{aligned}$$

Next, solve $2xy = 4$ for either variable. Let us solve for y.

$$2xy = 4$$
$$y = \frac{2}{x} \quad (10)$$

3. Solve each system.

(a) $x^2 + y^2 = 41$
$x^2 - y^2 = 9$

(b) $x^2 - 5y^2 = 4$
$x^2 - 3y^2 = 6$

(c) $x^2 + 3y^2 = 40$
$4x^2 - y^2 = 4$

ANSWERS
3. (a) $\{(5, 4), (5, -4), (-5, 4), (-5, -4)\}$
(b) $\{(3, 1), (-3, 1), (-3, -1), (3, -1)\}$
(c) $\{(2, 2\sqrt{3}), (2, -2\sqrt{3}), (-2, 2\sqrt{3}), (-2, -2\sqrt{3})\}$

4. Solve each system.

   (a) $x^2 + xy + y^2 = 3$
   $x^2 + y^2 = 5$

   (b) $x^2 + 7xy - 2y^2 = -8$
   $-2x^2 + 4y^2 = 16$

Now substitute $y = \dfrac{2}{x}$ into one of the original equations. It is easier to do this with (9).

$$x^2 - y^2 = 3 \qquad (9)$$

$$x^2 - \left(\dfrac{2}{x}\right)^2 = 3$$

$$x^2 - \dfrac{4}{x^2} = 3$$

To clear the equation of fractions, multiply both sides by x^2.

$$x^4 - 4 = 3x^2$$
$$x^4 - 3x^2 - 4 = 0$$
$$(x^2 - 4)(x^2 + 1) = 0$$

$x^2 - 4 = 0$ or $x^2 + 1 = 0$
$x^2 = 4$ $x^2 = -1$
$x = 2$ or $x = -2$ $x = i$ or $x = -i$

By substituting the four values of x from above into equation (10), we get the corresponding values for y.

If $x = 2$, then $y = 1$.   If $x = i$, then $y = -2i$.
If $x = -2$, then $y = -1$.   If $x = -i$, then $y = 2i$.

There are four solutions, two real and two imaginary, in the solution set: $\{(2, 1), (-2, -1), (i, -2i), (-i, 2i)\}$. ■

WORK PROBLEM 4 AT THE SIDE.

ANSWERS
4. (a) $\{(1, -2), (-1, 2), (2, -1), (-2, 1)\}$
   (b) $\{(0, 2), (0, -2), (2i\sqrt{2}, 0), (-2i\sqrt{2}, 0)\}$

name date hour

9.5 EXERCISES

Solve the following systems by the substitution method. See Examples 1 and 2.

1. $y = x^2 + 2x$
   $y = x$

2. $y = 2x^2 - x$
   $2y = 4x$

3. $y = x^2 - 4x + 4$
   $x + y = 2$

4. $y = x^2 + 2x + 1$
   $x - y = -7$

5. $x^2 + y^2 = 1$
   $x + 2y = 1$

6. $x^2 + 3y^2 = 3$
   $x = 3y$

7. $xy = 2$
   $x + 3y = -5$

8. $xy = -12$
   $x + 3y = 5$

CHAPTER 9 GRAPHS OF FUNCTIONS AND CONIC SECTIONS

9. $xy = -6$
$x + y = -1$

10. $xy = 4$
$x + y = 5$

11. $y = 2x^2 + 4x$
$y = x^2 - 3x - 10$

12. $y = 2x^2$
$y = 8x^2 - 2x - 4$

13. $4x^2 - y^2 = 7$
$y = 2x^2 - 3$

14. $y = x^2 - 3$
$x^2 + y^2 = 9$

15. $x^2 - xy + y^2 = 0$
$x - 2y = 1$

16. $x^2 - 3x + y^2 = 4$
$2x - y = 3$

Solve the following systems by the elimination method or by a combination of the elimination and substitution methods. See Examples 3 and 4.

17. $2x^2 + 3y^2 = 6$
$x^2 + 3y^2 = 3$

18. $3x^2 + y^2 = 13$
$4x^2 - 3y^2 = 13$

9.5 EXERCISES

19. $5x^2 - 2y^2 = -13$
$3x^2 + 4y^2 = 39$

20. $6x^2 + y^2 = 9$
$3x^2 + 4y^2 = 36$

21. $xy = 5$
$2x^2 - y^2 = 5$

22. $xy = 12$
$3x^2 + y^2 = 57$

23. $4x^2 + 4y^2 = 16$
$2x^2 + 3y^2 = 5$

24. $3x^2 + 3y^2 = 12$
$x^2 - y^2 = 6$

25. $x^2 + 2xy - y^2 = 7$
$x^2 - y^2 = 3$

26. $2x^2 + 3xy + 2y^2 = 21$
$x^2 + y^2 = 6$

27. $3x^2 + 2xy - 3y^2 = 5$
$-x^2 - 3xy + y^2 = 3$

28. $-2x^2 + 7xy - 3y^2 = 4$
$2x^2 - 3xy + 3y^2 = 4$

Write a nonlinear system of equations and solve.

29. The sum of the squares of two numbers is 106. The difference of the squares of the same two numbers is 56. Find the numbers.

30. The sum of the squares of two numbers is 8. The product of the two numbers is 4. Find the numbers.

31. Find the length and width of the rectangle whose perimeter is 50 centimeters and whose area is 100 square centimeters.

32. The area of a rectangle is 84 square inches and its perimeter is 38 inches. Find the length and width of the rectangle.

SKILL SHARPENERS

Graph each inequality. See Section 7.4.

33. $2x - y \leq 4$

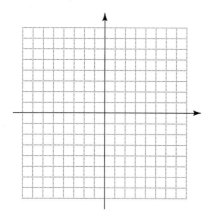

34. $-x + 3y > 9$

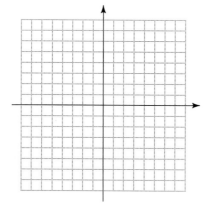

35. $4x + 2y > 8$

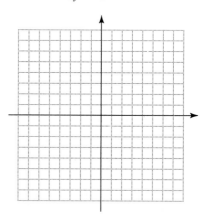

9.6 SECOND-DEGREE INEQUALITIES; SYSTEMS OF INEQUALITIES

OBJECTIVES

1. Graph second-degree inequalities.
2. Graph a system of inequalities.

1 Recall from Section 7.4 that a linear inequality such as $3x + 2y \leq 5$ is graphed by first graphing the boundary line $3x + 2y = 5$. **Second-degree inequalities** such as $x^2 + y^2 \leq 36$ are graphed in much the same way. The boundary of $x^2 + y^2 \leq 36$ is the graph of the equation $x^2 + y^2 = 36$, a circle with radius 6 and center at the origin, as shown in Figure 28. As with linear inequalities, the inequality $x^2 + y^2 \leq 36$ will include either the points outside the circle or the points inside the circle. Decide which region to shade by substituting any point not on the circle, such as $(0, 0)$, into the inequality. Since $0^2 + 0^2 < 36$ is a true statement, the inequality includes the points inside the circle, the shaded region in Figure 28.

1. Graph
$y \geq (x + 1)^2 - 5.$

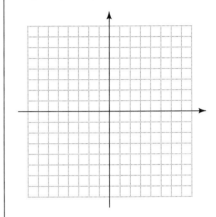

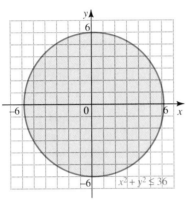

FIGURE 28

EXAMPLE 1 Graphing a Second-Degree Inequality
Graph $y < -2(x - 4)^2 - 3$.

The boundary, $y = -2(x - 4)^2 - 3$, is a parabola opening downward with vertex at $(4, -3)$. Using the point $(0, 0)$ as a test point gives

$$0 < -2(0 - 4)^2 - 3$$
$$0 < -32 - 3$$
$$0 < -35. \quad \text{False}$$

This is a false statement, so the points in the region containing $(0, 0)$ do not satisfy the inequality. Figure 29 shows the final graph; the parabola is drawn with a dashed line since the points of the parabola itself do not satisfy the inequality. ■

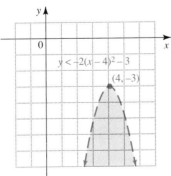

FIGURE 29

WORK PROBLEM 1 AT THE SIDE.

ANSWERS
1.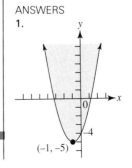

2. Graph
$x^2 + 4y^2 > 36$.

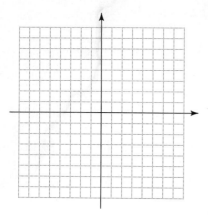

EXAMPLE 2 Graphing a Second-Degree Inequality

Graph $16y^2 \leq 144 + 9x^2$.

First rewrite the inequality as follows.

$$16y^2 - 9x^2 \leq 144$$

$$\frac{y^2}{9} - \frac{x^2}{16} \leq 1 \qquad \text{Divide both sides by 144.}$$

This form of the inequality shows that the boundary is the hyperbola

$$\frac{y^2}{9} - \frac{x^2}{16} = 1.$$

Since the test point (0, 0) satisfies the inequality $16y^2 \leq 144 + 9x^2$, the region containing (0, 0) is shaded, as shown in Figure 30. ∎

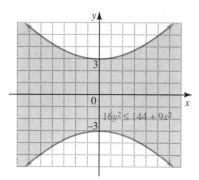

FIGURE 30

WORK PROBLEM 2 AT THE SIDE.

The solution set of a **system of inequalities,** such as

$$2x + 3y > 6$$
$$x^2 + y^2 < 16,$$

is the intersection of the solution sets of the individual inequalities.

The graph of the solution set of this system is the set of all points on the plane that belong to the graphs of both inequalities in the system. Graph this system by graphing both inequalities on the same coordinate axes, as shown in Figure 31. The heavily shaded region containing those points that belong to both graphs is the graph of the system. In this case the points of the boundary lines are not included.

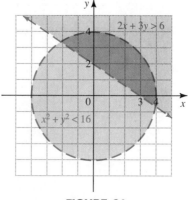

FIGURE 31

ANSWER
2.
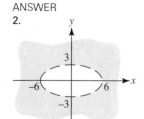

EXAMPLE 3 Graphing a System of Inequalities
Graph the solution of the system

$$x + y < 1$$
$$y \le 2x + 3$$
$$y \ge -2.$$

Graph each inequality separately, on the same axes. The graph of the system is the triangular region enclosed by the three lines in Figure 32. It contains all points that satisfy all three inequalities. One boundary line ($x + y = 1$) is dashed, while the other two are solid. ∎

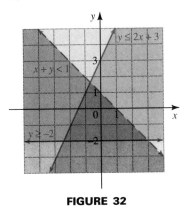

FIGURE 32

WORK PROBLEM 3 AT THE SIDE.

EXAMPLE 4 Graphing a System of Inequalities
Graph the solution of the system

$$y \ge x^2 - 2x + 1$$
$$2x^2 + y^2 > 4$$
$$y < 4.$$

The graph of $y = x^2 - 2x + 1$ is a parabola with vertex at (1, 0). Those points in the interior of the parabola satisfy the condition $y > x^2 - 2x + 1$. Thus points on the parabola or in the interior are in the solution of $y \ge x^2 - 2x + 1$. The graph of $2x^2 + y^2 = 4$ is an ellipse. To satisfy the inequality $2x^2 + y^2 > 4$, a point must lie outside the ellipse. The graph of $y < 4$ includes all points below the line $y = 4$. Finally, the graph of the system is the shaded region in Figure 33 that lies outside the ellipse, inside or on the boundary of the parabola, and below the line $y = 4$. ∎

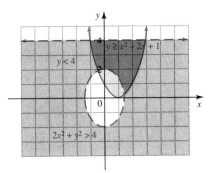

FIGURE 33

WORK PROBLEM 4 AT THE SIDE.

3. Graph the solution of the system

$$2x - 5y \ge 10$$
$$x + 3y \ge 6$$
$$y \le 2.$$

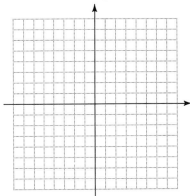

4. Graph the solution of the system

$$y \ge x^2 + 1$$
$$\frac{x^2}{9} + \frac{y^2}{4} \ge 1$$
$$y \le 5.$$

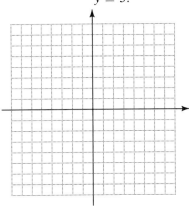

ANSWERS

3.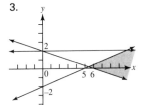

4.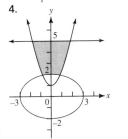

Historical Reflections

WOMEN IN MATHEMATICS:
Maria Gaetana Agnesi (1718–1799)

Maria Agnesi was born in Milan, Italy, in 1718, the first of her father's twenty-one children. She grew up in a scholarly atmosphere; her father was a mathematician on the faculty of the University of Bologna. She was fluent in several languages by the age of 13, but she chose mathematics over literature as her field of study. Much of her scholarly work was done in analytic geometry, but she also published in many other fields, including logic, gravitation, chemistry, botany, and zoology. The curve shown here, which is studied in analytic geometry, is known as the *Witch of Agnesi*.

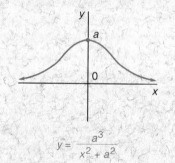

$$y = \frac{a^3}{x^2 + a^2}$$

Art: (Top) The Bettman Archive

9.6 EXERCISES

Graph each of the following inequalities. See Examples 1 and 2.

1. $y < x^2$

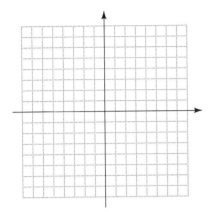

2. $y^2 < 36 - x^2$

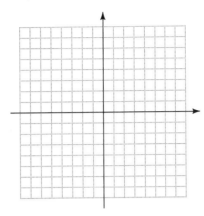

3. $x^2 \geq 4 - 2y^2$

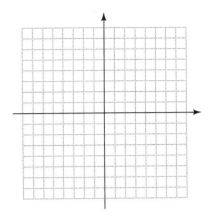

4. $y \geq x^2 - 4$

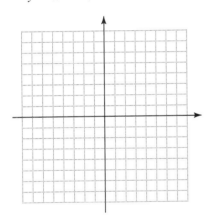

5. $y^2 \leq 16 + x^2$

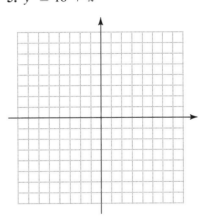

6. $x^2 \leq 36 + 4y^2$

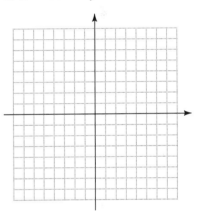

7. $y \leq x^2 + 4x + 6$

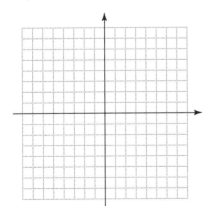

8. $y > 2x^2 - 4x + 3$

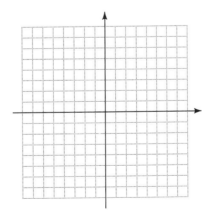

9. $9x^2 > 4y^2 - 36$

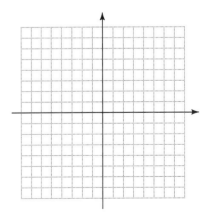

9.6 EXERCISES 573

10. $9x^2 \leq 4y^2 + 36$

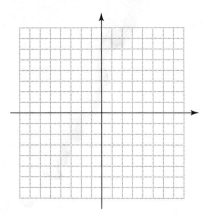

11. $y^2 \leq 4 - 2x^2$

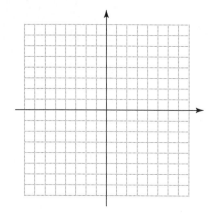

12. $x^2 + 2y^2 < 9$

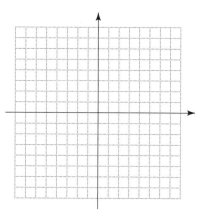

13. $2x^2 + 2y^2 \geq 98$

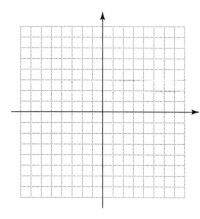

14. $4y^2 \leq 64 - 16x^2$

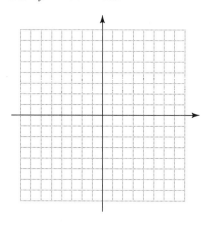

15. $x \leq y^2 - 4y + 1$

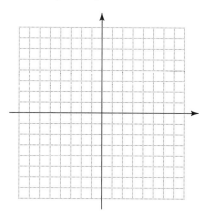

16. $x \geq 2y^2 + 6y - 1$

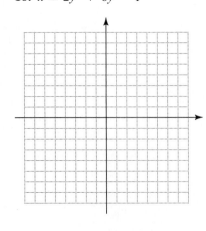

17. $y^2 - 9x^2 \geq 9$

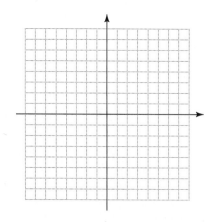

18. $16x^2 < 9y^2 + 144$

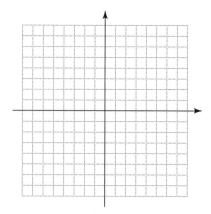

9.6 EXERCISES

Graph the following systems of inequalities. See Examples 3 and 4.

19. $3x + 2y < 12$
$x - 3y < 6$

20. $x - 2y > -4$
$2x + 3y < 8$

21. $2x - 3y \leq 6$
$4x + 5y \geq 10$

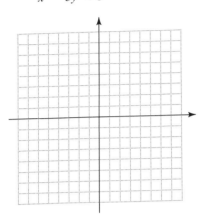
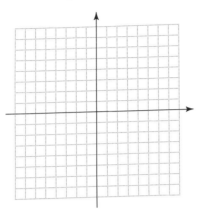
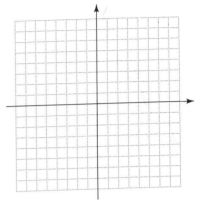

22. $3x - y \leq 0$
$x - 2y \leq 8$

23. $x \leq 4$
$y \leq 6$

24. $x \geq -3$
$y \leq 2$

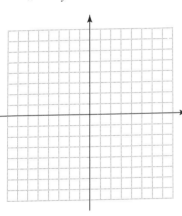
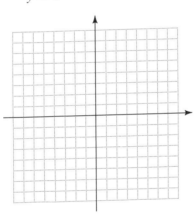
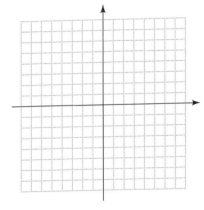

25. $2x - y > 0$
$y > x^2$

26. $y > x^2 - 2$
$y < -x^2 + 2$

27. $x^2 - y^2 \geq 1$
$\dfrac{x^2}{9} + \dfrac{y^2}{4} \leq 1$

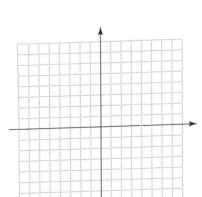
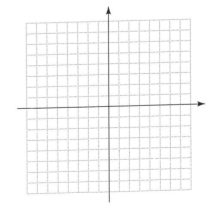

575

CHAPTER 9 GRAPHS OF FUNCTIONS AND CONIC SECTIONS

28. $-x^2 + \dfrac{y^2}{4} \geq 1$
$-4 \leq y \leq 4$

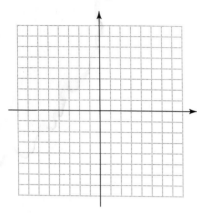

29. $x > 1$
$y < 2$
$x^2 + y^2 < 4$

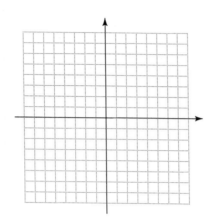

30. $x \geq 0$
$y \geq 0$
$x^2 + y^2 \geq 1$
$x + y \leq 4$

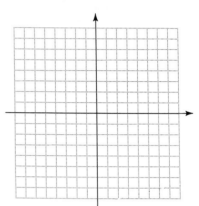

31. $y \leq -x^2$
$y \geq x - 3$
$y \leq -1$
$x < 1$

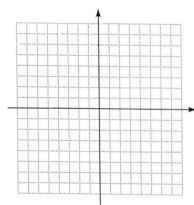

32. $y < x^2$
$y > -2$
$x + y < 3$
$3x - 2y > -6$

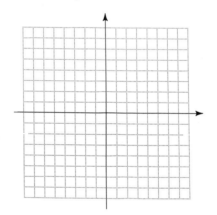

SKILL SHARPENERS
Graph each of the following equations. See Section 7.1.

33. $y = 2x - 3$

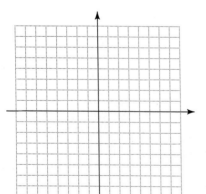

34. $2y = x + 3$

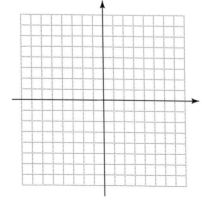

35. $3x + 5y = 15$

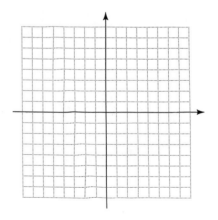

CHAPTER 9 SUMMARY

KEY TERMS

9.1 **conic sections** — Graphs that result from cutting an infinite cone with a plane are called conic sections.

parabola — The graph of a second-degree function is a parabola.

vertex — The point on a parabola that has the smallest y-value (if the parabola opens upward) or the largest y-value (if the parabola opens downward) is called the vertex of the parabola.

axis — The vertical or horizontal line through the vertex of a parabola is its axis.

quadratic function — A function that can be written in the form $f(x) = ax^2 + bx + c$, for real numbers a, b, and c, with $a \neq 0$, is a quadratic function.

9.3 **circle** — A circle is the set of all points in a plane that lie a fixed distance from a fixed point.

center — The fixed point discussed in the definition of a circle is the center of the circle.

radius — The radius of a circle is the fixed distance between the center and any point on the circle.

9.4 **ellipse** — An ellipse is the set of all points in a plane the sum of whose distances from two fixed points is constant.

hyperbola — A hyperbola is the set of all points in a plane the difference of whose distances from two fixed points is constant.

asymptotes of a hyperbola — The two intersecting lines that the branches of a hyperbola approach are called asymptotes of the hyperbola.

9.5 **nonlinear equation** — Any equation that cannot be written in the form $ax + by = c$, for real numbers a, b, and c, is a nonlinear equation.

nonlinear system of equations — A nonlinear system of equations is a system with at least one nonlinear equation.

9.6 **second-degree inequality** — A second-degree inequality is an inequality with at least one variable of degree two and no variable with degree greater than two.

system of inequalities — A system of inequalities consists of two or more inequalities to be solved at the same time.

QUICK REVIEW

Section	Concepts	Examples
9.1 Graphing Parabolas	1. The graph of the quadratic function $$f(x) = a(x - h)^2 + k, \quad a \neq 0$$ is a parabola with vertex at (h, k) and the vertical line $x = h$ as axis. 2. The graph opens upward if a is positive and downward if a is negative. 3. The graph is wider than $f(x) = x^2$ if $0 < |a| < 1$ and narrower if $|a| > 1$.	Graph $f(x) = -(x + 3)^2 + 1$.

CHAPTER 9 GRAPHS OF FUNCTIONS AND CONIC SECTIONS

Section	Concepts	Examples
9.2 More About Parabolas	The vertex of the graph of $f(x) = ax^2 + bx + c$, $a \neq 0$, may be found by completing the square. The vertex has coordinates $$\left(-\frac{b}{2a}, f\left(-\frac{b}{2a}\right)\right).$$ Steps in graphing a quadratic function: *Step 1* Find the y-intercept by evaluating $f(0)$. *Step 2* Find any x-intercepts by solving $f(x) = 0$. *Step 3* Find the vertex either by using the formula or by completing the square. *Step 4* Find and plot any additional points as needed, using the symmetry about the axis. *Step 5* Verify that the graph opens upward (if $a > 0$) or opens downward (if $a < 0$). If the discriminant, $b^2 - 4ac$, is positive, the graph of $f(x) = ax^2 + bx + c$ has two x-intercepts; if zero, one x-intercept; if negative, no x-intercepts. The graph of $x = ay^2 + by + c$ is a horizontal parabola, opening to the right if $a > 0$, or to the left if $a < 0$.	Graph $f(x) = x^2 + 4x + 3$. The vertex is $(-2, -1)$. Since $f(0) = 3$, the y-intercept is $(0, 3)$. The solutions of $x^2 + 4x + 3 = 0$ are -1 and -3, so the x-intercepts are $(-1, 0)$ and $(-3, 0)$. Graph $x = 2y^2 + 6y + 5$.

CHAPTER 9 SUMMARY

Section	Concepts	Examples
9.3 The Distance Formula and the Circle	The distance between (x_1, y_1) and (x_2, y_2) is $$d = \sqrt{(x_2 - x_1)^2 + (y_2 - y_1)^2}.$$ The circle with radius r and center at (h, k) has an equation of the form $$(x - h)^2 + (y - k)^2 = r^2.$$	The distance between $(3, -2)$ and $(-1, 1)$ is given by $$\sqrt{(-1 - 3)^2 + [1 - (-2)]^2}$$ $$= \sqrt{(-4)^2 + 3^2} = \sqrt{16 + 9}$$ $$= \sqrt{25} = 5.$$ The circle $(x + 2)^2 + (y - 3)^2 = 25$ has center $(-2, 3)$ and radius 5.
9.4 The Ellipse and the Hyperbola	The ellipse whose x-intercepts are $(a, 0)$ and $(-a, 0)$ and whose y-intercepts are $(0, b)$ and $(0, -b)$ has an equation of the form $$\frac{x^2}{a^2} + \frac{y^2}{b^2} = 1.$$ A hyperbola with x-intercepts $(a, 0)$ and $(-a, 0)$ has an equation of the form $$\frac{x^2}{a^2} - \frac{y^2}{b^2} = 1$$ and a hyperbola with y-intercepts $(0, b)$ and $(0, -b)$ has an equation of the form $$\frac{y^2}{b^2} - \frac{x^2}{a^2} = 1.$$ The extended diagonals of the fundamental rectangle with corners at the points (a, b), $(-a, b)$, $(-a, -b)$, and $(a, -b)$ are the asymptotes of these hyperbolas.	Graph $$\frac{x^2}{9} + \frac{y^2}{4} = 1.$$ Graph $$\frac{x^2}{4} - \frac{y^2}{4} = 1.$$ The fundamental rectangle has corners at $(2, 2)$, $(-2, 2)$, $(-2, -2)$, and $(2, -2)$.

CHAPTER 9 GRAPHS OF FUNCTIONS AND CONIC SECTIONS

Section	Concepts	Examples
9.5 Nonlinear Systems of Equations	Nonlinear systems can be solved by the substitution method, the elimination method, or a combination of the two.	Solve the system $$x^2 + 2xy - y^2 = 14$$ $$x^2 \quad\quad - y^2 = -16. \quad (*)$$ Multiply equation (*) by -1 and use elimination. $$\begin{aligned} x^2 + 2xy - y^2 &= 14 \\ -x^2 \quad\quad + y^2 &= 16 \\ \hline 2xy &= 30 \\ xy &= 15 \end{aligned}$$ Solve for y to obtain $y = \frac{15}{x}$, and substitute into equation (*). $$x^2 - \left(\frac{15}{x}\right)^2 = -16$$ This simplifies to $$x^2 - \frac{225}{x^2} = -16.$$ Multiply by x^2 and get one side equal to 0. $$x^4 + 16x^2 - 225 = 0$$ Factor and solve. $$(x^2 - 9)(x^2 + 25) = 0$$ $$x = \pm 3 \quad\quad x = \pm 5i$$ Find corresponding y values to get the solution set $\{(3, 5), (-3, -5), (5i, -3i), (-5i, 3i)\}$.
9.6 Second-Degree Inequalities; Systems of Inequalities	To graph a second-degree inequality, graph the corresponding equation as a boundary and use test points to determine which region(s) form the solution. Shade the appropriate region(s).	Graph $y \geq x^2 - 2x + 3$.
	The solution set of a system of inequalities is the intersection of the solution sets of the individual inequalities.	Graph the solution set of the system $$3x - 5y > -15$$ $$x^2 + y^2 \leq 25.$$

580

CHAPTER 9 REVIEW EXERCISES

[9.1] *Sketch the graph of each quadratic function. Identify the vertex.*

1. $f(x) = -5x^2$

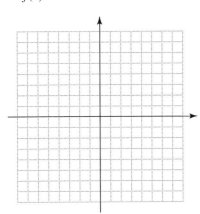

2. $f(x) = 3x^2 - 2$

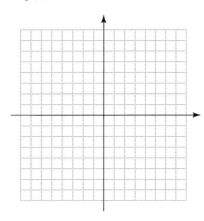

3. $f(x) = -x^2 + 4$

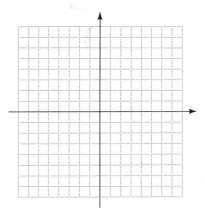

4. $f(x) = 6 - 2x^2$

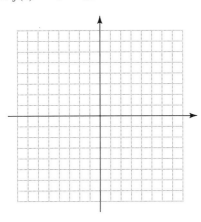

5. $f(x) = (x + 2)^2$

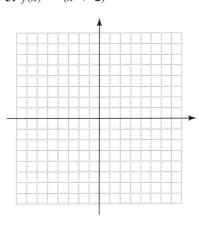

6. $f(x) = (x - 3)^2 - 7$

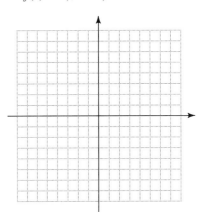

[9.2] *Identify the vertex of each parabola.*

7. $y = x^2 - 6x + 7$

8. $x = y^2 + 3y + 2$

9. $x = -2y^2 + 8y - 1$

10. $y = -3x^2 + 4x - 2$

A projectile is fired into the air. Its height (in feet) above the ground t seconds after firing is $s = -16t^2 + 160t$.

11. Find the number of seconds required for the projectile to reach maximum height.

12. Find the maximum height above the ground that the projectile reaches.

13. Find the length and width of a rectangle having a perimeter of 100 meters if the area is to be a maximum.

14. Find the two numbers whose sum is 10 and whose product is a maximum.

[9.3] *Find the distance between the points in each pair.*

15. $(1, 5)$ and $(-2, 3)$

16. $(-4, -2)$ and $(3, 0)$

17. $(7, -4)$ and $(7, 6)$

Write an equation for each of the following circles.

18. Center at $(-3, 4)$; radius 5

19. Center at $(6, -1)$; radius 3

20. Center at the origin; radius 8

Find the center and radius of each circle.

21. $x^2 + y^2 + 8x - 8y + 23 = 0$

22. $x^2 + y^2 - 10x - 4y - 7 = 0$

23. $3x^2 + 3y^2 - 6x - 12y - 21 = 0$

24. $5x^2 + 5y^2 - 5x + 10y = 8$

CHAPTER 9 REVIEW EXERCISES

[9.4] *Graph the following.*

25. $\dfrac{x^2}{9} + \dfrac{y^2}{25} = 1$

26. $\dfrac{x^2}{64} + \dfrac{y^2}{9} = 1$

27. $\dfrac{x^2}{16} - \dfrac{y^2}{36} = 1$

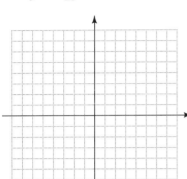
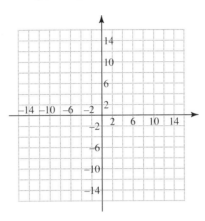
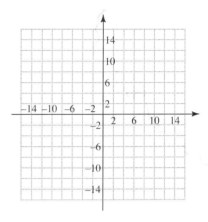

28. $\dfrac{x^2}{25} - \dfrac{y^2}{81} = 1$

29. $3x^2 + y^2 = 36$

30. $x^2 - y^2 = 16$

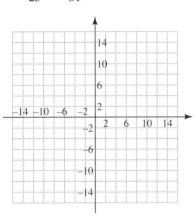
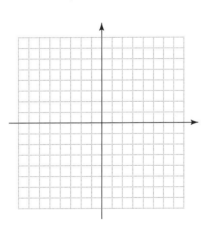
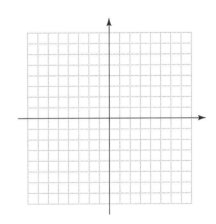

Identify each of the following as a straight line, parabola, circle, ellipse, *or* hyperbola.

31. $y = x^2 - 2$

32. $x^2 + y^2 = 16$

33. $8x - 9y = 2$

34. $y = x^2 - 4x + 2$

35. $x^2 - y^2 = 16$

36. $x^2 + 9y^2 = 18$

CHAPTER 9 GRAPHS OF FUNCTIONS AND CONIC SECTIONS

[9.5] *Solve each system.*

37. $y = 2x - x^2$
$x + 2y = -3$

38. $2y - 2 = x^2 + 3x$
$x - y = -3$

39. $3x^2 + y^2 = 19$
$2x - y = -2$

40. $xy = 15$
$2x - y = 1$

41. $x^2 + y^2 = 8$
$2x^2 - y^2 = 4$

42. $4x^2 + 3y^2 = 24$
$2x^2 - y^2 = 12$

[9.6] *Graph each of the following inequalities.*

43. $y \leq (x - 2)^2$

44. $4y^2 \geq 100 + 25x^2$

45. $x^2 + y^2 \leq 25$

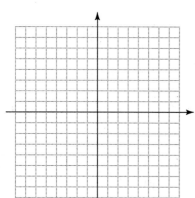
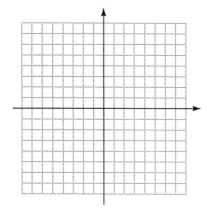
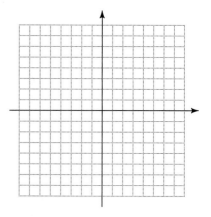

CHAPTER 9 REVIEW EXERCISES

Graph each of the following systems of inequalities.

46. $3x + 2y \geq 10$
$x - 2y \geq 5$

47. $y > 2x^2 - 3$
$y < -x^2 + 4$

48. $|x| \leq 2$
$|y| > 1$
$4x^2 + 9y^2 \leq 36$

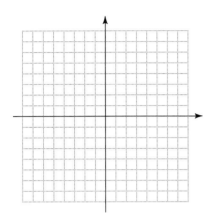
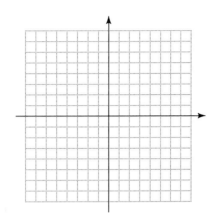
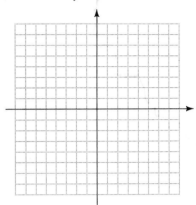

MIXED REVIEW EXERCISES
Graph.

49. $f(x) = -(x + 2)^2 - 1$

50. $x^2 + y^2 = 4$

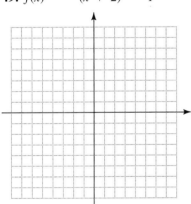
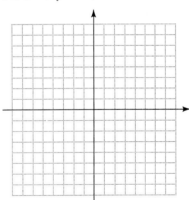

51. $\dfrac{x^2}{4} = 1 + \dfrac{y^2}{9}$

52. $\dfrac{x^2}{81} + \dfrac{y^2}{16} = 1$

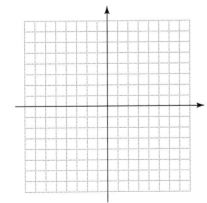
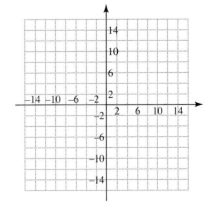

585

53. $f(x) = -\sqrt{2-x}$

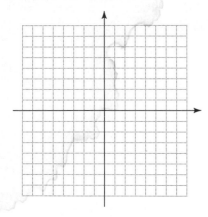

54. $f(x) = -\sqrt{25-x^2}$

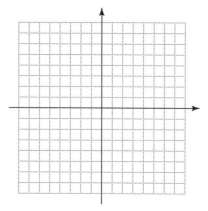

55. $x + 2y \geq 0$
$x \geq -2$
$y \leq 3$

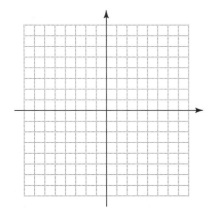

56. $x^2 + y^2 < 25$
$-x + y > 4$

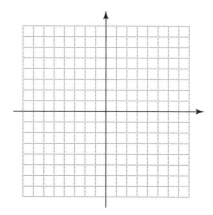

CHAPTER 9 TEST

1. Graph the quadratic function $f(x) = -x^2 + 4$. Identify the vertex.

1. _____

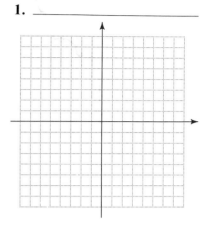

2. Identify the vertex of the graph of $f(x) = 2x^2 + 4x + 1$. Sketch the graph.

2. _____

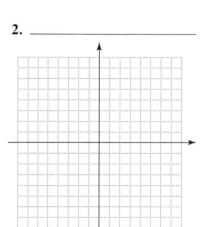

3. The distance in feet an object moves in t seconds is given by
$$d = -2t^2 + 12t + 32.$$
Find the maximum distance it reaches and the time it takes to reach that distance.

3.

4. Find the distance between the points $(-3, -4)$ and $(2, -1)$.

4. _____

5. _____

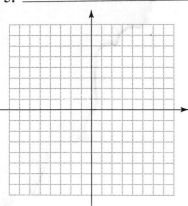

5. Find the center and radius of the circle whose equation is
$$(x + 1)^2 + (y - 2)^2 = 9.$$
Sketch its graph.

6. _____

6. Find the center and the radius of the circle whose equation is
$$x^2 - 2x + y^2 - 6y = 6.$$

7.

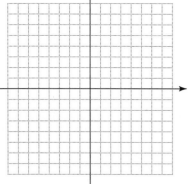

7. Graph the function $f(x) = \sqrt{16 - x^2}$.

8.

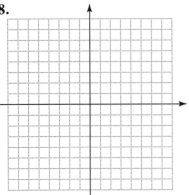

Graph each of the following equations.

8. $9x^2 + 16y^2 = 144$

CHAPTER 9 TEST

9. $4x^2 - 25y^2 = 100$

9.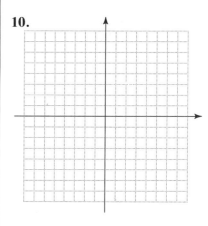

10. $x^2 + y^2 - 6x - 8y = 0$

10.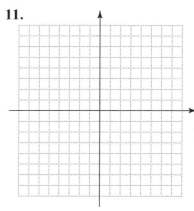

11. $x = (y + 1)^2 - 2$

11.

Identify each as the equation of a parabola, hyperbola, ellipse, *or* circle.

12. $2x^2 + 2y^2 = 36$ **13.** $6x^2 - 4y^2 = 12$

12. _____

13. _____

14. $3y^2 + 3x = 8$

15. $9x^2 + 25y^2 = 225$

Solve each nonlinear system.

16. $x + 4y = 10$
    $xy = 4$

17. $y + 2 = 3x$
    $x^2 + y^2 = 2$

18. $x^2 + y^2 = 16$
    $2x^2 - y^2 = 8$

19. Graph the inequality $x^2 + 16y^2 < 16$.

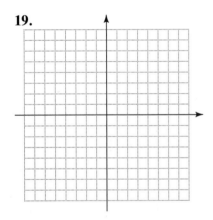

20. Graph the system
    $y \geq x^2$
    $x^2 + y^2 < 4$.

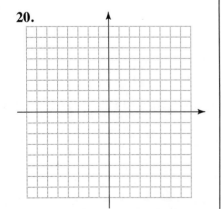

10 EXPONENTIAL AND LOGARITHMIC FUNCTIONS

10.1 INVERSE FUNCTIONS

A calculator with the following keys will be very helpful in this chapter.

$$y^x, \quad 10^x \text{ or } \log x, \quad e^x \text{ or } \ln x$$

We will explain how these keys are used at appropriate places in the chapter.

1 Suppose that G is the function $\{(-2, 2), (-1, 1), (0, 0), (1, 3), (2, 5)\}$. Another set of ordered pairs can be formed from G by exchanging the x- and y-values of each pair in G. Call this set F, with

$$F = \{(2, -2), (1, -1), (0, 0), (3, 1), (5, 2)\}.$$

To show that these two sets are related, F is called the *inverse* of G. For a function to have an inverse function, the given function must be *one-to-one*. In a **one-to-one function** each x-value corresponds to only one y-value and each y-value corresponds to just one x-value. The *inverse* of any function f is found by exchanging the components of the ordered pairs of f. The inverse of f is written f^{-1}. Read f^{-1} as "the inverse of f" or "f-inverse." The definition of the inverse of a function follows.

OBJECTIVES

1 Decide whether a function is one-to-one and, if it is, find its inverse.

2 Use the horizontal line test to determine whether a function is one-to-one.

3 Find the equation of the inverse of a function.

4 Graph the inverse f^{-1} from the graph of f.

The **inverse** of a one-to-one function f, written f^{-1}, is the set of all ordered pairs of the form (y, x), where (x, y) belongs to f.

CAUTION The symbol $f^{-1}(x)$ does not represent $\dfrac{1}{f(x)}$.

Since the inverse is formed by interchanging x and y, the domain of f becomes the range of f^{-1} and the range of f becomes the domain of f^{-1}.

EXAMPLE 1 Deciding Whether a Function Is One-to-One
Decide whether each function is one-to-one. If it is, find its inverse.

(a) $F = \{(-2, 1), (-1, 0), (0, 1), (1, 2), (2, 2)\}$

Each x-value in F corresponds to just one y-value. However, the y-value 2 corresponds to two x-values, 1 and 2. Also, the y-value 1 corre-

1. Decide whether or not each function is one-to-one. If it is, find the inverse.

   (a) {(1, 2), (2, 4), (3, 3), (4, 5)}

   (b) {(0, 3), (−1, 2), (1, 3)}

   (c) {(2, 5), (3, 6), (4, 8), (8, 7)}

sponds to both −2 and 0. Because some y-values correspond to more than one x-value, F is not one-to-one and does not have an inverse.

(b) $G = \{(3, 1), (0, 2), (2, 3), (4, 0)\}$

Every x-value in G corresponds to only one y-value, and every y-value corresponds to only one x-value, so G is a one-to-one function. The inverse function is found by exchanging the numbers in each ordered pair.

$$G^{-1} = \{(1, 3), (2, 0), (3, 2), (0, 4)\}.$$

■ WORK PROBLEM 1 AT THE SIDE.

2 It may be difficult to decide whether a function is one-to-one just by looking at the equation that defines the function. However, by graphing the function and observing the graph, we can use the following *horizontal line test* to tell whether it is one-to-one.

HORIZONTAL LINE TEST

A function is one-to-one if every horizontal line intersects the graph of the function at most once.

The horizontal line test follows from the definition of a one-to-one function. Any two points that lie on the same horizontal line have the same y-coordinate. No two ordered pairs that belong to a one-to-one function may have the same y-coordinate, and therefore no horizontal line will intersect the graph of a one-to-one function more than once.

EXAMPLE 2 Using the Horizontal Line Test
Use the horizontal line test to determine whether the graphs in Figures 1 and 2 are graphs of one-to-one functions.

(a)

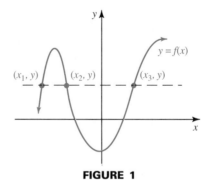

FIGURE 1

Because the horizontal line shown in Figure 1 intersects the graph in more than one point (actually three points in this case), the function is not one-to-one.

ANSWERS
1. (a) {(2, 1), (4, 2), (3, 3), (5, 4)}
   (b) not a one-to-one function
   (c) {(5, 2), (6, 3), (8, 4), (7, 8)}

(b)

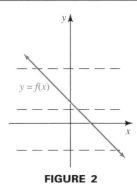

FIGURE 2

Every horizontal line will intersect the graph in Figure 2 in exactly one point. This function is one-to-one. ∎

WORK PROBLEM 2 AT THE SIDE.

3 By definition, the inverse of a function is found by exchanging the x- and y-values of the ordered pairs of the function. The equation of the inverse of a function defined by $y = f(x)$ is found in the same way.

INVERSE OF $y = f(x)$

For a one-to-one function f defined by an equation $y = f(x)$, find the defining equation of the inverse as follows.

Step 1 Exchange x and y.

Step 2 Solve for y.

Step 3 Replace y with $f^{-1}(x)$.

This procedure is illustrated in the following example.

EXAMPLE 3 Finding the Equation of the Inverse

Decide whether each of the following defines a one-to-one function. If so, find the equation of the inverse.

(a) $f(x) = 2x + 5$

Use the horizontal line test. This linear function has a straight line graph, so it is a one-to-one function. Use the steps given above to find the inverse. Let $y = f(x)$ so that

$$y = 2x + 5.$$

$x = 2y + 5$	Exchange x and y.
$2y = x - 5$	Subtract 5.
$y = \dfrac{x - 5}{2}$	Solve for y.

From the last equation,

$$f^{-1}(x) = \frac{x - 5}{2}.$$

2. Use the horizontal line test to determine whether each graph is the graph of a one-to-one function.

(a)

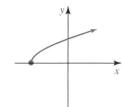

(b)

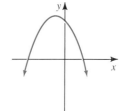

ANSWERS
2. (a) one-to-one
   (b) not one-to-one

CHAPTER 10 EXPONENTIAL AND LOGARITHMIC FUNCTIONS

3. Decide whether each equation defines a one-to-one function. If so, find the equation of the inverse.

   (a) $f(x) = 3x - 4$

   (b) $f(x) = x^3 + 1$

   (c) $f(x) = (x - 3)^2$

4. Find the following for $f(x) = \sqrt[3]{1 - x}$.

   (a) $f(2)$

   (b) $f(9)$

   (c) $f^{-1}(-1)$

   (d) $f^{-1}(-2)$

In the function defined by $y = 2x + 5$, the value of y is found by starting with a value of x, multiplying by 2, and adding 5. The inverse function has us *subtract* 5, and then *divide* by 2. This shows how an inverse is used to "undo" what a function does to the variable x.

(b) $f(x) = (x - 2)^3$

Because of the cube, each value of x produces just one value of y, so this is a one-to-one function. Find the inverse by replacing $f(x)$ with y and then exchanging x and y.

$$y = (x - 2)^3$$
$$x = (y - 2)^3$$

Take the cube root on each side to solve for y.

$$\sqrt[3]{x} = \sqrt[3]{(y - 2)^3}$$
$$\sqrt[3]{x} = y - 2$$
$$\sqrt[3]{x} + 2 = y$$

Replace y with $f^{-1}(x)$: $\quad f^{-1}(x) = \sqrt[3]{x} + 2$.

(c) $y = x^2 + 2$

As shown in Section 9.1, the graph of this equation is a parabola that opens upward. By the horizontal line test, there are many pairs of x-values that correspond to the same y-value. This means that the function defined by $y = x^2 + 2$ is not one-to-one. ∎

WORK PROBLEM 3 AT THE SIDE.

EXAMPLE 4 Finding $f(a)$ and $f^{-1}(a)$ for a Constant a
Let $f(x) = x^3$. Find the following.

(a) $f(-2) = (-2)^3 = -8$

(b) $f(3) = 3^3 = 27$

(c) $f^{-1}(-8)$

From part (a), $(-2, -8)$ belongs to the function f. Since f is one-to-one, $(-8, -2)$ belongs to f^{-1}, with $f^{-1}(-8) = -2$.

(d) $f^{-1}(27)$

Since $(3, 27)$ belongs to f, it follows that $(27, 3)$ belongs to f^{-1} and $f^{-1}(27) = 3$. ∎

WORK PROBLEM 4 AT THE SIDE.

4 Suppose the point (a, b) shown in Figure 3 belongs to a one-to-one function f. Then the point (b, a) belongs to f^{-1}. The line segment connecting (a, b) and (b, a) is perpendicular to, and cut in half by, the line $y = x$. The points (a, b) and (b, a) are "mirror images" of each other with respect to $y = x$. For this reason the graph of f^{-1} can be found from the graph of f by locating the mirror image of each point of f with respect to the line $y = x$.

ANSWERS

3. (a) $f^{-1}(x) = \dfrac{x + 4}{3}$

   (b) $f^{-1}(x) = \sqrt[3]{x - 1}$

   (c) not a one-to-one function

4. (a) -1 (b) -2 (c) 2 (d) 9

10.1 INVERSE FUNCTIONS

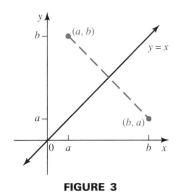

FIGURE 3

EXAMPLE 5 Graphing Inverses

Graph the inverses of the functions shown in Figure 4.

In Figure 4 the graphs of two functions are shown as solid lines. Their inverses are shown as dashed lines. ■

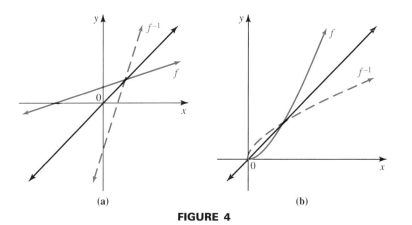

FIGURE 4

WORK PROBLEM 5 AT THE SIDE.

5. Use the given graphs to graph each inverse.

(a)
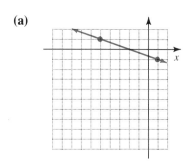

(b)
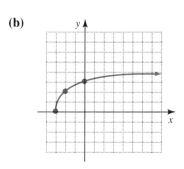

(c)
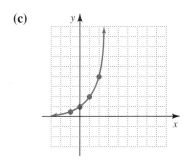

ANSWERS
5. (a)   (b)
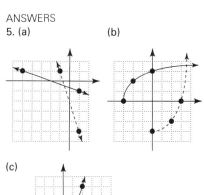

(c)
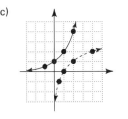

Historical Reflections

Forerunner of the Computer: Babbage's Difference Engine

Shown here is a model of the Difference Engine built in 1822 by Charles Babbage (1792–1871), a British mathematician. For a time, his research received financial assistance from the British government. Unfortunately, machine shops in his day could not produce parts of sufficient precision, the financial aid was discontinued, and his project had to be abandoned. The Difference Engine was the forerunner of today's electronic calculators.

On a vaster scale, Babbage worked on an Analytic Engine, which he began to design in 1833. It was to operate using punched cards as did early computers. His conception was grand, but it was too advanced for its time. The *IBM Automatic Sequence Controlled Calculator (ASCC)* was an outgrowth of the ideas of Babbage's Analytic Engine, and was completed in 1944 at Harvard University.

It is said that Babbage was inspired to use punched cards after he saw an incredible portrait of Joseph Marie Jacquard (1752–1823) woven in silk—so fine it appeared to be an etching. Punched cards had been programmed to weave the portrait on the Jacquard loom. Ada Augusta, Countess of Lovelace, said of Babbage's Analytic Engine ". . . the Analytic Engine weaves algebraical patterns just as the Jacquard loom weaves flowers and leaves."

Art: Courtesy of IBM

10.1 EXERCISES

For each of the following that defines a one-to-one function, find the inverse. See Examples 1–3.

1. $\{(3, 5), (2, 9), (4, 7)\}$

2. $\{(-1, 4), (0, 5), (1, -2), (2, -4), (3, 3)\}$

3. $\{(-2, 4), (-1, 1), (0, 0), (1, 1), (2, 4)\}$

4. $\{(-3, -1), (-2, 0), (-1, 1), (0, 2)\}$

5. $f(x) = 2x$

6. $f(x) = 3x - 1$

7. $y = \sqrt{x}$

8. $y = \sqrt{x + 2}$

9. $2y + 1 = 3x$

10. $5y + 6x = 30$

11. $f(x) = x^3 + 1$

12. $f(x) = x^3 - 2$

13. $2y = x^2 + 1$

14. $y = 2x^2 + 3$

15. $x - 1 = y^3$

16. $x = y^3 + 3$

The graphs of some functions are given below. **(a)** *Use the horizontal line test to determine whether the function is one-to-one.* **(b)** *If the function is one-to-one, graph the inverse of the function with dashed lines or curves on the same axes as the functions. See Examples 2 and 5.*

17.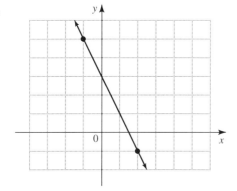

18.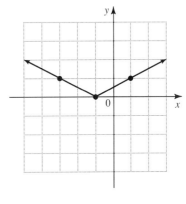

19.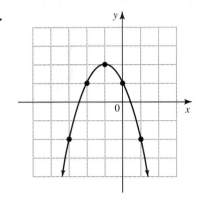

20.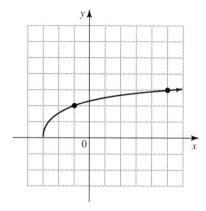

21.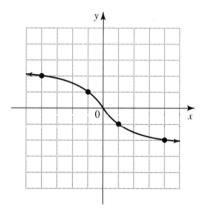

22.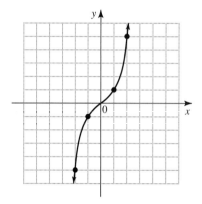

Graph each one-to-one function as a solid line or curve and its inverse as a dashed line or curve on the same axes. (The inverses of most of these functions were found in Exercises 5–16.) In Exercises 29–30 you should find a number of points to get the graph. See Example 5.

23. $f(x) = 2x$

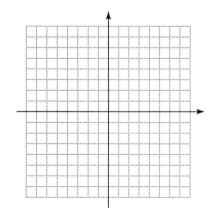

24. $f(x) = 3x - 1$

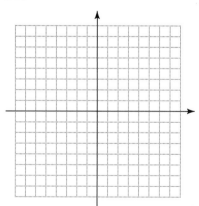

10.1 EXERCISES

25. $2y + 1 = 3x$

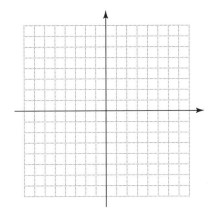

26. $5y + 6x = 30$

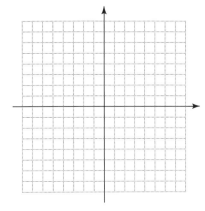

27. $2y = x + 1$

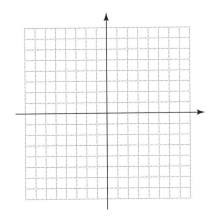

28. $y = 2x + 3$

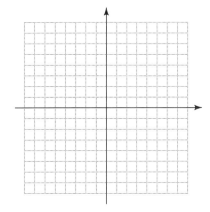

29. $x - 1 = y^3$

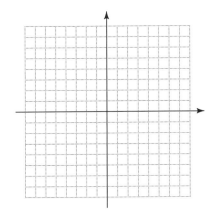

30. $y = \sqrt{x + 2}$

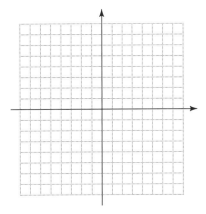

CHAPTER 10 EXPONENTIAL AND LOGARITHMIC FUNCTIONS

Let $f(x) = 2^x$. Find each of the following. See Example 4.

31. $f(3)$

32. $f(-2)$

33. $f^{-1}(8)$

34. $f^{-1}\left(\dfrac{1}{4}\right)$

35. $f(0)$

36. $f(1)$

37. $f^{-1}(1)$

38. $f^{-1}\left(\dfrac{1}{2}\right)$

SKILL SHARPENERS

Let $f(x) = 3^x$. Evaluate each of the following. See Section 7.5.

39. $f(3)$

40. $f(0)$

41. $f(-2)$

Evaluate each of the following. See Section 5.2.

42. $8^{2/3}$

43. $100^{-3/2}$

44. $81^{-5/4}$

10.2 EXPONENTIAL FUNCTIONS

1 In Section 5.2 expressions such as 2^x were evaluated for rational values of x. For example,

$$2^3 = 8$$
$$2^{-1} = \frac{1}{2}$$
$$2^{1/2} = \sqrt{2}$$
$$2^{3/4} = \sqrt[4]{2^3} = \sqrt[4]{8}.$$

In more advanced courses it is shown that 2^x exists for all real-number values of x, both rational and irrational. (Later in the chapter, we will see how to approximate the value of 2^x for irrational x.) With this assumption, we can now define an exponential function.

OBJECTIVES

1 Identify exponential functions.

2 Graph exponential functions.

3 Solve exponential equations of the form $a^x = a^k$ for x.

4 Use exponential functions in applications.

For $a > 0$, $a \neq 1$, and all real numbers x,

$$f(x) = a^x$$

defines an **exponential function**.

In this definition, we assume $a > 0$, since for negative values of a, a^x would not be a real number for all x. For example, with $a = -2$, $(-2)^{1/2}$, $(-2)^{1/4}$, $(-2)^{1/6}$, and so on are not real. If $a = 0$ or $a = 1$, the resulting function is a horizontal line, since $f(x) = 0^x = 0$ for all x (except 0) and $f(x) = 1^x = 1$ for all x. These functions are linear functions, not exponential functions.

2 Like other functions, exponential functions can be graphed by finding several ordered pairs that belong to the function. Plotting these points and connecting them with a smooth curve gives the graph.

EXAMPLE 1 Graphing an Exponential Function
Graph the exponential function $y = 2^x$.

Choose values of x and find the corresponding values of y.

x	-3	-2	-1	0	1	2	3	4
$y = 2^x$	$\frac{1}{8}$	$\frac{1}{4}$	$\frac{1}{2}$	1	2	4	8	16

Plotting these points and drawing a smooth curve through them gives the graph shown in Figure 5. This graph is typical of the graphs of exponential functions of the form $y = a^x$, where $a > 1$. The larger the value of a, the faster the graph rises. To see this, compare the graph of $y = 5^x$ with the graph of $y = 2^x$ in Figure 5 on the next page.

CHAPTER 10 EXPONENTIAL AND LOGARITHMIC FUNCTIONS

1. Graph.

(a) $y = 10^x$

(b) $y = \left(\dfrac{1}{4}\right)^x$

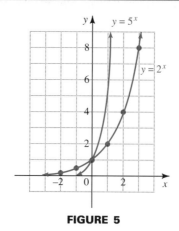

FIGURE 5

By the vertical line test, the graphs in Figure 5 represent functions. As these graphs suggest, the domain of an exponential function includes all real numbers. Since y is always positive, the range is $(0, \infty)$. Figure 5 also shows an important characteristic of exponential functions where $a > 1$; as x gets larger, y increases at a faster and faster rate. ∎

EXAMPLE 2 Graphing an Exponential Function

Graph $y = \left(\dfrac{1}{2}\right)^x$.

Again, find some sample points of the graph.

x	-3	-2	-1	0	1	2	3
$y = \left(\dfrac{1}{2}\right)^x$	8	4	2	1	$\dfrac{1}{2}$	$\dfrac{1}{4}$	$\dfrac{1}{8}$

The graph, shown in Figure 6, is the graph of a function. The graph is very similar to that of $y = 2^x$, shown in Figure 5, with the same domain and range, except that here as x gets larger, y decreases. This graph is typical of the graph of a function of the form $y = a^x$, where $0 < a < 1$. ∎

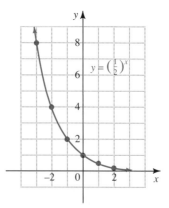

FIGURE 6

■ WORK PROBLEM 1 AT THE SIDE.

ANSWERS
1. (a) 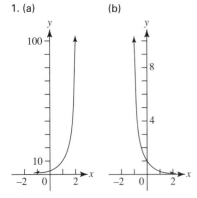 (b)

10.2 EXPONENTIAL FUNCTIONS

EXAMPLE 3 Graphing an Exponential Function

Graph $y = 3^{2x-4}$.

Find some ordered pairs. For example, if $x = 0$,

$$y = 3^{2(0)-4} = 3^{-4} = \frac{1}{81}.$$

Also, for $x = 2$,

$$y = 3^{2(2)-4} = 3^0 = 1.$$

The points $\left(0, \frac{1}{81}\right)$, $(2, 1)$, $\left(1, \frac{1}{9}\right)$, and $(3, 9)$ are shown on the graph in Figure 7. The graph is similar to the graph of $y = 2^x$ except that it is shifted to the right and rises more rapidly. ■

2. Graph $y = 2^{4x-3}$.

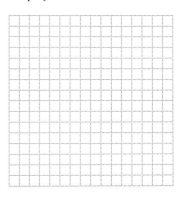

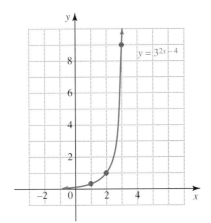

FIGURE 7

WORK PROBLEM 2 AT THE SIDE.

3 The following property of exponential expressions is useful for solving **exponential equations,** where the variable is an exponent.

For $a > 0$ and $a \neq 1$, if $a^x = a^y$ then $x = y$.

Notice that this property would not necessarily hold true for $a = 1$. For example, if $a = 1$,

$$1^4 = 1^5 \text{ is true}$$

but

$$4 \neq 5.$$

The following examples illustrate how this property is used to solve exponential equations.

ANSWER
2.
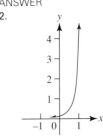

603

CHAPTER 10 EXPONENTIAL AND LOGARITHMIC FUNCTIONS

3. Solve. Check your answers.

(a) $25^x = 125$

(b) $4^x = 32$

(c) $81^p = 27$

EXAMPLE 4 Solving an Exponential Equation

Solve $5^x = 125$.

To use the property given above, first rewrite the equation so that both sides are exponential expressions with the same base. Since $5^3 = 125$, write the equation as

$$5^x = \mathbf{5^3},$$

and then use the property for exponential expressions given above to get

$$x = 3. \qquad \text{Set the exponents equal.}$$

Check by substituting in the given equation:

$$5^x = 125$$
$$5^3 = 125$$
$$125 = 125. \qquad \text{True}$$

The solution set is $\{3\}$. ∎

EXAMPLE 5 Solving an Exponential Equation

Solve $9^x = 27$.

Change both sides to the same base. Since $9 = 3^2$ and $27 = 3^3$, the equation $9^x = 27$ becomes

$$\mathbf{(3^2)^x = 3^3}, \qquad \text{or} \qquad 3^{2x} = 3^3.$$

Set the exponents equal and solve for x.

$$2x = 3$$
$$x = \frac{3}{2}.$$

Check that the solution set is $\left\{\frac{3}{2}\right\}$ by substituting $\frac{3}{2}$ for x in the given equation. ∎

■ WORK PROBLEM 3 AT THE SIDE.

4 Exponential equations frequently occur in applications describing growth or decay of some quantity. In particular, they are used to describe the growth and decay of populations.

EXAMPLE 6 Solving a Growth and Decay Problem

The air pollution, y, in appropriate units, in a large industrial city has been growing according to the equation

$$y = 1000(2)^{.3x}$$

where x is time in years from 1985. That is, $x = 0$ represents 1985, $x = 2$ represents 1987, and so on.

(a) Find the amount of pollution in 1985.
Let $x = 0$, and solve for y.

$$y = 1000(2)^{.3x}$$
$$= 1000(2)^{(.3)(\mathbf{0})} \qquad \text{Let } x = 0.$$
$$= 1000(2)^0$$
$$= 1000(1)$$
$$= 1000$$

The pollution in 1985 was 1000 units.

ANSWERS

3. (a) $\left\{\frac{3}{2}\right\}$ (b) $\left\{\frac{5}{2}\right\}$ (c) $\left\{\frac{3}{4}\right\}$

10.2 EXPONENTIAL FUNCTIONS

(b) Assuming that growth continues at the same rate, estimate the pollution in 1995.

Let $x = 10$ represent 1995.

$$y = 1000(2)^{.3x}$$
$$= 1000(2)^{(.3)(10)} \quad \text{Let } x = 10.$$
$$= 1000(2)^3$$
$$= 1000(8)$$
$$= 8000$$

In 1995 the pollution will be about 8000 units.

(c) Graph the equation $y = 1000(2)^{.3x}$.

The scale on the y-axis must be quite large to allow for the very large y-values. A calculator can be used to find a few more ordered pairs. The y^x (or x^y) key on the calculator is used to find values of numbers to a variable power. For example, to find y if $x = 15$, the equation gives $y = 1000(2)^{4.5}$. To evaluate $2^{4.5}$ with a calculator, touch 2, then the y^x key, then touch 4.5, then the "enter" or "=" key. You should get $2^{4.5} \approx 22.6$, so $y \approx 1000(22.6) = 22,600$. The graph is shown in Figure 8. ■

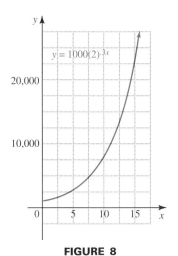

FIGURE 8

WORK PROBLEM 4 AT THE SIDE.

4. The amount of a radioactive substance, in grams, present at time t is $A = 100(3)^{-.5t}$, where t is in months. Find the amount present at

(a) $t = 0$;

(b) $t = 2$;

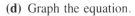

(c) $t = 10$.

(d) Graph the equation.

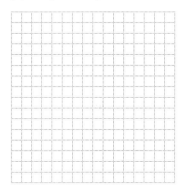

ANSWERS

4. (a) 100 grams **(b)** $33\frac{1}{3}$ grams
 (c) .41 grams
 (d)

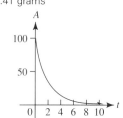

Historical Reflections

The Power of Exponents: A Story of a Chessboard

Once upon a time, a Persian king wished to please his executive officer, the Grand Vizier, with a gift of his choice. The Grand Vizier explained that he would like to be able to use his chessboard to accumulate wheat. A single grain of wheat would be received for the first square, two grains for the second square, four for the third, and so on. In each case, the number of grains would be twice the number on the preceding square. This doubling procedure is an example of exponential growth, and is defined by the exponential function

$$f(x) = 2^x$$

where x corresponds to the number of the chessboard square and $f(x)$ represents the number of grains of wheat corresponding to that square.

How many grains of wheat would be accumulated? As unlikely as it may seem, the number of grains would total 18.5 quintillion! The Grand Vizier evidently knew his mathematics.

10.2 EXERCISES

Graph the following. See Examples 1–3.

1. $y = 3^x$

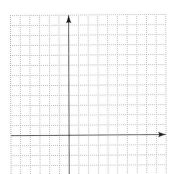

2. $y = 5^x$

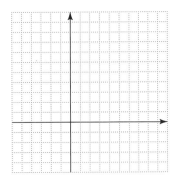

3. $y = \left(\dfrac{1}{4}\right)^x$

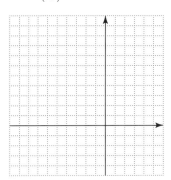

4. $y = \left(\dfrac{1}{3}\right)^x$

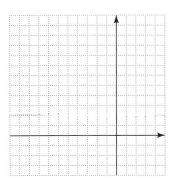

5. $y = 2^{2x}$

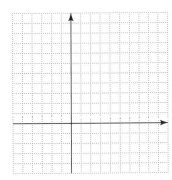

6. $y = 2^{-x-1}$

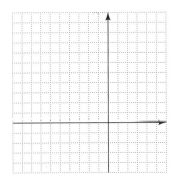

Solve the following equations. See Examples 4 and 5.

7. $5^x = 25$

8. $4^x = 64$

9. $4^x = 8$

10. $25^x = 125$

11. $16^x = 64$

12. $16^{-x+1} = 8$

13. $25^{-2x} = 3125$

14. $3^x = \dfrac{1}{9}$

15. $2^x = \dfrac{1}{8}$

CHAPTER 10 EXPONENTIAL AND LOGARITHMIC FUNCTIONS

16. $5^{-x} = \dfrac{1}{5}$

17. $\left(\dfrac{1}{2}\right)^x = 8$

18. $\left(\dfrac{3}{4}\right)^x = \dfrac{16}{9}$

Work each word problem. See Example 6.

19. The production of an oil well, in barrels, is decreasing according to the equation

$$y = 1,000,000(2)^{-.4t}$$

where t is time in years. Find the production at the following times.

(a) $t = 0$

(b) $t = 5$ (c) $t = 10$

(d) Graph the equation.

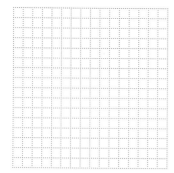

20. The number of bacteria, y, in a shipment of fresh milk is growing according to

$$y = 1000(3^x)$$

where x is the number of days since the shipment was made. Find the number of bacteria present

(a) when the shipment was made;

(b) after 2 days;   (c) after 5 days.

(d) Graph the equation.

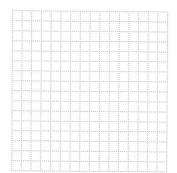

SKILL SHARPENERS

Evaluate each expression. Write the answers with simplified radicals in Exercises 25–28. See Sections 3.1 and 5.2.

21. 2^{-3}

22. $\left(\dfrac{1}{2}\right)^{-4}$

23. $36^{1/2}$

24. $\left(\dfrac{27}{8}\right)^{1/3}$

25. $3^{5/2}$

26. $5^{3/2}$

27. $2^{-4/3}$

28. $6^{-2/5}$

10.3 LOGARITHMIC FUNCTIONS

The graph of $y = 2^x$ is the dashed curve shown in Figure 9. Since $y = 2^x$ is a one-to-one function, it has an inverse. Interchanging x and y gives $x = 2^y$, the inverse of $y = 2^x$. As we saw in Section 10.1, the graph of the inverse is found by reflecting the graph of $y = 2^x$ about the line $y = x$. The graph of $x = 2^y$ is shown as a solid curve in Figure 9.

OBJECTIVES

1. Define a logarithm.
2. Write exponential statements in logarithmic form and logarithmic statements in exponential form.
3. Solve logarithmic equations of the form $\log_a b = k$ for a, b, or k.
4. Graph logarithmic functions.
5. Use logarithmic functions in applications.

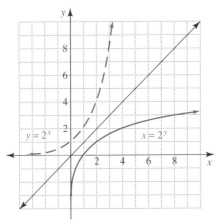

FIGURE 9

1 The equation $x = 2^y$ cannot be solved for the dependent variable y with the methods presented up to now. The following definition is used to solve $x = 2^y$ for y. For all positive numbers a and x, where $a \neq 1$,

$$y = \log_a x \quad \text{means the same as} \quad x = a^y.$$

The abbreviation **log** is used for **logarithm**. Read $\log_a x$ as "the logarithm of x to the base a." This key statement should be memorized. It is helpful to remember the location of the base and exponent in each form.

$$\text{Logarithmic form: } \mathbf{y} = \log_{\mathbf{a}} x$$

(Exponent ↑ above y; Base ↑ below a)

$$\text{Exponential form: } x = \mathbf{a}^{\mathbf{y}}$$

(Exponent ↑ above y; Base ↑ below a)

This discussion suggests the following definition of logarithm.

A **logarithm** is an exponent; $\log_a x$ is the exponent on the base a that yields the number x.

2 The definition of logarithm can be used to write exponential statements in logarithmic form and logarithmic statements in exponential form.

CHAPTER 10 EXPONENTIAL AND LOGARITHMIC FUNCTIONS

1. Complete the chart.

Exponential form	Logarithmic form
$2^5 = 32$	
$100^{1/2} = 10$	
	$\log_8 4 = \dfrac{2}{3}$
	$\log_6 \dfrac{1}{1296} = -4$

EXAMPLE 1 Converting Between Exponential and Logarithmic Form

The list below shows several pairs of equivalent statements. The same statement is written in both exponential and logarithmic form.

Exponential form	Logarithmic form
$3^2 = 9$	$\log_3 9 = 2$
$\left(\dfrac{1}{5}\right)^{-2} = 25$	$\log_{1/5} 25 = -2$
$10^5 = 100{,}000$	$\log_{10} 100{,}000 = 5$
$4^{-3} = \dfrac{1}{64}$	$\log_4 \dfrac{1}{64} = -3$ ∎

■ **WORK PROBLEM 1 AT THE SIDE.**

2. Solve each equation.

(a) $\log_3 27 = x$

(b) $\log_5 p = 2$

(c) $\log_m \dfrac{1}{16} = -4$

3 A **logarithmic equation** is an equation with a logarithm in at least one term. Logarithmic equations of the form $\log_a b = k$ can be solved for any of the three variables by first writing the equation in exponential form.

EXAMPLE 2 Solving Logarithmic Equations
Solve the following equations.

(a) $\log_4 x = -2$

By the definition of logarithm, $\log_4 x = -2$ is equivalent to $4^{-2} = x$. Then

$$x = 4^{-2} = \dfrac{1}{4^2} = \dfrac{1}{16}.$$

The solution set is $\left\{\dfrac{1}{16}\right\}$.

(b) $\log_{1/2} 16 = y$

Write the statement in exponential form.

$\log_{1/2} 16 = y$ is equivalent to $\left(\dfrac{1}{2}\right)^y = 16$.

Now write both sides of the equation with the same base. Since $\dfrac{1}{2} = 2^{-1}$ and $16 = 2^4$,

$$\left(\dfrac{1}{2}\right)^y = 16$$

becomes

$$(\mathbf{2^{-1}})^y = \mathbf{2^4}$$
$$2^{-y} = 2^4$$
$$-y = 4$$
$$y = -4.$$

The solution set is $\{-4\}$. ∎

■ **WORK PROBLEM 2 AT THE SIDE.**

ANSWERS

1. $\log_2 32 = 5$; $\log_{100} 10 = \dfrac{1}{2}$; $8^{2/3} = 4$; $6^{-4} = \dfrac{1}{1296}$

2. (a) $\{3\}$ (b) $\{25\}$ (c) $\{2\}$

For any positive real number b, we know that $b^1 = b$ and $b^0 = 1$. Writing these two statements in logarithmic form gives the following two properties of logarithms.

For any positive real number b, $b \neq 1$,

$$\log_b b = 1 \quad \text{and} \quad \log_b 1 = 0.$$

EXAMPLE 3 Using Properties of Logarithms

(a) $\log_7 7 = 1$

(b) $\log_{\sqrt{2}} \sqrt{2} = 1$

(c) $\log_9 1 = 0$

(d) $\log_{.2} 1 = 0$ ■

WORK PROBLEM 3 AT THE SIDE.

Now we can define the logarithmic function with base a as follows.

If a and x are positive numbers, with $a \neq 1$, then

$$f(x) = \log_a x$$

defines the **logarithmic function with base a**.

---

4 To graph a logarithmic function, it is helpful to write it in exponential form first. Then plot selected ordered pairs to determine the graph.

EXAMPLE 4 Graphing a Logarithmic Function
Graph $y = \log_{1/2} x$.

Writing $y = \log_{1/2} x$ in its exponential form as $x = \left(\frac{1}{2}\right)^y$ helps to identify ordered pairs that satisfy the equation. Here it is easier to choose values for y and find the corresponding values of x. Doing this gives the following pairs.

x	$\frac{1}{4}$	$\frac{1}{2}$	1	2	4	8
y	2	1	0	-1	-2	-3

Plotting these points (be careful to get them in the right order) and connecting them with a smooth curve gives the graph in Figure 10. This graph is typical of logarithmic functions with base $0 < a < 1$. The graph of $x = 2^y$ in Figure 9, which is equivalent to $y = \log_2 x$, is typical of graphs of logarithmic functions with base $a > 1$. ■

3. Find the value of each of the following.

(a) $\log_{2/5} \dfrac{2}{5}$

(b) $\log_{.4} 1$

ANSWERS
3. (a) 1 (b) 0

4. Graph.

(a) $y = \log_3 x$

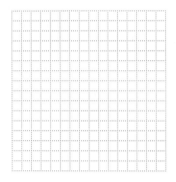

(b) $y = \log_{1/10} x$

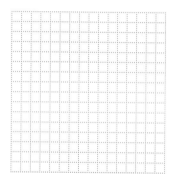

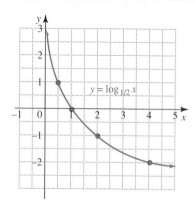

FIGURE 10

WORK PROBLEM 4 AT THE SIDE.

The graphs in Figures 9 and 10 suggest that the domain of the functions defined by $y = \log_2 x$ and $y = \log_{1/2} x$ is $(0, \infty)$. In both cases, the range is $(-\infty, \infty)$. This is typical of graphs of logarithmic functions.

5 Logarithmic functions, like exponential functions, are used in applications to describe growth and decay. As mentioned in the last section, a quantity that grows exponentially grows at a faster and faster rate. On the other hand, logarithmic functions describe growth that is taking place at a slower and slower rate.

EXAMPLE 5 Solving a Logarithmic Growth Problem

A population of mites in a laboratory is growing according to the equation

$$P = 80 \log_{10} (t + 10),$$

where t is the number of days after a study is begun.

(a) Find the number of mites at the beginning of the study.

Let $t = 0$ and find P.

$$\begin{aligned} P &= 80 \log_{10} (t + 10) \\ &= 80 \log_{10} (\mathbf{0} + 10) & \text{Let } t = 0. \\ &= 80 \log_{10} 10 \\ &= 80(\mathbf{1}) & \log_{10} 10 = 1 \\ &= 80 \end{aligned}$$

There were 80 mites at the beginning of the study.

(b) Find the number present after 90 days.

Substitute $t = 90$ in the expression for P.

$$\begin{aligned} P &= 80 \log_{10} (t + 10) \\ &= 80 \log_{10} (\mathbf{90} + 10) & \text{Let } t = 90. \\ &= 80 \log_{10} 100 \\ &= 80(\mathbf{2}) & \log_{10} 100 = 2 \\ &= 160 \end{aligned}$$

After 90 days, the number has doubled to 160.

ANSWERS
4. (a)   (b)

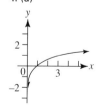 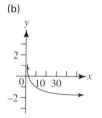

10.3 LOGARITHMIC FUNCTIONS

(c) Graph P.

Use the two ordered pairs $(0, 80)$ and $(90, 160)$ found in (a) and (b). Check that $(990, 240)$ also satisfies the equation. Use these three points, along with a knowledge of the general shape of the graph of a logarithmic function, to get the graph in Figure 11. ■

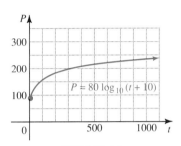

FIGURE 11

WORK PROBLEM 5 AT THE SIDE.

5. Sales (in thousands) of a new product are approximated by $S = 100 + 30 \log_3 (2t + 1)$, where t is the number of years after the product is introduced. Find S for the following values of t.

(a) $t = 0$

(b) $t = 13$

(c) $t = 40$

(d) Graph S.

ANSWERS
5. (a) 100 (b) 190 (c) 220
   (d) S

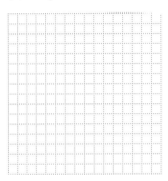

Historical Reflections

James Joseph Sylvester (1814–1897) and Arthur Cayley (1821–1895)

James Joseph Sylvester was working as an actuary in London in the early 1850s while Arthur Cayley was practicing law there. They met, and while not collaborators as such, they worked side-by-side on the algebra of *invariants*. (Invariants are algebraic expressions that remain virtually unaltered during certain "transformations," such as flips, rotations, and stretches.) Sylvester is credited with the term *matrix;* matrices provide, among other things, a systematic way of solving systems of equations.

Sylvester taught at several institutions during his lifetime, including the University of Virginia and Johns Hopkins. While at Johns Hopkins he helped to establish mathematical research at the university level. He founded the *American Journal of Mathematics*. Cayley spent most of his professorial time at Cambridge, and his work extended to nearly every branch of mathematics. He helped to open the horizons of algebra to include systems that worked even where the commutative law did not hold.

Art: Library of Congress

10.3 EXERCISES

Write in logarithmic form. See Example 1.

1. $3^4 = 81$
2. $5^2 = 25$
3. $2^5 = 32$

4. $6^3 = 216$
5. $\left(\dfrac{1}{2}\right)^{-2} = 4$
6. $\left(\dfrac{2}{3}\right)^{-3} = \dfrac{27}{8}$

7. $\left(\dfrac{3}{2}\right)^5 = \dfrac{243}{32}$
8. $\left(\dfrac{5}{6}\right)^{-2} = \dfrac{36}{25}$
9. $2^{-4} = \dfrac{1}{16}$

Write in exponential form. See Example 1.

10. $\log_2 8 = 3$
11. $\log_3 27 = 3$
12. $\log_{10} 10 = 1$

13. $\log_5 125 = 3$
14. $\log_4 16 = 2$
15. $\log_2 32 = 5$

16. $\log_2 \dfrac{1}{8} = -3$
17. $\log_3 \dfrac{1}{9} = -2$
18. $\log_{1/2} 4 = -2$

Evaluate the following. (Hint: Begin by letting the expression equal y.) See Example 2(b).

19. $\log_{10} 1000$
20. $\log_8 64$
21. $\log_3 27$

22. $\log_2 \dfrac{1}{2}$
23. $\log_{10} .1$
24. $\log_{10} .0001$

CHAPTER 10 EXPONENTIAL AND LOGARITHMIC FUNCTIONS

25. $\log_{36} 6$

26. $\log_{64} 8$

27. $\log_5 \dfrac{1}{25}$

28. $\log_7 \dfrac{1}{49}$

29. $\log_3 \sqrt{3^5}$

30. $\log_5 \sqrt{5^5}$

31. $\log_2 \sqrt{8}$

32. $\log_6 \sqrt{6^3}$

33. $\log_9 \sqrt{3^3}$

Solve each equation. See Examples 2 and 3.

34. $\log_x 9 = \dfrac{1}{2}$

35. $\log_m 125 = -3$

36. $\log_x 64 = -6$

37. $\log_4 x = \dfrac{5}{2}$

38. $\log_x \dfrac{1}{8} = -3$

39. $\log_p .1 = 1$

40. $\log_m 3 = \dfrac{1}{2}$

41. $\log_{1/2} x = -3$

42. $\log_2 x = 0$

43. $\log_k 4 = 1$

44. $\log_{12} P = 1$

45. $\log_b 1 = 0$

10.3 EXERCISES

Complete the tables of ordered pairs and then graph the given functions. See Example 4.

46. $y = \log_3 x$

x	y
27	
9	
3	
1	
$\frac{1}{3}$	
$\frac{1}{9}$	
$\frac{1}{27}$	

47. $y = \log_{10} x$

x	y
100	
10	
1	
$\frac{1}{10}$	
$\frac{1}{100}$	
$\frac{1}{1000}$	

48. $y = \log_{1/4} x$

x	y
64	
16	
4	
1	
$\frac{1}{4}$	
$\frac{1}{16}$	
$\frac{1}{64}$	

49. Graph $y = \log_{2.718} x$.

Solve the following word problems. See Example 5.

50. The number of fish in an aquarium is given by
$$f = 8 \log_5 (2t + 5)$$
where t is time in months. Find the number of fish present when

(a) $t = 0$; (b) $t = 10$; (c) $t = 60$.

51. A company analyst has found that total sales, in thousands of dollars, after a major advertising campaign, are
$$S = 100 \log_2 (x + 2)$$
where x is time in weeks after the campaign was introduced. Find the sales for

(a) $x = 0$; (b) $x = 2$; (c) $x = 6$.

(d) Graph f.

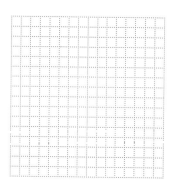

(d) Graph S.

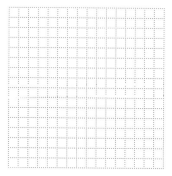

SKILL SHARPENERS

Use the properties of exponents to simplify the following expressions. See Sections 3.1 and 3.2.

52. $2^4 \cdot 2^5$

53. $3^{-2} \cdot 3^8$

54. $\dfrac{4^{-1}}{4^5}$

55. $\dfrac{5^8}{5^{-2}}$

56. $(2^4)^3$

57. $(5x^2)^4$

58. $\dfrac{p^5 p^{-3}}{(p^3)^2}$ $(p \neq 0)$

59. $\dfrac{2^3 (m^4)^3}{m^5}$ $(m \neq 0)$

10.4 PROPERTIES OF LOGARITHMS

Logarithms have been used as an aid to numerical calculation for several hundred years. Today the widespread use of calculators has made the use of logarithms for calculation obsolete. However, logarithms are very important in applications and in further work in mathematics. The properties that make logarithms so useful are given in this section.

1 One way in which logarithms simplify problems is by changing a problem of multiplication into one of addition. This is done with the following property of logarithms.

OBJECTIVES

1 Use the multiplication property for logarithms.

2 Use the division property for logarithms.

3 Use the power property for logarithms.

4 Use the properties of logarithms to write logarithmic expressions in alternative forms.

MULTIPLICATION PROPERTY FOR LOGARITHMS

If x, y, and b are positive real numbers, where $b \neq 1$, then

$$\log_b xy = \log_b x + \log_b y.$$

To prove this rule, recall that

$\log_b x = m$ is equivalent to $b^m = x$
$\log_b y = n$ is equivalent to $b^n = y$.

Then, by substitution,

$$xy = b^m \cdot b^n = b^{m+n},$$

using the product rule for exponents. Now, go back to logarithmic form.

$xy = b^{m+n}$ is equivalent to $\log_b xy = m + n.$

Substituting for $m + n$ gives

$$\log_b xy = \log_b x + \log_b y,$$

which is the desired result.

EXAMPLE 1 Using the Multiplication Property
Use the multiplication property to rewrite the following (assume $x > 0$).

(a) $\log_5 (6 \cdot 9)$

By the multiplication property,

$$\log_5 (\mathbf{6 \cdot 9}) = \log_5 \mathbf{6} + \log_5 \mathbf{9}.$$

(b) $\log_7 8 + \log_7 12$

$$\log_7 8 + \log_7 12 = \log_7 (8 \cdot 12) = \log_7 96$$

(c) $\log_3 (3x)$

$$\log_3 (3x) = \log_3 3 + \log_3 x$$

Since $\log_3 3 = 1$,

$$\log_3 (3x) = 1 + \log_3 x.$$

CHAPTER 10 EXPONENTIAL AND LOGARITHMIC FUNCTIONS

1. Use the multiplication property to rewrite the following.

(a) $\log_6 (5 \cdot 8)$

(b) $\log_4 3 + \log_4 7$

(c) $\log_8 8k \quad (k > 0)$

(d) $\log_5 m^2$

2. Use the division property to rewrite the following.

(a) $\log_7 \dfrac{49}{4}$

(b) $\log_3 \dfrac{p}{q} \quad (p > 0, q > 0)$

(c) $\log_4 \dfrac{3}{16}$

ANSWERS
1. (a) $\log_6 5 + \log_6 8$ (b) $\log_4 21$
   (c) $1 + \log_8 k$ (d) $2 \log_5 m \ (m > 0)$
2. (a) $2 - \log_7 4$
   (b) $\log_3 p - \log_3 q$ (c) $\log_4 3 - 2$

(d) $\log_4 x^3$

Since $x^3 = x \cdot x \cdot x$,

$$\log_4 x^3 = \log_4 (x \cdot x \cdot x)$$
$$= \log_4 x + \log_4 x + \log_4 x$$
$$= 3 \log_4 x. \quad \blacksquare$$

■ **WORK PROBLEM 1 AT THE SIDE.**

2 The rule for division is similar to the rule for multiplication.

DIVISION PROPERTY FOR LOGARITHMS

If x, y, and b are positive real numbers, where $b \neq 1$, then

$$\log_b \dfrac{x}{y} = \log_b x - \log_b y.$$

The proof of this rule is very similar to the proof of the multiplication property.

EXAMPLE 2 Using the Division Property
Use the division property to rewrite the following.

(a) $\log_4 \dfrac{7}{9} = \log_4 7 - \log_4 9$

(b) If $x > 0$, then $\log_5 6 - \log_5 x = \log_5 \dfrac{6}{x}$.

(c) $\log_3 \dfrac{27}{5} = \log_3 27 - \log_3 5$
$= 3 - \log_3 5 \qquad \log_3 27 = 3 \quad \blacksquare$

■ **WORK PROBLEM 2 AT THE SIDE.**

3 The next rule gives a method for evaluating powers and roots such as

$$2^{\sqrt{2}}, \quad (\sqrt{2})^{3/4}, \quad (.032)^{5/8}, \quad \text{and} \quad \sqrt[5]{12}.$$

This rule makes it possible to find approximations for numbers that could not be evaluated before. By the multiplication property for logarithms,

$$\log_5 2^3 = \log_5 (2 \cdot 2 \cdot 2)$$
$$= \log_5 2 + \log_5 2 + \log_5 2$$
$$= 3 \log_5 2.$$

Also, $\log_2 7^4 = \log_2 (7 \cdot 7 \cdot 7 \cdot 7)$
$$= \log_2 7 + \log_2 7 + \log_2 7 + \log_2 7$$
$$= 4 \log_2 7.$$

Furthermore, we saw in Example 1(d) that $\log_4 x^3 = 3 \log_4 x$. These examples suggest the following generalization.

620

10.4 PROPERTIES OF LOGARITHMS

POWER PROPERTY FOR LOGARITHMS

If x and b are positive real numbers, where $b \neq 1$, and r is any real number, then

$$\log_b x^r = r(\log_b x).$$

As examples of this result,

$$\log_b m^5 = 5 \log_b m \quad \text{and} \quad \log_3 5^{3/4} = \frac{3}{4} \log_3 5.$$

To prove the power property, let $\log_b x = m$. Then changing to exponential form gives

$$b^m = x.$$

Now raise both sides to the power r.

$$(b^m)^r = x^r$$
$$b^{mr} = x^r$$

Change back to logarithmic form.

$$\log_b x^r = mr = rm$$

Substitute $\log_b x$ for m to get the desired result,

$$\log_b x^r = r \log_b x.$$

As a special case of this rule, let $r = \frac{1}{p}$, so that $\log_b \sqrt[p]{x} = \log_b x^{1/p} = \frac{1}{p} \log_b x$. For example, using this result with $x > 0$,

$$\log_b \sqrt[5]{x} = \frac{1}{5} \log_b x \quad \text{and} \quad \log_b \sqrt[3]{x^4} = \frac{4}{3} \log_b x.$$

WORK PROBLEM 3 AT THE SIDE.

Two special properties involving both exponential and logarithmic expressions come directly from the fact that logarithmic and exponential functions are inverses of each other.

If $b > 0$ and $b \neq 1$, then

$$b^{\log_b x} = x \quad (x > 0) \quad \text{and} \quad \log_b b^x = x.$$

To prove the first statement, let $y = \log_b x$, which can be written in exponential form as $b^y = x$. Now replace y with $\log_b x$ to get $b^{\log_b x} = x$. The proof of the second statement is similar.

3. Use the power rule to rewrite the following. Assume $a > 0$, $b > 0$, $a \neq 1$, and $b \neq 1$.

(a) $\log_3 5^2$

(b) $\log_a x^4$

(c) $\log_b \sqrt{8}$

(d) $\log_2 \sqrt[3]{2}$

ANSWERS
3. (a) $2 \log_3 5$ (b) $4 \log_a x \; (x > 0)$
(c) $\frac{1}{2} \log_b 8$ (d) $\frac{1}{3}$

621

4. Find the value of each expression.

   (a) $\log_{10} 10^3$

   (b) $\log_2 8$

   (c) $5^{\log_5 3}$

5. Write as a sum or difference of logarithms. Assume all variable expressions are positive.

   (a) $\log_6 36m^5$

   (b) $\log_2 \sqrt{9z}$

   (c) $\log_q \dfrac{8r}{m-1}$ $(m \neq 1, q \neq 1)$

   (d) $\log_4 (3x + y)$

ANSWERS
4. (a) 3 (b) 3 (c) 3
5. (a) $2 + 5 \log_6 m$
   (b) $\log_2 3 + \dfrac{1}{2} \log_2 z$
   (c) $\log_q 8 + \log_q r - \log_q (m - 1)$
   (d) cannot be rewritten

EXAMPLE 3 Using the Special Properties
Find the value of the following logarithmic expressions.

(a) $\log_5 5^4 = 4$, by the first property.

(b) $\log_3 9$

Since $9 = 3^2$,
$$\log_3 \mathbf{9} = \log_3 \mathbf{3^2} = 2.$$

The second property was used in the last step.

(c) $4^{\log_4 10} = 10$ ∎

WORK PROBLEM 4 AT THE SIDE.

4 The properties of logarithms are useful for writing expressions in an alternative form. This use of logarithms is important in calculus.

EXAMPLE 4 Writing Logarithms in Alternative Forms
Use the properties of logarithms to rewrite each expression. Assume all variables represent positive real numbers.

(a) $\log_4 4x^3 = \log_4 4 + \log_4 x^3$ Multiplication property

$\qquad\qquad\quad\; = 1 + 3 \log_4 x$ Power property: $\log_4 4 = 1$

(b) $\log_7 \sqrt{\dfrac{m}{n}} = \log_7 \left(\dfrac{m}{n}\right)^{1/2}$

$\qquad\qquad\;\; = \dfrac{1}{2} \log_7 \dfrac{m}{n}$ Power property

$\qquad\qquad\;\; = \dfrac{1}{2} (\log_7 m - \log_7 n)$ Division property

(c) $2 \log_{10} m + \log_{10} m = \log_{10} m^2 + \log_{10} m$ Power property

$\qquad\qquad\qquad\qquad\; = \log_{10} m^2 \cdot m$ Multiplication property

$\qquad\qquad\qquad\qquad\; = \log_{10} m^3.$

(d) $\log_8 (2p + 3r)$ cannot be rewritten by the properties of logarithms. ∎

CAUTION   Remember that there is no property of logarithms to rewrite the logarithm of a *sum*. That is, we *cannot* write $\log_b (x + y)$ in terms of $\log_b x$ and $\log_b y$.

WORK PROBLEM 5 AT THE SIDE.

10.4 PROPERTIES OF LOGARITHMS

EXAMPLE 5 Using the Properties of Logarithms
Given that $\log_2 5 = 2.3219$ and $\log_2 3 = 1.5850$, evaluate the following.

(a) $\log_2 15$

Since 15 is $3 \cdot 5$, use the multiplication property.

$$\begin{aligned}\log_2 15 &= \log_2 3 \cdot 5 \\ &= \log_2 3 + \log_2 5 \\ &= 1.5850 + 2.3219 = 3.9069\end{aligned}$$

(b) $\log_2 .6$

Since $.6 = \frac{6}{10} = \frac{3}{5}$, use the division property to get

$$\begin{aligned}\log_2 .6 &= \log_2 \frac{3}{5} \\ &= \log_2 3 - \log_2 5 \\ &= 1.5850 - 2.3219 = -.7369.\end{aligned}$$

(c) $\log_2 27$

Use the power property and the fact that $27 = 3^3$.

$$\begin{aligned}\log_2 27 &= \log_2 3^3 \\ &= 3 \log_2 3 \\ &= 3(1.5850) = 4.7550 \quad \blacksquare\end{aligned}$$

WORK PROBLEM 6 AT THE SIDE.

EXAMPLE 6 Deciding Whether Statements about Logarithms Are True
Decide whether each of the following statements is true or false.

(a) $\log_2 8 - \log_2 4 = \log_2 4$

Since $\log_2 8 = 3$ and $\log_2 4 = 2$, the statement becomes

$$\log_2 8 - \log_2 4 = \log_2 4$$
$$3 - 2 = 2,$$

which is false.

(b) $\log_3 (\log_2 8) = \dfrac{\log_7 49}{\log_8 64}$

Evaluate both sides: $\log_2 8 = 3$, so

$$\log_3 (\log_2 8) = \log_3 (3) = 1;$$

also,

$$\dfrac{\log_7 49}{\log_8 64} = \dfrac{2}{2} = 1.$$

The statement is true. $\blacksquare$

WORK PROBLEM 7 AT THE SIDE.

6. Use the values given in Example 5 to find the following.

(a) $\log_2 \dfrac{5}{3}$

(b) $\log_2 \sqrt{5}$

7. Decide whether the following statements are true or false.

(a) $\dfrac{\log_5 10}{\log_5 20} = \dfrac{1}{2}$

(b) $\log_6(\log_2 16) = \dfrac{\log_6 6}{\log_6 36}$

ANSWERS
6. (a) .7369 (b) 1.1610
7. (a) false (b) false

Historical Reflections

Euclidean Tools and the Three Famous Problems of Antiquity

Renaissance sculptor Antonio del Pollaiolo (1433–1498) depicted geometry in this bronze relief *Geometria*, found on the sloping base of the tomb of Pope Sixtus IV. The important feature in this photo is the inclusion of compasses, one of the two instruments allowed in classical (Euclidean) geometry for constructions. The other Euclidean tool is the unmarked straightedge. These two tools proved to be sufficient for Greek geometers to accomplish a great number of geometric constructions. Basic constructions such as copying an angle or constructing the perpendicular bisector of a segment are easily performed and verified.

There were, however, three constructions that the Greeks were not able to accomplish with these tools. Now known as the *three famous problems of antiquity*, they consisted of the following constructions:

1. To trisect an arbitrary angle;
2. To construct the length of the edge of a cube having twice the volume of a given cube;
3. To construct a square having the same area as that of a given circle.

In the nineteenth century it was learned why the Greeks had not been able to perform these constructions. They are, in fact, impossible to accomplish with Euclidean tools. Over the years other methods have been devised to accomplish these constructions. For example, trisecting an arbitrary angle can be accomplished if one is allowed the luxury of a single mark on the straightedge! However, this violates the rules of the game of the Greeks; the straightedge may not be so marked.

Despite the fact that these problems have been *proven* to be impossible to solve with compasses and straightedge, there are, nevertheless, people who claim to have found solutions. It seems that mathematics will always be plagued with these "angle trisectors" and "circle squarers."

Art: "Geometria" by Pollaiolo/Art Resource, NY

10.4 EXERCISES

Use the properties of logarithms to express each of the following as a sum, difference, or product of logarithms, or as a single number if possible. Assume all variables represent positive numbers. See Examples 1–4.

1. $\log_6 \dfrac{3}{4}$

2. $\log_6 \dfrac{5}{8}$

3. $\log_2 8^{1/2}$

4. $\log_3 9^{3/2}$

5. $\log_5 \dfrac{5x}{y}$

6. $\log_4 \dfrac{16p}{3q}$

7. $\log_2 \dfrac{5\sqrt{7}}{3}$

8. $\log_2 \sqrt{mn}$

9. $\log_3 \sqrt{\dfrac{ma}{b}}$

10. $\log_5 \dfrac{\sqrt[3]{x} \cdot \sqrt[4]{y}}{z}$

11. $\log_5 (9x + 4y)$

12. $\log_3 (2x + 3y)$

Use the properties of logarithms to write each of the following as a single logarithm. Assume all variables represent positive numbers not equal to 1. See Examples 1–4.

13. $\log_b x + \log_b y$

14. $\log_b a - \log_b c$

15. $4 \log_b m - \log_b n$

16. $x \log_b 3 + \log_b 2$

17. $(\log_b m - \log_b n) + \log_b r$

18. $\log_b 2 + \log_b 3 - \log_b 5$

19. $3 \log_b 4 - 2 \log_b 3$

20. $5 \log_b 2 + \log_b 3$

21. $\log_{10} (x + 1) + \log_{10} (x - 1)$ $(x > 1)$

22. $2 \log_p x + \dfrac{1}{2} \log_p y - \dfrac{3}{2} \log_p z - 3 \log_p a$

23. $\log_p \dfrac{5}{2} + \log_p \dfrac{5}{4} - \left(\log_p \dfrac{3}{4} + \log_p 2\right)$

24. $\dfrac{1}{3} \log_b x + \dfrac{1}{3} \log_b x^2 - \dfrac{1}{6} \log_b x^3$

Given that $\log_{10} 2 \approx .3010$ *and* $\log_{10} 3 \approx .4771$, *evaluate the following. See Example 5.*

25. $\log_{10} 4$

26. $\log_{10} 5$

$$\left(\text{Hint: } 5 = \dfrac{10}{2}.\right)$$

27. $\log_{10} 9$

28. $\log_{10} 12$

29. $\log_{10} 16$

30. $\log_{10} 180$

Answer true *or* false *for each of the following. See Example 6.*

31. $\log_{10} 2 + \log_{10} 3 = \log_{10} (2 + 3)$

32. $\log_6 24 - \log_6 4 = 1$

33. $\dfrac{\log_{10} 8}{\log_{10} 16} = \dfrac{1}{2}$

34. $\dfrac{\log_{10} 10}{\log_{100} 10} = 2$

SKILL SHARPENERS

Write each logarithmic statement in exponential form and each exponential statement in logarithmic form. See Section 10.3.

35. $\log_{10} 9 = .9542$

36. $\log_{10} 8 = .9031$

37. $\log_{10} 140 = 2.1461$

38. $10^{.6990} = 5$

39. $10^{1.4314} = 27$

40. $10^{-.2358} = .581$

10.5 EVALUATING LOGARITHMS

As mentioned earlier, logarithms are important in many applications of mathematics to everyday problems, particularly in biology, engineering, economics, and social science. In this section we show how to find numerical approximations for logarithms. Traditionally base 10 logarithms have been used most extensively, since our number system is base 10. Logarithms to base 10 are called **common logarithms** and $\log_{10} x$ is abbreviated as simply $\log x$, where the base is understood to be 10. Extensive tables of values are available for evaluating common logarithms. Appendix A gives a brief explanation of the use of the table of common logarithms given at the back of this book.

OBJECTIVES

1. Evaluate common logarithms by using a calculator.
2. Find the antilogarithm of a common logarithm.
3. Use common logarithms in applied problems.
4. Evaluate natural logarithms using a calculator.
5. Find the antilogarithm of a natural logarithm.
6. Use natural logarithms in applied problems.
7. Use the change-of-base rule.

1 Most people today use a calculator to evaluate common logarithms. The next example gives the results of evaluating some common logarithms using a calculator with a log key. (This may be a second function key on some calculators.) Just enter the number, then touch the log key. We will give all logarithms to four decimal places.

EXAMPLE 1 Evaluating Common Logarithms
Evaluate each logarithm.

(a) $\log 327.1 = 2.5147$

(b) $\log 437{,}000 = 5.6405$

(c) $\log .0615 = -1.2111$

In Example 1(c), $\log .0615 = -1.2111$, a negative result. The common logarithm of a number between 0 and 1 is always negative because the logarithm is the exponent on 10 that produces the number. For example,

$$10^{-1.2111} = .0615.$$

If the exponent (the logarithm) were positive, the result would be greater than 1, since $10^0 = 1$. ∎

WORK PROBLEM 1 AT THE SIDE.

2 In Example 1(a), the number 327.1, whose common logarithm is 2.5147, is called the common **antilogarithm** (abbreviated *antilog*) of 2.5147. In the same way 437,000, in Example 1(b), is the common antilogarithm of 5.6405. In general, in the expression $y = \log_{10} x$, x is the antilog.

NOTE From Example 1,

$$\underbrace{\log 327.1}_{\text{antilogarithm}} = \underbrace{2.5147}_{\text{logarithm}}$$

In exponential form, this is written as

$$10^{\overbrace{2.5147}^{\text{logarithm}}} = \overbrace{327.1}^{\text{antilogarithm}},$$

which shows that the logarithm 2.5147 is just an exponent. A common logarithm is the exponent on 10 that produces the antilogarithm.

1. Find the following.

   (a) $\log 41{,}600$

   (b) $\log 43.5$

   (c) $\log .442$

ANSWERS
1. (a) 4.6191   (b) 1.6385
   (c) $-.3546$

2. Find the antilogarithm of each logarithm.

   (a) 3.6503

   (b) −1.3696

   (c) .6064

3. Find the pH of solutions with the following hydronium ion concentrations.

   (a) 3.7×10^{-8}

   (b) 1.2×10^{-3}

ANSWERS
2. (a) 4470 (b) .0427 (c) 4.04
3. (a) 7.4 (b) 2.9

EXAMPLE 2 Finding Common Antilogarithms
Find the common antilogarithm of each of the following.

(a) 2.6454

Use the fact that for an antilogarithm N, $10^{2.6454} = N$. Some calculators have a 10^x key. With such a calculator, enter 2.6454, then touch the 10^x key to get the antilogarithm:

$$N \approx 442.$$

On other calculators, antilogarithms are found using the INV and log keys: enter 2.6454, touch the INV key, then touch the log key to get 442. INV indicates "inverse," so the INV key followed by the log key gives the inverse of the logarithm.

(b) −2.3686

The antilogarithm is N, where

$$10^{-2.3686} = N.$$

Using the 10^x key or the INV and log keys gives

$$N \approx .00428.$$

(c) 1.5203

The antilogarithm is 33.14 to four significant figures. ■

■ **WORK PROBLEM 2 AT THE SIDE.**

Most calculators will work by one of the methods we have described. There are a few that may not. If yours is one of those, see your instruction booklet. Your teacher also may be able to help you.

3 In chemistry, the **pH** of a solution is defined as follows.

$$\text{pH} = -\log [H_3O^+],$$

where $[H_3O^+]$ is the hydronium ion concentration in moles per liter.

The pH is a measure of the acidity or alkalinity of a solution, with water, for example, having a pH of 7. In general, acids have pH numbers less than 7, and alkaline solutions have pH values greater than 7.

EXAMPLE 3 Finding pH
Find the pH of grapefruit with a hydronium ion concentration of 6.3×10^{-4}.

Use the definition of pH.

$$\begin{aligned}
\text{pH} &= -\log (6.3 \times 10^{-4}) \\
&= -(\log 6.3 + \log 10^{-4}) \quad \text{Multiplication property} \\
&= -[.7993 - 4(1)] \\
&= -.7993 + 4 \approx 3.2
\end{aligned}$$

It is customary to round pH values to the nearest tenth. ■

■ **WORK PROBLEM 3 AT THE SIDE.**

10.5 EVALUATING LOGARITHMS

EXAMPLE 4 Finding Hydronium Ion Concentration
Find the hydronium ion concentration of drinking water with a pH of 6.5.

$$pH = 6.5 = -\log [H_3O^+]$$
$$\log [H_3O^+] = -6.5$$

Since the antilogarithm of -6.5 is 3.2×10^{-7},

$$[H_3O^+] = 3.2 \times 10^{-7}. \blacksquare$$

WORK PROBLEM 4 AT THE SIDE.

4 The most important logarithms used in applications are **natural logarithms**, which have as base the number e. The number e is irrational, like π: $e \approx 2.7182818$. Logarithms to base e are called natural logarithms because they occur in biology and the social sciences in natural situations that involve growth or decay. The base e logarithm of x is written $\ln x$ (read "el en x"). A graph of $y = \ln x$, the natural logarithm function, is given in Figure 12.

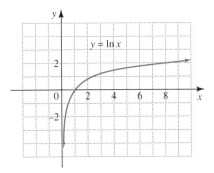

FIGURE 12

The next example shows how to find natural logarithms with a calculator that has a key labeled $\ln x$.

EXAMPLE 5 Finding Natural Logarithms
Find each of the following logarithms to four significant digits.

(a) $\ln .5841$

Enter .5841 and touch the ln key to get

$$\ln .5841 \approx -.5377.$$

As with common logarithms, a number between 0 and 1 has a negative natural logarithm.

(b) $\ln 192.7 \approx 5.261$

(c) $\ln 10.84 \approx 2.383$ $\blacksquare$

If your calculator has an e^x key, but not a key labeled $\ln x$, find natural logarithms by entering the number, touching the INV key, and then touching the e^x key. This works because $y = e^x$ is the inverse function of $y = \ln x$ (or $y = \log_e x$).

WORK PROBLEM 5 AT THE SIDE.

Tables of natural logarithms are available, but we do not include one in this book.

4. Find the hydronium ion concentrations of solutions with the following pHs.

(a) 4.6

(b) 7.5

5. Find each logarithm.

(a) $\ln .01$

(b) $\ln 27$

(c) $\ln 529$

ANSWERS
4. (a) 2.5×10^{-5} (b) 3.2×10^{-8}
5. (a) -4.6052 (b) 3.2958 (c) 6.2710

CHAPTER 10 EXPONENTIAL AND LOGARITHMIC FUNCTIONS

6. Find each natural antilogarithm.

   (a) 5.123

   (b) .0492

   (c) −1.468

5 Antilogarithms of natural logarithms are found as for common logarithms.

$$\text{If} \quad \underset{\downarrow}{\ln x} = y, \quad \text{then} \quad \underset{\downarrow}{e^y} = x,$$

$$\text{(logarithm)} \qquad \text{(antilogarithm)}$$

since the base of natural logarithms is e. Some calculators have a key labeled e^x. With others, use the INV key and the ln x key to evaluate a natural antilogarithm.

EXAMPLE 6 Finding Natural Antilogarithms

Find the natural antilogarithm of each of the following.

(a) 2.5017

Let N be the antilogarithm. Then

$$\ln N = 2.5017.$$

Find N by entering 2.5017, then touching the INV and ln x keys, getting

$$N = 12.20.$$

Alternatively, to use the e^x key, enter 2.5017 and touch the e^x key to get 12.20.

(b) −1.429

The antilogarithm is .2395.

(c) .0053

The antilogarithm is 1.0053. ∎

■ WORK PROBLEM 6 AT THE SIDE.

6 One of the most common applications of exponential functions depends on the fact that in many situations involving growth or decay of a population, the amount or number of some quantity present at time t can be closely approximated by

$$y = y_0 e^{kt}$$

where y_0 is the amount or number present at time $t = 0$, k is a constant, and e is the base of natural logarithms mentioned earlier.

EXAMPLE 7 Applying Natural Logarithms

Suppose that the population of a small town in the Southwest is

$$P = 10,000 e^{.04t},$$

where t represents time, measured in years. The population at time $t = 0$ is

$$P = 10,000 e^{(.04)(0)}$$
$$= 10,000 e^0$$
$$= 10,000(1) \qquad e^0 = 1$$
$$= 10,000.$$

ANSWERS

6. (a) 167.8 (b) 1.0504 (c) .2304

The population of the town is 10,000 at time $t = 0$, that is, at present. The population of the town 5 years from now, at $t = 5$, is

$$P = 10,000e^{(.04)(5)}$$
$$= 10,000e^{.2}.$$

The number $e^{.2}$ can be found with the e^x key of a calculator. Enter .2, then touch e^x to get 1.22. Alternatively, use the INV and ln x keys for the same result. Use this result to evaluate P.

$$P \approx (10,000)(1.22)$$
$$= 12,200$$

In 5 years the population of the town will be about 12,200. ∎

WORK PROBLEM 7 AT THE SIDE.

7. In Example 7, find the population 10 years from now.

EXAMPLE 8 Applying Natural Logarithms

The number of years, $N(r)$, since two independently evolving languages split off from a common ancestral language is approximated by

$$N(r) = -5000 \ln r$$

where r is the percent of words from the ancestral language common to both languages now. Find N if $r = 70\%$.

Write 70% as .7 and find $N(.7)$.

$$N(.7) = -5000 \ln .7$$
$$\approx -5000(-.3567)$$
$$\approx 1783$$

Approximately 1800 years have passed since the two languages separated. ∎

WORK PROBLEM 8 AT THE SIDE.

8. Find $N(.1)$ in Example 8.

7 A calculator (or a table) can be used to approximate the values of common logarithms (base 10) or natural logarithms (base e). However, sometimes it is convenient to use logarithms to other bases. The following rule is used to convert logarithms from one base to another.

CHANGE-OF-BASE RULE

If $a > 0$, $a \neq 1$, $b > 0$, $b \neq 1$, and $x > 0$, then

$$\log_a x = \frac{\log_b x}{\log_b a}.$$

NOTE As an aid in remembering the change-of-base rule, notice that x is "above" a on both sides of the equation.

Any positive number other than 1 can be used for base b in the change of base rule, but usually the only practical bases are e and 10, since calculators (or tables) give logarithms only for these two bases.

ANSWERS
7. about 15,000
8. about 12,000 years

9. Find each logarithm.

(a) $\log_3 17$
(Use common logarithms.)

(b) $\log_9 121$
(Use natural logarithms.)

EXAMPLE 9 Using the Change-of-Base Rule

Find each logarithm.

(a) $\log_5 12$

Use common logarithms and the rule for change of base.

$$\log_5 12 = \frac{\log 12}{\log 5} \approx \frac{1.0792}{.6990}$$

Now evaluate this quotient.

$$\log_5 12 \approx \frac{1.0792}{.6990} \approx 1.5439$$

(b) $\log_2 134$

Use natural logarithms and the change-of-base rule.

$$\log_2 134 = \frac{\ln 134}{\ln 2}$$
$$\approx \frac{4.8978}{.6931}$$
$$= 7.0665 \quad \blacksquare$$

WORK PROBLEM 9 AT THE SIDE.

The change-of-base rule can be used to find natural logarithms with a calculator that has the $\log x$ key, but not the $\ln x$ key. For example,

$$\ln 5 = \log_e 5 = \frac{\log 5}{\log e} \approx \frac{\log 5}{\log 2.718},$$

since $e \approx 2.718$. Similarly, if a calculator has only the $\ln x$ key, then the change-of-base rule could be used to find common logarithms.

To prove the formula for change of base, let $\log_a x = m$.

$\log_a x = m$	
$a^m = x$	Change to exponential form.
$\log_b (a^m) = \log_b x$	Take logs on both sides.
$m \log_b a = \log_b x$	Use the power property.
$(\log_a x)(\log_b a) = \log_b x$	Substitute for m.
$\log_a x = \dfrac{\log_b x}{\log_b a}$	Divide both sides by $\log_b a$.

ANSWERS
9. (a) 2.5789
   (b) 2.1827

10.5 EXERCISES

You will need to use a calculator for most of the problems in this exercise set.

Find each logarithm. See Examples 1 and 5. (Hint: Use the properties of logarithms first in Exercises 20–24.)

1. log 278
2. log 3460
3. log 3.28

4. log 57.3
5. log 9.83
6. log 589

7. log .327
8. log .0763
9. log .000672

10. log .00382
11. log 675,000
12. log 371,000,000

13. ln 5
14. ln 6
15. ln 697

16. ln 4.83
17. ln 1.72
18. ln 38.5

19. ln 97.6
20. ln e^2
21. ln $e^{3.1}$

22. ln $5e^3$
23. ln $4e^2$
24. ln $3e^{-1}$

Find the common antilogarithm of each of the following numbers. Give answers to three significant digits. See Example 2.

25. .5340
26. .9309
27. 2.6571

28. −.0691
29. −1.118
30. −2.3893

Find the natural antilogarithm of each of the following numbers. Give answers to three significant digits. See Example 6.

31. 3.8494　　　　　　　　**32.** 1.7938　　　　　　　　**33.** 1.3962

34. −.0259　　　　　　　　**35.** −.4168　　　　　　　　**36.** −1.2190

Use common logarithms or natural logarithms to find each of the following to the nearest hundredth. See Example 9.

37. $\log_5 11$　　　　　　　**38.** $\log_6 15$　　　　　　　**39.** $\log_3 1.89$

40. $\log_4 7.21$　　　　　　**41.** $\log_8 9.63$　　　　　　**42.** $\log_2 3.42$

43. $\log_{11} 47.3$　　　　　**44.** $\log_{18} 51.2$　　　　　**45.** $\log_{50} 31.3$

Use the formula pH = $-\log[H_3O^+]$ to find the pH of substances with the given hydronium ion concentrations. See Example 3.

46. Milk, 4×10^{-7}　　　　　　　　**47.** Sodium hydroxide (lye), 3.2×10^{-14}

48. Limes, 1.6×10^{-2}　　　　　　　**49.** Crackers, 3.9×10^{-9}

Use the formula for pH to find the hydronium ion concentrations of substances with the given pH values. See Example 4.

50. Shampoo, 5.5　　　**51.** Beer, 4.8　　　**52.** Soda pop, 2.7　　　**53.** Wine, 3.4

10.5 EXERCISES

Solve the following problems. See Example 7.

54. Suppose the quantity, measured in grams, of a radioactive substance present at time t is given by
$$Q(t) = 500e^{-.05t}$$
where t is measured in days. Find the quantity present at the following times.

(a) $t = 0$ **(b)** $t = 4$

(c) $t = 8$ **(d)** $t = 20$

55. Suppose that the population of a city is given by
$$P(t) = 1{,}000{,}000 e^{.02t}$$
where t represents time measured in years. Find the following values.

(a) $P(0)$ **(b)** $P(2)$

(c) $P(4)$ **(d)** $P(10)$

56. When a bactericide is introduced into a certain culture, the number of bacteria present is given by
$$D(t) = 50{,}000 e^{-.01t}$$
where t is time measured in hours. Find the number of bacteria present at the following times.

(a) $t = 0$ **(b)** $t = 5$

(c) $t = 20$ **(d)** $t = 50$

57. Let the number of bacteria present in a certain culture be given by
$$B(t) = 25{,}000 e^{.2t}$$
where t is measured in hours, and $t = 0$ corresponds to noon. Find the number of bacteria present at the following times.

(a) Noon **(b)** 1 P.M.

(c) 2 P.M. **(d)** 5 P.M.

58. Suppose a certain collection of termites is growing according to the relationship
$$y = 3000 e^{.04t}$$
where t is measured in months. If there are 3000 termites present on January 1, how many will be present on July 1?

59. The number of ants in an anthill grows according to the relationship
$$y = 300 e^{.4t}$$
where t is measured in days. Find the time it will take for the number of ants to triple.

CHAPTER 10 EXPONENTIAL AND LOGARITHMIC FUNCTIONS

For Exercises 60–64, refer to Example 8 and use the function $N(r) = -5000 \ln r$.

60. Find $N(.9)$.

61. Find $N(.3)$.

62. Find $N(.5)$.

63. How many years have elapsed since the split if 80% of the words of the ancestral language are common to both languages today?

64. If the equation for $N(r)$ is solved for r, the result is

$$r = e^{-N(r)/5000}.$$

Find r if the split occurred 2000 years ago.

SKILL SHARPENERS
Solve the following equations. See Sections 4.5, 5.6, and 6.1.

65. $\dfrac{x + 2}{x} = 5$

66. $\dfrac{3z}{2z - 1} = 4$

67. $\sqrt{k + 2} = 7$

68. $(2x + 3)^2 = 16$

69. $x^2 = 48$

70. $3p^2 = 15$

10.6 EXPONENTIAL AND LOGARITHMIC EQUATIONS

As mentioned at the beginning of this chapter, exponential and logarithmic functions are important in many applications of mathematics. Using these functions in applications requires solving exponential and logarithmic equations. Some simple equations were solved in Sections 10.2 and 10.3. More general methods for solving these equations depend on the following properties.

OBJECTIVES

1. Solve equations involving exponents.
2. Solve equations involving logarithms.
3. Solve applied problems involving exponential equations.

For all real numbers $b > 0$, $b \neq 1$, and any real numbers x and y:

1. If $x = y$, then $b^x = b^y$.
2. If $b^x = b^y$, then $x = y$.
3. If $x = y$, and $x > 0$, $y > 0$, then $\log_b x = \log_b y$.
4. If $x > 0$, $y > 0$, and $\log_b x = \log_b y$, then $x = y$.

Property 2 was used to solve exponential equations in Section 10.2.

1 The first examples illustrate a general method for solving exponential equations using property 3.

EXAMPLE 1 Solving an Exponential Equation

Solve the equation $3^m = 12$.

Use property 3 to write

$$\log 3^m = \log 12.$$

By the power property for logarithms, $\log 3^m = m \log 3$, so

$$m \log 3 = \log 12$$
$$m = \frac{\log 12}{\log 3}.$$

This quotient is the exact solution. To get a decimal approximation for the solution, use a calculator or table of logarithms. A calculator gives

$$m \approx \frac{1.0792}{.4771}$$
$$m \approx 2.262, \quad \text{Divide 1.0792 by .4771.}$$

and the solution set is $\{2.262\}$. ∎

CAUTION Be careful: $\frac{\log 12}{\log 3}$ is *not* equal to $\log 4$, since $\log 4 \approx .6021$, but $\frac{\log 12}{\log 3} \approx 2.262$.

1. Give decimal approximations for the solutions.

   (a) $2^p = 9$

   (b) $10^k = 4$

WORK PROBLEM 1 AT THE SIDE.

ANSWERS
1. (a) $\{3.170\}$ (b) $\{.6021\}$

CHAPTER 10 EXPONENTIAL AND LOGARITHMIC FUNCTIONS

2. Solve each of the following in two ways. Use the method illustrated in Example 2 and then use the method discussed in Section 10.2.

(a) $9^{3x-1} = \left(\dfrac{1}{27}\right)^{x/2}$

(b) $\left(\dfrac{1}{16}\right)^{2x-5} = 32^{x+4}$

ANSWERS
2. (a) $\left\{\dfrac{4}{15}\right\}$ (b) $\{0\}$

EXAMPLE 2 Solving an Exponential Equation

Solve $\left(\dfrac{1}{16}\right)^{2x+3} = 8^{(5x+1)/3}$

Since both 8 (or 2^3) and $\dfrac{1}{16}$ (or 2^{-4}) are powers of 2, take base 2 logarithms of both sides.

$$\log_2 \left(\dfrac{1}{16}\right)^{2x+3} = \log_2 8^{(5x+1)/3}$$

$$(2x+3)\log_2\left(\dfrac{1}{16}\right) = \left(\dfrac{5x+1}{3}\right)\log_2 8 \quad \text{Power property}$$

Since $\log_2 \dfrac{1}{16} = -4$ and $\log_2 8 = 3$,

$$(2x+3)(-4) = \left(\dfrac{5x+1}{3}\right)(3)$$

$$-8x - 12 = 5x + 1 \quad \text{Distributive property}$$

$$-13x = 13$$

$$x = -1.$$

A check will show that -1 is the solution of the original equation, so the solution set is $\{-1\}$. This equation could also be solved using base 10 logarithms as in Example 1. Another method would be to write each side as a power of 2 and then use the second property in Section 10.2. ∎

■ **WORK PROBLEM 2 AT THE SIDE.**

In summary, exponential equations can be solved by one of the following methods. (The method used depends upon the form of the equation.)

SOLVING AN EXPONENTIAL EQUATION

1. Using property 3, take logarithms to the same base on each side; then use the power property of logarithms on both sides.

2. Using property 2, write both sides as exponentials with the same base; then set the exponents equal.

2 The next three examples illustrate ways to solve equations with logarithms. The properties of logarithms from Section 10.4 are useful here, as is using the definition of a logarithm to change to exponential form.

EXAMPLE 3 Solving a Logarithmic Equation

Solve: $\log_2 (x+5)^3 = 4$.

Using the definition of a logarithm gives $(x+5)^3 = 2^4$, or

$$(x+5)^3 = 16$$

$$x + 5 = \sqrt[3]{16}$$

$$x = -5 + \sqrt[3]{16}.$$

Verify that the solution satisfies the equation, so the solution set is $\{-5 + \sqrt[3]{16}\}$. ∎

10.6 EXPONENTIAL AND LOGARITHMIC EQUATIONS

CAUTION Recall that the domain of $y = \log_b x$ is $(0, \infty)$. For this reason, *it is always necessary to check that the solution of an equation with logarithms is in the domain.*

3. Solve.

(a) $\log_3 (2x + 6)^4 = 8$

WORK PROBLEM 3 AT THE SIDE.

EXAMPLE 4 Solving a Logarithmic Equation
Solve $\log_2 (x + 1) - \log_2 x = \log_2 8$.

By the division property for logarithms, $\log_b m - \log_b n = \log_b \frac{m}{n}$,

$$\log_2 (x + 1) - \log_2 x = \log_2 \frac{x+1}{x}.$$

(b) $\log_5 \sqrt{x - 7} = 1$

The original equation becomes

$$\log_2 \frac{x+1}{x} = \log_2 8,$$

which gives

$$\frac{x+1}{x} = 8, \quad \text{Property 4}$$

or

$$8x = x + 1 \quad \text{Multiply by } x.$$

$$x = \frac{1}{7}. \quad \text{Subtract } x; \text{ divide by 7.}$$

Check this solution by substitution in the given equation. Here, both $x + 1$ and x must be positive. If $x = \frac{1}{7}$, this condition is satisfied, and the solution set is $\left\{\frac{1}{7}\right\}$. ∎

4. Solve.

(a) $\log_8 (2x + 5) + \log_8 3 = \log_8 33$

WORK PROBLEM 4 AT THE SIDE.

EXAMPLE 5 Solving a Logarithmic Equation
Solve $\log x + \log \frac{3x}{2} = 5$.

Use the multiplication property for logarithms to write

$$\log x + \log \frac{3x}{2} = \log \frac{3x^2}{2}.$$

Also, the log notation indicates that the base is 10, giving the equation

$$\log_{10} \left(\frac{3x^2}{2}\right) = 5.$$

Using the definition of logarithm to switch to exponential form gives

$$\frac{3x^2}{2} = 10^5,$$

(b) $\log_{10} 5x - \log_{10} (2x - 1) = \log_{10} 4$

ANSWERS
3. (a) $\left\{\frac{3}{2}\right\}$ (b) {32}

4. (a) {3} (b) $\left\{\frac{4}{3}\right\}$

639

CHAPTER 10 EXPONENTIAL AND LOGARITHMIC FUNCTIONS

5. Solve.

(a) $\log(2x - 1) + \log 10x = 1$

(b) $\log_3 2x - \log_3 (3x + 15) = -2$

or

$$\frac{3x^2}{2} = 100{,}000 \quad\quad 10^5 = 100{,}000$$

$$3x^2 = 200{,}000 \quad\quad \text{Multiply both sides by 2.}$$

$$x^2 = \frac{200{,}000}{3} \quad\quad \text{Divide both sides by 3.}$$

$$x \approx 258.2. \quad\quad \text{Take the square root.}$$

The negative square root, -258.2, cannot be used as a solution, since x must be positive in this equation. The solution set is $\{258.2\}$. ∎

■ **WORK PROBLEM 5 AT THE SIDE.**

In summary, use the following steps to solve a logarithmic equation.

SOLVING A LOGARITHMIC EQUATION

Step 1 Use the multiplication or division properties of logarithms to get a single logarithm on one side.

Step 2 (a) Use property 3: If $\log_b x = \log_b y$, then $x = y$. (See Example 4.)
(b) Use the definition of logarithm to write the equation in exponential form. (See Example 3 and Example 5.)

3 The final two examples show applications of exponential equations.

EXAMPLE 6 Solving a Compound Interest Problem

When interest is compounded (interest is paid on interest), P dollars deposited at a rate of interest i compounded annually for t years becomes

$$A = P(1 + i)^t.$$

For example, $1000 deposited at 12% compounded annually for 7 years becomes

$$A = \boxed{1000}(1 + \boxed{.12})^7.$$

Logarithms can be used to find an approximate value for this amount.

$$\log 1000 (1 + .12)^7 = \log 1000 + 7 \log 1.12$$
$$= 3 + 7(.0492)$$
$$= 3 + .3444 = 3.3444$$

Use a calculator to find that the antilogarithm of 3.3444 is 2210. Thus, $1000 would become about $2210 in 7 years. ∎

6. Find the value of $2000 deposited at 5% compounded annually for 10 years.

■ **WORK PROBLEM 6 AT THE SIDE.**

EXAMPLE 7 Solving an Inflation Problem

Suppose that over several years the average annual rate of inflation is 7% compounded annually. How long would it take for the average level of prices to double?

Use the formula from Example 6,

$$A = P(1 + i)^t.$$

ANSWERS
5. (a) {1} (b) {1}
6. about $3260.00

10.6 EXPONENTIAL AND LOGARITHMIC EQUATIONS

If prices double, then P will grow to $2P$, so let $A = 2P$ and $i = .07$ in the formula, and divide each side by P to get

$$2P = P(1 + .07)^t$$
$$2 = (1.07)^t.$$

Take common logarithms on each side, and use the power property on the right.

$$\log 2 = t \log 1.07$$

$$t = \frac{\log 2}{\log 1.07} \quad \text{Divide both sides by log 1.07.}$$

$$\approx \frac{.3010}{.0294} \quad \text{Evaluate the logarithms.}$$

$$= 10.2 \text{ years, or about 10 years, 2 months}$$

At this rate, a person retiring at age 60 would find the purchasing power of his or her dollars cut in half at around age 70. ■

WORK PROBLEM 7 AT THE SIDE.

7. (a) In Example 7, how long would it take for prices to double if the annual rate of inflation were 6% compounded annually?

(b) How long at 12% compounded annually?

ANSWERS
7. (a) 11.9 years **(b)** 6.1 years

Historical Reflections

John Napier (1550–1617)

Until the advent of hand-held calculators, logarithms were used extensively for computational purposes. Products, quotients, powers, and roots could all be found using tables of common logarithms. Today's technology has made computing with logarithms a thing of the past.

John Napier's most significant mathematical contribution, developed over a period of at least twenty years, was the concept of logarithms. Despite the importance of his work, he regarded mathematics as a recreation and devoted much of his time to political and religious causes. He was a supporter of John Knox and James I, and he published a widely read anti-Catholic work that analyzed the book of Revelations and concluded that the Pope was the Antichrist and that the Creator would end the world between 1688 and 1700.

Art: Courtesy of IBM

name   date   hour

10.6 EXERCISES

Solve the following equations. Round solutions to the nearest hundredth if necessary. See Examples 1 and 2.

1. $25^x = 125$
2. $16^y = 64$
3. $5^m = 10$

4. $8^{-p} = 12$
5. $6^{y+1} = 8$
6. $2^{-3+y} = 4.5$

7. $\left(\dfrac{1}{2}\right)^x = 10$
8. $3^{y+1} = 2$
9. $7^{2y-1} = 1$

10. $8^m = 3^{m+1}$
11. $2^{-x} = 27$
12. $7^{x^2+2x} = \dfrac{1}{7}$

13. $(1 + .03)^n = 90$
14. $100(1 + .02)^{3+n} = 150$

Solve the following equations. See Examples 3–5.

15. $\log (x + 2) = \log 6$
16. $\log x = \log (1 - x)$

17. $\log_5 (3x + 2) - \log_5 x = \log_5 4$
18. $\log_2 (x + 5) - \log_2 (x - 1) = \log_2 3$

19. $\log 4x = \log 2 + \log (x - 3)$

20. $\log (-x) + \log 3 = \log (2x - 15)$

21. $\log_2 x = 3$

22. $\log_x 10 = 3$

23. $\log_y 11 = 2$

24. $\log_m 4 = \dfrac{3}{2}$

25. $\log_a 5 = -\dfrac{3}{4}$

26. $2 + \log x = 0$

27. $\log_3 x + \log_3 (2x + 5) = 1$

28. $\log_2 x + \log_2 (x - 7) = 3$

Solve the following problems. See Example 6.

29. Find the amount of money in the bank after 12 years if $5000 is deposited at 6% compounded annually.

30. How much money will be in the bank at the end of 8 years if $4500 is deposited at 7% compounded annually?

10.6 EXERCISES

31. How much money must be deposited in a bank today to amount to $1000 in 10 years, at 7% compounded annually?

32. Find the amount of money that must be deposited today at 8% compounded annually to amount to $8000 in 5 years.

How long would it take for the average price level to double if the average rate of inflation is as follows? (Check your answers by using the rule of 70: the time for prices to double is given by $\frac{70}{x}$, where x is the percent of annual inflation.) See Example 7.

33. 3%

34. 5%

35. 6%

36. 8%

A machine purchased for business use depreciates, or wears out, over a period of years. The value of the machine at the end of its useful life is called its scrap value. By one method of depreciation (where it is assumed a constant percentage of the value depreciates annually), the scrap value, S, is given by

$$S = C(1 - r)^n$$

where C is the original cost, n is the useful life in years, and r is the constant percentage of depreciation.

37. Find the scrap value of a machine costing $30,000, having a useful life of 12 years and a constant annual rate of depreciation of 15%.

38. What is the scrap value of a machine that cost $50,000 new, with a useful life of 8 years and a constant annual depreciation of 10%?

645

39. If a machine has its value cut in half in 6 years, find the constant annual rate of depreciation.

40. What is the constant annual rate of depreciation for a piece of equipment that retains $\frac{1}{3}$ of its value after 8 years?

Solve each problem.

41. Sales of a new product (in hundreds) are given by $S = 80 \log_5 (t + 1)$, where t is time in years after the product is introduced. When will the sales reach 80 (hundreds)?

42. An average worker on a certain production line produces $p = 10 + \log_3 t$ items per day, where t is the number of days the worker has been on the job. When will such a worker produce 15 items a day?

43. The population of buffalo in an area t years after they were introduced into the area is $P = 15 + 10 \ln (2t + 1)$. How many buffalo were in the area after 3 years?

44. The number of fish in a stream t months after being planted there is $P = 25 \ln (3t + 20)$. What was the fish population after 2 months?

Historical Reflections

Leonhard Euler (1707–1783)

Leonhard Euler (pronounced "oiler") of Switzerland was the most prolific mathematician of his time, despite blindness that forced him to dictate from memory. He did not view his ailment as a handicap; his comment was "I'll have fewer distractions."

The stamp shown here features one of the various "Euler equations," this one from trigonometry. Euler wrote in all mathematical fields, creating new results as well as organizing several fields, most notably calculus and analysis. Much of the symbolism used in mathematics today was originated by Euler; his contributions include the function notation $f(x)$, the imaginary unit i, and the number e that serves as the base for natural logarithms.

One Euler invention, the Beta function, provided a major breakthrough in 1968 in the study of physics. It proved to be a critical step in the development of the *superstring theory*, a promising new model encompassing all known forces of nature. It is based on the idea that matter consists of tiny strings rather than point particles.

CHAPTER 10 SUMMARY

KEY TERMS

10.1 **one-to-one function**   A one-to-one function is a function in which each x-value corresponds to just one y-value and each y-value corresponds to just one x-value.

inverse of a function f   If f is a one-to-one function, the inverse of f is the set of all ordered pairs of the form (y, x), where (x, y) belongs to f.

10.2 **exponential equation**   An equation involving an exponential, where the variable is in the exponent, is an exponential equation.

10.3 **logarithm**   A logarithm is an exponent; $\log_a x$ is the exponent on the base a that gives the number x.

logarithmic equation   A logarithmic equation is an equation with a logarithm in at least one term.

10.5 **common logarithm**   A common logarithm is a logarithm to the base 10.

antilogarithm   An antilogarithm is the number that corresponds to a given logarithm.

natural logarithm   A natural logarithm is a logarithm to the base e.

NEW SYMBOLS

$f^{-1}(x)$	the inverse of f
$\log_a x$	the logarithm of x to the base a
$\log x$	common (base 10) logarithm of x
$\ln x$	natural (base e) logarithm of x
e	a constant, approximately 2.7182818

QUICK REVIEW

Section	Concepts	Examples
10.1 Inverse Functions	If a horizontal line intersects the graph of a function in no more than one point, then the function is one-to-one. For a one-to-one function f defined by an equation $y = f(x)$, the defining equation of the inverse function f^{-1} is found by exchanging x and y, solving for y, and replacing y with $f^{-1}(x)$.	Find f^{-1} if $f(x) = 2x - 3$. The graph of f is a straight line, so f is one-to-one by the horizontal line test. Exchange x and y in the equation $y = 2x - 3$. $$x = 2y - 3$$ Solve for y to get $$y = \frac{1}{2}x + \frac{3}{2}.$$ Therefore, $f^{-1}(x) = \frac{1}{2}x + \frac{3}{2}$.

CHAPTER 10 SUMMARY

Section	Concepts	Examples
	The graph of f^{-1} is a mirror image of the graph of f with respect to the line $y = x$.	The graphs of a function f and its inverse f^{-1} are given below.
10.2 Exponential Functions	For $a > 0$, $a \neq 1$, $f(x) = a^x$ is an exponential function with base a.	$f(x) = 10^x$ is an exponential function with base 10.
10.3 Logarithmic Functions	$y = \log_a x$ has the same meaning as $a^y = x$. For $b > 0$, $b \neq 1$, $\quad \log_b b = 1$ and $\log_b 1 = 0$. For $a > 0$, $a \neq 1$, $x > 0$, $f(x) = \log_a x$ is the logarithmic function with base a.	$y = \log_2 x$ means $x = 2^y$. $\log_3 3 = 1$ $\log_5 1 = 0$ $f(x) = \log_6 x$ is the logarithmic function with base 6.
10.4 Properties of Logarithms	**Multiplication Property** $\quad \log_a xy = \log_a x + \log_a y$ **Division Property** $\quad \log_a \dfrac{x}{y} = \log_a x - \log_a y$ **Power Property** $\quad \log_a x^r = r \log_a x$ **Special Properties** $b^{\log_b x} = x$ and $\log_b b^x = x$	$\log_2 3m = \log_2 3 + \log_2 m$ $\log_5 \dfrac{9}{4} = \log_5 9 - \log_5 4$ $\log_{10} 2^3 = 3 \log_{10} 2$ $6^{\log_6 10} = 10 \quad \log_3 3^4 = 4$
10.5 Evaluating Logarithms	**Change-of-Base Rule** If $a > 0$, $a \neq 1$, $b > 0$, $b \neq 1$, $x > 0$, then $\quad \log_a x = \dfrac{\log_b x}{\log_b a}.$	$\log_3 17 = \dfrac{\ln 17}{\ln 3}$

CHAPTER 10 EXPONENTIAL AND LOGARITHMIC FUNCTIONS

Section	Concepts	Examples
10.6 Exponential and Logarithmic Equations	To solve exponential equations, use these properties ($b > 0$, $b \neq 1$). **1.** If $b^x = b^y$, then $x = y$.	Solve: $2^{3x} = 2^5$. $3x = 5$ $x = \dfrac{5}{3}$
	2. If $x = y$, ($x > 0$, $y > 0$), then $\log_b x = \log_b y$.	Solve: $5^m = 8$. $\log 5^m = \log 8$ $m \log 5 = \log 8$ $m = \dfrac{\log 8}{\log 5}$
	To solve logarithmic equations, use these properties, where $b > 0$, $b \neq 1$, $x > 0$, $y > 0$. First use the properties of 10.4, if necessary, to get the equation in the proper form.	
	1. If $\log_b x = \log_b y$, then $x = y$.	Solve: $\log_3 2x = \log_3 (x + 1)$. $2x = x + 1$ $x = 1$
	2. If $\log_b x = y$, then $b^y = x$.	Solve: $\log_2 (3a - 1) = 4$. $3a - 1 = 2^4 = 16$ $3a = 17$ $a = \dfrac{17}{3}$

CHAPTER 10 REVIEW EXERCISES

[10.1] *For each equation of a one-to-one function, find the equation of the inverse.*

1. $3y = x - 1$ **2.** $f(x) = x^2 - 4$ **3.** $f(x) = \sqrt{2x + 5}$

Determine whether the graph is the graph of a one-to-one function.

4.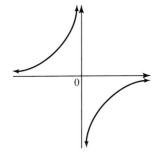

5.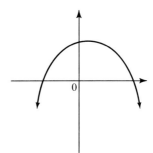

Graph the inverse of each one-to-one function on the same axes as the function.

6.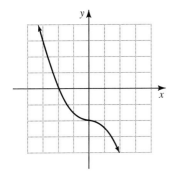

7.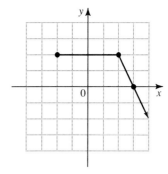

[10.2] *Graph the following.*

8. $y = 4^x$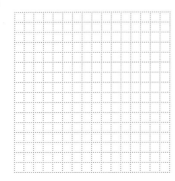

9. $y = \left(\dfrac{1}{2}\right)^x$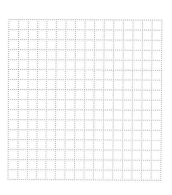

10. $y = 3^{2x}$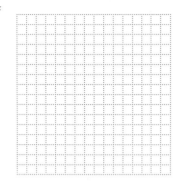

11. $y = 2^{-x+1}$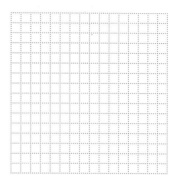

[10.3]

12. $y = \log_4 x$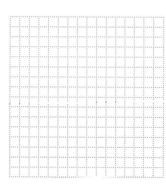

13. $y = \log_{1/2} x$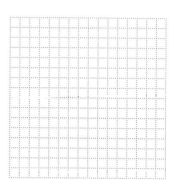

Complete the following charts.

	Logarithmic form	Exponential form
14.	$\log_3 27 = 3$	
16.	$\log_8 2 = \dfrac{1}{3}$	
18.		$9^{1/2} = 3$

	Logarithmic form	Exponential form
15.	$\log_5 \dfrac{1}{5} = -1$	
17.		$4^3 = 64$
19.		$2^{-3} = \dfrac{1}{8}$

CHAPTER 10 REVIEW EXERCISES

[10.4] *Express each of the following as a sum, difference, or product of logarithms. Assume all variables represent positive numbers.*

20. $\log_5 \dfrac{3}{10} x$

21. $\log_3 m^2 p \sqrt{g}$

22. $\log_2 \dfrac{5k}{3r^3}$

23. $\log_{10} (2x + 3)(5x)$

Write each of the following as a single logarithm. Assume all variable expressions represent positive numbers.

24. $\log_a 2x + \log_a 3$

25. $3 \log_b p - 2 \log_b q$

26. $\log_5 (x - 1) - \log_5 (x^2 - 1) \quad (x > 1)$

27. $4 \log_2 x - 3 \log_2 x$

[10.5] *Evaluate each logarithm. A calculator may be needed for most of the remaining exercises in this set.*

28. $\log 2.95$ **29.** $\log 432$ **30.** $\log .0714$ **31.** $\log .16$

32. $\log 10^{.2}$ **33.** $\log 1$ **34.** $\ln 8$ **35.** $\ln 25$

36. $\ln e$ **37.** $\ln e^{.1}$ **38.** $\ln e^4$ **39.** $\ln 1$

Find each of the following.

40. $\log_6 10$ **41.** $\log_3 2.51$ **42.** $\ln 7$ **43.** $\ln 100$

44. $\log 4e^2$ **45.** $\log 2e^{1.3}$ **46.** $\ln 4e^{1.2}$ **47.** $\ln 3e^{5.2}$

653

CHAPTER 10 EXPONENTIAL AND LOGARITHMIC FUNCTIONS

[10.6] *Solve each equation.*

48. $27^x = 81$

49. $2^{2y-3} = 8$

50. $4^{m+1} = 5$

51. $\log(3x - 1) = \log 10$

52. $\log_3(p + 2) - \log_3 p = \log_3 2$

53. $\log_3 k = -2$

54. $\log_m 125 = 3$

55. $\log(2x + 3) = \log x + 1$

Solve the following problems.

56. How much will $10,000 invested at 7% compounded annually amount to in 5 years?

57. How much money must be deposited today to amount to $12,000 in 18 years at 6% compounded annually?

58. Use the formula $S = C(1 - r)^n$ to find the scrap value of equipment costing $50,000, with a useful life of 10 years, if the annual depreciation is 10%.

59. Use the formula $S = C(1 - r)^n$ to find the constant annual depreciation rate of a machine that has lost $\frac{1}{4}$ of its value in 2 years.

CHAPTER 10 TEST

1. True or false: $y = x^2 + 4$ defines a one-to-one function.

    1. _____

2. Find the inverse of the one-to-one function $f(x) = \sqrt[3]{4 - x}$.

    2. _____

3. Graph the inverse of f, given the graph of f below.

    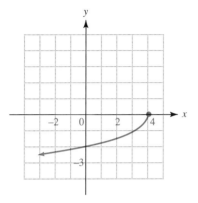

    3.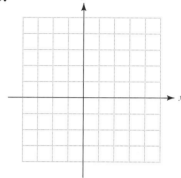

Graph the following functions.

4. $y = \left(\dfrac{1}{3}\right)^x$

    4.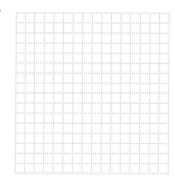

5. $y = \log_3 x$

    5.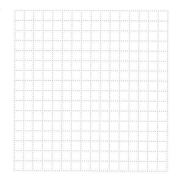

CHAPTER 10 EXPONENTIAL AND LOGARITHMIC FUNCTIONS

Given that $\log M = 2.3156$ *and* $\log N = .1827$, *evaluate each of the following.*

6. _____   6. $\log MN$

7. _____   7. $\log \dfrac{M}{N}$

8. _____   8. $\log M^3$

9. _____   9. $\log \sqrt{N}$

Solve each equation.

10. _____   10. $3^x = 81$

11. _____   11. $\log_6 k = 2$

Find the following.

12. _____   12. $\log_6 36$

13. _____   13. $\log_2 \left(\dfrac{1}{64}\right)$

CHAPTER 10 TEST

14. $\log_9 9^{1.4}$

14. _____

15. $\log 246$

15. _____

16. $\log .000317$

16. _____

17. $\log 10$

17. _____

18. $\ln 89.1$

18. _____

19. $\ln .43$

19. _____

20. $\log_2 18$

20. _____

21. $\log_{15} 9$

21. _____

Solve the following equations.

22. $2^m = 14$

22. _____

23. $\log_2 (x + 3) + \log_2 (x - 1) = \log_2 5$

23. _____

24. $\log_x 5 = 6$

24. _____

Solve the following word problem.

25. Suppose a culture of bacteria grows according to the function
$$C(t) = 8000e^{.4t}$$
where $C(t)$ is the number of bacteria present at time t (measured in hours).

(a) How many bacteria will be present at time $t = 5$?

25. (a) _____

(b) When will the culture contain 16,000 bacteria?

(b) _____

FINAL EXAMINATION

Let $S = \left\{-\frac{9}{4}, -2, -\sqrt{2}, 0, .6, \sqrt{11}, \sqrt{-8}, 6, \frac{30}{3}\right\}$. List the elements of S that are members of the following sets.

1. Integers

2. Rational numbers

3. Irrational numbers

4. Real numbers

Simplify the following.

5. $|-8| + 6 - |-2| - (-6 + 2)$

6. $-12 - |-3| - 7 - |-5|$

7. $2(-5) + (-8)(4) - (-3)$

Solve the following.

8. $7 - (3 + 4a) + 2a = -5(a - 1) - 3$

9. $2m + 2 \leq 5m - 1$

10. $|2x - 5| = 9$

11. $|3p| - 4 = 12$

FINAL EXAMINATION

12. $|3k - 8| \leq 1$

13. $|4m + 2| > 10$

Perform the indicated operations.

14. $(2p + 3)(3p - 1)$

15. $(4k - 3)^2$

16. $(3m^3 + 2m^2 - 5m) - (8m^3 + 2m - 4)$

17. Divide $6t^4 + 17t^3 - 4t^2 + 9t + 4$ by $3t + 1$.

Factor as completely as possible.

18. $8x + x^3$

19. $24y^2 - 7y - 6$

20. $5z^3 - 19z^2 - 4z$

21. $16a^2 - 25b^4$

22. $8c^3 + d^3$

FINAL EXAMINATION

23. $16r^2 + 56rq + 49q^2$

23. _____

Simplify as much as possible in Exercises 24–27.

24. $\dfrac{(5p^3)^4(-3p^7)}{2p^2(4p^4)}$

24. _____

25. $\dfrac{x^2 - 9}{x^2 + 7x + 12} \div \dfrac{x - 3}{x + 5}$

25. _____

26. $\dfrac{2}{k + 3} - \dfrac{5}{k - 2}$

26. _____

27. $\dfrac{3}{p^2 - 4p} - \dfrac{4}{p^2 + 2p}$

27. _____

28. Candy worth $1.00 per pound is to be mixed with 10 pounds of candy worth $1.96 per pound to get a mixture that will be sold for $1.60 per pound. How many pounds of the $1.00 candy should be used?

28. _____

Simplify in Exercises 29–31.

29. $\left(\dfrac{5}{4}\right)^{-2}$

29. _____

30. $\dfrac{6^{-3}}{6^2}$

30. _____

31. $2\sqrt{32} - 5\sqrt{98}$

31. _____

32. Multiply: $(5 + 4i)(5 - 4i)$.

32. _____

661

Solve the equations or inequalities in Exercises 33–35.

33. $10p^2 + p - 2 = 0$

34. _____

34. $a^2 + 3a = 10$

35. _____

35. $k^2 + 2k - 8 > 0$

36. _____

36. Find the slope of the line through $(-3, 2)$ and $(8, 4)$.

37. _____

37. Find the equation of the line through $(5, -1)$ and parallel to the line with equation $3x - 4y = 12$.

Graph the following.

38.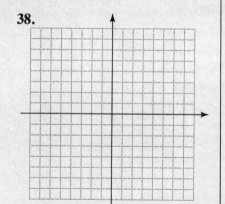

38. $5x + 2y = 10$

39.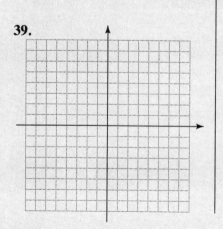

39. $-4x + y \leq 5$

40. $f(x) = \frac{1}{3}(x-1)^2 + 2$

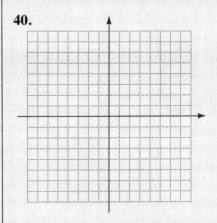

40.

41. $\dfrac{x^2}{9} + \dfrac{y^2}{16} = 1$

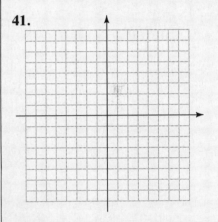

41.

42. $25x^2 - 16y^2 = 400$

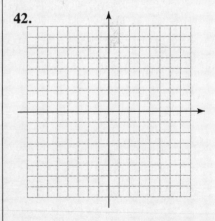

42.

Solve each system of equations.

43. $5x - 3y = 14$
$2x + 5y = 18$

43. _____

44.
$$x + 2y + 3z = 11$$
$$3x - y + z = 8$$
$$2x + 2y - 3z = -12$$

Evaluate the following.

45. $\begin{vmatrix} -2 & -1 \\ 5 & 3 \end{vmatrix}$

46. $\begin{vmatrix} 2 & 4 & 5 \\ 1 & 3 & 0 \\ 0 & -1 & -2 \end{vmatrix}$

47. Convert $5^3 = 125$ to logarithmic form.

48. Convert to exponential form: $\log_3 9 = 2$.

49. Solve $7^{2x} = 49$.

50. Solve for x: $\log_5 x + \log_5 (x + 4) = 1$.

APPENDICES

APPENDIX A: USING THE TABLE OF COMMON LOGARITHMS

1 This appendix gives a brief explanation of the table of common logarithms given on pages 670–71. To find logarithms using the table, first write the number in scientific notation (see Section 3.2). For example, to find log 423, first write 423 in scientific notation.

$$\log 423 = \log (4.23 \times 10^2)$$

By the multiplication property for logarithms,

$$\log 423 = \log 4.23 + \log 10^2.$$

The logarithm of 10^2 is 2 (the exponent). To find log 4.23, use the table of common logarithms. (A portion of that table is reproduced here.) Read down on the left to the row headed 4.2. Then read across to the column headed by 3 to get .6263. From the table,

$$\log 423 \approx .6263 + 2 = 2.6263.$$

x	0	1	2	3	4	5	6	7	8	9
4.0	.6021	.6031	.6042	.6053	.6064	.6075	.6085	.6096	.6107	.6117
4.1	.6128	.6138	.6149	.6160	.6170	.6180	.6191	.6201	.6212	.6222
4.2	.6232	.6243	.6253	.6263	.6274	.6284	.6294	.6304	.6314	.6325
4.3	.6335	.6345	.6355	.6365	.6375	.6385	.6395	.6405	.6415	.6425
4.4	.6435	.6444	.6454	.6464	.6474	.6484	.6493	.6503	.6513	.6522

The decimal part of the logarithm, .6263, is called the **mantissa**. The integer part, 2, is called the **characteristic**.

WORK PROBLEMS 1 AND 2 AT THE SIDE.

EXAMPLE 1 Using the Table to Find Common Logarithms
Find the following common logarithms.

(a) log 437,000

Since $437,000 = 4.37 \times 10^5$, the characteristic is 5. From the portion of the logarithm table given above, the mantissa of log 437,000 is .6405, and log 437,000 = 5 + .6405 = 5.6405.

(b) log .0415

Express .0415 as 4.15×10^{-2}. The characteristic is -2. The mantissa, from the table, is .6180, and

$$\log .0415 = -2 + .6180 = -1.3820.$$

To retain the mantissa, the answer may be given as $-2 + .6180$, or $.6180 - 2$. A calculator with a logarithm key gives the result as -1.3820. ∎

OBJECTIVES

1 Evaluate base 10 logarithms by using a logarithm table.

2 Find the antilogarithm of a common logarithm using a table.

3 Interpolate the values in a logarithm table.

4 Interpolate when finding antilogarithms.

1. Write each number in scientific notation. Then give the characteristic of its logarithm.

   (a) 793

   (b) 105,000

   (c) .0674

   (d) 2.96

2. Find *just the mantissa* of the logarithm of each number.

   (a) 421

   (b) 41,500

   (c) .0439

ANSWERS
1. (a) 7.93×10^2; 2
   (b) 1.05×10^5; 5
   (c) 6.74×10^{-2}; -2
   (d) 2.96×10^0; 0
2. (a) .6243 (b) .6180 (c) .6425

APPENDIX A: USING THE TABLE OF COMMON LOGARITHMS

3. Find the following.

(a) log 41,600

(b) log 43.5

(c) log .442

4. Find the antilogarithm of each logarithm.

(a) 3.6503

(b) .6304 − 2

(c) .6064

WORK PROBLEM 3 AT THE SIDE.

2 In Example 1(a), the number 437,000, whose logarithm is 5.6405, is called the **antilogarithm** (abbreviated *antilog*) of 5.6405. In the same way, in Example 1(b), .0415 is the antilogarithm of the logarithm .6180 − 2.

From Example 1,

$$\log 437{,}000 = 5.6405.$$

In exponential form, this is equivalent to

$$437{,}000 = 10^{5.6405}.$$

This shows that the antilogarithm, 437,000, is just an exponential.

EXAMPLE 2 Using the Table to Find Antilogarithms
Find the antilogarithm of each of the following logarithms.

(a) 2.6454

To find the number whose logarithm is 2.6454, first look in the table for a mantissa of .6454. The number whose mantissa is .6454 is 4.42. Since the characteristic is 2, the antilogarithm is $4.42 \times 10^2 = 442$.

(b) .6314 − 3

The mantissa is .6314 and the characteristic is −3. In the table, the number with a mantissa of .6314 is 4.28, so the antilogarithm is $4.28 \times 10^{-3} = .00428$. ■

WORK PROBLEM 4 AT THE SIDE.

3 The table of logarithms contains logarithms to four decimal places. This table can be used to find the logarithm of any positive number containing three significant digits. **Linear interpolation** is used to find logarithms of numbers containing four significant digits. In Example 3 we show how this is done.

EXAMPLE 3 Using Linear Interpolation to Find a Logarithm
Find log 4238.

Since $\qquad 4230 < 4238 < 4240,$

$$\log 4230 < \log 4238 < \log 4240.$$

Find log 4230 and log 4240 in the table.

$$10 \left\{ 8 \left\{ \begin{array}{l} \log 4230 = 3.6263 \\ \log 4238 = \\ \log 4240 = 3.6274 \end{array} \right. \right\} .0011$$

Since 4238 is $\frac{8}{10}$ of the way from 4230 to 4240, take $\frac{8}{10}$ of .0011, the difference between the two logarithms.

$$\frac{8}{10}(.0011) \approx .0009$$

Now add .0009 to log 4230 to get

$$\log 4238 = 3.6263 + .0009$$
$$= 3.6272.$$

ANSWERS
3. (a) 4.6191 (b) 1.6385
   (c) .6454 − 1 or −.3546
4. (a) 4470 (b) .0427 (c) 4.04

APPENDIX A: USING THE TABLE OF COMMON LOGARITHMS

Check the answer by consulting a more accurate logarithm table or use a calculator to find that log 4238 = 3.62716 (to five places). ■

EXAMPLE 4 Using Linear Interpolation to Find a Logarithm
Find log .02386.

Arrange the work as follows.

$$10\left\{6\left\{\begin{array}{l}\log .0238 = .3766 - 2\\ \log .02386 = \\ \log .0239 = .3784 - 2\end{array}\right\}.0018\right.$$

$$\frac{6}{10}(.0018) \approx .0011$$

Now add .3766 − 2 and .0011.

$$\begin{array}{r}.3766 - 2\\ .0011\\ \hline \log .02386 = .3777 - 2\end{array}$$ ■

WORK PROBLEM 5 AT THE SIDE.

4 Interpolation can also be used to find antilogarithms with four significant digits.

EXAMPLE 5 Using Linear Interpolation to Find an Antilogarithm
Find the antilogarithm of 3.5894.

From the table, the mantissas closest to .5894 are .5888 and .5899. Find the antilogarithms of 3.5888 and 3.5899.

$$.0011\left\{.0006\left\{\begin{array}{l}3.5888 = \log 3880\\ 3.5894 = \\ 3.5899 = \log 3890\end{array}\right\}10\right.$$

$$\left(\frac{.0006}{.0011}\right)10 \approx .5(10) = 5$$

Add the 5 to 3880, giving

$$\log 3885 = 3.5894.$$

The antilogarithm of 3.5894 is 3885. ■

WORK PROBLEM 6 AT THE SIDE.

5. Use interpolation to find each logarithm.

   (a) 47.23

   (b) .1968

6. Find the antilogarithm of each number.

   (a) 2.6315

   (b) .3842 − 2

ANSWERS
5. (a) 1.6742 (b) .2941 − 1
6. (a) 428.1 (b) .02422

name   date   hour

APPENDIX A EXERCISES

Use the table on pages 670–71 to find each of the following logarithms. See Example 1.

1. log 2.37   **2.** log 4.69   **3.** log 194

4. log 83   **5.** log 12   **6.** log 1870

7. log 25,000   **8.** log 36,400   **9.** log .6

10. log .05   **11.** log .000211   **12.** log .00432

Use the table to find each of the following antilogarithms. See Example 2.

13. 2.5366   **14.** 1.8407   **15.** .6599   **16.** 3.9258

17. 1.3979   **18.** 4.8716   **19.** .0792 − 3   **20.** .7259 − 1

21. .8727 − 2   **22.** .4843 − 2   **23.** .2041 − 4   **24.** .9586 − 2

Use the table and linear interpolation to find the following logarithms. See Examples 3 and 4.

25. log .8973   **26.** log 2.635   **27.** log 53.89   **28.** log 218.4

29. log 5248   **30.** log 19,040   **31.** log .01253   **32.** log .6529

Use the table and interpolation to find the antilogarithms of the following logarithms to four significant digits. See Example 5.

33. 2.7138   **34.** 1.9146   **35.** .8342 − 1

36. .7138 − 2   **37.** .2008 − 4   **38.** .2413 − 3

TABLE OF COMMON LOGARITHMS

n	0	1	2	3	4	5	6	7	8	9
1.0	.0000	.0043	.0086	.0128	.0170	.0212	.0253	.0294	.0334	.0374
1.1	.0414	.0453	.0492	.0531	.0569	.0607	.0645	.0682	.0719	.0755
1.2	.0792	.0828	.0864	.0899	.0934	.0969	.1004	.1038	.1072	.1106
1.3	.1139	.1173	.1206	.1239	.1271	.1303	.1335	.1367	.1399	.1430
1.4	.1461	.1492	.1523	.1553	.1584	.1614	.1644	.1673	.1703	.1732
1.5	.1761	.1790	.1818	.1847	.1875	.1903	.1931	.1959	.1987	.2014
1.6	.2041	.2068	.2095	.2122	.2148	.2175	.2201	.2227	.2253	.2279
1.7	.2304	.2330	.2355	.2380	.2405	.2430	.2455	.2480	.2504	.2529
1.8	.2553	.2577	.2601	.2625	.2648	.2672	.2695	.2718	.2742	.2765
1.9	.2788	.2810	.2833	.2856	.2878	.2900	.2923	.2945	.2967	.2989
2.0	.3010	.3032	.3054	.3075	.3096	.3118	.3139	.3160	.3181	.3201
2.1	.3222	.3243	.3263	.3284	.3304	.3324	.3345	.3365	.3385	.3404
2.2	.3424	.3444	.3464	.3483	.3502	.3522	.3541	.3560	.3579	.3598
2.3	.3617	.3636	.3655	.3674	.3692	.3711	.3729	.3747	.3766	.3784
2.4	.3802	.3820	.3838	.3856	.3874	.3892	.3909	.3927	.3945	.3962
2.5	.3979	.3997	.4014	.4031	.4048	.4065	.4082	.4099	.4116	.4133
2.6	.4150	.4166	.4183	.4200	.4216	.4232	.4249	.4265	.4281	.4298
2.7	.4314	.4330	.4346	.4362	.4378	.4393	.4409	.4425	.4440	.4456
2.8	.4472	.4487	.4502	.4518	.4533	.4548	.4564	.4579	.4594	.4609
2.9	.4624	.4639	.4654	.4669	.4683	.4698	.4713	.4728	.4742	.4757
3.0	.4771	.4786	.4800	.4814	.4829	.4843	.4857	.4871	.4886	.4900
3.1	.4914	.4928	.4942	.4955	.4969	.4983	.4997	.5011	.5024	.5038
3.2	.5051	.5065	.5079	.5092	.5105	.5119	.5132	.5145	.5159	.5172
3.3	.5185	.5198	.5211	.5224	.5237	.5250	.5263	.5276	.5289	.5302
3.4	.5315	.5328	.5340	.5353	.5366	.5378	.5391	.5403	.5416	.5428
3.5	.5441	.5453	.5465	.5478	.5490	.5502	.5514	.5527	.5539	.5551
3.6	.5563	.5575	.5587	.5599	.5611	.5623	.5635	.5647	.5658	.5670
3.7	.5682	.5694	.5705	.5717	.5729	.5740	.5752	.5763	.5775	.5786
3.8	.5798	.5809	.5821	.5832	.5843	.5855	.5866	.5877	.5888	.5899
3.9	.5911	.5922	.5933	.5944	.5955	.5966	.5977	.5988	.5999	.6010
4.0	.6021	.6031	.6042	.6053	.6064	.6075	.6085	.6096	.6107	.6117
4.1	.6128	.6138	.6149	.6160	.6170	.6180	.6191	.6201	.6212	.6222
4.2	.6232	.6243	.6253	.6263	.6274	.6284	.6294	.6304	.6314	.6325
4.3	.6335	.6345	.6355	.6365	.6375	.6385	.6395	.6405	.6415	.6425
4.4	.6435	.6444	.6454	.6464	.6474	.6484	.6493	.6503	.6513	.6522
4.5	.6532	.6542	.6551	.6561	.6571	.6580	.6590	.6599	.6609	.6618
4.6	.6628	.6637	.6646	.6656	.6665	.6675	.6684	.6693	.6702	.6712
4.7	.6721	.6730	.6739	.6749	.6758	.6767	.6776	.6785	.6794	.6803
4.8	.6812	.6821	.6830	.6839	.6848	.6857	.6866	.6875	.6884	.6893
4.9	.6902	.6911	.6920	.6928	.6937	.6946	.6955	.6964	.6972	.6981
5.0	.6990	.6998	.7007	.7016	.7024	.7033	.7042	.7050	.7059	.7067
5.1	.7076	.7084	.7093	.7101	.7110	.7118	.7126	.7135	.7143	.7152
5.2	.7160	.7168	.7177	.7185	.7193	.7202	.7210	.7218	.7226	.7235
5.3	.7243	.7251	.7259	.7267	.7275	.7284	.7292	.7300	.7308	.7316
5.4	.7324	.7332	.7340	.7348	.7356	.7364	.7372	.7380	.7388	.7396
n	0	1	2	3	4	5	6	7	8	9

TABLE OF COMMON LOGARITHMS (CONTINUED)

n	0	1	2	3	4	5	6	7	8	9
5.5	.7404	.7412	.7419	.7427	.7435	.7443	.7451	.7459	.7466	.7474
5.6	.7482	.7490	.7497	.7505	.7513	.7520	.7528	.7536	.7543	.7551
5.7	.7559	.7566	.7574	.7582	.7589	.7597	.7604	.7612	.7619	.7627
5.8	.7634	.7642	.7649	.7657	.7664	.7672	.7679	.7686	.7694	.7701
5.9	.7709	.7716	.7723	.7731	.7738	.7745	.7752	.7760	.7767	.7774
6.0	.7782	.7789	.7796	.7803	.7810	.7818	.7825	.7832	.7839	.7846
6.1	.7853	.7860	.7868	.7875	.7882	.7889	.7896	.7903	.7910	.7917
6.2	.7924	.7931	.7938	.7945	.7952	.7959	.7966	.7973	.7980	.7987
6.3	.7993	.8000	.8007	.8014	.8021	.8028	.8035	.8041	.8048	.8055
6.4	.8062	.8069	.8075	.8082	.8089	.8096	.8102	.8109	.8116	.8122
6.5	.8129	.8136	.8142	.8149	.8156	.8162	.8169	.8176	.8182	.8189
6.6	.8195	.8202	.8209	.8215	.8222	.8228	.8235	.8241	.8248	.8254
6.7	.8261	.8267	.8274	.8280	.8287	.8293	.8299	.8306	.8312	.8319
6.8	.8325	.8331	.8338	.8344	.8351	.8357	.8363	.8370	.8376	.8382
6.9	.8388	.8395	.8401	.8407	.8414	.8420	.8426	.8432	.8439	.8445
7.0	.8451	.8457	.8463	.8470	.8476	.8482	.8488	.8494	.8500	.8506
7.1	.8513	.8519	.8525	.8531	.8537	.8543	.8549	.8555	.8561	.8567
7.2	.8573	.8579	.8585	.8591	.8597	.8603	.8609	.8615	.8621	.8627
7.3	.8633	.8639	.8645	.8651	.8657	.8663	.8669	.8675	.8681	.8686
7.4	.8692	.8698	.8704	.8710	.8716	.8722	.8727	.8733	.8739	.8745
7.5	.8751	.8756	.8762	.8768	.8774	.8779	.8785	.8791	.8797	.8802
7.6	.8808	.8814	.8820	.8825	.8831	.8837	.8842	.8848	.8854	.8859
7.7	.8865	.8871	.8876	.8882	.8887	.8893	.8899	.8904	.8910	.8915
7.8	.8921	.8927	.8932	.8938	.8943	.8949	.8954	.8960	.8965	.8971
7.9	.8976	.8982	.8987	.8993	.8998	.9004	.9009	.9015	.9020	.9025
8.0	.9031	.9036	.9042	.9047	.9053	.9058	.9063	.9069	.9074	.9079
8.1	.9085	.9090	.9096	.9101	.9106	.9112	.9117	.9122	.9128	.9133
8.2	.9138	.9143	.9149	.9154	.9159	.9165	.9170	.9175	.9180	.9186
8.3	.9191	.9196	.9201	.9206	.9212	.9217	.9222	.9227	.9232	.9238
8.4	.9243	.9248	.9253	.9258	.9263	.9269	.9274	.9279	.9284	.9289
8.5	.9294	.9299	.9304	.9309	.9315	.9320	.9325	.9330	.9335	.9340
8.6	.9345	.9350	.9355	.9360	.9365	.9370	.9375	.9380	.9385	.9390
8.7	.9395	.9400	.9405	.9410	.9415	.9420	.9425	.9430	.9435	.9440
8.8	.9445	.9450	.9455	.9460	.9465	.9469	.9474	.9479	.9484	.9489
8.9	.9494	.9499	.9504	.9509	.9513	.9518	.9523	.9528	.9533	.9538
9.0	.9542	.9547	.9552	.9557	.9562	.9566	.9571	.9576	.9581	.9586
9.1	.9590	.9595	.9600	.9605	.9609	.9614	.9619	.9624	.9628	.9633
9.2	.9638	.9643	.9647	.9652	.9657	.9661	.9666	.9671	.9675	.9680
9.3	.9685	.9689	.9694	.9699	.9703	.9708	.9713	.9717	.9722	.9727
9.4	.9731	.9736	.9741	.9745	.9750	.9754	.9759	.9763	.9768	.9773
9.5	.9777	.9782	.9786	.9791	.9795	.9800	.9805	.9809	.9814	.9818
9.6	.9823	.9827	.9832	.9836	.9841	.9845	.9850	.9854	.9859	.9863
9.7	.9868	.9872	.9877	.9881	.9886	.9890	.9894	.9899	.9903	.9908
9.8	.9912	.9917	.9921	.9926	.9930	.9934	.9939	.9943	.9948	.9952
9.9	.9956	.9961	.9965	.9969	.9974	.9978	.9983	.9987	.9991	.9996
n	0	1	2	3	4	5	6	7	8	9

APPENDIX B: FORMULAS

FORMULAS FROM GEOMETRY

Square
Perimeter: $P = 4s$
Area: $A = s^2$

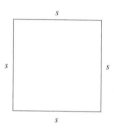

Rectangle
Perimeter: $P = 2L + 2W$
Area: $A = LW$

Triangle
Perimeter: $P = a + b + c$
Area: $A = \dfrac{1}{2}bh$

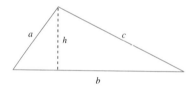

Isosceles Triangle
Two sides equal

Equilateral Triangle
All sides equal

APPENDIX B: FORMULAS

Right Triangle
One 90° (right) angle

Pythagorean Formula (for right triangles)
$c^2 = a^2 + b^2$

Sum of the Angles
$A + B + C = 180°$

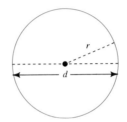

Circle
Diameter: $d = 2r$
Circumference: $C = 2\pi r = \pi d$
Area: $A = \pi r^2$

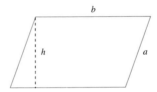

Parallelogram
Area: $A = bh$

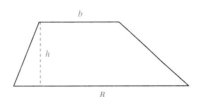

Trapezoid
Area: $A = \dfrac{1}{2}(B + b)h$

Sphere
Volume: $V = \dfrac{4}{3}\pi r^3$
Surface area: $S = 4\pi r^2$

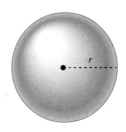

APPENDIX B: FORMULAS

Cone
Volume: $V = \dfrac{1}{3}\pi r^2 h$
Surface area: $S = \pi r \sqrt{r^2 + h^2}$

Rectangular Solid
Volume: $V = LWH$
Surface area: $A = 2HW + 2LW + 2LH$

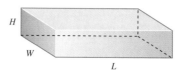

Right Circular Cylinder
Volume: $V = \pi r^2 h$
Surface area: $S = 2\pi rh + 2\pi r^2$

Right Pyramid
Volume: $V = \dfrac{1}{3}Bh$
$B = $ area of the base

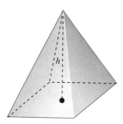

OTHER FORMULAS

Distance: $d = rt$; $r = $ rate or speed, $t = $ time

Percent: $p = br$; $p = $ percentage
$b = $ base
$r = $ rate

Temperature: $F = \dfrac{9}{5}C + 32$
$C = \dfrac{5}{9}(F - 32)$

Simple Interest: $I = prt$; $p = $ principal or amount invested
$r = $ rate or percent
$t = $ time in years

ANSWERS TO SELECTED EXERCISES

The solutions to selected odd-numbered exercises are given in the section beginning on page A-29.

In this section we provide the answers that we think most students will obtain when they work the exercises using the methods explained in the text. If your answer does not look exactly like the one given here, it is not necessarily wrong. In many cases there are equivalent forms of the answer that are correct. For example, if the answer section shows $\frac{3}{4}$ and your answer is .75, you have obtained the right answer but written it in a different (yet equivalent) form. Unless the directions specify otherwise, .75 is just as valid an answer as $\frac{3}{4}$.

In general, if your answer does not agree with the one given in the text, see whether it can be transformed into the other form. If it can, then it is the correct answer. If you still have doubts, talk with your instructor.

Diagnostic Pretest (page xiii)

1. -23 **2.** $\frac{91}{11}$ **3.** -5 **4.** -8
5. $-|-2|$ **6.** $-10x$ **7.** $2x + 26$ **8.** $\frac{16}{3}$
9. $-\frac{3}{4}$ **10.** $y > -2$ **11.** $12x^5y^3 - 18x^4y^4$
12. $k^3 + 3k^2 - k + 1$ **13.** $6x^2 + 17x - 14$
14. $25x^2 - 20xy + 4y^2$ **15.** $x^2 + 2x + 7$
16. $(p + 2q)(p - 2q)$ **17.** $(x + 5)(x - 3)$
18. $(2y + 3)(3y - 1)$ **19.** $\frac{m + 3}{m}$
20. $\frac{a^2(a + b)}{3b(a - b)}$ **21.** $\frac{r + 2k - 3}{rk}$ **22.** $-6, -8$
23. $1, -\frac{7}{2}$ **24.** $(1, 2)$ **25.** $r = \frac{pD}{t}$
26. $6k^5, -6k^5$ **27.** $4\sqrt{14}$ **28.** $-3\sqrt{2}$

29.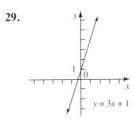

30. length: 20 meters; width: 15 meters

CHAPTER 1

Section 1.1 (page 7)

1.

3.

5.

7. $\{1, 2, 3, 4, 5, 6\}$ **9.** $\{12, 13, 14, \ldots\}$
11. $\{12, 14, 16, 18, \ldots\}$ **13.** $\emptyset$ **15.** $\{3, -3\}$
17. $\{0, 5, 10, 15, \ldots\}$ **19.** 9 **21.** -6
23. -3 **25.** -16 **27.** 5 **29.** 5 **31.** 4
33. 15 **35.** 8 **37.** -8 **39.** 9 **41.** 0
43. -3 **45.** -10 **47.** 8 **49.** 1 **51.** 18
53. -5 **55.** $2, 3, \frac{10}{2}$ (or 5)
57. $-6, 0, 2, 3, \frac{10}{2}$ **59.** $-\sqrt{3}, \sqrt{2}$ **61.** false
63. false **65.** true **67.** false **69.** true
71. false **73.** 0 **75.** -1

Section 1.2 (page 13)

1. $6 < 10$ **3.** $2 > x$ **5.** $r \neq 4$
7. $2p + 1 \leq 9$ **9.** $3 \geq 7y$ **11.** $-6 \leq -6$
13. $5 < x < 9$ **15.** $-8 \leq 3k \leq 4$
17. $1 \leq a < 5$ **19.** false **21.** true **23.** false
25. true **27.** false **29.** true **31.** true
33. true **35.** true **37.** true **39.** false
41. false
43. $(-1, \infty)$ **45.** $(-\infty, 4]$

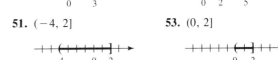

47. $(0, 3)$ **49.** $[2, 5]$

51. $(-4, 2]$ **53.** $(0, 2]$

55. yes, $-4 \le x \le 4$ **57.** $-1 < x < 0$ or $x > 1$

Section 1.3 (page 21)

1. 9 **3.** -5 **5.** -20 **7.** -6 **9.** $-\dfrac{19}{12}$
11. -2 **13.** -12 **15.** -11 **17.** $-\dfrac{5}{24}$
19. $-\dfrac{13}{12}$ **21.** $-\dfrac{15}{28}$ **23.** 5.6 **25.** -25.82
27. -57.6 **29.** 13 **31.** -12 **33.** 29
35. 7 **37.** -42 **39.** -33 **41.** -63
43. 8 **45.** 216 **47.** -168 **49.** -2
51. -6 **53.** -25 **55.** -12 **57.** $\dfrac{6}{5}$
59. 1 **61.** -18.706 **63.** -1.0766 **65.** $-\dfrac{1}{8}$
67. $\dfrac{8}{5}$ **69.** $-\dfrac{13}{4}$ **71.** -2 **73.** -9
75. 12 **77.** 20 **79.** undefined **81.** $-\dfrac{7}{8}$
83. $\dfrac{8}{69}$ **85.** -1.9 **87.** $-.09$
89. false; $3 - (1 - 5) = 7$, but $(3 - 1) - 5 = -3$
91. false; $|6 - 9| = 3$, but $|6| - |9| = -3$

Section 1.4 (page 31)

1. 8 **3.** 1000 **5.** $\dfrac{1}{25}$ **7.** $\dfrac{729}{1000}$ **9.** -27
11. 256 **13.** -8 **15.** -125 **17.** .0256
19. 9.2416 **21.** exp: 7; base: 5
23. exp: 5; base: 12 **25.** exp: 4; base: -9
27. 16 **29.** 28 **31.** 18 **33.** 19 **35.** 5
37. 10 **39.** 3 **41.** 5 **43.** 3 **45.** 3
47. 18 **49.** 4 **51.** 14 **53.** 43 **55.** -48
57. 39 **59.** 2 **61.** 5 **63.** -2
65. undefined **67.** 7 **69.** -1 **71.** $-\dfrac{4}{5}$
73. -100 **75.** 40 **77.** 24 **79.** $-\dfrac{39}{4}$
81. -88 **83.** 496 **85.** $4(5) - 5 = 15$ and $15 \ne 4$; multiply before subtracting

Section 1.5 (page 41)

1. $8k$ **3.** $16r$ **5.** cannot be simplified **7.** $6a$
9. $15c$ **11.** $3a + 3b$ **13.** $-10d - 5f$
15. $-20k + 12$ **17.** $-3k - 7$ **19.** $7x + 16$
21. $-6y - 3$ **23.** $p + 1$ **25.** $-2k + 9$
27. $m - 14$ **29.** $-6y + 29$ **31.** $5p + 7$
33. $-6z - 39$ **35.** $-4m - 10$

37. $-24x - 12$ **39.** $(2 + 3)x = 5x$
41. $(2 \cdot 4)x = 8x$ **43.** 0 **45.** 7
47. $(3 + 5 + 6)a = 14a$ **49.** -2
51. $8(2) + 8(3) = 40$ **53.** 0
55. no; for example, $7 + (5 \cdot 3) \ne (7 + 5)(7 + 3)$
57. yes; any different nonzero numbers a and b that have the same absolute value—for example, 3 and -3

Chapter 1 Review Exercises (page 47)

1.    **2.**

3. 12 **4.** 31 **5.** -4 **6.** -5
7. $\dfrac{12}{3}$ (or 4) **8.** $-9, 0, \dfrac{12}{3}$ (or 4)
9. $-9, -\dfrac{4}{3}, 0, \dfrac{5}{3}, \dfrac{12}{3}$ **10.** $-\sqrt{10}, \sqrt{7}$
11. $-2 < x < 1$ **12.** $0 < x \le 3$ **13.** true
14. false **15.** false **16.** true **17.** true
18. true

19. $(-\infty, -4)$ **20.** $(-2, 5]$

21. $\dfrac{41}{24}$ **22.** $-\dfrac{1}{2}$ **23.** 4 **24.** -17.09
25. -39 **26.** 6 **27.** $\dfrac{23}{20}$ **28.** $-\dfrac{5}{18}$
29. -35 **30.** 30 **31.** $\dfrac{2}{3}$ **32.** 17.94 **33.** 5
34. undefined **35.** -2.9 **36.** 9 **37.** 10,000
38. $\dfrac{27}{343}$ **39.** -125 **40.** -125 **41.** 2.89
42. 20 **43.** 3 **44.** 7 **45.** 3 **46.** 2
47. -3 **48.** 12 **49.** $\dfrac{7}{3}$ **50.** -2
51. -29 **52.** -19 **53.** -54 **54.** $\dfrac{10}{3}$
55. $-\dfrac{34}{5}$ **56.** 1 **57.** $11q$ **58.** $-14z$
59. $5m$ **60.** $4p$ **61.** $-2k + 6$ **62.** $6r + 18$
63. $18m + 27n$ **64.** $-3k + 4h$ **65.** $p - 6q$
66. $-2x + 6$ **67.** $-7y + 1$ **68.** $a + 2$
69. $4k - 2$ **70.** $-6m + 4$ **71.** $-7p - 10$
72. $-6k - 9$ **73.** $(-2 \cdot 5)y$ **74.** $9 \cdot 3$
75. 0 **76.** 0 **77.** $\dfrac{256}{625}$ **78.** 20 **79.** 43
80. 9 **81.** 0 **82.** 2 **83.** $-\dfrac{4}{3}$ **84.** -6.16
85. -9

Chapter 1 Test (page 51)

[1.1] 1. (number line with points at $\frac{3}{4}$ between markers from -3 to 4)

2. (number line with points at $\frac{-5}{8}$ and $\frac{5}{2}$ from -2 to 3)

3. $-\sqrt{7}, \sqrt{11}$

4. $-2, -\frac{1}{2}, \frac{0}{2}$ (or 0), 7.5, 3, $\frac{24}{2}$ (or 12)

5. $-2, \frac{0}{2}$ (or 0), 3, $\frac{24}{2}$ (or 12)

[1.2] 6. $(2, \infty)$ 7. $[-2, 5)$

(number line graphs for 6 and 7)

[1.3] 8. -4 9. 0 **[1.4]** 10. -19 11. 5
12. 1 13. $\frac{16}{7}$ 14. $\frac{11}{23}$ 15. exp: 4; base: y
16. -8 17. $\frac{81}{16}$ 18. 144 19. 0 20. 4
21. 8 22. 173 23. undefined 24. -36
25. $-\frac{14}{3}$ **[1.5]** 26. $6x - 3y$ 27. $2k - 10$
28. $2p + 1$ 29. $\frac{3}{2}$ 30. 0

CHAPTER 2

Section 2.1 (page 59)

1. $\{3\}$ 3. $\{3\}$ 5. $\{-3\}$ 7. $\{-7\}$ 9. $\{1\}$
11. $\left\{\frac{5}{3}\right\}$ 13. $\left\{-\frac{10}{3}\right\}$ 15. $\left\{-\frac{1}{2}\right\}$ 17. $\{2\}$
19. $\{4\}$ 21. $\{7\}$ 23. $\{0\}$ 25. $\{-5\}$
27. $\left\{-\frac{18}{5}\right\}$ 29. $\left\{-\frac{5}{6}\right\}$ 31. $\{-20\}$
33. $\{15\}$ 35. $\{0\}$ 37. $\{-4\}$ 39. $\{-5\}$
41. identity; {all real numbers} 43. contradiction; ∅
45. conditional; $\{0\}$ 47. identity; {all real numbers}
49. contradiction; ∅ 51. -7 53. -34
55. $-\frac{37}{5}$ 57. equivalent 59. not equivalent
61. not equivalent 63. 135 65. 1920
67. 212 69. $\frac{105}{2}$

Section 2.2 (page 67)

1. $r = \frac{d}{t}$ 3. $b = \frac{A}{h}$ 5. $a = P - b - c$
7. $h = \frac{2A}{b}$
9. $h = \frac{S - 2\pi r^2}{2\pi r}$ or $h = \frac{S}{2\pi r} - r$

11. $F = \frac{9}{5}C + 32$ 13. $b = \frac{2A}{h}$
15. $W = \frac{V}{LH}$
17. $B = \frac{2A}{h} - b$ or $B = \frac{2A - bh}{h}$
19. 4 hours 21. 14.5 meters 23. 10 meters
25. 5° 27. 104° 29. 120 inches 31. 5 inches
33. 10% 35. 25% 37. $45 39. $82.50
41. $x = \frac{7y + 3}{4 + b}$ 43. $x = 32y - 25$
45. $k = \frac{y}{1 - r}$ or $k = \frac{-y}{r - 1}$ 47. $\{5\}$
49. $\{200\}$ 51. $\{2000\}$

Section 2.3 (page 77)

1. $x - 9$ 3. $13 + x$ 5. $-2x$ 7. $x - 7$
9. $-1 + 6x$ 11. $\frac{x}{-9}$ 13. $\frac{x}{12}$ 15. $\frac{2}{3}x$
17. $\frac{x + 3}{6}$ 19. $x + (-4) = -23$; -19
21. $7 - \frac{5}{12}x = 17$; -24 23. equation
25. expression 27. equation
29. length: 12 meters; width: 7 meters 31. 10 feet
33. 50 35. $560 37. $21.20
39. $3000 at 8%; $9000 at 9%
41. $2000 at 8%; $3000 at 14% 43. $87,000
45. $2\frac{1}{2}$ liters 47. $\frac{1}{2}$ liter 49. 2 liters
51. length: 30 meters; width: 15 meters
53. 5 liters 55. $5000 57. 14 toll calls
59. 150 61. 50 63. 40

Section 2.4 (page 87)

1. 16 nickels; 14 pennies
3. 25 half dollars; 20 quarters
5. 30 quarters; 5 dimes; 5 pennies
7. 30 three-cent pieces; 10 two-cent pieces
9. 120 students; 180 non-students
11. 20 $2 tickets; 14 $3 tickets 13. 3 hours
15. 5.5 hours 17. 8 hours 19. $\frac{5}{6}$ hour
21. 15 miles per hour 23. 18 miles 25. $\frac{1}{2}$ hour

27. (number line graph, -2 to 0) 29. (number line graph, 0 to 3)
31. (number line graph, -2 to 5) 33. (number line graph, 0 to 3 to 9)

Section 2.5 (page 97)

1. $(2, \infty)$
3. $(-\infty, -3]$
5. $[5, \infty)$
7. $(3, \infty)$
9. $(-3, \infty)$
11. $(-\infty, -28]$
13. $[3, \infty)$
15. $(-\infty, -3)$
17. $\left(-\infty, \dfrac{1}{2}\right)$
19. $(2, \infty)$
21. $[1, \infty)$
23. $(2, \infty)$
25. $\left(-\infty, \dfrac{23}{5}\right)$
27. $\left[-\dfrac{5}{6}, \infty\right)$
29. $\left(-\infty, \dfrac{36}{11}\right)$
31. $(2, 11)$
33. $(4, 8)$
35. $[-5, 4]$
37. $\left[-\dfrac{14}{3}, -\dfrac{4}{3}\right]$
39. $\left[-\dfrac{19}{2}, \dfrac{35}{2}\right]$
41. $(1, \infty)$
43. $\left(-\infty, -\dfrac{1}{5}\right)$
45. $(-\infty, 5]$
47. $\left(-\dfrac{7}{6}, \dfrac{5}{6}\right)$
49. $\left[\dfrac{5}{9}, \dfrac{2}{3}\right]$
51. at least 74
53. at least 5 red lights
55. 4 green pills
57. {all real numbers}
59. ∅
61.
63.

Section 2.6 (page 105)

1. {a, c, e} or B
3. {d} or D
5. {a}
7. {a, b, c, d, e, f} or A
9. {a, c, e, f}
11. {a, d, f}
13. $(-1, 2)$
15. $(-\infty, 3]$
17. ∅
19. $[2, \infty)$
21. $[-1, 2]$
23. $[4, 8]$
25. $(-2, -1)$
27. ∅
29. $(-\infty, 4]$
31. $(-\infty, -2) \cup (3, \infty)$
33. $(-\infty, 1] \cup [4, \infty)$
35. $(-\infty, 7]$
37. $(-\infty, 1] \cup [7, \infty)$
39. $[-2, \infty)$
41. {all real numbers}
43. $(-\infty, -4) \cup (4, \infty)$
45. $(-\infty, 1) \cup (2, \infty)$
47. $(-4, -1)$
49. $(-\infty, 3)$
51. $[4, 12]$
53. $(-\infty, 0] \cup [2, \infty)$
55. 5
57. 1
59. 11

Section 2.7 (page 115)

1. $\{-8, 8\}$
3. $\{-3, 3\}$
5. $\{-5, 11\}$
7. $\{-5, 4\}$
9. $\left\{-2, \dfrac{9}{2}\right\}$
11. $\left\{-\dfrac{17}{2}, \dfrac{7}{2}\right\}$
13. $\{-14, 2\}$
15. $\left\{-\dfrac{8}{3}, \dfrac{16}{3}\right\}$
17. $\left\{-1, \dfrac{1}{3}\right\}$

19. $(-\infty, -2) \cup (2, \infty)$ **21.** $(-\infty, -3] \cup [3, \infty)$

23. $(-\infty, -10) \cup (6, \infty)$ **25.** $\left(-\infty, -\dfrac{4}{3}\right] \cup [2, \infty)$

27. $(-\infty, -1) \cup (7, \infty)$ **29.** $[-2, 2]$

31. $(-3, 3)$ **33.** $[-10, 6]$

35. $\left(-\dfrac{4}{3}, 2\right)$ **37.** $[-1, 7]$

39. $(-\infty, -13) \cup (5, \infty)$ **41.** $\{-5, -2\}$

43. $\left[-\dfrac{7}{3}, 3\right]$ **45.** $\left(-\dfrac{7}{6}, -\dfrac{5}{6}\right)$

47. $\{-8, 8\}$ **49.** $\{-6, -2\}$

51. $\left(-\infty, -\dfrac{3}{2}\right) \cup \left(\dfrac{1}{2}, \infty\right)$ **53.** $[-9, -1]$

55. $\left\{-8, \dfrac{6}{5}\right\}$ **57.** $\left\{-3, \dfrac{5}{3}\right\}$

59. $\left\{-\dfrac{5}{3}, -\dfrac{1}{3}\right\}$ **61.** $\left\{\dfrac{1}{4}\right\}$ **63.** $\emptyset$ **65.** $\emptyset$

67. $\left\{\dfrac{1}{4}\right\}$ **69.** $\emptyset$ **71.** {all real numbers}

73. $\left\{\dfrac{3}{7}\right\}$ **75.** {all real numbers}

77. $\left(-\infty, -\dfrac{9}{10}\right) \cup \left(-\dfrac{9}{10}, \infty\right)$ **79.** -13 or 3

81. all numbers less than or equal to -4, or greater than or equal to 4

83. all numbers between -12 and 4, inclusive   **85.** 8

87. -64 **89.** -25 **91.** $\dfrac{8}{27}$

Summary on Solving Linear Equations and Inequalities (page 119)

1. $\{13\}$ **3.** $\{3\}$ **5.** $\{-10, 4\}$ **7.** $[-2, \infty)$

9. $\{-2\}$ **11.** $(-\infty, 4]$ **13.** $\{0\}$ **15.** $\left[\dfrac{8}{5}, \infty\right)$

17. $(-\infty, -1) \cup (1, \infty)$ **19.** $\left\{\dfrac{96}{5}\right\}$

21. $(-\infty, -12)$ **23.** $\left\{\dfrac{11}{6}\right\}$

25. {all real numbers}   **27.** $(-\infty, 1) \cup (2, \infty)$

29. $\left(-\dfrac{7}{5}, 1\right)$ **31.** $\left[-\dfrac{1}{3}, 2\right]$ **33.** $\{0, 1\}$

35. $(-\infty, -1] \cup \left[\dfrac{5}{3}, \infty\right)$ **37.** $\left\{\dfrac{7}{2}\right\}$

39. $\left[-\dfrac{1}{2}, \dfrac{15}{2}\right]$ **41.** {all real numbers}

43. {all real numbers}   **45.** $\{-1\}$

47. $\left(-1, \dfrac{1}{5}\right)$

Chapter 2 Review Exercises (page 125)

1. $\{2\}$ **2.** $\{-1\}$ **3.** $\left\{-\dfrac{6}{5}\right\}$ **4.** $\left\{-\dfrac{2}{3}\right\}$

5. $\left\{\dfrac{4}{3}\right\}$ **6.** $\left\{-\dfrac{31}{5}\right\}$ **7.** $\emptyset$ **8.** $\{13\}$

9. $\{0\}$ **10.** {all real numbers}   **11.** $x = \dfrac{9}{4yz}$

12. $x = \dfrac{Q - 2y - 4z}{3}$ **13.** $x = 4M - 2y$

14. $x = \dfrac{3P + 18}{2}$ **15.** 30 kilometers per hour

16. 20 meters   **17.** 15 centimeters   **18.** $1500

19. 18 millimeters   **20.** 194° F   **21.** $5 + \dfrac{1}{2}x$

22. $8 - 4x$ **23.** length: 13 meters; width: 8 meters

24. 17 inches, 17 inches, 19 inches   **25.** 12 kilograms

26. 30 liters   **27.** $18,000 at 12%; $14,000 at 14%

28. $25,000 at 15%; $17,000 at 12%

29. 850 reserved; 246 general admission

30. 249 students; 62 non-students   **31.** 2.2 hours

32. 50 kilometers per hour; 65 kilometers per hour

33. 1 hour   **34.** 46 miles per hour

35. $(-14, \infty)$ **36.** $[-12, \infty)$

37. $(-6, \infty)$ **38.** $\left(-\infty, -\dfrac{14}{5}\right]$

39. $\left(\frac{5}{6}, \infty\right)$ **40.** $\left(-\infty, -\frac{11}{9}\right)$

41. $(-2, \infty)$ **42.** $[-3, 6]$

43. $[3, 5)$ **44.** $\left(-\frac{1}{2}, 6\right)$

45. $\left(\frac{59}{31}, \infty\right)$ **46.** $\left(-\infty, \frac{14}{17}\right)$

47. $\{4, 8, 12\}$ **48.** $\{2, 4, 6, 8, 10, 12, 16\}$

49. $(6, 8)$ **50.** $\emptyset$

51. $(-\infty, -1] \cup (5, \infty)$ **52.** {all real numbers}

53. {all real numbers}   **54.** $(-\infty, -2] \cup [7, \infty)$

55. $\{-4, 4\}$ **56.** $\{-8, 4\}$ **57.** $\left\{1, \frac{11}{3}\right\}$

58. $\emptyset$ **59.** $\{1, 6\}$ **60.** $\emptyset$ **61.** $\{-2, 3\}$

62. $\left\{-4, -\frac{2}{3}\right\}$

63. $(-4, 4)$ **64.** $[1, 11]$

65. $[-4, -1]$ **66.** $(-\infty, -10] \cup [8, \infty)$

67. $\left(-\infty, -\frac{13}{5}\right) \cup (3, \infty)$ **68.** {all real numbers}

69. $\left(-\infty, \frac{4}{3}\right)$ **70.** $x = \frac{4R + 48}{3}$ **71.** $[-1, 5)$

72. $\left\{-\frac{10}{13}\right\}$ **73.** 15 inches   **74.** $(-\infty, 6]$

75. $\left(-\infty, -\frac{3}{7}\right) \cup (1, \infty)$ **76.** {all real numbers}

77. $\{-1, 11\}$ **78.** $[-4, -2]$

79. 540 votes; 675 votes

80. 15 $5 bills; 19 $20 bills

Chapter 2 Test (page 131)

[2.1]   **1.** $\{1\}$ **2.** $\{-8\}$ **3.** {all real numbers}

[2.2]   **4.** $v = \dfrac{S + 16t^2}{t}$ **[2.3–2.4]**   **5.** 4 years

6. $400   **7.** 4 meters

8. $1500 at 7%; $4500 at 6%

[2.5]   **9.** $(-\infty, 30)$ **10.** $(-6, \infty)$

11. $[-3, 3]$ **12.** 78

[2.6]   **13.** $(-5, 4)$ **14.** $[8, \infty)$

15. $(-\infty, 3) \cup [5, \infty)$

[2.7]   **16.** $\left\{-\dfrac{7}{3}, \dfrac{11}{3}\right\}$ **17.** $\left\{-\dfrac{3}{7}, 3\right\}$ **18.** $\emptyset$

19. $\left[-\dfrac{7}{2}, 2\right]$ **20.** $(-\infty, -6) \cup (1, \infty)$

CHAPTER 3

Section 3.1 (page 139)

1. exp: 7; base: 5   **3.** exp: 4; base: -9

5. exp: 4; base: 9   **7.** exp: -7; base: p

9. exp: -4; base: q **11.** exp: 3; base: $-m + z$

13. 625   **15.** $\dfrac{25}{9}$ **17.** $\dfrac{1}{49}$ **19.** $-\dfrac{1}{8}$

21. $-\dfrac{1}{81}$ **23.** $-\dfrac{1}{32}$ **25.** 25   **27.** -128

29. $\dfrac{9}{8}$ **31.** 8   **33.** $\dfrac{9}{4}$ **35.** $\dfrac{5}{6}$ **37.** 2^{16}

39. 7^4 **41.** 3^3 **43.** $\dfrac{1}{6^2}$ **45.** $\dfrac{1}{3^3}$ **47.** $\dfrac{1}{9^2}$

49. $\dfrac{1}{t^7}$ **51.** r^5 **53.** $\dfrac{1}{a^5}$ **55.** x^{11} **57.** r^6

59. $-\dfrac{56}{k^2}$ **61.** 1   **63.** 1   **65.** -1 **67.** 0

69. $(2 + 3)^{-1} = \dfrac{1}{5}$, while $2^{-1} + 3^{-1} = \dfrac{5}{6}$
71. $(2 + 3)^2 = 25$, while $2^2 + 3^2 = 13$
73. $(-2)^6 = 2^6$ **75.** $\left(\dfrac{3}{4}\right)^4$

Section 3.2   (page 147)

1. $\left(\dfrac{4}{3}\right)^2$ **3.** $\dfrac{5}{6}$ **5.** $\dfrac{1}{3^6 \cdot 4^3}$ **7.** $4^6 \cdot 7^{10}$
9. $\dfrac{1}{z^4}$ **11.** $-\dfrac{3}{r^7}$ **13.** $\dfrac{3^3}{a^{18}}$ **15.** $\dfrac{x^5}{y^2}$
17. $\dfrac{2^2 \cdot 4^2}{p^2 q^4}$ **19.** $\dfrac{1}{5p^{10}}$ **21.** $\dfrac{4}{a^2}$ **23.** $\dfrac{1}{6y^{13}}$
25. $\dfrac{2^2 k^5}{m^2}$ **27.** $\dfrac{z^5}{2^2 y^5}$ **29.** $\dfrac{2^2 k^{17}}{5^3}$ **31.** $\dfrac{2k^5}{3}$
33. $\dfrac{2^3}{3pq^{10}}$ **35.** $\dfrac{y^9}{2^3}$ **37.** 2.30×10^2
39. 2×10^{-2} **41.** 3.27×10^{-6}
43. -4.72×10^4 **45.** 6500   **47.** .0152
49. .005   **51.** $-.000568$ **53.** 3×10^{-4} or .0003
55. 3×10^{-5} or .00003   **57.** 6×10^5 or 600,000
59. 1.5×10^2 or 150   **61.** 2×10^5 or 200,000
63. 5.8657×10^{12} miles (rounded)
65. 2×10^5 seconds or 55.6 hours (rounded)
67. $-3p$ **69.** $2x - 5$

Section 3.3   (page 155)

1. 9   **3.** -11 **5.** -1 **7.** 1
9. $3x^3 + 9x^2 + 8x$ **11.** $4y^4 - 6y^3 + 8y^2 - 3$
13. binomial; 1   **15.** trinomial; 2   **17.** trinomial; 4
19. monomial; 0   **21.** none of these; 3   **23.** $4k$
25. $20z^5$ **27.** $4p^5$ **29.** 0   **31.** $8a^2 - 2a$
33. $-2r^2 + 6r$ **35.** $-x^2 + 5x$
37. (a) 0 (b) 9   **39.** (a) -7 (b) -1
41. (a) -4 (b) 8   **43.** $-9p^2 + 11p - 9$
45. $5a + 18$ **47.** $14m^2 - 13m + 6$
49. $13z^2 + 10z - 3$ **51.** $27p^5 + 7p^3 + 2p$
53. $12k^5 - k^4 + 5k - 6$
55. $-5a^4 - 6a^3 + 9a^2 - 11$
57. $6x^5 - 11x^4 + 4x^3 - x$ **59.** $5x - 9$
61. $-3z + 9$ **63.** $5y^3 - 6y + 7$
65. $-5m^3 + 7m^2 - 11$ **67.** $-5p^3 - 2p^2 - 16p$
69. $-6z^3 - 3z^2 - 8z - 4$ **71.** $y^4 - 4y^2 - 4$
73. $-3x^2 + 12x - 6$ **75.** $-3y^3 + 9y^2 + y$
77. $48r^6$ **79.** $12x^2 y^7$ **81.** $36m^2 n^4$

Section 3.4   (page 165)

1. $15p^2$ **3.** $-45a^5$ **5.** $15a - 27$
7. $-10r^2 + 4r$ **9.** $8a^3 + 40a^2$
11. $-18r^4 + 12r^3$ **13.** $-6z^4 - 15z^3 + 3z^2$
15. $6b^5 - 3b^4 + 24b^3$
17. $12y^5 - 8y^6 + 16y^4 - 10y^3$ **19.** $15k^2 - 7k - 2$
21. $16a^2 - 2a - 5$ **23.** $-3y^3 + 11y^2 - 7y - 5$
25. $6k^2 - 11kz - 10z^2$ **27.** $49y^2 - 121z^2$

29. $10k^4 + 29k^3 + 4k^2 - 23k - 20$
31. $16y^5 + 22y^4 - 39y^3 + 25y^2 + 37y + 30$
33. $-2k^4 - 2k^3 + 7k^2 + 10k$
35. $m^2 - 3m - 40$ **37.** $6a^2 - a - 2$
39. $10y^2 + 23y - 42$ **41.** $18a^2 - 3a - 10$
43. $21a^2 + 22a + 5$ **45.** $2r^2 - rs - s^2$
47. $24m^2 - 49mn + 15n^2$ **49.** $k^2 + \dfrac{1}{6}k - \dfrac{1}{3}$
51. $3w^2 - \dfrac{23}{4}wz - \dfrac{1}{2}z^2$ **53.** $36a^2 - 1$
55. $25x^2 - 4y^2$ **57.** $25r^2 - 9w^2$ **59.** $64x^2 - y^2$
61. $r^2 - \dfrac{9}{16}$ **63.** $4y^6 - 1$ **65.** $k^2 - 16k + 64$
67. $9r^2 + 6rt + t^2$ **69.** $36p^2 - 60pq + 25q^2$
71. $9k^2 - \dfrac{2}{3}k + \dfrac{1}{81}$ **73.** $r^2 + \dfrac{4}{3}rs + \dfrac{4}{9}s^2$
75. $9m^2 - 12m + 4 + 6mp - 4p + p^2$
77. $16k^2 + 8kh + h^2 - 32k - 8h + 16$
79. $m^2 + 2mp + p^2 - 25$
81. $9m^2 - 6my + y^2 - z^2$
83. $(2 + 3)^3 = 125$, while $2^3 + 3^3 = 35$; $(x + y)^3 = x^3 + 3x^2 y + 3xy^2 + y^3$
85. $2p^4$ **87.** $\dfrac{4}{3s^4}$ **89.** $-7p^2 - 10p - 4$

Section 3.5   (page 173)

1. $a^2 - 2a + 1$ **3.** $m + 2 + \dfrac{3}{m}$
5. $3 + \dfrac{2}{a} - \dfrac{5}{2a^2}$ **7.** $\dfrac{3}{4q} + \dfrac{3}{2p} - \dfrac{1}{2pq}$
9. $\dfrac{c}{5} - \dfrac{2b}{5} + \dfrac{3a}{5}$ **11.** $x + 3$ **13.** $r + 3$
15. $3b^2 + b + \dfrac{1}{2b - 3}$
17. $a^2 - 5a - 9 + \dfrac{-30}{a - 3}$
19. $2x - 5 + \dfrac{-4x + 5}{2x^2 - 4x + 3}$
21. $4p + 1 + \dfrac{-11p + 15}{2p^2 + 3}$ **23.** $2y^2 + 3y - 1$
25. $9t^2 - 4t + 1$ **27.** $\dfrac{2}{3}x - 1$
29. $\dfrac{3}{4}a - 2 + \dfrac{1}{4a + 3}$ **31.** $3k + 2$
33. $m^2 + 3m - 7$ kilometers per hour   **35.** -6
37. -6

Section 3.6   (page 179)

1. $x + 7$ **3.** $3m - 8$ **5.** $3a + 4 + \dfrac{3}{a + 2}$
7. $p - 4 + \dfrac{9}{p + 1}$ **9.** $4a^2 + a + 3 + \dfrac{4}{a - 1}$
11. $6x^4 + 12x^3 + 22x^2 + 40x + 83 + \dfrac{164}{x - 2}$

13. $-4r^5 - 7r^4 - 10r^3 - 5r^2 - 11r - 8 + \dfrac{-5}{r-1}$
15. $-3y^4 + 8y^3 - 21y^2 + 36y - 72 + \dfrac{143}{y+2}$
17. $y^2 - y + 1 + \dfrac{-2}{y+1}$ 19. 7   21. 202
23. 0   25. 69   27. yes   29. yes   31. no
33. yes   35. no   37. $9(6 + r)$
39. $7(2x - 3y)$ 41. $(x + 1)(x + y)$

Section 3.7 (page 185)

1. $15(k + 2)$ 3. $2p(3p + 2)$ 5. $mx(4 - 5x)$
7. $2m(4m + 3)$ 9. $3x(4x - 1)$
11. $16m^3(9m + m^2 - 2)$ 13. $3a(5a^2 + 4a - 1)$
15. $7a^2b(2ab + 1 - 3a^3b^2)$
17. $-3mp^3(5m^2 + 2m^2p + 3)$
19. $(m - 9)(2m + 3)$ 21. $(3k - 7)(2k + 7)$
23. $m^5(r + s + t + u)$
25. $(3 - x)(6 + 2x - x^2)$
27. $5(m + p)^2(m + p - 2 - 3m^2 - 6mp - 3p^2)$
29. $36y(y - 2)$ or $-36y(-y + 2)$
31. $2x^2(-x^3 + 3x + 2)$ or $-2x^2(x^3 - 3x - 2)$
33. $16a^2m^3(-2a^2m^2 - 1 - 4a^3m^3)$ or
    $-16a^2m^3(2a^2m^2 + 1 + 4a^3m^3)$
35. $(a + 2b)(x + y)$ 37. $(b + c)(2 + a)$
39. $(p + q)(p - 3y)$ 41. $(x - 3)(x + 2)$
43. $(3r - 2)(r + 5)$ 45. $(8p + 3q)(-2p + q)$
47. $(a^2 + b^2)(-3a + 2b)$ 49. $(1 - a)(1 - b)$
51. $(4 - 3y^3)(2 - 3y)$ 53. $m^{-5}(3 + m^2)$
55. $p^{-3}(3 + 2p)$ 57. $p^2 - 2p - 35$
59. $21y^2 + 29y - 10$ 61. $64p^2 - 9q^2$

Section 3.8 (page 193)

1. $(m + 2)(m + 3)$ 3. $(a - 4)(a + 3)$
5. $(r - 5)(r + 4)$ 7. $(a - 8)(a + 2)$
9. $(a + 6b)(a - 3b)$ 11. $(p - 5q)(p + 3q)$
13. $(yw + 7)(yw - 3)$ 15. $(8y - 3)(y + 2)$
17. $(6x - 5)(3x + 2)$ 19. $(7p - 5)(5p + 3)$
21. $(6a - 5b)(2a + 3b)$
23. $(2k - 3a)(2k - 3a)$ or $(2k - 3a)^2$
25. $(7x + 3y)(5x - 8y)$ 27. $(3kp + 2)(2kp + 3)$
29. $2(3m - 4)(2m + 5)$ 31. $3(2a - 3)(3a + 2)$
33. $6a(a + 5)(a - 3)$ 35. $13y(y + 4)(y - 1)$
37. $2xy^3(x - 12y)(x - 12y)$ or $2xy^3(x - 12y)^2$
39. $(3x^2 + 1)(x^2 - 5)$ 41. $(z^2 - 10)(z^2 + 3)$
43. $(3x^2 - 5)(2x^2 + 5)$ 45. $(6p^2 - r)(2p^2 + 5r)$
47. $(2p + 7)(3p + 14)$
49. $(2z + 2k + 1)(3z + 3k - 5)$
51. $(a + b)^2(a - 3b)(a + 2b)$
53. $(p + q)^2(p + 3q)$ 55. $4m^2 - 25$
57. $x^2 + 12xy + 36y^2$ 59. $y^3 - 8$

Section 3.9 (page 199)

1. $(x + 5)(x - 5)$ 3. $(6m + 1)(6m - 1)$
5. $(4y + 9q)(4y - 9q)$ 7. $3(3x^2 + y^2)(3x^2 - y^2)$
9. $(x + y + 4)(x + y - 4)$
11. $(5 + r + 3s)(5 - r - 3s)$ 13. $4ab$
15. $(x + 2)^2$ 17. $(3r - s)^2$
19. $(x - 2 - w)(x - 2 + w)$
21. $(3m - 2 + n)(3m - 2 - n)$
23. $(a + b - 1)(a - b + 1)$
25. $(5xy - 2)^2$ 27. $2(6m - 5p)^2$
29. $(a + b + 1)^2$ 31. $(m - p + 2)^2$
33. $(2a + 1)(4a^2 - 2a + 1)$
35. $(3x - 4y)(9x^2 + 12xy + 16y^2)$
37. $2(5m - p)(25m^2 + 5mp + p^2)$
39. $(4y^2 + 1)(16y^4 - 4y^2 + 1)$
41. $(x + y - 3)(x^2 + 2xy + y^2 + 3x + 3y + 9)$
43. $r(r^2 + 3r + 3)$ 45. $(2m + 3)(m^2 + 3m + 9)$
47. $2q(3p^2 + q^2)$ 49. 3^2 51. 2   53. $\{2\}$
55. $\left\{-\dfrac{1}{2}\right\}$

Summary of Factoring Methods (page 201)

1. $(10a + 3b)(10a - 3b)$ 3. $6p^3(3p^2 - 4 + 2p^3)$
5. $(x + 7)(x - 5)$ 7. $(7z + 4)(7z - 4)$
9. $(x - 10)(x^2 + 10x + 100)$
11. $(k - 8)(k + 2)$ 13. $(3t - 7u)(2t + 11u)$
15. $8p(5p - 4)$ 17. $(2k + 7r)^2$
19. $(m - 2)(n + 5)$
21. $(3m - 5n + p)(3m - 5n - p)$
23. $7(2k - 5)(4k^2 + 10k + 25)$
25. $16z^2x(zx - 2)$
27. $(m - 2)(m + 2)(m - 2)$ or $(m + 2)(m - 2)^2$
29. $3(3m + 8n)^2$
31. $(5m^2 + 6)(25m^4 - 30m^2 + 36)$
33. $(2m + 5n)(m - 3n)$ 35. $4y(y - 2)$
37. $(7z + 2k)(2z - k)$ 39. $16(4b + 5c)(4b - 5c)$
41. $8(5z + 4)(25z^2 - 20z + 16)$
43. $(5r - s)(2r + 5s)$ 45. $8x^2(4 + 2x - 3x^3)$
47. $(7x + 5q)(2x - 5q)$ 49. $(y + 5)(y - 2)$
51. $2a(a^2 + 3a - 2)$ 53. $(9p - 5r)(2p + 7r)$
55. $(x - 2y + 2)(x - 2y - 2)$
57. $(5r + 2s - 3)^2$ 59. $(z + 2)(z - 2)(z^2 - 5)$

Section 3.10 (page 209)

1. $\{2, 3\}$ 3. $\left\{-\dfrac{6}{5}, \dfrac{8}{3}\right\}$ 5. $\left\{4, -\dfrac{5}{2}\right\}$
7. $\{-3, 4\}$ 9. $\{-7, -2\}$ 11. $\left\{-3, \dfrac{1}{2}\right\}$
13. $\left\{-\dfrac{2}{3}, \dfrac{1}{5}\right\}$ 15. $\left\{-2, -\dfrac{1}{4}\right\}$
17. $\left\{-\dfrac{4}{3}, 1\right\}$ 19. $\{-6, -5\}$
21. $\left\{-\dfrac{4}{3}, 5\right\}$ 23. $\{0, 9\}$ 25. $\{-2, 2\}$
27. $\{-3, 3\}$ 29. 3 and 9
31. -12 and -11, or 11 and 12   33. 14 feet
35. 10 meters   37. $\{2, 4\}$ 39. $\dfrac{3y^4}{2}$
41. $-\dfrac{2}{3r^3s^2}$ 43. $\dfrac{35}{63}$ 45. $\dfrac{42}{30}$

Chapter 3 Review Exercises (page 215)

1. exp: 3; base: 9
2. exp: 4; base: -5
3. exp: 2; base: 8
4. exp: -2; base: z
5. 243
6. $\dfrac{1}{16}$
7. -64
8. 147
9. $\dfrac{64}{27}$
10. $\dfrac{25}{16}$
11. 9^8
12. 2^6
13. y^{13}
14. $\dfrac{216}{p^4}$
15. 5^3
16. m^5
17. 3.450×10^3
18. 7.6×10^{-8}
19. 1.3×10^{-1}
20. 210,000
21. .0038
22. .378
23. $\dfrac{1}{3^8}$
24. 1
25. -1
26. $\dfrac{1}{z^{15}}$
27. $\dfrac{5^2}{m^{18}}$
28. $3^2 r^{11}$
29. $\dfrac{5}{z^2}$
30. $\dfrac{1}{96m^7}$
31. $\dfrac{5^2 3^4}{2^3 r^4}$
32. 2^2 or 4
33. 2×10^{-4} or .0002
34. 5×10^3 or 5000
35. 4×10^{-5} or .00004
36. 2.7×10^{-2} or .027
37. 14
38. -1
39. 29
40. 104
41. (a) $11k^3 - 3k^2 + 9k$ (b) trinomial (c) 3
42. (a) $-9m^7 + 14m^6$ (b) binomial (c) 7
43. (a) $-5y^4 + 3y^3 + 7y^2 - 2y$ (b) none of these (c) 4
44. (a) 16 (b) -4
45. (a) -11 (b) -7
46. $x^2 + 11x + 6$
47. $12y - 11$
48. $-4y^3 - 2y^2 + 16y - 14$
49. $1 + 16w$
50. $-k + 1$
51. $6a^3 - 15a^2 + 7a$
52. $5y^2 - 2y + 6$
53. $-8a^3 - 22a$
54. $2y^2 + 5y - 12$
55. $24y^2 + 19y - 35$
56. $2p^3 + 13p^2 + 17p - 12$
57. $10r^2 - 9rs - 9s^2$
58. $12k^4 + 20k^3 - 18k^2 - 30k$
59. $6z^4 - 8z^3 + 10z^2 + 9z - 2$
60. $9r^4 - 25$
61. $z^2 - \dfrac{4}{9}$
62. $y^2 + 4y + 4$
63. $9p^2 - 42p + 49$
64. $\dfrac{3r}{2} - \dfrac{8}{3}$
65. $m^2 - \dfrac{11m}{5} + 2$
66. $8x + 1 + \dfrac{5}{x-3}$
67. $5p + 7 + \dfrac{-3}{3p-2}$
68. $k^2 - 7k + 6$
69. $m^2 + 3m - 6 + \dfrac{-m+14}{5m^2-3}$
70. $p^2 + 6p + 9 + \dfrac{54}{2p-3}$
71. $4y^2 + 1 + \dfrac{-2y}{3y^2+1}$
72. $3p + 2$
73. $10k - 23 + \dfrac{31}{k+2}$
74. $2k^2 + k + 3 + \dfrac{21}{k-3}$
75. $3z^3 - 4z^2 + 2z + 2 + \dfrac{6}{z+6}$
76. $9a^4 - a^3 + 4a^2 + 3a + 6 + \dfrac{-9}{a+4}$
77. $y^4 + y^3 + y^2 + y + 1$
78. yes
79. no
80. -13
81. -5
82. $k(11 + 12k)$
83. $5y(3y + 4)$
84. $6m(2m - 1)$
85. $6ab(2a + 3b^2 - 4a^2b)$
86. $9mn(n^2 - 8mn + 6m^2)$
87. $(x - 4)(-2x - 1)$
88. $(a + 2)(2a - 3)$
89. $(x + y)(4 + m)$
90. $(x + y)(x + 5)$
91. $(x - 8)(x - 3)$
92. $(a + 18)(a - 2)$
93. $(4p - 1)(p + 1)$
94. $(2m + 3)(3m - 1)$
95. $(4p - 5)(3p + 4)$
96. $(3r + 2)(6r - 5)$
97. $(4a - b)(5a - 2b)$
98. prime
99. $a(5 + m)(6 - m)$
100. $z(2r - 3)(r + 1)$
101. $(r^2 + 3)(r^2 - 2)$
102. $(2k^2 + 1)(k^2 - 3)$
103. $(3p + 7)(3p - 7)$
104. $(4z + 11)(4z - 11)$
105. $9(4r + 3s + 7)(4r - 3s + 1)$
106. $(p + 7)^2$
107. $(3k - 2)^2$
108. $(5z - 3m)^2$
109. $(2 - a)(4 + 2a + a^2)$
110. $2(r + 3)(r^2 - 3r + 9)$
111. $3(5x - 1)(25x^2 + 5x + 1)$
112. $\left\{\dfrac{2}{5}, -1\right\}$
113. $\{2, 3\}$
114. $\{-4, 2\}$
115. $\left\{\dfrac{10}{3}, -\dfrac{5}{2}\right\}$
116. $\left\{-\dfrac{3}{2}, \dfrac{1}{3}\right\}$
117. $\left\{-\dfrac{3}{2}, -\dfrac{1}{4}\right\}$
118. $-9, 4$
119. $2, 4$
120. $-\dfrac{20}{3}, 5$
121. 10 meters
122. 9 feet
123. 6 inches
124. $8x^2 - 10x - 3$
125. $\dfrac{1}{125}$
126. $\dfrac{1}{2^4 y^{18}}$
127. $-14 + 16w - 8w^2$
128. $m - 3$
129. $-\dfrac{1}{5z^9}$
130. -9
131. $2500z^6$
132. $-3k^2 + 4k - 7$
133. $21p^9 + 7p^8 + 14p^7$
134. $2y^2x + \dfrac{3y^3}{2x} + \dfrac{5x^2}{2}$
135. $(5m - 3)(2m + 1)$
136. $12z(1 - 6z)$
137. $(4 - p)(16 + 4p + p^2)$
138. $(k + 3q)(k - 2q)$
139. $4c^2(4c + 7)$
140. $(4d + 3)^2$

Chapter 3 Test (page 221)

[3.1] 1. $\dfrac{8}{27}$ 2. 0 [3.2] 3. $\dfrac{1}{8^3}$
4. $\dfrac{a^6}{3^3 b^9}$ 5. $\dfrac{r^3}{5 \cdot 3^2}$
[3.3] 6. $-k^3 - 3k^2 - 8k - 9$
[3.4] 7. $8x^2 + 17x - 21$
8. $12m^3 - 26m^2 + 27m - 10$ 9. $9x^2 - 4y^2$
10. $4p^2 + 4pq + q^2$ [3.5] 11. $z - 2 + \dfrac{3}{z}$
12. $3y^3 - 3y^2 + 4y + 1 + \dfrac{-10}{2y+1}$

[3.6] 13. yes   [3.7] 14. $12f(f-1)$
15. $(g-1)(3g+2)$ 16. $(x+y)(a+b)$
[3.8] 17. $(2p-3q)(p-q)$ 18. prime
[3.9] 19. $(x+10)(x^2-10x+100)$
20. $(10d+3m^2)(10d-3m^2)$
21. $(3k^2+4)(6k^2-5)$ [3.10] 22. $\left\{-3,-\dfrac{5}{2}\right\}$
23. $\left\{\dfrac{1}{4},\dfrac{3}{2}\right\}$ 24. $\left\{-\dfrac{2}{5},1\right\}$
25. length: 9 meters; width: 7 meters

CHAPTER 4

Section 4.1 (page 227)

1. 4   3. $\dfrac{7}{3}$ 5. 0
7. Any real number can replace p.   9. $-3, 4$
11. $-2, \dfrac{1}{2}$ 13. Any real number can replace z.
15. $\dfrac{3m}{2n^2}$ 17. $\dfrac{4p}{q}$ 19. $\dfrac{x-5}{x-1}$ 21. $\dfrac{1}{3}$
23. already in lowest terms   25. $\dfrac{2}{3}$ 27. $\dfrac{z}{2}$
29. $\dfrac{b+1}{2}$ 31. $\dfrac{2-y}{4-y}$ 33. $\dfrac{a-2}{a+2}$
35. $\dfrac{r+4}{r+6}$ 37. $\dfrac{c+6d}{c-d}$ 39. $\dfrac{z^2-zx+x^2}{z-x}$
41. $\dfrac{x-w}{x+w}$ 43. -1 45. $\dfrac{-1}{y+x}$
47. $\dfrac{-(4-n)}{4+n}$ or $\dfrac{n-4}{4+n}$ 49. $-\dfrac{2}{3}$ 51. $-x$
53. $-\dfrac{2}{3}$ 55. $\dfrac{5}{2}$

Section 4.2 (page 231)

1. 2   3. $\dfrac{a^4}{2}$ 5. $\dfrac{2y}{x^3}$ 7. $\dfrac{4}{3a^3b^2}$
9. $\dfrac{7r}{6}$ 11. $\dfrac{35}{8}$ 13. $\dfrac{m(m+3)}{m-3}$
15. $\dfrac{3(2k+3)}{8k(2k-3)}$ 17. $\dfrac{-(z+1)}{2z}$ or $\dfrac{-z-1}{2z}$
19. $\dfrac{-p(p+6)}{p+1}$ 21. -2 23. $\dfrac{a+3}{a-3}$
25. $\dfrac{2a+3b}{2a-3b}$ 27. $\dfrac{3x-y}{3x+2y}$ 29. $\dfrac{2k+5r}{k+5r}$
31. $(k-1)(k-2)$ 33. $\dfrac{(a+5)(a-1)}{3a^2-2a+1}$
35. 10   37. 36

Section 4.3 (page 239)

1. $8x$ 3. $5k^2$ 5. $35a$ 7. m^2n^2
9. $z(z-1)$ 11. $2(a+3)$ 13. $40(m+2)$

15. $(a-4)^2(a+4)$ 17. $(x+y)(x-y)$
19. $r(r-1)(r+1)$ 21. $p(p-4)(p+1)$
23. $(m+5)(m-2)(2m-3)$ 25. $\dfrac{7}{p}$ 27. $\dfrac{4}{5k}$
29. 1   31. 6   33. $a-b$ 35. $\dfrac{1}{p-2}$
37. $\dfrac{23}{2r}$ 39. $\dfrac{4z+3}{3z^2}$ 41. $\dfrac{29}{12x}$ 43. $-\dfrac{17}{18k}$
45. $\dfrac{2m-1}{m(m-1)}$ 47. $\dfrac{2t-16}{(t+2)(t-2)}$
49. $\dfrac{3}{m-4}$ or $\dfrac{-3}{4-m}$ 51. $\dfrac{x+y}{x-y}$
53. $\dfrac{13}{12(a+3)}$ 55. $\dfrac{x+9}{(x-3)(x-2)(x+1)}$
57. $\dfrac{2(2x-1)}{x-1}$ 59. $\dfrac{7}{y}$ 61. $\dfrac{2(w+5)}{w-2}$
63. $\dfrac{5r^2-17r+4}{(r-2)(r+1)(r-3)}$
65. $\dfrac{2x(x+12y)}{(x+2y)(x-y)(x+6y)}$
67. $\dfrac{7y^2+11yz-y-2z}{(y+z)(y-z)(y+2z)}$ 69. $\dfrac{15}{4}$ 71. $\dfrac{17}{7}$

Section 4.4 (page 245)

1. $\dfrac{x-1}{2x}$ 3. $\dfrac{8p}{3(p+5)}$ 5. $\dfrac{m-1}{4(m+1)}$
7. $\dfrac{5}{27z}$ 9. $\dfrac{1}{4x}$ 11. $\dfrac{7(5a-b)}{4}$
13. $\dfrac{2-m}{m+2}$ 15. $\dfrac{2r-3q}{2qr+1}$ 17. $\dfrac{a}{a-2b}$
19. $\dfrac{2}{y(2y+3x)}$ 21. $\dfrac{b^2+a^2}{a^2b^2}$ 23. $\dfrac{xy}{y-x}$
25. $\dfrac{pq}{q-p}$ 27. $\{3\}$ 29. $\left\{-\dfrac{2}{3},\dfrac{1}{2}\right\}$

Section 4.5 (page 249)

1. $\{16\}$ 3. $\{-3\}$ 5. $\left\{\dfrac{19}{4}\right\}$ 7. $\{1\}$
9. $\{5\}$ 11. $\{-3\}$ 13. $\{-3\}$ 15. $\{5\}$
17. $\{7\}$ 19. $\{0\}$ 21. $\{2\}$ 23. $\{2\}$
25. $\{-6, 4\}$ 27. $\{-6\}$ 29. $\emptyset$ 31. $\left\{\dfrac{4}{3}\right\}$
33. $\emptyset$ 35. $\{-3\}$ 37. $\emptyset$ 39. $\emptyset$
41. $\left\{-\dfrac{27}{2}\right\}$ 43. $r=\dfrac{d}{t}$ 45. $h=\dfrac{2A}{b}$

Summary of Operations Involving Rational Expressions (page 253)

1. $\{10\}$ 3. $\dfrac{-2y(y+2)}{5(y-2)}$ 5. $\dfrac{y-x}{y+x}$
7. $\{7\}$ 9. $\left\{\dfrac{1}{2}\right\}$ 11. $\dfrac{5}{4(q+2)}$

13. $\dfrac{12p}{p+2}$ 15. $\{0\}$ 17. $\dfrac{3q}{5}$ 19. $\{2\}$

21. $\dfrac{m}{3m+5k}$ 23. $\dfrac{p+2}{p-1}$ 25. $\left\{\dfrac{5}{4}\right\}$

27. $\dfrac{5m+3}{(m+2)(m-1)(m+3)}$ 29. $\{-10\}$

31. $\dfrac{r^2+3pr-r-p}{(r+p)(r+2p)(r+3p)}$ 33. $\dfrac{x-3y}{x+3y}$

Section 4.6 (page 257)

1. $\dfrac{2}{5}$ 3. $\dfrac{25}{4}$ 5. 24   7. $M=\dfrac{Fd^2}{Gm}$

9. $b=\dfrac{ac}{c-a}$ or $b=\dfrac{-ac}{a-c}$ 11. $T=\dfrac{PVt}{pv}$

13. $V=at+v$ 15. $n=\dfrac{2S}{(a+l)d}$

17. $n=\dfrac{t-a+d}{d}$ or $n=\dfrac{t-a}{d}+1$

19. $r=\dfrac{eR}{E-e}$ 21. 1   23. $\dfrac{6}{5}$

Section 4.7 (page 263)

1. -10 3. 3   5. -10 or $-\dfrac{9}{19}$ 7. $\dfrac{1}{15}$

9. $\dfrac{30}{11}$ or $2\dfrac{8}{11}$ hours   11. 12 hours

13. 20 hours   15. $1\dfrac{1}{5}$ hours   17. $2\dfrac{4}{5}$ hours

19. 3 miles per hour   21. $12\dfrac{2}{3}$ miles per hour

23. 240 miles   25. 150 miles   27. 2
29. 2   31. 2

Chapter 4 Review Exercises (page 271)

1. -11 2. -2 3. 2, 5   4. $\dfrac{5n^2}{2m}$ 5. $\dfrac{x}{2}$

6. $\dfrac{y+5}{y-3}$ 7. $\dfrac{5m+n}{5m-n}$ 8. p^2-pq+q^2

9. $\dfrac{-1}{2+r}$ 10. $\dfrac{3y}{2}$ 11. $\dfrac{10p^3}{3q}$ 12. $\dfrac{3}{10}$

13. $\dfrac{3y^2(2y+3)}{2y-3}$ 14. $\dfrac{-3(w+4)}{w}$

15. $\dfrac{y(y+5)}{y-5}$ 16. $\dfrac{z(z+2)}{z+5}$ 17. $\dfrac{2p+3}{6}$

18. 1   19. $60x$ 20. $96b^5$ 21. $p(p+6)$
22. $9r^2(3r+1)$ 23. $45(2k+1)$

24. $(3x-1)(2x+5)(3x+4)$ 25. $-\dfrac{3}{k}$ 26. $\dfrac{13}{8y}$

27. $\dfrac{16z-3}{2z^2}$ 28. $\dfrac{8}{t-2}$ or $\dfrac{-8}{2-t}$

29. $\dfrac{8y+4}{(y+1)(y-1)}$ or $\dfrac{4(2y+1)}{(y+1)(y-1)}$

30. $\dfrac{71}{30(a+2)}$ 31. $\dfrac{-7zy}{(2z+y)(z-y)(3z-2y)}$

32. $\dfrac{13r^2+5rs}{(5r+s)(2r-s)(r+s)}$ 33. $\dfrac{3(p+5)}{4p}$

34. $\dfrac{mn^4}{2}$ 35. $\dfrac{1}{16}$ 36. $\dfrac{3-5x}{6x+1}$

37. $\dfrac{1}{3q+2p}$ 38. $\dfrac{4(a^2-5)}{5(12-a)}$ 39. $\{1\}$

40. $\{-3\}$ 41. $\{-2\}$ 42. $\{0\}$ 43. $\{-3\}$

44. $\left\{\dfrac{1}{3}\right\}$ 45. $\dfrac{15}{2}$ 46. 30   47. $m=\dfrac{Fd^2}{GM}$

48. $t=\dfrac{vpT}{VP}$ 49. $\ell=\dfrac{2S}{n}-a$ or $\ell=\dfrac{2S-na}{n}$

50. $d=\dfrac{t-a}{n-1}$ or $d=\dfrac{a-t}{1-n}$ 51. 1

52. $\dfrac{18}{5}$ or $3\dfrac{3}{5}$ hours   53. $\dfrac{24}{5}$ or $4\dfrac{4}{5}$ minutes

54. 60 minutes   55. 16 kilometers per hour
56. bus: 50 miles per hour; train: 60 miles per hour

57. $\dfrac{6z-5}{3z^2}$ 58. $\dfrac{p-3}{36p^2+6p+1}$ 59. $\dfrac{y^2-6}{4y+2}$

60. $\dfrac{y(9y+1)}{3y+1}$ 61. $\dfrac{2}{5}$ 62. $\dfrac{11}{x-3}$ or $\dfrac{-11}{3-x}$

63. $\dfrac{13}{3(x+2)}$ 64. $\dfrac{4x^2+6xy+12y^2}{(x+3y)(x-2y)(x+y)}$

65. $r=\dfrac{AR}{R-A}$ or $r=\dfrac{-AR}{A-R}$ 66. $\{1,4\}$

67. $\{0\}$ 68. $\dfrac{60}{7}$ or $8\dfrac{4}{7}$ minutes

Chapter 4 Test (page 275)

[4.1] 1. $-3, \dfrac{1}{2}$ 2. $\dfrac{p}{p-4}$ [4.2] 3. $\dfrac{4}{9k^2}$

4. $\dfrac{x-5}{x+4}$ 5. $\dfrac{y}{y+5}$ [4.3] 6. $36z^3$

7. $m(m+3)(m-2)$ 8. $\dfrac{21}{5k}$

9. $\dfrac{4a+2b}{(a+b)(a-b)}$ 10. $\dfrac{28-5t}{2(t+4)(t-4)}$

[4.4] 11. $\dfrac{9}{4}$ 12. $\dfrac{-1}{s+t}$ [4.5] 13. $\{6\}$

14. $\emptyset$ 15. $\left\{\dfrac{1}{2}\right\}$ [4.6] 16. $m=\dfrac{5zy}{2y-z}$

17. $B=\dfrac{2A}{h}-b$ or $B=\dfrac{2A-bh}{h}$

[4.7] 18. $-\dfrac{3}{2}$ 19. $\dfrac{45}{14}$ or $3\dfrac{3}{14}$ hours

20. 15 miles per hour

CHAPTER 5

Section 5.1 (page 281)

1. $-5, 5$ **3.** $-19, 19$ **5.** $-25, 25$
7. $-47, 47$ **9.** $79, -79$ **11.** $27.537, -27.537$
13. 2 **15.** 16 **17.** -45 **19.** not a real number
21. x^2 **23.** 3.317 **25.** -7.483 **27.** -9.539
29. 16.733 **31.** 9 **33.** 4 **35.** -6 **37.** -8
39. 3 **41.** -5 **43.** not a real number **45.** -9
47. 2 **49.** 4 **51.** -2 **53.** 4 **55.** $-z^3$
57. m^5 **59.** (a) 110 miles per hour (b) 49 miles per hour (c) 70 miles per hour **61.** x^3 **63.** m^2
65. s^{15}

Section 5.2 (page 287)

1. 11 **3.** 8 **5.** 4 **7.** -9 **9.** $\dfrac{1}{2}$
11. $\dfrac{4}{5}$ **13.** $\dfrac{2}{5}$ **15.** 1000 **17.** 64
19. -1728 **21.** 144 **23.** $\dfrac{8}{27}$ **25.** $-\dfrac{243}{32}$
27. $\dfrac{5}{2}$ **29.** 2^2 **31.** 9^2 **33.** $81^{1/2}$ **35.** x
37. $\dfrac{1}{k^{2/3}}$ **39.** $\dfrac{1}{n^{2/3}}$ **41.** $4p^6q^4$ **43.** $\dfrac{1}{r^2}$
45. $\dfrac{z^{5/2}}{x^3}$ **47.** $m^{1/4}h^{3/4}$ **49.** p^2 **51.** $\dfrac{a^{5/2}}{m^{23/12}}$
53. 2^5 **55.** 6^4 **57.** $-11\sqrt[3]{11^2}$ or $-11\sqrt[3]{121}$
59. $x\sqrt[4]{x^3}$ **61.** $a\sqrt[14]{a^{11}}$ **63.** $\sqrt[6]{x}$
65. 50 centimeters **67.** p^{11} **69.** $\dfrac{1}{24x^8}$ **71.** $\dfrac{s}{16r}$

Section 5.3 (page 293)

1. $\sqrt{77}$ **3.** $\sqrt{42}$ **5.** $\sqrt[3]{14}$ **7.** $\sqrt[4]{24}$
9. $\dfrac{4}{5}$ **11.** $\dfrac{\sqrt{7}}{5}$ **13.** $\dfrac{\sqrt{r}}{10}$ **15.** $\dfrac{m^4}{5}$
17. $-\dfrac{3}{4}$ **19.** $\dfrac{\sqrt[3]{k^2}}{5}$ **21.** $3\sqrt{2}$ **23.** $6\sqrt{2}$
25. $-4\sqrt{3}$ **27.** $-2\sqrt{6}$ **29.** $2\sqrt{15}$ **31.** $5\sqrt{6}$
33. $-2\sqrt[3]{2}$ **35.** $2\sqrt[3]{3}$ **37.** $5\sqrt[3]{3}$ **39.** $2\sqrt[4]{2}$
41. $-5\sqrt[4]{2}$ **43.** $2\sqrt[5]{4}$ **45.** $5k$ **47.** $\dfrac{2\sqrt[3]{4}}{5}$
49. $10y^5$ **51.** $-2k^3$ **53.** $-12m^5z$ **55.** $5m^3b^6c^8$
57. $\dfrac{1}{2}m^3x^4$ **59.** $5y\sqrt{3y}$ **61.** $10m^4\sqrt{10m}$
63. $x^2y^3\sqrt{7x}$ **65.** $2z^3r^4$ **67.** $-2zt^3\sqrt[3]{2zt^2}$
69. $2a^2b^3$ **71.** $2km^2\sqrt[4]{2km^2}$ **73.** $\dfrac{m^4\sqrt{m}}{4}$
75. $\dfrac{y^3\sqrt[3]{y}}{3}$ **77.** $2\sqrt{3}$ **79.** $m\sqrt{m}$ **81.** $\sqrt[6]{675}$
83. $\sqrt[15]{864}$ **85.** $\dfrac{5\sqrt{21}}{3}$ feet **87.** $6x^2$
89. $11q - 6q^2$ **91.** cannot be simplified further

Section 5.4 (page 299)

1. 16 **3.** $23\sqrt{3}$ **5.** $-13\sqrt{3}$ **7.** $12\sqrt{2}$
9. $5\sqrt{7}$ **11.** $-16\sqrt{5}$ **13.** $10\sqrt{10}$
15. $-5\sqrt{2x}$ **17.** $17m\sqrt{2}$ **19.** $-\sqrt[3]{2}$
21. $10\sqrt[3]{x}$ **23.** $19\sqrt[4]{2}$ **25.** $-\sqrt[3]{x^2y}$
27. $-x\sqrt[3]{xy^2}$ **29.** $19\sqrt[4]{2}$ **31.** $9\sqrt[4]{2a^3}$
33. $x\sqrt[4]{xy}$ **35.** $12\sqrt{5} + 5\sqrt{3}$ centimeters
37. $37\sqrt{2} + 10\sqrt{3}$ yards
39. $\sqrt{3} \approx 1.732$, $\sqrt{12} \approx 3.464$; yes
41. 5 **43.** 2 **45.** $-12p^3 + 6p$
47. $15a^2 + 11ab - 14b^2$

Section 5.5 (page 307)

1. $6 + 2\sqrt{3}$ **3.** $6 + \sqrt{6}$ **5.** 4 **7.** 4
9. 15 **11.** 2 **13.** $\sqrt{6} - 2\sqrt{2} + 3\sqrt{3} - 6$
15. $4p - 36\sqrt{p} + \sqrt{7p} - 9\sqrt{7}$ **17.** $k - 16m$
19. $36a - 60\sqrt{a} + 25$ **21.** $16t + 8\sqrt{t} + 1$
23. $15\sqrt[3]{3z^2} - 3\sqrt[3]{9z} - 14$ **25.** $4\sqrt{3}$
27. $\dfrac{\sqrt{35}}{5}$ **29.** $\dfrac{9\sqrt{15}}{5}$ **31.** $\dfrac{3\sqrt{22}}{11}$
33. $\dfrac{3\sqrt{2}}{4}$ **35.** $\dfrac{9\sqrt{5}}{10}$ **37.** $-\sqrt{15}$
39. $\dfrac{8\sqrt{3k}}{k}$ **41.** $\dfrac{5\sqrt{2my}}{y^2}$ **43.** $\dfrac{3\sqrt{2}}{2}$
45. $-\dfrac{3\sqrt{14}}{7}$ **47.** $\dfrac{\sqrt{6}}{10}$ **49.** $\dfrac{5\sqrt{y}}{y}$
51. $\dfrac{6\sqrt{2r}}{r}$ **53.** $-\dfrac{5m\sqrt{3mn}}{n}$ **55.** $\dfrac{6x^4\sqrt{2y}}{y^2}$
57. $\dfrac{10r^3\sqrt{10rs}}{s^3}$ **59.** $\dfrac{\sqrt[3]{5}}{5}$ **61.** $\dfrac{\sqrt[3]{30}}{3}$
63. $\dfrac{m^3\sqrt[3]{n^2}}{n}$ **65.** $\dfrac{p^4\sqrt[3]{q}}{q^4}$ **67.** $2(3 + \sqrt{7})$
69. $\dfrac{3(\sqrt{3} + 1)}{2}$ **71.** $\dfrac{5(\sqrt{3} + \sqrt{7})}{4}$
73. $3\sqrt{3}(2 - \sqrt{3})$ **75.** $\dfrac{\sqrt{15} - 3 + \sqrt{5} - \sqrt{3}}{2}$
77. $7\sqrt{3} + 8\sqrt{2}$ **79.** $\dfrac{5\sqrt{k}(2\sqrt{k} - \sqrt{q})}{4k - q}$
81. $\dfrac{6p + 10\sqrt{10ps} + 15s}{2p - 45s}$ **83.** $\dfrac{3 - 2\sqrt{5}}{4}$
85. $-1 + \sqrt{2}$ **87.** $\dfrac{2 - 9\sqrt{2}}{3}$ **89.** $\dfrac{1 - y\sqrt{2y}}{2}$
91. $\{-1\}$ **93.** $\left\{\dfrac{2}{5}, \dfrac{1}{3}\right\}$ **95.** $5 - 3x^2$

Section 5.6 (page 317)

1. {4} **3.** {11} **5.** $\left\{\dfrac{1}{3}\right\}$ **7.** ∅ **9.** {9}
11. {5} **13.** {18} **15.** {4} **17.** {5} **19.** {5}
21. {4} **23.** {5} **25.** {−7, −2} **27.** ∅
29. {8} **31.** {3} **33.** {5} **35.** {6} **37.** {−4}
39. {−4} **41.** {5} **43.** {−2} **45.** {−1, −2}
47. {0, 3} **49.** 3.2 feet **51.** $15 + 5y$
53. $\dfrac{2(4 - \sqrt{3})}{13}$ **55.** $\dfrac{\sqrt{10} - \sqrt{6} + \sqrt{35} - \sqrt{21}}{2}$

Section 5.7 (page 327)

1. $12i$ **3.** $-15i$ **5.** $i\sqrt{3}$ **7.** $5i\sqrt{3}$
9. $10i\sqrt{5}$ **11.** $26i\sqrt{2}$ **13.** 0 **15.** -3
17. -18 **19.** $\sqrt{2}$ **21.** $3i$ **23.** 6
25. $10 + 5i$ **27.** $-1 + 3i$ **29.** $0 + 0i$
31. $8 - i$ **33.** $1 - i$ **35.** $-2 - 3i$
37. $-6 + 4i$ **39.** $6 + 4i$ **41.** -10
43. -28 **45.** -30 **47.** $6 + 6i$ **49.** $-38 - 8i$
51. $-7 + i$ **53.** $-8 - 6i$ **55.** 13 **57.** 2
59. $1 - i$ **61.** $\dfrac{1}{5} + \dfrac{2}{5}i$ **63.** $-2i$ **65.** $-i$
67. -1 **69.** 1 **71.** -1 **73.** 1
75. $\dfrac{37}{10} - \dfrac{19}{10}i$ **77.** $\dfrac{1}{2} + \dfrac{1}{2}i$
79. $(1 - 5i)^2 - 2(1 - 5i) + 26$
$= (1 - 10i + 25i^2) - 2 + 10i + 26$
$= 1 - 10i - 25 - 2 + 10i + 26 = 0$
81. $-5, 5$ **83.** $-3\sqrt{5}, 3\sqrt{5}$ **85.** {5, −8}

Chapter 5 Review Exercises (page 335)

1. 42 **2.** -17 **3.** not a real number
4. 6 **5.** -5 **6.** $-3z^4$ **7.** -2
8. $2r^2s^4$ **9.** 6.325 **10.** 8.775
11. 17.607 **12.** 32 **13.** -4 **14.** $-\dfrac{216}{125}$
15. -32 **16.** $\dfrac{1000}{27}$ **17.** 25 **18.** 96
19. p^4 **20.** $\dfrac{k^{17/12}}{2}$ **21.** 3^9 **22.** $7^4\sqrt{7}$
23. $m^4\sqrt[3]{m}$ **24.** $k^2\sqrt[4]{k}$ **25.** $\sqrt{78}$
26. $\sqrt{5r}$ **27.** $\sqrt[3]{30}$ **28.** $\sqrt[4]{21}$ **29.** $2\sqrt{5}$
30. $5\sqrt{3}$ **31.** $5\sqrt{5}$ **32.** $3\sqrt[3]{4}$
33. $10y^3\sqrt{y}$ **34.** $4p^2q\sqrt[3]{p^2q^2}$ **35.** $\dfrac{7}{9}$
36. $\dfrac{y\sqrt{y}}{12}$ **37.** $\dfrac{m^5}{3}$ **38.** $\dfrac{\sqrt[3]{r^2}}{2}$
39. $3a^2b\sqrt[3]{4a^2b^2}$ **40.** $2r^2t\sqrt[3]{79r^2t}$
41. $-11\sqrt{2}$ **42.** $23\sqrt{5}$ **43.** $7\sqrt{3y}$
44. $26m\sqrt{6m}$ **45.** $19\sqrt[3]{2}$ **46.** $-8\sqrt[4]{2}$
47. $1 - \sqrt{3}$ **48.** 2 **49.** $9 - 7\sqrt{2}$
50. $86 + 8\sqrt{55}$ **51.** $15 - 2\sqrt{26}$
52. $12 - 2\sqrt{35}$ **53.** $\dfrac{2\sqrt{7}}{7}$ **54.** $\dfrac{\sqrt{30}}{5}$
55. $-3\sqrt{6}$ **56.** $\dfrac{3\sqrt{7py}}{y}$ **57.** $\dfrac{\sqrt{22}}{4}$
58. $-\dfrac{\sqrt[3]{45}}{5}$ **59.** $\dfrac{3m\sqrt[3]{4n}}{n^2}$ **60.** $\dfrac{-2 - \sqrt{7}}{3}$
61. $\dfrac{-5(\sqrt{6} + \sqrt{3})}{3}$ **62.** {2} **63.** {6}
64. ∅ **65.** {0, 5} **66.** {9} **67.** {3}
68. {7} **69.** $\left\{-\dfrac{1}{2}\right\}$ **70.** {6} **71.** $5i$
72. $10i\sqrt{2}$ **73.** $-8i\sqrt{2}$ **74.** $-10 - 2i$
75. $14 + 7i$ **76.** $-\sqrt{35}$ **77.** -45
78. 3 **79.** $5 + i$ **80.** $42 - 32i$ **81.** $1 - i$
82. $-\dfrac{3}{13} - \dfrac{28}{13}i$ **83.** $-i$ **84.** 1
85. $-i$ **86.** -4 **87.** $-13ab^2$ **88.** $\dfrac{1}{100}$
89. $\dfrac{1}{y^{1/2}}$ **90.** $\dfrac{x^{3/4}}{z^{3/4}}$ **91.** k^6 **92.** $3z^3t^2\sqrt[3]{2t^2}$
93. $57\sqrt{2}$ **94.** $6x\sqrt[3]{y^2}$
95. $\sqrt{35} + \sqrt{15} - \sqrt{21} - 3$ **96.** $-\dfrac{\sqrt{3}}{6}$
97. $\dfrac{\sqrt[3]{60}}{5}$ **98.** $\dfrac{2\sqrt{z}(\sqrt{z} + 2)}{z - 4}$ **99.** $7i$
100. $3 - 7i$ **101.** $-5i$ **102.** 1
103. $\dfrac{13}{10} - \dfrac{11}{10}i$ **104.** {5} **105.** {−4}

Chapter 5 Test (page 339)

[5.1] **1.** $-15, 15$ **2.** -41 **3.** -4
[5.2] **4.** $\dfrac{1}{81}$ **5.** $3^{3/4}$ **6.** $\dfrac{1}{x^{1/12}}$
[5.3] **7.** $6x^2\sqrt{3x}$ **8.** $2ab^2\sqrt[4]{ab^3}$ **9.** $3\sqrt[3]{2}$
[5.4] **10.** $8\sqrt{5}$ [5.5] **11.** $9\sqrt{5} - 8$
12. $93 - 30\sqrt{6}$ **13.** $\dfrac{-9\sqrt{10}}{20}$ **14.** $\dfrac{3\sqrt[3]{2}}{2}$
15. $-4(\sqrt{7} + \sqrt{5})$ [5.6] **16.** {−1}
17. {−4} [5.7] **18.** $-14 + 21i$
19. $-2 + 16i$ **20.** $\dfrac{1}{10} + \dfrac{7}{10}i$

CHAPTER 6

Section 6.1 (page 347)

1. $\{7, -7\}$ 3. $\{\sqrt{13}, -\sqrt{13}\}$
5. $\{2\sqrt{3}, -2\sqrt{3}\}$ 7. $\{-9, -3\}$
9. $\left\{\dfrac{1+\sqrt{3}}{3}, \dfrac{1-\sqrt{3}}{3}\right\}$
11. $\left\{\dfrac{-1+2\sqrt{3}}{4}, \dfrac{-1-2\sqrt{3}}{4}\right\}$
13. $\{6i\sqrt{2}, -6i\sqrt{2}\}$ 15. $\{5+i, 5-i\}$
17. $\left\{\dfrac{1+i\sqrt{3}}{2}, \dfrac{1-i\sqrt{3}}{2}\right\}$ 19. $\{5, -3\}$
21. $\left\{\dfrac{5}{2}, -3\right\}$ 23. $\left\{\dfrac{-5+\sqrt{41}}{4}, \dfrac{-5-\sqrt{41}}{4}\right\}$
25. $\{1+\sqrt{2}, 1-\sqrt{2}\}$
27. $\{-3+\sqrt{2}, -3-\sqrt{2}\}$
29. $\left\{\dfrac{1+\sqrt{3}}{2}, \dfrac{1-\sqrt{3}}{2}\right\}$
31. $\left\{\dfrac{4+\sqrt{3}}{3}, \dfrac{4-\sqrt{3}}{3}\right\}$ 33. $\{-3+i, -3-i\}$
35. $\left\{\dfrac{2+\sqrt{5}}{5}, \dfrac{2-\sqrt{5}}{5}\right\}$ 37. $\sqrt{13}$ 39. 29
41. $\sqrt{165}$

Section 6.2 (page 355)

1. $\{-3, -5\}$ 3. $\left\{\dfrac{-2+\sqrt{2}}{2}, \dfrac{-2-\sqrt{2}}{2}\right\}$
5. $\{5+\sqrt{7}, 5-\sqrt{7}\}$
7. $\left\{\dfrac{-1+\sqrt{2}}{3}, \dfrac{-1-\sqrt{2}}{3}\right\}$
9. $\left\{\dfrac{-1+\sqrt{2}}{2}, \dfrac{-1-\sqrt{2}}{2}\right\}$
11. $\left\{\dfrac{1+\sqrt{29}}{2}, \dfrac{1-\sqrt{29}}{2}\right\}$
13. $\left\{\dfrac{-2+i\sqrt{2}}{3}, \dfrac{-2-i\sqrt{2}}{3}\right\}$
15. $\{3+i\sqrt{5}, 3-i\sqrt{5}\}$
17. $\left\{\dfrac{1+i\sqrt{6}}{2}, \dfrac{1-i\sqrt{6}}{2}\right\}$
19. $\left\{\dfrac{-1+i\sqrt{3}}{2}, \dfrac{-1-i\sqrt{3}}{2}\right\}$
21. $\left\{\dfrac{1+i\sqrt{47}}{6}, \dfrac{1-i\sqrt{47}}{6}\right\}$ 23. $\{8.589, .078\}$
25. 1.3 seconds 27. $3.92 29. (c) 31. (a)
33. (c) 35. (a) 37. (d) 39. (b) 41. (c)
43. (c) 45. $(2r+3)(4r-7)$
47. $(4k-3)(2k+11)$ 49. cannot be factored
51. $\left\{\dfrac{1+\sqrt{13}}{2}\right\}$ 53. $\{8\}$
55. $4r^2 - 7r - 2$; $(4r+1)(r-2)$
57. $2t^2 - 5t + 2$; $(2t-1)(t-2)$

Section 6.3 (page 365)

1. $\left\{-\dfrac{1}{3}, 2\right\}$ 3. $\{-7, 4\}$ 5. $\left\{-\dfrac{2}{3}, 1\right\}$
7. $\{-1+\sqrt{2}, -1-\sqrt{2}\}$
9. $\left\{\dfrac{2+\sqrt{6}}{2}, \dfrac{2-\sqrt{6}}{2}\right\}$ 11. $\left\{-2, -\dfrac{3}{4}\right\}$
13. $\{16\}$ 15. $\emptyset$ 17. $\left\{-\dfrac{2}{3}\right\}$
19. $\{-2, 2, -5, 5\}$ 21. $\left\{\dfrac{4}{3}, -\dfrac{4}{3}, 1, -1\right\}$
23. $\{-2, 7\}$ 25. $\left\{\dfrac{1}{2}, -\dfrac{4}{3}\right\}$ 27. $\left\{-2, -\dfrac{16}{3}\right\}$
29. $\left\{\dfrac{2}{3}, 2\right\}$ 31. 50 miles per hour
33. 80 miles per hour 35. 9 minutes 37. $\{1, 27\}$
39. $\left\{\dfrac{1}{4}, 9\right\}$ 41. $h = \dfrac{2A}{b}$
43. $h = \dfrac{S - 2\pi r^2}{2\pi r}$ 45. $r^2 = 3q^2 + 6$

Section 6.4 (page 373)

1. $a = \pm\sqrt{c^2 - b^2}$ 3. $d = \pm\dfrac{\sqrt{skw}}{kw}$
5. $d = \pm\dfrac{\sqrt{kR}}{R}$ 7. $h = \pm\dfrac{d^2\sqrt{kL}}{L}$
9. $r = \pm\dfrac{\sqrt{3\pi Vh}}{\pi h}$
11. $t = \dfrac{-B \pm \sqrt{B^2 - 4AC}}{2A}$ 13. $h = \dfrac{D^2}{k}$
15. $\ell = \dfrac{p^2 g}{k}$ 17. 9.3 seconds
19. 24 feet 21. 16 meters
23. width: 12 inches; length: 20 inches
25. 1 foot 27. 5.5 hours 29. 10 minutes
31. about 34.1 miles per hour
33. 100 miles per hour 35. 10% 37. $11.93
39.  41.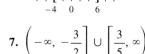
43. ◄———)| | |(———►
       -2 0 1

Section 6.5 (page 383)

1. $\left(-\infty, -\dfrac{1}{2}\right) \cup (5, \infty)$ 3. $[-4, 6]$

5. $(1, 3)$

7. $\left(-\infty, -\dfrac{3}{2}\right] \cup \left[\dfrac{3}{5}, \infty\right)$

9. $\left[-4, \frac{2}{3}\right]$

11. $\left(-3, \frac{5}{2}\right)$

13. $\left(-1, -\frac{3}{4}\right)$

15. $\left[\frac{3-\sqrt{3}}{3}, \frac{3+\sqrt{3}}{3}\right]$

17. $\left(-\infty, -\frac{5}{4}\right) \cup \left(-\frac{1}{2}, \frac{2}{3}\right)$

19. $\left[-\frac{5}{2}, -2\right] \cup \left[\frac{3}{4}, \infty\right)$

21. $\left(-\infty, -\frac{5}{3}\right) \cup \left(-\frac{1}{2}, 3\right)$

23. $(-\infty, 1) \cup [21, \infty)$

25. $(-\infty, 1) \cup [4, \infty)$

27. $\left(-\infty, \frac{1}{2}\right) \cup \left(\frac{5}{4}, \infty\right)$

29. $(-1, 5)$

31. $(-5, \infty)$

33. $\left(-\infty, \frac{4}{7}\right) \cup \left(\frac{2}{3}, \infty\right)$

35. $(-\infty, -2) \cup \left[-\frac{3}{2}, 0\right]$

37. -6 39. -9

Chapter 6 Review Exercises (page 387)

1. $\{4, -4\}$ 2. $\{\sqrt{7}, -\sqrt{7}\}$ 3. $\{-6, 1\}$

4. $\left\{\frac{2+4i}{3}, \frac{2-4i}{3}\right\}$

5. $\{2+\sqrt{19}, 2-\sqrt{19}\}$ 6. $\left\{1, \frac{1}{2}\right\}$

7. $\left\{\frac{7}{2}, -3\right\}$ 8. $\left\{\frac{-5+\sqrt{53}}{2}, \frac{-5-\sqrt{53}}{2}\right\}$

9. $\left\{\frac{-3+\sqrt{89}}{4}, \frac{-3-\sqrt{89}}{4}\right\}$

10. $\left\{\frac{7}{3}\right\}$ 11. $\left\{\frac{2+i\sqrt{2}}{3}, \frac{2-i\sqrt{2}}{3}\right\}$

12. $\left\{\frac{-7+\sqrt{37}}{2}, \frac{-7-\sqrt{37}}{2}\right\}$

13. $\{\sqrt{3}, -\sqrt{3}, i, -i\}$

14. $\left\{1, -1, \frac{i\sqrt{6}}{3}, \frac{-i\sqrt{6}}{3}\right\}$ 15. $\left\{\sqrt{2}, \frac{\sqrt{2}}{2}\right\}$

16. 5.5 hours and 6.5 hours   17. 4.6 hours

18. (c)   19. (a)   20. (d)   21. (d)   22. (b)

23. (c)   24. $(6x+5)(4x+9)$

25. $(12x-5)(3x+7)$ 26. $\left\{3, -\frac{5}{2}\right\}$

27. $\left\{-\frac{3}{4}, 1\right\}$ 28. $\{\sqrt{2}, -\sqrt{2}, 2\sqrt{2}, -2\sqrt{2}\}$

29. $\left\{-\frac{11}{8}, -\frac{7}{4}\right\}$ 30. $\emptyset$ 31. $\{4+\sqrt{51}\}$

32. $\emptyset$ 33. 40 kilometers per hour

34. Bill: 6 hours; Linda: 8 hours   35. $d = \frac{\pm\sqrt{SkI}}{I}$

36. $v = \frac{\pm\sqrt{rFkw}}{kw}$

37. $r = \frac{-\pi h \pm \sqrt{\pi^2 h^2 + 2\pi S}}{2\pi}$

38. $t = \frac{3m \pm \sqrt{9m^2 + 20m}}{2m}$ 39. .5 second

40. 3.3 seconds   41. $\sqrt{561}$ or about 23.7 meters

42. 9 feet, 12 feet, and 15 feet

43. length: 13 inches; width: 2 inches   44. 1 inch

45. Janet: 4 hours; Laurie: 8 hours

46. Ben: 2.6 hours; Arnold: 3.6 hours

47. 40 miles per hour

48. $10 + 10\sqrt{2}$ or about 24.1 miles per hour

49. $.50   50. 8%   51. 4.7 seconds

52. .3 second

53. $\left(-\infty, -\frac{3}{2}\right) \cup (1, \infty)$ 54. $(-4, 3)$

55. $\left[-\frac{1}{2}, 3\right]$ 56. $(-\infty, -1) \cup \left(\frac{1}{2}, 1\right)$

57. $\{2\}$ 58. $R = \frac{\pm\sqrt{Vh - r^2 h}}{h}$

59. $\left\{\frac{3+i\sqrt{3}}{3}, \frac{3-i\sqrt{3}}{3}\right\}$ 60. $\{4, -2, 3, -1\}$

61. $\left[-7, -\frac{3}{2}\right] \cup [1, \infty)$ 62. 3 hours and 6 hours

63. 7 miles per hour   64. $\{4+\sqrt{6}, 4-\sqrt{6}\}$

65. $y = \frac{3p^2}{x}$ 66. $\left\{1, \frac{3}{5}\right\}$ 67. $\left\{-\frac{5}{3}, -\frac{3}{2}\right\}$

68. $\left(-5, -\frac{23}{5}\right]$ 69. $\left\{-\frac{3}{2}, -\frac{1}{2}\right\}$

Chapter 6 Test (page 393)

[6.1] **1.** $\{2\sqrt{2}, -2\sqrt{2}\}$ **2.** $\left\{0, -\dfrac{4}{3}\right\}$
3. $\{1 + \sqrt{2}, 1 - \sqrt{2}\}$ [6.2] **4.** $\left\{1, \dfrac{1}{2}\right\}$
5. $\left\{\dfrac{-2 + i\sqrt{11}}{3}, \dfrac{-2 - i\sqrt{11}}{3}\right\}$ **6.** $\left\{\dfrac{2}{3}\right\}$
7. Mario: 11.1 hours; Luis: 9.1 hours
8. two imaginary solutions
9. two irrational solutions
10. two rational solutions
[6.3] **11.** $\left\{\dfrac{-1 + \sqrt{13}}{2}, \dfrac{-1 - \sqrt{13}}{2}\right\}$
12. $\left\{2, -2, \dfrac{\sqrt{10}}{2}, -\dfrac{\sqrt{10}}{2}\right\}$ **13.** $\left\{2, -\dfrac{3}{2}\right\}$
14. 15 miles per hour   [6.4] **15.** $r = \dfrac{\pm\sqrt{\pi S}}{2\pi}$
16. $p = \dfrac{k \pm \sqrt{k^2 - 4kr}}{2k}$ **17.** 2 feet
18. 120 meters

[6.5] **19.** $\left(-5, \dfrac{3}{2}\right)$ **20.** (1, 5]

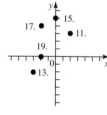   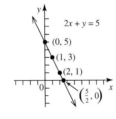

CHAPTER 7

Section 7.1 (page 401)

1. I   **3.** III   **5.** IV   **7.** II   **9.** none

11–19.   **21.** $(0, 5), \left(\dfrac{5}{2}, 0\right), (1, 3), (2, 1)$

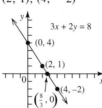

23. $(0, -4), (4, 0),$
$(2, -2), (3, -1)$ **25.** $(0, 4), (5, 0),$
$\left(3, \dfrac{8}{5}\right), \left(\dfrac{5}{2}, 2\right)$

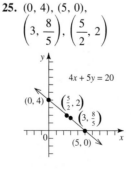

27. $(0, 4), \left(\dfrac{8}{3}, 0\right),$
$(2, 1), (4, -2)$

29. $\left(0, \dfrac{7}{3}\right), \left(\dfrac{7}{4}, 0\right),$
$(1, 1), (4, -3)$

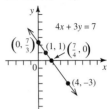

31. $(5, 0); (0, 2)$ **33.** $(6, 0); (0, 2)$

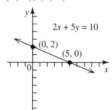
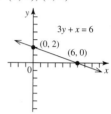

35. $\left(\dfrac{8}{7}, 0\right); \left(0, -\dfrac{8}{3}\right)$ **37.** $(5, 0); \left(0, \dfrac{5}{2}\right)$

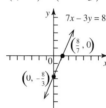
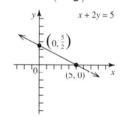

39. (a) 233.4 million   (b) 238.5 million
(c) 247 million   (d) 2030   (e)

41. $\dfrac{1}{2}$ **43.** $\dfrac{2}{5}$

45. $-\dfrac{1}{5}$

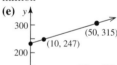

Section 7.2 (page 411)

1. slope: -8 **3.** slope: -1

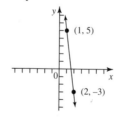
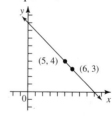

5. slope: 0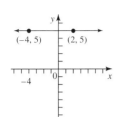

7. slope: $-\dfrac{1}{2}$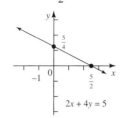

9. slope: 1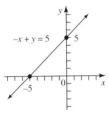

11. slope: $\dfrac{6}{5}$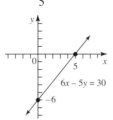

13. slope: $\dfrac{2}{5}$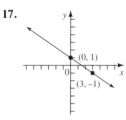

15. slope: -3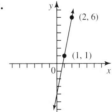

17.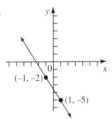

19.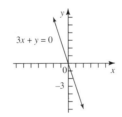

21.

23. not parallel
25. parallel
27. parallel
29. perpendicular
31. not perpendicular
33. perpendicular
35. yes; yes

37. $y = 5x - 12$
39. $2x + 3y = 27$
41. $5x - 3y = -17$

Section 7.3 (page 421)

1. $y = -x + 8$; -1; $(0, 8)$
3. $y = -\dfrac{5}{2}x + 5$; $-\dfrac{5}{2}$; $(0, 5)$
5. $y = \dfrac{2}{3}x - \dfrac{5}{3}$; $\dfrac{2}{3}$; $\left(0, -\dfrac{5}{3}\right)$

7. $y = -\dfrac{5}{3}x - \dfrac{4}{3}$; $-\dfrac{5}{3}$; $\left(0, -\dfrac{4}{3}\right)$
9. $y = -\dfrac{8}{11}x + \dfrac{9}{11}$; $-\dfrac{8}{11}$; $\left(0, \dfrac{9}{11}\right)$
11. $y = 5x + 4$ **13.** $y = -\dfrac{2}{3}x + \dfrac{1}{2}$
15. $y = \dfrac{2}{5}x - 1$ **17.** $y = 4$
19. $y = -1.573x + 4.209$ **21.** $5x + 6y = 2$
23. $x - y = -5$ **25.** $x - 4y = 9$ **27.** $x = 1$
29. $5x + y = -10$ **31.** $x = -4$ **33.** $x = 3$
35. $3x + 8y = -1$
37. $2x - 13y = -6$
39. $y = 5$ **41.** $\sqrt{5}x - 2y = -\sqrt{5}$
43. $3x - y = -24$ **45.** $x - 2y = 2$
47. $x + 2y = 18$ **49.** $y = 7$
51. $4400x - 3y = 94{,}100$
53. $\left(-\infty, \dfrac{1}{2}\right)$ **55.** $[-2, \infty)$ **57.** $\left(-\infty, -\dfrac{8}{5}\right)$

Section 7.4 (page 431)

1.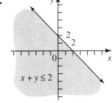

3.

5.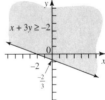

7.

9.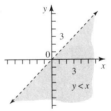

11.

13.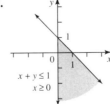

15.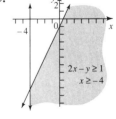

17. 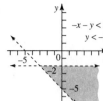 **19.**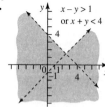

21. 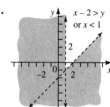 **23.**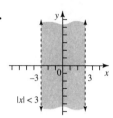

25. 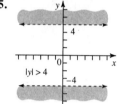 **27.**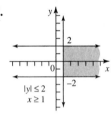

29. 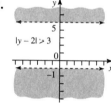 **31.**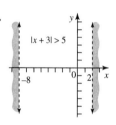

33. no solution **35.** $w = \dfrac{V}{lh}$ **37.** $140

Section 7.5 (page 441)

1. domain: {5, 3, 4, 7}; range: {1, 2, 9, 6}; function
3. domain: {2, 0}; range: {4, 2, 5}; not a function
5. domain: {−3, −2, 4}; range: {1, 7}; function
7. domain: {1, 2, 3, 5}; range: {10, 15, 19, 27}; not a function
9. domain: {2, 5, 11, 17, 3}; range: {1, 7, 20}; function
11. function; domain: $(-\infty, \infty)$
13. function; domain: $(-\infty, \infty)$; linear
15. function; domain: $(-\infty, \infty)$
17. not a function; domain: $(-\infty, \infty)$
19. not a function; domain: $(-\infty, \infty)$
21. function; domain: $(-\infty, 0) \cup (0, \infty)$
23. not a function; domain: $(-\infty, \infty)$
25. function; domain: $[0, \infty)$ **27.** 4 **29.** $\dfrac{1}{2}$
31. $3 + \dfrac{p}{2}$ or $\dfrac{6+p}{2}$ **33.** $\dfrac{3}{4}$ **35.** 39 **37.** $\dfrac{3}{4}$
39. −13 **41.** 21 **43.** function **45.** function

47. $2.25 **49.** $2.50 **51.** yes **53.** $\left\{-\dfrac{25}{11}\right\}$
55. $\left\{\dfrac{1}{3}\right\}$ **57.** {18}

Section 7.6 (page 451)

1. $\dfrac{16}{9}$ **3.** $\dfrac{5}{16}$ **5.** $\dfrac{1000}{9}$ **7.** 850 ohms
9. $333\dfrac{1}{3}$ cycles per second **11.** 126.4 pounds
13. 3.57×10^{-10} meter **15.** about 8.5
17. (0, 4), (5, −11), (−2, 10)
19. (−1, 1), (0, 0), (1, 1)
21. (3, 3), (4, 4), (4, −4), (3, −3)

Chapter 7 Review Exercises (page 457)

1. (0, 5), $\left(\dfrac{10}{3}, 0\right)$, **2.** (2, −6), (5, −3), (3, −5), (6, −2)

(2, 2), $\left(\dfrac{14}{3}, -2\right)$

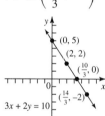 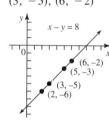

3. (3, 0); (0, −4) **4.** $\left(\dfrac{28}{5}, 0\right)$; (0, 4)

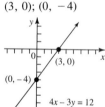 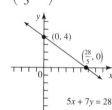

5. $-\dfrac{7}{5}$ **6.** $-\dfrac{1}{2}$ **7.** 2 **8.** $\dfrac{3}{4}$

9. undefined **10.** $\dfrac{2}{3}$ **11.** $-\dfrac{1}{3}$

12. One pair of opposite sides has slope 0 and the other pair of opposite sides has undefined slope. Since opposite sides are parallel and adjacent sides are perpendicular, the figure is a rectangle.

13. $y = \dfrac{3}{5}x + 2$ **14.** $y = -\dfrac{1}{3}x - 1$
15. $y = -2$ **16.** $4x + 3y = 29$
17. $3x - y = -7$ **18.** $x = 2$ **19.** $9x + y = 13$
20. $7x - 5y = -16$ **21.** $4x - y = 29$
22. $5x + 2y = 26$

23. 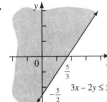 **24.**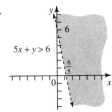

25. 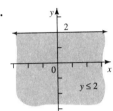 **26.**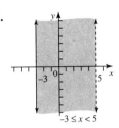

27. 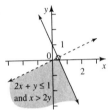 **28.**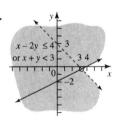

29.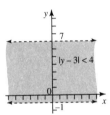

30. domain: {1, 2, 3, 4}; range: {5, 7, 9, 11}; function
31. domain: {−4, 1}; range: {2, −2, 5, −5}; not a function
32. domain: {14, 91, 17, 75, 23}; range: {9, 12, 18, 70, 56, 5}; not a function
33. domain: {A, B, C, D, E}; range: {x, y, z}; function
34. domain: $(-\infty, \infty)$; function
35. domain: $\left[\frac{1}{2}, \infty\right)$; not a function
36. domain: $(-\infty, \infty)$; function
37. domain: $[0, \infty)$; not a function
38. domain: $(-\infty, \infty)$; not a function
39. domain: $(-\infty, 1) \cup (1, \infty)$; function **40.** −2
41. −1 **42.** 14 **43.** $-\frac{7}{4}$ **44.** $k^2 + 2k - 1$
45. −2 **46.** not a function **47.** function
48. function **49.** 48 **50.** 12
51. 430 millimeters **52.** 60 kilometers
53. $20\sqrt{2} \approx 28.3$ **54.** 5556 pounds (rounded)
55. .8 million dollars
56. $\frac{45}{8}$ or 5.625 kilometers per second

Chapter 7 Test (page 463)

[7.2] **1.** $\frac{2}{7}$

[7.1–7.2] **2.** $\frac{3}{2}$; $\left(\frac{20}{3}, 0\right)$; $(0, -10)$

3. 0; no x-intercept; (0, 3)
4. undefined; (3, 0); no y-intercept
[7.3] **5.** $y = 4$ **6.** $3x + y = 11$
7. $3x + 5y = -11$

[7.1] **8.** 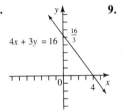 **9.**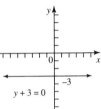

[7.4] **10.** 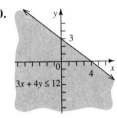 **11.**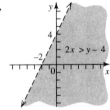

[7.5] **12.** domain: $[-1, \infty)$; not a function
13. domain: $[0, \infty)$; function **14.** −19
15. $-\frac{1}{m^2} + \frac{4}{m} - 7$ [7.6] **16.** 800 pounds

CHAPTER 8

Section 8.1 (page 473)

1. yes **3.** no **5.** {(5, 5)}

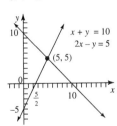

7. {(5, 0)}
9. $\{(x, y) | 7x + 2y = 3\}$; dependent equations
11. {(11, 7)} **13.** {(2, −3)} **15.** {(2, −4)}
17. ∅; inconsistent system **19.** {(2, 1)}
21. {(−2, −10)} **23.** {(2, 3)} **25.** {(5, 4)}
27. $\left\{\left(-5, -\frac{10}{3}\right)\right\}$ **29.** {(2, 6)} **31.** {(2, 4)}
33. {(4, −5)} **35.** $\left\{\left(\frac{1}{a}, \frac{1}{b}\right)\right\}$

37. $\left\{\left(-\dfrac{3}{5a}, \dfrac{7}{5}\right)\right\}$

39. (a) about $1000 for A and $550 for B
(b) year 3.5; about $800
(c) about 7 years for A; about 1.5 years for B

41. (a) $x = 8$ or 800 items; $3000 (b) about $-$$500
(c) $1000   43. $-6x + 12y - 3z = -24$

45. $2x - 6y + 4z = -36$ 47. $3x + 11y = -3$

Section 8.2 (page 483)

1. $\{(-1, 2, 3)\}$ 3. $\{(2, 1, -4)\}$
5. $\{(3, -1, 4)\}$ 7. $\left\{\left(-\dfrac{7}{3}, \dfrac{22}{3}, 7\right)\right\}$
9. $\{(4, 5, 3)\}$ 11. $\{(2, 2, 2)\}$ 13. $\{(-1, 2, -2)\}$
15. ∅   17. $\left\{\left(\dfrac{20}{11}, \dfrac{10}{11}, -\dfrac{41}{11}\right)\right\}$
19. $\{(x, y, z) | x + 4y - z = 3\}$
21. $\{(x, y, z) | 2x + y - z = 6\}$
23. $\{(x, y, z) | -x + 4y - z = 5\}$ 25. $\{(2, 1, 5, 3)\}$
27. 20 centiliters   29. 50 inches, 53 inches, 60 inches
31. $-4, 8, 12$

Section 8.3 (page 493)

1. 13 centimeters, 19 centimeters, 19 centimeters
3. square: 10 centimeters; triangle: 6 centimeters
5. small: $1.50; large: $2.50
7. 6 units of yarn; 2 units of thread
9. dark clay: $5 per kilogram; light clay: $1 per kilogram
11. 5 pounds of oranges; 1 pound of apples
13. 5 twenties; 30 fifties
15. $2000 at 8%; $4000 at 4%
17. $24,000 at 14%; $32,000 at 7%
19. 1 liter of pure acid; 8 liters of 10% acid
21. 4 liters of pure antifreeze
23. 28 kilograms of nuts; 32 kilograms of cereal
25. express train: 80 kilometers per hour; freight train: 50 kilometers per hour
27. plane: 160 kilometers per hour; train: 60 kilometers per hour
29. first: 9; second: 10; third: 13
31. largest: 84°; middle: 70°; smallest: 26°
33. shortest: 16 inches; medium: 18 inches; longest: 22 inches   35. 7 nickels; 17 dimes; 20 quarters
37. subcompact: 9; compact: 5; full size: 6
39. up close: $10; in the middle: $8; far out: $7
41. $20,000 in the video rental firm; $30,000 each in the bond and the money market fund
43. 60 ounces of Orange Pekoe; 30 ounces of Irish Breakfast; 10 ounces of Earl Grey   45. -4
47. 16   49. -3

Section 8.4 (page 503)

1. 13   3. -19 5. 0   7. 0   9. 20
11. 6   13. 1   15. -11 17. 0   19. 0
21. 0   23. -5 25. $8m$ 27. -55 29. 0

31. $\{11\}$ 33. The line has the equation $y - y_1 = \dfrac{y_2 - y_1}{x_2 - x_1}(x - x_1)$. Simplify this equation and simplify the given determinant equation. Each simplifies to $x_2y - x_1y - x_2y_1 - xy_2 + x_1y_2 + xy_1 = 0$.

35. $\dfrac{3}{5}$ 37. 19

Section 8.5 (page 509)

1. $\{(4, 6)\}$ 3. $\{(3, -4)\}$ 5. $\{(1, 0)\}$
7. $\{(3, 1)\}$ 9. $\{(-2, -3)\}$ 11. $\{(3, 1, 2)\}$
13. Cramer's rule does not apply.   15. $\{(0, 0, 0)\}$
17. $\left\{\left(\dfrac{20}{59}, -\dfrac{33}{59}, \dfrac{35}{59}\right)\right\}$ 19. $\{(20, -13, -12)\}$
21. $\left\{\left(-\dfrac{22}{5}, \dfrac{17}{15}, 1\right)\right\}$ 23. $\{(-1, 2, 5, 1)\}$
25. $f(3) = -9; f(-2) = 11$
27. $f(3) = -26; f(-2) = 9$
29. $f(3) = -\dfrac{3}{29}; f(-2) = \dfrac{1}{2}$

Chapter 8 Review Exercises (page 517)

1. $\{(2, -3)\}$ 2. $\{(1, 4)\}$

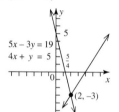
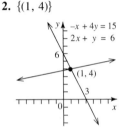

3. $\{(-1, 2)\}$ 4. $\{(-3, 4)\}$ 5. $\left\{\left(-\dfrac{8}{9}, -\dfrac{4}{3}\right)\right\}$
6. ∅   7. $\{(2, 4)\}$ 8. $\{(5, 2)\}$ 9. $\{(1, -5, 3)\}$
10. $\{(1, 2, 3)\}$ 11. ∅   12. ∅
13. length: 12 meters; width: 9 meters
14. 6 weekend days; 4 weekdays
15. 15 pounds of $2-a-pound candy; 35 pounds of $1-a-pound candy
16. plane: 300 miles per hour; wind: 20 miles per hour
17. $20,000 at 10%; $50,000 at 6%; $70,000 at 5%
18. 85°; 60°; 35°
19. 5 liters of 8%; none of 10%; 3 liters of 20%
20. A: 141.5 pounds; B: 170.3 pounds; C: 48.6 pounds
21. -6 22. 17   23. 20   24. 49   25. -19
26. -5 27. -231 28. 0
29. 2   30. $abc - b$ 31. $\left\{\left(\dfrac{37}{11}, \dfrac{14}{11}\right)\right\}$
32. $\left\{\left(\dfrac{39}{10}, \dfrac{6}{5}\right)\right\}$ 33. $\left\{\left(\dfrac{34}{13}, -\dfrac{20}{39}\right)\right\}$

34. $\{(3, -2, 1)\}$ **35.** $\left\{\left(\dfrac{172}{67}, -\dfrac{14}{67}, -\dfrac{87}{67}\right)\right\}$
36. $\left\{\left(\dfrac{49}{9}, -\dfrac{155}{9}, \dfrac{136}{9}\right)\right\}$ **37.** $\{(5, 3)\}$ **38.** $\emptyset$
39. $\{(2, 1, 1)\}$ **40.** $\{(0, 4)\}$ **41.** $\{(0, 0, 0)\}$
42. $\{(3, -1)\}$ **43.** $\{(12, 9)\}$
44. $\left\{\left(\dfrac{82}{23}, -\dfrac{4}{23}\right)\right\}$

Chapter 8 Test (page 521)

[8.1] **1.** $\{(-1, 2)\}$ **2.** $\{(-2, 1)\}$
3. $\{(-1, 4)\}$ **4.** $\emptyset$ **5.** $\{(x, y) | 12x - 5y = 8\}$
[8.2] **6.** $\{(3, -1, 2)\}$ **7.** $\{(-3, -2, -4)\}$
[8.3] **8.** 2 liters of 20% alcohol; 4 liters of 50% alcohol **9.** 45 miles per hour; 75 miles per hour
10. shortest: 5 centimeters; medium: 6 centimeters; longest: 7 centimeters
[8.4] **11.** -10 **12.** 0 **13.** 8
[8.5] **14.** $\left\{\left(-\dfrac{9}{4}, \dfrac{5}{4}\right)\right\}$
15. $\{(-1, 2, 3)\}$

CHAPTER 9

Section 9.1 (page 529)

1. $(0, -5)$ **3.** $(0, 0)$ **5.** $(-4, 0)$
7. $(9, 12)$ **9.** downward; narrower
11. upward; wider

13. 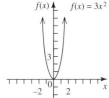 **15.**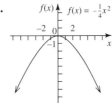

17. 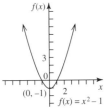 **19.**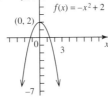

21. 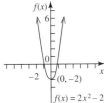 **23.**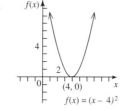

25. 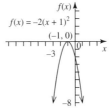 **27.**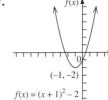

29. 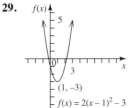 **31.**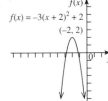

33. 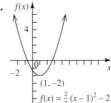 **35.** $\{-3 \pm 2\sqrt{3}\}$
37. $\left\{\dfrac{6 \pm \sqrt{46}}{2}\right\}$
39. $\left\{\dfrac{-3 \pm \sqrt{17}}{2}\right\}$

Section 9.2 (page 539)

1. $(-4, -21)$; upward; same; two x-intercepts
3. $\left(\dfrac{3}{2}, \dfrac{21}{4}\right)$; downward; same; two x-intercepts
5. $\left(\dfrac{3}{2}, -\dfrac{3}{2}\right)$; upward; narrower; two x-intercepts
7. $(-2, -1)$; to the right; narrower
9. $(1, 2)$; to the right; narrower

11. 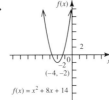 **13.**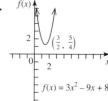

15. 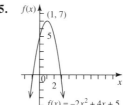 **17.**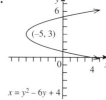

19.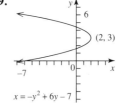
21. 7 units of tacos; $62
23. 2500 feet; 12.5 seconds
25. 40 and 40
27. (a) 855 (b) 480 (c) 0
29. 5 **31.** 10

Section 9.3 (page 547)

1. $\sqrt{34}$ 3. $\sqrt{89}$ 5. $2\sqrt{61}$
7. $\sqrt{a^2 - 2ab + 5b^2}$ 9. $x^2 + y^2 = 64$
11. $(x - 2)^2 + (y - 4)^2 = 25$
13. $x^2 + (y - 3)^2 = 2$
15. $x^2 + (y + 1)^2 = 16$ 17. $(-3, 1)$; 2
19. $(-4, -2)$; 7 21. $(1, -3)$; 3

23.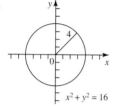
25.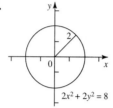
27.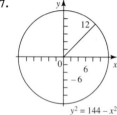
29.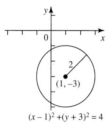
31.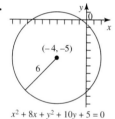
33.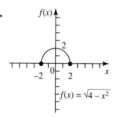
35.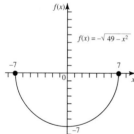
37.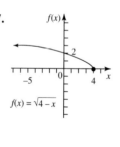

39. (a) $|x + p|$ (b) The distance from the point to the origin should equal the distance from the line to the origin. (c) $\sqrt{(x - p)^2 + y^2}$ (d) $y^2 = 4px$

41. x-intercept: $\left(\dfrac{3}{5}, 0\right)$; y-intercept: $\left(0, -\dfrac{3}{8}\right)$

43. x-intercept: $(2, 0)$; y-intercept: $(0, 3)$

Section 9.4 (page 557)

1.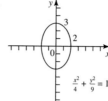
3.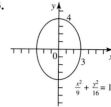
5.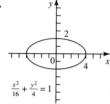
7.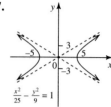
9.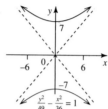
11.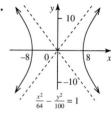

13. circle
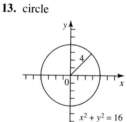

15. circle
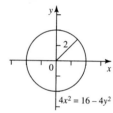

17. parabola
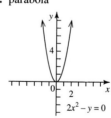

19. circle
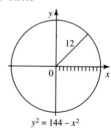

21. ellipse
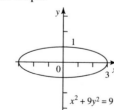

23. hyperbola
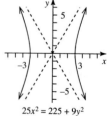

25.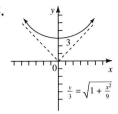

29. (a) 10 meters **(b)** 36 meters

31. $\left\{\sqrt{3}, -\sqrt{3}, \dfrac{i\sqrt{2}}{2}, -\dfrac{i\sqrt{2}}{2}\right\}$

33. $\{2, -2, 2i, -2i\}$

Section 9.5 (page 565)

1. $\{(0, 0), (-1, -1)\}$ **3.** $\{(2, 0), (1, 1)\}$
5. $\left\{(1, 0), \left(-\dfrac{3}{5}, \dfrac{4}{5}\right)\right\}$ **7.** $\left\{\left(-3, -\dfrac{2}{3}\right), (-2, -1)\right\}$
9. $\{(2, -3), (-3, 2)\}$ **11.** $\{(-2, 0), (-5, 30)\}$
13. $\{(\sqrt{2}, 1), (-\sqrt{2}, 1)\}$
15. $\left\{\left(\dfrac{-i\sqrt{3}}{3}, \dfrac{-3 - i\sqrt{3}}{6}\right),\right.$
$\left.\left(\dfrac{i\sqrt{3}}{3}, \dfrac{-3 + i\sqrt{3}}{6}\right)\right\}$
17. $\{(\sqrt{3}, 0), (-\sqrt{3}, 0)\}$
19. $\{(1, 3), (1, -3), (-1, 3), (-1, -3)\}$
21. $\left\{\left(\dfrac{i\sqrt{10}}{2}, -i\sqrt{10}\right), \left(\dfrac{-i\sqrt{10}}{2}, i\sqrt{10}\right),\right.$
$\left.(\sqrt{5}, \sqrt{5}), (-\sqrt{5}, -\sqrt{5})\right\}$
23. $\{(\sqrt{7}, i\sqrt{3}), (\sqrt{7}, -i\sqrt{3}), (-\sqrt{7}, i\sqrt{3}),$
$(-\sqrt{7}, -i\sqrt{3})\}$
25. $\{(2, 1), (-2, -1), (i, -2i), (-i, 2i)\}$
27. $\{(-2, 1), (2, -1), (i, 2i), (-i, -2i)\}$
29. 9 and 5, 9 and -5, -9 and 5, or -9 and -5
31. length: 20 centimeters; width: 5 centimeters
33. 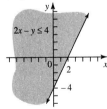   **35.**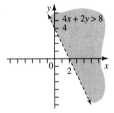

Section 9.6 (page 573)

1. 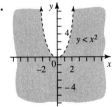   **3.**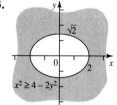

5. 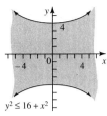   **7.**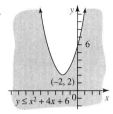

9. 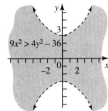   **11.**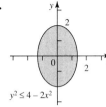

13. 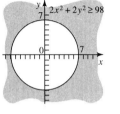   **15.**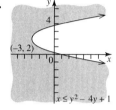

17.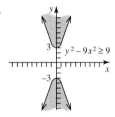

In the graphs for systems of inequalities we give the equations of the boundaries instead of the inequalities of the system.

19.   **21.**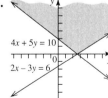

23. 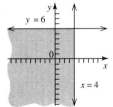   **25.**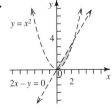

27. 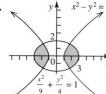 **29.**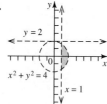

31. 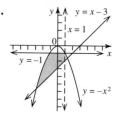 **33.**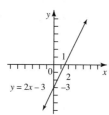

35.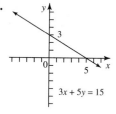

10. $\left(\dfrac{2}{3}, -\dfrac{2}{3}\right)$ **11.** 5 seconds   **12.** 400 feet
13. length: 25 meters; width: 25 meters   **14.** 5 and 5
15. $\sqrt{13}$ **16.** $\sqrt{53}$ **17.** 10
18. $(x + 3)^2 + (y - 4)^2 = 25$
19. $(x - 6)^2 + (y + 1)^2 = 9$ **20.** $x^2 + y^2 = 64$
21. $(-4, 4)$; 3   **22.** $(5, 2)$; 6
23. $(1, 2)$; $2\sqrt{3}$ **24.** $\left(\dfrac{1}{2}, -1\right)$; $\dfrac{\sqrt{285}}{10}$

25. 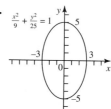 **26.**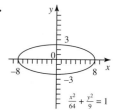

27. 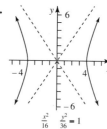 **28.**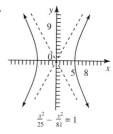

29. 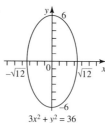 **30.**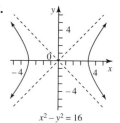

31. parabola   **32.** circle   **33.** straight line
34. parabola   **35.** hyperbola   **36.** ellipse
37. $\left\{\left(-\dfrac{1}{2}, -\dfrac{5}{4}\right), (3, -3)\right\}$
38. $\left\{\left(\dfrac{-1 + \sqrt{17}}{2}, \dfrac{5 + \sqrt{17}}{2}\right), \left(\dfrac{-1 - \sqrt{17}}{2}, \dfrac{5 - \sqrt{17}}{2}\right)\right\}$
39. $\left\{(1, 4), \left(-\dfrac{15}{7}, -\dfrac{16}{7}\right)\right\}$
40. $\left\{\left(-\dfrac{5}{2}, -6\right), (3, 5)\right\}$
41. $\{(2, 2), (2, -2), (-2, 2), (-2, -2)\}$
42. $\{(\sqrt{6}, 0), (-\sqrt{6}, 0)\}$

Chapter 9 Review Exercises (page 581)

1. $(0, 0)$ **2.** $(0, -2)$

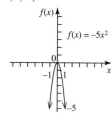   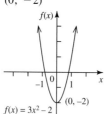

3. $(0, 4)$ **4.** $(0, 6)$

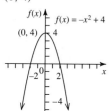   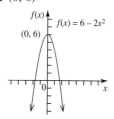

5. $(-2, 0)$ **6.** $(3, -7)$

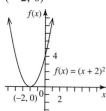   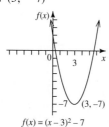

7. $(3, -2)$ **8.** $\left(-\dfrac{1}{4}, -\dfrac{3}{2}\right)$ **9.** $(7, 2)$

43.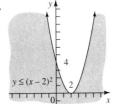
44.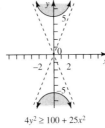
45.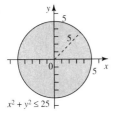
46.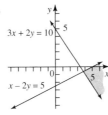
47.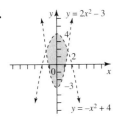
48.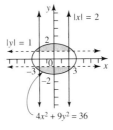
49.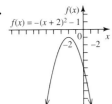
50.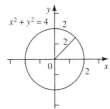
51.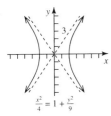
52.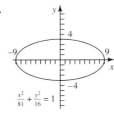
53.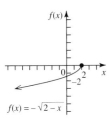
54.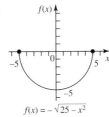
55.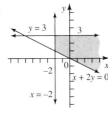
56.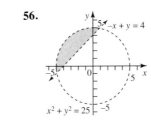

Chapter 9 Test (page 587)

[9.1] **1.** $(0, 4)$ 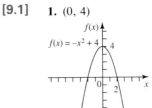 [9.2] **2.** $(-1, -1)$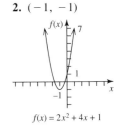

3. 50 feet; 3 seconds [9.3] **4.** $\sqrt{34}$

5. $(-1, 2); 3$ **6.** $(1, 3); \ 4$

7. 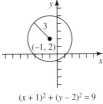 [9.4] **8.**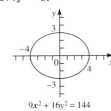

9. 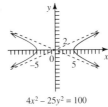 [9.3] **10.**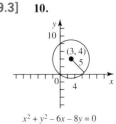

[9.2] **11.**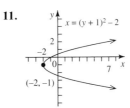

[9.4] **12.** circle **13.** hyperbola **14.** parabola

15. ellipse [9.5] **16.** $\left\{\left(8, \dfrac{1}{2}\right), (2, 2)\right\}$

17. $\left\{(1, 1), \left(\dfrac{1}{5}, -\dfrac{7}{5}\right)\right\}$

18. $\{(2\sqrt{2}, 2\sqrt{2}), (2\sqrt{2}, -2\sqrt{2}), (-2\sqrt{2}, 2\sqrt{2}), (-2\sqrt{2}, -2\sqrt{2})\}$

[9.6] 19.
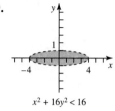

20.
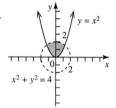

CHAPTER 10

Section 10.1 (page 597)

1. {(5, 3), (9, 2), (7, 4)}   3. not one-to-one
5. $f^{-1}(x) = \dfrac{x}{2}$ 7. $f^{-1}(x) = x^2,\ x \geq 0$
9. $f^{-1}(x) = \dfrac{2x + 1}{3}$ 11. $f^{-1}(x) = \sqrt[3]{x - 1}$
13. not one-to-one   15. $f^{-1}(x) = x^3 + 1$
17. (a) one-to-one   19. (a) not one-to-one
(b)
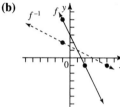

21. (a) one-to-one
(b)
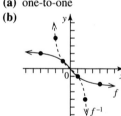

23.
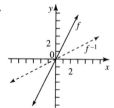
25.
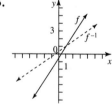

27.
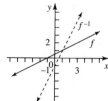
29.
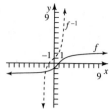

31. 8   33. 3   35. 1   37. 0   39. 27
41. $\dfrac{1}{9}$ 43. $\dfrac{1}{1000}$

Section 10.2 (page 607)

1.
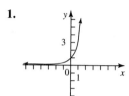
3.
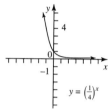

5.
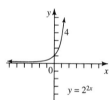
7. {2}   9. $\left\{\dfrac{3}{2}\right\}$
11. $\left\{\dfrac{3}{2}\right\}$ 13. $\left\{-\dfrac{5}{4}\right\}$
15. {−3}   17. { 3}

19. (a) 1,000,000 barrels
(b) 250,000 barrels
(c) 62,500 barrels
(d)
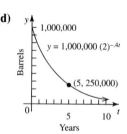

21. $\dfrac{1}{8}$ 23. 6   25. $9\sqrt{3}$ 27. $\dfrac{\sqrt[3]{4}}{4}$

Section 10.3 (page 615)

1. $\log_3 81 = 4$ 3. $\log_2 32 = 5$
5. $\log_{1/2} 4 = -2$ 7. $\log_{3/2}\left(\dfrac{243}{32}\right) = 5$
9. $\log_2\left(\dfrac{1}{16}\right) = -4$ 11. $3^3 = 27$
13. $5^3 = 125$ 15. $2^5 = 32$ 17. $3^{-2} = \dfrac{1}{9}$
19. 3   21. 3   23. −1   25. $\dfrac{1}{2}$ 27. −2
29. $\dfrac{5}{2}$ 31. $\dfrac{3}{2}$ 33. $\dfrac{3}{4}$ 35. $\left\{\dfrac{1}{5}\right\}$ 37. {32}
39. $\left\{\dfrac{1}{10}\right\}$ 41. {8}   43. {4}
45. {$b \mid b$ is a positive real number, $b \neq 1$}

47. $(100, 2), (10, 1),$ $(1, 0), \left(\frac{1}{10}, -1\right),$ $\left(\frac{1}{100}, -2\right), \left(\frac{1}{1000}, -3\right)$

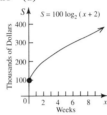

49.

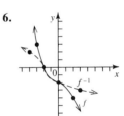

51. (a) 100 thousand dollars (b) 200 thousand dollars (c) 300 thousand dollars (d)

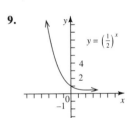

53. 3^6 **55.** 5^{10}
57. $625x^8$ **59.** $8m^7$

Section 10.4 (page 625)

1. $\log_6 3 - \log_6 4$ **3.** $\frac{1}{2} \log_2 8 = \frac{3}{2}$
5. $1 + \log_5 x - \log_5 y$
7. $\log_2 5 + \frac{1}{2} \log_2 7 - \log_2 3$
9. $\frac{1}{2} \log_3 m + \frac{1}{2} \log_3 a - \frac{1}{2} \log_3 b$
11. cannot be rewritten using the properties given in Section 10.4   **13.** $\log_b (xy)$ **15.** $\log_b \left(\frac{m^4}{n}\right)$
17. $\log_b \left(\frac{mr}{n}\right)$ **19.** $\log_b \left(\frac{64}{9}\right)$
21. $\log_{10} (x^2 - 1)$ **23.** $\log_p \left(\frac{25}{12}\right)$ **25.** .6020
27. .9542   **29.** 1.2040   **31.** false   **33.** false
35. $10^{.9542} = 9$ **37.** $10^{2.1461} = 140$
39. $\log_{10} 27 = 1.4314$

Section 10.5 (page 633)

1. 2.4440   **3.** .5159   **5.** .9926
7. −.4855   **9.** −3.1726   **11.** 5.8293
13. 1.6094   **15.** 6.5468   **17.** .5423
19. 4.5809   **21.** 3.1   **23.** 3.3863   **25.** 3.42
27. 454   **29.** .0762   **31.** 47.0   **33.** 4.04
35. .659   **37.** 1.49   **39.** .58   **41.** 1.09
43. 1.61   **45.** .88   **47.** 13.5   **49.** 8.4
51. 1.6×10^{-5} **53.** 4.0×10^{-4}
55. (a) 1,000,000 (b) about 1,040,800 (c) about 1,083,300 (d) about 1,221,400
57. (a) 25,000 (b) about 30,500 (c) about 37,300 (d) about 68,000   **59.** about 2.75 days

61. about 6020 years   **63.** about 1120 years
65. $\left\{\frac{1}{2}\right\}$ **67.** $\{47\}$ **69.** $\{-4\sqrt{3}, 4\sqrt{3}\}$

Section 10.6 (page 643)

1. $\left\{\frac{3}{2}\right\}$ **3.** $\{1.43\}$ **5.** $\{.16\}$ **7.** $\{-3.32\}$
9. $\left\{\frac{1}{2}\right\}$ **11.** $\{-4.75\}$ **13.** $\{152.23\}$ **15.** $\{4\}$
17. $\{2\}$ **19.** $\emptyset$ **21.** $\{8\}$ **23.** $\{\sqrt{11}\}$
25. $\left\{\frac{\sqrt[3]{25}}{25}\right\}$ **27.** $\left\{\frac{1}{2}\right\}$ **29.** about $10,061
31. about $508   **33.** about 23 years
35. about 12 years   **37.** about $4270
39. about 11%   **41.** in 4 years   **43.** about 34

Chapter 10 Review Exercises (page 651)

1. $y = 3x + 1$ **2.** not one-to-one
3. $f^{-1}(x) = \frac{x^2 - 5}{2}, x \geq 0$ **4.** one-to-one
5. not one-to-one   **6.**
7. not one-to-one

8.   **9.**

10.   **11.**

12.   **13.**

14. $3^3 = 27$ **15.** $5^{-1} = \frac{1}{5}$ **16.** $8^{1/3} = 2$

17. $\log_4 64 = 3$ **18.** $\log_9 3 = \frac{1}{2}$
19. $\log_2 \frac{1}{8} = -3$ **20.** $\log_5 3 + \log_5 x - \log_5 10$
21. $2\log_3 m + \log_3 p + \frac{1}{2}\log_3 g$
22. $\log_2 5 + \log_2 k - \log_2 3 - 3\log_2 r$
23. $\log_{10}(2x+3) + \log_{10} 5 + \log_{10} x$ **24.** $\log_a 6x$
25. $\log_b \left(\frac{p^3}{q^2}\right)$ **26.** $\log_5 \left(\frac{1}{x+1}\right)$ **27.** $\log_2 x$
28. .4698 **29.** 2.6355 **30.** -1.1463
31. $-.7959$ **32.** .2 **33.** 0 **34.** 2.0794
35. 3.2189 **36.** 1 **37.** .1 **38.** 4 **39.** 0
40. 1.2851 **41.** .8377 **42.** 1.9459
43. 4.6052 **44.** 1.4706 **45.** .8656
46. 2.5863 **47.** 6.2986 **48.** $\left\{\frac{4}{3}\right\}$ **49.** $\{3\}$
50. $\{.161\}$ **51.** $\left\{\frac{11}{3}\right\}$ **52.** $\{2\}$ **53.** $\left\{\frac{1}{9}\right\}$
54. $\{5\}$ **55.** $\left\{\frac{3}{8}\right\}$ **56.** about $14,026
57. about $4204 **58.** about $17,430
59. about 13.4%

[2.1] **8.** $\left\{-\frac{2}{3}\right\}$ [2.5] **9.** $[1, \infty)$ [2.7] **10.** $\{-2, 7\}$
11. $\left\{-\frac{16}{3}, \frac{16}{3}\right\}$ **12.** $\left[\frac{7}{3}, 3\right]$ **13.** $(-\infty, -3) \cup (2, \infty)$
[3.4] **14.** $6p^2 + 7p - 3$ **15.** $16k^2 - 24k + 9$
[3.3] **16.** $-5m^3 + 2m^2 - 7m + 4$
[3.5] **17.** $2t^3 + 5t^2 - 3t + 4$ [3.7] **18.** $x(8 + x^2)$
[3.8] **19.** $(3y - 2)(8y + 3)$ **20.** $z(5z + 1)(z - 4)$
[3.9] **21.** $(4a + 5b^2)(4a - 5b^2)$
22. $(2c + d)(4c^2 - 2cd + d^2)$ **23.** $(4r + 7q)^2$
[4.1] **24.** $-\frac{1875p^{13}}{8}$ **25.** $\frac{x+5}{x+4}$
[4.3] **26.** $\frac{-3k-19}{(k+3)(k-2)}$ **27.** $\frac{22-p}{p(p-4)(p+2)}$
[2.3] **28.** 6 pounds [3.1] **29.** $\frac{16}{25}$ **30.** $\frac{1}{6^5}$
[5.4] **31.** $-27\sqrt{2}$ [5.7] **32.** 41
[6.1–6.2] **33.** $\left\{-\frac{1}{2}, \frac{2}{5}\right\}$ **34.** $\{2, -5\}$
[6.5] **35.** $(-\infty, -4) \cup (2, \infty)$
[7.2] **36.** $\frac{2}{11}$ [7.3] **37.** $3x - 4y = 19$

[7.1] **38.** [7.4] **39.**

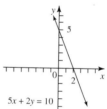
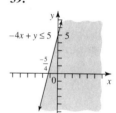

[9.1] **40.** [9.4] **41.**

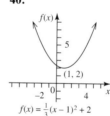
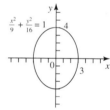

42.

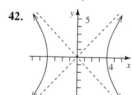

[8.1] **43.** $\{(4, 2)\}$
[8.2] **44.** $\{(1, -1, 4)\}$
[8.4] **45.** -1 **46.** -9
[10.3] **47.** $\log_5 125 = 3$
48. $3^2 = 9$
[10.6] **49.** $\{1\}$ **50.** $\{1\}$

Chapter 10 Test (page 655)

[10.1] **1.** false **2.** $f^{-1}(x) = 4 - x^3$

3. [10.2] **4.**

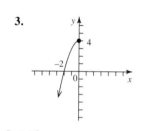
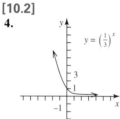

[10.3]
5.

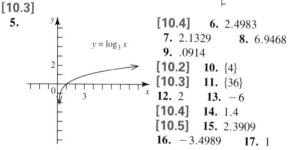

[10.4] **6.** 2.4983
7. 2.1329 **8.** 6.9468
9. .0914
[10.2] **10.** $\{4\}$
[10.3] **11.** $\{36\}$
12. 2 **13.** -6
[10.4] **14.** 1.4
[10.5] **15.** 2.3909
16. -3.4989 **17.** 1
18. 4.4898 **19.** $-.8440$ **20.** 4.1699 **21.** .8114
[10.6] **22.** $\{3.81\}$ **23.** $\{2\}$ **24.** $\{\sqrt[6]{5}\}$ or $\{1.31\}$
25. (a) about 59,100 (b) at 1.7 hours

Final Examination (page 659)

[1.1] **1.** $-2, 0, 6, \frac{30}{3}$ (or 10)
2. $-\frac{9}{4}, -2, 0, .6, 6, \frac{30}{3}$ **3.** $-\sqrt{2}, \sqrt{11}$
4. all except $\sqrt{-8}$ [1.3] **5.** 16 **6.** -27
7. -39

Appendix A (page 669)

1. .3747 **3.** 2.2878 **5.** 1.0792 **7.** 4.3979
9. .7782 − 1 **11.** .3243 − 4 **13.** 344 **15.** 4.57
17. 25.0 **19.** .00120 **21.** .0746 **23.** .000160
25. .9530 − 1 **27.** 1.7315 **29.** 3.7200 **31.** .0980 − 2
33. 517.4 **35.** .6827 **37.** .0001588

SOLUTIONS TO SELECTED EXERCISES

For the answers to all odd-numbered section exercises, all chapter review exercises, and all chapter tests, see the section beginning on page A-1.

If you would like to see more solutions, you may order the *Student's Solutions Manual for Intermediate Algebra* from your college bookstore. It contains solutions to the odd-numbered exercises that do not appear in this section, as well as solutions to all chapter review exercises and chapter tests.

As you are looking at these solutions, remember that many algebraic exercises and problems can be solved in a variety of ways. In this section we provide only one method of solving selected exercises from each section; space does not permit showing other methods that may be equally correct.

If you work the exercise differently but obtain the same answer, then *as long as your steps are mathematically valid,* your work is correct. Mathematical thinking is a creative process, and solving problems in more than one way is an example of this creativity.

CHAPTER 1

Section 1.1 (page 7)

1. Place dots at 2, 3, 4, and 5. See the graph in the answer section.
5. Place dots as needed to indicate the points $-\frac{1}{2}, \frac{3}{4}, \frac{5}{3}$, and $\frac{7}{2}$. See the graph in the answer section.
9. The set of integers is $\{\ldots, -3, -2, -1, 0, 1, 2, 3, \ldots\}$, so the set of integers greater than 11 is $\{12, 13, 14, \ldots\}$.
13. There are no irrational numbers that are also rational, so the set is $\emptyset$.
17. $\{0 \cdot 5, 1 \cdot 5, 2 \cdot 5, 3 \cdot 5, \ldots\} = \{0, 5, 10, 15, \ldots\}$
21. $-|6| = -(6) = -6$
25. $-|16| = -(16) = -16$
29. $|-8| - |3| = 8 - 3 = 5$
33. $|-12| + |-3| = 12 + 3 = 15$
37. Additive inverse of 8 is -8 (change the sign).
41. Additive inverse of 0 is $-0 = 0$.
45. Additive inverse of $|10|$ is $-|10| = -10$.
49. Additive inverse of $-|-1|$ is $|-1| = 1$ (change the sign).
53. The additive inverse of $|-2| + |-3|$ is found by first evaluating the expression $|-2| + |-3| = 2 + 3 = 5$, then changing the sign to get -5.
57. The integers are the set $\{\ldots, -3, -2, -1, 0, 1, 2, 3, \ldots\}$ so the integers in S are $-6, 0, 2, 3,$ and $\frac{10}{2}$ (or 5).
61. Since $\frac{1}{2}$ is a rational number but not an integer, the statement is false.

65. Since the whole numbers are also real numbers, the statement is true.
69. True; $5 = \frac{10}{2}$ is a rational number that is also an integer.
73. $|x| = -|x|$ only if $x = 0$. If x is not zero, then $|x|$ is positive and $-|x|$ is negative and thus $|x|$ cannot equal $-|x|$.

Section 1.2 (page 13)

1. $6 < 10$
5. $r \neq 4$
9. $3 \geq 7y$
13. $5 < x < 9$
17. $1 \leq a < 5$
21. $4 \leq 4$ is read "4 is less than or equal to 4." We know 4 is not less than 4 but it is equal to 4; so the statement is true.
25. True, because $5 + 2 = 7$ and $6 \neq 7$.
29. True, because $5 + 2 = 7$ and $-6 \leq 7$.
33. True, because $3(10) = 30, 6 - 5 = 1,$ and $30 > 1$.
37. True, because $|-5| = 5, |5| = 5,$ and "$\leq$" includes "$=$".
41. False, because $9 \cdot |-6| = 9 \cdot 6 = 54$
    $18 + |36| = 18 + 36 = 54,$ and $54 \not< 54$.

For Exercises 45–53 see the graphs in the answer section.

45. $\{x | x \leq 4\}$ includes all numbers that are less than or equal to 4, written $(-\infty, 4]$. Use a square bracket to show that 4 is included.
49. $\{x | 2 \leq x \leq 5\}$ includes all numbers between 2 and 5 and the numbers 2 and 5, written $[2, 5]$. Use square brackets to show that 2 and 5 are included.

53. $\{x|0 < x \leq 2\}$ includes all numbers greater than 0 and less than or equal to 2, written $(0, 2]$. The parenthesis indicates that 0 is not included and the square bracket indicates that 2 is included.

57. When $-1 < x < 0$ or when $x > 1$, then $\frac{1}{x} < x$.

For example, if $x = -\frac{1}{2}$, then $\frac{1}{-\frac{1}{2}} < -\frac{1}{2}$ is

equivalent to $-2 < -\frac{1}{2}$, which is true. On the

other hand, if $x = \frac{1}{2}$, then $\frac{1}{\frac{1}{2}} < \frac{1}{2}$ is equivalent to

$2 < \frac{1}{2}$, which is false. If $x = 2$, then $\frac{1}{2} < 2$ is true.

Section 1.3 (page 21)

1. $12 + (-3) = 12 - 3 \quad |-3| < |12|$
$= 9$

5. $-12 + (-8) = -(12 + 8)$ Both are negative.
$= -20$

9. $-\frac{7}{3} + \frac{3}{4} = -\frac{28}{12} + \frac{9}{12} = \frac{-28 + 9}{12} = -\frac{19}{12}$

13. $-3 - 9 = -3 + (-9) = -12$

17. $-\frac{3}{8} - \left(-\frac{1}{6}\right) = -\frac{3}{8} + \left(\frac{1}{6}\right)$
$= -\frac{9}{24} + \frac{4}{24} = -\frac{5}{24}$

21. $-\frac{9}{7} + \frac{3}{4} = -\frac{36}{28} + \frac{21}{28} = -\frac{15}{28}$

25. $-18.31 - 7.51 = -18.31 + (-7.51)$
$= -(18.31 + 7.51)$
$= -(25.82)$
$= -25.82$

29. $|-5| + 8 = 5 + 8 = 13$

33. $|-20| + |-9| = 20 + 9 = 29$

37. $6(-7) = -42$
The product of 6 and -7 is negative, because 6 is positive and -7 is negative.

41. $-9(7) = -63$

45. $(-3)(-12)(6) = (36)(6) = 216$

49. $-8\left(\frac{1}{4}\right) = -2$

53. $\frac{5}{8}(-40) = 5\left(\frac{1}{8}\right)(-40) = 5(-5) = -25$

57. $\left(-\frac{5}{2}\right)\left(-\frac{12}{25}\right) = \left(\frac{1}{2}\right)(-5)\left(\frac{1}{25}\right)(-12)$
$= \left(\frac{1}{2}\right)\left(-\frac{1}{5}\right)(-12)$
$= \left(-\frac{1}{5}\right)\left(\frac{1}{2}\right)(-12)$
$= \left(-\frac{1}{5}\right)(-6) = \frac{6}{5}$

61. $-4.7(3.98) = -[(4.7)(3.98)]$
$= -(18.706)$
$= -18.706$

65. The reciprocal of -8 is $\frac{1}{-8}$ or $-\frac{1}{8}$ since
$(-8)\left(-\frac{1}{8}\right) = 1.$

69. The reciprocal of $-\frac{4}{13}$ is $-\frac{13}{4}$ since
$\left(-\frac{4}{13}\right)\left(-\frac{13}{4}\right) = 1.$

73. $\frac{18}{-2} = -9$

77. $\frac{-360}{-18} = 20$

81. $-\frac{3}{4} \div \frac{6}{7} = \frac{-3}{4} \cdot \frac{7}{6} = \frac{-21}{24} = -\frac{7}{8}$

85. $\frac{-1.71}{.9} = -\frac{1.71}{.9} = -1.9$

89. $a - (b - c) = (a - b) - c$
For example, if $a = 3, b = 2, c = 1$, we get
$a - (b - c) = 3 - (2 - 1)$
$= 3 - 1$
$= 2$
and $(a - b) - c = (3 - 2) - 1$
$= 1 - 1$
$= 0.$
Since $2 \neq 0$, the statement is false.

Section 1.4 (page 31)

1. $2^3 = 2 \cdot 2 \cdot 2$ 2 is a factor 3 times.
$= 8$

5. $\left(\frac{1}{5}\right)^2 = \left(\frac{1}{5}\right)\left(\frac{1}{5}\right) = \frac{1}{25}$

9. $(-3)^3 = (-3)(-3)(-3) = -27$

13. $-2^3 = -(2^3) = -(2)(2)(2) = -8$

17. $(.4)^4 = (.4)(.4)(.4)(.4) = .0256$

21. In 5^7 the exponent is 7, and the base is 5.

25. In $(-9)^4$ the exponent is 4, and the base is -9.

29. $\sqrt{784} = 28$, because $784 > 0$ and $28^2 = 784$.

33. $\sqrt{361} = 19$, because $19^2 = 361$.

37. $\sqrt[3]{1000} = 10$, because $10^3 = 1000$.

41. $\sqrt[4]{625} = 5$, because $5^4 = 625$.

45. $\sqrt[6]{729} = 3$, because $3^6 = 729$.

49. $2[-5 - (-7)] = 2(-5 + 7) = 2(2) = 4$

53. $-6 - 5(-8) + 3^2 = -6 + 40 + 9 = 43$

57. $(-8 - 5)(-\sqrt[3]{8} - 1) = (-8 - 5)(-2 - 1)$
  Find the root.
$= (-13)(-3)$
  Work in parentheses.
$= 39$ Multiply.

61. $\frac{-8 + (-7)}{-3} = \frac{-15}{-3}$ Add in numerator.
$= 5$ Divide.

65. $\frac{2(-5) + (-3)(-4)}{-6 + 5 + 1} = \frac{-10 + 12}{0}$
Stop here. This expression is undefined.

69. $\dfrac{-3(-4) - 5(2^2)}{2 - 4 + 10} = \dfrac{-3(-4) - 5(4)}{2 - 4 + 10}$ Use exponents.

$\phantom{\dfrac{-3(-4) - 5(2^2)}{2 - 4 + 10}} = \dfrac{12 - 20}{2 - 4 + 10}$ Multiply in numerator.

$\phantom{\dfrac{-3(-4) - 5(2^2)}{2 - 4 + 10}} = \dfrac{-8}{8}$ Add and subtract.

$\phantom{\dfrac{-3(-4) - 5(2^2)}{2 - 4 + 10}} = -1$ Divide.

73. $5\left[1 + \dfrac{3}{4}(-12) - 8 \cdot \dfrac{3}{2}\right]$

$= 5(1 - 9 - 12)$ Multiply.
$= 5(-20)$ Subtract.
$= -100$ Multiply.

77. $-2(a^2 + 4c)$
$= -2[(-4)^2 + 4(-7)]$ Substitute.
$= -2[16 + 4(-7)]$ Use exponents.
$= -2[16 - 28]$ Multiply inside parentheses.
$= -2(-12)$ Subtract.
$= 24$ Multiply.

81. $-6a^2 + b = -6(-4)^2 + 8$ Substitute.
$= -6(16) + 8$ Use exponents.
$= -96 + 8$ Multiply.
$= -88$ Add.

85. $4(5) - 5 = 15$ and $15 \ne 4$. Multiply before subtracting.

Section 1.5 (page 41)

1. $5k + 3k = (5 + 3)k$ Distributive property
$= 8k$

5. $-8z + 3w$ cannot be simplified since there is no common factor to use with the distributive property.

9. $14c + c = 14c + 1c$ Identity property
$= (14 + 1)c$ Distributive property
$= 15c$ Combine terms.

13. $-5(2d + f) = (-5)(2d) + (-5)(f)$
Distributive property
$= -10d - 5f$ Multiply.

17. $-(3k + 7) = -1(3k + 7)$ Identity property
$= (-1)(3k) + (-1)(7)$
Distributive property
$= -3k - 7$ Multiply.

21. $-12y + 4y - 3 + 2y$
$= -12y + 4y + 2y - 3$
$= -6y - 3$

25. $3(k + 2) - 5k + 6 - 3$
$= 3k + 6 - 5k + 6 - 3$
$= 3k - 5k + 6 + 6 - 3$
$= -2k + 9$

29. $3(2y + 3) - 4(3y - 5)$
$= 6y + 9 - 12y + 20$
$= 6y - 12y + 9 + 20$
$= -6y + 29$

33. $2 + 3(2z - 5) - 3(4z + 6) - 8$
$= 2 + 6z - 15 - 12z - 18 - 8$
$= 2 - 15 - 18 - 8 + 6z - 12z$
$= -6z - 39$

37. $4(-2x - 5) + 6(2 - 3x) - (4 - 2x)$
$= -8x - 20 + 12 - 18x - 4 + 2x$
$= -8x - 18x + 2x - 20 + 12 - 4$
$= -24x - 12$

41. $2(4x) = (2 \cdot 4)x = 8x$

45. $0 + 7 = 7$

49. $0 = 2 + (-2)$

53. $0(3 - 8) = 0$

57. Yes. Any two different nonzero numbers a and b that have the same absolute value satisfy. For example if $a = 3$ and $b = -3$, $\dfrac{3}{-3} = \dfrac{-3}{3} = -1$.

CHAPTER 2

Section 2.1 (page 59)

1. $2k + 6 = 12$
$2k = 12 - 6$
$2k = 6$
$k = 3$
The solution set is $\{3\}$.

5. $3 - 2r = 9$
$3 - 9 = 2r$
$-6 = 2r$
$-\dfrac{6}{2} = r$
$-3 = r$
The solution set is $\{-3\}$.

9. $7m - 2m + 4 - 5 = 3m - 5 + 6$
$5m - 1 = 3m + 1$
$5m - 3m = 1 + 1$
$2m = 2$
$m = 1$
The solution set is $\{1\}$.

13. $2(k - 4) = 5k + 2$
$2k - 8 = 5k + 2$
$2k - 5k = 2 + 8$
$-3k = 10$
$k = -\dfrac{10}{3}$
The solution set is $\left\{-\dfrac{10}{3}\right\}$.

17. $2y + 3(y - 4) = 2(y - 3)$
$2y + 3y - 12 = 2y - 6$
$5y - 12 = 2y - 6$
$5y - 2y = 12 - 6$
$3y = 6$
$y = 2$
The solution set is $\{2\}$.

21. $-[z - (4z + 2)] = 2 + (2z + 7)$
$-z + (4z + 2) = 2 + 2z + 7$
$-z + 4z + 2 = 9 + 2z$
$3z + 2 = 9 + 2z$
$3z - 2z = 9 - 2$
$z = 7$
The solution set is $\{7\}$.

25. $-(9 - 3a) - (4 + 2a) - 3 = -(2 - 5a) + (-a) + 1$
$-9 + 3a - 4 - 2a - 3 = -2 + 5a - a + 1$
$3a - 2a - 9 - 4 - 3 = -2 + 4a + 1$
$a - 16 = 4a - 1$
$1 - 16 = 4a - a$
$-15 = 3a$
$-5 = a$
The solution set is $\{-5\}$.

29. $\dfrac{6x}{5} = -1$
$\dfrac{5}{6} \cdot \dfrac{6x}{5} = \dfrac{5}{6} \cdot (-1)$
$x = -\dfrac{5}{6}$
The solution set is $\left\{-\dfrac{5}{6}\right\}$.

33. $\dfrac{8r}{3} - \dfrac{6r}{5} = 22$
$15\left(\dfrac{8r}{3} - \dfrac{6r}{5}\right) = 15(22)$
$5(8r) - 3(6r) = 330$
$40r - 18r = 330$
$22r = 330$
$r = 15$
The solution set is $\{15\}$.

37. $\dfrac{2p - 3}{4} + \dfrac{p - 1}{4} = \dfrac{p - 4}{2}$
$4\left(\dfrac{2p - 3}{4} + \dfrac{p - 1}{4}\right) = 4\left(\dfrac{p - 4}{2}\right)$
$2p - 3 + p - 1 = 2(p - 4)$
$3p - 4 = 2p - 8$
$3p - 2p = -8 + 4$
$p = -4$
The solution set is $\{-4\}$.

41. $-7m + 8 + 4m = -3(m - 3) - 1$
$-7m + 8 + 4m = -3m + 9 - 1$ Distributive property
$-3m + 8 = -3m + 8$ Combine like terms.
$-3m + 3m + 8 = -3m + 3m + 8$
$8 = 8$
The equation is an identity.
The solution set is {all real numbers}.

45. $3p - 5(p + 4) + 9 = -11 + 15p$
$3p - 5p - 20 + 9 = -11 + 15p$ Distributive property
$-2p - 11 = -11 + 15p$ Combine like terms.
$-2p - 11 + 11 = -11 + 15p + 11$ Add 11.
$-2p = 15p$ Simplify.
$-2p - 15p = 15p - 15p$ Subtract 15p.
$-17p = 0$ Simplify.
$\dfrac{-17p}{-17} = \dfrac{0}{-17}$ Divide by −17.
$p = 0$ Simplify.
The equation is conditional.
The solution set is $\{0\}$.

49. $-3[-5 - (-9 + 2r)] = 2(3r - 1)$
$-3[-5 + 9 - 2r] = 6r - 2$ Distributive property
$-3[4 - 2r] = 6r - 2$ Simplify.
$-12 + 6r = 6r - 2$ Distribute.
$-12 + 6r - 6r = 6r - 2 - 6r$ Subtract 6r.
$-12 = -2$ False
The equation is a contradiction.
The solution set is ∅.

53. $8x - 9 = k - 7$
$8(-4) - 9 = k - 7$ Substitute −4 for x.
$-32 - 9 = k - 7$
$-41 = k - 7$
$k = -34$

57. $x + 1 = 9$
$\dfrac{x + 1}{8} = \dfrac{9}{8}$
Solve the first equation.
$x + 1 = 9$
$x + 1 - 1 = 9 - 1$
$x = 8$
The solution set is $\{8\}$.
Solve the second equation.
$\dfrac{x + 1}{8} = \dfrac{9}{8}$
$8\left(\dfrac{x + 1}{8}\right) = 8\left(\dfrac{9}{8}\right)$
$x + 1 = 9$
$x = 8$
The solution set is $\{8\}$.
Since the solution sets are the same, the equations are said to be equivalent.

61. $p^2 = 25$
$p = 5$
Since $(-5)^2 = 25$ and $5^2 = 25$, the solution set of the first equation is $\{-5, 5\}$, but the solution set of the second equation is $\{5\}$, so the two equations are not equivalent.

65. $prt = (8000)(.12)(2)$
   Substitute 8000 for p, .12 for r, and 2 for t.
$= 1920$

69. $\dfrac{1}{2}bh = \dfrac{1}{2}(21)(5)$ Substitute 21 for b and 5 for h.
$= \dfrac{105}{2}$

Section 2.2 (page 67)

1. $d = rt$ for r
$\dfrac{1}{t} \cdot d = \dfrac{1}{t} \cdot rt$
$\dfrac{d}{t} = r\left(\dfrac{1}{t} \cdot t\right)$
$\dfrac{d}{t} = r$

5. $P = a + b + c$
$P - b - c = a$

9. $$S = 2\pi rh + 2\pi r^2$$
$$S - 2\pi rh = 2\pi h$$
$$\frac{S - 2\pi r^2}{2\pi r} = h \quad \text{or} \quad h = \frac{S}{2\pi r} - r$$

13. $$A = \frac{1}{2}bh$$
$$2A = bh$$
$$\frac{2A}{h} = b$$

17. $$A = \frac{1}{2}h(B + b)$$
$$2A = h(B + b)$$
$$\frac{2A}{h} = B + b$$
$$B = \frac{2A}{h} - b \quad \text{or} \quad B = \frac{2A - bh}{h}$$

21. The formula is $P = 2l + 2w$. Solve for l.
$$P = 2w + 2l$$
$$P - 2w = 2l$$
$$\frac{P - 2w}{2} = l$$
Let $P = 36$ and $w = 3.5$.
$$l = \frac{36 - 2(3.5)}{2} = \frac{36 - 7}{2} = \frac{29}{2} = 14.5$$
The length is 14.5 meters.

25. Use the formula relating Celsius and Fahrenheit that has C alone on one side. It is
$$C = \frac{5}{9}(F - 32).$$
Let $F = 41$ in the formula.
$$C = \frac{5}{9}(41 - 32) = \frac{5}{9}(9) = 5$$
The Celsius temperature is 5°.

29. The formula is $C = 2\pi r$. Solve this formula for r.
$$C = 2\pi r$$
$$\frac{C}{2\pi} = \frac{2\pi r}{2\pi}$$
$$r = \frac{C}{2\pi}$$
Let $C = 240\pi$ in the formula.
$$r = \frac{240\pi}{2\pi} = 120$$
The radius is 120 inches.

33. Use the formula
$$\text{Percent} = \frac{\text{Percentage}}{\text{Amount}}.$$
The percentage is 2 (liters) and the amount is 20 (liters).
$$\text{Percent} = \frac{2}{20} = .10 = 10\%$$
The alcohol concentration is 10%.

37. Use the formula
$$\text{Percentage} = \text{Percent} \times \text{Amount}.$$
The percent is 10% and the amount is $450.
$$\text{Percentage} = 10\%(450) = .10(450) = 45$$
The agent earns a commission of $45 each month.

41. $$4x = 7y - bx + 3$$
$$4x + bx = 7y + 3 \quad \text{Add } bx \text{ to both sides.}$$
$$(4 + b)x = 7y + 3 \quad \text{Factor out } x \text{ on left side.}$$
$$\frac{(4 + b)x}{4 + b} = \frac{7y + 3}{4 + b} \quad \text{Divide both sides by } 4 + b.$$
$$x = \frac{7y + 3}{4 + b}$$

45. $$r = \frac{k - y}{k}$$
$$kr = k - y \quad \text{Multiply both sides by } k.$$
$$kr - k = -y \quad \text{Subtract } k \text{ from both sides.}$$
$$(r - 1)k = -y \quad \text{Factor out } k \text{ on left side.}$$
$$\frac{(r - 1)k}{r - 1} = \frac{-y}{r - 1} \quad \text{Divide both sides by } r - 1.$$
$$k = \frac{-y}{r - 1} \quad \text{or} \quad k = \frac{y}{1 - r}$$

49. $$x + .10x = 220$$
$$100(x + .10x) = 100(220) \quad \text{Multiply both sides by 100.}$$
$$100x + 10x = 22{,}000 \quad \text{Distributive property}$$
$$110x = 22{,}000 \quad \text{Combine terms.}$$
$$x = 200 \quad \text{Divide by 100.}$$
The solution set is {200}.

Section 2.3 (page 77)

1. The phrase "decreased by" indicates subtraction. Since a number is decreased by 9, 9 is subtracted from the number. The phrase translates as $x - 9$.

5. The word "product" indicates multiplication. The phrase translates as $-2x$.

9. The words "increased by" indicate addition, and the word "times" indicates multiplication. The phrase translates as $-1 + 6x$.

13. The word "quotient" indicates division. The unknown number is mentioned first, so it is the numerator. 12 is mentioned second, so it is the denominator. The phrase translates as $\frac{x}{12}$.

17. The word "quotient" indicates division. The numerator is 3 more than a number, or $x + 3$. The denominator is 6. The phrase translates as $\frac{x + 3}{6}$.

21. Let x represent the unknown number. $\frac{5}{12}x$ is subtracted from 7 to give a result of 17, so the equation is
$$7 - \frac{5}{12}x = 17.$$
To solve the equation, start by multiplying both sides by 12.
$$12\left(7 - \frac{5}{12}x\right) = 12(17)$$
$$84 - 5x = 204$$
$$-5x = 204 - 84$$
$$-5x = 120$$
$$x = -24$$
The number is -24.

25. Because there is no equals sign, we have an expression.

29. Let w = the width of the rectangle. Then the length is $2w - 2$. Since $P = 2l + 2w$,
$$38 = 2(2w - 2) + 2w$$
$$38 = 4w - 4 + 2w$$
$$38 = 6w - 4$$
$$42 = 6w$$
$$7 = w.$$
The width is 7 meters and the length is $2(7) - 2 = 12$ meters.

33. Let x represent the number. Then $.12x$ is 12% of x.
$$x - .12x = 44$$
$$.88x = 44$$
$$x = \frac{44}{.88} = 50$$
The number is 50.

37. Let x be the total due the motel. Then $.08x$ represents the amount of the tax. Add these two amounts to get \$286.20.
$$x + .08x = 286.20$$
$$1.08x = 286.20$$
$$x = \frac{286.20}{1.08} = 265$$
The amount due the motel is \$265. Therefore, the amount of the tax is 8% of this, or $.08(\$265) = \$21.20$.

41. Let x represent the amount Evelyn invested at 8%. Then the amount she invested at 14% is represented by $2x - 1000$. Since $I = prt$ and $t = 1$,
Interest at 8% = $.08x$ and
Interest at 14% = $.14(2x - 1000)$.
The total interest is \$580.
$$.08x + .14(2x - 1000) = 580$$
$$.08x + .28x - 140 = 580$$
$$.36x - 140 = 580$$
$$.36x = 720$$
$$x = 2000$$
She invested \$2000 at 8% and $2(\$2000) - \$1000 = \$3000$ at 14%.

45. Let x be the number of liters of 10% solution. Make a table.

Strength	Amount of Solution	Amount of Pure Acid
4%	5	$.04(5) = .2$
10%	x	$.10x$
6%	$5 + x$	$.06(5 + x)$

The amount of pure acid from the 4% solution plus the amount of pure acid from the 10% solution must equal the amount of pure acid in the final (6%) solution. So
$$.2 + .10x = .06(5 + x)$$
$$.2 + .10x = .3 + .06x$$
$$.2 + .04x = .3$$
$$.04x = .1$$
$$x = 2.5 \text{ or } 2\frac{1}{2}$$

$2\frac{1}{2}$ liters of the 10% solution are needed.

49. Let x be the number of liters of 20% solution. Make a table.

Strength	Amount of Solution	Amount of Pure Alcohol
20%	x	$.20x$
12%	6	$.12(6) = .72$
14%	$x + 6$	$.14(x + 6)$

The amount of pure alcohol from the 20% solution plus the amount of pure alcohol from the 12% solution must equal the amount of pure alcohol in the final (14%) solution.
$$.20x + .72 = .14(x + 6)$$
$$.20x + .72 = .14x + .84$$
$$.06x = .12$$
$$x = 2$$
2 liters of the 20% solution are needed.

53. Let x be the number of liters drained from the radiator. Make a table.

Strength	Amount of Solution	Amount of Pure Antifreeze
20%	$20 - x$	$.20(20 - x)$
100%	x	$1x = x$
40%	20	$.40(20)$

The amount of pure antifreeze from the 20% solution plus the amount of pure antifreeze added must equal the amount in the final (40%) solution.
$$.20(20 - x) + x = .40(20)$$
$$4 - .20x + x = 8$$
$$4 + .80x = 8$$
$$.80x = 4$$
$$x = 5$$
5 liters of pure antifreeze should be used.

57. Let x be the number of toll calls. Since each of the toll calls costs 50¢, the total cost of the toll calls is $.5x$ dollars. The cost of local calls plus toll calls is $.5x + 10$ dollars. The tax on this is $.05(.5x + 10) = .025x + .5$ dollars, and so the total bill is $.5x + 10 + .025x + .5$ dollars $= .525x + 10.5$ dollars. Now
$$.525x + 10.5 = 17.85$$
$$.525x = 7.35$$
$$x = 14.$$
There were 14 toll calls.

61. $d = rt$
$1000 = 20t$ Let $d = 1000$ and $r = 20$.
$50 = t$ Divide by 20.

Section 2.4 (page 87)

1. Let p represent the number of pennies in Mary Ann's purse. Then the number of nickels is $30 - x$. Then
   monetary value of pennies $= .01x$, and
   monetary value of nickels $= .05(30 - x)$.
   Since the total monetary value is $.94, add the value of the pennies and the value of the nickels to get .94.
   $$.01x + .05(30 - x) = .94$$
   $$.01x + 1.5 - .05x = .94$$
   $$-.04x = .94 - 1.5$$
   $$-.04x = -.56$$
   $$x = 14$$
   She has 14 pennies and $30 - 14 = 16$ nickels.

5. Let x represent the number of pennies. Then x must also represent the number of dimes. The number of quarters is $40 -$ (total number of other coins), or $40 - 2x$. Then
   monetary value of pennies $= .01x$
   monetary value of dimes $= .10x$
   monetary value of quarters $= .25(40 - 2x)$.
   Add these values to get 8.05.
   $$.01x + .10x + .25(40 - 2x) = 8.05$$
   $$.01x + .10x + 10 - .50x = 8.05$$
   $$-.39x + 10 = 8.05$$
   $$-.39x = 8.05 - 10$$
   $$-.39x = -1.95$$
   $$x = 5$$
   Gary has 5 pennies, 5 dimes, and $40 - 2(5) = 30$ quarters in the box.

9. Let x represent the number of students that attended. Then $300 - x$ represents the number of non-students that attended. Since student tickets cost $1.50 each and non-student tickets cost $3.50 each, the amount of money collected for students was $1.50x$, and the amount collected for non-students was $3.50(300 - x)$. Add these two amounts to get 810.
   $$1.50x + 3.50(300 - x) = 810$$
   $$1.50x + 1050 - 3.50x = 810$$
   $$-2x + 1050 = 810$$
   $$-2x = -240$$
   $$x = 120$$
   Therefore 120 students attended, and $300 - 120 = 180$ non-students attended.

13. Let x be the number of hours it takes for them to be 315 miles apart. Build a table based on $d = rt$.

	d	r	t
Eastbound car	$50x$	50	x
Westbound car	$55x$	55	x

    Since the total distance is to be 315 miles, add $50x$ and $55x$ to get 315.
    $$50x + 55x = 315$$
    $$105x = 315$$
    $$x = 3$$
    It will take them 3 hours to be 315 miles apart.

17. Let x be the time for each part.
    Then $2x$ is the time for the entire trip.
    Build a table based on $d = rt$.

	d	r	t
BR to WC	$10x$	10	x
WC to NO	$15x$	15	x
Combined	100 mi		$2x$

    The distance from   and
    Baton Rouge to
    White Castle
    $\downarrow$ $\downarrow$
    $10x$ $+$

    the distance from   is   the length
    White Castle to   of the entire
    New Orleans   trip.
    $\downarrow$ $\downarrow$ $\downarrow$
    $15x$ $=$ 100

    Solve the equation.
    $$10x + 15x = 100$$
    $$25x = 100$$
    $$x = 4$$
    The entire trip took $2(4) = 8$ hours.

21. Let x be the speed when Maria rides, so that her speed when walking is $x - 10$. Make a chart. Complete the d column from the formula $d = rt$.

	d	r	t
Walking	$\frac{3}{4}(x - 10)$	$x - 10$	$\frac{3}{4}$
Riding	$\frac{1}{4}x$	x	$\frac{1}{4}$

    Since the distance is the same by either method,
    $$\frac{3}{4}(x - 10) = \frac{1}{4}x.$$
    Multiply both sides by 4.
    $$4 \cdot \frac{3}{4}(x - 10) = 4 \cdot \frac{1}{4}x$$
    $$3(x - 10) = 1x$$
    $$3x - 30 = x$$
    $$2x = 30$$
    $$x = 15$$
    Her speed riding is 15 miles per hour.

25. Let x represent Steve's time in hours. Complete a chart.

	d	r	t
Steve	$50x$	50	x
David	$60\left(x - \frac{1}{2}\right)$	60	$x - \frac{1}{2}$
	80		

Since the total distance is 80,
$$50x + 60\left(x - \frac{1}{2}\right) = 80.$$
$$50x + 60x - 30 = 80$$
$$110x - 30 = 80$$
$$110x = 110$$
$$x = 1$$

Steve's time is 1 hour, so David's time is $\frac{1}{2}$ hour and they meet after $\frac{1}{2}$ hour.

See the graphs in the answer section for Exercises 29 and 33.

29. In order to graph $(3, \infty)$, use (at 3 since 3 is not included, and shade to the right on the number line.

33. In order to graph $[3, 9)$, use [at 3 since 3 is included, and use) at 9 since 9 is not included. Shade in between 3 and 9 on the number line.

Section 2.5 (page 97)

See the graphs in the answer section for Exercises 1–49.

1. $4x > 8$
   $\frac{4x}{4} > \frac{8}{4}$
   $x > 2$
   Solution set: $(2, \infty)$

5. $3r + 1 \geq 16$
   $3r \geq 15$
   $r \geq 5$
   Solution set: $[5, \infty)$

9. $-4x < 12$
   $\frac{-4x}{-4} > \frac{12}{-4}$ Reverse the inequality symbol ($<$ becomes $>$).
   $x > -3$
   Solution set: $(-3, \infty)$

13. $-\frac{3}{2}y \leq -\frac{9}{2}$
    $-\frac{2}{3}\left(-\frac{3}{2}y\right) \geq -\frac{2}{3}\left(-\frac{9}{2}\right)$ Change $\leq$ to $\geq$.
    $y \geq 3$
    Solution set: $[3, \infty)$

17. $\frac{2k - 5}{-4} > 1$
    $-4\left(\frac{2k - 5}{-4}\right) < -4(1)$ Change $>$ to $<$.
    $2k - 5 < -4$
    $2k < 1$
    $k < \frac{1}{2}$
    Solution set: $\left(-\infty, \frac{1}{2}\right)$

21. $y + 4(2y - 1) \geq 5y$
    $y + 8y - 4 \geq 5y$
    $9y \geq 5y + 4$
    $9y - 5y \geq 4$
    $4y \geq 4$
    $y \geq 1$
    Solution set: $[1, \infty)$

25. $-3(z - 6) > 2z - 5$
    $-3z + 18 > 2z - 5$
    $18 > 2z + 3z - 5$
    $18 > 5z - 5$
    $18 + 5 > 5z$
    $23 > 5z$
    $\frac{23}{5} > z$
    Solution set: $\left(-\infty, \frac{23}{5}\right)$

29. $-\frac{1}{4}(p + 6) + \frac{3}{2}(2p - 5) < 0$
    Multiply by 4 to clear fractions.
    $(4)\left(-\frac{1}{4}\right)(p + 6) + (4)\left(\frac{3}{2}\right)(2p - 5) < 4(0)$
    $(-1)(p + 6) + 2(3)(2p - 5) < 0$
    $-p - 6 + 6(2p - 5) < 0$
    $-p - 6 + 12p - 30 < 0$
    $11p - 36 < 0$
    $11p < 36$
    $p < \frac{36}{11}$
    Solution set: $\left(-\infty, \frac{36}{11}\right)$

33. $9 < k + 5 < 13$
    $9 - 5 < k + 5 - 5 < 13 - 5$
    $4 < k < 8$
    Solution set: $(4, 8)$

37. $-19 \leq 3x - 5 \leq -9$
    $-14 \leq 3x \leq -4$
    $-\frac{14}{3} \leq x \leq -\frac{4}{3}$
    Solution set: $\left[-\frac{14}{3}, -\frac{4}{3}\right]$

41. $3x - 8 > -5$
    $3x > 3$
    $x > 1$
    Solution set: $(1, \infty)$

45. $3 - 2x \geq -7$
    $-2x \geq -10$
    $x \leq 5$ Divide by -2 and reverse inequality symbol.
    Solution set: $(-\infty, 5]$

49. $-1 \leq 5 - 9x \leq 0$
    $-6 \leq -9x \leq -5$
    $\frac{-6}{-9} \geq \frac{-9x}{-9} \geq \frac{-5}{-9}$
    $\frac{2}{3} \geq x \geq \frac{5}{9}$
    Solution set: $\left[\frac{5}{9}, \frac{2}{3}\right]$

53. Let x represent the number of red lights. Then $2x$ represents the number of green lights. "At least" indicates greater than or equal to, so
$$x + 2x \geq 15$$
$$3x \geq 15$$
$$x \geq 5.$$
The tree has at least 5 red lights.

57. $3(2x - 4) - 4x < 2x + 3$
$$6x - 12 - 4x < 2x + 3$$
$$2x - 12 < 2x + 3$$
$$-12 < 3$$
The final inequality is a true statement. The solution set is {all real numbers}.

See the answer section for the graph in Exercise 61.

61. In order to graph $[-2, \infty)$, use [at -2 since -2 is included, and shade to the right on the number line.

Section 2.6 (page 105)

1. The intersection of sets A and B contains only those elements that are in both A and B: {a, c, e} or set B.

5. The intersection of sets B and C contains the elements that are in both B and C, so a is the only element in $B \cap C$ and $B \cap C = \{a\}$.

9. The union of B and C contains all elements in either B or C or both B and C, so $B \cup C = \{a, c, e, f\}$.

See the graphs in the answer section for Exercises 13–53.

13. $x < 2$ and $x > -1$
The solution set is $(-1, 2)$.

17. $x \leq 3$ and $x \geq 5$ is impossible, so the solution set is $\emptyset$.

21. $x \geq -1$ and $x \leq 2$ is true whenever $-1 \leq x \leq 2$. The solution set is $[-1, 2]$.

25. $3x < -3$ and $x + 2 > 0$
$x < -1$ and $x > -2$
This is true only when $-2 < x < -1$.
The solution set is $(-2, -1)$.

29. $6x - 8 \leq 16$ and $4x - 1 \leq 15$
$6x \leq 24$ and $4x \leq 16$
$x \leq 4$ and $x \leq 4$
The solution set is $(-\infty, 4]$.

33. $x \leq 1$ or $x \geq 4$
The solution set is $(-\infty, 1] \cup [4, \infty)$.

37. $x \leq 1$ or $x \geq 7$
The solution set is $(-\infty, 1] \cup [7, \infty)$.

41. $x \geq -2$ or $x \leq 3$
$[-2, \infty) \cup (-\infty, 3]$ consists of all real numbers.
The solution set is {all real numbers}.

45. $2x + 2 > 6$ or $4x - 1 < 3$
$2x > 4$ or $4x < 4$
$x > 2$ or $x < 1$
The solution set is $(-\infty, 1) \cup (2, \infty)$.

49. $x < 3$ or $x < -2$
$(-\infty, 3) \cup (-\infty, -2) = (-\infty, 3)$
The solution set is $(-\infty, 3)$.

53. $-3x \leq -6$ or $-5x \geq 0$
$x \geq 2$ or $x \leq 0$
The solution set is $(-\infty, 0] \cup [2, \infty)$.

57. $|-8| - |4| - 3 = 8 - 4 - 3$
$= 4 - 3$
$= 1$

Section 2.7 (page 115)

1. $|x| = 8$ gives the two equations $x = 8$ and $x = -8$.
The solution set is $\{8, -8\}$.

5. $|y - 3| = 8$
$y - 3 = 8$ or $y - 3 = -8$
$y = 11$ or $y = -5$
The solution set is $\{-5, 11\}$.

9. $|4r - 5| = 13$
$4r - 5 = 13$ or $4r - 5 = -13$
$4r = 18$ $\qquad$ $4r = -8$
$r = \dfrac{18}{4}$ $\qquad$ $r = -\dfrac{8}{4}$
$r = \dfrac{9}{2}$ $\qquad$ $r = -2$
The solution set is $\left\{-2, \dfrac{9}{2}\right\}$.

13. $\left|\dfrac{1}{2}x + 3\right| = 4$
$\dfrac{1}{2}x + 3 = 4$ or $\dfrac{1}{2}x + 3 = -4$
$\dfrac{1}{2}x = 1$ $\qquad$ $\dfrac{1}{2}x = -7$
$x = 2$ or $x = -14$
The solution set is $\{-14, 2\}$.

17. $|2(3x + 1)| = 4$
$|6x + 2| = 4$
$6x + 2 = 4$ or $6x + 2 = -4$
$6x = 2$ $\qquad$ $6x = -6$
$x = \dfrac{2}{6}$ $\qquad$ $x = -1$
$x = \dfrac{1}{3}$
The solution set is $\left\{-1, \dfrac{1}{3}\right\}$.

See the graphs in the answer section for Exercises 21–53.

21. $|k| \geq 3$
$k \geq 3$ or $k \leq -3$
The solution set is $(-\infty, -3] \cup [3, \infty)$.

25. $|3x - 1| \geq 5$
$3x - 1 \geq 5$ or $3x - 1 \leq -5$
$3x \geq 6$ $\qquad$ $3x \leq -4$
$x \geq 2$ $\qquad$ $x \leq -\dfrac{4}{3}$
The solution set is $\left(-\infty, -\dfrac{4}{3}\right] \cup [2, \infty)$.

29. $|x| \leq 2$
$-2 \leq x \leq 2$
The solution set is $[-2, 2]$.

33. $|t + 2| \leq 8$
$-8 \leq t + 2 \leq 8$
$-8 - 2 \leq t + 2 - 2 \leq 8 - 2$
$-10 \leq t \leq 6$
The solution set is $[-10, 6]$.

37. $|3 - x| \leq 4$
$-4 \leq 3 - x \leq 4$
$-4 - 3 \leq 3 - x - 3 \leq 4 - 3$
$-7 \leq -x \leq 1$
$7 \geq x \geq -1$ Multiply by -1 and reverse inequality.
The solution set is $[-1, 7]$.

41. $|7 + 2z| = 3$
$7 + 2z = 3 \quad$ or $\quad 7 + 2z = -3$
$2z = -4 \qquad\qquad 2z = -10$
$z = -2 \qquad\qquad z = -5$
The solution set is $\{-5, -2\}$.

45. $|6(x + 1)| < 1$
$|6x + 6| < 1$
$-1 < 6x + 6 < 1$
$-1 - 6 < 6x + 6 - 6 < 1 - 6$
$-7 < 6x < -5$
$-\dfrac{7}{6} < x < -\dfrac{5}{6}$
The solution set is $\left(-\dfrac{7}{6}, -\dfrac{5}{6}\right)$.

49. $|x + 4| + 1 = 3$
$|x + 4| = 2 \quad$ Subtract 1 from both sides.
$x + 4 = 2 \quad$ or $\quad x + 4 = -2$
$x = -2 \qquad\qquad x = -6$
The solution set is $\{-6, -2\}$.

53. $|x + 5| - 6 \leq -2$
$|x + 5| \leq 4 \quad$ Add 6 to both sides.
$-4 \leq x + 5 \leq 4$
$-9 \leq x \leq -1$
The solution set is $[-9, -1]$.

57. $\left|m - \dfrac{1}{2}\right| = \left|\dfrac{1}{2}m - 2\right|$
$m - \dfrac{1}{2} = \dfrac{1}{2}m - 2$
$2m - 1 = m - 4 \quad$ Multiply both sides by 2.
$2m - m = -4 + 1$
$m = -3$

or $\quad m - \dfrac{1}{2} = -\left(\dfrac{1}{2}m - 2\right)$
$m - \dfrac{1}{2} = -\dfrac{1}{2}m + 2$
$2m - 1 = -m + 4 \quad$ Multiply both sides by 2.
$2m + m = 4 + 1$
$3m = 5$
$m = \dfrac{5}{3}$
The solution set is $\left\{-3, \dfrac{5}{3}\right\}$.

61. $|2p - 6| = |2p + 5|$
$2p - 6 = 2p + 5 \quad$ or $\quad 2p - 6 = -(2p + 5)$
$-6 = 5 \qquad\qquad\qquad 2p - 6 = -2p - 5$
Subtract $2p$. $\qquad\qquad\qquad 2p + 2p = -5 + 6$
This part has no $\qquad\qquad\qquad 4p = 1$
solution, as indicated by $\qquad\qquad p = \dfrac{1}{4}$
the false statement
$-6 = 5$.
The solution set is $\left\{\dfrac{1}{4}\right\}$.

65. $|12t - 3| = -4$
Since the absolute value of an expression can never be negative, this equation has no solution. The solution set is $\emptyset$.

69. $|2q - 1| < -5$
The absolute value of an expression can never be negative, and thus can never be less than -5. The solution set is $\emptyset$.

73. $|7x - 3| \leq 0$
The absolute value of $7x - 3$ can never be negative. However, it can be 0; we must solve the following equation:
$7x - 3 = 0$
$7x = 3$
$x = \dfrac{3}{7}$.
The solution set is $\left\{\dfrac{3}{7}\right\}$.

77. $|10z + 9| > 0$
The absolute value of $10z + 9$ will be greater than 0 in all cases *except* when $10x + 9 = 0$.
$10x + 9 = 0$
$10x = -9$
$x = -\dfrac{9}{10}$
The solution set consists of all real numbers except $-\dfrac{9}{10}$, or $\left(-\infty, -\dfrac{9}{10}\right) \cup \left(-\dfrac{9}{10}, \infty\right)$.

81. Let x be any such number.
..., the result is at least 4.
$\qquad |x| \qquad \geq \qquad 4$
$|x| \geq 4$ means $x \leq -4$ or $x \geq 4$. The numbers that satisfy the statement are all numbers less than or equal to -4, or greater than or equal to 4.

85. $2^3 = (2)(2)(2) = 8$

89. $-5^2 = -(5 \cdot 5) = -25$

Summary on Solving Linear Equations and Inequalities (page 119)

1. $4z + 1 = 53$
$4z = 52$
$z = 13$
Solution set: $\{13\}$

5. $|a + 3| = 7$
$a + 3 = 7 \quad$ or $\quad a + 3 = -7$
$a = 4 \qquad$ or $\qquad a = -10$
Solution set: $\{-10, 4\}$

9. $2q - 1 = -5$
$2q = -4$
$q = -2$
Solution set: $\{-2\}$

13. $9y - 3(y + 1) = 8y - 3$
$9y - 3y - 3 = 8y - 3$
$6y - 3 = 8y - 3$
$-3 = 2y - 3$
$0 = 2y$
$0 = y$
Solution set: $\{0\}$

17. $|q| > 1$ is equivalent to $q < -1$ or $q > 1$.
Solution set: $(-\infty, -1) \cup (1, \infty)$

21. $\frac{1}{4}p < -3$
$4\left(\frac{1}{4}p\right) < 4(-3)$
$p < -12$
Solution set: $(-\infty, -12)$

25. $3r + 9 + 5r = 4(3 + 2r) - 3$
$8r + 9 = 12 + 8r - 3$
$8r + 9 = 8r + 9$
Solution set: {all real numbers}

29. $|5a + 1| < 6$ is equivalent to $-6 < 5a + 1 < 6$.
$-7 < 5a < 5$
$-\frac{7}{5} < a < 1$
Solution set: $\left(-\frac{7}{5}, 1\right)$

33. $|7z - 1| = |5z + 1|$
$7z - 1 = 5z + 1$ or $7z - 1 = -(5z + 1)$
$2z - 1 = 1$ $\qquad\qquad 7z - 1 = -5z - 1$
$2z = 2$ $\qquad\qquad\quad 12z - 1 = -1$
$z = 1$ $\qquad\qquad\qquad 12z = 0$
$\qquad\qquad\qquad\qquad\quad z = 0$
Solution set: $\{0, 1\}$

37. $-(m - 4) + 2 = 3m - 8$
$-m + 4 + 2 = 3m - 8$
$-m + 6 = 3m - 8$
$-4m + 6 = -8$
$-4m = -14$
$m = \frac{7}{2}$
Solution set: $\left\{\frac{7}{2}\right\}$

41. $|y - 1| \geq -2$
Since absolute value is always nonnegative, $|y - 1| \geq -2$ is true for any real number y.
Solution set: {all real numbers}

45. $|r - 5| = |r + 7|$
$r - 5 = r + 7$ or $r - 5 = -(r + 7)$
$-5 = 7$ $\qquad\qquad r - 5 = -r - 7$
No solution, as $\qquad 2r - 5 = -7$
indicated by $\qquad\qquad 2r = -2$
the false statement $\qquad r = -1$
Solution set: $\{-1\}$

CHAPTER 3

Section 3.1 (page 139)

1. In 5^7, the exponent is 7 and the base is 5.

5. In -9^4, the exponent is 4 and since -9 is not in parentheses, the base is 9.

9. $-3q^{-4}$ has exponent -4 and since $-3q$ is not in parentheses, the exponent affects only q. Thus, the base is q.

13. $5^4 = 5 \cdot 5 \cdot 5 \cdot 5 = 625$

17. $7^{-2} = \frac{1}{7^2} = \frac{1}{49}$

21. $-(-3)^{-4} = -\frac{1}{(-3)^4} = -\frac{1}{81}$

25. $\frac{1}{5^{-2}} = \frac{1}{\frac{1}{5^2}} = \frac{1}{\frac{1}{25}} = 25$

29. $\frac{2^{-3}}{3^{-2}} = \frac{3^2}{2^3} = \frac{3 \cdot 3}{2 \cdot 2 \cdot 2} = \frac{9}{8}$

33. $\left(\frac{2}{3}\right)^{-2} = \frac{1}{\left(\frac{2}{3}\right)^2} = \frac{1}{\frac{2^2}{3^2}} = \frac{1}{\frac{4}{9}} = \frac{9}{4}$

37. $2^6 \cdot 2^{10} = 2^{6+10}$ Product rule
$= 2^{16}$ Add exponents.

41. $\frac{3^5}{3^2} = 3^{5-2}$ Quotient rule
$= 3^3$

45. $\frac{3^{-5}}{3^{-2}} = 3^{-5-(-2)}$ Quotient rule
$= 3^{-5+2} = 3^{-3} = \frac{1}{3^3}$

49. $t^5 t^{-12} = t^{5+(-12)}$ Product rule
$= t^{-7} = \frac{1}{t^7}$

53. $a^{-3} a^2 a^{-4} = a^{-3+2+(-4)}$ Product rule
$= a^{-5} = \frac{1}{a^5}$

57. $\frac{r^3 r^{-4}}{r^{-2} r^{-5}} = \frac{r^{3+(-4)}}{r^{-2+(-5)}}$ Product rule
$= \frac{r^{-1}}{r^{-7}}$
$= r^{-1-(-7)}$
$= r^{-1+7} = r^6$ Quotient rule

61. $8^0 = 1$ Zero exponent

65. $-2^0 = -1$
In -2^0, the absence of parentheses means that 2^0 is evaluated first, then the negative symbol is applied to the result.

69. $(x + y)^{-1} = x^{-1} + y^{-1}$
$(2 + 3)^{-1} = 2^{-1} + 3^{-1}$ Substitute 2 for x, 3 for y.
$5^{-1} = \frac{1}{2} + \frac{1}{3}$
$\frac{1}{5} \neq \frac{5}{6}$
The statement is false.

73. $(-2)^3 \cdot (-2)^3 = (-2)(-2)(-2) \cdot (-2)(-2)(-2)$
$= (-2)^6$
$= 2^6$

Section 3.2 (page 147)

1. $\left(\dfrac{3}{4}\right)^{-2} = \left(\dfrac{4}{3}\right)^2 \quad \dfrac{4}{3}$ is reciprocal of $\dfrac{3}{4}$.

5. $(3^{-2} \cdot 4^{-1})^3 = 3^{(-2)(3)} 4^{(-1)(3)}$
$\qquad\qquad\qquad$ Power rule
$= 3^{-6} \cdot 4^{-3}$
$= \dfrac{1}{3^6} \cdot \dfrac{1}{4^3}$ Negative exponents
$= \dfrac{1}{3^6 \cdot 4^3}$

9. $(z^3)^{-2} z^2 = z^{(3)(-2)} z^2$ Power rule
$= z^{-6} z^2$
$= z^{-6+2}$ Product rule
$= z^{-4}$
$= \dfrac{1}{z^4}$ Negative exponent

13. $(3a^{-2})^3 (a^3)^{-4} = 3^3 a^{-6} a^{-12}$ Power rule
$= 3^3 a^{-18} = \dfrac{3^3}{a^{18}}$ Negative exponent

17. $(2^1 p^2 q^{-3})^2 (4^1 p^{-3} q^1)^2$
$= 2^{(1)(2)} p^{(2)(2)} q^{(-3)(2)} \cdot 4^{(1)(2)} p^{(-3)(2)} q^{(1)(2)}$
$\qquad\qquad\qquad$ Power rule
$= 2^2 p^4 q^{-6} \cdot 4^2 p^{-6} q^2$
$= (2^2 \cdot 4^2)(p^4 \cdot p^{-6})(q^{-6} q^2)$
$= (2^2 \cdot 4^2)(p^{4+(-6)})(q^{-6+2})$
$\qquad\qquad\qquad$ Product rule
$= 2^2 \cdot 4^2 \cdot p^{-2} q^{-4}$
$= 2^2 \cdot 4^2 \cdot \dfrac{1}{p^2} \cdot \dfrac{1}{q^4}$ Negative exponents
$= \dfrac{2^2 \cdot 4^2}{p^2 q^4}$

21. $\dfrac{4a^5 (a^{-1})^3}{(a^{-2})^{-2}} = \dfrac{4a^5 \cdot a^{(-1)(3)}}{a^{(-2)(-2)}}$ Power rule
$= \dfrac{4a^5 \cdot a^{-3}}{a^4}$
$= \dfrac{4a^{5+(-3)}}{a^4}$ Product rule
$= \dfrac{4a^2}{a^4}$
$= 4a^{2-4}$ Quotient rule
$= 4a^{-2}$
$= \dfrac{4}{a^2}$

25. $\dfrac{(2k)^2 m^{-5}}{(km)^{-3}} = \dfrac{2^2 k^2 m^{-5}}{k^{-3} m^{-3}}$ Power rule
$= 2^2 k^{2-(-3)} m^{-5-(-3)}$
$\qquad\qquad$ Quotient rule
$= 2^2 k^5 m^{-2}$
$= \dfrac{2^2 k^5}{m^2}$

29. $\dfrac{(2k)^2 k^3}{k^{-1} k^{-5}} (5k^{-2})^{-3} = \dfrac{2^2 k^2 k^3}{k^{-1} k^{-5}} \cdot \dfrac{5^{-3} k^6}{1}$
$\qquad\qquad\qquad$ Power rule
$= \dfrac{2^2 k^{2+3+6}}{5^3 k^{(-1)+(-5)}}$ Product rule
$= \dfrac{2^2 k^{11}}{5^3 k^{-6}}$
$= \dfrac{2^2 k^{11-(-6)}}{5^3}$ Quotient rule
$= \dfrac{2^2 k^{17}}{5^3}$

33. $\left(\dfrac{2p}{q^2}\right)^3 \left(\dfrac{3p^4}{q^{-4}}\right)^{-1} = \dfrac{2^3 p^3}{q^6} \cdot \dfrac{3^{-1} p^{-4}}{q^4}$ Power rule
$= \dfrac{2^3 p^{-1}}{3^1 q^{10}}$ Product rule
$= \dfrac{2^3}{3 p q^{10}}$

37. $230 = 2_\wedge 30 \quad$ Place caret to right of 2, count 2 places (left to right) to decimal point.
$= 2.30 \times 10^2 \quad$ Number gets smaller, exponent positive.

41. $.000003_\wedge 27 \quad$ Place caret to the right of 3, count 6 places to decimal point.
$3.27 \times 10^{-6} \quad$ Number gets larger, exponent negative.

45. $6.5 \times 10^3 = 6500 \quad$ Move decimal point 3 places to the right for a positive exponent.

49. $5 \times 10^{-3} = 005 \quad$ Move decimal point 3 places to the left for a negative exponent.

53. $\dfrac{9 \times 10^2}{3 \times 10^6} = \dfrac{9}{3} \times 10^{2-6} = 3 \times 10^{-4} = .0003$

57. $\dfrac{.002 \times 3900}{.000013} = \dfrac{(2 \times 10^{-3}) \times (3.9 \times 10^3)}{1.3 \times 10^{-5}}$
$= \dfrac{7.8 \times 10^0}{1.3 \times 10^{-5}} = 6 \times 10^5$
$= 600,000$

61. $\dfrac{840,000 \times .03}{0.00021 \times 600} = \dfrac{(8.4 \times 10^5) \times (3 \times 10^{-2})}{(2.1 \times 10^{-4}) \times (6 \times 10^2)}$
$= \dfrac{(8.4)(3)}{(2.1)(6)} \times \dfrac{10^{5+(-2)}}{10^{-4+2}}$
$= 2 \times \dfrac{10^3}{10^{-2}}$
$= 2 \times 10^5$
$= 200,000$

65. Multiply 2×10^{-7} by 10^{12}:
$2 \times 10^{-7} \times 10^{12} = 2 \times 10^5$ seconds.
Since there are 3600 seconds in 1 hour, divide 2×10^5 by 3600 to get the answer in hours.
$\dfrac{2 \times 10^5}{3600} = \dfrac{2 \times 10^5}{3.6 \times 10^3} = .556 \times 10^2$
$= 55.6$ hours (rounded)

69. $7x - (5 + 5x)$
$= 7x - 5 - 5x$ Distributive property
$= 2x - 5$

Section 3.3 (page 155)

1. The numerical factor in $9k$ is 9, so the coefficient is 9.

5. The numerical factor in $-y = -1 \cdot y$ is -1, so the coefficient is -1.

9. $8x + 9x^2 + 3x^3 = 3x^3 + 9x^2 + 8x$
The largest exponent is 3 and the smallest is 1, since $8x = 8x^1$.

13. In $4k - 9$ there are two terms, $4k$ and -9. Thus it is a binomial. The exponent of the variable $k = k^1$ is 1, so the degree of the polynomial is 1.

17. In $a^3 - a^2 + a^4$ there are three terms, a^3, $-a^2$, and a^4. So it is a trinomial. The degrees of the terms are 4, 3, and 2, so the degree of the polynomial is 4.

21. In $-5x^3 + 8x^2 - 5x + 2$ there are four terms. So the polynomial is not a monomial, binomial, or trinomial. The degrees of the terms are 3, 2, 1, and 0, so the degree of the polynomial is 3.

25. $12z^5 + 8z^5 = (12 + 8)z^5 = 20z^5$

29. $z - z + z - z = (1 - 1 + 1 - 1)z = 0z = 0$

33. $8r + 3r^2 - 5r^2 - 2r$
$= 8r - 2r + 3r^2 - 5r^2$ Group terms
$= (8 - 2)r + (3 - 5)r^2$ of same degree.
$= -2r^2 + 6r$

37. $P(x) = 3x + 3$
(a) Substitute -1 for x to get
$P(-1) = 3(-1) + 3 = -3 + 3 = 0$.
(b) Substitute 2 for x to get
$P(2) = 3(2) + 3 = 6 + 3 = 9$.

41. $P(x) = x^4 - 3x^2 + 2x$
(a) Substitute -1 for x to get
$P(-1) = (-1)^4 - 3(-1)^2 + 2(-1)$
$= 1 - 3 \cdot 1 - 2$
$= 1 - 3 - 2 = -4$.
(b) Substitute 2 for x to get
$P(2) = (2)^4 - 3(2)^2 + 2 \cdot 2$
$= 16 - 3 \cdot 4 + 4$
$= 16 - 12 + 4 = 8$.

45. Subtract.
$12a + 15$
$\underline{7a - 3}$
Add the opposite.
$12a + 15$
$\underline{-7a + 3}$ All signs changed
$5a + 18$ Answer

49. $12z^2 - 11z + 8$ Add
$5z^2 + 16z - 2$ vertically.
$\underline{-4z^2 + 5z - 9}$
$13z^2 + 10z - 3$

53. $12k^5 - 9k^4$ Add
$\underline{8k^4 + 5k - 6}$ vertically.
$12k^5 - k^4 + 5k - 6$

57. $6x^5 - 9x^4 + x^3$ Subtraction;
$\underline{-(2x^4 - 3x^3 + x)}$ change signs
becomes   in bottom row.
$6x^5 - 9x^4 + x^3$
$\underline{ - 2x^4 + 3x^3 - x}$
$6x^5 - 11x^4 + 4x^3 - x$

61. $(5z + 6) - (8z - 3)$
$= 5z + 6 - 8z + 3$
$= 5z - 8z + 6 + 3$ Group like terms.
$= -3z + 9$

65. $(-2m^3 + 5m^2 - 6) - (3m^3 - 2m^2 + 5)$
Subtraction; change signs in second polynomial and add.
$-2m^3 + 5m^2 - 6 - 3m^3 + 2m^2 - 5$
$= -2m^3 - 3m^3 + 5m^2 + 2m^2 - 6 - 5$
$= -5m^3 + 7m^2 - 11$

69. $(-2z^3 - 5z^2 - 8z) + (-4z^3 + 2z^2 - 4)$
Group like terms and add.
$-2z^3 - 4z^3 - 5z^2 + 2z^2 - 8z - 4$
$= -6z^3 - 3z^2 - 8z - 4$

73. $-2x^2 + 5x - 4$
$\underline{-(x^2 - 7x + 2)}$ Change signs and add.
becomes
$-2x^2 + 5x - 4$
$\underline{-x^2 + 7x - 2}$
$-3x^2 + 12x - 6$

77. $8r^2(6r^4) = 8 \cdot 6 \cdot r^2 \cdot r^4$
$= 48r^{2+4}$ Product rule
$= 48r^6$

81. $-9m(-4mn^4) = -9 \cdot (-4)m^1 m^1 n^4 = 36m^2 n^4$

Section 3.4 (page 165)

1. $3p(5p) = 3 \cdot 5 \cdot p \cdot p = 15p^{1+1} = 15p^2$

5. $3(5a - 9) = 3 \cdot 5a - 3 \cdot 9 = 15a - 27$

9. $8a^2(a + 5) = 8a^2 \cdot a + 8a^2 \cdot 5$
$= 8a^3 + 40a^2$

13. $-3z^2(2z^2 + 5z - 1)$
$= -3(2)z^2 \cdot z^2 + (-3)(5)z^2 \cdot z$
$ + (-3)(-1)z^2$
$= -6z^4 - 15z^3 + 3z^2$

17. $2y^3(6y^2 - 4y^3 + 8y - 5)$
$= 2(6)y^3 \cdot y^2 + 2(-4)y^3 \cdot y^3$
$ + 2(8)y^3 \cdot y + 2(-5)y^3$
$= 12y^5 - 8y^6 + 16y^4 - 10y^3$

21. $8a - 5$
$\underline{2a + 1}$
$8a - 5$ $\leftarrow 1 \cdot (8a - 5)$
$\underline{16a^2 - 10a}$ $\leftarrow 2a \cdot (8a - 5)$
$16a^2 - 2a - 5$ $\leftarrow$ Add.

25. $2k - 5z$
$\underline{3k + 2z}$
$4kz - 10z^2$ $\leftarrow 2z \cdot (2k - 5z)$
$\underline{6k^2 - 15kz}$ $\leftarrow 3k \cdot (2k - 5z)$
$6k^2 - 11kz - 10z^2$ $\leftarrow$ Add.

29. $5k^3 + 2k^2 - 3k - 4$
$\underline{2k + 5}$
$25k^3 + 10k^2 - 15k - 20$
$\underline{10k^4 + 4k^3 - 6k^2 - 8k}$
$10k^4 + 29k^3 + 4k^2 - 23k - 20$

33.
$$\begin{array}{r} 2k^2 + 6k + 5 \\ -\ k^2 + 2k \\ \hline 4k^3 + 12k^2 + 10k \\ -2k^4 - 6k^3 - 5k^2 \\ \hline -2k^4 - 2k^3 + 7k^2 + 10k \end{array}$$

37. $(3a - 2)(2a + 1)$
$= 6a^2 + 3a - 4a - 2$
$= 6a^2 - a - 2$

41. $(6a - 5)(3a + 2)$
$= 18a^2 + 12a - 15a - 10$
$= 18a^2 - 3a - 10$

45. $(r - s)(2r + s)$
$= 2r^2 + rs - 2rs - s^2$
$= 2r^2 - rs - s^2$

49. $\left(k - \dfrac{1}{2}\right)\left(k + \dfrac{2}{3}\right)$
$= k^2 + \dfrac{2}{3}k - \dfrac{1}{2}k - \dfrac{1}{3}$
$= k^2 + \dfrac{1}{6}k - \dfrac{1}{3}$
since $\dfrac{2}{3} - \dfrac{1}{2} = \dfrac{4}{6} - \dfrac{3}{6} = \dfrac{1}{6}$

53. $(6a - 1)(6a + 1) = (6a)^2 - (1)^2$
$= 36a^2 - 1$

57. $(5r - 3w)(5r + 3w) = (5r)^2 - (3w)^2$
$= 25r^2 - 9w^2$

61. $\left(r - \dfrac{3}{4}\right)\left(r + \dfrac{3}{4}\right) = (r)^2 - \left(\dfrac{3}{4}\right)^2$
$= r^2 - \dfrac{9}{16}$

65. $(k - 8)^2 = k^2 - 2(k)(8) + 8^2 = k^2 - 16k + 64$

69. $(6p - 5q)^2$
$= (6p)^2 - 2(6p)(5q) + (5q)^2$
$= 36p^2 - 60pq + 25q^2$

73. $\left(r + \dfrac{2}{3}s\right)^2 = (r)^2 + 2\left(\dfrac{2}{3}\right)(r)(s) + \left(\dfrac{2}{3}s\right)^2$
$= r^2 + \dfrac{4}{3}rs + \dfrac{4}{9}s^2$

77. $[(4k + h) - 4]^2$
$= (4k + h)^2 - 2(4k + h)(4) + (4)^2$
$= (4k)^2 + 2(4k)(h) + h^2 - 8(4k + h)$
$\quad + 16$
$= 16k^2 + 8kh + h^2 - 32k - 8h + 16$

81. $[(3m - y) + z][(3m - y) - z]$
$= (3m - y)^2 - z^2$
$= [(3m)^2 - 2(3m)(y) + y^2] - z^2$
$= 9m^2 - 6my + y^2 - z^2$

85. $\dfrac{12p^7}{6p^3} = \dfrac{12}{6}p^{7-3} = 2p^4$

89. $-4p^2 - 8p + 5$
$-(3p^2 + 2p + 9)$
Change signs in the bottom row and add.
$$\begin{array}{r} -4p^2 - 8p + 5 \\ -3p^2 - 2p - 9 \\ \hline -7p^2 - 10p - 4 \end{array}$$

Section 3.5 (page 173)

1. $\dfrac{5a^2 - 10a + 5}{5} = \dfrac{5a^2}{5} - \dfrac{10a}{5} + \dfrac{5}{5}$
$= a^2 - 2a + 1$

5. $\dfrac{12a^3 + 8a^2 - 10a}{4a^3} = \dfrac{12a^3}{4a^3} + \dfrac{8a^2}{4a^3} - \dfrac{10a}{4a^3}$
$= 3 + \dfrac{2}{a} - \dfrac{5}{2a^2}$

9. $\dfrac{3abc^2 - 6ab^2c + 9a^2bc}{15abc}$
$= \dfrac{3abc^2}{15abc} - \dfrac{6ab^2c}{15abc} + \dfrac{9a^2bc}{15abc}$
$= \dfrac{c}{5} - \dfrac{2b}{5} + \dfrac{3a}{5}$

13.
$$\begin{array}{r} r + 3 \\ 2r + 7\overline{\smash{)}2r^2 + 13r + 21} \\ \underline{2r^2 + 7r} \\ 6r + 21 \\ \underline{6r + 21} \\ 0 \end{array}$$

17.
$$\begin{array}{r} a^2 - 5a - 9 \\ a - 3\overline{\smash{)}a^3 - 8a^2 + 6a - 3} \\ \underline{a^3 - 3a^2} \\ -5a^2 + 6a \\ \underline{-5a^2 + 15a} \\ -9a - 3 \\ \underline{-9a + 27} \\ -30 \end{array}$$
Answer: $a^2 - 5a - 9 + \dfrac{-30}{a - 3}$

21.
$$\begin{array}{r} 4p + 1 \\ 2p^2 + 3\overline{\smash{)}8p^3 + 2p^2 + p + 18} \\ \underline{8p^3 + 12p} \\ 2p^2 - 11p + 18 \\ \underline{2p^2 + 3} \\ -11p + 15 \end{array}$$
Answer: $4p + 1 + \dfrac{-11p + 15}{2p^2 + 3}$

25.
$$\begin{array}{r} 9t^2 - 4t + 1 \\ t^2 - t + 2\overline{\smash{)}9t^4 - 13t^3 + 23t^2 - 9t + 2} \\ \underline{9t^4 - 9t^3 + 18t^2} \\ -4t^3 + 5t^2 - 9t \\ \underline{-4t^3 + 4t^2 - 8t} \\ t^2 - t + 2 \\ \underline{t^2 - t + 2} \\ 0 \end{array}$$
Answer: $9t^2 - 4t + 1$

29.
$$\begin{array}{r} (3/4)a - 2 \\ 4a + 3\overline{\smash{)}3a^2 - (23/4)a - 5} \\ \underline{3a^2 + (9/4)a} \\ -8a - 5 \\ \underline{-8a - 6} \\ 1 \end{array}$$

A-42 SOLUTIONS TO SELECTED EXERCISES

Answer: $(3/4)a - 2 + 1/(4a + 3)$

33. $2m + 9 \overline{\smash{\big)}\begin{array}{r} m^2 + 3m - 7 \\ 2m^3 + 15m^2 + 13m - 63 \end{array}}$ Use $\dfrac{d}{t} = r$.
$$\begin{array}{r} \underline{2m^3 + 9m^2} \\ 6m^2 + 13m \\ \underline{6m^2 + 27m} \\ -14m - 63 \\ \underline{-14m - 63} \\ 0 \end{array}$$

Answer: $m^2 + 3m - 7$ kilometers per hour.

37. $2x^3 + x^2 - 5x - 4; x = -2$
Substitute -2 for x; use parentheses.
$2(-2)^3 + (-2)^2 - 5(-2) - 4$
$= 2(-8) + 4 + 10 - 4$
$= -16 + 4 + 10 - 4$
$= -6$

Section 3.6 (page 179)

1. $\dfrac{x^2 + 6x - 7}{x - \boxed{1}}$

$1\overline{\smash{\big)}\,1 \quad +6 \quad -7}$
Bring down $\downarrow \quad +1 \quad +7$ Copy coefficients of numerator.
$\boxed{1} \quad 7 \quad 0$

Coefficient of x

Answer: $\dfrac{x^2 + 6x - 7}{x - 1} = x + 7$

5. $\dfrac{3a^2 + 10a + 11}{a + 2}$

$\boxed{-2}\overline{\smash{\big)}\,3 \quad +10 \quad +11}$
$\,\, \downarrow \quad \boxed{-6} \quad -8$
$\,\, \boxed{3} \quad +4 \quad +3$
Multiply. Add. ↑ Remainder

Thus, $\dfrac{3a^2 + 10a + 11}{a + 2} = 3a + 4 + \dfrac{3}{a + 2}$.

9. $\dfrac{4a^3 - 3a^2 + 2a + 1}{a - 1}$: $1\overline{\smash{\big)}\,4 \quad -3 \quad +2 \quad +1}$
$\, \downarrow \quad 4 \quad 1 \quad 3$
$\,4 \quad +1 \quad +3 \quad 4$

$\dfrac{4a^3 - 3a^2 + 2a + 1}{a - 1} = 4a^2 + a + 3 + \dfrac{4}{a - 1}$

13. $1\overline{\smash{\big)}\,-4 \quad -3 \quad -3 \quad +5 \quad -6 \quad +3 \quad +3}$
$\, \downarrow \quad -4 \quad -7 \quad -10 \quad -5 \quad -11 \quad -8$
$\, \boxed{-4} \quad \boxed{-7} \quad -10 \quad -5 \quad -11 \quad -8 \quad -5$

$\dfrac{-4r^6 - 3r^5 - 3r^4 + 5r^3 - 6r^2 + 3r + 3}{r - 1}$
$= \boxed{-4}r^5 \boxed{-7}r^4 - 10r^3 - 5r^2 - 11r - 8$
$+ \dfrac{-5}{r - 1}$

17. $\dfrac{y^3 - 1}{y + 1} = \dfrac{y^3 + 0y^2 + 0y - 1}{y + 1}$

$-1\overline{\smash{\big)}\,1 \quad +0 \quad +0 \quad -1}$
$\, \underline{ \quad -1 \quad +1 \quad -1}$
$\,1 \quad -1 \quad +1 \quad -2$

$\dfrac{y^3 - 1}{y + 1} = y^2 - y + 1 + \dfrac{-2}{y + 1}$

21. $-4\overline{\smash{\big)}\,-1 \quad 8 \quad -3 \quad -2}$
$\, \underline{ \quad 4 \quad -48 \quad 204}$
$\,-1 \quad 12 \quad -51 \quad 202$

The remainder is 202, so $P(-4) = 202$.

25. $-2\overline{\smash{\big)}\,1 \quad -3 \quad 5 \quad -2 \quad 5}$
$\, \underline{ \quad -2 \quad 10 \quad -30 \quad 64}$
$\,1 \quad -5 \quad 15 \quad -32 \quad 69$

The remainder is 69, so $P(-2) = 69$.

29. $-2\overline{\smash{\big)}\,1 \quad -3 \quad -2 \quad 16}$
$\, \underline{ \quad -2 \quad 10 \quad -16}$
$\,1 \quad -5 \quad 8 \quad 0$ Remainder is 0.

Yes, -2 is a solution.

33. $-3\overline{\smash{\big)}\,2 \quad -1 \quad -13 \quad 24}$
$\, \underline{ \quad -6 \quad 21 \quad -24}$
$\,2 \quad -7 \quad 8 \quad 0$ Remainder is 0.

Yes, -3 is a solution.

37. $9 \cdot 6 + 9 \cdot r = 9(6 + r)$ The common factor is 9.

41. $x(x + 1) + y(x + 1) = (x + 1)(x + y)$
The common factor is $x + 1$.

Section 3.7 (page 185)

1. $15k + 30 = 15 \cdot k + 15 \cdot 2$
15 is a common factor.
$= 15(k + 2)$

5. $4mx - 5mx^2 = mx \cdot 4 - mx \cdot 5x$
mx is a common factor.
$= mx(4 - 5x)$

9. $12x^2 - 3x = 3x(4x - 1)$
$3x$ is a common factor.

13. $15a^3 + 12a^2 - 3a = 3a(5a^2 + 4a - 1)$
$3a$ is a common factor.

17. $-15m^3p^3 - 6m^3p^4 - 9mp^3$
$-3mp^3$ is a common factor.
$= -3mp^3(5m^2 + 2m^2p + 3)$

21. $(3k - 7)(k + 2) + (3k - 7)(k + 5)$
$3k - 7$ is a common factor.
$= (3k - 7)(k + 2 + k + 5)$
$= (3k - 7)(2k + 7)$

25. $4(3 - x)^2 - (3 - x)^3 + 3(3 - x)$
$(3 - x)$ is a common factor.
$= (3 - x)[4(3 - x) - (3 - x)^2 + 3]$
$= (3 - x)(12 - 4x - 9 + 6x - x^2 + 3)$
$= (3 - x)(6 + 2x - x^2)$

29. $36y^2 - 72y$
Factor out $36y$.
$= 36y(y - 2)$
Now factor out $-36y$.
$= -36y(-y + 2)$

33. $-32a^4m^5 - 16a^2m^3 - 64a^5m^6$
First factor out $16a^2m^3$.
$= 16a^2m^3(-2a^2m^2 - 1 - 4a^3m^3)$
Now factor out $-16a^2m^3$.
$= -16a^2m^3(2a^2m^2 + 1 + 4a^3m^3)$

37. $2b + 2c + ab + ac = 2(b + c) + a(b + c)$
Factor out $(b + c)$.
$= (b + c)(2 + a)$

41. $x^2 - 3x + 2x - 6$
$= (x^2 - 3x) + (2x - 6)$
$= x(x - 3) + 2(x - 3)$
$= (x - 3)(x + 2)$

45. $-16p^2 - 6pq + 8pq + 3q^2$
Group the first two terms and the last two terms.
$= (-16p^2 - 6pq) + (8pq + 3q^2)$
$-2p$ is a common factor for the first group. q is a common factor for the second group. Factor them out.
$= -2p(8p + 3q) + q(8p + 3q)$
Now $8p + 3q$ is a common factor.
$= (8p + 3q)(-2p + q)$.

49. $1 - a + ab - b$
$= 1 - a - b - (-ab)$
$= 1(1 - a) - b(1 - a)$
$= (1 - a)(1 - b)$

53. $3m^{-5} + m^{-3}$
The smaller exponent is -5.
Factor out m^{-5}.
$= 3 \cdot m^{-5} + m^2 \cdot m^{-5}$
$= m^{-5}(3 + m^2)$
To find m^2, divide m^{-3} by m^{-5}.

57. $(p + 5)(p - 7) = p^2 - 7p + 5p - 35$
$= p^2 - 2p - 35$

61. $(8p + 3q)(8p - 3q) = 64p^2 - 24pq + 24pq - 9q^2$
$= 64p^2 - 9q^2$

Section 3.8 (page 193)

1. To factor $m^2 + 5m + 6$, we need two integers whose sum is 5 and whose product is 6. 3 and 2 are the numbers. Thus, we have
$m^2 + 5m + 6 = (m + 2)(m + 3)$.

5. To factor $r^2 - r - 20$, find two integers whose product is -20 and whose sum is -1. The numbers are -5 and 4.
$r^2 - r - 20 = (r - 5)(r + 4)$

9. To factor $a^2 + 3ab - 18b^2$, find two integers whose sum is 3 (the coefficient of the middle term) and whose product is -18. By inspection, we have 6 and -3, since $6 + (-3) = 3$ and $6 \cdot (-3) = -18$. Thus, we have $a^2 + 3ab - 18b^2 = (a + 6b) \cdot (a - 3b)$. (Note the second terms contain b to get b^2 in the product.)

13. $y^2w^2 + 4yw - 21 = (yw)^2 + 4(yw) - 21$
To factor, find two integers whose sum is 4 and whose product is -21. 7 and -3 are the numbers, since $7 \cdot (-3) = -21$ and $7 + (-3) = 4$. Thus, $(yw)^2 + 4(yw) - 21 = (yw + 7)(yw - 3)$.

17. In $18x^2 - 3x - 10$, use 3 and 6 or 2 and 9 for the first terms and 2 and -5 or -2 and 5 for the second terms. By trial,
$18x^2 - 3x - 10 = (6x - 5)(3x + 2)$.

21. In $12a^2 + 8ab - 15b^2$, use 12 and 1, 3 and 4, or 2 and 6 for the first terms, and 3 and -5 or -3 and 5 for the second terms. By trial,
$12a^2 + 8ab - 15b^2 = (6a - 5b)(2a + 3b)$.

25. To factor $35x^2 - 41xy - 24y^2$, write $35x^2$ as $7x \cdot 5x$ and $-24y^2$ as $3y(-8y)$. Use these factors in the binomial factors to get
$35x^2 - 41xy - 24y^2 = (7x + 3y)(5x - 8y)$.

29. $12m^2 + 14m - 40 = 2(6m^2 + 7m - 20)$
$= 2(3m - 4)(2m + 5)$

33. $6a^3 + 12a^2 - 90a = 6a(a^2 + 2a - 15)$
$= 6a(a + 5)(a - 3)$

37. $2x^3y^3 - 48x^2y^4 + 288xy^5$
$= 2xy^3(x^2 - 24xy + 144y^2)$
$= 2xy^3(x - 12y)(x - 12y)$
$= 2xy^3(x - 12y)^2$

41. $z^4 - 7z^2 - 30$ Let $m = z^2$.
$m^2 - 7m - 30 = (m - 10)(m + 3)$
Substitute z^2 for m.
$= (z^2 - 10)(z^2 + 3)$

45. $12p^4 + 28p^2r - 5r^2$ Let $x = p^2$.
$12x^2 + 28xr - 5r^2$
$= (6x - r)(2x + 5r)$
$= (6p^2 - r)(2p^2 + 5r)$ Substitute p^2 for x.

49. $6(z + k)^2 - 7(z + k) - 5$
Let $x = z + k$ to get $6x^2 - 7x - 5$.
Factor to get $(2x + 1)(3x - 5)$.
Replace x with $z + k$.
$[2(z + k) + 1][3(z + k) - 5]$
$= (2z + 2k + 1)(3z + 3k - 5)$

53. $p^2(p + q) + 4pq(p + q) + 3q^2(p + q)$
Let $x = p + q$ and substitute to get
$= p^2x + 4pqx + 3q^2x$.
Factor out the common factor of x.
$= x(p^2 + 4pq + 3q^2)$
Factor the trinomial.
$= x(p + q)(p + 3q)$
Replace x with $p + q$ to get
$= (p + q)(p + q)(p + 3q)$
$= (p + q)^2(p + 3q)$.

57. $(x + 6y)^2$
Use the pattern for a perfect square.
$= x^2 + 2(x)(6y) + (6y)^2$
$= x^2 + 12xy + 36y^2$

Section 3.9 (page 199)

1. $x^2 - 25 = x^2 - 5^2 = (x + 5)(x - 5)$

5. $16y^2 - 81q^2 = (4y)^2 - (9q)^2$
$= (4y + 9q)(4y - 9q)$

9. $(x + y)^2 - 16$
$= (x + y)^2 - (4)^2$
$= (x + y + 4)(x + y - 4)$

13. $(a + b)^2 - (a - b)^2$
Let $x = a + b$, $y = a - b$.
$(a + b)^2 - (a - b)^2$
$= x^2 - y^2$
$= (x - y)(x + y)$
$= [(a + b) - (a - b)][(a + b) + (a - b)]$
$= (2b)(2a) = 4ab$

17. $9r^2 - 6rs + s^2$
$= (3r)^2 - 2 \cdot 3r \cdot s + (s)^2$
$= (3r - s)^2$

21. $9m^2 - 12m + 4 - n^2$
Group the first three terms.
$= (9m^2 - 12m + 4) - n^2$
$= [(3m)^2 - 2(3m)(2) + 2^2] - n^2$
$= (3m - 2)^2 - n^2$
$= [(3m - 2) + n][(3m - 2) - n]$
$= (3m - 2 + n)(3m - 2 - n)$

25. $25x^2y^2 - 20xy + 4 = (5xy)^2 - 2(5xy)(2) + 2^2$
$= (5xy - 2)^2$

29. $(a + b)^2 + 2(a + b) + 1$ Let $x = a + b$
$(a + b)^2 + 2(a + b) + 1 = x^2 + 2x + 1$
$= (x + 1)^2$
$= (a + b + 1)^2$

33. $8a^3 + 1 = (2a)^3 + 1^3$
$= (2a + 1)[(2a)^2 - (2a)(1) + 1^2]$
$= (2a + 1)(4a^2 - 2a + 1)$

37. $250m^3 - 2p^3$
$= 2(125m^3 - p^3)$
$= 2[(5m)^3 - p^3]$
$= 2[(5m - p)((5m)^2 + (5m)(p) + p^2)]$
$= 2[(5m - p)(25m^2 + 5mp + p^2)]$
$= 2(5m - p)(25m^2 + 5mp + p^2)$

41. $(x + y)^3 - 27$
$= [(x + y) - 3][(x + y)^2 + 3(x + y) + 3^2]$
$= (x + y - 3)(x^2 + 2xy + y^2 + 3x + 3y + 9)$

45. $m^3 + (m + 3)^3$
$= [m + (m + 3)][m^2 - m(m + 3) + (m + 3)^2]$
$= (m + m + 3)(m^2 - m^2 - 3m + m^2 + 6m + 9)$
$= (2m + 3)(m^2 + 3m + 9)$

49. $\dfrac{3^6 \cdot 3^{-1}}{3^3} = 3^{6-1-3} = 3^2$

53. $5r - 2 = 8$
$5r = 10$ Add 2.
$r = 2$ Divide by 5.
Solution set: $\{2\}$

Summary of Factoring Methods   (page 201)

1. $100a^2 - 9b^2 = (10a)^2 - (3b)^2$
Difference of squares
$= (10a + 3b)(10a - 3b)$

5. $x^2 + 2x - 35 = (x + 7)(x - 5)$
The outer product, $-5x$, plus the inner product, $7x$, equals $2x$, the middle term.

9. $x^3 - 1000 = (x)^3 - (10)^3$
$= (x - 10)(x^2 + 10x + 100)$

13. $6t^2 + 19tu - 77u^2 = (3t - 7u)(2t + 11u)$
Since $2t(-7u) + 11u(3t) = -14tu + 33tu = 19tu$

17. $4k^2 + 28kr + 49r^2$
$= (2k)^2 + 2(2k)(7r) + (7r)^2$ A perfect square
$= (2k + 7r)^2$

21. $9m^2 - 30mn + 25n^2 - p^2$
$= (9m^2 - 30mn + 25n^2) - p^2$
$= [(3m)^2 - 2(3m)(5n) + (5n)^2] - p^2$
$= (3m - 5n)^2 - p^2$
$= [(3m - 5n) + p][(3m - 5n) - p]$
$= (3m - 5n + p)(3m - 5n - p)$

25. $16z^3x^2 - 32z^2x = 16z^2x(zx - 2)$
Factor out $16z^2x$ from each term.

29. $27m^2 + 144mn + 192n^2$
$= 3(9m^2 + 48mn + 64n^2)$
$= 3[(3m)^2 + 2(3m)(8n) + (8n)^2]$
Factor the square of a sum.
$= 3(3m + 8n)^2$

33. $2m^2 - mn - 15n^2 = (2m + 5n)(m - 3n)$
Since $5n(m) + 2m(-3n) = 5nm - 6nm$
$= -mn$, the middle term

37. $14z^2 - 3zk - 2k^2 = (7z + 2k)(2z - k)$
Since $7z(-1k) + 2k(2z) = -7zk + 4zk$
$= -3zk$, the middle term

41. In $1000z^3 + 512$, first factor out a common factor of 8 to get
$$8(125z^3 + 64).$$
Now, $125z^3 + 64 = (5z)^3 + (4)^3$.
Factor the sum of two cubes.
$= (5z + 4)[(5z)^2 - (5z)(4) + (4)^2]$
$= (5z + 4)(25z^2 - 20z + 16)$
Putting 8 back out in front of this last result, we get the complete factorization as
$$8(5z + 4)(25z^2 - 20z + 16).$$

45. $32x^2 + 16x^3 - 24x^5 = 8x^2(4 + 2x - 3x^3)$
Factor out $8x^2$ from each term.

49. $y^2 + 3y - 10 = (y + 5)(y - 2)$
Since $5 - 2 = 3$ and $5(-2) = -10$

53. $18p^2 + 53pr - 35r^2$
Write $18p^2$ as $9p \cdot 2p$ and $-35r^2$ as $-5r \cdot 7r$ to get $(9p - 5r)(2p + 7r)$. The outer product, $63pr$, plus the inner product, $-10pr$, equals the middle term, $53pr$, so this is correct.

57. $(5r + 2s)^2 - 6(5r + 2s) + 9$
Let $5r + 2s = x$. Then the expression becomes $x^2 - 6x + 9$, which factors as $(x - 3)(x - 3)$.
Now replace x with $5r + 2s$ to get
$(5r + 2s - 3)(5r + 2s - 3)$
$= (5r + 2s - 3)^2$.

Section 3.10   (page 209)

1. $(x - 2)(x - 3) = 0$
$x - 2 = 0$ or $x - 3 = 0$
$x = 2$ or $x = 3$
Solution set: $\{2, 3\}$

5. $(r - 4)(2r + 5) = 0$
$r - 4 = 0$ or $2r + 5 = 0$
$r = 4$ or $r = -\dfrac{5}{2}$
Solution set: $\left\{4, -\dfrac{5}{2}\right\}$

9. $y^2 + 9y + 14 = 0$
$(y + 2)(y + 7) = 0$
$y + 2 = 0$ or $y + 7 = 0$
$y = -2$ or $y = -7$
Solution set: $\{-2, -7\}$

13. $15r^2 + 7r = 2$
$15r^2 + 7r - 2 = 0$
$(3r + 2)(5r - 1) = 0$ Factor.
$3r + 2 = 0$ or $5r - 1 = 0$
$3r = -2$ or $5r = 1$
$r = -\dfrac{2}{3}$ or $r = \dfrac{1}{5}$
The solution set is $\left\{-\dfrac{2}{3}, \dfrac{1}{5}\right\}$.

17. $3n^2 + n - 4 = 0$ implies
$(3n + 4)(n - 1) = 0$.   Thus,
$3n + 4 = 0$ or $n - 1 = 0$
$3n = -4$ or $n = 1$
$n = -\dfrac{4}{3}$
The solution set is $\left\{1, -\dfrac{4}{3}\right\}$.

21. $(3x + 1)(x - 3) = 2 + 3(x + 5)$
$3x^2 - 8x - 3 = 2 + 3x + 15$ Multiply.
$3x^2 - 8x - 3 = 17 + 3x$ Combine terms.
$3x^2 - 11x - 20 = 0$ Subtract 17 + 3x on both sides.
$(3x + 4)(x - 5) = 0$ Factor.
$3x + 4 = 0$ or $x - 5 = 0$ Set each factor equal to 0.
$3x = -4$ $x = 5$
$x = -\dfrac{4}{3}$
The solution set is $\left\{-\dfrac{4}{3}, 5\right\}$.

25. $-2a^2 + 8 = 0$
$-2(a^2 - 4) = 0$
$-2(a + 2)(a - 2) = 0$
$a + 2 = 0$ or $a - 2 = 0$
$a = -2$ or $a = 2$
The solution set is $\{-2, 2\}$.

29. Let x be one of the numbers. The other, then, is $12 - x$.
$x^2 + (12 - x)^2 = 90$
$x^2 + 144 - 24x + x^2 = 90$
$2x^2 - 24x + 54 = 0$
$2(x^2 - 12x + 27) = 0$
$2(x - 9)(x - 3) = 0$
Thus,
$x - 9 = 0$ or $x - 3 = 0$
$x = 9$ or $x = 3$
If $x = 9$, the other number is $12 - 9 = 3$. If $x = 3$, the other number is $12 - 3 = 9$. The two numbers are 9 and 3 in either case.

33. Let x be the height of the triangle. Then the base is $x + 2$. Since Area $= \dfrac{1}{2} \times$ Base $\times$ Height,
$$112 = \dfrac{1}{2} \cdot (x + 2) \cdot x$$
$224 = (x + 2)x$
$224 = x^2 + 2x$
$0 = x^2 + 2x - 224$
$0 = (x - 14)(x + 16)$
$x - 14 = 0$ or $x + 16 = 0$
$x = 14$ $x = -16$
The height cannot be negative, so the only answer is 14 feet.

37. $(x - 1)^2 - (2x - 5)^2 = 0$
This is a difference of squares. Factor on the left.
$[(x - 1) + (2x - 5)][(x - 1) - (2x - 5)] = 0$
$(3x - 6)(-x + 4) = 0$
$3x - 6 = 0$ or $-x + 4 = 0$
$3x = 6$ $-x = -4$
$x = 2$ $x = 4$
The solution set is $\{2, 4\}$.

41. $\dfrac{-32r^3s^5}{48r^6s^7} = \dfrac{-2}{3}r^{3-6}s^{5-7}$
$= \dfrac{-2}{3}r^{-3}s^{-2} = -\dfrac{2}{3r^3s^2}$

45. $\dfrac{7}{5} = \dfrac{7}{5} \cdot \dfrac{6}{6} = \dfrac{42}{30}$

CHAPTER 4

Section 4.1 (page 227)

1. $\dfrac{m}{m - 4}$ is not defined when the denominator $m - 4 = 0$. When $m = 4$, this rational expression is undefined.

5. The expression $\dfrac{6 + z}{z}$ is not defined when the denominator is equal to 0. So when $z = 0$, the rational expression is undefined.

9. $\dfrac{m + 5}{m^2 - m - 12}$ is not defined when the denominator is equal to 0.
$m^2 - m - 12 = 0$
$(m - 4)(m + 3) = 0$
$m = 4$ or $m = -3$
The rational expression is undefined for $m = 4$ or $m = -3$.

13. $\dfrac{8z - 2}{z^2 + 16}$
$z^2 + 16 = 0$ Set denominator equal to 0.
$z^2 = -16$
The square root of -16 is not a real number, so there are no real numbers which make this expression undefined. Any real number can replace z.

17. $\dfrac{64p^4q^6}{16p^3q^7} = \dfrac{4p}{q}$ Divide numerator and denominator by $16p^3q^6$.

21. $\dfrac{p(p-4)}{3p(p-4)} = \dfrac{1}{3}$ Divide numerator and denominator by $p(p-4)$.

25. $\dfrac{2x-2}{3x-3} = \dfrac{2(x-1)}{3(x-1)} = \dfrac{2}{3}$

Factor numerator and denominator, and divide each by $x-1$.

29. $\dfrac{2b^2-2}{4b-4} = \dfrac{2(b^2-1)}{4(b-1)}$ Factor numerator and denominator.

$= \dfrac{2(b+1)(b-1)}{4(b-1)}$ Factor numerator again.

$= \dfrac{b+1}{2}$ Divide by $2(b-1)$.

33. $\dfrac{a^2-4}{a^2+4a+4}$

$= \dfrac{(a+2)(a-2)}{(a+2)(a+2)}$ Factor.

$= \dfrac{a-2}{a+2}$ Divide by $a+2$.

37. $\dfrac{c^2+cd-30d^2}{c^2-6cd+5d^2}$

$= \dfrac{(c+6d)(c-5d)}{(c-5d)(c-d)}$ Factor.

$= \dfrac{c+6d}{c-d}$ Divide by $c-5d$.

41. $\dfrac{xy-yw+xz-zw}{xy+yw+xz+zw} = \dfrac{(xy-yw)+(xz-zw)}{(xy+yw)+(xz+zw)}$

$= \dfrac{y(x-w)+z(x-w)}{y(x+w)+z(x+w)}$

$= \dfrac{(x-w)(y+z)}{(x+w)(y+z)}$

$= \dfrac{x-w}{x+w}$

45. $\dfrac{x-y}{y^2-x^2} = \dfrac{x-y}{(y-x)(y+x)}$

$= \dfrac{-1(y-x)}{(y-x)(y+x)}$ Factor -1 from numerator.

$= \dfrac{-1}{y+x}$ Divide numerator and denominator by $y-x$.

49. $\dfrac{2k-4}{6-3k} = \dfrac{2(k-2)}{-3(k-2)}$ Factor out 2 in numerator and -3 in denominator.

$= -\dfrac{2}{3}$ Divide numerator and denominator by $k-2$.

53. $-\dfrac{3}{8} \cdot \dfrac{16}{9} = -\dfrac{48}{72} = -\dfrac{2 \cdot 24}{3 \cdot 24} = -\dfrac{2}{3}$

Section 4.2 (page 231)

1. $\dfrac{m^3}{2} \cdot \dfrac{4m}{m^4} = \dfrac{4m^4}{2m^4} = 2$

5. $\dfrac{6y^5x^6}{y^3x^4} \cdot \dfrac{y^4x^2}{3y^5x^7} = \dfrac{6y^9x^8}{3y^8x^{11}} = \dfrac{6}{3} \cdot \dfrac{y^9}{y^8} \cdot \dfrac{x^8}{x^{11}}$

$= 2 \cdot y \cdot \dfrac{1}{x^3} = \dfrac{2y}{x^3}$

9. $\dfrac{2r}{8r+4} \cdot \dfrac{14r+7}{3} = \dfrac{2r}{4(2r+1)} \cdot \dfrac{7(2r+1)}{3}$

$= \dfrac{14r(2r+1)}{12(2r+1)} = \dfrac{7r}{6}$

13. $\dfrac{(m+3)^2}{m} \cdot \dfrac{m^2}{m^2-9}$

$= \dfrac{(m+3)(m+3)}{m} \cdot \dfrac{m \cdot m}{(m+3)(m-3)}$

$= \dfrac{m(m+3)}{m-3}$

17. $\dfrac{z^2-1}{2z} \cdot \dfrac{1}{1-z} = \dfrac{(z-1)(z+1)}{2z} \cdot \dfrac{-1}{z-1}$

Rewrite $\dfrac{1}{1-z}$ as $\dfrac{-1}{z-1}$.

$= \dfrac{-(z+1)}{2z}$ or $\dfrac{-z-1}{2z}$

21. $\dfrac{6r-5s}{3r+2s} \cdot \dfrac{6r+4s}{5s-6r} = \dfrac{-(5s-6r)}{3r+2s} \cdot \dfrac{2(3r+2s)}{5s-6r}$

$= -2$

25. $\dfrac{6a^2+5ab-6b^2}{12a^2-11ab+2b^2} \cdot \dfrac{8a^2-14ab+3b^2}{4a^2-12ab+9b^2}$

$= \dfrac{(3a-2b)(2a+3b)}{(3a-2b)(4a-b)} \cdot \dfrac{(4a-b)(2a-3b)}{(2a-3b)(2a-3b)}$

$= \dfrac{2a+3b}{2a-3b}$

29. $\dfrac{6k^2+kr-2r^2}{6k^2-5kr+r^2} \div \dfrac{3k^2+17kr+10r^2}{6k^2+13kr-5r^2}$

$= \dfrac{(3k+2r)(2k-r)}{(3k-r)(2k-r)} \div \dfrac{(3k+2r)(k+5r)}{(3k-r)(2k+5r)}$

$= \dfrac{(3k+2r)(2k-r)}{(3k-r)(2k-r)} \cdot \dfrac{(3k-r)(2k+5r)}{(3k+2r)(k+5r)}$

$= \dfrac{2k+5r}{k+5r}$

33. $\dfrac{a^2(2a+b)+6a(2a+b)+5(2a+b)}{3a^2(2a+b)-2a(2a+b)+(2a+b)} \div \dfrac{a+1}{a-1}$

$= \dfrac{a^2(2a+b)+6a(2a+b)+5(2a+b)}{3a^2(2a+b)-2a(2a+b)+1(2a+b)} \cdot \dfrac{a-1}{a+1}$

$= \dfrac{(2a+b)(a^2+6a+5)}{(2a+b)(3a^2-2a+1)} \cdot \dfrac{a-1}{a+1}$

$= \dfrac{(2a+b)(a+5)(a+1)(a-1)}{(2a+b)(3a^2-2a+1)(a+1)}$

$= \dfrac{(a+5)(a-1)}{3a^2-2a+1}$

37. $12 = 2^2 \cdot 3$ and $18 = 3^2 \cdot 2$
To find the least common denominator, multiply all prime factors that appear in *any* factorization, choosing the largest exponent for each prime.
Least Common Denominator $= 2^2 \cdot 3^2 = 36$

Section 4.3 (page 239)

1. $\dfrac{3}{8}, \dfrac{5}{2x}$

Factor: $8 = 2 \cdot 2 \cdot 2 = 2^3$ and $2x = 2 \cdot x$. The least common denominator (LCD) is $2^3 \cdot x = 8x$ (Take the highest power of each factor.)

5. $\dfrac{3}{7a}, \dfrac{2}{5a}$

The LCD $= 7 \cdot 5 \cdot a = 35a$.

9. $\dfrac{3}{z}, \dfrac{2}{z-1}$

The LCD $= z(z-1)$.

13. $\dfrac{5}{8m+16}, \dfrac{3}{5m+10}$

Factor each denominator: $8m + 16 = 8(m+2)$ and $5m + 10 = 5(m+2)$. The LCD $= 5 \cdot 8 \cdot (m+2) = 40(m+2)$.

17. $\dfrac{9}{x+y}, \dfrac{2}{x-y}$

The LCD $= (x+y)(x-y)$.

21. $\dfrac{3p+2}{p^2-3p-4}, \dfrac{2}{p^2+p}$

Factor both denominators:
$p^2 - 3p - 4 = (p-4)(p+1)$;
$p^2 + p = p(p+1)$.
The LCD $= p(p-4)(p+1)$.

25. $\dfrac{2}{p} + \dfrac{5}{p} = \dfrac{2+5}{p} = \dfrac{7}{p}$

29. $\dfrac{1}{m+1} + \dfrac{m}{m+1} = \dfrac{1+m}{m+1} = 1$

33. $\dfrac{a^2}{a+b} - \dfrac{b^2}{a+b} = \dfrac{a^2-b^2}{a+b} = \dfrac{(a+b)(a-b)}{a+b} = a-b$

37. $\dfrac{9}{r} + \dfrac{5}{2r} = \dfrac{2 \cdot 9}{2 \cdot r} + \dfrac{5}{2r} = \dfrac{18+5}{2r} = \dfrac{23}{2r}$

41. $\dfrac{3}{4x} + \dfrac{5}{3x} = \dfrac{3 \cdot 3}{3 \cdot 4x} + \dfrac{4 \cdot 5}{4 \cdot 3x} = \dfrac{9}{12x} + \dfrac{20}{12x} = \dfrac{9+20}{12x} = \dfrac{29}{12x}$

45. $\dfrac{1}{m-1} + \dfrac{1}{m} = \dfrac{m \cdot 1}{m(m-1)} + \dfrac{1 \cdot (m-1)}{m(m-1)} = \dfrac{m+m-1}{m(m-1)} = \dfrac{2m-1}{m(m-1)}$

49. $\dfrac{5}{m-4} + \dfrac{2}{4-m} = \dfrac{5}{m-4} + \dfrac{(-1)2}{(-1)(4-m)} = \dfrac{5}{m-4} + \dfrac{-2}{-4+m} = \dfrac{5}{m-4} + \dfrac{-2}{m-4} = \dfrac{5-2}{m-4} = \dfrac{3}{m-4}$

By using LCD $= 4 - m$, the form of the final answer would be $(-3)/(4-m)$, which is equivalent to $3/(m-4)$.]

53. $\dfrac{7}{3a+9} - \dfrac{5}{4a+12} = \dfrac{7}{3(a+3)} - \dfrac{5}{4(a+3)}$
$= \dfrac{4 \cdot 7}{4 \cdot 3(a+3)} - \dfrac{3 \cdot 5}{3 \cdot 4(a+3)}$
$= \dfrac{28}{12(a+3)} - \dfrac{15}{12(a+3)}$
$= \dfrac{28-15}{12(a+3)} = \dfrac{13}{12(a+3)}$

57. $\dfrac{4x}{x-1} - \dfrac{2}{x+1} - \dfrac{4}{x^2-1}$

The LCD is $(x+1)(x-1)$.
$= \dfrac{4x(x+1)}{(x+1)(x-1)} - \dfrac{2(x-1)}{(x+1)(x-1)} - \dfrac{4}{(x+1)(x-1)}$
$= \dfrac{4x^2 + 4x - 2x + 2 - 4}{(x+1)(x-1)}$
$= \dfrac{4x^2 + 2x - 2}{(x+1)(x-1)} = \dfrac{2(2x^2 + x - 1)}{(x+1)(x-1)}$
$= \dfrac{2(2x-1)(x+1)}{(x+1)(x-1)} = \dfrac{2(2x-1)}{x-1}$

61. $\dfrac{2w+3}{w-2} + \dfrac{7}{w} + \dfrac{14}{w^2-2w}$
$= \dfrac{w(2w+3)}{w(w-2)} + \dfrac{7(w-2)}{w(w-2)} + \dfrac{14}{w(w-2)}$
$= \dfrac{2w^2 + 3w + 7w - 14 + 14}{w(w-2)}$
$= \dfrac{2w^2 + 10w}{w(w-2)} = \dfrac{2w(w+5)}{w(w-2)} = \dfrac{2(w+5)}{w-2}$

65. $\dfrac{5x}{x^2+xy-2y^2} - \dfrac{3x}{x^2+5xy-6y^2}$
$= \dfrac{5x}{(x+2y)(x-y)} - \dfrac{3x}{(x+6y)(x-y)}$
$= \dfrac{5x(x+6y)}{(x+2y)(x-y)(x+6y)}$
$- \dfrac{3x(x+2y)}{(x+2y)(x-y)(x+6y)}$
$= \dfrac{5x^2 + 30xy - 3x^2 - 6xy}{(x+2y)(x-y)(x+6y)}$
$= \dfrac{2x^2 + 24xy}{(x+2y)(x-y)(x+6y)}$
$= \dfrac{2x(x+12y)}{(x+2y)(x-y)(x+6y)}$

69. $\dfrac{\dfrac{2}{3} + \dfrac{1}{6}}{\dfrac{5}{9} - \dfrac{1}{3}} = \dfrac{\dfrac{4}{6} + \dfrac{1}{6}}{\dfrac{5}{9} - \dfrac{3}{9}} = \dfrac{\dfrac{5}{6}}{\dfrac{2}{9}}$
$= \dfrac{5}{6} \div \dfrac{2}{9} = \dfrac{5}{6} \cdot \dfrac{9}{2} = \dfrac{45}{12} = \dfrac{15}{4}$

Section 4.4 (page 245)

1. $\dfrac{\dfrac{3}{x}}{\dfrac{6}{x-1}} \cdot \dfrac{x(x-1)}{x(x-1)}$

Multiply numerator and denominator by the LCD of $\dfrac{3}{x}$ and $\dfrac{6}{x-1}$, which is $x(x-1)$. This gives

$$\dfrac{\dfrac{3}{x}x(x-1)}{\dfrac{6}{(x-1)}x(x-1)} = \dfrac{3(x-1)}{6x} = \dfrac{x-1}{2x}.$$

5. $\dfrac{\dfrac{m-1}{4m}}{\dfrac{m+1}{m}} \cdot \dfrac{4m}{4m} = \dfrac{m-1}{4(m+1)}$

9. $\dfrac{\dfrac{x-3y}{5x}}{\dfrac{8x-24y}{10}}$

$= \dfrac{x-3y}{5x} \div \dfrac{8x-24y}{10}$

$= \dfrac{x-3y}{5x} \cdot \dfrac{10}{8x-24y}$ Definition of division

$= \dfrac{x-3y}{5x} \cdot \dfrac{10}{8(x-3y)}$ Factor.

$= \dfrac{10(x-3y)}{5(8)x(x-3y)}$ Multiply.

$= \dfrac{1}{4x}$ Simplify.

13. $\dfrac{\dfrac{2}{m} - 1}{1 + \dfrac{2}{m}}$

Multiply both the numerator and the denominator by m.

$= \dfrac{\left(\dfrac{2}{m} - 1\right)m}{\left(1 + \dfrac{2}{m}\right)m} = \dfrac{\dfrac{2}{m}(m) - 1(m)}{1(m) + \dfrac{2}{m}(m)}$

$= \dfrac{2-m}{m+2}$

17. $\dfrac{\dfrac{a}{b} + 2}{\dfrac{a}{b} - \dfrac{4b}{a}} = \dfrac{\left(\dfrac{a}{b} + 2\right)ab}{\left(\dfrac{a}{b} - \dfrac{4b}{a}\right)ab} = \dfrac{a^2 + 2ab}{a^2 - 4b^2}$

$= \dfrac{a(a+2b)}{(a+2b)(a-2b)} = \dfrac{a}{a-2b}$

21. $a^{-2} + b^{-2} = \dfrac{1}{a^2} + \dfrac{1}{b^2} = \dfrac{b^2}{a^2b^2} + \dfrac{a^2}{a^2b^2}$

$= \dfrac{b^2 + a^2}{a^2b^2}$

25. $(p^{-1} - q^{-1})^{-1} = \dfrac{1}{p^{-1} - q^{-1}}$

$= \dfrac{1}{\dfrac{1}{p} - \dfrac{1}{q}}$

$= \dfrac{(1) \cdot pq}{\left(\dfrac{1}{p} - \dfrac{1}{q}\right) \cdot pq} = \dfrac{pq}{q-p}$

29. $6r^2 + r = 2$
$6r^2 + r - 2 = 0$
$(3r+2)(2r-1) = 0$
$3r + 2 = 0$ or $2r - 1 = 0$
$3r = -2$ $2r = 1$
$r = -\dfrac{2}{3}$ or $r = \dfrac{1}{2}$

Solution set: $\{-2/3, 1/2\}$

Section 4.5 (page 249)

1. $\dfrac{p}{4} - \dfrac{p}{8} = 2$

The least common denominator is 8, so multiply both sides of the equation by 8.

$8\left(\dfrac{p}{4} - \dfrac{p}{8}\right) = 8(2)$
$2p - p = 16$
$p = 16$

Solution set: $\{16\}$

5. $\dfrac{3y-6}{2} = \dfrac{5y+1}{6}$

$3(3y-6) = 5y + 1$ Multiply both sides by 6.
$9y - 18 = 5y + 1$
$4y = 19$
$y = \dfrac{19}{4}$

Solution set: $\left\{\dfrac{19}{4}\right\}$

9. $\dfrac{1}{a} + \dfrac{2}{3a} = \dfrac{1}{3}$

$3a \cdot \dfrac{1}{a} + 3a \cdot \dfrac{2}{3a} = 3a \cdot \dfrac{1}{3}$ Multiply by $3a$.
$3 + 2 = a$
$5 = a$

Solution set: $\{5\}$

13. $\dfrac{1}{r-1} + \dfrac{2}{3r-3} = \dfrac{-5}{12}$

$\dfrac{1}{r-1} + \dfrac{2}{3(r-1)} = \dfrac{-5}{12}$

$12 + 8 = -5(r-1)$ Multiply by $12(r-1)$.
$20 = -5r + 5$
$5r = -15$
$r = -3$

Solution set: $\{-3\}$

17. $\dfrac{5}{6q+14} - \dfrac{2}{3q+7} = \dfrac{1}{56}$
Multiply by LCD, $56(3q+7)$.
$28(5) - 2(56) = 3q + 7$
$140 - 112 = 3q + 7$
$28 = 3q + 7$
$21 = 3q$
$q = 7$
Solution set: $\{7\}$

21. $\dfrac{2}{x-1} + \dfrac{3}{x+1} = \dfrac{9}{x^2-1}$
$2(x+1) + 3(x-1) = 9$ Multiply by LCD, $x^2 - 1$.
$2x + 2 + 3x - 3 = 9$
$5x = 10$
$x = 2$
Solution set: $\{2\}$

25. $y + 2 = \dfrac{24}{y}$
$y^2 + 2y = 24$ Multiply by y.
$y^2 + 2y - 24 = 0$
$(y+6)(y-4) = 0$ Factor.
$y + 6 = 0$ or $y - 4 = 0$
$y = -6$ or $y = 4$
Solution set: $\{-6, 4\}$

29. $\dfrac{a+3}{a} - \dfrac{a+4}{a+5} = \dfrac{15}{a^2+5a}$
$(a+3)(a+5) - a(a+4) = 15$ Multiply by $a(a+5)$.
$a^2 + 8a + 15 - a^2 - 4a = 15$
$4a = 0$
$a = 0$
When 0 is replaced for a in the original equation, we obtain
$\dfrac{3}{0} - \dfrac{4}{5} = \dfrac{15}{0}$.
Since 0 cannot appear in the denominator of a fraction, the equation has no solution.
Solution set: $\emptyset$

33. $\dfrac{5}{x-4} - \dfrac{3}{x-1} = \dfrac{x+11}{x^2-5x+4}$
$\dfrac{5}{x-4} - \dfrac{3}{x-1} = \dfrac{x+11}{(x-1)(x-4)}$
$5(x-1) - 3(x-4) = x + 11$ Multiply by $(x-1)\cdot(x-4)$.
$5x - 5 - 3x + 12 = x + 11$
$2x + 7 = x + 11$
$2x = x + 4$
$x = 4$
When x is replaced with 4, the original equation becomes
$\dfrac{5}{4-4} - \dfrac{3}{4-1} = \dfrac{4+11}{(4)^2 - 5(4) + 4}$
$\dfrac{5}{0} - \dfrac{3}{3} = \dfrac{15}{0}$,
which is not possible since we can't divide by 0. The equation has no solution, so the solution set is $\emptyset$.

37. $\dfrac{1}{x+2} - \dfrac{5}{x^2+9x+14} = \dfrac{-3}{x+7}$
$\dfrac{1}{x+2} - \dfrac{5}{(x+2)(x+7)} = \dfrac{-3}{x+7}$
$x + 7 - 5 = -3(x+2)$ Multiply by $(x+2)\cdot(x+7)$.
$x + 2 = -3x - 6$
$4x + 2 = -6$
$4x = -8$
$x = -2$

Check: $\dfrac{1}{-2+2} - \dfrac{5}{(-2)^2 + 9(-2) + 14}$
$= \dfrac{-3}{-2+7}$
$\dfrac{1}{0} - \dfrac{5}{4 - 18 + 14} = \dfrac{-3}{5}$

Since -2 leads to a zero denominator, there is no solution, and the solution set is $\emptyset$.

41. $\dfrac{-4x}{6x-3} + \dfrac{2}{6x-3} = \dfrac{9}{x}$
Multiply by $x(6x-3)$.
$-4x(x) + 2x = 9(6x - 3)$
$-4x^2 + 2x = 54x - 27$
$4x^2 + 52x - 27 = 0$
$(2x - 1)(2x + 27) = 0$
$2x - 1 = 0$ or $2x + 27 = 0$
$2x = 1$ or $2x = -27$
$x = \dfrac{1}{2}$ or $x = -\dfrac{27}{2}$
But $x = \dfrac{1}{2}$ produces undefined fractions.
Solution set: $\left\{-\dfrac{27}{2}\right\}$

45. $A = \dfrac{1}{2}bh$
$2A = bh$
$\dfrac{2A}{b} = \dfrac{bh}{b}$
$h = \dfrac{2A}{b}$

Summary of Operations Involving Rational Expressions (page 253)

1. $\dfrac{a}{2} - \dfrac{a}{5} = 3$
$10\left(\dfrac{a}{2} - \dfrac{a}{5}\right) = 10 \cdot 3$
$5a - 2a = 30$
$3a = 30$
$a = 10$
Solution set: $\{10\}$

5. $\dfrac{\dfrac{1}{x} - \dfrac{1}{y}}{\dfrac{1}{x} + \dfrac{1}{y}} = \dfrac{\dfrac{y}{y} \cdot \dfrac{1}{x} - \dfrac{1}{y} \cdot \dfrac{x}{x}}{\dfrac{y}{y} \cdot \dfrac{1}{x} + \dfrac{1}{y} \cdot \dfrac{x}{x}}$

$= \dfrac{\dfrac{y}{yx} - \dfrac{x}{yx}}{\dfrac{y}{yx} + \dfrac{x}{yx}}$

$= \dfrac{\dfrac{y - x}{yx}}{\dfrac{y + x}{yx}}$

$= \dfrac{y - x}{yx} \cdot \dfrac{yx}{y + x} = \dfrac{y - x}{y + x}$

9. $\dfrac{1}{x} - \dfrac{1}{2x} = 1$

$2x\left(\dfrac{1}{x} - \dfrac{1}{2x}\right) = 2x \cdot 1$

$2 - 1 = 2x$

$1 = 2x$

$x = \dfrac{1}{2}$

Solution set: $\left\{\dfrac{1}{2}\right\}$

13. $\dfrac{3p(p - 2)}{p + 5} \div \dfrac{p^2 - 4}{4p + 20}$

$= \dfrac{3p(p - 2)}{p + 5} \cdot \dfrac{4p + 20}{p^2 - 4}$

$= \dfrac{3p(p - 2)}{p + 5} \cdot \dfrac{4(p + 5)}{(p + 2)(p - 2)}$

$= \dfrac{12p(p - 2)(p + 5)}{(p + 5)(p - 2)(p + 2)}$

$= \dfrac{12p}{p + 2}$

17. $\dfrac{2q^2 + 5q - 3}{q^2 + q - 6} \div \dfrac{10q^2 - 5q}{3q^3 - 6q^2}$

$= \dfrac{2q^2 + 5q - 3}{q^2 + q - 6} \cdot \dfrac{3q^3 - 6q^2}{10q^2 - 5q}$

$= \dfrac{(2q - 1)(q + 3)}{(q + 3)(q - 2)} \cdot \dfrac{3q^2(q - 2)}{5q(2q - 1)}$

$= \dfrac{3q}{5}$

21. $\dfrac{\dfrac{3}{k} - \dfrac{5}{m}}{\dfrac{9m^2 - 25k^2}{km^2}} = \dfrac{\left(\dfrac{3}{k} - \dfrac{5}{m}\right)km^2}{\left(\dfrac{9m^2 - 25k^2}{km^2}\right)km^2}$

$= \dfrac{3m^2 - 5km}{9m^2 - 25k^2}$

$= \dfrac{m(3m - 5k)}{(3m + 5k)(3m - 5k)}$

$= \dfrac{m}{3m + 5k}$

25. $\dfrac{8}{3p + 9} - \dfrac{2}{5p + 15} = \dfrac{8}{15}$

$\dfrac{8}{3(p + 3)} - \dfrac{2}{5(p + 3)} = \dfrac{8}{15}$

$15(p + 3)\left[\dfrac{8}{3(p + 3)} - \dfrac{2}{5(p + 3)}\right]$

$= 15(p + 3) \cdot \dfrac{8}{15}$

$40 - 6 = 8(p + 3)$

$34 = 8p + 24$

$10 = 8p$

$p = \dfrac{10}{8} = \dfrac{5}{4}$

Solution set: $\left\{\dfrac{5}{4}\right\}$

29. $\dfrac{2}{q + 1} - \dfrac{3}{q^2 - q - 2} = \dfrac{3}{q - 2}$

$\dfrac{2}{q + 1} - \dfrac{3}{(q + 1)(q - 2)} = \dfrac{3}{q - 2}$

$(q + 1)(q - 2)\left[\dfrac{2}{q + 1} - \dfrac{3}{(q + 1)(q - 2)}\right]$

$= (q + 1)(q - 2) \cdot \dfrac{3}{q - 2}$

$2(q - 2) - 3 = 3(q + 1)$

$2q - 4 - 3 = 3q + 3$

$2q - 7 = 3q + 3$

$-10 = q$

Solution set: $\{-10\}$

33. $\dfrac{2x^2 - 11xy + 15y^2}{2x^2 + 5xy - 3y^2} \cdot \dfrac{8x^2 - 2xy - y^2}{8x^2 - 18xy - 5y^2}$

$= \dfrac{(2x - 5y)(x - 3y)}{(2x - y)(x + 3y)} \cdot \dfrac{(4x + y)(2x - y)}{(4x + y)(2x - 5y)}$

$= \dfrac{(2x - 5y)(4x + y)(2x - y)(x - 3y)}{(2x - 5y)(4x + y)(2x - y)(x + 3y)}$

$= \dfrac{x - 3y}{x + 3y}$

Section 4.6 (page 257)

1. $I = \dfrac{E}{R}$

$20 = \dfrac{8}{R}$ Let $I = 20$, $E = 8$.

$20R = 8$

$R = \dfrac{8}{20} = \dfrac{2}{5}$

5. $\dfrac{1}{a} = \dfrac{1}{b} + \dfrac{1}{c}$

$\dfrac{1}{8} = \dfrac{1}{b} + \dfrac{1}{12}$ Let $a = 8$, $c = 12$.

$3b = 24 + 2b$ Multiply by $24b$.

$b = 24$

9. $\dfrac{1}{a} = \dfrac{1}{b} + \dfrac{1}{c}$
Multiply both sides by abc.
$bc = ac + ab$
$bc - ab = ac$
$b(c - a) = ac$
$b = \dfrac{ac}{c - a}$ or $-\dfrac{ac}{a - c}$

13. $a = \dfrac{V - v}{t}$
$at = V - v$ Multiply by t.
$V = at + v$

17. $\quad t = a + (n - 1)d$
$\quad t = a + nd - d$
$t - a + d = nd$
$\dfrac{t - a + d}{d} = n$ or $n = \dfrac{t - a}{d} + 1$

21. Let x = the number. The reciprocal is $\dfrac{1}{x}$.
$\dfrac{1}{x} + 3 = 4$
$\dfrac{1}{x} = 1$
$1 = x$
The number is 1.

Section 4.7 (page 263)

1. Let x be the number.
$\dfrac{12}{13 + x} = 4$
$4(13 + x) = 12$
$52 + 4x = 12$
$4x = -40$
$x = -10$
The number is -10.

5. Let x be the number.
$\dfrac{1}{x} + \dfrac{1}{x + 1} = -\dfrac{19}{90}$
Multiply both sides by the LCD, $90x(x + 1)$.
$90(x + 1) + 90x = -19x(x + 1)$
$90x + 90 + 90x = -19x^2 - 19x$
$19x^2 + 199x + 90 = 0$
$(19x + 9)(x + 10) = 0$
$19x + 9 = 0$ or $x + 10 = 0$
$x = -\dfrac{9}{19}$ $\qquad x = -10$
The number is -10 or $-\dfrac{9}{19}$.

9. Let x represent the number of hours it will take them if they work together.
$\dfrac{1}{6}$ = part of the job done by A.J. alone in 1 hour
$\dfrac{1}{5}$ = part of the job done by Audrey alone in 1 hour
$\dfrac{1}{x}$ = part of the job done in 1 hour when they work together
$\dfrac{1}{6} + \dfrac{1}{5} = \dfrac{1}{x}$
$30x\left(\dfrac{1}{6}\right) + 30x\left(\dfrac{1}{5}\right) = 30x\left(\dfrac{1}{x}\right)$
$5x + 6x = 30$
$11x = 30$
$x = \dfrac{30}{11}$
It will take them $\dfrac{30}{11}$ or $2\dfrac{8}{11}$ hours to do the job if they work together.

13. Let x represent the amount of time (in hours) it will take to fill the vat if both pipes are left open. Because work is being "un-done" by the outlet pipe, subtraction must be used in the equation.
$\dfrac{1}{10} - \dfrac{1}{20} = \dfrac{1}{x}$
$20x\left(\dfrac{1}{10}\right) - 20x\left(\dfrac{1}{20}\right) = 20x\left(\dfrac{1}{x}\right)$
$2x - x = 20$
$x = 20$
It will take 20 hours to fill the vat if both pipes are open.

17. In x hours Mimi can mess up $\dfrac{x}{2}$ of the house. In x hours Arlene and Mort can clean up $\dfrac{x}{7}$ of the house.
So to completely mess up the house we have
$\dfrac{x}{2} - \dfrac{x}{7} = 1$
$14\left(\dfrac{x}{2} - \dfrac{x}{7}\right) = 14$
$7x - 2x = 14$
$5x = 14$
$x = \dfrac{14}{5}$
$x = 2\dfrac{4}{5}.$
The house will be a shambles in $2\dfrac{4}{5}$ hours.

21. Let x = speed of the boat in still water.

	d	r	t
Downstream	22	$x + 2$	$\dfrac{22}{x + 2}$
Upstream	16	$x - 2$	$\dfrac{16}{x - 2}$

The times upstream and downstream are equal. So
$$\frac{22}{x+2} = \frac{16}{x-2}$$
$$22(x-2) = 16(x+2)$$
$$22x - 44 = 16x + 32$$
$$6x = 76$$
$$x = 12\frac{2}{3}.$$

The speed of the boat is $12\frac{2}{3}$ miles per hour.

25. If it takes him t hours via the interstate, then it takes him $t + 2$ hours via the old highway. The distances, $50t$ and $30(t + 2)$, are the same.
$$30(t+2) = 50t$$
$$30t + 60 = 50t$$
$$20t = 60$$
$$t = 3$$
The distance is $50 \cdot 3$, or 150 miles.

29. Because $2^3 = 8$, $\sqrt[3]{8} = 2$.

CHAPTER 5

Section 5.1 (page 281)

1. The square roots of 25 are 5 and -5 since $5^2 = 25$ and $(-5)^2 = 25$.
5. The square roots of 625 are 25 and -25.
9. The square roots of 6241 are 79 and -79.
13. $\sqrt{4} = 2$ since $2^2 = 4$.
17. $-\sqrt{2025} = -(45) = -45$
21. $\sqrt{x^4} = x^2$, since $(x^2)^2 = x^4$.
25. $-\sqrt{56} \approx -7.483$
29. $\sqrt{280} \approx 16.733$
33. $\sqrt[3]{64} = 4$ since $4^3 = 64$.
37. $-\sqrt[3]{512} = -8$ since $8^3 = 512$.
41. $-\sqrt[4]{625} = -5$ since $5^4 = 625$. The negative sign outside the radical stays with the answer.
45. $-\sqrt[4]{6561} = -9$ since $9^4 = 6561$.
49. $-\sqrt[5]{-1024} = 4$ since $(-4)^5 = -1024$ and $-(-4) = 4$.
53. $\sqrt[6]{4096} = 4$ since $4^6 = 4096$.
57. $\sqrt[3]{m^{15}} = \sqrt[3]{(m^5)^3} = m^5$
61. $x^{-3} \cdot x^6 = x^{-3+6} = x^3$
65. $(s^3)^5 = s^{3\cdot 5} = s^{15}$

Section 5.2 (page 287)

1. $121^{1/2} = \sqrt{121} = 11$
5. $256^{1/4} = \sqrt[4]{256} = 4$
9. $\left(\frac{1}{32}\right)^{1/5} = \frac{1^{1/5}}{32^{1/5}} = \frac{1}{\sqrt[5]{32}} = \frac{1}{2}$

13. $\left(\frac{8}{125}\right)^{1/3} = \frac{8^{1/3}}{125^{1/3}} = \frac{\sqrt[3]{8}}{\sqrt[3]{125}} = \frac{2}{5}$

17. $32^{6/5} = (\sqrt[5]{32})^6 = 2^6 = 64$

21. $(-1728)^{2/3} = (\sqrt[3]{-1728})^2 = (-12)^2 = 144$

25. $-\left(\frac{9}{4}\right)^{5/2} = -\frac{(\sqrt{9})^5}{(\sqrt{4})^5} = -\frac{3^5}{2^5} = -\frac{243}{32}$

29. $2^{1/2} \cdot 2^{3/2} = 2^{1/2 + 3/2} = 2^{4/2} = 2^2$

33. $\frac{81^{5/4}}{81^{3/4}} = 81^{5/4 - 3/4} = 81^{2/4} = 81^{1/2}$

37. $\frac{k^{2/3}k^{-1}}{k^{1/3}} = k^{2/3 + (-1) - 1/3}$
$= k^{2/3 - (4/3)} = k^{-2/3}$
$= \frac{1}{k^{2/3}}$

41. $(8p^9q^6)^{2/3} = 8^{2/3}p^6q^4$
$= (\sqrt[3]{8})^2 p^6 q^4 = 4p^6q^4$

45. $\left(\frac{z^{10}}{x^{12}}\right)^{1/4} = \frac{z^{10/4}}{x^{12/4}} = \frac{z^{5/2}}{x^3}$

49. $\frac{p^{1/5}p^{7/10}p^{1/2}}{(p^3)^{-1/5}} = \frac{p^{1/5 + (7/10) + (1/2)}}{p^{-3/5}}$
$= p^{7/5 - (-3/5)}$
$= p^2$

53. $\sqrt{2^{10}} = (2^{10})^{1/2} = 2^{10(1/2)} = 2^5$

57. $-\sqrt[3]{11^5} = -(11^5)^{1/3} = -11^{5/3} = -11^{3/3} \cdot 11^{2/3}$
$= -11\sqrt[3]{11^2}$ or $-11\sqrt[3]{121}$

61. $\sqrt[7]{a^9} \cdot \sqrt{a} = a^{9/7} \cdot a^{1/2} = a^{9/7 + 1/2} = a^{25/14}$
$= a^{14/14 + 11/14}$
$= a^{14/14} \cdot a^{11/14} = a\sqrt[14]{a^{11}}$

65. $L = \left(\frac{S}{a}\right)^{1/2}$
$= \left(\frac{1000}{\frac{2}{5}}\right)^{1/2}$ Substitute given numbers.
$= \left(\frac{1000 \cdot 5}{2}\right)^{1/2}$ Multiply by reciprocal of denominator.
$= (2500)^{1/2}$
$= 50$ centimeters

69. $(6x^2)^{-1}(2x^3)^{-2} = 6^{-1}x^{-2}2^{-2}x^{-6}$
$= \frac{x^{-2+(-6)}}{6 \cdot 2^2}$
$= \frac{x^{-8}}{24}$
$= \frac{1}{24x^8}$

Section 5.3 (page 293)

1. $\sqrt{7} \cdot \sqrt{11} = \sqrt{7 \cdot 11} = \sqrt{77}$
5. $\sqrt[3]{2} \cdot \sqrt[3]{7} = \sqrt[3]{2 \cdot 7} = \sqrt[3]{14}$

9. $\sqrt{\dfrac{16}{25}} = \dfrac{\sqrt{16}}{\sqrt{25}} = \dfrac{4}{5}$

13. $\sqrt{\dfrac{r}{100}} = \dfrac{\sqrt{r}}{\sqrt{100}} = \dfrac{\sqrt{r}}{10}$

17. $\sqrt[3]{-\dfrac{27}{64}} = \dfrac{\sqrt[3]{-27}}{\sqrt[3]{64}} = \dfrac{-3}{4} = -\dfrac{3}{4}$

21. $\sqrt{18} = \sqrt{9 \cdot 2} = \sqrt{9} \cdot \sqrt{2} = 3\sqrt{2}$

25. $-\sqrt{48} = -\sqrt{16 \cdot 3} = -\sqrt{16} \cdot \sqrt{3} = -4\sqrt{3}$

29. $\sqrt{60} = \sqrt{4 \cdot 15} = \sqrt{4} \cdot \sqrt{15} = 2\sqrt{15}$

33. $\sqrt[3]{-16} = \sqrt[3]{-8 \cdot 2} = \sqrt[3]{-8} \cdot \sqrt[3]{2} = -2\sqrt[3]{2}$

37. $\sqrt[3]{375} = \sqrt[3]{125 \cdot 3} = \sqrt[3]{125} \cdot \sqrt[3]{3} = 5\sqrt[3]{3}$

41. $-\sqrt[4]{1250} = -\sqrt[4]{625 \cdot 2}$
$= -\sqrt[4]{625} \cdot \sqrt[4]{2}$
$= -5\sqrt[4]{2}$

45. $\sqrt{25k^2} = 5k$
No absolute value bars are needed because $k > 0$.

49. $\sqrt{100y^{10}} = 10y^5$
No absolute value bars are needed because $y > 0$.

53. $-\sqrt{144m^{10}z^2} = -12m^5 z$
No absolute value bars are needed because $m > 0$ and $z > 0$.

57. $\sqrt[4]{\dfrac{1}{16}m^{12}x^{16}} = \dfrac{1}{2}m^3 x^4$
No absolute value bars are needed because $m > 0$.

61. $\sqrt{1000m^9} = \sqrt{100m^8 \cdot 10m} = 10m^4 \sqrt{10m}$

65. $\sqrt[3]{8z^9 r^{12}} = 2z^3 r^4$

69. $\sqrt[4]{16a^8 b^{12}} = 2a^2 b^3$

73. $\sqrt{\dfrac{m^9}{16}} = \dfrac{\sqrt{m^8 \cdot m}}{\sqrt{16}} = \dfrac{m^4 \sqrt{m}}{4}$

77. $\sqrt[4]{12^2} = 12^{2/4} = 12^{1/2} = \sqrt{12}$
$= \sqrt{4} \cdot \sqrt{3} = 2\sqrt{3}$

81. $\sqrt[3]{5} \cdot \sqrt{3}$ The common index of 3 and 2 is 6.
$\sqrt[3]{5} = 5^{1/3} = 5^{2/6} = \sqrt[6]{5^2}$
$\sqrt{3} = 3^{1/2} = 3^{3/6} = \sqrt[6]{3^3}$
So,
$\sqrt[3]{5} \cdot \sqrt{3} = \sqrt[6]{5^2} \cdot \sqrt[6]{3^3}$
$= \sqrt[6]{5^2 \cdot 3^3} = \sqrt[6]{675}$

85. $d = \sqrt{\dfrac{k}{l}}$
$d = \sqrt{\dfrac{700}{12}} = \dfrac{\sqrt{700}}{\sqrt{12}} = \dfrac{\sqrt{100 \cdot 7}}{\sqrt{4 \cdot 3}}$
$= \dfrac{10\sqrt{7}}{2\sqrt{3}} \cdot \dfrac{\sqrt{3}}{\sqrt{3}} = \dfrac{10\sqrt{21}}{6} = \dfrac{5\sqrt{21}}{3}$
The distance is $5\sqrt{21}/3$ feet.

89. $9q + 2q - 5q^2 - q^2 = 11q - 6q^2$

Section 5.4 (page 299)

1. $\sqrt{36} + \sqrt{100} = 6 + 10 = 16$

5. $4\sqrt{12} - 7\sqrt{27} = 4\sqrt{4 \cdot 3} - 7\sqrt{9 \cdot 3}$
$= 4(2\sqrt{3}) - 7(3\sqrt{3})$
$= 8\sqrt{3} - 21\sqrt{3}$
$= -13\sqrt{3}$

9. $2\sqrt{63} - 2\sqrt{28} + 3\sqrt{7}$
$= 2\sqrt{9 \cdot 7} - 2\sqrt{4 \cdot 7} + 3\sqrt{7}$
$= 6\sqrt{7} - 4\sqrt{7} + 3\sqrt{7} = 5\sqrt{7}$

13. $2\sqrt{40} + 6\sqrt{90} - 3\sqrt{160}$
$= 2\sqrt{4 \cdot 10} + 6\sqrt{9 \cdot 10} - 3\sqrt{16 \cdot 10}$
$= 4\sqrt{10} + 18\sqrt{10} - 12\sqrt{10} = 10\sqrt{10}$

17. $3\sqrt{72m^2} + 2\sqrt{32m^2} - 3\sqrt{18m^2}$
$= 3\sqrt{36m^2 \cdot 2} + 2\sqrt{16m^2 \cdot 2} - 3\sqrt{9m^2 \cdot 2}$
$= 3(6m)\sqrt{2} + 2(4m)\sqrt{2} - 3(3m)\sqrt{2}$
$= 18m\sqrt{2} + 8m\sqrt{2} - 9m\sqrt{2} = 17m\sqrt{2}$

21. $2\sqrt[3]{27x} + 2\sqrt[3]{8x} = 2\sqrt[3]{27 \cdot x} + 2\sqrt[3]{8 \cdot x}$
$= 2\sqrt[3]{27} \cdot \sqrt[3]{x} + 2\sqrt[3]{8} \cdot \sqrt[3]{x}$
$= 2 \cdot 3\sqrt[3]{x} + 2 \cdot 2\sqrt[3]{x}$
$= 6\sqrt[3]{x} + 4\sqrt[3]{x}$
$= 10\sqrt[3]{x}$

25. $\sqrt[3]{x^2 y} - \sqrt[3]{8x^2 y} = \sqrt[3]{x^2 y} - 2\sqrt[3]{x^2 y} = -\sqrt[3]{x^2 y}$

29. $5\sqrt[4]{32} + 3\sqrt[4]{162} = 5\sqrt[4]{16} \cdot \sqrt[4]{2} + 3\sqrt[4]{81} \cdot \sqrt[4]{2}$
$= 5 \cdot 2\sqrt[4]{2} + 3 \cdot 3\sqrt[4]{2}$
$= 10\sqrt[4]{2} + 9\sqrt[4]{2}$
$= 19\sqrt[4]{2}$

33. $3\sqrt[4]{x^5 y} - 2x\sqrt[4]{xy}$
$= 3\sqrt[4]{x^4}\sqrt[4]{xy} - 2x\sqrt[4]{xy}$
$= 3x\sqrt[4]{xy} - 2x\sqrt[4]{xy}$
$= x\sqrt[4]{xy}$

37. Use $P = a + b + c + d$ with $a = 3\sqrt{18}$, $b = 2\sqrt{32}$, $c = 4\sqrt{50}$, and $d = 5\sqrt{12}$.
$P = 3\sqrt{18} + 2\sqrt{32} + 4\sqrt{50} + 5\sqrt{12}$ Substitution.
$= 3\sqrt{9}\sqrt{2} + 2\sqrt{16}\sqrt{2} + 4\sqrt{25}\sqrt{2}$
$\quad + 5\sqrt{4}\sqrt{3}$
$= 9\sqrt{2} + 8\sqrt{2} + 20\sqrt{2} + 10\sqrt{3}$
$= 37\sqrt{2} + 10\sqrt{3}$
The perimeter is $37\sqrt{2} + 10\sqrt{3}$ yards.

41. $\sqrt{5} \cdot \sqrt{5} = \sqrt{25} = 5$

45. $-6p(2p^2 - 1) = -6p(2p^2) - 6p(-1)$
$= -12p^3 + 6p$

Section 5.5 (page 307)

1. $\sqrt{3}(\sqrt{12} + 2) = \sqrt{3}\sqrt{12} + \sqrt{3} \cdot 2$
$= \sqrt{36} + 2\sqrt{3} = 6 + 2\sqrt{3}$

5. $(\sqrt{5} - 1)(\sqrt{5} + 1) = (\sqrt{5})^2 - (1)^2$
$= 5 - 1 = 4$

9. $(\sqrt{20} - \sqrt{5})(\sqrt{20} + \sqrt{5})$
$= (\sqrt{20})^2 - (\sqrt{5})^2$
$= 20 - 5 = 15$

13. $(\sqrt{2} + 3)(\sqrt{3} - 2)$
$= \sqrt{2}\sqrt{3} + \sqrt{2}(-2) + 3\sqrt{3} + 3(-2)$
$= \sqrt{6} - 2\sqrt{2} + 3\sqrt{3} - 6$

17. $(\sqrt{k} + 4\sqrt{m})(\sqrt{k} - 4\sqrt{m})$
$= (\sqrt{k})^2 - (4\sqrt{m})^2 = k - 16m$

21. $(4\sqrt{t} + 1)^2 = (4\sqrt{t})^2 + 2(4\sqrt{t}) + 1^2$
$= 16t + 8\sqrt{t} + 1$

25. $\dfrac{12}{\sqrt{3}} = \dfrac{12 \cdot \sqrt{3}}{\sqrt{3} \cdot \sqrt{3}} = \dfrac{12\sqrt{3}}{3} = 4\sqrt{3}$

29. $\dfrac{9\sqrt{3}}{\sqrt{5}} = \dfrac{9\sqrt{3} \cdot \sqrt{5}}{\sqrt{5} \cdot \sqrt{5}} = \dfrac{9\sqrt{15}}{5}$

33. $\dfrac{3}{\sqrt{8}} = \dfrac{3}{2\sqrt{2}} = \dfrac{3 \cdot \sqrt{2}}{2\sqrt{2} \cdot \sqrt{2}} = \dfrac{3\sqrt{2}}{2 \cdot 2} = \dfrac{3\sqrt{2}}{4}$

37. $\dfrac{-6\sqrt{5}}{\sqrt{12}} = \dfrac{-6\sqrt{5}}{2\sqrt{3}} = \dfrac{-6\sqrt{5} \cdot \sqrt{3}}{2\sqrt{3} \cdot \sqrt{3}} = \dfrac{-6\sqrt{15}}{2 \cdot 3}$
$= \dfrac{-6\sqrt{15}}{6} = -\sqrt{15}$

41. $\dfrac{5\sqrt{2m}}{\sqrt{y^3}} = \dfrac{5\sqrt{2m} \cdot \sqrt{y}}{\sqrt{y^3} \cdot \sqrt{y}} = \dfrac{5\sqrt{2my}}{\sqrt{y^4}} = \dfrac{5\sqrt{2my}}{y^2}$

45. $-\sqrt{\dfrac{18}{7}} = -\dfrac{\sqrt{9} \cdot \sqrt{2}}{\sqrt{7}} = -\dfrac{3\sqrt{2} \cdot \sqrt{7}}{\sqrt{7} \cdot \sqrt{7}}$
$= -\dfrac{3\sqrt{14}}{7}$

49. $\sqrt{\dfrac{25}{y}} = \dfrac{\sqrt{25}}{\sqrt{y}} = \dfrac{5 \cdot \sqrt{y}}{\sqrt{y} \cdot \sqrt{y}} = \dfrac{5\sqrt{y}}{y}$

53. $-\sqrt{\dfrac{75m^3}{n}} = -\dfrac{\sqrt{75m^3}}{\sqrt{n}}$
$= -\dfrac{\sqrt{25m^2} \cdot \sqrt{3m} \cdot \sqrt{n}}{\sqrt{n} \cdot \sqrt{n}}$
$= -\dfrac{5m\sqrt{3mn}}{n}$

57. $\sqrt{\dfrac{1000r^7}{s^5}} = \dfrac{\sqrt{1000r^7}}{\sqrt{s^5}} = \dfrac{\sqrt{100r^6} \cdot \sqrt{10r}}{\sqrt{s^4} \cdot \sqrt{s}}$
$= \dfrac{10r^3\sqrt{10r} \cdot \sqrt{s}}{s^2\sqrt{s} \cdot \sqrt{s}}$
$= \dfrac{10r^3\sqrt{10rs}}{s^2 \cdot s} = \dfrac{10r^3\sqrt{10rs}}{s^3}$

61. $\sqrt[3]{\dfrac{10}{9}} = \dfrac{\sqrt[3]{10}}{\sqrt[3]{9}} = \dfrac{\sqrt[3]{10} \cdot \sqrt[3]{3}}{\sqrt[3]{9} \cdot \sqrt[3]{3}} = \dfrac{\sqrt[3]{30}}{\sqrt[3]{27}} = \dfrac{\sqrt[3]{30}}{3}$

65. $\sqrt[3]{\dfrac{p^{12}}{q^{11}}} = \dfrac{\sqrt[3]{p^{12}}}{\sqrt[3]{q^{11}}} = \dfrac{\sqrt[3]{(p^4)^3}}{\sqrt[3]{q^9}\sqrt[3]{q^2}}$
$= \dfrac{p^4}{\sqrt[3]{(q^3)^3}\sqrt[3]{q^2}} = \dfrac{p^4}{q^3\sqrt[3]{q^2}}$
$= \dfrac{p^4 \cdot \sqrt[3]{q}}{q^3\sqrt[3]{q^2} \cdot \sqrt[3]{q}} = \dfrac{p^4\sqrt[3]{q}}{q^3\sqrt[3]{q^2 \cdot q}}$
$= \dfrac{p^4\sqrt[3]{q}}{q^3\sqrt[3]{q^3}} = \dfrac{p^4\sqrt[3]{q}}{q^3 \cdot q} = \dfrac{p^4\sqrt[3]{q}}{q^4}$

69. $\dfrac{3}{\sqrt{3} - 1} = \dfrac{3(\sqrt{3} + 1)}{(\sqrt{3} - 1)(\sqrt{3} + 1)}$
$= \dfrac{3(\sqrt{3} + 1)}{(\sqrt{3})^2 - (1)^2}$
$= \dfrac{3(\sqrt{3} + 1)}{3 - 1}$
$= \dfrac{3(\sqrt{3} + 1)}{2}$

73. $\dfrac{\sqrt{27}}{2 + \sqrt{3}} = \dfrac{\sqrt{9}\sqrt{3}(2 - \sqrt{3})}{(2 + \sqrt{3})(2 - \sqrt{3})}$
$= \dfrac{3\sqrt{3}(2 - \sqrt{3})}{(2)^2 - (\sqrt{3})^2}$
$= \dfrac{3\sqrt{3}(2 - \sqrt{3})}{4 - 3}$
$= \dfrac{3\sqrt{3}(2 - \sqrt{3})}{1}$
$= 3\sqrt{3}(2 - \sqrt{3})$

77. $\dfrac{5 + \sqrt{6}}{\sqrt{3} - \sqrt{2}} = \dfrac{(5 + \sqrt{6})(\sqrt{3} + \sqrt{2})}{(\sqrt{3} - \sqrt{2})(\sqrt{3} + \sqrt{2})}$
$= \dfrac{5\sqrt{3} + 5\sqrt{2} + \sqrt{6}\sqrt{3} + \sqrt{6}\sqrt{2}}{(\sqrt{3})^2 - (\sqrt{2})^2}$
$= \dfrac{5\sqrt{3} + 5\sqrt{2} + \sqrt{18} + \sqrt{12}}{3 - 2}$
$= \dfrac{5\sqrt{3} + 5\sqrt{2} + \sqrt{9}\sqrt{2} + \sqrt{4}\sqrt{3}}{1}$
$= 5\sqrt{3} + 5\sqrt{2} + 3\sqrt{2} + 2\sqrt{3}$
$= 7\sqrt{3} + 8\sqrt{2}$

81. $\dfrac{3\sqrt{2p} + \sqrt{5s}}{\sqrt{2p} - 3\sqrt{5s}}$
$= \dfrac{(3\sqrt{2p} + \sqrt{5s})(\sqrt{2p} + 3\sqrt{5s})}{(\sqrt{2p} - 3\sqrt{5s})(\sqrt{2p} + 3\sqrt{5s})}$
$= \dfrac{3\sqrt{2p}\sqrt{2p} + 3\sqrt{2p}\cdot 3\sqrt{5s} + \sqrt{5s}\sqrt{2p} + \sqrt{5s}\cdot 3\sqrt{5s}}{(\sqrt{2p})^2 - (3\sqrt{5s})^2}$
$= \dfrac{3 \cdot 2p + 9\sqrt{10ps} + \sqrt{10ps} + 3 \cdot 5s}{2p - 9 \cdot 5s}$
$= \dfrac{6p + 10\sqrt{10ps} + 15s}{2p - 45s}$

85. $\dfrac{-5 + 5\sqrt{2}}{5} = \dfrac{5(-1 + \sqrt{2})}{5}$ Factor.
$= -1 + \sqrt{2}$

89. $\dfrac{11y - \sqrt{242y^5}}{22y}$
$= \dfrac{11y - \sqrt{121y^4}\sqrt{2y}}{22y}$
$= \dfrac{11y - 11y^2\sqrt{2y}}{22y}$
$= \dfrac{11y(1 - y\sqrt{2y})}{22y} = \dfrac{1 - y\sqrt{2y}}{2}$

93.
$$15a^2 + 2 = 11a$$
$$15a^2 - 11a + 2 = 0$$
$$(3a - 1)(5a - 2) = 0$$
$$3a - 1 = 0 \quad \text{or} \quad 5a - 2 = 0$$
$$3a = 1 \qquad\qquad 5a = 2$$
$$a = \frac{1}{3} \quad \text{or} \quad a = \frac{2}{5}$$
Solution set: $\left\{\frac{1}{3}, \frac{2}{5}\right\}$

Section 5.6 (page 317)

1. $\sqrt{k} = 2$ Square both sides: $k = 4$.
Solution set: $\{4\}$

5. $\sqrt{6k - 1} = 1$
$6k - 1 = 1$ Square both sides.
$6k = 2$
$k = \frac{1}{3}$
Check this solution. Solution set: $\left\{\frac{1}{3}\right\}$

9. $\sqrt{w} - 3 = 0$
$\sqrt{w} = 3$
$w = 9$ Square both sides.
Check this solution. Solution set: $\{9\}$

13. $4 - \sqrt{x - 2} = 0$
$\sqrt{x - 2} = 4$
$x - 2 = 16$ Square both sides.
$x = 18$
Check this solution. Solution set: $\{18\}$

17. $\sqrt{9a - 4} = \sqrt{8a + 1}$
$9a - 4 = 8a + 1$ Square both sides.
$a = 5$
Check: $\sqrt{9(5) - 4} = \sqrt{45 - 4} = \sqrt{41}$
$\sqrt{8(5) + 1} = \sqrt{40 + 1} = \sqrt{41}$
Solution set: $\{5\}$

21. $2\sqrt{k} = \sqrt{3k + 4}$
$4k = 3k + 4$ Square both sides.
$k = 4$
Check this solution. Solution set: $\{4\}$

25. $\sqrt{7z + 50} = z + 8$
$(\sqrt{7z + 50})^2 = (z + 8)^2$
$7z + 50 = z^2 + 16z + 64$
$z^2 + 9z + 14 = 0$
$(z + 7)(z + 2) = 0$
$z + 7 = 0 \quad \text{or} \quad z + 2 = 0$
$z = -7 \quad \text{or} \quad z = -2$
Check by substituting back in the original equation.
The solution set is $\{-7, -2\}$.

29. $\sqrt{m^2 + 3m + 12} - m - 2 = 0$
Get the radical alone on one side.
$\sqrt{m^2 + 3m + 12} = m + 2$
$(\sqrt{m^2 + 3m + 12})^2 = (m + 2)^2$
$m^2 + 3m + 12 = m^2 + 4m + 4$
$8 = m$
Check by substituting back in the original equation. The solution set is $\{8\}$.

33. $\sqrt{y + 4} + \sqrt{y - 4} = 4$
$\sqrt{y + 4} = 4 - \sqrt{y - 4}$
$(\sqrt{y + 4})^2 = (4 - \sqrt{y - 4})^2$
$y + 4 = 16 - 8\sqrt{y - 4} + y - 4$
$-8 = -8\sqrt{y - 4}$
$1 = \sqrt{y - 4}$
$(1)^2 = (\sqrt{y - 4})^2$
$1 = y - 4$
$5 = y$
Check by substituting back in the original equation.
The solution set is $\{5\}$.

37. Cube both sides.
$\sqrt[3]{r^2 + 2r + 8} = \sqrt[3]{r^2}$
$(\sqrt[3]{r^2 + 2r + 8})^3 = (\sqrt[3]{r^2})^3$
$r^2 + 2r + 8 = r^2$
$2r = -8$
$r = -4$
Solution set: $\{-4\}$

41. Raise both sides to the fourth power.
$\sqrt[4]{z + 11} = \sqrt[4]{2z + 6}$
$(\sqrt[4]{z + 11})^4 = (\sqrt[4]{2z + 6})^4$
$z + 11 = 2z + 6$
$5 = z$
Check by substituting back in the original equation.
The solution set is $\{5\}$.

45. $(3y + 7)^{1/2} - (y + 2)^{1/2} = 1$
$\sqrt{3y + 7} - \sqrt{y + 2} = 1$
$\sqrt{3y + 7} = 1 + \sqrt{y + 2}$
$(\sqrt{3y + 7})^2 = (1 + \sqrt{y + 2})^2$
$3y + 7 = 1^2 + 2 \cdot 1 \cdot \sqrt{y + 2}$
$\qquad\qquad + (\sqrt{y + 2})^2$
$3y + 7 = 1 + 2\sqrt{y + 2} + y + 2$
$2y + 4 = 2\sqrt{y + 2}$
$y + 2 = \sqrt{y + 2}$
$(y + 2)^2 = (\sqrt{y + 2})^2$
$y^2 + 4y + 4 = y + 2$
$y^2 + 3y + 2 = 0$
$(y + 1)(y + 2) = 0$
$y + 1 = 0 \quad \text{or} \quad y + 2 = 0$
$y = -1 \qquad\qquad y = -2$
Both solutions check, so the solution set is $\{-1, -2\}$.

49. Use $P = 2\pi\sqrt{\dfrac{L}{32}}$ with $P = 2$ and $\pi \approx 3.14$.

$2 = 2(3.14)\sqrt{\dfrac{L}{32}}$

$2 = 6.28\sqrt{\dfrac{L}{32}}$

$2^2 = \left(6.28\sqrt{\dfrac{L}{32}}\right)^2$

$4 = 39.44\left(\dfrac{L}{32}\right)$

$4 = 1.23L$
$L = 3.2$
The pendulum is 3.2 feet long.

53. $\dfrac{2}{4+\sqrt{3}} \cdot \dfrac{4-\sqrt{3}}{4-\sqrt{3}} = \dfrac{2(4-\sqrt{3})}{4^2-(\sqrt{3})^2}$
$= \dfrac{2(4-\sqrt{3})}{16-3}$
$= \dfrac{2(4-\sqrt{3})}{13}$

Section 5.7 (page 327)

1. $\sqrt{-144} = \sqrt{144 \cdot -1} = \sqrt{144} \cdot \sqrt{-1} = 12i$
5. $\sqrt{-3} = \sqrt{-1 \cdot 3} = \sqrt{-1} \cdot \sqrt{3} = i\sqrt{3}$
9. $\sqrt{-500} = \sqrt{-1 \cdot 100 \cdot 5} = \sqrt{-1} \cdot \sqrt{100} \cdot \sqrt{5}$
   $= i \cdot 10\sqrt{5} = 10i\sqrt{5}$
13. $-3\sqrt{-28} + 2\sqrt{-63}$
    $= -3\sqrt{-1 \cdot 4 \cdot 7} + 2\sqrt{-1 \cdot 9 \cdot 7}$
    $= -3 \cdot 2i\sqrt{7} + 2 \cdot 3i\sqrt{7}$
    $= -6i\sqrt{7} + 6i\sqrt{7}$
    $= 0$
17. $\sqrt{-4} \cdot \sqrt{-81} = 2i \cdot 9i = 18i^2 = 18(-1)$
    $= -18$
21. $\dfrac{\sqrt{-54}}{\sqrt{6}} = \dfrac{3i\sqrt{6}}{\sqrt{6}} = 3i$
25. $(6 + 2i) + (4 + 3i)$
    $= (6 + 4) + (2i + 3i)$
    $= 10 + 5i$
29. $(5 - i) + (-5 + i)$
    $= (5 - 5) + (-i + i)$
    $= 0 + 0i$
33. $(4 + i) + (-3 - 2i)$
    $= (4 - 3) + (i - 2i)$
    $= 1 - i$
37. $(-4 + 11i) + (-2 - 4i) + (-3i)$
    $= (-4 - 2) + (11i - 4i - 3i)$
    $= -6 + 4i$
41. $(2i)(5i) = 10i^2 = -10$
45. $(-6i)(-5i) = 30i^2 = -30$
49. $(7 + 3i)(-5 + i)$
    $= -35 + 7i - 15i + 3i^2$
    $= -35 - 8i - 3$
    $= -38 - 8i$
53. $(-1 + 3i)^2 = 1 - 6i + 9i^2$
    $= 1 - 6i - 9$
    $= -8 - 6i$
57. $(1 + i)(1 - i) = 1 - i^2$
    $= 1 + 1$
    $= 2$
61. $\dfrac{i}{2+i} \cdot \dfrac{2-i}{2-i} = \dfrac{2i - i^2}{4 - i^2} = \dfrac{2i - (-1)}{4 - (-1)}$
    $= \dfrac{1 + 2i}{5} = \dfrac{1}{5} + \dfrac{2}{5}i$
65. $\dfrac{1-i}{1+i} = \dfrac{(1-i)(1-i)}{(1+i)(1-i)}$
    $= \dfrac{1 - 2i + i^2}{1 - i^2}$
    $= \dfrac{1 - 2i - 1}{1 - (-1)}$
    $= \dfrac{-2i}{2} = -i$

69. $i^{36} = (i^4)^9 = 1^9 = 1$
73. $i^{-4} = (i^4)^{-1} = 1^{-1} = 1$
77. $I = \dfrac{E}{R + (X_L - X_c)i}$
    $I = \dfrac{2 + 3i}{5 + (4 - 3)i}$
    $= \dfrac{2 + 3i}{5 + i}$
    $= \dfrac{2 + 3i}{5 + i} \cdot \dfrac{5 - i}{5 - i}$
    $= \dfrac{10 - 2i + 15i - 3i^2}{25 - i^2}$
    $= \dfrac{13 + 13i}{26}$
    $= \dfrac{13}{26} + \dfrac{13i}{26}$
    $= \dfrac{1}{2} + \dfrac{1}{2}i$
81. The square roots of 25 are -5 and 5.
85. $x^2 + 3x - 40 = 0$
    $(x - 5)(x + 8) = 0$
    $x - 5 = 0$ or $x + 8 = 0$
    $x = 5$ $x = -8$
    Solution set: $\{5, -8\}$

CHAPTER 6

Section 6.1 (page 347)

1. $b^2 = 49$, $b = \pm\sqrt{49} = \pm 7$
   Solution set: $\{7, -7\}$
5. $k^2 = 12$
   $k = \sqrt{12}$ or $k = -\sqrt{12}$
   $k = 2\sqrt{3}, -2\sqrt{3}$
   Solution set: $\{2\sqrt{3}, -2\sqrt{3}\}$
9. $(3a - 1)^2 = 3$
   $3a - 1 = \sqrt{3}$ or $3a - 1 = -\sqrt{3}$
   $3a = 1 + \sqrt{3}$ $3a = 1 - \sqrt{3}$
   $a = \dfrac{1 + \sqrt{3}}{3}$ or $a = \dfrac{1 - \sqrt{3}}{3}$
   Solution set: $\left\{\dfrac{1 + \sqrt{3}}{3}, \dfrac{1 - \sqrt{3}}{3}\right\}$
13. $m^2 = -72$
    $m = \sqrt{-72}$ or $m = -\sqrt{-72}$
    $m = i\sqrt{72}$ $m = -i\sqrt{72}$
    $m = 6i\sqrt{2}$ or $m = -6i\sqrt{2}$
    Solution set: $\{6i\sqrt{2}, -6i\sqrt{2}\}$.
17. $(2m - 1)^2 = -3$
    $2m - 1 = i\sqrt{3}$ or $2m - 1 = -i\sqrt{3}$
    $2m = 1 + i\sqrt{3}$ $2m = 1 - i\sqrt{3}$
    $m = \dfrac{1 + i\sqrt{3}}{2}$ or $m = \dfrac{1 - i\sqrt{3}}{2}$
    Solution set: $\left\{\dfrac{1 + i\sqrt{3}}{2}, \dfrac{1 - i\sqrt{3}}{2}\right\}$

21.
$$2y^2 + y = 15$$
$$y^2 + \left(\frac{1}{2}\right)y = \frac{15}{2} \quad \text{Divide by 2.}$$

Half of $\frac{1}{2}$ is $\frac{1}{4}$; square $\frac{1}{4}$ to get $\frac{1}{16}$. Add $\frac{1}{16}$ to *both* sides to get

$$y^2 + \frac{1}{2}y + \frac{1}{16} = \frac{15}{2} + \frac{1}{16}$$
$$\left(y + \frac{1}{4}\right)^2 = \frac{121}{16}$$
$$y + \frac{1}{4} = \frac{11}{4} \quad \text{or} \quad y + \frac{1}{4} = -\frac{11}{4}$$
$$y = \frac{5}{2} \quad \text{or} \quad y = -3$$

Solution set: $\left\{\frac{5}{2}, -3\right\}$

25. $x^2 - 2x - 1 = 0$
$$x^2 - 2x = 1$$

Half of -2, squared, is 1. Add 1 to *both* sides to get
$$x^2 - 2x + 1 = 2$$
$$(x - 1)^2 = 2$$
$$x - 1 = \sqrt{2} \quad \text{or} \quad x - 1 = -\sqrt{2}$$
$$x = 1 + \sqrt{2} \quad \text{or} \quad x = 1 - \sqrt{2}$$

Solution set: $\{1 + \sqrt{2},\ 1 - \sqrt{2}\}$

29.
$$x^2 - x = \frac{1}{2}$$
$$x^2 - x + \frac{1}{4} = \frac{1}{2} + \frac{1}{4}$$
$$\left(x - \frac{1}{2}\right)^2 = \frac{3}{4}$$
$$x - \frac{1}{2} = \frac{\sqrt{3}}{2} \quad \text{or} \quad x - \frac{1}{2} = -\frac{\sqrt{3}}{2}$$
$$x = \frac{1 + \sqrt{3}}{2} \qquad x = \frac{1 - \sqrt{3}}{2}$$

Solution set: $\left\{\dfrac{1 + \sqrt{3}}{2}, \dfrac{1 - \sqrt{3}}{2}\right\}$

33. $m^2 + 6m + 10 = 0$
$$m^2 + 6m = -10$$
$$m^2 + 6m + 9 = -10 + 9$$
$$(m + 3)^2 = -1$$
$$m + 3 = i \quad \text{or} \quad m + 3 = -i$$
$$m = -3 + i \quad \text{or} \quad m = -3 - i$$

Solution set: $\{-3 + i, -3 - i\}$

37. $\sqrt{b^2 - 4ac} = \sqrt{(1)^2 - 4(3)(-1)}$
$a = 3, b = 1, c = -1$
$$= \sqrt{1 + 12}$$
$$= \sqrt{13}$$

41. $\sqrt{b^2 - 4ac} = \sqrt{(5)^2 - 4(7)(-5)}$
$a = 7, b = 5, c = -5$
$$= \sqrt{25 + 140}$$
$$= \sqrt{165}$$

Section 6.2 (page 355)

1. $m^2 + 8m + 15 = 0 \quad a = 1, b = 8, c = 15$
$$m = \frac{-b \pm \sqrt{b^2 - 4ac}}{2a}$$
$$= \frac{-8 \pm \sqrt{8^2 - 4(1)(15)}}{2(1)}$$
$$= \frac{-8 \pm \sqrt{64 - 60}}{2}$$
$$= \frac{-8 \pm \sqrt{4}}{2} = \frac{-8 \pm 2}{2}$$
$$m = \frac{-8 + 2}{2} = -3 \quad \text{or} \quad m = \frac{-8 - 2}{2} = -5$$

Solution set: $\{-3, -5\}$

5. $m^2 + 18 = 10m$
$m^2 - 10m + 18 = 0 \quad a = 1, b = -10, c = 18$
$$m = \frac{-b \pm \sqrt{b^2 - 4ac}}{2a}$$
$$m = \frac{-(-10) \pm \sqrt{(-10)^2 - 4(1)(18)}}{2(1)}$$
$$= \frac{10 \pm \sqrt{100 - 72}}{2}$$
$$= \frac{10 \pm \sqrt{28}}{2} = \frac{10 \pm 2\sqrt{7}}{2} = 5 \pm \sqrt{7}$$

Solution set: $\{5 + \sqrt{7}, 5 - \sqrt{7}\}$

9. Put $4x(x + 1) = 1$ into $ax^2 + bx + c = 0$ form.
$$4x(x + 1) = 1$$
$$4x^2 + 4x = 1$$
$$4x^2 + 4x - 1 = 0$$

Use $a = 4$, $b = 4$, and $c = -1$ in the quadratic formula.
$$x = \frac{-b \pm \sqrt{b^2 - 4ac}}{2a}$$
$$x = \frac{-4 \pm \sqrt{(4)^2 - 4(4)(-1)}}{2(4)}$$
$$= \frac{-4 \pm \sqrt{16 + 16}}{8}$$
$$= \frac{-4 \pm \sqrt{32}}{8}$$
$$= \frac{-4 \pm 4\sqrt{2}}{8}$$
$$= \frac{4(-1 \pm \sqrt{2})}{8}$$
$$= \frac{-1 \pm \sqrt{2}}{2}$$

Solution set: $\left\{\dfrac{-1 + \sqrt{2}}{2}, \dfrac{-1 - \sqrt{2}}{2}\right\}$

13. $3x^2 + 4x + 2 = 0$
$a = 3, b = 4, c = 2$
$$x = \frac{-b \pm \sqrt{b^2 - 4ac}}{2a}$$
$$x = \frac{-4 \pm \sqrt{4^2 - 4(3)(2)}}{2(3)}$$
$$= \frac{-4 \pm \sqrt{-8}}{6}$$
$$= \frac{-4 \pm 2i\sqrt{2}}{6} = \frac{-2 \pm i\sqrt{2}}{3}$$
Solution set: $\left\{\dfrac{-2 + i\sqrt{2}}{3}, \dfrac{-2 - i\sqrt{2}}{3}\right\}$

17. $4z^2 = 4z - 7$
$4z^2 - 4z + 7 = 0$
$a = 4, b = -4, c = 7$
$$z = \frac{-b \pm \sqrt{b^2 - 4ac}}{2a}$$
$$z = \frac{-(-4) \pm \sqrt{(-4)^2 - 4(4)(7)}}{2(4)}$$
$$= \frac{4 \pm \sqrt{-96}}{8}$$
$$= \frac{4 \pm \sqrt{-16 \cdot 6}}{8} = \frac{4 \pm 4i\sqrt{6}}{8}$$
$$= \frac{1 \pm i\sqrt{6}}{2}$$
Solution set: $\left\{\dfrac{1 + i\sqrt{6}}{2}, \dfrac{1 - i\sqrt{6}}{2}\right\}$

21. $3 - \dfrac{1}{w} + \dfrac{4}{w^2} = 0$
$w^2\left(3 - \dfrac{1}{w} + \dfrac{4}{w^2}\right) = w^2 \cdot 0$
$3w^2 - w + 4 = 0$
$a = 3, b = -1, c = 4$
$$w = \frac{-(-1) \pm \sqrt{(-1)^2 - 4(3)(4)}}{2(3)}$$
$$= \frac{1 \pm \sqrt{1 - 48}}{6} = \frac{1 \pm i\sqrt{47}}{6}$$
Solution set: $\left\{\dfrac{1 + i\sqrt{47}}{6}, \dfrac{1 - i\sqrt{47}}{6}\right\}$

25. Let $d = 10$ in the equation $d = t^2 + 5t + 2$.
$10 = t^2 + 5t + 2$
$0 = t^2 + 5t - 8$
$a = 1, b = 5, c = -8$
$$t = \frac{-5 \pm \sqrt{(5)^2 - 4(1)(-8)}}{2(1)}$$
$$= \frac{-5 \pm \sqrt{57}}{2}$$
Reject $\dfrac{-5 - \sqrt{57}}{2}$, since t cannot be negative.
The object will be 10 feet from its starting point at
$t = \dfrac{-5 + \sqrt{57}}{2} \approx 1.3$ seconds.

29. $x^2 + 4x + 1 = 0 \quad a = 1, b = 4, c = 1$
$b^2 - 4ac = 4^2 - 4(1)(1) = 16 - 4 = 12 > 0$
12 is not a perfect square; two distinct irrational number solutions, choice (c)

33. $4y^2 - y - 10 = 0$; then $a = 4, b = -1$, and $c = -10$.
$b^2 - 4ac = (-1)^2 - 4(4)(-10)$
$= 1 + 160$
$= 161$
Since 161 is not a perfect square, the answer is (c).

37. $9y^2 - 30y + 45 = 0$; then $a = 9, b = -30$, and $c = 45$.
$b^2 - 4ac = (-30)^2 - 4(9)(45)$
$= 900 - 1620$
$= -720$
Since $-720 < 0$, the answer is (d).

41. Rewrite the equation with integer coefficients.
$$\frac{w^2}{2} + 5w - 1 = 0$$
$$2\left(\frac{w^2}{2} + 5w - 1\right) = 2(0)$$
$$w^2 + 10w - 2 = 0$$
Then $a = 1, b = 10$, and $c = -2$.
$b^2 - 4ac = 10^2 - 4(1)(-2)$
$= 100 + 8$
$= 108$
Since 108 is not a perfect square, the answer is (c).

45. $8r^2 - 2r - 21$
$a = 8, b = -2, c = -21$
$b^2 - 4ac = (-2)^2 - 4(8)(-21)$
$= 4 + 672$
$= 676$
Since $676 = 26^2$, the polynomial can be factored.
$8r^2 - 2r - 21 = (2r + 3)(4r - 7)$

49. $15m^2 - 17m - 12$
$a = 15, b = -17, c = -12$
$b^2 - 4ac = (-17)^2 - 4(15)(-12)$
$= 289 + 720$
$= 1009$
Since 1009 is not a perfect square, the polynomial cannot be factored.

53. $\sqrt{y + 1} = -1 + \sqrt{2y}$
$y + 1 = 1 - 2\sqrt{2y} + 2y$
$2\sqrt{2y} = y$
$4(2y) = y^2$
$0 = y^2 - 8y$
$0 = y(y - 8)$
$y = 0 \quad$ or $\quad y = 8$
Check: $\sqrt{0 + 1} = -1 + \sqrt{2 \cdot 0}$ Let $y = 0$.
$1 = -1$ False
$\sqrt{8 + 1} = -1 + \sqrt{2(8)}$ Let $y = 8$.
$\sqrt{9} = -1 + \sqrt{16}$
$3 = -1 + 4$
$3 = 3$ True
Solution set: $\{8\}$

57. $2(r + 3)^2 - 5(r + 3) + 2; \; r + 3 = t$
$2t^2 - 5t + 2 = (2t - 1)(t - 2)$

Section 6.3 (page 365)

1.
$$3b - 5 = \frac{2}{b}$$
$3b^2 - 5b = 2$ Multiply by b.
$3b^2 - 5b - 2 = 0$ Add -2.
$(3b + 1)(b - 2) = 0$ Factor.
$3b + 1 = 0$ or $b - 2 = 0$
$b = -\frac{1}{3}$ or $b = 2$

Solution set: $\left\{-\frac{1}{3}, 2\right\}$

5.
$$3 = \frac{1}{y} + \frac{2}{y^2}$$
$3y^2 = y + 2$ Multiply by y^2.
$3y^2 - y - 2 = 0$
$(3y + 2)(y - 1) = 0$
$y = -\frac{2}{3}, 1$

Solution set: $\left\{-\frac{2}{3}, 1\right\}$

9. $1 - \frac{2}{m} - \frac{1}{2m^2} = 0$
$2m^2 - 4m - 1 = 0$ Multiply by $2m^2$.
Use the quadratic formula.
$$m = \frac{-(-4) \pm \sqrt{(-4)^2 - 4(2)(-1)}}{2(2)}$$
$$= \frac{4 \pm \sqrt{16 + 8}}{4} = \frac{4 \pm 2\sqrt{6}}{4} = \frac{2 \pm \sqrt{6}}{2}$$
$$m = \frac{2 + \sqrt{6}}{2}, \frac{2 - \sqrt{6}}{2}$$

Solution set: $\left\{\frac{2 + \sqrt{6}}{2}, \frac{2 - \sqrt{6}}{2}\right\}$

13.
$$p = 8 + 2\sqrt{p}$$
$p - 8 = 2\sqrt{p}$
$(p - 8)^2 = (2\sqrt{p})^2$
$p^2 - 16p + 64 = 4p$
$p^2 - 20p + 64 = 0$
$(p - 16)(p - 4) = 0$
$p = 16$ or $p = 4$
Check: $p = 16$ | $p = 4$
$16 = 8 + 2\sqrt{16}$ | $4 = 8 + 2\sqrt{4}$
$16 = 8 + 2(4)$ | $4 = 8 + 2(2)$
$16 = 16$ True | $4 = 12$ False

Solution set: $\{16\}$

17.
$$3k + 1 = -\sqrt{3k + 3}$$
$9k^2 + 6k + 1 = 3k + 3$
$9k^2 + 3k - 2 = 0$
$(3k + 2)(3k - 1) = 0$
$3k + 2 = 0$ or $3k - 1 = 0$
$3k = -2$ or $3k = 1$
$k = \frac{-2}{3}$ or $k = \frac{1}{3}$

Check: $k = -\frac{2}{3}$
$3\left(-\frac{2}{3}\right) + 1 = -\sqrt{3\left(-\frac{2}{3}\right) + 3}$
$-2 + 1 = -\sqrt{-2 + 3}$
$-1 = -1$ True

$k = \frac{1}{3}$
$3\left(\frac{1}{3}\right) + 1 = -\sqrt{3\left(\frac{1}{3}\right) + 3}$
$1 + 1 = -\sqrt{1 + 3}$
$2 = -2$ False

Solution set: $\left\{-\frac{2}{3}\right\}$

21. Let $x = m^2$; then $x^2 = m^4$.
So $9m^4 - 25m^2 + 16 = 0$ becomes
$9x^2 - 25x + 16 = 0$.
$(9x - 16)(x - 1) = 0$ Factor.
$9x - 16 = 0$ or $x - 1 = 0$
$9x = 16$ or $x = 1$
$x = \frac{16}{9}$

If $m^2 = \frac{16}{9}$, $m = \pm\frac{4}{3}$, and if $m^2 = 1$,
$m = \pm 1$.

Solution set: $\left\{\frac{4}{3}, -\frac{4}{3}, 1, -1\right\}$

25. Let $x = 2p + 2$. Then $3(2p + 2)^2 - 7(2p + 2) - 6 = 0$ becomes
$3x^2 - 7x - 6 = 0$
$(3x + 2)(x - 3) = 0$
$3x + 2 = 0$ or $x - 3 = 0$
$3x = -2$ or $x = 3$
$x = -\frac{2}{3}$

Solve $x = 2p + 2$ to get $p = \frac{1}{2}(x - 2)$.

When $x = 3$, $p = \frac{1}{2}(3 - 2) = \frac{1}{2}$

When $x = -\frac{2}{3}$, $p = \frac{1}{2}\left(-\frac{2}{3} - 2\right) = \frac{1}{2}\left(-\frac{8}{3}\right)$
$= -\frac{4}{3}$.

Solution set: $\left\{\frac{1}{2}, -\frac{4}{3}\right\}$

29. Let $x = p - 1$. Then
$3 = 2(p - 1)^{-1} + (p - 1)^{-2}$ becomes
$3 = 2x^{-1} + x^{-2}$
or $3 = \frac{2}{x} + \frac{1}{x^2}$.

Multiply each side by the LCD, x^2.
$$3x^2 = 2x + 1$$
$$3x^2 - 2x - 1 = 0$$
$$(3x + 1)(x - 1) = 0 \quad \text{Factor.}$$
$$3x + 1 = 0 \quad \text{or} \quad x - 1 = 0$$
$$x = -\frac{1}{3} \quad \text{or} \quad x = 1$$
$p = x + 1$, so
$$p = \frac{2}{3} \quad \text{or} \quad p = 2.$$
Solution set: $\left\{\frac{2}{3}, 2\right\}$

33. Let x be Steve's average speed.
Then $x - 20$ is Paula's speed.

Steve's time from Albuquerque to Amarillo is $\dfrac{300}{x}$
and Paula's time is $\dfrac{300}{x - 20}$. Since it takes Paula $\dfrac{5}{4}$ hours longer,
$$\frac{300}{x - 20} - \frac{300}{x} = \frac{5}{4}.$$
$$\frac{60}{x - 20} - \frac{60}{x} = \frac{1}{4} \quad \text{Divide by 5.}$$
Multiply each side by the LCD, $4x(x - 20)$.
$$4x(x - 20)\left(\frac{60}{x - 20} - \frac{60}{x}\right) = 4x(x - 20)\left(\frac{1}{4}\right)$$
$$240x - 240(x - 20) = x(x - 20)$$
$$4800 = x^2 - 20x$$
$$x^2 - 20x - 4800 = 0$$
$$(x - 80)(x + 60) = 0$$
$$x - 80 = 0 \quad \text{or} \quad x + 60 = 0$$
$$x = 80 \quad \text{or} \quad x = -60$$
$x = -60$ is not reasonable. Steve's average speed is 80 miles per hour.

37. Let $z^{1/3} = x$. Then $z^{2/3} = 4z^{1/3} - 3$ becomes
$$x^2 = 4x - 3.$$
$$x^2 - 4x + 3 = 0$$
$$(x - 3)(x - 1) = 0$$
$$x = 3 \quad \text{or} \quad x = 1$$
$$z^{1/3} = 3 \qquad z^{1/3} = 1$$
$$z = 3^3 = 27 \qquad z = 1^3 = 1$$
Solution set: $\{1, 27\}$

41. $A = \dfrac{1}{2}bh$ for h
$$2A = bh$$
$$\frac{2A}{b} = h$$

45. $\dfrac{r^2 - 6}{3} = q^2$ for r^2
$$r^2 - 6 = 3q^2$$
$$r^2 = 3q^2 + 6$$

Section 6.4 (page 373)

1. $c^2 = a^2 + b^2$ for a
$$a^2 = c^2 - b^2$$
$$a = \pm\sqrt{c^2 - b^2}$$

5. $R = \dfrac{k}{d^2}$ for d
$$d^2 = \frac{k}{R} \quad \begin{array}{l}\text{Multiply by } d^2; \\ \text{divide by } R.\end{array}$$
$$d = \pm\sqrt{\frac{k}{R}} = \pm\frac{\sqrt{kR}}{R}$$

9. $V = \left(\dfrac{1}{3}\right)\pi r^2 h$ for r
$$3V = \pi r^2 h$$
$$r^2 = \frac{3V}{\pi h}$$
$$r = \pm\sqrt{\frac{3V}{\pi h}} = \pm\frac{\sqrt{3\pi V h}}{\pi h}$$

13. $D = \sqrt{kh}$ for h
$$D^2 = kh$$
$$h = \frac{D^2}{k}$$

17. $D = 13t^2 - 100t$
$$200 = 13t^2 - 100t \quad D = 200 \text{ feet}$$
$$0 = 13t^2 - 100t - 200$$
$$t = \frac{100 \pm \sqrt{(-100)^2 - 4(13)(-200)}}{2(13)}$$
$$= \frac{100 \pm \sqrt{10{,}000 + 10{,}400}}{26}$$
$$= \frac{100 \pm \sqrt{20{,}400}}{26}$$
$t \approx 9.3 \quad \text{or} \quad t \approx -1.6$
Since time is positive, the skid would take about 9.3 seconds.

21. Let h be the height of the tower. Then the distance to the top of the tower from a point 30 meters from the base is $2h + 2$. This configuration forms a right triangle, and so
$$h^2 + (30)^2 = (2h + 2)^2.$$
$$h^2 + 900 = 4h^2 + 8h + 4$$
$$3h^2 + 8h - 896 = 0$$
$$(3h + 56)(h - 16) = 0$$
$$h = \frac{-56}{3} \quad \text{or} \quad h = 16$$
$\dfrac{-56}{3}$ cannot be a solution of the problem. The tower is 16 meters high.

25. Let x be the width of the strip. The dimensions of the rug are $20 - 2x$ by $15 - 2x$, so the area of the rug is
$$(20 - 2x)(15 - 2x) = 234.$$

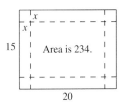

Solve the equation.
$$300 - 70x + 4x^2 = 234$$
$$4x^2 - 70x + 66 = 0$$
$$2(2x - 33)(x - 1) = 0$$
$$x = \frac{33}{2} \text{ or } x = 1$$

Reject $\frac{33}{2}$ as too large. The strip should be 1 foot wide.

29. Let m be the time the faucet would take to fill the sink. Then $m + 5$ minutes is the time the drain takes to empty the sink. The rate of water coming into the sink is $\frac{1}{m}$, and the rate of water going out is $\frac{1}{m+5}$. With the faucet and the drain both open, the rate of filling the sink is
$$\frac{1}{m} - \frac{1}{m+5} = \frac{1}{30}.$$
$$30m(m+5)\left(\frac{1}{m} - \frac{1}{m+5}\right) = \frac{1}{30}(30m)(m+5)$$
$$30(m+5) - 30m = m^2 + 5m$$
$$30m + 150 - 30m = m^2 + 5m$$
$$0 = m^2 + 5m - 150$$
$$m = \frac{-5 \pm \sqrt{5^2 - 4(1)(-150)}}{2(1)}$$
$$= \frac{-5 \pm \sqrt{25 + 600}}{2}$$
$$= \frac{-5 \pm \sqrt{625}}{2} = \frac{-5 \pm 25}{2}$$
$$m = 10 \text{ or } m = -15$$

The time must be positive, so $m = 10$ is the solution. The faucet can fill the sink in 10 minutes.

33. Let x be the plane's speed in still air. Then $x - 20$ is the speed against the wind and $x + 20$ is the speed with the wind. The time flown against the wind is $\frac{600}{x-20}$ and the time with the wind is $\frac{600}{x+20}$.

Since the time flying with the wind was 2.5 hours less,
$$\frac{600}{x+20} = \frac{600}{x-20} - 2.5.$$
$$(x+20)(x-20)\left(\frac{600}{x+20}\right)$$
$$= \left(\frac{600}{x-20} - 2.5\right)(x+20) \cdot$$
$$(x - 20)$$
$$(x - 20)(600) = 600(x+20) - 2.5(x+20) \cdot$$
$$(x - 20)$$
$$600x - 12{,}000 = 600x + 12{,}000 - 2.5(x^2 - 400)$$
$$-12{,}000 = 12{,}000 - 2.5x^2 + 1000$$
$$2.5x^2 - 25{,}000 = 0$$
$$x^2 - 10{,}000 = 0$$
$$(x+100)(x-100) = 0$$
$$x = -100 \text{ or } x = 100$$

Choose the positive solution. The speed of the plane in still air is 100 miles per hour.

37. For supply to equal demand, we must have
$$5P - 1 = \frac{700}{P}.$$
Now solve for P.
$$5P^2 - P = 700$$
$$5P^2 - P - 700 = 0$$
$$P = \frac{-(-1) \pm \sqrt{(-1)^2 - 4(5)(-700)}}{2(5)}$$
$$= \frac{1 \pm \sqrt{14{,}001}}{10}$$
$$\approx 11.93, -11.73$$

Choose the positive value. Supply equals demand at $11.93.

41. In order to graph $(-1, 5]$, use (at -1 since -1 is not included, and use] at 5 since 5 is included. Shade the graph between -1 and 5. See the answer section for the graph.

Section 6.5 (page 383)

See the answer section for all final graphs in this section.

1. $(2x + 1)(x - 5) > 0$
Solve $(2x + 1)(x - 5) = 0$.
$2x + 1 = 0$ or $x - 5 = 0$
$x = -\frac{1}{2}$ or $x = 5$

Locate these numbers on a number line.

```
        T     F     T
     ───┼───────┼───→
       -½      5
```

Let $x = -1$:
$(2 \cdot -1 + 1)(-1 - 5) = -1 \cdot -6$
$\qquad = 6 > 0.$ True

Let $x = 0$:
$(2 \cdot 0 + 1)(0 - 5) = 1 \cdot (-5)$
$\qquad = -5 > 0.$ False

Let $x = 6$:
$(2 \cdot 6 + 1)(6 - 5) = 13 \cdot 1$
$\qquad = 13 > 0.$ True

The solution set is $\left(-\infty, -\frac{1}{2}\right) \cup (5, \infty)$.

5. $x^2 - 4x + 3 < 0$
$x^2 - 4x + 3 = 0$
$(x - 3)(x - 1) = 0$
$x - 3 = 0$ or $x - 1 = 0$
$x = 3$ or $x = 1$

Locate these numbers on a number line.

```
        F     T     F
     ───┼───────┼───→
        1      3
```

Let $x = 0$: $0^2 - 4 \cdot 0 + 3 = 3 < 0.$ False
Let $x = 2$: $2^2 - 4 \cdot 2 + 3 = -1 < 0.$ True
Let $x = 4$: $4^2 - 4 \cdot 4 + 3 = 3 < 0.$ False

The solution set is $(1, 3)$.

9. $3r^2 + 10r \leq 8$
$3r^2 + 10r = 8$
$3r^2 + 10r - 8 = 0$
$(3r - 2)(r + 4) = 0$
$3r - 2 = 0$ or $r + 4 = 0$
$r = \dfrac{2}{3}$ or $r = -4$

Locate these numbers on a number line.

```
   F        T        F
───┼────────┼────────→
  -4       2/3
```

Let $r = -5$:
$3(-5)^2 + 10(-5) = 75 - 50$
$= 25 \leq 8.$ False
Let $r = 0$: $3(0)^2 + 10(0) = 0 \leq 8.$ True
Let $r = 1$: $3(1)^2 + 10(1) = 13 \leq 8.$ False

The solution set is $\left[-4, \dfrac{2}{3}\right]$.

13. $4y^2 + 7y + 3 < 0$
$4y^2 + 7y + 3 = 0$
$(4y + 3)(y + 1) = 0$
$4y + 3 = 0$ or $y + 1 = 0$
$y = -\dfrac{3}{4}$ or $y = -1$

Locate these numbers on a number line.

```
   F        T        F
───┼────────┼────────→
  -1      -3/4
```

Let $y = -2$:
$4(-2)^2 + 7(-2) + 3 = 16 - 14 + 3$
$= 5 < 0.$ False.
Let $y = -\dfrac{7}{8}$: $4\left(-\dfrac{7}{8}\right)^2 + 7\left(-\dfrac{7}{8}\right) + 3$
$= 4\left(\dfrac{49}{64}\right) + 7\left(-\dfrac{7}{8}\right) + 3$
$= \dfrac{49}{16} - \dfrac{49}{8} + 3$
$= \dfrac{49}{16} - \dfrac{98}{16} + \dfrac{48}{16}$
$= -\dfrac{1}{16} < 0.$ True

Let $y = 0$: $4(0)^2 + 7(0) + 3 = 3 < 0.$
False

The solution set is $\left(-1, -\dfrac{3}{4}\right)$.

17. $(2r + 1)(3r - 2)(4r + 5) < 0$

The numbers $-\dfrac{1}{2}, \dfrac{2}{3}$, and $-\dfrac{5}{4}$ are solutions to the cubic equation $(2r + 1)(3r - 2)(4r + 5) = 0$. They divide the number line into 4 regions, as shown in the figure.

```
   A     B     C     D
───┼─────┼─────┼─────→
  -5/4  -1/2  2/3
```

Test a point from each region.

A: If $r = -2$, then
$(2r + 1)(3r - 2)(4r + 5) = (-3)(-8)(-3)$
$= -72 < 0$ True
B: If $r = -1$, then
$(2r + 1)(3r - 2)(4r + 5) = (-1)(-5)(1)$
$= 5 < 0$ False
C: If $r = 0$, then
$(2r + 1)(3r - 2)(4r + 5) = (1)(-2)(5)$
$= -10 < 0$ True
D: If $r = 1$, then
$(2r + 1)(3r - 2)(4r + 5) = (3)(1)(9)$
$= 27 < 0$ False

Based on these results, the solution set includes points in both Regions A and C.

The solution set is $\left(-\infty, -\dfrac{5}{4}\right) \cup \left(-\dfrac{1}{2}, \dfrac{2}{3}\right)$.

21. $(3m + 5)(2m + 1)(m - 3) < 0$

The numbers $-\dfrac{5}{3}, -\dfrac{1}{2}$, and 3 are solutions to the cubic equation $(3m + 5)(2m + 1)(m - 3) = 0$. These numbers divide the number line into 4 regions, as shown in the figure.

```
   A     B     C     D
───┼─────┼─────┼─────→
  -5/3  -1/2   3
```

Test a point from each region.
A: If $m = -2$, then
$(3m + 5)(2m + 1)(m - 3) =$
$(-1)(-3)(-5) = -15 < 0$ True
B: If $m = -1$, then
$(3m + 5)(2m + 1)(m - 3) = (2)(-1)(-4)$
$= 8 < 0$ False
C: If $m = 0$, then
$(3m + 5)(2m + 1)(m - 3) = (5)(1)(-3)$
$= -15 < 0$ True
D: If $m = 4$, then
$(3m + 5)(2m + 1)(m - 3) = (17)(9)(1)$
$= 153 < 0$ False

Based on these results, the solution set includes points in both Regions A and C.

The solution set is $\left(-\infty, -\dfrac{5}{3}\right) \cup \left(-\dfrac{1}{2}, 3\right)$.

25. $\dfrac{3}{m - 1} \leq 1$

$\dfrac{3}{m - 1} = 1$ and $m - 1 = 0$
$3 = m - 1$ $m = 1$
$m = 4$

These two values, 1 and 4, divide the number line into 3 regions.

```
       A     B     C
    ───┼─────┼─────→
       1     4
```

Test a point from each region.

A: If $m = 0$, $\dfrac{3}{0 - 1} \leq 1$
$-3 \leq 1$ True

B: If $m = 2$, $\dfrac{3}{2 - 1} \leq 1$
$3 \leq 1$ False

C: If $m = 5$, $\dfrac{3}{5 - 1} \leq 1$
$\dfrac{3}{4} \leq 1$ True

Based on these results, the solution set includes points in regions A and C. The endpoint 1 is not included, since it makes the denominator 0; the endpoint 4 is included since the symbol $\leq$ includes the equality. The solution set is $(-\infty, 1) \cup [4, \infty)$.

29. $\dfrac{x + 1}{x - 5} < 0$

$\dfrac{x + 1}{x - 5} = 0$ and $x - 5 = 0$

$x + 1 = 0$
$x = -1$ and $x = 5$

These two values, -1 and 5, divide the number line into 3 regions.

```
    A   B   C
  ──┼───┼──→
   -1   5
```

Test a point from each region.

A: $x = -2$ $\dfrac{-2 + 1}{-2 - 5} < 0$
$\dfrac{1}{7} < 0$ False

B: $x = 0$ $\dfrac{0 + 1}{0 - 5} < 0$
$-\dfrac{1}{5} < 0$ True

C: $x = 6$ $\dfrac{6 + 1}{6 - 5} < 0$
$7 < 0$ False

Based on these results, only the points in region B satisfy the inequality. Neither endpoint is included. The solution set is $(-1, 5)$.

33. $\dfrac{y}{3y - 2} > -2$

$\dfrac{y}{3y - 2} = -2$ and $3y - 2 = 0$

$y = -2(3y - 2)$ $3y = 2$
$y = -6y + 4$
$7y = 4$
$y = \dfrac{4}{7}$ and $y = \dfrac{2}{3}$

These two values, $\dfrac{4}{7}$ and $\dfrac{2}{3}$, divide the number line into 3 regions.

```
    A   B   C
  ──┼───┼──→
   4/7  2/3
```

Test a point from each region.

A: $y = 0$ $\dfrac{0}{3(0) - 2} > -2$
$0 > -2$ True

B: $y = .6$ $\dfrac{.6}{3(.6) - 2} > -2$
$-3 > -2$ False

C: $y = 1$ $\dfrac{1}{3(1) - 2} > -2$
$1 > -2$ True

Based on these results, the points in both regions A and C are in the solution set. Neither endpoint is included. The solution set is $\left(-\infty, \dfrac{4}{7}\right) \cup \left(\dfrac{2}{3}, \infty\right)$.

37. $3x - 2y = 12$; $x = 0$
$3(0) - 2y = 12$
$-2y = 12$
$y = -6$

CHAPTER 7

Section 7.1 (page 401)

1. $(1, 5)$ is in quadrant I since both coordinates are positive.

5. $(2, -3)$ is in quadrant IV since the x-coordinate is positive and the y-coordinate is negative.

9. $(2, 0)$ is on the positive x-axis, thus not in any quadrant.

13. Plot $(-3, -2)$ on the coordinate system by going three units to the left from zero along the x-axis and two units down parallel to the y-axis.

17. Plot $(-2, 4)$ by going two units to the left from zero along the x-axis and four units up parallel to the y-axis.

See the graphs in the answer section of your text for Exercises 21–37.

21.
If $x = 0$,   If $y = 0$,
$2x + y = 5$ $2x + y = 5$
$2(0) + y = 5$ $2x + 0 = 5$
$0 + y = 5$ $2x = 5$
$y = 5.$ $x = \dfrac{5}{2}.$

If $x = 1$,   If $y = 1$,
$2x + y = 5$ $2x + y = 5$
$2(1) + y = 5$ $2x + 1 = 5$
$2 + y = 5$ $2x = 4$
$y = 3.$ $x = 2.$

The ordered pairs are $(0, 5)$, $(5/2, 0)$, $(1, 3)$, and $(2, 1)$.

25.
If $x = 0$,
$4x + 5y = 20$
$4(0) + 5y = 20$
$0 + 5y = 20$
$5y = 20$
$y = 4$.

If $x = 3$,
$4x + 5y = 20$
$4(3) + 5y = 20$
$12 + 5y = 20$
$5y = 8$
$y = \dfrac{8}{5}$.

If $y = 0$,
$4x + 5y = 20$
$4x + 5(0) = 20$
$4x + 0 = 20$
$4x = 20$
$x = 5$.

If $y = 2$,
$4x + 5y = 20$
$4x + 5(2) = 20$
$4x + 10 = 20$
$4x = 10$
$x = \dfrac{10}{4} = \dfrac{5}{2}$.

The ordered pairs are $(0, 4)$, $(5, 0)$, $(3, 8/5)$, and $(5/2, 2)$.

29.
If $x = 0$,
$4x + 3y = 7$
$4(0) + 3y = 7$
$0 + 3y = 7$
$3y = 7$
$y = \dfrac{7}{3}$.

If $x = 1$,
$4x + 3y = 7$
$4(1) + 3y = 7$
$4 + 3y = 7$
$3y = 3$
$y = 1$.

If $y = 0$,
$4x + 3y = 7$
$4x + 3(0) = 7$
$4x + 0 = 7$
$4x = 7$
$x = \dfrac{7}{4}$.

If $y = -3$,
$4x + 3y = 7$
$4x + 3(-3) = 7$
$4x - 9 = 7$
$4x = 16$
$x = 4$.

The ordered pairs are $(0, 7/3)$, $(7/4, 0)$, $(1, 1)$, and $(4, -3)$.

33. For the x-intercept, let $y = 0$ in $3y + x = 6$ to get $(6, 0)$; for the y-intercept, let $x = 0$ to get $(0, 2)$. See the graph of the equation in the answer section.

37. Let $y = 0$ in $x + 2y = 5$ to get x-intercept $(5, 0)$; $x = 0$ gives y-intercept $(0, 5/2)$. See the graph in the answer section.

41. $\dfrac{5 - 3}{6 - 2} = \dfrac{2}{4} = \dfrac{1}{2}$

45. $\dfrac{-2 - (-3)}{-1 - 4} = \dfrac{1}{-5} = -\dfrac{1}{5}$

Section 7.2 (page 411)

See the graphs in the answer section of your text for Exercises 1–21.

1. $(2, -3)$ and $(1, 5)$
$m = \dfrac{\text{change in } y}{\text{change in } x} = \dfrac{-3 - 5}{2 - 1} = \dfrac{-8}{1} = -8$

5. $(2, 5)$ $(-4, 5)$
$m = \dfrac{5 - 5}{-4 - 2}$
$= \dfrac{0}{-6} = 0$ Horizontal line

9. To find the slope, find two points and draw the line $-x + y = 5$. Use the points to graph the line.
Let $x = 0$; then $y = 5$; $(0, 5)$
Let $y = 0$; then $x = -5$; $(-5, 0)$.
The slope is $m = \dfrac{0 - 5}{-5 - 0} = \dfrac{-5}{-5} = 1$.

13. Replace x with 0 in $2x - 5y = 10$ to get
$2(0) - 5y = 10$ or $y = -2$; $(0, -2)$.
Replace y with 0 in $2x - 5y = 10$ to get
$2x - 5(0) = 10$ or $x = 5$; $(5, 0)$.
Using these results, $(5, 0)$ and $(0, -2)$, in the slope formula gives
$$m = \dfrac{0 - (-2)}{5 - 0} = \dfrac{2}{5}.$$

17. Locate $(0, 1)$. The slope is $m = -\dfrac{2}{3}$. So go from $(0, 1)$ 2 units down and 3 units right to get to $(3, -1)$. Draw the line through $(0, 1)$ and $(3, -1)$.

21. Locate $(-1, -2)$. The slope is $m = -\dfrac{3}{2}$. So from $(-1, -2)$ go 3 units down and 2 units right to get to $(1, -5)$. Draw the line through $(-1, -2)$ and $(1, -5)$.

25. Two points on the line $x = 6$ are $(6, 0)$ and $(6, 1)$. The slope, $m = \dfrac{0 - 1}{6 - 6} = -\dfrac{1}{0}$, is undefined. Two points on the line $6 - x = 8$ are $(-2, 0)$ and $(-2, 1)$. The slope, $m = \dfrac{0 - 1}{-2 - (-2)} = -\dfrac{1}{0}$, is undefined.
Since the two lines have undefined slopes they are both vertical and so they are parallel.

29. To determine if the lines $2x = y + 3$ and $2y + x = 3$ are perpendicular, show that the product of their slopes is -1. Find two points on the line $2x = y + 3$.
Let $x = 1$; then $y = -1$; $(1, -1)$.
Let $x = 2$; then $y = 1$; $(2, 1)$.
The slope $m = \dfrac{-1 - 1}{1 - 2} = \dfrac{-2}{-1} = 2$.
Find two points on the line $2y + x = 3$.
Let $x = 1$; then $y = 1$; $(1, 1)$.
Let $x = 2$; then $y = \dfrac{1}{2}$; $\left(2, \dfrac{1}{2}\right)$.
The slope $m = \dfrac{1 - \dfrac{1}{2}}{1 - 2} = \dfrac{\dfrac{1}{2}}{-1} = -\dfrac{1}{2}$.
The product of the slopes is $(2)\left(-\dfrac{1}{2}\right) = -1$; therefore the lines are perpendicular.

33. The points $(-3, 0)$ and $(-1, 1)$ are on the line $2y - x = 3$ and give slope $m = \dfrac{0 - 1}{-3 - (-1)} = \dfrac{-1}{-2} = \dfrac{1}{2}$. The points $\left(\dfrac{1}{2}, 0\right)$ and $(0, 1)$ are on the line $y + 2x = 1$ and give slope
$$m = \dfrac{1 - 0}{0 - \dfrac{1}{2}} = \dfrac{1}{-\dfrac{1}{2}} = -2.$$
The product $\left(\dfrac{1}{2}\right)(-2) = -1$ so the lines are perpendicular.

37. $5x - y = 12$
$-y = 12 - 5x$
$y = 5x - 12$

41. $y - (-1) = \dfrac{5}{3}[x - (-4)]$

$y + 1 = \dfrac{5}{3}(x + 4)$

$3(y + 1) = 5(x + 4)$ Multiply both sides by 3.

$3y + 3 = 5x + 20$
$3y = 5x + 17$
$5x - 3y = -17$ Standard form

Section 7.3 (page 421)

1. To get slope-intercept form, solve $x + y = 8$ for y to get $y = -x + 8$. Thus, $m = -1$ (the coefficient of x) and $b = 8$, and the y-intercept is $(0, 8)$.

5. $2x - 3y = 5$
Solve for y.
$2x = 5 + 3y$
$3y = 2x - 5$
$y = \dfrac{2}{3}x - \dfrac{5}{3}; \; m = \dfrac{2}{3}; \; b = -\dfrac{5}{3}$

The y-intercept is $\left(0, -\dfrac{5}{3}\right)$.

9. $8x + 11y = 9$
$11y = -8x + 9$
$y = -\dfrac{8}{11}x + \dfrac{9}{11}$
$m = -\dfrac{8}{11}; \; b = \dfrac{9}{11}$

The y-intercept is $\left(0, \dfrac{9}{11}\right)$.

13. $m = -\dfrac{2}{3}; \; b = \dfrac{1}{2}$

Use $y = mx + b$. Here, $m = -\dfrac{2}{3}$ and $b = \dfrac{1}{2}$,

so $y = -\dfrac{2}{3}x + \dfrac{1}{2}$.

17. Slope 0; y-intercept $(0, 4)$
Use $y = mx + b$. Let $m = 0$ and $b = 4$ to get $y = 0x + 4$ or $y = 4$.

21. Through $(4, -3)$; $m = -\dfrac{5}{6}$

Use $y - y_1 = m(x - x_1)$ with $m = -\dfrac{5}{6}, x_1 = 4$,
and $y_1 = -3$ to get
$$y + 3 = -\dfrac{5}{6}(x - 4)$$
$$6y + 18 = -5x + 20 \quad \text{Multiply by 6 and clear parentheses.}$$
or $5x + 6y = 2$.

25. Through $(1, -2)$; $m = \dfrac{1}{4}$

Use $y - y_1 = m(x - x_1)$ with $m = \dfrac{1}{4}, x_1 = 1$, and $y_1 = -2$ to get
$$y + 2 = \dfrac{1}{4}(x - 1)$$
$$4y + 8 = x - 1 \quad \text{Multiply by 4 and clear parentheses.}$$
or $x - 4y = 9$.

29. x-intercept $(-2, 0)$ means that the line goes through $(-2, 0)$. Use $y - y_1 = m(x - x_1)$ with $m = -5$, $x_1 = -2, y_1 = 0$ to get
$$y - 0 = -5(x + 2)$$
$$y = -5x - 10 \quad \text{Clear parentheses.}$$
or $5x + y = -10$.

33. The vertical line through $(3, -9)$ is $x = 3$.

37. The slope of the line through $\left(-\dfrac{2}{5}, \dfrac{2}{5}\right)$ and $\left(\dfrac{4}{3}, \dfrac{2}{3}\right)$ is
$$m = \dfrac{\dfrac{2}{5} - \dfrac{2}{3}}{-\dfrac{2}{5} - \dfrac{4}{3}} = \dfrac{\dfrac{6 - 10}{15}}{\dfrac{-6 - 20}{15}} = \dfrac{-\dfrac{4}{15}}{-\dfrac{26}{15}} = \dfrac{2}{13}.$$

Use $m = \dfrac{2}{13}, x_1 = -\dfrac{2}{5}$, and $y_1 = \dfrac{2}{5}$

in $y - y_1 = m(x - x_1)$. Then
$$y - \dfrac{2}{5} = \dfrac{2}{13}\left(x + \dfrac{2}{5}\right)$$
$$65\left(y - \dfrac{2}{5}\right) = \left(\dfrac{2}{13}x + \dfrac{4}{65}\right)65$$
$$65y - 26 = 10x + 4$$
$$10x - 65y = -30 \quad \text{or} \quad 2x - 13y = -6.$$

41. The slope of the line through $(1, \sqrt{5})$ and $(3, 2\sqrt{5})$ is
$$m = \dfrac{2\sqrt{5} - \sqrt{5}}{3 - 1} = \dfrac{\sqrt{5}}{2}.$$
Thus, the equation of the line is
$$y - \sqrt{5} = \dfrac{\sqrt{5}}{2}(x - 1)$$
$$2(y - \sqrt{5}) = \sqrt{5}(x - 1)$$
$$2y - 2\sqrt{5} = \sqrt{5}x - \sqrt{5}$$
$$-\sqrt{5}x + 2y = \sqrt{5}$$
or $\sqrt{5}x - 2y = -\sqrt{5}$.

45. Find the slope of $-x + 2y = 3$.
$$2y = x + 3$$
$$y = \frac{1}{2}x + \frac{3}{2}$$
The slope is $\frac{1}{2}$, so the parallel lines each have slope $\frac{1}{2}$. Let $m = \frac{1}{2}$, $x_1 = -2$, and $y_1 = -2$. Then
$$y - (-2) = \frac{1}{2}[x - (-2)]$$
$$y + 2 = \frac{1}{2}(x + 2)$$
$$2(y + 2) = x + 2$$
$$2y + 4 = x + 2$$
$$2y - x = -2 \quad \text{or} \quad x - 2y = 2.$$

49. The line $y = 4$ is a horizontal line whose slope $= 0$, so a line parallel to it will also have a slope of 0. Thus, let $m = 0$, $x_1 = -2$, and $y_1 = 7$.
$$y - 7 = 0[x - (-2)]$$
$$y - 7 = 0$$
$$y = 7$$

53. $2x + 5 < 6$
$$2x < 1$$
$$x < \frac{1}{2}$$
Solution set: $\left(-\infty, \frac{1}{2}\right)$

57. $-5x - 5 > 3$
$$-5x > 8$$
$$x < -\frac{8}{5} \quad \text{Divide by } -5 \text{ and reverse inequality.}$$
Solution set: $\left(-\infty, -\frac{8}{5}\right)$

Section 7.4 (page 431)

See the graphs in the answer section of your text for Exercises 1–33.

1. To graph $x + y \leq 2$, first graph $x + y = 2$. Write in slope-intercept form: $y = -x + 2$. Slope $= -1$, y-intercept is $(0, 2)$. Solve the original inequality for y:
$$x + y \leq 2$$
$$y \leq -x + 2.$$
Because of $\leq$, shade the region *below* the line.

5. $x + 3y \geq -2$
Graph $x + 3y = -2$. Solve the original inequality for y:
$$x + 3y \geq -2$$
$$3y \geq -x - 2$$
$$y \geq -\frac{1}{3}x - \frac{2}{3}.$$
Because of $\geq$, shade the region *above* the line.

9. Graph $y = x$. Two points on the line are $(0, 0)$ and $(1, 1)$. Draw a dashed line through these points. Test a point not on the line, say $(5, 4)$, in $y < x$. $4 < 5$ is true, so shade the side of the line containing $(5, 4)$.

13. Graph $x + y = 1$, which contains $(1, 0)$ and $(0, 1)$. Draw a solid line through these points. Notice $(0, 0)$ gives $0 \leq 1$, a true statement, so shade the side of the line containing the origin. Graph the vertical line $x = 0$ as a solid line and shade to the right of it. The intersection is the part where the shaded regions overlap, as shown in the figure.

17. Graph $-x - y = 5$, which contains $(-5, 0)$ and $(0, -5)$. Draw a dashed line through these points. Notice $(0, 0)$ gives $0 < 5$, a true statement, so shade the side of the line containing the origin. Graph $y = -2$ as a dashed horizontal line and shade below it. Where the shaded regions overlap is the intersection, which is shown in the figure.

21. Graph $x - 2 = y$, which contains $(2, 0)$ and $(0, -2)$. Draw a dashed line through these points. Notice $(0, 0)$ gives $-2 > 0$, a false statement, so shade the side of the line not containing the origin. Graph $x = 1$, which is a vertical line containing $(1, 0)$. Draw a vertical dashed line through this point and shade to the left of the line. All shaded regions are included in the union, which is shown in the figure.

25. $|y| > 4$ becomes $y > 4$ or $y < -4$. The graph of $y > 4$ consists of the dashed horizontal line $y = 4$ and shading above it. The graph of $y < -4$ consists of the dashed horizontal line $y = -4$ and shading below it.

29. $|y - 2| > 3$ means $y - 2 > 3$ or $y - 2 < -3$, so $y > 5$ or $y < -1$. The graph consists of the two dashed horizontal lines, $y = 5$ and $y = -1$, with shading above $y = 5$ and below $y = -1$.

33. $|y - 1| < x$ Sketch the boundary lines $y - 1 = x$ and $y - 1 = -x$. Use test points to decide which of the four regions formed satisfy $|y - 1| < x$. For example, the point $(3, 0)$ satisfies, while $(-3, 0)$, $(0, 3)$ and $(0, -3)$ do not. Since x must be less than or equal to 0, there are no points that satisfy both inequalities. There is no solution.

37. Use the formula for simple interest, $I = prt$.
$I = 84$, $r = .10$, $t = 6$, and p is to be found.
$$84 = p(.10)(6)$$
$$84 = .60p$$
$$\frac{84}{.60} = p$$
$$p = 140$$
$140 must be invested.

Section 7.5 (page 441)

1. $\{(5, 1), (3, 2), (4, 9), (7, 6)\}$
Domain: $\{3, 4, 5, 7\}$
Range: $\{1, 2, 6, 9\}$
This is a set of ordered pairs in which no first components are repeated. It is a function.

5. $\{(-3, 1), (4, 1), (-2, 7)\}$
Domain: $\{-3, -2, 4\}$
Range: $\{1, 7\}$
Each domain value is assigned to exactly one range value; ordered pairs *can* have the same y-value in a function. Thus, it is a function.

9. Domain: {2, 5, 11, 17, 3}
   Range: {1, 7, 20}
   Each domain value corresponds to exactly one range value. This is a function.

13. $3x = 5 - 2y$
    Each value of x from the domain $(-\infty, \infty)$ corresponds to exactly one y-value. For example, if $x = 5$,
    $$3(5) = 5 - 2y$$
    $$15 = 5 - 2y$$
    $$10 = -2y$$
    $$y = -5.$$
    Thus it is a function. The equation can be written as $f(x) = y = -\dfrac{3}{2}x + \dfrac{5}{2}$, so it is a linear function.

17. $x \ne y$
    The domain is $(-\infty, \infty)$. Any element from the domain corresponds to more than one element from the range. For example, $x = 2$ corresponds to any y-value except $y = 2$. Thus, this equation does not define a function.

21. $x = \dfrac{1}{y}$
    Here $x \ne 0$, since the numerator of $\dfrac{1}{y}$ is 1. The domain is $(-\infty, 0) \cup (0, \infty)$. Each x-value corresponds to exactly one y-value, so the equation defines a function. This function is *not* linear, because it cannot be written in the form $f(x) = y = mx + b$.

25. $y = \sqrt{x}$
    Each x-value from the domain $[0, \infty)$ corresponds to just one y-value. (Recall $\sqrt{x}$ is the principal square root and is always nonnegative.) For example, $x = 4$ corresponds to $y = 2$. Thus the equation defines a function. It is not a linear function because $\sqrt{x} = x^{1/2}$. In order to be a linear function the highest power of x must be 1, and in this case it is $\dfrac{1}{2}$.

29. $f(x) = 3 - \dfrac{1}{2}x$
    $f(5) = 3 - \dfrac{1}{2}(5) = \dfrac{6-5}{2} = \dfrac{1}{2}$

33. $f(x) = 3 - \dfrac{1}{2}x$
    $f(-3) = 3 - \dfrac{1}{2}(-3) = \dfrac{6+3}{2} = \dfrac{9}{2}$
    $f[f(-3)]; f\left(\dfrac{9}{2}\right) = 3 - \dfrac{1}{2}\left(\dfrac{9}{2}\right) = 3 - \dfrac{9}{4}$
    $= \dfrac{12-9}{4} = \dfrac{3}{4}$

37. If $g(x) = -x^2 + 4x - 1$, then
    $g\left(\dfrac{1}{2}\right) = -\left(\dfrac{1}{2}\right)^2 + 4\left(\dfrac{1}{2}\right) - 1$
    $= -\dfrac{1}{4} + 2 - 1 = \dfrac{3}{4}.$

41. Using both $f(x) = -9x + 12$ and $g(x) = -x^2 + 4x - 1$, first find $g(0) = -0^2 + 4(0) - 1 = -1$. Then
    $$f[g(0)] = f(-1)$$
    $$= -9(-1) + 12 = 21.$$

45. Using the vertical line test, the line will cross the graph at most once, anywhere. Thus, the graph represents a function.

49. Let $f(x) = .50x$. Then
    $f(5) = .50(5) = 2.50$.
    The cost is $2.50.

53. $-5(2m + 3) - m = 10$
    $-10m - 15 - m = 10$
    $-11m = 25$
    $m = -\dfrac{25}{11}$
    Solution set: $\left\{-\dfrac{25}{11}\right\}$

57. $-2\left(\dfrac{3n+1}{5}\right) + n = -4$
    $5(-2)\left(\dfrac{3n+1}{5}\right) + 5 \cdot n = 5(-4)$
    $-2(3n+1) + 5n = -20$
    $-6n - 2 + 5n = -20$
    $-n - 2 = -20$
    $-n = -18$
    $n = 18$
    Solution set: $\{18\}$

Section 7.6 (page 451)

1. Since a varies directly as the square of b.
   $a = kb^2$, for some constant k.
   We know that $a = 4$ when $b = 3$. Substituting these values into the equation $a = kb^2$ gives
   $$a = kb^2$$
   $$4 = k(3)^2$$
   $$4 = 9k$$
   $$k = \dfrac{4}{9}.$$
   Let $b = 2$, and now solve for a, using $k = \dfrac{4}{9}$.
   $$a = \dfrac{4}{9}(2)^2$$
   $$a = \dfrac{4}{9}(4) = \dfrac{16}{9}$$

5. Since p varies jointly as q and r^2, $p = kqr^2$, for some constant k.
   Given that $p = 100$ when $q = 2$ and $r = 3$, solve for k.
   $$100 = k(2)(3)^2$$
   $$100 = 18k$$
   $$k = \dfrac{100}{18} = \dfrac{50}{9}$$
   Using $k = \dfrac{50}{9}$, find p when $q = 5$ and $r = 2$.
   $$p = \left(\dfrac{50}{9}\right)(5)(2)^2 = \dfrac{50}{9}(5)(4) = \dfrac{1000}{9}$$

9. Let f represent the frequency and L the length of the string. Then
$$f = \frac{k}{L},$$
where k is the constant of variation. Find k by substituting 2 for L and 250 for f.
$$250 = \frac{k}{2}$$
$$k = 500$$
Now substitute $k = 500$ and $L = 1.5$ in the variation equation to get
$$f = \frac{500}{1.5} = 333\frac{1}{3}$$
cycles per second.

13. $F = \dfrac{k}{d^2}$

$F = 100$, when $d = 5 \times 10^{-10}$, so
$$100 = \frac{k}{(5 \times 10^{-10})^2}$$
$$k = 2.5 \times 10^{-17}.$$
Then $d^2 = \dfrac{k}{F} = \dfrac{2.5 \times 10^{-17}}{F}$. When $F = 196$,
$$d^2 = \frac{2.5 \times 10^{-17}}{196}$$
$$d^2 = 1.2755 \times 10^{-19}$$
$$d \approx 3.57 \times 10^{-10}$$
So $d \approx 3.57 \times 10^{-10}$ meter.

17. $3x + y = 4$
Solve for y: $y = -3x + 4$.
If $x = 0$, $y = -3(0) + 4 = 4$.
If $x = 5$, $y = -3(5) + 4 = -11$.
If $x = -2$, $y = -3(-2) + 4 = 10$.
The ordered pairs are $(0, 4)$, $(5, -11)$, and $(-2, 10)$.

21. $x = |y|$
If $y = 3$, $x = |3| = 3$.
If $y = 4$, $x = |4| = 4$.
If $y = -4$, $x = |-4| = 4$.
If $y = -3$, $x = |-3| = 3$.
The ordered pairs are $(3, 3)$, $(4, 4)$, $(4, -4)$, and $(3, -3)$.

CHAPTER 8

Section 8.1 (page 473)

To decide whether a given ordered pair is a solution, see if it satisfies both equations.

1. $x + y = 5 \quad\quad 4 + 1 = 5$
   $x - y = 3 \quad\quad 4 - 1 = 3$
   Yes; the ordered pair satisfies both equations, so $(4, 1)$ is a solution.

5. (See the graph in the answer section of your textbook.) Solve the following system by adding the equations. Solve for x.
$$\begin{aligned} x + y &= 10 \\ 2x - y &= 5 \\ \hline 3x &= 15 \\ x &= 5 \end{aligned}$$
Substitute $x = 5$ into $x + y = 10$.
$$5 + y = 10$$
$$y = 5$$
Solution set: $\{(5, 5)\}$

9. Multiply first equation by 2: $\quad 14x + 4y = 6.$
   Leave second unchanged: $\quad\quad -14x - 4y = -6.$
   Add: $\quad\quad\quad\quad\quad\quad\quad\quad\quad\quad\quad 0 = 0.$
   Since we get a true statement, the equations are dependent. The solution set is $\{(x, y) | 7x + 2y = 3\}$.

13. Multiply first equation by 4: $\quad 20x + 12y = 4.$
    Multiply second equation by 3: $\quad -9x - 12y = 18.$
    Add: $\quad\quad\quad\quad\quad\quad\quad\quad\quad 11x = 22.$
    $\quad\quad\quad\quad\quad\quad\quad\quad\quad\quad x = 2$
    Substitute 2 for x in either equation to get $y = -3$.
    Solution set: $\{(2, -3)\}$

17. Leave the first equation unchanged: $\quad 2x - 3y = 2$
    Multiply the second equation by -6: $\quad -2x + 3y = 2$
    Add: $\quad\quad\quad\quad\quad\quad\quad\quad\quad 0 = 4 \quad$ False
    The false statement indicates an inconsistent system. The solution set is $\emptyset$.

21. Substitute $5x$ for y in the first equation.
$$2x = 5x + 6$$
$$-3x = 6$$
$$x = -2$$
Since $y = 5x$, we have $y = 5(-2)$, or $y = -10$.
Solution set: $\{(-2, -10)\}$

25. Solve second equation for x: $x = 2y - 3$. Substitute into first equation.
$$5(2y - 3) - 4y = 9$$
$$10y - 15 - 4y = 9$$
$$6y = 24$$
$$y = 4$$
Since $x = 2y - 3$, we have $x = 2(4) - 3$ or $x = 5$.
Solution set: $\{(5, 4)\}$

29. Substitute $3x$ for y in the first equation.
$$\frac{x}{2} + \frac{3x}{3} = 3$$
$$\frac{x}{2} + x = 3$$
$$\frac{3}{2}x = 3$$
$$x = 2$$
Then $y = 6$, and the solution set is $\{(2, 6)\}$.

33. $\dfrac{2}{x} - \dfrac{5}{y} = \dfrac{3}{2}$

$\dfrac{4}{x} + \dfrac{1}{y} = \dfrac{4}{5}$

Let $p = \dfrac{1}{x}$, $q = \dfrac{1}{y}$.

$2p - 5q = \dfrac{3}{2}$ (1)

$4p + q = \dfrac{4}{5}$ (2)

$4p - 10q = 3$ (3) Multiply (1) by 2.
$20p + 5q = 4$ (4) Multiply (2) by 5.

$4p - 10q = 3$
$40p + 10q = 8$ Multiply (4) by 2.
$44p = 11$ Add.

$p = \dfrac{1}{4}$

Substitute $p = \dfrac{1}{4}$ in (2).

$4\left(\dfrac{1}{4}\right) + q = \dfrac{4}{5}$

$1 + q = \dfrac{4}{5}$

$q = -\dfrac{1}{5}$

$x = \dfrac{1}{p} = \dfrac{1}{\tfrac{1}{4}} = 4$, $y = \dfrac{1}{q} = \dfrac{1}{-\tfrac{1}{5}} = -5$

Solution set: $\{(4, -5)\}$

37. $3ax + 2y = 1$ (1)
$-ax + y = 2$ (2)

$3ax + 2y = 1$ (1)
$-3ax + 3y = 6$ Multiply (2) by 3.
$5y = 7$ Add the equations.

$y = \dfrac{7}{5}$

Substitute $y = \dfrac{7}{5}$ in (2).

$-ax + \dfrac{7}{5} = 2$

$-ax = \dfrac{3}{5}$

$x = -\dfrac{3}{5a}$ Solution set: $\left\{\left(-\dfrac{3}{5a}, \dfrac{7}{5}\right)\right\}$

41. (a) The point where the graphs cross is (8, 3000). At $x = 8$ when 800 parts are produced, the revenue is $3000.

(b) When $x = 4$, 400 parts are sold. At $x = 4$, the graph shows the cost is $2000 and the revenue is $1500. The profit is $1500 - $2000 = -$500, or a loss of $500.

(c) When $x = 0$, the cost is $1000, so the fixed cost is $1000.

45. $-x + 3y - 2z = 18$
$2x - 6y + 4z = -36$ Multiply by -2.

Section 8.2 (page 483)

1. $2x + y + z = 3$
$3x - y + z = -2$
$4x - y + 2z = 0$

Add. $2x + y + z = 3$ Add. $2x + y + z = 3$
$ 3x - y + z = -2$ $ 4x - y + 2z = 0$
$ 5x + 2z = 1$ $ 6x + 3z = 3$

Solve the system $5x + 2z = 1$
$6x + 3z = 3$.

Multiply first equation by 3, second by -2.
Add. $15x + 6z = 3$
$-12x - 6z = -6$
$3x = -3$
$x = -1$

Substitute back to get $z = 3$ and $y = 2$, so the solution set is $\{(-1, 2, 3)\}$.

5. $2x + 5y + 2z = 9$
$4x - 7y - 3z = 7$
$3x - 8y - 2z = 9$

Add. $2x + 5y + 2z = 9$
$ 3x - 8y - 2z = 9$
$ 5x - 3y = 18$

Multiply first equation of original system by 3, the second by 2, then add.

$6x + 15y + 6z = 27$
$8x - 14y - 6z = 14$
$14x + y = 41$

Solve the system $5x - 3y = 18$
$14x + y = 41$.

Multiply the second equation by 3, and add to the first.
$5x - 3y = 18$
$42x + 3y = 123$
$47x = 141$
$x = 3$

Substitute back to get $y = -1$, $z = 4$.
Solution set is $\{(3, -1, 4)\}$.

9. $2x - 3y + 2z = -1$ (1)
$x + 2y = 14$ (2)
$x - 3z = -5$ (3)

Multiply equation (1) by 2.
Multiply equation (2) by 3. Add.
$4x - 6y + 4z = -2$
$3x + 6y = 42$
$7x + 4z = 40$ (4)

Multiply equation (3) by -7. Add.
$7x + 4z = 40$ (4)
$-7x + 21z = 35$
$25z = 75$
$z = 3$

Substitute $z = 3$ into (3). $x - 3(3) = -5$ (3)
$x - 9 = -5$
$x = 4$

Substitute $x = 4$ into (2). $4 + 2y = 14$ (2)
$2y = 10$
$y = 5$

Solution set: $\{(4, 5, 3)\}$

13. $x + y \phantom{{}+2z} = 1$ (1)
   $2x \phantom{{}+y} - z = 0$ (2)
   $y + 2z = -2$ (3)

Multiply (1) by -1.   Add to (3).
   $-x - y \phantom{{}+2z} = -1$
   $y + 2z = -2$ (3)
   $\overline{-x \phantom{{}+y} + 2z = -3}$ (4)

Multiply equation (4) by 2.   Add to (2).
   $-2x + 4z = -6$
   $2x - z = 0$ (2)
   $\overline{3z = -6}$
   $z = -2$

Substitute $z = -2$ into (3).
   $y + 2(-2) = -2$ (3)
   $y - 4 = -2$
   $y = 2$

Substitute $z = -2$ into (2).
   $2x - (-2) = 0$ (2)
   $2x + 2 = 0$
   $2x = -2$
   $x = -1$ Solution set: $\{(-1, 2, -2)\}$

17. $x - 2y \phantom{{}+z} = 0$ (1)
   $3y + z = -1$ (2)
   $4x \phantom{{}+2y} - z = 11$ (3)

   $3y + z = -1$ (2)
   $\underline{4x \phantom{{}+3y} - z = 11}$ (3)
   $4x + 3y \phantom{{}-z} = 10$ (4)

   $-4x + 8y = 0$ Multiply (1) by -4.
   $\underline{4x + 3y = 10}$ (4)
   $11y = 10$
   $y = \dfrac{10}{11}$ Substitute $y = \dfrac{10}{11}$ in (2).

   $3\left(\dfrac{10}{11}\right) + z = -1$

   $z = -1 - \dfrac{30}{11} = -\dfrac{11}{11} - \dfrac{30}{11} = -\dfrac{41}{11}$

Substitute $y = \dfrac{10}{11}$ in (1).

   $x - 2\left(\dfrac{10}{11}\right) = 0$

   $x = \dfrac{20}{11}$

Solution set: $\left\{\left(\dfrac{20}{11}, \dfrac{10}{11}, -\dfrac{41}{11}\right)\right\}$

21. $2x + y - z = 6$ (1)
   $4x + 2y - 2z = 12$ (2)
   $-x - \dfrac{1}{2}y + \dfrac{1}{2}z = -3$ (3)

   $4x + 2y - 2z = 12$ 2 times (1)
   $4x + 2y - 2z = 12$ -4 times (3)
   $4x + 2y - 2z = 12$ (2)

The equations are dependent.
Solution set: $\{(x, y, z) | 2x + y - z = 6\}$

25. $x + y + z - w = 5$ (1)
   $2x + y - z + w = 3$ (2)
   $x - 2y + 3z + w = 18$ (3)
   $-x - y + z + 2w = 8$ (4)

   $x + y + z - w = 5$ (1)
   $\underline{2x + y - z + w = 3}$ (2)
   $3x + 2y \phantom{{}+3z+w} = 8$ (5)

   $2x + y - z + w = 3$ (2)
   $\underline{-x - y + z + 2w = 8}$ (4)
   $x \phantom{{}+y-z} + 3w = 11$ (6)

   $x + y + z - w = 5$ (1)
   $\underline{-x - y + z + 2w = 8}$ (4)
   $2z + w = 13$ (7)

   $x + y + z - w = 5$ (1)
   $\underline{x - 2y + 3z + w = 18}$ (3)
   $2x - y + 4z \phantom{{}+w} = 23$ (8)

   $2x - 4y + 6z + 2w = 36$ 2 times (3).
   $\underline{x + y - z - 2w = -8}$ -1 times (4).
   $3x - 3y + 5z \phantom{{}+w} = 28$ (9)

   $10x - 5y + 20z = 115$ 5 times (8).
   $\underline{-12x + 12y - 20z = -112}$ -4 times (9).
   $-2x + 7y \phantom{{}+20z} = 3$ (10)

   $6x + 4y = 16$ 2 times (5)
   $\underline{-6x + 21y = 9}$ 3 times (10).
   $25y = 25$
   $y = 1$

   $3x + 2(1) = 8$ Substitute 1 for y in (5).
   $3x = 6$
   $x = 2$

   $2 + 3w = 11$ Substitute 2 for x in (6).
   $3w = 9$
   $w = 3$

   $2z + 3 = 13$ Substitute 3 for w in (7).
   $2z = 10$
   $z = 5$

Solution set: $\{(x, y, z, w) = (2, 1, 5, 3)\}$

29. Let x represent the length of the longest side in inches. Then $\dfrac{5}{6}x$ is the length of the shortest side, and $x - 7$ is the length of the medium side. Use the formula $P = a + b + c$.

   $163 = x + \dfrac{5}{6}x + (x - 7)$
   $978 = 6x + 5x + 6x - 42$ Multiply by 6.
   $978 = 17x - 42$
   $1020 = 17x$
   $60 = x$

The shortest side is $\dfrac{5}{6}(60) = 50$ inches, the medium side is $60 - 7 = 53$ inches, and the longest side is 60 inches.

Section 8.3 (page 493)

1. Let x represent the length of the two equal sides, and let y represent the length of the remaining side. From the perimeter formula,
$$2x + y = 51.$$
The other equation that can be obtained is
$$y = 2x - 25.$$
The system to be solved is
$$2x + y = 51$$
$$y = 2x - 25.$$
Substitute $2x - 25$ for y in the first equation.
$$2x + (2x - 25) = 51$$
$$4x = 76$$
$$x = 19$$
Since $x = 19$, $y = 2(19) - 25 = 13$.
The side lengths are 13 centimeters, 13 centimeters, and 19 centimeters.

5. Let x represent the price of a small box and let y represent the price of a large box.
$$10x + 20y = 65$$
$$6x + 10y = 34$$
Multiply the second equation by -2.
$$10x + 20y = 65$$
$$-12x - 20y = -68$$
$$-2x = -3$$
$$x = 1.50$$
Substitute 1.50 for x in either equation to find $y = 2.50$. A small box costs \$1.50 and a large box costs \$2.50.

9. Let $x =$ the price of the dark clay per kilogram. Let $y =$ the price of the light clay per kilogram. Then $2x + 3y = 13$ since 2 kilograms of dark clay and 3 kilograms of light clay cost \$13. Since 1 kilogram of dark clay and 2 kilograms of light clay cost \$7, $x + 2y = 7$.
Solve the system.
$$2x + 3y = 13 \quad (1)$$
$$x + 2y = 7 \quad (2)$$
Multiply (2) by -2. Add.
$$2x + 3y = 13 \quad (1)$$
$$-2x - 4y = -14$$
$$-y = -1$$
$$y = 1$$
Substitute $y = 1$ into (1).
$$2x + 3(1) = 13$$
$$2x = 10$$
$$x = 5$$
Dark clay is \$5 per kilogram and light clay is \$1 per kilogram.

13. Let $x =$ number of twenties and $y =$ number of fifties.
Since there is a total of 35 bills,
$$x + y = 35.$$
Since the value of the money is \$1600,
$$20x + 50y = 1600.$$
We must solve the system
$$x + y = 35$$
$$20x + 50y = 1600.$$
Multiply the first equation by -20 and add.
$$-20x - 20y = -700$$
$$20x + 50y = 1600$$
$$30y = 900$$
$$y = 30$$
Substitute 30 for y in either equation to find $x = 5$.
There were 5 twenties and 30 fifties deposited.

17. Let $x =$ amount invested at 14% and
$y =$ amount invested at 7%.
Since the total amount invested was \$56,000,
$$x + y = 56,000.$$
Since the annual interest was the same as if she had invested the entire amount at 10%,
$$.14x + .07y = .10(x + y).$$
The system to be solved is
$$x + y = 56,000$$
$$.14x + .07y = .10(x + y).$$
Solve the first equation for y, and substitute into the second equation.
$$y = 56,000 - x$$
$$.14x + .07(56,000 - x) = .10[x + (56,000 - x)]$$
$$14x + 7(56,000 - x) = 10(56,000)$$
$$14x + 392,000 - 7x = 560,000$$
$$7x = 168,000$$
$$x = 24,000$$
She invested \$24,000 at 14% and \$56,000 $-$ \$24,000 $=$ \$32,000 at 7%.

21.

Kind	Amount	Pure Antifreeze
100%	x	$1.00x$
4%	y	$.04y$
20%	24	$.20(24) = 4.8$

Since the total amount of antifreeze is 24 liters, one equation is
$$x + y = 24.$$
The amount of pure antifreeze in the final mixture must be 4.8 liters, so the other equation is
$$1.00x + .04y = 4.8.$$
The system to be solved is
$$x + y = 24$$
$$x + .04y = 4.8.$$
Multiply the second equation by -1 and add.
$$x + y = 24$$
$$-x - .04y = -4.8$$
$$.96y = 19.2$$
$$y = 20$$
Since $x + y = 24$, $x + 20 = 24$ or $x = 4$.
Four liters of pure antifreeze are needed.

25. Let $x =$ speed of express train.
Let $y =$ speed of freight train.

	d	r	t
Express	$3x$	x	3
Freight	$3y$	y	3

Since the express is 30 kilometers per hour faster than the freight, $x = y + 30$. After 3 hours the express has traveled $3x$ kilometers and the freight $3y$ kilometers. The total distance is 390, so $3x + 3y = 390$ or $x + y = 130$.
Solve the system.
$$x = y + 30 \quad (1)$$
$$x + y = 130 \quad (2)$$
Substitute $y + 30$ for x into (2).
$$(y + 30) + y = 130 \quad (2)$$
$$2y + 30 = 130$$
$$2y = 100$$
$$y = 50$$
Substitute $y = 50$ into (1).
$$x = 50 + 30 \quad (1)$$
$$x = 80$$
The express train is traveling 80 kilometers per hour and the freight train is traveling 50 kilometers per hour.

29. Let x = first number
y = second number
z = third number.
Since their sum is 32,
$$x + y + z = 32.$$
Since the second is 1 more than the first,
$$y = x + 1.$$
Since the first is 4 less than the third,
$$x = z - 4.$$
The system to be solved is
$$x + y + z = 32 \quad (1)$$
$$y = x + 1 \quad (2)$$
$$x = z - 4. \quad (3)$$
From (2), $y = x + 1$ and from (3), $z = x + 4$. Substitute these expressions for y and z in (1) and solve.
$$x + (x + 1) + (x + 4) = 32$$
$$3x + 5 = 32$$
$$3x = 27$$
$$x = 9$$
From (2), $y = 9 + 1 = 10$. From (3), $z = 9 + 4 = 13$.
The first number is 9, the second number is 10, and the third number is 13.

33. Let x = longest side, y = medium side, and z = shortest side. The three equations are
$$x + y + z = 56$$
$$x = y + z - 12$$
$$3z = x + 26.$$
The second equation can be rewritten as $x - y - z = -12$. Add this to the first equation to get $2x = 44$, or $x = 22$. From the third equation, $3z = 22 + 26$, which leads to $z = 16$. From the first equation, find that $y = 18$. The longest side measures 22 inches, the medium side measures 18 inches, and the shortest side measures 16 inches.

37. Let x = the number of sub-compact cars he must sell,
y = the number of compact cars he must sell, and
z = the number of full size cars he must sell.
Since he must sell a total of 20 new cars, one equation is
$$x + y + z = 20.$$
Since he must sell 4 more sub-compact cars than compact cars, another equation is
$$x = 4 + y.$$
Since he must sell 1 more full size car than compact cars, a third equation is $z = 1 + y$.
The system to be solved is as follows.
$$x + y + z = 20 \quad (1)$$
$$x = 4 + y \quad (2)$$
$$z = 1 + y \quad (3)$$
Substitute the expressions for x and z found in (2) and (3) into (1).
$$(4 + y) + y + (1 + y) = 20$$
$$3y + 5 = 20$$
$$3y = 15$$
$$y = 5$$
From (2) $x = 4 + 5 = 9$ and from (3) $z = 1 + 5 = 6$. He must sell 9 sub-compacts, 5 compacts, and 6 full size cars.

41. Let x = amount invested at 7% per year in the video rental firm,
y = amount invested at 6% per year in a tax-free bond, and
z = amount invested at 12% per year in a money market fund.
Since she invests a total of $80,000, one equation is
$$x + y + z = 80,000.$$
Since she invests equal amounts at 6% and 12%, a second equation is
$$y = z.$$
Since her annual return on these investments is $6800, a third equation is
$$.07x + .06y + .12z = 6800$$
or
$$7x + 6y + 12z = 680,000.$$
The system to be solved is as follows.
$$x + y + z = 80,000 \quad (1)$$
$$y = z \quad (2)$$
$$7x + 6y + 12z = 680,000 \quad (3)$$
Since $y = z$, replace y with z in equations (1) and (3).
$$x + 2z = 80,000 \quad (4)$$
$$7x + 18z = 680,000 \quad (5)$$
Multiply (4) by -7 and add to (5).
$$-7x - 14z = -560,000$$
$$\underline{7x + 18z = 680,000}$$
$$4z = 120,000$$
$$z = 30,000$$
From (2), $y = 30,000$. From (1),
$$x + 30,000 + 30,000 = 80,000$$
$$x = 20,000.$$
She invested $20,000 in the video rental firm, and $30,000 each in the tax-free bond and the money market fund.

45. $(-4)(-2) + 2(-3) - (-1)(-6)$
$= 8 - 6 - 6$
$= -4$

49. $[(5)(-4) + (-1)(-6)] + [(-2)(-4) - (-1)(3)]$
$= [-20 + 6] + [8 + 3]$
$= [-14] + [11]$
$= -3$

Section 8.4 (page 503)

1. $\begin{vmatrix} 2 & 5 \\ -1 & 4 \end{vmatrix} = 2 \cdot 4 - (-1) \cdot 5 = 8 + 5 = 13$

5. $\begin{vmatrix} -2 & 2 \\ 2 & -2 \end{vmatrix} = (-2)(-2) - 2 \cdot 2 = 4 - 4 = 0$

9. $\begin{vmatrix} 1 & 2 & 3 \\ -3 & -2 & -1 \\ 2 & -1 & 1 \end{vmatrix}$

$= 1 \begin{vmatrix} -2 & -1 \\ -1 & 1 \end{vmatrix} - (-3) \begin{vmatrix} 2 & 3 \\ -1 & 1 \end{vmatrix}$

$\quad + 2 \begin{vmatrix} 2 & 3 \\ -2 & -1 \end{vmatrix}$

$= 1[(-2)(1) - (-1)(-1)]$
$\quad + 3[2(1) - (-1)(3)]$
$\quad + 2[2(-1) - (-2)(3)]$
$= 1[-3] + 3[5] + 2[4] = 20$

13. $\begin{vmatrix} 1 & 0 & 0 \\ 0 & 1 & 0 \\ 0 & 0 & 1 \end{vmatrix} = 1 \begin{vmatrix} 1 & 0 \\ 0 & 1 \end{vmatrix} - 0 + 0$

$= 1[(1)(1) - (0)(0)] = 1$

17. $\begin{vmatrix} 2 & 0 & 3 \\ -1 & 0 & 3 \\ 5 & 0 & 7 \end{vmatrix}$ Expand about column 2.

$-0 \begin{vmatrix} -1 & 3 \\ 5 & 7 \end{vmatrix} + 0 \begin{vmatrix} 2 & 3 \\ 5 & 7 \end{vmatrix} - 0 \begin{vmatrix} 2 & 3 \\ -1 & 3 \end{vmatrix} = 0$

21. $\begin{vmatrix} 1 & 1 & 2 \\ 5 & 5 & 7 \\ 3 & 3 & 1 \end{vmatrix}$ Expand about row 1.

$1 \begin{vmatrix} 5 & 7 \\ 3 & 1 \end{vmatrix} - 1 \begin{vmatrix} 5 & 7 \\ 3 & 1 \end{vmatrix} + 2 \begin{vmatrix} 5 & 5 \\ 3 & 3 \end{vmatrix}$

$= 1[5(1) - (3)(7)]$
$\quad - 1[5(1) - (3)(7)]$
$\quad + 2[5(3) - (3)(5)]$
$= -16 + 16 + 0 = 0$

25. $\begin{vmatrix} 1 & 3m & 2 \\ 0 & 2m & 4 \\ 1 & 10m & 20 \end{vmatrix}$ Expand about column 2.

$-3m \begin{vmatrix} 0 & 4 \\ 1 & 20 \end{vmatrix} + 2m \begin{vmatrix} 1 & 2 \\ 1 & 20 \end{vmatrix} - 10m \begin{vmatrix} 1 & 2 \\ 0 & 4 \end{vmatrix}$

$= -3m[0(20) - (1)(4)]$
$\quad + 2m[1(20) - (1)(2)]$
$\quad - 10m[1(4) - (0)(2)]$
$= -3m(-4) + 2m(18) - 10m(4) = 8m$

29. Expand using second row, since it contains two zeros.

$\begin{vmatrix} 3 & 5 & 1 & 9 \\ 0 & 5 & 2 & 0 \\ 2 & -1 & -1 & -1 \\ -4 & 2 & 2 & 2 \end{vmatrix}$

$= 5 \begin{vmatrix} 3 & 1 & 9 \\ 2 & -1 & -1 \\ -4 & 2 & 2 \end{vmatrix} - 2 \begin{vmatrix} 3 & 5 & 9 \\ 2 & -1 & -1 \\ -4 & 2 & 2 \end{vmatrix}$

$= 5 \cdot 0 - 2 \cdot 0 = 0$

33. See answer section for this proof.

37. $\dfrac{-5(-5) - 2(3)}{-7(-3) - (-4)(-5)} = \dfrac{25 - 6}{21 - 20} = \dfrac{19}{1} = 19$

Section 8.5 (page 509)

1. $8x - 4y = 8$
$x + 3y = 22$

$D = \begin{vmatrix} 8 & -4 \\ 1 & 3 \end{vmatrix} = 24 + 4 = 28$

$D_x = \begin{vmatrix} 8 & -4 \\ 22 & 3 \end{vmatrix} = 24 + 88 = 112$

$D_y = \begin{vmatrix} 8 & 8 \\ 1 & 22 \end{vmatrix} = 176 - 8 = 168$

$x = \dfrac{D_x}{D} = \dfrac{112}{28} = 4$ and $y = \dfrac{D_y}{D} = \dfrac{168}{28} = 6$

Solution set: $\{(4, 6)\}$

5. $3x + 8y = 3$
$2x - 4y = 2$

$D = \begin{vmatrix} 3 & 8 \\ 2 & -4 \end{vmatrix} = -12 - 16 = -28$

$D_x = \begin{vmatrix} 3 & 8 \\ 2 & -4 \end{vmatrix} = -12 - 16 = -28$

(coefficients of x replaced by constants)

$D_y = \begin{vmatrix} 3 & 3 \\ 2 & 2 \end{vmatrix} = 6 - 6 = 0$

$x = \dfrac{D_x}{D} = \dfrac{-28}{-28} = 1$

and

$y = \dfrac{D_y}{D} = \dfrac{0}{-28} = 0$

Solution set: $\{(1, 0)\}$

9. $3x + 5y = -21$
$-4x - 2y = 14$

$D = \begin{vmatrix} 3 & 5 \\ -4 & -2 \end{vmatrix} = -6 + 20 = 14$

$D_x = \begin{vmatrix} -21 & 5 \\ 14 & -2 \end{vmatrix} = 42 - 70 = -28$

$D_y = \begin{vmatrix} 3 & -21 \\ -4 & 14 \end{vmatrix} = 42 - 84 = -42$

$$x = \frac{D_x}{D} = -\frac{28}{14} = -2$$

$$y = \frac{D_y}{D} = -\frac{42}{14} = -3$$

Solution set: $\{(-2, -3)\}$

13. $2x + 2y + z = 1$
$x + 3y + 2z = 5$
$x - y - z = 6$

$$D = \begin{vmatrix} 2 & 2 & 1 \\ 1 & 3 & 2 \\ 1 & -1 & -1 \end{vmatrix} \quad \text{Expand by row 1.}$$

$$= 2\begin{vmatrix} 3 & 2 \\ -1 & -1 \end{vmatrix} - 2\begin{vmatrix} 1 & 2 \\ 1 & -1 \end{vmatrix} + 1\begin{vmatrix} 1 & 3 \\ 1 & -1 \end{vmatrix}$$

$D = 2(-1) - 2(-3) + 1(-4) = 0.$
$D = 0$, so Cramer's rule does not apply.

17. $x + 2y + 3z = 1$
$-x - y + 3z = 2$
$6x - y - z = 2$

$$D = \begin{vmatrix} 1 & 2 & 3 \\ -1 & -1 & 3 \\ 6 & -1 & -1 \end{vmatrix} \quad \text{Expand by row 1.}$$

$$= 1\begin{vmatrix} -1 & 3 \\ -1 & -1 \end{vmatrix} - 2\begin{vmatrix} -1 & 3 \\ 6 & -1 \end{vmatrix}$$

$$+ 3\begin{vmatrix} -1 & -1 \\ 6 & -1 \end{vmatrix}$$

$$= 1(4) - 2(-17) + 3(7) = 59$$

$$D_x = \begin{vmatrix} 1 & 2 & 3 \\ 2 & -1 & 3 \\ 2 & -1 & -1 \end{vmatrix} \quad \text{Expand by row 1.}$$

$$= 1\begin{vmatrix} -1 & 3 \\ -1 & -1 \end{vmatrix} - 2\begin{vmatrix} 2 & 3 \\ 2 & -1 \end{vmatrix}$$

$$+ 3\begin{vmatrix} 2 & -1 \\ 2 & -1 \end{vmatrix}$$

$$= 1(4) - 2(-8) + 3(0) = 20$$

$$D_y = \begin{vmatrix} 1 & 1 & 3 \\ -1 & 2 & 3 \\ 6 & 2 & -1 \end{vmatrix} \quad \text{Expand by row 1.}$$

$$= 1\begin{vmatrix} 2 & 3 \\ 2 & -1 \end{vmatrix} - 1\begin{vmatrix} -1 & 3 \\ 6 & -1 \end{vmatrix}$$

$$+ 3\begin{vmatrix} -1 & 2 \\ 6 & 2 \end{vmatrix}$$

$$= 1(-8) - 1(-17) + 3(-14)$$
$$= -33$$

$$D_z = \begin{vmatrix} 1 & 2 & 1 \\ -1 & -1 & 2 \\ 6 & -1 & 2 \end{vmatrix} \quad \text{Expand by row 1.}$$

$$= 1\begin{vmatrix} -1 & 2 \\ -1 & 2 \end{vmatrix} - 2\begin{vmatrix} -1 & 2 \\ 6 & 2 \end{vmatrix}$$

$$+ 1\begin{vmatrix} -1 & -1 \\ 6 & -1 \end{vmatrix}$$

$$= 1(0) - 2(-14) + 1(7) = 35$$

$$x = \frac{D_x}{D} = \frac{20}{59}, \ y = \frac{D_y}{D} = \frac{-33}{59}$$

$$z = \frac{D_z}{D} = \frac{35}{59}$$

Solution set: $\left\{\left(\dfrac{20}{59}, -\dfrac{33}{59}, \dfrac{35}{59}\right)\right\}$

21. $5x + 2z = -20$
$x + 3y = -1$
$ z = 1$

$$D = \begin{vmatrix} 5 & 0 & 2 \\ 1 & 3 & 0 \\ 0 & 0 & 1 \end{vmatrix} = 5\begin{vmatrix} 3 & 0 \\ 0 & 1 \end{vmatrix} - 0\begin{vmatrix} 1 & 0 \\ 0 & 1 \end{vmatrix}$$

$$+ 2\begin{vmatrix} 1 & 3 \\ 0 & 0 \end{vmatrix} \quad \text{Expand by Row 1.}$$

$$= 5(3) - 0 + 0 = 15$$

$$D_x = \begin{vmatrix} -20 & 0 & 2 \\ -1 & 3 & 0 \\ 1 & 0 & 1 \end{vmatrix} = -20\begin{vmatrix} 3 & 0 \\ 0 & 1 \end{vmatrix} - 0\begin{vmatrix} -1 & 0 \\ 1 & 1 \end{vmatrix}$$

$$+ 2\begin{vmatrix} -1 & 3 \\ 1 & 0 \end{vmatrix}$$

$$= -20(3) - 0 + 2(-3)$$
$$= -66$$

$$D_y = \begin{vmatrix} 5 & -20 & 2 \\ 1 & -1 & 0 \\ 0 & 1 & 1 \end{vmatrix} = 5\begin{vmatrix} -1 & 0 \\ 1 & 1 \end{vmatrix} + 20\begin{vmatrix} 1 & 0 \\ 0 & 1 \end{vmatrix}$$

$$+ 2\begin{vmatrix} 1 & -1 \\ 0 & 1 \end{vmatrix}$$

$$= 5(-1) + 20(1) + 2(1)$$
$$= 17$$

$$x = \frac{D_x}{D} = \frac{-66}{15} = -\frac{22}{5}$$

$$y = \frac{D_y}{D} = \frac{17}{15}$$

$z = 1$ (from third equation of the system)

Solution set: $\left\{\left(-\dfrac{22}{5}, \dfrac{17}{15}, 1\right)\right\}$

25. $f(x) = -4x + 3$
$f(3) = -4(3) + 3 = -12 + 3 = -9$
$f(-2) = -4(-2) + 3 = 8 + 3 = 11$

29. $f(x) = \dfrac{-3}{x^3 + 2}$

$$f(3) = \frac{-3}{3^3 + 2}$$

$$= \frac{-3}{27 + 2} = -\frac{3}{29}$$

$$f(-2) = \frac{-3}{(-2)^3 + 2}$$

$$= \frac{-3}{-8 + 2}$$

$$= \frac{-3}{-6} = \frac{1}{2}$$

CHAPTER 9

Section 9.1 (page 529)

1. Write $f(x) = x^2 - 5$ in $f(x) = a(x - h)^2 + k$ form to get $f(x) = 1(x - 0)^2 - 5$. Thus, the vertex $(h, k) = (0, -5)$.

5. The equation $f(x) = (x + 4)^2$ in $f(x) = a(x - h)^2 + k$ form is $f(x) = (x + 4)^2 + 0$. There is a shift 4 units to the left. So the vertex (h, k) is $(-4, 0)$.

9. Write $f(x) = -5x^2$ in $f(x) = a(x - h)^2 + k$ form to get $f(x) = -5(x - 0)^2 + 0$. Since $a < 0$ $(a = -5)$, the graph opens downward, and is narrower than $y = x^2$ since $|a| > 1$.

See the answer section in your text for the graphs in Exercises 13–33.

13. The graph of $f(x) = 3x^2$ opens upward since $3 > 0$ and is narrower than $y = x^2$ since $3 > 1$. Vertex is $(0, 0)$. Two other points are $(-1, 3)$ and $(1, 3)$.

17. Vertex of $f(x) = x^2 - 1$ is at $(0, -1)$; opens upward; same shape as $y = x^2$. Two other points are $(-2, 3)$ and $(2, 3)$.

21. Vertex of $f(x) = 2x^2 - 2$ is at $(0, -2)$; opens upward; narrower than $y = x^2$. Two other points are $(-2, 6)$ and $(2, 6)$.

25. Vertex of $f(x) = -2(x + 1)^2$ is at $(-1, 0)$; opens downward; narrower than $y = x^2$. Two other points are $(1, -8)$ and $(-3, -8)$.

29. Vertex of $f(x) = 2(x - 1)^2 - 3$ is at $(1, -3)$; opens upward; narrower than $y = x^2$. Two other points are $(-1, 5)$ and $(3, 5)$.

33. Vertex of $f(x) = \frac{2}{3}(x - 1)^2 - 2$ is at $(1, -2)$; opens upward; wider than $y = x^2$. Two other points are $(-2, 4)$ and $(4, 4)$.

37. $2x^2 - 12x = 5$

$x^2 - 6x = \dfrac{5}{2}$ Divide each side by 2.

$x^2 - 6x + 9 = \dfrac{5}{2} + 9$ Add $\left[\dfrac{1}{2}(6)\right]^2 = 9$ to each side.

$(x - 3)^2 = \dfrac{23}{2}$ Factor on the left; add on the right.

$x - 3 = \pm\sqrt{\dfrac{23}{2}}$ Square root on each side.

$x - 3 = \dfrac{\pm\sqrt{23}}{\sqrt{2}}$

$x - 3 = \dfrac{\pm\sqrt{46}}{2}$ Rationalize the denominator.

$x = 3 \pm \dfrac{\sqrt{46}}{2}$ Add 3 to each side.

$x = \dfrac{6 \pm \sqrt{46}}{2}$ Add on the right side.

Solution set: $\left\{\dfrac{6 \pm \sqrt{46}}{2}\right\}$

Section 9.2 (page 539)

1. $y = x^2 + 8x - 5$
$y = x^2 + 8x + (16 - 16) - 5$ Complete the square in x.
$y = (x + 4)^2 - 21$
Vertex is at $(-4, -21)$; opens upward; same shape as $y = x^2$. The x-intercepts are the points where $y = 0$—that is, where $x^2 + 8x - 5 = 0$. To find the number of x-intercepts, use the discriminant to find the number of *real* solutions. The discriminant is $b^2 - 4ac$. Here $a = 1$, $b = 8$, and $c = -5$. The discriminant is $8^2 - 4(1)(-5) = 64 + 20 = 84$. Since 84 has *two real* square roots, there are two x-intercepts.

5. $y = 2x^2 - 6x + 3$
$y = 2(x^2 - 3x) + 3$
$y = 2\left(x^2 - 3x + \dfrac{9}{4} - \dfrac{9}{4}\right) + 3$
$= 2\left[\left(x - \dfrac{3}{2}\right)^2 - \dfrac{9}{4}\right] + 3$
$= 2\left(x - \dfrac{3}{2}\right)^2 - \dfrac{9}{2} + 3$
$= 2\left(x - \dfrac{3}{2}\right)^2 - \dfrac{3}{2}$

Vertex is at $\left(\dfrac{3}{2}, -\dfrac{3}{2}\right)$; opens upward; narrower than $y = x^2$. Set $y = 0$, so $2x^2 - 6x + 3 = 0$. To find the number of x-intercepts, use the discriminant, $b^2 - 4ac$. $a = 2$, $b = -6$, $c = 3$. The discriminant is $(-6)^2 - 4(2)(3) = 36 - 24 = 12$. Since 12 has two real square roots, there are two x-intercepts.

9. $x = 2y^2 - 8y + 9$
This is a horizontal parabola.
Complete the square on y:
$x = 2(y^2 - 4y) + 9$
$= 2(y^2 - 4y + 4 - 4) + 9$
$= 2(y^2 - 4y + 4) - 8 + 9$
$= 2(y - 2)^2 + 1$

Vertex is at $(1, 2)$; opens to the right since $a = 2$ is positive; narrower than $x = y^2$ since $|a| = 2$ is greater than 1.

See the graphs in the answer section of your text in Exercises 13–17.

13. $f(x) = 3x^2 - 9x + 8$
Find the y-intercept:
$f(0) = 3(0)^2 - 9(0) + 8 = 8$
The y-intercept is $(0, 8)$.
Find the x-intercepts:
$0 = 3x^2 - 9x + 8$
$x = \dfrac{9 \pm \sqrt{81 - 4(3)(8)}}{2(3)} = \dfrac{9 \pm \sqrt{-15}}{6}$

Both solutions are imaginary, so there are no x-intercepts.

Find the vertex:
$$y = 3(x^2 - 3x) + 8$$
$$= 3\left(x^2 - 3x + \frac{9}{4} - \frac{9}{4}\right) + 8$$
$$= 3\left(x^2 - 3x + \frac{9}{4}\right) - \frac{27}{4} + 8$$
$$= 3\left(x - \frac{3}{2}\right)^2 + \frac{5}{4}$$

The vertex is $\left(\frac{3}{2}, \frac{5}{4}\right)$.

Find additional points:
If $x = 2$, $y = 2$; if $x = 1$, $y = 2$; if $x = 3$, $y = 8$.
Plot these points and the vertex and y-intercept.

17. $x = y^2 - 6y + 4$
This is a horizontal parabola opening to the right.
Find any intercepts. If $y = 0$,
$$x = 0^2 - 6(0) + 4 = 4,$$
giving $(4, 0)$. If $x = 0$,
$$0 = y^2 - 6y + 4$$
$$y = \frac{6 \pm \sqrt{36 - 16}}{2} = \frac{6 \pm \sqrt{20}}{2} \approx 5.2 \text{ or } .8,$$
giving $(0, 5.2)$ and $(0, .8)$.
Find the vertex:
$$x = (y^2 - 6y + 9 - 9) + 4$$
$$= (y^2 - 6y + 9) - 9 + 4$$
$$= (y - 3)^2 - 5$$
The vertex is $(-5, 3)$. Plot the points $(4, 0)$, $(0, 5.2)$, $(0, .8)$, and $(-5, 3)$, and use symmetry to complete the graph.

21. The number of units of tacos to sell to produce the lowest cost will occur at the vertex.
$$c = 2x^2 - 28x + 160$$
$$= 2(x^2 - 14x) + 160$$
$$= 2(x^2 - 14x + 49 - 49) + 160$$
$$= 2(x^2 - 14x + 49) - 98 + 160$$
$$= 2(x - 7)^2 + 62$$
Since x is the number of units of tacos and c the corresponding cost, she should produce 7 units of tacos, which will cost $62.

25. Let x be one number. Then $80 - x$ is the other number. The product is $P = x(80 - x)$. Find the vertex of $P = x(80 - x)$ by putting it into $P = a(x - h)^2 + k$ form.
$$P = 80x - x^2$$
$$P = -1(x^2 - 80x)$$
$$P = -1(x^2 - 80x + 1600) - 1600(-1)$$
Complete the square.
$$P = -1(x - 40)^2 + 1600$$
The vertex is $(40, 1600)$ and since $a < 0$ ($a = -1$), the graph opens downward. Thus the maximum occurs where $x = 40$, so $80 - x = 80 - 40 = 40$. The pair of numbers is 40 and 40.

29. $\sqrt{3^2 + 4^2} = \sqrt{9 + 16} = \sqrt{25} = 5$
(Note $\sqrt{3^2 + 4^2} \neq 3 + 4 = 7$.)

Section 9.3 (page 547)

1. Let $(x_1, y_1) = (-2, 1)$ and $(x_2, y_2) = (3, -2)$.
$$d = \sqrt{(x_2 - x_1)^2 + (y_2 - y_1)^2}$$
$$= \sqrt{(3 + 2)^2 + (-2 - 1)^2}$$
$$= \sqrt{25 + 9} = \sqrt{34}$$

5. Let $(x_1, y_1) = (-8, 7)$ and $(x_2, y_2) = (4, -3)$.
$$d = \sqrt{(x_2 - x_1)^2 + (y_2 - y_1)^2}$$
$$= \sqrt{[4 - (-8)]^2 + (-3 - 7)^2}$$
$$= \sqrt{12^2 + (-10)^2}$$
$$= \sqrt{144 + 100}$$
$$= \sqrt{244} = \sqrt{4 \cdot 61} = 2\sqrt{61}$$

9. Center $(0, 0)$; radius 8
Let $(h, k) = (0, 0)$ and $r = 8$.
$$(x - h)^2 + (y - k)^2 = r^2$$
$$(x - 0)^2 + (y - 0)^2 = 8^2$$
$$x^2 + y^2 = 64$$

13. With center $(0, 3)$ and radius $\sqrt{2}$, let $(h, k) = (0, 3)$ and $r = \sqrt{2}$.
$$(x - h)^2 + (y - k)^2 = r^2$$
$$(x - 0)^2 + (y - 3)^2 = (\sqrt{2})^2$$
$$x^2 + (y - 3)^2 = 2$$

17. Begin by rearranging terms.
$$x^2 + y^2 + 6x - 2y + 6 = 0$$
$$x^2 + 6x + y^2 - 2y = -6$$
$$(x^2 + 6x + 9) + (y^2 - 2y + 1) = -6 + 9 + 1$$
Complete the squares by adding 10 to both sides.
$$(x + 3)^2 + (y - 1)^2 = 4$$
The center is $(-3, 1)$ and the radius is 2.

21. $\quad 2x^2 + 2y^2 - 4x + 12y + 2 = 0$
$$x^2 + y^2 - 2x + 6y + 1 = 0$$
Divide through by 2.
$$x^2 - 2x + y^2 + 6y = -1$$
Rearrange terms.
$$(x^2 - 2x + 1) + (y^2 + 6y + 9) = -1 + 1 + 9$$
Complete the squares.
$$(x - 1)^2 + (y + 3)^2 = 9$$
The center is $(1, -3)$ and the radius is 3.

For Exercises 25–37, refer to the graphs in the answer section of your textbook.

25. Put $2x^2 + 2y^2 = 8$ in the form
$$(x - h)^2 + (y - k)^2 = r^2.$$
$$2x^2 + 2y^2 = 8$$
$$x^2 + y^2 = 4$$
$$(x - 0)^2 + (y - 0)^2 = 4$$
The center is $(0, 0)$ and the radius is 2.

29. Given $(x - 1)^2 + (y + 3)^2 = 4$. The center is $(1, -3)$ and the radius is 2.

33. Let $f(x) = y$ and square both sides.
$$y = \sqrt{4 - x^2}$$
$$y^2 = 4 - x^2$$
$$x^2 + y^2 = 4.$$
This is a circle with center $(0, 0)$ and $r = 2$. But $y = \sqrt{4 - x^2}$ is satisfied only if $y \geq 0$. So the graph is the semicircle above the x-axis with endpoints at $(-2, 0)$ and $(2, 0)$.

37. Let $y = f(x)$. Square both sides of $y = \sqrt{4-x}$ to get
$$y^2 = 4 - x$$
$$x = -y^2 + 4,$$
which is the equation of a horizontal parabola, opening to the left, with vertex $(4, 0)$. If $x = 0$, $y = \sqrt{4-x} = \sqrt{4-0} = 2$. Also, if $x = -5$, $y = \sqrt{4-x} = \sqrt{4+5} = \sqrt{9} = 3$. The graph is the half of the horizontal parabola above the x-axis with an endpoint at $(4, 0)$ and going through the points $(0, 2)$ and $(-5, 3)$.

41. $5x - 8y = 3$
If $x = 0$, $5(0) - 8y = 3$
$$-8y = 3$$
$$y = -\frac{3}{8}.$$
If $y = 0$, $5x - 8(0) = 3$
$$5x = 3$$
$$x = \frac{3}{5}.$$
The x-intercept is $\left(\frac{3}{5}, 0\right)$; the y-intercept is $\left(0, -\frac{3}{8}\right)$.

Section 9.4 (page 557)

See the answer section in your text for the graphs for these exercises.

1. $\dfrac{x^2}{4} + \dfrac{y^2}{9} = 1$
x-intercepts are $(2, 0), (-2, 0)$
y-intercepts are $(0, 3), (0, -3)$

5. $\dfrac{x^2}{16} + \dfrac{y^2}{4} = 1$
x-intercepts are $(4, 0), (-4, 0)$
y-intercepts are $(0, 2), (0, -2)$

9. $\dfrac{y^2}{49} - \dfrac{x^2}{36} = 1$
The y-intercepts are $(0, 7)$ and $(0, -7)$.
The asymptotes go through $(6, 7)$ and $(-6, -7)$, and through $(-6, 7)$ and $(6, -7)$.

13. $x^2 + y^2 = 16$
Graph is a circle with center at the origin and radius 4.

17. $2x^2 - y = 0$
Graph is a parabola with vertex at $(0, 0)$, going through $(-2, 8)$ and $(2, 8)$.

21. $x^2 + 9y^2 = 9$
$\dfrac{x^2}{9} + \dfrac{y^2}{1} = 1$
Graph is an ellipse, with x-intercepts $(3, 0), (-3, 0)$ and y-intercepts $(0, 1), (0, -1)$.

25. $\dfrac{y}{3} = \sqrt{1 + \dfrac{x^2}{9}}$
$\dfrac{y^2}{9} = 1 + \dfrac{x^2}{9} \rightarrow -\dfrac{x^2}{9} + \dfrac{y^2}{9} = 1$
Graph is the upper branch of a hyperbola with y-intercept $(0, 3)$. The asymptotes go through $(3, 3)$ and $(0, 0)$, and through $(-3, 3)$ and $(0, 0)$.

29. (a) Divide each term of $100x^2 + 324y^2 = 32,400$ by 32,400 to get $\dfrac{x^2}{324} + \dfrac{y^2}{100} = 1$ in $\dfrac{x^2}{a^2} + \dfrac{y^2}{b^2} = 1$ form. The height in the center is the y-intercept, or $(0, b)$. Thus, the height is 10 meters.

(b) The width of the ellipse is the distance between the x-intercepts, which are $x = \pm 18$ [since $(18)^2 = 324$]. Thus the width across the bottom is 36 meters.

33. $x^4 - 16 = 0$
$(x^2 - 4)(x^2 + 4) = 0$ Difference of two squares
$x^2 - 4 = 0$ or $x^2 + 4 = 0$
$x^2 = 4$ $x^2 = -4$
$x = \pm 2$ or $x = \pm 2i$
Solution set: $\{2, -2, 2i, -2i\}$

Section 9.5 (page 565)

1. By substitution,
$$x = x^2 + 2x$$
$$0 = x^2 + x$$
$$0 = x(x + 1)$$
$$x = 0 \text{ or } x = -1.$$
If $x = 0$ then $y = 0$; if $x = -1$ then $y = -1$.
Solution set is $\{(0, 0), (-1, -1)\}$.

5. Solve the second equation for x:
$x = 1 - 2y$. Substitute.
$$(1 - 2y)^2 + y^2 = 1$$
$$1 - 4y + 4y^2 + y^2 = 1$$
$$5y^2 - 4y = 0$$
$$y(5y - 4) = 0$$
$$y = 0 \text{ or } y = \frac{4}{5}$$
If $y = 0$ then $x = 1$, giving $(1, 0)$; if $y = \dfrac{4}{5}$ then $x = -\dfrac{3}{5}$, giving $\left(-\dfrac{3}{5}, \dfrac{4}{5}\right)$. Solution set is $\left\{(1, 0), \left(-\dfrac{3}{5}, \dfrac{4}{5}\right)\right\}$.

9. Solve the first equation for x: $x = -\dfrac{6}{y}$. Substitute.
$$-\frac{6}{y} + y = -1$$
$$-6 + y^2 = -y$$
$$y^2 + y - 6 = 0$$
$$(y + 3)(y - 2) = 0$$
$$y = -3 \text{ or } y = 2$$

Substitute into $x = -\dfrac{6}{y}$ to find x.

If $y = -3$, $x = \dfrac{-6}{-3} = 2$.

If $y = 2$, $x = -\dfrac{6}{2} = -3$.

Solution set: $\{(2, -3), (-3, 2)\}$

13. $4x^2 - y^2 = 7$ (1)
$y = 2x^2 - 3$ (2)

Substitute (2) into (1).
$$4x^2 - (2x^2 - 3)^2 = 7$$
$$4x^2 - [4x^4 - 12x^2 + 9] = 7$$
$$-4x^4 + 16x^2 - 9 = 7$$
$$4x^4 - 16x^2 + 16 = 0$$
$$x^4 - 4x^2 + 4 = 0$$

Factor
$$(x^2 - 2)(x^2 - 2) = 0$$
$$x^2 = 2$$
$$x = \pm\sqrt{2}$$

If $x = \pm\sqrt{2}$, then $y = 2(\pm\sqrt{2})^2 - 3 = 1$. The solution set is $\{(\sqrt{2}, 1), (-\sqrt{2}, 1)\}$.

17. Subtract $x^2 + 3y^2 = 3$ from $2x^2 + 3y^2 = 6$ to get $x^2 = 3$ or $x = \pm\sqrt{3}$. Substitute into either equation to find y. Solution set is $\{(\sqrt{3}, 0), (-\sqrt{3}, 0)\}$.

21. Substitute $y = \dfrac{5}{x}$ in $2x^2 - y^2 = 5$.

$$2x^2 - \left(\dfrac{5}{x}\right)^2 = 5$$
$$2x^2 - \dfrac{25}{x^2} = 5$$
$$2x^4 - 25 = 5x^2$$
$$2x^4 - 5x^2 - 25 = 0$$
$$(2x^2 + 5)(x^2 - 5) = 0$$
$$x = \dfrac{\pm i\sqrt{10}}{2} \quad \text{or} \quad x = \pm\sqrt{5}$$

Substitute into $y = \dfrac{5}{x}$ to find y.

Solution set is $\left\{\left(\dfrac{i\sqrt{10}}{2}, -i\sqrt{10}\right),\right.$ $\left.\left(\dfrac{-i\sqrt{10}}{2}, i\sqrt{10}\right), (\sqrt{5}, \sqrt{5}), (-\sqrt{5}, -\sqrt{5})\right\}$.

25. Subtract $x^2 - y^2 = 3$ from $x^2 + 2xy - y^2 = 7$ to get
$$2xy = 4$$
$$xy = 2$$
$$y = \dfrac{2}{x}.$$

Substitute. $x^2 - \left(\dfrac{2}{x}\right)^2 = 3$
$$x^2 - \dfrac{4}{x^2} = 3$$
$$x^4 - 4 = 3x^2$$
$$x^4 - 3x^2 - 4 = 0$$
$$(x^2 - 4)(x^2 + 1) = 0$$
$$x = \pm 2 \quad \text{or} \quad \pm i$$

Substitute into $y = \dfrac{2}{x}$ to find y.

Solution set is $\{(2, 1), (-2, -1), (i, -2i), (-i, 2i)\}$.

29. Let the two numbers be x and y. Then
$$x^2 + y^2 = 106 \quad (1)$$
$$x^2 - y^2 = 56. \quad (2)$$

Add the equations to get
$$2x^2 = 162$$
$$x^2 = 81$$
$$x = \pm 9.$$

Let $x = \pm 9$ in equation (1). Then
$$(\pm 9)^2 + y^2 = 106$$
$$y^2 = 25$$
$$y = \pm 5.$$

This gives $(9, 5)$, $(9, -5)$, $(-9, 5)$, and $(-9, -5)$. Thus the two numbers are 9 and 5, 9 and -5, -9 and 5, or -9 and -5.

33. Graph $2x - y = 4$ with a solid line through the points $(2, 0)$ and $(0, -4)$. Test $(0, 0)$ in $2x - y \le 4$ to get $0 \le 4$, a true statement. So shade the side of the line containing the origin. See the answer graph in the answer section of this textbook.

Section 9.6 (page 573)

In Exercises 1–33, see the answer graphs in the textbook.

1. $y < x^2$

To graph $y < x^2$, first graph the parabola $y = x^2$ which opens upward, with a dashed curve. The parabola divides the plane into two parts. Test a point not on the parabola, say $(3, 0)$, in $y < x^2$ to get $0 < 3^2$ or $0 < 9$. Since this is a true statement, shade the region of the plane that includes $(3, 0)$, that is, the region outside of the parabola.

5. Treat as an equation and simplify $y^2 \le 16 + x^2$.
$$y^2 = 16 + x^2$$
$$y^2 - x^2 = 16$$
$$\dfrac{y^2}{16} - \dfrac{x^2}{16} = 1 \quad \text{Divide each term by 16.}$$

This is a hyperbola of the form
$$\dfrac{y^2}{a^2} - \dfrac{x^2}{b^2} = 1$$

intersecting the y-axis at ± 4. The graph is a solid curve, because of the $\le$ sign. Now test a point not on the curve, such as $(0, 0)$, in $y^2 \le 16 + x^2$.
$$0^2 \le 16 + 0^2$$
or $\quad 0 \le 16 \quad$ True

Because $(0, 0)$ is between the two branches of the hyperbola, and $(0, 0)$ produced a true statement above, the shading is between the two branches.

9. $9x^2 > 4y^2 - 36$

Simplify the equation
$$9x^2 = 4y^2 - 36.$$
$$36 = 4y^2 - 9x^2$$
$$1 = \dfrac{y^2}{9} - \dfrac{x^2}{4} \quad \text{Divide each term by 36.}$$

or $\quad \dfrac{y^2}{9} - \dfrac{x^2}{4} = 1.$

This is a hyperbola of the form $\dfrac{y^2}{a^2} - \dfrac{x^2}{b^2} = 1$, intersecting the y-axis at ± 3. The graph is a dashed curve because of the $>$ sign. Now, test a point not on the curve, such as $(0, 0)$,
$$9x^2 > 4y^2 - 36.$$
$$9 \cdot 0^2 > 4 \cdot 0^2 - 36$$
$$0 > -36 \quad \text{True}$$
Because $(0, 0)$ is between the two branches of the hyperbola, and $(0, 0)$ produced a true statement above, shade the region between the two branches.

13. $2x^2 + 2y^2 \geq 98$
Simplify
$$2x^2 + 2y^2 = 98.$$
$$x^2 + y^2 = 49 \quad \text{Divide by 2.}$$
This is a solid circle of radius 7 with center at the origin. Test a point not on the circle, say $(0, 0)$, in the inequality to get $0 \geq 98$, a false statement. Since $(0, 0)$ is inside the circle and produces a false statement, shade the region outside the circle.

17. $y^2 - 9x^2 \geq 9$
Simplify the equation
$$y^2 - 9x^2 = 9.$$
$$\dfrac{y^2}{9} - \dfrac{x^2}{1} = 1 \quad \begin{array}{l}\text{Divide each}\\\text{term by 9.}\end{array}$$
This is a hyperbola of the form
$$\dfrac{y^2}{a^2} - \dfrac{x^2}{b^2} = 1,$$
intersecting the y-axis at ± 3. The graph is a solid curve because of the $\geq$ sign. Now, test a point not on the curve, such as $(0, 0)$, in
$$y^2 - 9x^2 \geq 9.$$
$$0^2 - 9 \cdot 0^2 \geq 9$$
$$0 \geq 9 \quad \text{False}$$
Because $(0, 0)$ is between the two branches of the hyperbola, and $(0, 0)$ produced a false statement, the shading is in the regions above and below the two branches.

21. $2x - 3y \leq 6$
$4x + 5y \geq 10$
Graph $2x - 3y = 6$ as a solid line which goes through $(3, 0)$ and $(0, -2)$. Shade above the line. Graph $4x + 5y = 10$ as a solid line which goes through $(2.5, 0)$ and $(0, 2)$. Shade above the line. The solution will be the intersection of the shaded regions.

25. $2x - y > 0$
$y > x^2$
Graph $2x - y = 0$ as a dashed line which goes through $(0, 0)$ and $(1, 2)$. Shade below the line. Graph $y = x^2$ as a dashed parabola with vertex $(0, 0)$, which opens upward. Shade the inside of the parabola. The solution will be the intersection of the shaded regions.

29. $\quad x > 1$
$\quad y < 2$
$x^2 + y^2 < 4$
$x > 1$ is the right side of the dashed vertical line $x = 1$. To graph $y < 2$, graph $y = 2$ as a dashed horizontal line and shade below. To graph $x^2 + y^2 < 4$, draw the dashed circle $x^2 + y^2 = 4$ with center $(0, 0)$ and radius 2. Test $(0, 0)$ in the inequality to get $0 < 4$, a true statement. Since $(0, 0)$ is inside the circle and yields a true statement, shade the inside of the circle. The region of overlap of all 3 shaded portions is the intersection.

33. $y = 2x - 3$
The intercepts are $(0, -3)$ and $\left(\dfrac{3}{2}, 0\right)$. Plot these points and draw a straight line through them.

CHAPTER 10

Section 10.1 (page 597)

1. Exchange x and y in each ordered pair to get the inverse, $\{(5, 3), (9, 2), (7, 4)\}$.

5. Replace $f(x)$ with y in $f(x) = 2x$ to get $y = 2x$. Exchange x and y, then solve for y. $x = 2y$ leads to $y = \dfrac{x}{2}$. The inverse is $f^{-1}(x) = \dfrac{x}{2}$.

9. Exchange x and y in $2y + 1 = 3x$ to get $2x + 1 = 3y$.
Solve for y: $2x + 1 = 3y$
$$y = \dfrac{2x + 1}{3}$$
The inverse is $f^{-1}(x) = \dfrac{2x + 1}{3}$.

13. $2y = x^2 + 1$ is not a one-to-one function because, for example, both $x = 1$ and $x = -1$ correspond to $y = 1$.

For Exercises 17–29, refer to the graphs in the answer section of your textbook.

17. (a) The graph passes the horizontal line test. It is the graph of a one-to-one function.
(b) Exchange the x's and y's of the points $(-1, 5)$ and $(2, -1)$ on the line to get $(5, -1)$ and $(-1, 2)$. Draw the line through these two points.

21. (a) The graph passes the horizontal line test. It is the graph of a one-to-one function.
(b) The given points on the curve are $(-4, 2)$, $(-1, 1)$, $(1, -1)$, and $(4, -2)$. Exchange the x's and y's, and join these new points: $(2, -4)$, $(1, -1)$, $(-1, 1)$, and $(-2, 4)$.

25. To graph $2y + 1 = 3x$, plot the points $\left(\dfrac{1}{3}, 0\right)$ and $\left(0, -\dfrac{1}{2}\right)$. The inverse function is also linear and passes through $\left(0, \dfrac{1}{3}\right)$ and $\left(-\dfrac{1}{2}, 0\right)$.

29. To graph $x - 1 = y^3$, plot the points $(0, -1)$, $(1, 0)$, $(2, 1)$, $(9, 2)$, $(-7, -2)$. The inverse function will pass through the points $(-1, 0)$, $(0, 1)$, $(1, 2)$, $(2, 9)$, $(-2, -7)$.

33. From Exercise 31, $f(3) = 2^3 = 8$. Therefore, by definition of inverse function, $f^{-1}(8) = 3$.

37. From Exercise 35, $f(0) = 2^0 = 1$. By definition of inverse function, $f^{-1}(1) = 0$.

41. $f(x) = 3^x$

$$f(-2) = 3^{-2} = \frac{1}{3^2} = \frac{1}{9}$$

Section 10.2 (page 607)

See the answer section for the graphs in Exercises 1–5.

1. Complete some ordered pairs.

x	-3	-2	-1	0	1	2	3
$y = 3^x$	$3^{-3} = \frac{1}{27}$	$3^{-2} = \frac{1}{9}$	$3^{-1} = \frac{1}{3}$	$3^{-0} = 1$	3	9	27

5. Complete some ordered pairs.

x	-2	-1	0	1	2
$y = 2^{2x}$	$2^{-4} = \frac{1}{16}$	$2^{-2} = \frac{1}{4}$	$2^0 = 1$	$2^2 = 4$	$2^4 = 16$

9. $\quad 4^x = 8$

$\quad (2^2)^x = 2^3 \quad$ Write each side using the same base.

$\quad 2^{2x} = 2^3$

$\quad 2x = 3$

Since the bases are equal, the exponents must be equal. Solve for x.

$$x = \frac{3}{2}$$

Solution set: $\left\{\dfrac{3}{2}\right\}$

13. $\quad 25^{-2x} = 3125$

$\quad (5^2)^{-2x} = 5^5$

$\quad 5^{-4x} = 5^5$

$\quad -4x = 5$

$$x = -\frac{5}{4}$$

Solution set: $\left\{-\dfrac{5}{4}\right\}$

17. $\quad \left(\dfrac{1}{2}\right)^x = 8$

$\quad (2^{-1})^x = 2^3$

$\quad 2^{-x} = 2^3$

$\quad -x = 3$

$\quad x = -3$

Solution set: $\{-3\}$

21. $2^{-3} = \dfrac{1}{2^3} = \dfrac{1}{8}$

25. $3^{5/2} = (\sqrt{3})^5$

$\quad 3^{5/2} = \sqrt{243}$

$\quad \quad \ = \sqrt{81 \cdot 3}$

$\quad \quad \ = 9\sqrt{3}$

Section 10.3 (page 615)

1. $3^4 = 81$ becomes $\log_3 81 = 4$

5. $\left(\dfrac{1}{2}\right)^{-2} = 4$ becomes $\log_{1/2} 4 = -2$.

9. $2^{-4} = \dfrac{1}{16}$ becomes $\log_2\left(\dfrac{1}{16}\right) = -4$.

13. $\log_5 125 = 3$ becomes $5^3 = 125$

17. $\log_3 \dfrac{1}{9} = -2$ becomes $3^{-2} = \dfrac{1}{9}$.

21. $3^3 = 27$, so $\log_3 27 = 3$.

25. $36^{1/2} = 6$, so $\log_{36} 6 = \dfrac{1}{2}$.

29. $\sqrt{3^5} = 3^{5/2}$, so $\log_3 \sqrt{3^5} = \log_3 3^{5/2} = \dfrac{5}{2}$.

33. $\log_9 \sqrt{3^3} = \log_9 3^{3/2} = \log_9 (3^2)^{3/4} = \log_9 9^{3/4} = \dfrac{3}{4}$.

37. Write $\log_4 x = \dfrac{5}{2}$ in exponential form as $4^{5/2} = x$.

Thus, $x = 4^{5/2} = (4^{1/2})^5 = 2^5 = 32$.

Solution set: $\{32\}$

41. Write $\log_{1/2} x = -3$ in exponential form as

$\left(\dfrac{1}{2}\right)^{-3} = x$.

Thus $x = \left(\dfrac{1}{2}\right)^{-3} = 2^3 = 8$.

Solution set: $\{8\}$

45. Write $\log_b 1 = 0$ in exponential form as $b^0 = 1$. Since for any nonzero base b, $b^0 = 1$, b can be any positive real number (except 1, since 1 is not used as a logarithmic base).

Solution set: $\{b \mid b$ is a positive real number, $b \neq 1\}$

49. Write $y = \log_{2.718} x$ in exponential form as $(2.718)^y = x$. Then choose values for y. With a calculator find the corresponding values of x and complete the table.

x	0.135	0.368	1	2.718	7.388
y	-2	-1	0	1	2

Plot these ordered pairs and draw a smooth curve through the points.

See the graph in the answer section of this textbook.

53. $3^{-2} \cdot 3^8 = 3^{-2+8} = 3^6$

57. $(5x^2)^4 = 5^4(x^2)^4 = 625x^8$

Section 10.4 (page 625)

1. $\log_6 \dfrac{3}{4} = \log_6 3 - \log_6 4$ Quotient rule for logarithms

5. $\log_5 \dfrac{5x}{y} = \log_5 (5x) - \log_5 y$ Quotient rule
   $= \log_5 5 + \log_5 x - \log_5 y$ Product rule
   $= 1 + \log_5 x - \log_5 y$ Since $\log_b b = 1$

9. $\log_3 \sqrt{\dfrac{ma}{b}} = \log_3 \left(\dfrac{ma}{b}\right)^{1/2}$
   $= \left(\dfrac{1}{2}\right) \log_3 \dfrac{ma}{b}$ Power rule
   $= \left(\dfrac{1}{2}\right) [\log_3 (ma) - \log_3 b]$ Quotient rule
   $= \left(\dfrac{1}{2}\right) (\log_3 m + \log_3 a - \log_3 b)$
        Product rule
   $= \dfrac{1}{2} \log_3 m + \dfrac{1}{2} \log_3 a - \dfrac{1}{2} \log_3 b$
        Distributive property

13. $\log_b x + \log_b y = \log_b (xy)$ Product rule

17. $(\log_b m - \log_b n) + \log_b r$
    $= \log_b \left(\dfrac{m}{n}\right) + \log_b r$ Quotient rule
    $= \log_b \left(\dfrac{mr}{n}\right)$ Product rule

21. $\log_{10}(x+1) + \log_{10}(x-1)$
    $= \log_{10}[(x+1)(x-1)]$ Product rule
    $= \log_{10}(x^2 - 1)$

25. $\log_{10} 4 = \log_{10} 2^2 = 2 \log_{10} 2 = 2(.3010) = .6020$

29. $\log_{10} 16 = \log_{10} 2^4 = 4 \log_{10} 2 = 4(.3010) = 1.2040$

33. $\dfrac{\log_{10} 8}{\log_{10} 16} = \dfrac{.9031}{1.2041} = .75 \neq \dfrac{1}{2}$
    The statement is false.

37. $10^{2.1461} = 140$

Section 10.5 (page 633)

1. Enter 278 into your calculator and touch the log key.
   $\log 278 = 2.4440$

5. Use your calculator to get $\log 9.83 = .9926$.

9. Use your calculator to get
   $\log .000672 = -3.1726$.

13. Enter 5 in your calculator and touch the ln key.
    $\ln 5 = 1.6094$.

17. Use your calculator to get
    $\ln 1.72 = .5423$

21. $\ln e^{3.1} = 3.1(\ln e) = 3.1(1) = 3.1$

25. Enter .5340 in your calculator, and touch INV LOG or 10^x to get 3.42.

29. Enter -1.118 in your calculator, and touch INV LOG or 10^x to get .0762.

33. Enter 1.3962 in your calculator and touch INV ln or e^x to get 4.04.

37. $\log_5 11 = \dfrac{\log 11}{\log 5} = \dfrac{1.0414}{.6990} = 1.49$

41. $\log_8 9.63 = \dfrac{\log 9.63}{\log 8} = 1.09$

45. $\log_{50} 31.3 = \dfrac{\log 31.3}{\log 50} = .88$

49. $\text{pH} = -\log [H_3O^+]$
    $= -\log (3.9 \times 10^{-9})$
    $= -[\log 3.9 + \log 10^{-9}]$
    $= -[.6 + (-9)]$
    $= -(-8.4) = 8.4$

53. $\quad\; 3.4 = -\log [H_3O^+]$
    $-3.4 = \log [H_3O^+]$
    $[H_3O^+] = 10^{-3.4} \approx 4.0 \times 10^{-4}$

57. $B(t) = 25{,}000\, e^{.2t}$
    (a) $B(0) = 25{,}000\, e^0 = 25{,}000$
    There are 25,000 bacteria.
    (b) $B(1) = 25{,}000 e^{.2(1)}$
    $\approx 25{,}000(1.2214)$
    $\approx 30{,}535$
    There are about 30,500 bacteria at 1 P.M.
    (c) $B(2) = 25{,}000 e^{.2(2)}$
    $\approx 25{,}000(1.4918)$
    $\approx 37{,}300$
    There are about 37,300 bacteria at 2 P.M.
    (d) $B(5) = 25{,}000 e^{.2(5)}$
    $\approx 25{,}000(2.7183)$
    $\approx 67{,}957$
    There are about 68,000 bacteria at 5 P.M.

61. $N(.3) = -5000 \ln (.3) \approx -5000(-1.204)$
    $= 6020$ years (rounded)

65. $\dfrac{x+2}{x} = 5$
    $5x = x + 2$
    $4x = 2$
    $x = \dfrac{1}{2}$
    Solution set: $\left\{\dfrac{1}{2}\right\}$

69. $x^2 = 48$
    $x = \sqrt{48}$ or $x = -\sqrt{48}$
    $x = 4\sqrt{3}$ $x = -4\sqrt{3}$
    Solution set: $\{-4\sqrt{3}, 4\sqrt{3}\}$

Section 10.6 (page 643)

1. $25^x = 125$
   $(5^2)^x = 5^3$
   $5^{2x} = 5^3$
   $2x = 3$
   $x = \dfrac{3}{2}$
   Solution set: $\left\{\dfrac{3}{2}\right\}$

5. $6^{y+1} = 8$
$(y+1)\log 6 = \log 8$
$y + 1 = \dfrac{\log 8}{\log 6}$
$y = \dfrac{\log 8}{\log 6} - 1$ Exact answer
$y \approx .16$
Solution set: $\{.16\}$

9. $7^{2y-1} = 1$
$7^{2y-1} = 7^0$
$2y - 1 = 0$
$y = \dfrac{1}{2}$
Solution set: $\left\{\dfrac{1}{2}\right\}$

13. $(1 + .03)^n = 90$
$1.03^n = 90$
$n \log 1.03 = \log 90$
$n = \dfrac{\log 90}{\log 1.03}$
$= \dfrac{1.9542}{.0128}$
≈ 152.23
Solution set: $\{152.23\}$

17. $\log_5 (3x + 2) - \log_5 x = \log_5 4$
$\log_5 \left(\dfrac{3x+2}{x}\right) = \log_5 4$
$\dfrac{3x+2}{x} = 4$
$3x + 2 = 4x$
$2 = x$
Solution set: $\{2\}$

21. $\log_2 x = 3$
Write in exponential form as
$2^3 = x$
$x = 8.$
Solution set: $\{8\}$

25. $\log_a 5 = -\dfrac{3}{4}$
Write in exponential form as
$a^{-3/4} = 5.$
Raising both sides to the $-\dfrac{4}{3}$ power, we have the following.
$a = 5^{-4/3}$
$= \dfrac{1}{\sqrt[3]{625}} = \dfrac{1}{\sqrt[3]{125 \cdot 5}}$
$= \dfrac{1}{5\sqrt[3]{5}} \cdot \dfrac{\sqrt[3]{25}}{\sqrt[3]{25}} = \dfrac{\sqrt[3]{25}}{25}$
Solution set: $\left\{\dfrac{\sqrt[3]{25}}{25}\right\}$

29. Use the formula $A = P(1 + i)^n$, where $P = 5000$, $i = .06$, and $n = 12$.
$A = 5000(1 + .06)^{12} = 5000(1.06)^{12}$
$(1.06)^{12} \approx 2.0122$ (with a calculator), so
$A \approx 5000(2.0122)$
$\approx 10{,}061$
The amount is about \$10,061.

33. Use the formula $A = P(1 + i)^t$ with $A = 2P$ and $i = .03$.
$2P = P(1.03)^t$
$2 = (1.03)^t$
$\log 2 = t \log 1.03$
$t = \dfrac{\log 2}{\log 1.03}$
$= \dfrac{.3010}{.0128} = 23.45$, or about 23 years

37. Solve $S = 30{,}000(1 - .15)^{12} = 30{,}000(.85)^{12}$.
$(.85)^{12} \approx .1422$ (with a calculator), so
$S \approx 30{,}000(.1422)$
≈ 4270
The scrap value is about \$4270.

41. Use $S = 80 \log_5 (t + 1)$ with $S = 80$. Solve for t.
$80 = 80 \log_5 (t + 1)$
$1 = \log_5 (t + 1)$ Divide both sides by 80.
$5^1 = t + 1$ Exponential form
$t = 4$
The sales will reach 80 (hundreds) in 4 years.

GLOSSARY

This glossary provides an alphabetized listing of all the entries found in the "Key Terms" section of each Chapter Summary. For reference or further study, the corresponding section number is given at the end of each entry.

A

absolute value The absolute value of a number is its distance from 0 or its magnitude without regard to sign.   **[1.1]**

absolute value equation; absolute value inequality Absolute value equations and inequalities are equations and inequalities that involve the absolute value of a variable expression.   **[2.7]**

addition method The addition method of solving a system of equations involves the elimination of a variable by adding the two equations.   **[8.1]**

addition property of equality This property states that the same number may be added to (or subtracted from) each side of an equation to obtain an equivalent equation.   **[2.1]**

addition property of inequality This property states that the same number may be added to (or subtracted from) each side of an inequality to obtain an equivalent inequality.   **[2.5]**

additive inverse (negative, opposite) The additive inverse of a number a is $-a$.   **[1.1]**

algebraic expression An algebraic expression is an expression indicating any combination of the following operations: addition, subtraction, multiplication, division (except by 0), and taking roots on any collection of variables and numbers.   **[2.1]**

antilogarithm An antilogarithm is the number that corresponds to a given logarithm.   **[10.5]**

asymptotes of a hyperbola The two intersecting lines that the branches of a hyperbola approach are called asymptotes of the hyperbola.   **[9.4]**

axis The line through the vertex of a parabola parallel to the x-axis or y-axis is the axis of the parabola.   **[9.1]**

B

base The base is a number that is a repeated factor in a product.   **[1.4]**

binomial A binomial is a polynomial with exactly two terms.   **[3.3]**

boundary line In the graph of a linear inequality, the boundary line separates the region that satisfies the inequality from the region that does not satisfy the inequality.   **[7.4]**

C

center The fixed point mentioned in the definition of circle is the center of the circle.   **[9.3]**

circle A circle is the set of all points in a plane that lie a fixed distance from a fixed point.   **[9.3]**

coefficient A coefficient (also known as a **numerical coefficient**) is the numerical factor of a term.   **[1.5]**

combining like terms Combining like terms is a method of adding or subtracting like terms by using the properties of real numbers.   **[1.5]**

common denominator A common denominator of several denominators is an expression that is divisible by each of these denominators.   **[4.3]**

common logarithm A common logarithm is a logarithm to the base 10.   **[10.5]**

complex fraction A fraction that has a fraction in the numerator, the denominator, or both, is called a complex fraction.   **[4.4]**

complex number A complex number is a number that can be written in the form $a + bi$, where a and b are real numbers.   **[5.7]**

compound inequality A compound inequality is formed by joining two inequalities with a connective word, such as *and* or *or*.   **[2.6]**

conditional equation An equation that has a finite number of elements in its solution set is called a conditional equation.   **[2.1]**

conic sections Graphs that result from cutting an infinite cone with a plane are called conic sections.   **[9.1]**

conjugate The conjugate of $a + bi$ is $a - bi$.   **[5.7]**

contradiction An equation that has no solutions (that is, its solution set is ∅) is called a contradiction.   **[2.1]**

coordinate The number that corresponds to a point on the number line is its coordinate.   **[1.1]**

coordinates The numbers in an ordered pair are called the coordinates of the corresponding point.   **[7.1]**

Cramer's rule Cramer's rule is a method of solving a system of equations by using determinants. **[8.5]**

cube root The cube root of a number r is the number that can be cubed to get r. **[1.4]**

D

degree of a polynomial The degree of a polynomial is the highest degree of any of the terms in the polynomial. **[3.3]**

degree of a term The degree of a term with one variable is the exponent on that variable. **[3.3]**

dependent equations Dependent equations are equations whose graphs are the same line. **[8.1]**

dependent variable If the quantity y depends on x in an equation relating x and y, then y is called the dependent variable. **[7.5]**

descending powers A polynomial is written in descending powers if the exponents on the variables in the terms decrease from left to right. **[3.3]**

determinant A determinant is a real number that is associated with a square matrix. **[8.4]**

difference The result of subtraction is called the difference. **[1.3]**

discriminant The discriminant is the quantity under the radical in the quadratic formula. **[6.2]**

domain The domain of a relation is the set of first components (x-values) of the ordered pairs of a relation. **[7.5]**

E

elimination method The elimination method of solving a system of equations involves the elimination of a variable by adding the two equations. **[8.1]**

ellipse An ellipse is the set of all points in a plane the sum of whose distances from two fixed points is constant. **[9.4]**

empty set A set with no elements is called the empty set. **[1.1]**

equation An equation is a statement that two algebraic expressions are equal. **[2.1]**

equivalent equations Equivalent equations are equations that have the same solution set. **[2.1]**

equivalent inequalities Equivalent inequalities are inequalities with the same solution set. **[2.5]**

exponent An exponent is a number that shows how many times a factor is repeated in a product. **[1.4]**

exponential A base with an exponent is called an exponential or a **power**. **[1.4]**

exponential equation An equation involving an exponential, where the variable is in the exponent, is an exponential equation. **[10.2]**

extraneous solution An extraneous solution of a radical equation is a solution of $x = a^2$ that is not a solution of $\sqrt{x} = a$. **[5.6]**

F

factors Two numbers whose product is a third number are factors of that third number. **[1.4]**

formula A formula is a mathematical statement in which more than one letter is used to express a relationship. **[2.2]**

fourth root A fourth root of a number a is a number b whose fourth power is a. **[5.1]**

function A function is a set of ordered pairs in which each value of the first component x corresponds to exactly one value of the second component y. **[7.5]**

G

graph The point on the number line that corresponds to a number is its graph. **[1.1]**

graph of a relation The graph of a relation is the graph of the ordered pairs of the relation. **[7.5]**

greatest common factor The product of the largest numerical factor and the variable factor of lowest degree of every term in a polynomial is the greatest common factor of the terms of the polynomial. **[3.7]**

H

half-plane Either of the two regions determined by a boundary line is called a half-plane. **[7.4]**

hyperbola A hyperbola is the set of all points in a plane the difference of whose distances from two fixed points is constant. **[9.4]**

hypotenuse The hypotenuse is the longest side in a right triangle. **[6.4]**

I

identity An equation that is satisfied by every number for which both sides are defined is called an identity. **[2.1]**

imaginary number A complex number $a + bi$ with $b \neq 0$ is called an imaginary number. **[5.7]**

imaginary part The imaginary part of a complex number $a + bi$ is b. **[5.7]**

inconsistent system A system is inconsistent if it has no solution. **[8.1]**

independent variable If the quantity y depends on x in an equation relating x and y, x is the independent variable. **[7.5]**

index In the expression $\sqrt[n]{a}$, n is the index. **[5.1]**

inequality An inequality is a mathematical statement that two quantities are not equal. **[1.2]**

intersection The intersection of two sets A and B is the set of elements that belong to both A and B. **[2.5]**

interval An interval is a portion of a number line. **[1.2]**

interval notation Interval notation uses symbols to describe an interval on the number line. **[1.2]**

inverse of a function If f is a one-to-one function, the inverse of f is the set of all ordered pairs of the form (y, x), where (x, y) belongs to f. **[10.1]**

L

leg The two shorter sides of a right triangle are called the legs. **[6.4]**

like fractions Like fractions are fractions with the same denominators. **[4.3]**

like terms Like terms are terms with the same variables raised to the same powers. **[1.5]**

linear equation or first-degree equation in one variable An equation is linear or first-degree in the variable x if it can be written in the form $ax + b = c$, where a, b, and c are real numbers, with $a \neq 0$. **[2.1]**

linear equation in two variables A first-degree equation with two variables is a linear equation in two variables. **[7.1]**

linear inequality in one variable An inequality is linear in the variable x if it can be written in the form $ax + b < c$, $ax + b \leq c$, $ax + b > c$, or $ax + b \geq c$, where a, b, and c are real numbers, with $a \neq 0$. **[2.5]**

linear system A linear system is a system of equations that contains only linear equations. **[8.1]**

logarithm A logarithm is an exponent; $\log_a x$ is the exponent on the base a that gives the number x. **[10.3]**

logarithmic equation A logarithmic equation is an equation with a logarithm in at least one term. **[10.3]**

M

matrix A matrix is an ordered array of numbers. **[8.4]**

monomial A monomial is a polynomial with exactly one term. **[3.3]**

multiplication property of equality This property states that the same nonzero number may be multiplied by or divided into each side of an equation to obtain an equivalent equation. **[2.1]**

multiplication property of inequality This property states that both sides of an inequality may be multiplied or divided by the same positive number without changing the direction of the inequality. If both sides are multiplied or divided by a negative number, the inequality symbol must be reversed. **[2.5]**

N

natural logarithm A natural logarithm is a logarithm to the base e. **[10.5]**

nonlinear equation Any equation that cannot be written in the form $ax + by = c$, for real numbers a, b, and c, is a nonlinear equation. **[9.5]**

nonlinear system of equations A nonlinear system of equations is a system with at least one nonlinear equation. **[9.5]**

non-strict inequality An inequality that involves $\geq$ or $\leq$ is called a non-strict inequality. **[2.5]**

number line A number line is a line with a scale to indicate the set of real numbers. **[1.1]**

O

one-to-one function A one-to-one function is a function in which each x-value corresponds to just one y-value and each y-value corresponds to just one x-value. **[10.1]**

origin When two number lines intersect at a right angle, the origin is the common zero point. **[7.1]**

P

parabola A parabola is a graph of a second-degree function. **[9.1]**

plot To plot an ordered pair is to locate it on a rectangular coordinate system. **[7.1]**

polynomial A polynomial is a finite sum of terms with only whole number exponents on the variables and no variable denominators. **[3.3]**

prime factored form An integer that has been factored as a product of prime numbers is in prime factored form. **[3.7]**

prime number A prime number is a positive integer greater than 1 that has only itself and 1 as factors. **[3.7]**

principal root For a positive number a and even value of n, the principal nth root of a is the positive nth root of a. **[5.1]**

product The result of multiplication is called the product. **[1.3]**

Pythagorean formula The Pythagorean formula states that in a right triangle the square of the length of the hypotenuse equals the sum of the squares of the lengths of the legs. **[6.4]**

Q

quadrant A quadrant is one of the four regions in the plane determined by a rectangular coordinate system. **[7.1]**

quadratic equation A quadratic equation is one that can be written in the form $ax^2 + bx + c = 0$ ($a \neq 0$). **[3.10]**

quadratic formula The quadratic formula is a formula for solving quadratic equations. **[6.2]**

quadratic inequality A quadratic inequality is an inequality that can be written in the form $ax^2 + bx + c < 0$ or $ax^2 + bx + c > 0$, or with $\leq$ or $\geq$. **[6.5]**

quadratic function A function that can be written in the form $f(x) = ax^2 + bx + c$, for real numbers a, b, and c, with $a \neq 0$, is a quadratic function. **[9.1]**

quadratic in form An equation that can be written as a quadratic equation is called quadratic in form. **[6.3]**

quotient The result of division is called the quotient. **[1.3]**

R

radical expression A radical expression is an algebraic expression that contains radicals. **[5.4]**

radicand In the expression $\sqrt[n]{a}$, a is the radicand. **[5.1]**

radius The radius of a circle is the fixed distance between the center and any point on the circle. **[9.3]**

range The range of a relation is the set of second components (y-values) of the ordered pairs of the relation. **[7.5]**

rational expression A rational expression (or algebraic fraction) is the quotient of two polynomials with denominator not 0. **[4.1]**

rationalizing the denominator The process of removing radicals from the denominator so that the denominator contains only rational quantities is called rationalizing the denominator. **[5.5]**

real part The real part of $a + bi$ is a. **[5.7]**

reciprocals Two numbers whose product is 1 are reciprocals. **[1.3]**

rectangular coordinate system Two number lines that intersect at a right angle at their zero points form a rectangular coordinate system. **[7.1]**

relation A relation is any set of ordered pairs. **[7.5]**

S

scientific notation In scientific notation, a number is written as the product of a number between 1 and 10 (or -1 and -10) and some integer power of 10. **[3.2]**

second-degree inequality A second-degree inequality is an inequality with at least one variable of degree two and no variable with degree greater than two. **[9.6]**

set A set is a collection of objects. **[1.1]**

set-builder notation Set-builder notation is used to describe a set of numbers without listing them. **[1.1]**

signed numbers Positive and negative numbers are signed numbers. **[1.1]**

slope The ratio of the change in y compared to the change in x along a line is the slope of the line. **[7.2]**

solution A solution of an equation is a number that makes the equation true when substituted for the variable. **[2.1]**

solution set The solution set of an equation is the set of all its solutions. **[2.1]**

square matrix A square matrix has the same number of rows as columns. **[8.4]**

square root The square root of a number r is the number that can be squared to get r. **[1.4]**

standard form A linear equation is in standard form when written as $ax + by = c$ with $a \geq 0$, and a, b, and c integers. **[7.1]**

strict inequality An inequality that involves $>$ or $<$ is called a strict inequality. **[2.5]**

substitution method The substitution method of solving a system involves substituting an expression for one variable in terms of another. **[8.1]**

sum The result of addition is called the sum. **[1.3]**

synthetic division Synthetic division is a shortcut procedure for dividing a polynomial by a polynomial of the form $x - k$. **[3.6]**

system of equations Two or more equations to be solved at the same time form a system of equations. **[8.1]**

system of inequalities A system of inequalities consists of two or more inequalities that are solved at the same time. **[9.6]**

T

term A term is a number or the product of a number and one or more variables. **[1.5]**

trinomial A trinomial is a polynomial with exactly three terms. **[3.3]**

U

union The union of two sets A and B is the set of elements that belong to either A or B (or both). **[2.6]**

unlike fractions Unlike fractions are fractions with different denominators. **[4.3]**

V

variable A variable is a letter used to represent a number or a set of numbers. **[1.1]**

vertex The point on a parabola that has the smallest y-value (if the parabola opens upward) or the largest y-value (if the parabola opens downward) is called the vertex of the parabola. **[9.1]**

vertical line test The vertical line test says that if a vertical line cuts the graph of a relation in more than one point, then the relation is not a function. **[7.5]**

X

***x*-axis** The horizontal number line in a rectangular coordinate system is called the *x*-axis. **[7.1]**

***x*-intercept** The point where a graph crosses the *x*-axis is the *x*-intercept. **[7.1]**

Y

***y*-axis** The vertical number line in a rectangular coordinate system is called the *y*-axis. **[7.1]**

***y*-intercept** The point where a graph crosses the *y*-axis is the *y*-intercept. **[7.1]**

INDEX

A

Absolute value, 5
Absolute value equation, 109
Absolute value inequality, 109, 428
Addition
  of complex numbers, 323
  of polynomials, 152
  of radicals, 297
  of rational expressions, 233
  of real numbers, 15
Addition method for systems, 467
Addition property
  of equality, 54
  of inequality, 91
Additive inverse, 4, 323
Agnesi, Maria, 572
Algebraic expression, 53, 149
Algebraic fraction, 223
Antilogarithm, 627, 630
Applications of linear systems, 487
Archimedes, 58
Area, formulas for. *See Appendix B*.
Array of signs, 501
Associative properties, 38
Asymptotes of a hyperbola, 553
Ada Augusta, Countess of Lovelace, 316
Axis
  of a parabola, 523
  x- and y-, 395

B

Babbage, Charles, 316, 596
Base, 25, 133
Between, 12
Binomial, 150
  square of, 163
Boundary line, 425

C

Cardano, Girolamo, 453
Cayley, Arthur, 614
Center of a circle, 544
Change of base rule, 631
Châtelet, Marquise du, 146
Circle, 544
  center of, 544
  equation of, 545
  radius of, 544
Coefficient, 37, 149
Combining like terms, 37

Common logarithms, 627
  evaluating, 627
  table. *See Appendix A*.
Commutative properties, 38
Completing the square, 343, 533
Complex fraction, 243
Complex numbers, 322, 400
  addition of, 323
  division of, 325
  multiplication of, 324
  subtraction of, 323
Components of an ordered pair, 395
Compound inequality, 101
Computers, 316, 596
Conditional equation, 57
Conic sections, 523
  summary of, 555
Conjugate, 304
  of a complex number, 324
Consecutive integers, 2–3
Constant of variation, 445
Contradiction equation, 57
Coordinate, 2
Coordinates of a point, 395
Coordinate system, 395
Counting numbers, 1
Cramer's rule, 506–7
Cube of a number, 25
Cube root, 27, 278
Cubes
  difference of two, 196
  sum of two, 197
Cubic equations, solving, 453

D

Degree of a polynomial, 150
Denominator
  least common, 234
  rationalizing, 301
Dependent equations, 467, 481
Dependent variable, 435
Descending powers, 149
Determinant, 499
  Cramer's rule, 506–7
  expansion of, 500
  minor of, 500
Determinant method for systems, 505
Difference, 16
  of two cubes, 196
  of two squares, 195
Direct variation, 445
Discriminant, 353, 536

Distance, 5, 109
Distance formula, 65, 543
Distributive property, 35
Division
  of complex numbers, 325
  of logarithms, 620
  of polynomials, 169
  of radicals, 290, 301
  of rational expressions, 230
  of real numbers, 19
  synthetic, 175
Domain
  of a function, 436
  of rational expressions, 233

E

e, 629
Element of a set, 1
Elimination method for systems, 467, 478
Ellipse, 551
  equation of, 551
  foci, 551
  intercepts of, 551
Empty set, 1
Equality
  addition property of, 54
  multiplication property of, 54
Equation, 9, 53
  absolute value, 109
  of a circle, 545
  conditional, 57
  contradiction, 57
  dependent, 467, 481
  equivalent, 53
  exponential, 603, 637
  first-degree, 53
  graph of, 397
  identity, 57
  linear, 53, 415
  logarithmic, 610, 638
  nonlinear, 561
  quadratic, 206, 341
  quadratic in form, 359
  with radicals, 311
  with rational expressions, 247
  solution of, 53
  solving, 53
  See also System of equations.
Equivalent equations, 53
Equivalent inequalities, 91
Euclid, 267

INDEX

Euclidean tools, 624
Euler, Leonhard, 647
Expansion of a determinant, 500
Exponent, 25, 133, 606
   integer, 133
   negative, 135, 285
   power rules for, 141, 285
   product rule for, 134, 285
   quotient rule for, 137, 285
   rational, 283
   zero, 138, 285
Exponential equation, 603, 637
Exponential function, 601
   graph of, 602
Extraneous solution, 311

F

Factor, 25
   greatest common, 181
Factoring
   difference of two cubes, 196
   difference of two squares, 195
   by grouping, 183
   perfect square trinomials, 195
   polynomials, 181
   solving equations by, 205
   substitution method for, 192
   sum of two cubes, 197
   trinomials, 187
Factors of a number, 25
Fermat, Pierre de, 482
Fermat's Last Theorem, 482
Finite function, 437
First-degree equation, 53
Focus of an ellipse, 551
FOIL method, 161, 324
Formulas, 63, 369
   distance, 65, 543
   Heron's, 282
   list of. *See* Appendix B.
   Pythagorean, 369, 543
   quadratic, 349
   with rational expressions, 255
   solving for a specified variable, 63
Fourth root, 27, 278
Fractions
   algebraic, 223
   complex, 243
   inequalities with, 379
   like, 233
   unlike, 233
Function, 435
   domain of, 436
   exponential, 601
   finite, 437
   graph of, 438
   infinite, 437
   inverse of, 591
   linear, 440
   logarithmic, 609
   one-to-one, 591
   quadratic, 524
   range of, 436
   square root, 546
   vertical line test for, 438
Functional notation, 439
Fundamental principle of rational numbers, 224
Fundamental rectangle, 553

G

Galilei, Galileo, 511
Gauss, Carl, 400
Geometric formulas. *See* Appendix B.
Geometry, Euclidean, 267
Graph
   of an equation, 397
   of an exponential function, 602
   of a function, 438
   of a horizontal line, 399
   of an inequality, 11
   of inverses, 595
   of a linear inequality, 425
   of a logarithmic function, 611
   of a number, 2
   of a relation, 438
   of a set, 11, 109
   of a system, 466
   See also Circle; Ellipse; Hyperbola; Parabola.
Greater than, 10
Greatest common factor, 181
Grouping, factoring by, 183
Growth and decay, 604

H

Heron's formula, 282
Hook's Law, 452
Horizontal line, graph of, 399
Horizontal line test, 592
Hyperbola, 552
   asymptotes of, 553
   equation of, 553
   intercepts of, 552
Hypotenuse, 369

I

i, 321
   powers of, 325
Identity equation, 57
Identity properties, 37
Imaginary numbers, 322
Inconsistent system, 467, 480
Independent variable, 435
Index of a radical, 278
Inequality, 9
   absolute value, 109, 428
   addition property of, 91
   compound, 101
   equivalent, 91
   fractional, 379
   interval notation, 11
   linear, 91, 425
   multiplication property of, 91
   nonlinear, 377
   non-strict, 96
   quadratic, 377
   second-degree, 569
   solving, 91
   strict, 96
   system of, 570
Inequality symbols, 10
Infinite function, 437
Integers, 2
Intercepts
   of an ellipse, 551
   of a hyperbola, 552
   x-, 397, 536
   y-, 397
Interest formula, 63
Intersection of sets, 101
Interval notation, 11
Invariants, 614
Inverse, additive, 4
Inverse of a function, 591
Inverse properties, 36
Inverse variation, 447
Investment problems, 75
Irrational numbers, 2, 280

J

Joint variation, 448

L

Least common denominator, 234
Legs of a triangle, 369
Less than, 10
Like fractions, 233
Like terms, 37, 151
Linear equation, 53, 415
   applications of, 83
   in one variable, 53
   standard form, 397
   in two variables, 397
Linear function, 440
Linear inequality, 91
   graph of, 425
   in one variable, 91
   in two variables, 91, 425

INDEX

Linear system of equations, 465
   applications of, 487
   inconsistent, 467
   methods of solving, 466–67, 470, 478, 505
   with three variables, 477
   with two variables, 465
Logarithm, 609, 642
   antilogarithm, 627, 630
   change of base, 631
   common, 627
   natural, 629
   properties of, 611, 619
   table. *See Appendix A.*
Logarithmic equation, 610, 638
Logarithmic function, 609
   graph of, 611
Logic puzzles, 420

M

Mapping, 436
Member of a set, 1
Minor of a determinant, 500
Mixture problems, 76
Monomial, 150
Motion problems, 84
Multiplication
   of complex numbers, 324
   of polynomials, 159
   of radicals, 289, 301
   of rational expressions, 229
   of real numbers, 17
Multiplication property
   of equality, 54
   of inequality, 91
   of logarithms, 619
   of zero, 40
Multiplicative inverse, 36

N

Napier, John, 642
Natural logarithm, 629
Natural numbers, 1
Negative exponent, 135, 285
Negative numbers, 2, 10
Negative of a number, 4
Newton, Sir Isaac, 146, 511
Noether, Amalie, 382
Nonlinear equation, 561
Nonlinear inequality, 377
Nonlinear system, 561
Non-strict inequality, 96
Notation, scientific, 143
nth root, 278
Number line, 2, 11, 109

Numbers
   complex, 322
   counting, 1
   graph of, 2
   imaginary, 322
   integers, 2
   irrational, 2, 280
   natural, 1
   rational, 2
   real, 2
   signed, 2
   whole, 2
Numerical coefficient, 37

O

One-to-one function, 591
   horizontal line test for, 592
Opposite of a number, 4
Ordered pair, 395
   components of, 395
Ordered triple, 477
Order of operations, 28
Origin, 395

P

Pair, ordered, 395
Parabola, 523
   appplications of, 536
   axis of, 523
   graph of, 524, 535
   horizontal shift of, 525
   vertex of, 523, 533
   vertical shift of, 524
Parallel lines, 409
Percent, 65
Perfect square trinomial, 195
Perpendicular lines, 409
pH, 628
Pi, computation of, 430
Plane, 477
Plotting points, 395
Point-slope form, 416
Polya, George, 86
Polynomial, 149
   addition of, 152
   degree of, 150
   descending powers, writing in, 149
   division of, 169
   factoring, 181
   in x, 149
   multiplication of, 159
   prime, 187
   subtraction of, 153
Positive numbers, 2, 10
Power, 133

Power rule
   for equations with radicals, 311
   for exponents, 141, 285
   for logarithms, 621
Powers of i, 325
Prime factored form, 181
Prime number, 181
Prime polynomial, 187
Principal root, 278
Product, 17
   of sum and difference, 163
Product rule
   for exponents, 134, 285
   for radicals, 289
Properties
   of equality, 54
   of inequalities, 91
   of logarithms, 611, 619
   of real numbers, 35
Proportionality, 445
Puzzles, 420
Pythagorean formula, 369, 543

Q

Quadrant, 396
Quadratic discriminant, 353, 536
Quadratic equation, 206, 341
Quadratic formula, 349
Quadratic function, 524
Quadratic inequality, 377
Quotient, 19
Quotient rule
   for exponents, 137, 285
   for radicals, 290

R

Radical expression, 297
   addition of, 297
   division of, 290, 301
   multiplication of, 289
   subtraction of, 297, 301
Radical sign, 277
Radicals
   equations containing, 311
   index of, 278
   power rule for, 311
   product rule for, 289
   quotient rule for, 290
   rationalizing denominators with, 301
   simplified form of, 290
Radicand, 278
Radius of a circle, 544
Ramanujan, Srinivasa, 450
Range of a function, 436
Rational exponent, 283

Rational expressions, 223
   addition of, 233
   division of, 230
   domain of, 233
   equations with, 247
   formulas with, 255
   in lowest terms, 224
   multiplication of, 229
   subtraction of, 233
Rationalizing denominators, 301
Rational numbers, 2
   as exponents, 283
   fundamental principle of, 224
Real numbers, 2
   addition of, 15
   division of, 19
   multiplication of, 17
   properties of, 35
   subtraction of, 16
Reciprocal, 18, 229
Rectangular coordinate system, 395
Relation, 435
   graph of, 438
Remainder theorem, 177
Root
   cube, 27, 278
   fourth, 27, 278
   nth, 278
   principal, 278
   square, 27, 277

S

Scientific notation, 143
Second-degree inequalities, 569
Set, 1
   elements of, 1
   empty, 1
   graph of, 11, 109
   members of, 1
   solution, 53
Set-builder notation, 1
Shanks, William, 430

Signed numbers, 2
Simplified form of a radical, 290
Slope of a line, 405
Slope-intercept form, 415
Smullyan, Raymond, 420
Solution of an equation, 53
Solution set, 53
Solving for a specified variable, 63
Square of a binomial, 163
Square of a number, 25
Square root, 27, 277
Square root function, 546
Square root property, 342
Standard form of a linear equation, 397
Strict inequality, 96
Substitution method for factoring, 192
Substitution method for systems, 470
Subtraction
   of complex numbers, 323
   of polynomials, 153
   of radicals, 297, 301
   of rational expressions, 233
   of real numbers, 16
Sum and difference, product of, 163
Sum of two cubes, 197
Superstring theory, 647
Sylvester, James, 614
Synthetic division, 175
System
   of equations, 465, 477
   graph for, 466
   of inequalities, 570
   of nonlinear equations, 561

T

Tartaglia, Nicolo, 453
Temperature formula. *See Appendix B.*
Term of a polynomial, 37, 149
Third-degree equations, 453
Translating from words to mathematical expressions, 71
Trinomial, 150
   factoring, 187

   perfect square, 195
Triple, ordered, 477

U

Union of sets, 103
Unlike fractions, 233

V

Variable, 1
   dependent, 435
   independent, 435
Variation, 445
   constant of, 445
   direct, 445
   inverse, 447
   joint, 448
Vertex of a parabola, 523, 533
Vertical line, graph of, 399
Vertical line test for a function, 438
Voltaire, 146
Volume, formulas for. *See Appendix B.*

W

Whole numbers, 2
Women in mathematics, 146, 316, 382, 572
Word problems, solving, 76, 84, 86

X

x-axis, 395
x-intercept, 397, 536

Y

y-axis, 395
y-intercept, 397

Z

Zero exponent, 138, 285
Zero-factor property, 205, 341